Developmental Biology

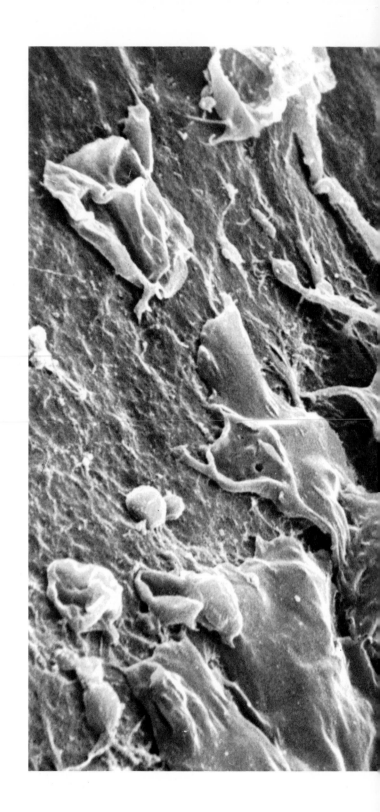

Migration of epithelium during the wound-healing
stage of regeneration in the amputated newt limb
(see Figure 16.3, page 353).

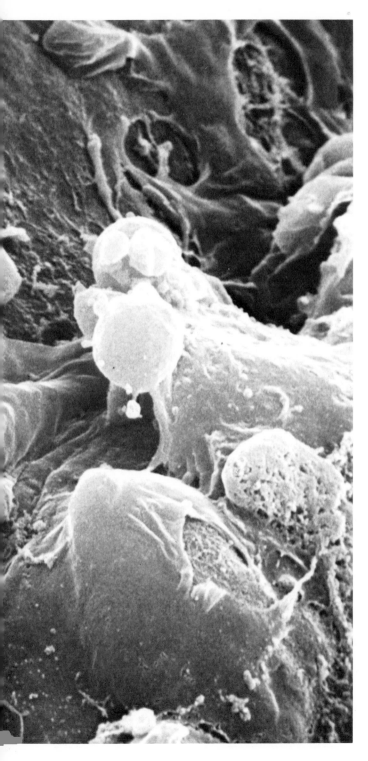

Developmental
BIOLOGY

PATTERNS /
PROBLEMS /
PRINCIPLES

John W. Saunders, Jr.
Department of Biological Sciences
State University of New York at Albany
and The Marine Biological Laboratory,
Woods Hole

Macmillan Publishing Co., Inc.
New York

Collier Macmillan Publishers
London

To the memory of
Benjamin Harrison Willier,
Teacher, Scholar.

Macmillan Publishing Co., Inc.
866 Third Avenue, New York, New York 10022

Collier Macmillan Canada, Ltd.

Library of Congress Cataloging in Publication Data

Saunders, John Warren
  Developmental biology.

  Includes bibliographies and index.
  1. Developmental biology. I. Title.
QH491.S29 1980   574.3   79-15853
ISBN 0-02-406370-3

Printing: 1 2 3 4 5 6 7 8   Year: 2 3 4 5 6 7 8

FRONT COVER: Photograph of a living chick embryo in an egg that had been incubated for 96 hours. A window was made in the egg shell, membranes covering the embryo were removed, and the embryo was lightly stained with a vital dye, Nile Blue Sulfate. The embryo lies on its left side, with the right eye, heart, and fore- and hindlimb buds clearly visible. Dark blue areas on the posterior and anterior sides, respectively, of fore- and hindlimb buds are zones of massive cellular degeneration, which occurs normally in a precise program during limb development.

BACK COVER: Photograph of a fixed and stained section cut through the ovary of *Lilium*. The haploid immature female gametophyte (center) is surrounded by sporogenous tissue, which is enveloped by the integuments (cf. Figure 6.8F on page 50). Three successive divisions of the haploid nucleus lead to formation of the functional female gametophyte, consisting of egg cell, synergids, polar nuclei, and antipodal cells (cf. Figure 6.8K on page 50). [Courtesy of Carolina Biological Supply Company.]

Preface

The concerns of this book are basically the same as those stated in the preface to a work that I published some years ago, *Patterns and Principles of Animal Development* (New York: Macmillan, 1970, 282 pages). In the present work, however, coverage is expanded to include plant development, and selected aspects of the factual and conceptual content of the field of developmental biology are pursued to greater depth. The text deals with the following issues: (1) developmental patterns, including the physiological patterns that lead to formation of the gametes and to breeding behavior and fertilization, and the patterns of morphological change that occur after fertilization in plants and animals, namely, cleavage, germ-layer formation, and the establishment of organ rudiments; (2) the problems presented by the events of development and the formulation of these problems in conceptual contexts that suggest experimental approaches to their solution; and (3) the emergence of the form and function of the organism as expressions of the genetic endowments and the epigenetic events that determine successive patterns of gene expression.

During the past three decades, especially, the field of developmental biology has expanded tremendously. To a large degree, this expansion has been made possible by, and has been enriched by, the entrance into this field of imaginative investigators whose basic training has been in areas such as biophysics, theoretical chemistry, mathematics, molecular biology, and physics. Because the study of development now has such an enormous scope, determining the content of a suitable modern course in developmental biology is a difficult task. It does seem clear to me, however, that the course should be so constructed that the students will acquire many of the more important insights that can be derived from the use of new analytical approaches that nontraditionalists have brought to the field. In addition, however, and perhaps more basically, a modern course should provide a background of fact and principle gained from a study of developmental patterns at various levels of organizational complexity in a variety of organisms. Without this background, the insights that have accrued by virtue of recent advances are of diminished significance.

This text has been written as a text for use in an upper division college course in developmental biology, but its approach and content do not require that students should have had a prior course at a lower level. There is sufficient advanced material in the work, however, that it could readily serve as the sole text or the second text in a two-course sequence. It is assumed that students using this book will have previously acquired a basic biological vocabulary. I have, therefore, not included a glossary of biological terms, but each term that may be new or unfamiliar is introduced in the text in boldface type and is immediately defined, or its definition emerges from the context in which it is used.

Part One presents the aims and scope of the study of development, emphasizing the recognition of unifying principles. The principle of differentiation, one of the central themes of developmental biology, is given special treatment. Part II provides an outline of the general features of meiosis with added emphasis on the events of meiotic prophase in preparation for the later extended discussion of spermatogenesis and oögenesis. A brief discussion of the significance of sexual reproduction is also included in Part Two.

Part Three is quite extensive, consisting of eight chapters. It deals with production of the gametes, fertilization, and early embryonic development. There is a somewhat protracted treatment of the basic reproductive physiology of plants and animals, an aspect of developmental biology that is often neglected in developmental biology textbooks. Yet this topic has significance for two major concerns: population control and the need to increase agricultural productivity. These two concerns also warrant a great deal of attention to the development of gametes and the control of fertilization. Part Three also deals with the partitioning of the egg, the formation of the organ rudiments in plants and of the germ layers in animals, and basic aspects of the formation of definitive organs in plants and animals. This part includes almost all of the developmental anatomy of plants and animals that is to be found in the book. The anatomical descriptions, in fact, could almost constitute a small textbook. Finally, Part Three includes, wherever possible, recent experimental studies relating to the emergence of form, and also the analysis of gene expression during various phases of development.

Part Four deals with the major principles of morphogenesis that have emerged from the classical period of the study of "Entwicklungsmechanik" which was initiated by the experiments of Wilhelm Roux and Hans Driesch in the latter part of the nineteenth century and further enriched by the experiments of Hans Spemann and his students in the early 1900's. Entwicklungsmechanik (or Developmental Physiology) continues to dominate much of what we call experimental embryology today. In this part, too, I deal with the problem of the apparent selectivity of cellular association, a phenomenon that is under investigation by many cellular and developmental biologists as well as by biochemists, immunologists, and neurobiologists. Part Four also treats of principles that emerge from the study of regenerative growth and tissue repair, both of which are areas of great importance to developmental biologists and have been the subjects of lifetime studies by numerous investigators.

Part Five consists of one chapter, which deals with metamorphosis, the study of larval animals and their transformation to the adult form. Metamorphosis is a topic that might readily be expanded into one or more volumes of some magnitude. Necessarily, I have had to confine the content of this chapter to a few organisms whose metamorphosis has been best studied. Emphasis is on the genetic programming of biochemical and morphological changes that characterize the transformation of larva to adult.

The final part of the book, Part Six, deals with the genetic and epigenetic control of developmental processes. Chapter 18 deals with the many points at which the ultimate expression of the gene, the synthesis of specific proteins, may be controlled. Chapter 19, however, presents the most difficult problem now facing developmental biologists, namely, how the characteristic form and function of an organism, its phenotype, emerge from the expression of its genes at the molecular level. New insights are clearly needed in order to solve this many-faceted problem.

Relatively comprehensive summaries are found at the end of each major chapter. Reviewers have suggested that perhaps the summary of each chapter should be read first, thus alerting the student to what the author, at least, finds most important in what he has written. The summary is followed by a list of questions which, if pursued, will lead students to recall important factual and conceptual matters contained in the chapter and will challenge them to put these matters into different contexts. Sources for additional reading about subjects introduced in the chapter are listed at the end of each chapter. These sources consist chiefly of reviews and of other articles that are of historical interest or that expand on important materials treated in the text. In addition, each chapter has a list of references to specific works, usually individual research papers. These references will enable students and teachers to pursue in depth matters that are, necessarily, reported only briefly in the text.

Writing a textbook is a labor of love and requires a love of learning. I hope that my readers will find what they learn from this text as rewarding as I have found the research and writing required to produce it.

Many people have helped in the production of this book and I am grateful to them. I have worked with a number of biology editors at the Macmillan Publishing Company during the overlong gestation period of this book. I mention especially Mr. Woodrow Chapman, whose patience and understanding during the early phases of the writing were of immeasurable help, and Mr. Gregory Payne whose persistent interest helped to bring the work to completion. My sincere thanks go to Mrs. Elisabeth

Belfer, Production Supervisor, who personally did much of the copyediting of the text and who exercised close supervision over all aspects of the work of her staff on this text. Mrs. Belfer's criticisms have been most constructive, and I have learned much about writing from her.

I am greatly indebted to my colleagues in the Department of Biological Sciences at the State University of New York in Albany and to my associates at the Marine Biological Laboratory, Woods Hole, Massachusetts, for many helpful discussions that have contributed to the content of this work. I mention, particularly, Dr. Joseph Mascarenhas, who has participated with me for many years in the teaching of a course in developmental biology. He generously provided references to important papers dealing with many aspects of plant development. Also, we have a seminar course together, largely organized by him, from which emerged a significant part of Chapter 18. To Dr. Mascarenhas I am greatly indebted. I would like, also, to acknowledge the kindness of Dr. Colin Izzard who shared with me his insights into mechanisms of cellular motility, which greatly assisted me in the writing of a part of Chapter 11.

A number of distinguished biologists have reviewed one or more chapters of the book during early phases of the writing and have provided most welcome criticisms. I acknowledge most gratefully the efforts of Drs. Susan Bryant, Tom Humphries, Hans Laufer, Richard Lockshin, William Sheridan, and Trygve Steen. They helped me to avoid some embarrassing mistakes. I am, however, fully responsible for the errors of fact and interpretation that may remain. These are likely to be many in a field that is changing as rapidly as developmental biology is today.

In its penultimate draft, the entire manuscript was read by Drs. George Malacinski, Steven Roth, and Fred Wilt. Because of the production schedule for the book, I was unable to react fully to the suggestions and criticisms of these referees. Happily, however, a number of their suggestions were anticipated in the final draft. I am grateful to these eminent biologists for their help.

My gratitude extends further to the many investigators who have provided photographs and drawings for reproduction in this book and to publishers who have generously permitted reproduction of previously published illustrations. The help of these kind people is acknowledged in the captions of the illustrations. I am also greatly indebted to Mrs. Mary Gasseling Lange, friend and collaborator in research. She supervises the day-to-day operation of my laboratory, and without her skilled assistance I could not have undertaken and finished this book. Most of the original drawings found in the text are the work of Ann Vancura, a distinguished artist and photographer in the Albany area. I thank her for her patience and for the originality that she infused into many of the illustrations. Early versions of many chapters were typed by Mrs. Linda Santandrea, who worked with diligence and accuracy from handwritten manuscript. Her help is greatly appreciated. Margaret A. (Saunders) Geist read both galley and page proof and aided in the production of the index. I am grateful for her diligent assistance. Finally, I wish to express my thanks to Lilyan C. Saunders, who did the final typing of this manuscript and who gave most constructive criticism based on her years of teaching biology at the Albany branch of Russell Sage College. Her loyalty and support have made this work possible.

J. W. S.
Woods Hole, Massachusetts

Contents

Part One
Aims and Scope of Developmental Biology

Part Four
Major Principles of Morphogenesis

xi

Contents

Developmental Biology

P A R T

O N E

Aims and Scope of Developmental Biology

All phenomena of life are parts of cyclic patterns, one cycle, or generation, emerging from and following a preceding one. A cell grows and divides, and its daughter cells do the same; a bacteriophage infects a bacterial cell and replicates, and a multitude of new infective units is released to start the cycle anew. Each generation is characterized to some degree by a phase during which higher levels of order or complexity emerge from systems of lesser order. Examples are the origin of a new individual from the fertilized egg of a plant or animal, the assembly of a randomly ordered population of protein subunits into the orderly pattern of the head of a T_2 bacteriophage. Biological cycles also involve the degeneration of systems from higher to lower levels of organization. Thus, degradation of order occurs progressively during the senescence of an individual or of a population, and it occurs cataclysmically when an organism dies.

The field of developmental biology is broadly concerned with all phases of generative cycles of whatever degree of complexity. Most research and pedadogy in this field, however, deal with those phases in the life cycles of organisms during which progressively higher levels of order emerge in rapid succession. This progression occurs, for example, in the germination of a plant seed, the development of a fertilized egg, the metamorphosis of an insect, or the regeneration of an amphibian limb. This book deals chiefly with the phase of emergent order in development, and the emphasis is on the factors that control the origin of the phenotype in plant and animal species.

In most living things a new generation begins with a fertilized egg, a **zygote,** a cell formed by the union of the egg, or **ovum,** from the female parent and a specialized reproductive cell, the **sperm** or **spermatozoon,** from the male. The word zygote, derived from the Greek, means "yoked" or "joined together," reminding us that each of the parents contributes to the characteristics of the offspring, including size, shape, physiology, behavior, and so on—in other words, its **phenotype.** The **development** of an organism embraces the totality of the processes whereby these attributes are achieved and whereby they are modified during growth, adulthood, and senescence, ending in death.

The early phases of development that an organism undergoes before it becomes capable of an independent self-sustaining existence are usually referred to as its **embryonic period.** The structural features that the organism acquires during its embryonic development collectively constitute the **form** of the organism. The study of form is called **morphology,** and the sum of the processes whereby form is achieved is called **morphogenesis.** Morphogenesis is studied **descriptively** by examining the sequence of morphogenetic changes that occur during development. It may also be studied **experimentally** by manipulating the organism in such a way that observations reveal the causality of morphogenetic change.

1 Patterns, Problems, and Principles

that is essential to the realization of the next stage, which has its own particular pattern of developmental processes.

1.2 Problems— Causal Analysis

This book is also about problems of development. These **problems** revolve about the **causality** of morphogenetic events. Experimental studies often reveal that in development there is a cause-and-effect relationship between one particular morphogenetic event and a subsequent one. It is important for the analysis of development that such causal relationships be localized in time and space, and that the mechanisms underlying such causal relationships be learned. For example, the development of the olfactory organ in the vertebrate embryo requires an interaction of the prospective nasal cells with the cells of the future forebrain. This is a causal relationship. What is the mechanism whereby the forebrain exercises its effect? Do its cells perhaps release certain molecules that enter the future nasal cells and elicit a particular pattern of gene activity?

1.1 Pattern

This book is about **patterns** of development. Of course, form is **pattern,** but the word pattern has much broader connotations. A structure may have the form of a hand, but the pattern may be one of a right hand or of a left hand (i.e., it may show right or left **asymmetry**). Also, one may speak of the pattern in which the cytoplasmic components of an egg are arrayed, and one may compare these patterns in different kinds of eggs. Moreover, the sequences of form changes that are shown in different developing systems occur in distinctive patterns. Finally, we note that the processes that underlie events at each developmental stage occur with a precise timing, intensity, and duration. The timing and duration of these processes constitute a pattern

The search for causal mechanisms in morphogenesis involves efforts to describe events at the molecular and cellular levels of organization. These organizational levels are less complex than those of the interacting tissues and organ rudiments. The goal of causal analysis is thus ultimately the same as that of morphological research, that is, a description of events. But the description at cellular and molecular levels is a good deal more challenging and eventually leads to the analysis of factors controlling genetic activity.

1.3 Principles—Making Knowledge Grow

Is the goal of developmental biology simply the accretion of data about events that occur in developing systems at progressively lower levels of complexity? Is the accumulation of information the only way in which knowledge grows? "Information is but the raw material, the precursor of knowledge," as Paul Weiss once pointed out. It is at best knowledge in unorganized form. True enlightenment about development, he further said, "emerges from the distilling, shaping and integrating of the raw materials into concepts and rules." Concepts, rules, and principles of science are terms used more or less interchangeably to refer to a higher order of knowledge that emerges from the integration of facts into rationally connected constructs. They are **generalizations** that are more meaningful than the facts from which they were induced. Thus, a major goal of the analysis of development must be to make knowledge grow through the discovery of new generalizations based on a solid groundwork of fact. How one arrives at a knowledge of principles of development and applies them to increase one's understanding of development is illustrated in the next chapter, which treats briefly the concept of **differentiation**.

1.4 Summary

The field of developmental biology embraces the entire area of cyclical generative phenomena. Its goals are chiefly restricted, however, to describing the sequence of events that marks the progressive emergence of higher order in living things as they proceed from egg to adult and to the causal analysis of the relationships between these events. The analysis of causality requires an experimental approach designed to reveal the conditions under which particular events occur—that is, their underlying mechanisms—at progressively lower organizational levels of the living matter.

Whether the approach is descriptive or experimental, its end result must be more than the compilation of data. Enlightment and understanding of development emerge from the integration of data into rational constructs—principles or concepts that constitute a higher order of knowledge than the mere accretion of information. The formulation of meaningful principles of development is thus a major goal of the analysis of development.

1.5 Questions for Thought and Review

1. What is meant by the level of organization of a biologic system? Is it reasonable to consider the fertilized egg to be at a lower level of organization than the reproductively functional plant or animal?

2. What is meant by a scientific generalization? Why is it important to be able to make general statements about developing organisms?

3. The word "causality" is used in the text. What does this word mean? How, in a general way, does the scientist go about investigating the causality of a phenomenon?

In the development of a new individual a single cell, the zygote, becomes an organism composed of tens of trillions of cells, as in the case of higher plants and animals, or perhaps of only one or two dozen cells, as in rotifers and some lower plants. Regardless of total number, however, the cells that are found in the new organism at the end of its embryonic period are of a variety of kinds, of finite proportions, and arranged in a species-specific geometry.

Cells of multicellular organisms are ordered as tissues, tissues as organs, and organs as organ systems. The different organs and their component cells and tissues are so ordered as to carry out different functions. Thus, the components of the digestive system of an animal are specialized for the internal transport and breakdown of ingested food. The stem of a tree is composed of tissues specialized for support of the branches and leaves, for the vertical transport of fluids and nutrients, and for growth in diameter.

From the observation that there are these differences and that they arise progressively during **embryogenesis** (development of the embryo), we arrive at the concept of **differentiation.** Within this concept are embraced all data relative to the fact of differentiation and to the means of its accomplishment during development. The major goal of the analysis of development is to discover how these differences emerge from the fertilized egg.

The magnitude of the egg's problem may be appreciated if one examines some of the kinds of cells that are differentiated during development of some higher animals and plants. In vertebrates (Figure 2.1) these include

1. Muscle cells, which produce the contractile proteins and thereby can move the skeletal framework of the animal, pump its blood, and propel material through its digestive tract.
2. Connective tissue cells, which synthesize collagen and other tough fibrils of protein that lace other kinds of cells together into tissues and organs.
3. Red blood cells, packages of hemoglobin, which is the chemical vehicle for transport of oxygen.
4. Liver cells, which, among other activities, manufacture serum proteins and bile.
5. Columnar cells, lining the digestive tract, some of which secrete mucoprotein while others pro-

2

Differentiation

duce and release the enzymes that catalyze the hydrolysis of food.
6. Cells of the various endocrine organs, which produce protein or steroid hormones that are transported by the blood and affect the activities of target organs elsewhere in the body.

Among the cell types in plants (Figure 2.2) are

1. Protective cells in the root, which form a cap over the rapidly dividing meristematic cells of the tip.
2. Epidermal root cells with long root hairs through which uptake of fluid and nutrients from the soil is effected.
3. Sclerenchyma cells, which provide structural support for roots, stems, and leaves.
4. Conducting cells, such as sieve tubes and tracheids, in roots, stems, and leaves.
5. Columnar cells, called palisade cells, laden with chloroplasts, which transduce radiant energy into chemical energy in the leaves.
6. Guard cells, which regulate the flow of gases through the pores connecting the interior environment of the leaf with the exterior.

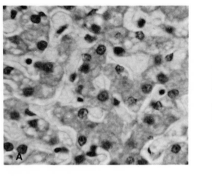
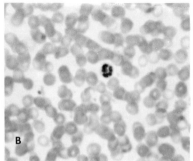

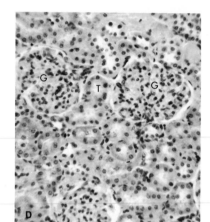
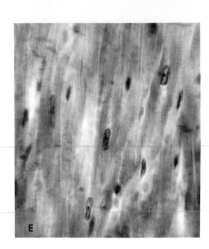

FIGURE 2.1 Representative tissues of the adult mammal: (A) liver; (B) blood; (C) bladder, showing transitional epithelium, E, and connective tissue, CT; (D) kidney, showing glomeruli, G, and tubules, T; (E) heart muscle.

Implicit in the foregoing is that during embryonic development the parts of the organism undergo a sequence of events leading to their integration into a morphologically and functionally differentiated array. It is appropriate, therefore, to speak of the parts of the organism as having achieved their final **state of differentiation** at the end of their development and to refer to the sequence of changes leading to the differentiated state as the **course of differentiation**. The events underlying sequences of differentiative events are properly called **mechanisms of differentiation**.

These terms are applied at different levels of organizational complexity. One may refer, for example, to the course of differentiation of an organ such as the liver (**organ differentiation**), to the final state of differentiation of a tissue (**histodifferentiation**), or to the mechanisms leading to the morphologically and functionally differentiated states of a cell (**cytodifferentiation**).

The following material deals with some generalizations that may be made about the state of differentiation and the course of differentiation. Mechanisms that underlie differentiative changes in cells and tissues ultimately reside in the genome, but gene activity is mediated by factors of the environment and by intracellular cytoplasmic factors that act on the genome. These matters are considered at some length in later chapters.

2.1 Properties of the Differentiated State

In any organism differentiation leads to the production of a finite number of discrete kinds of cells, and each kind has a distinctive repertory, or catalog, of

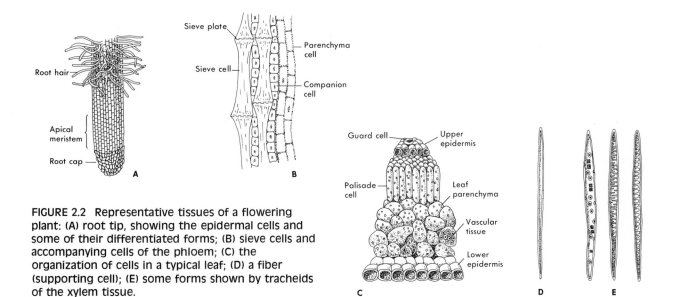

FIGURE 2.2 Representative tissues of a flowering plant: (A) root tip, showing the epidermal cells and some of their differentiated forms; (B) sieve cells and accompanying cells of the phloem; (C) the organization of cells in a typical leaf; (D) a fiber (supporting cell); (E) some forms shown by tracheids of the xylem tissue.

biosynthetic activities, as well as regular maintenance or "housekeeping" activities engaged in by almost all cells. Also, different kinds of cells show different structural properties. A nerve cell has its particular character and so does a cartilage cell, and one does not find intergradations between. Likewise, the palisade cell of a leaf is distinctive from a sieve tube in a vascular bundle of the plant, and hybrid forms between them do not occur. From these and other examples we recognize a subordinate principle under the general concept of differentiation, namely, the **discreteness of differentiation** of types.

Moreover, we find that cells that have achieved a distinctive state of cytodifferentiation do not, as a rule, transform into cells of another kind. Muscle cells, once differentiated, do not transform into mucus-secreting cells like those of the intestinal lining, for example, nor do epithelial cells of the skin become nerve cells. Differentiation thus leads to a stable set of cellular properties. From observations of the relative stability of differentiated cells we recognize as a general principle the **stability of the differentiated state.** There are many exceptions to this rule, but they do not detract from its general applicability and utility. The exceptions are often dramatic, however. For example, dissociated individual cells of carrot root, growing in liquid culture, can form miniature plant embryos that, when transferred to an appropriate substratum, can produce a complete carrot plant (see Chapter 16).

Instability of the differentiated state of cells is less common in animals, but it does occur. For example, when the lens of the eye of a newt is removed, a new lens forms from cells of the iris, and in the clawed toad, *Xenopus,* a missing lens can regenerate from the cornea (see Chapter 16). These exceptions notwithstanding, it is a general rule that cells differentiate into a stable condition. The loss of a cell's differentiated properties is called **dedifferentiation.** Dedifferentiation of a cell, followed by its redifferentiation as another cellular type, is unusual, especially in animals.

2.2 The Course of Differentiation

Not yet knowing the facts of differentiation, the reader may feel that a discussion of generalizations about the course of differentiation is premature.

7

Nevertheless, it will prove helpful in the construction of a conceptual framework for the ordering of facts and ideas that are discussed later. The generalizations are prcsented here in terms of simple cases that can be grasped without a detailed knowledge of development; later, they reappear in more sophisticated contexts.

Most experiments on the course and causality of differentiative events have been carried out on embryos of animals, rather than on those of plants, because the early events of plant differentiation occur hidden under the coats of the seed, whereas those of most animals take place under more exposed conditions. Moreover, cells of plant embryos are rather rigidly locked in place by heavy cellulose walls, whereas those of animal embryos are free of heavy matrix materials and are easily shifted about. Thus, most examples that we bring to bear on the course of differentiative events in this section are taken from animal sources.

2.2.1 PROGRESSIVENESS. A great deal of simple observation about the course of development and much that has been revealed by experimentation can be categorized under the heading of the general concept **progressiveness of differentiation.** Watching a fertilized chick egg at intervals after the start of incubation shows us that the parts known to be present at the time of hatching are not at hand when incubation begins. Rather, they appear in rudimentary form and in regular sequence during the early stages of incubation. Early on the future brain can be seen, then the rudiments of the nose, eyes, and ears. Parts begin to appear in sequence from head to tail until the future organs of the entire chick are laid out in a prescribed pattern.

Similarly, one can follow the progressiveness of differentiation in a flowering plant. Soon after fertilization the dividing egg forms many cells. Among them there can be identified groups that will form the future stem, root, and leaves, all encased within the seed coats. Thereafter, one may follow visually the germination of the seed and the progressive appearance and differentiation of the parts of the mature plant.

2.2.2 IRREVERSIBILITY. Observation of the progress of differentiation also reveals a corollary of the principle of progressiveness, namely, the **irreversi-bility of development.** The dividing egg does not become a zygote once more; the leaf does not revert to a part of the twig; when the limb begins to emerge as a bud from the body wall of a frog embryo, it does not revert to body wall; the mature woman does not become a baby girl again.

2.2.3 EACH STEP IN DIFFERENTIATION IS DEPENDENT ON PRIOR EVENTS. Embryonic development is not simply the unfolding and growth of preexisting tissues and organs. Rather, it involves the progressive appearance, in successive stages, of structures not present in an earlier stage. Thus, the embryo proceeds from a condition of greater homogeneity to one of lesser homogeneity and specialization. This was recognized at the time of Aristotle, and proposed as the doctrine of **epigenesis** by Von Baer in 1828, based on his observations of the developing chick embryo. Implicit in the notion of epigenesis is that the events that occur at each stage of development do so by virtue of conditions created at the immediately prior stage. This implication is readily verified by observation and experiment. The simplest case is, of course, fertilization. Fertilization of the egg or other appropriate stimulus is the antecedent event for development in most plant and animal eggs. In the absence of this event the egg dies. Subsequent steps in development, however, offer more interesting and challenging opportunities for analysis of epigenesis.

It was noted in Chapter 1 that the future brain forms in advance of the nasal rudiment of the early embryo. In the early frog embryo it is a relatively simple matter to remove the brain-forming tissues microsurgically or to prevent their origin by appropriate chemical treatment. When this is done, the nasal organ as well as other sense organs of the head fail to form, even though the cells that should give rise to them are still present. Thus, the formation of the sense organs of the head requires the normal antecedent step, formation of the future brain. This example also suggests another principle, which is considered in some detail in Chapter 13, namely, the principle of dependent differentiation.

2.2.4 SELF-DIFFERENTIATION AND THE LOSS OF DEVELOPMENTAL OPTIONS. Whereas at each stage in development new structures arise from conditions created in the preceding stages, nevertheless,

8

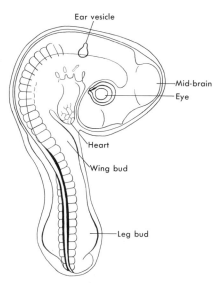

FIGURE 2.3 General form of the chick embryo after 3 days of incubation.

during the succeeding stage, these parts tend to become independent of the conditions that created them. That is to say, they become capable of continuing their differentiation in the absence of the conditions that made their original differentiation possible. Moreover, having acquired the power of self-differentiation, they lose the option of forming anything different. These ideas can be illustrated for the case of the developing limb of the chick embryo.

Consider first that the future limbs arise from **limb buds** that initially appear as thickenings of the body wall. The bud that forms the wing arises posterior to the future neck. The leg bud appears immediately anterior to the future tail (Figure 2.3). Between the two limb-forming areas is the future flank region. By marking cells of the body wall before the limb buds appear, and following their distribution, one can map at an early stage where they subsequently will appear. Having done so, one may remove the prospective wing or leg region of the body wall and graft it to a foreign location on another embryo. There it will grow and form a wing or a leg, depending on the position from which it was taken. The limb-forming areas thus apparently have no developmental options other than making wing or leg, respectively, from the earliest stages at

which one can identify their source. At present it is not clear when and under what circumstances the individual limb-forming areas acquire the power of self-differentiation.

Nevertheless, within each limb bud during its early development cells from different parts of the bud retain for a while the option of forming other parts than they normally would. For example, at the stage shown in Figure 2.3, that part of the leg bud nearest the body wall would normally form tissues of the thigh, adorned with distinctive thigh feathers. The furthest from the body wall would normally form foot parts armed with scales and claws. But if one cuts out a bit of prospective thigh tissue from the leg bud of an embryo at the stage shown in Figure 2.3, and grafts it to the tip of the bud, it will form scale-covered, clawed toes. At this early stage, therefore, future thigh tissues have a **prospective potency** that is greater than their normal **prospective fate**. If, however, the same operation is carried out at a later stage of development, such as shown in Figure 11.10E, for example, the prospective thigh tissue forms only thigh structures (Cairns and Saunders, 1954). These and similar experiments show that as limb development progresses, patterning factors (see Chapter 19) within the limb bud confer two concomitant effects on individual parts of the limb bud: each portion of the bud (1) acquires the ability to form its specific part of the structural pattern even when grafted to a new location and (2) loses a prior option to form any other part of the limb. In other words, as the limb parts acquire the power of self-differentiation, each has its prospective potency limited to its prospective fate.

Once an embryonic part is capable of realizing its prospective fate in the absence of the conditions that established that capability, it is said to be **determined**. Determination is thus a differentiative step in the progressive realization of the phenotypic fate of a part of the embryo. It is a step that limits the subsequent development of the part to a specific course of tissue and cellular differentiation.

The concept of determination is explored extensively in Chapter 13. Meanwhile, this and other concepts relative to the course of development are invoked from time to time in the context of specific differentiative events that are taken up in intervening sections of the book.

9

2.3 Summary

The fertilized egg gives rise to an organism with a multiplicity of kinds of cells, tissues, and organs. The state of their being different and the course of events leading to their differentness are subservient to the concept of differentiation, which is the concept under which most principles of developmental biology may be ordered.

The end of embryonic development finds cells, tissues, and organs in their final differentiated state, which is the species-specific morphological and functional condition. Differentiated cells are of discrete types, each with its peculiar repertory of biochemical activities and possible morphological configurations. We, therefore, recognize the principle of discreteness of differentiation. Also, the final state of differentiation is relatively stable; one differentiated type of tissue or cell does not change to another type, as a general rule. This attribute and related factors are considered under the general principle of stability of the differentiated state.

Examination of the course of differentiation leads to the recognition of several other principles, also. Outstanding is the principle of progressiveness, with its corollary irreversibility. During differentiation each step is requisite to and establishes a basis for the following one. Development is, therefore, epigenetic. Epigenetic development results in individual parts of the embryo acquiring the power to continue their differentiation in the absence of extrinsic morphogenetic influences. Systems that are thus emancipated are said to have the power of self-differentiation and are in a condition referred to as determined. Prior to determination in the course of differentiation the prospective potency of each part of the embryo is generally greater than its normal fate. Once determined, however, an embryonic part loses its options for alternative developmental routes. Its prospective potency is restricted to its prospective fate.

2.4 Questions for Thought and Review

1. What is meant by differentiation? Make a rational distinction between the concept of differentiation and the course of differentiation.

2. What does one imply by stating that a cell is in its final state of cytodifferentiation?

3. Define the principle of discreteness of differentiation. What is meant by the stability of the differentiated state?

4. In the text a number of generalizations are made about the course of differentiation. The principal generalization is that of progressiveness. What does this mean? Recall the other generalizations that were made concerning the course of differentiation. What are they? Define them. How are they related to the idea of progressiveness?

2.5 Suggestions for Further Reading

BERRILL, N. J., and G. KARP. 1976. *Development.* New York: McGraw-Hill, ch. 17.

EBERT, J., and I. M. SUSSEX. 1970. *Interacting Systems in Development.* New York: Holt, Rinehart and Winston, ch. 12.

WEISS, P. 1939. *Principles of Development.* New York: Holt.

2.6 Reference

CAIRNS, J. M., and J. W. SAUNDERS. 1954. The influence of embryonic mesoderm in the regional specification of epidermal derivatives in the chick. *J Exp Zool* 127:221–48.

T W O

Meiosis and the Significance of Sexual Reproduction

There is no discontinuity of life from one generation to the next in a species. In sexually reproducing organisms the next generation is produced by the fusion of living **germ cells**, or **gametes**, produced by the bodies of the parental generation. In organisms that propagate by nonsexual means, some employ continuous vegetative growth, as when a plant or sessile animal extends itself by means of stolons or runners; others, such as certain flatworms and annelids, produce a new generation by the fragmentation of the parent. Alternation of sexual and asexual generations occurs in some animals and in most plants. Alternation of generations is treated in Chapters 6 and 17, which deal respectively, with gametogenesis and fertilization in certain plants and with the larval forms and metamorphosis of certain animals. Asexual reproduction is, to some degree, a part of the life cycle of essentially all plants. It is closely allied to regenerative processes in animals. Asexual reproduction in animals is also considered briefly in connection with studies of their regeneration in Chapter 16.

For higher plants and animals there is usually a morphologically distinguishable male parent or male gamete-producing tissue. The female gamete, or egg, is usually much larger than, and considerably different in appearance from, the male gamete, or sperm. Higher plants and animals thus produce **heterogametes**, a term derived from the Greek, referring to their different appearance. But many kinds of organisms produce **isogametes**, that is, gametes that are morphologically alike. This book deals entirely with heterogametic sexual reproduction.

The first chapter in this section is concerned with the nuclear phenomena that are involved in **meiosis** and that are common to plants and animals. Meiosis, as most readers already know, brings about the reduction of chromosome numbers in the gametes to one half the number usually found in cells of the body. After this, a brief chapter treats some of the advantages that accrue to a species by virtue of its utilizing the sexual method of reproduction.

Plants and animals differ in the manner in which meiosis is related to the life cycle and to the production of gametes. These differences are treated in some detail in Chapters 5 and 7. For the present, however, let us realize that the cell populations from which the gametes arise are derived by mitotic divisions. Mitosis is the mechanism whereby each daughter nucleus obtains a copy of each chromosome of the parent cell. The cells of each kind of plant and animal have a characteristic number of chromosomes. For instance, there are 46 chromosomes in human cells, 8 in the cells of the common fruit fly, 16 in the common onion, and 28 in the tiger salamander. The nuclear changes involved in mitosis and the accompanying divisions of the cytoplasm (cytokinesis) are probably well known to most readers. For those who do not recall them, a brief outline of the salient features of the mitotic cycle is provided in Appendix A.

When the metaphase chromosomes of a cell are arranged in order according to size and shape, they fall naturally into pairs. Although the human being has 46 chromosomes in each cell, there are only 23 different kinds, or pairs, of chromosomes. The two chromosomes of each kind constitute a homologous pair; if the number of homologous pairs is n, then the number of chromosomes in the cell is $2n$. The n number of human chromosomes is 23. The shapes and sizes of chromosomes (i.e., their **karyotype**) are relatively constant in a species and enable cytologists to compare different kinds of plants and animals, or sometimes, genetically aberrant individ-

3

Meiosis

uals. In Figure 3.1 the full complement of chromosomes of the grass frog, *Rana pipiens,* is arranged in distinctive pairs.

Because the new generation in sexually reproducing organisms is formed by the union of two gametes, it is evident that if each brought the $2n$ number of chromosomes to the union, the resulting individual would have $4n$ chromosomes in each

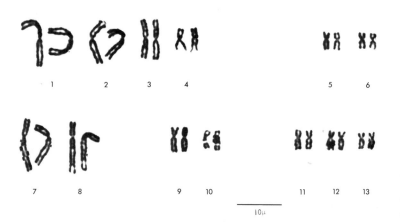

FIGURE 3.1 Karyotype of *Rana pipiens* prepared from a photograph of chromosomes in a tail cell from a young larva. The 26 chromosomes may be arranged in 13 distinctive pairs. [From M. A. DiBerardino, *Dev Biol* 5:101–26 (1962), by courtesy of the author and permission of Academic Press, Inc., New York.]

cell; in the next generation there would be offspring with $8n$ chromosomes, and so on, in geometric progression. Therefore, each gamete can contribute only one copy of each kind of chromosome at fertilization, if the species-specific number of chromosomes is to be maintained. Reduction of the chromosome number to n in the sexual cells is accomplished by two distinctive nuclear divisions, which constitute the process called meiosis. When the gametes unite at fertilization, the $2n$ number is restored, one chromosome of each kind coming from the male gamete and another set from the female gamete. The $2n$ number is referred to as the diploid chromosome number, and the n number as haploid. Each haploid set of chromosomes contributes one full complement of the hereditary material, deoxyribonucleic acid, or DNA. The amount of DNA in a sperm or egg is thus referred to as the $1C$ amount. The $2C$ quantity of DNA is usually present in each cell of the body.

The gross changes that occur during meiosis are treated in most elementary textbooks of biology. They are, nevertheless, briefly reviewed here in order to form the basis for a more extended discussion of the prophase of the first meiotic division. The analysis of the first meiotic prophase has been productive of a number of insights into the control of early developmental stages and thus deserves special consideration.

3.1 An Outline of the Stages of Meiosis

Meiosis consists of two specialized nuclear divisions (Figure 3.2). Prospective meiotic cells are derived from a population of cells that was produced by mitotic division. After mitotic division ceases, a period of premeiotic DNA synthesis begins (**premeiotic S**). During this period the DNA of each chromosome is replicated, just as in the S phase of the cell cycle, so that each chromosome comprises **sister chromatids,** which share the single **centromere,** or spindle fiber attachment. Each premeiotic cell now contains the $4C$ quantity of DNA.

During meiotic prophase condensation of the chromosomes begins, as in mitosis, but then, in contrast to mitosis, each chromosome comes to be alongside its homologue. The chromosomes thus lie in pairs. But because each member of each pair is composed of two chromatids (formed after premeiotic S), each of the pairs now is made up of four chromatids, although the centromeres are undivided. Except for the centromere regions, therefore, the paired homologues are quadripartite structures. Paired chromosomes in this condition were formerly called tetrads; now they are usually referred to as **bivalents.**

Shortly after pairing occurs the interchange of segments between chromatids of the paired homologues takes place. This is genetic **crossing over.** As the prophase condensation continues, crossing over causes the bivalents to assume rather bizarre shapes.

At metaphase each bivalent shows the original two centromeres, one provided by each homologue. At anaphase of the first meiotic division, these centromeres move apart, carrying the homologous chromosomes, including segments exchanged during crossing over, to opposite poles of the cell. At telophase the daughter nuclei of the first division thus receive as many chromosomes as there were bivalents at the metaphase plate. Each chromosome contains two chromatids, however, and the total DNA in each cell is the $2C$ amount, which is reduced in the next division.

After telophase the daughter cells seldom form a true interphase nucleus; rather, their chromosomes move directly to the metaphase spindle for the second meiotic division. In preparation for this division the centromeres divide so that each sister chromosome has its own centromere as the chromosomes separate during the second anaphase. Thus the two divisions of meiosis result in the formation of four nuclei, each with the n, or haploid, number of chromosomes and the $1C$ complement of DNA.

The subsequent history of these nuclei is quite different insofar as the formation of male and female gametes is concerned. At this point, however, we turn to a somewhat more detailed account of the first meiotic prophase, emphasizing several of its aspects that have contributed insights into the processes underlying early development at the molecular level.

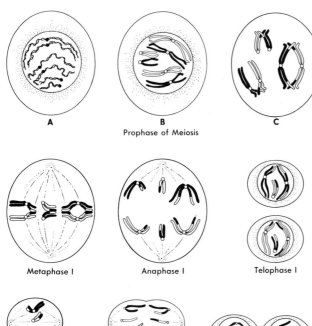

A   B   C

Prophase of Meiosis

Metaphase I   Anaphase I   Telophase I

Metaphase II   Anaphase II   Telophase II

FIGURE 3.2 Meiosis: a summary of nuclear events in a hypothetical cell containing three pairs of chromosomes. Homologous chromosomes, already having replicated their DNA, pair during early prophase (A) and then show the phenomenon of crossing over (B, C). Bivalents, which become visible at prophase, are separated, in two successive divisions into four cells each of which contains one representative of each kind of chromosome.

15

3.2 Salient Features of Meiotic Prophase

Prophase is divided into five phases, largely on the basis of observations made by means of the light microscope. The first phase is called **leptotene** (Gr., *leptos,* thin). It is marked by the appearance of long thin filaments of chromatin that show enlargements here and there along their length. In organisms that have only a few chromosomes, the threads may be well separated, and it can be recognized that their number corresponds to the number of diploid chromosomes.

The arrangement of the leptotene chromosomes is irregular in many species. In many kinds of plants

the leptotene stage is marked by the dense clumping of chromosomes in a pattern called **synezesis** (Figure 3.3A). In a number of animal species the ends of the leptotene chromosomes are directed toward a common region of the nuclear membrane. The central loops of the chromosomes spray out into the nucleoplasm from this point in a "bouquet" arrangement. The term **bouquet stage** is sometimes applied to this aspect of leptotene, although the bouquet configuration may persist into the next stage (zygotene), as shown in Figure 3.4B. During the leptotene stage the $4C$ complement of DNA is present, each chromosome having replicated prior to the onset of meiosis, but the double character of the chromatin threads is not visible at this time.

The onset of side-by-side pairing, or **synapsis,** of the homologues marks the beginning of the next

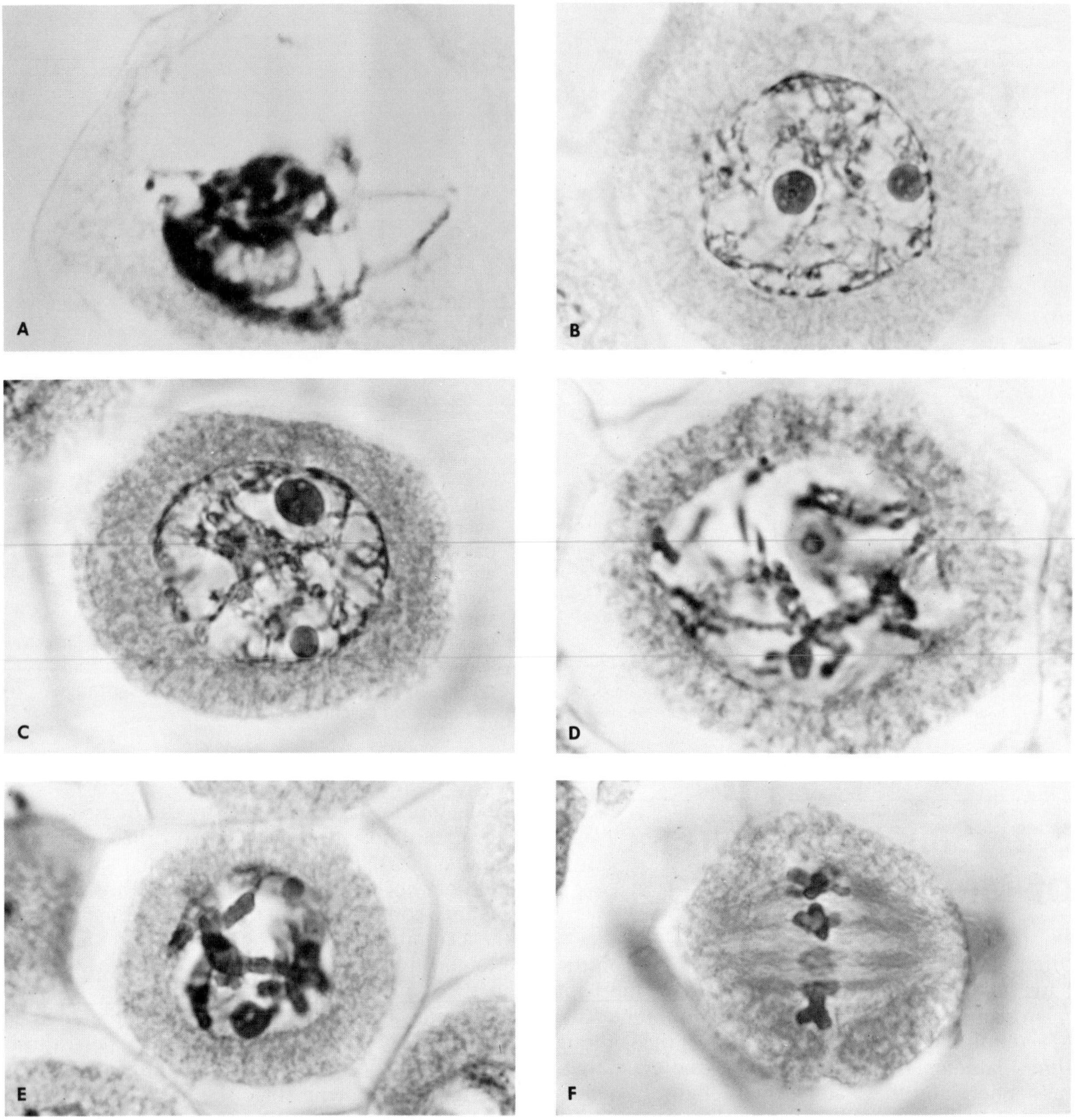

FIGURE 3.3 Stages of prophase (A–E) and metaphase (F) in the first meiotic division of the microspore mother cell in *Lilium:* (A) leptotene, showing the dense clumping of chromatin in the configuration of synizesis; (B) zygotene; (C) pachytene; (D) diplotene; (E) diakinesis; (F) metaphase. [Courtesy of Carolina Biological Supply Company.]

stage, **zygotene** (Gr., *zygos,* yoked or paired). Pairing of the homologous strands is quite precise, so that chromomeres of the partners come to lie in exact register. The process begins at any point along each chromosome and, once initiated, proceeds to "zipper" up the homologues along their entire length.

Examination of meiotic cells with the electron microscope reveals the development during zygotene of ribbons of electron-dense structures in the nucleus. Serial sections of nuclei show these ribbons to be equal in number to the haploid number of chromosomes. Usually the two ends of a strand are seen to be inserted in the nuclear membrane, either at some distance from each other (Figure 3.4A) or adjacent to each other and near the insertion points of the other ribbons (Figure 3.4B). These structures can also be pictured by electron microscopy of whole mounts of nuclei, selectively stained and spread by surface tension on an aqueous interphase (Moses, 1977). They are now clearly recognized as the paired homologous chromosomes, and in this configuration they are referred to as the **synaptonemal complex.** Each synaptonemal complex is composed of two longitudinally parallel lateral elements (Figure 3.5) whose cores are about 120 nm (1 nm = 10^{-9} m) apart, as measured in *Locusta* (Moens, 1969). These elements are connected to a median longitudinal element by transverse fibers.

The metabolic factors involved in the formation of the synaptonemal complex have been analyzed during male meiosis most intensively in the lily, but also in mice and rats. These factors were reviewed in detail by Stern and Hotta (1978, 1980). During the preleptotene synthesis of DNA approximately 0.3–0.4% (in the lily) of the chromosome remains unreplicated; it replicates at the time of zygotene. This DNA consists of stretches of about 10,000 nucleotide base pairs interdispersed among the bulk DNA on all the chromosomes. It has been designated as **Z-DNA.** When its replication is prevented by means of drugs, the synaptonemal complex fails to form and meiotic progress ceases. Likewise requisite to pairing are components of a heavy lipoprotein complex that appears at zygotene and persists through pachytene. Interference with lipid synthesis inhibits Z-DNA replication and also meiotic progress. Also associated with the heavy lipoprotein fraction extracted from cells at zygotene is a

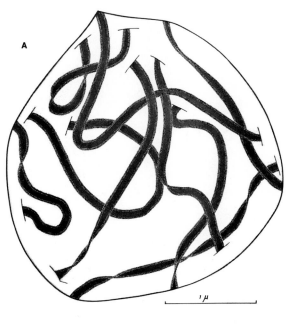

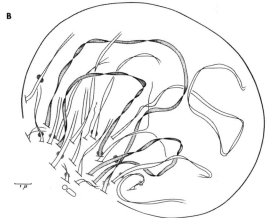

17

FIGURE 3.4 The disposition of paired homologues in the synaptonemal complex as seen in the marine mycetozoan *Labrynthula* (A) and in the grasshopper *Locusta migratoria* (B). Note the insertion of paired homologues in the nuclear membrane in each case and the "bouquet" arrangement of the chromosomes in *Locusta*. [A from P. B. Moens and F. O. Perkins, *Science*, 166:1289–91 (1969), by courtesy of Dr. Moens and with permission. Copyright 1969 by the American Association for the Advancement of Science. B from P. B. Moens, *Chromosoma* 28:1–25 (1969), by courtesy of Dr. Moens and permission of Springer-Verlag.]

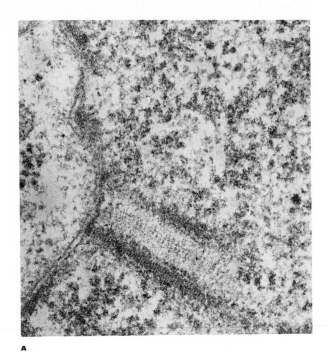

A

18

B

FIGURE 3.5 A: The ultrastructure of the synaptonemal complex as revealed by the electron microscope. Not all of the elements of the complex identified in the diagram (B) can be discerned in the photograph. [A from P. B. Moens, *Chromosoma*, 28:1–25 (1969), by courtesy of Dr. Moens and permission of Springer-Verlag. B redrawn with modifications from P. B. Moens in R. F. Grell, ed., *Mechanisms in Recombination*, Plenum, New York, 1974, p. 378.]

protein that facilitates strand matching, a primary requirement for chromosome alignment. This has been designated **r-protein,** or reassociation protein, by virtue of its ability to catalyze the reassociation of single-stranded DNA.

The synaptonemal complex persists into the next stage, which is marked by the visible thickening of the paired structures and is called **pachytene** (Gr, *pachy,* thick) (Figure 3.3C). The dual nature of each member of the pair is now clearly visible by means of the electron microscope. Now, there is also visible evidence of crossing over. This involves first a rupture of a chromatid from each homologous pair and then transposition of and fusion of the segments. Each of the chromatids thus acquires a corresponding piece of the other. This may occur at more than one point along the paired homologues.

Biochemical analysis of lily and of mouse male meiotic cells reveals that the onset of pachytene is marked by the appearance of an endonuclease that differs from endonucleases found in somatic cells. This enzyme catalyzes the breaking of internucleotide links in one strand of the DNA double helix and is specific for double-stranded DNA (Hotta and Stern, 1974; Stern and Hotta, 1974). The activity of this enzyme is apparently responsible for a period of DNA repair synthesis that occurs during pachytene and that is involved in crossing over. The repair synthesis of DNA has been shown to occur preferentially in regions of DNA containing base sequences that are moderately repetitive throughout the genome in both the lily and the mouse (Hotta et al., 1977).

As the pachytene stage comes to an end, the synaptonemal complex is no longer revealed by the electron microscope. This marks the beginning of **diplotene** (Gr., *diplos,* double), a stage during which the homologous chromosomes condense somewhat more and separate slightly. Separation is complete except at the points where crossing over has occurred. There the homologues are held together because sister chromatids tend to remain together even though segments are interchanged in crossovers (Figure 3.3D). The points of interchange are seen with ease during diplotene and are called **chiasmata** (sing., **chiasma;** Gr., *chiazlin,* to mark with a *chi,* i.e., X). The chromosomes progressively shorten and thicken during the later phases of diplotene, and this brings about the progressive displacement of chiasmata toward the ends of the bivalents, a process known as the terminalization of chiasmata.

During formation of the female gamete in many kinds of animals, diplotene is a prolonged phase, and the cell is essentially quiescent until proper hormonal conditions elicit the completion of meiosis. When a human female is born, most of her future egg cells (oöcytes) are in diplotene. Their number is on the order of a million. Most of these cells die without developing further. By the end of a woman's approximately 40 years of reproductive life, all her oöcytes are gone. Only about 500 will have developed into fertilizable eggs (Baker, 1963).

Most notably in animals, but possibly in some plants, diplotene includes a phase during which loops of chromatin project out into the nucleoplasm from the chromosomes, apparently exposing considerable segments of the genome for possible genetic activity. In this condition, chromosomes are referred as to as **lampbrush chromosomes** (see Figure 7.26). A lampbrush phase is seen in the development of many kinds of egg cells and, to a lesser extent, in the formation of sperm. The significance of this phase of meiosis is treated in more detail in Chapter 7.

Diplotene grades into the stage of **diakinesis** (Gr., *dia-*, apart; *kinesis*, moving). The homologues continue to shorten and thicken, usually showing a fuzzy appearance when stained for light microscopy, and their centromeres move farther apart. The homologues tend to be joined only at their ends, as chiasmata become terminalized (Figure 3.3E).

Diakinesis is the last stage of the first meiotic prophase. At the end of this stage the nuclear membrane breaks down and the chromosomes move to the metaphase plate (Figure 3.3F). Thereafter, the first division of meiosis is completed and the second proceeds as previously described.

ried the diploid number of chromosomes into the zygote. What is a geometric progression?

2. Recall the stages in the meiotic cycle in plants and animals. If you do not know these, consult elementary biology textbooks. At what stage in the cycle does DNA synthesis occur?

3. What do the terms haploid and diploid refer to with respect to chromosome number in a cell? With respect to its content of genetic information?

4. What is a karyotype? Of what value might it be to examine the karyotype of a prospective human parent? Of an infant? Consider the fact that genes are on the chromosomes and that they determine the characteristic of the individual.

5. As applied to the nucleus distinguish between the use of n and C, as in such terms as $2n$ and $2C$, $4n$ and $4C$, and so on.

6. Review the five stages in the first meiotic prophase. What are the features that distinguish each stage?

7. Of what significance might be the synthesis of one or more special proteins during zygotene and pachytene? How would you detect the occurrence of such specific synthesis? You may be better able to answer this question if you have studied biochemistry and molecular biology.

8. What is the synaptonemal complex? How is it related to the chromosomes?

19

3.4 Suggestions for Further Reading

CHANDLEY, A. C., Y. HOTTA, and H. STERN. 1977. Biochemical analysis of meiosis in the male mouse. *Chromosoma* 62:243–53.

HOTTA, Y., and H. STERN. 1978. Absence of satellite DNA synthesis during meiotic prophase in mouse and human spermatocytes. *Chromosoma* 69: 323–30.

STERN, H., and Y. HOTTA. 1977. Biochemistry of mitosis. *Philos Trans R Soc Lond* [B]277:277–94.

3.3 Questions for Thought and Review

1. Chromosome numbers would increase in geometric progression from generation to generation if gametes car-

3.5 References

BAKER, T. G. 1963. A quantitative and qualitative study of germ cells in human ovaries. *Proc R Soc Lond* [B] 158:417–33.

HOTTA, Y., A. C. CHANDLEY, and H. STERN, 1977. Biochemical analysis of meiosis in the male mouse. II. DNA metabolism at pachytene. *Chromosoma* 62:255–68.

HOTTA, Y., and H. STERN. 1974. DNA scission and repair during pachytene in *Lilium. Chromosoma* 46:279–96.

MOENS, P. B. 1969. The fine structure of meiotic chromosome polarization and pairing in *Locusta migratoria* spermatocytes. *Chromosoma* 28:1–25.

MOSES, M. J. 1977. Synaptonemal complex karyotyping in spermatocytes of the Chinese hamster *(Cricetulus grisens)*. I. Morphology of the autosomal complex in spread preparations. *Chromosoma* 60:99–125.

STERN, H., and Y. HOTTA. 1974. Biochemical controls of meiosis. *Annu Rev Genet* 7:37–66.

STERN, H., and Y. HOTTA. 1978. Regulatory mechanisms in meiotic crossing-over. *Annu Rev Plant Physiol* 29: 415–36.

STERN, H., and Y. HOTTA. 1980. The organization of DNA metabolism during the recombinational phase of meiosis with special reference to humans. *Mol Cell Biochem* 29:145–48.

The significance of sexual reproduction is treated at length in textbooks of genetics and evolution and therefore merits only very brief attention here. But, because sexual and asexual forms of reproduction have such markedly different effects on the genetic endowment of successive generations, and because this book places considerable emphasis on the action of genes during development, it is appropriate to devote some remarks to the significance of sexual reproduction as contrasted to that of asexual means of propagating species.

The survival of a species requires not only that it be suited to its environment but also that it be capable of adapting to new environmental circumstances, such as changing climatic conditions, modified predator-prey relationships, and so on. This capability requires that there be present within the various breeding populations of the species a reserve of genes of appropriate frequencies and combinations on which natural selection can operate. The gene pool available for natural selection depends ultimately on the frequency of gene mutation at various loci on the chromosomes.

4

Significance of Sexual Reproduction

at anaphase along with a centromere and sequence of genes largely of maternal origin. Moreover, at anaphase of the first meiotic division, chromosomes are not sorted into sets according to their maternal or paternal origin. Rather, centromeres go randomly to either pole of the spindle regardless of their parental origin. Therefore, each haploid gamete will have a total genetic constitution different from the gene sets possessed by either of its parents, and it will unite with another gamete, equally distinctive as to its genome. Variability of gene combinations in succeeding generations is assured, and, therefore, the chances for a population (assuming it to be large enough and randomly breeding) to have favorable gene frequencies and combinations for survival of the species are enhanced.

4.1 Meiosis and Fertilization Enhance Genetic Variability

The origin of successive new generations from gametes produced through meiosis by randomly breeding parents assures that the pool of genetic variability in a population will be passed on to the next generation. Each zygote receives two complete species-specific sets of matched gene loci, one set from each parent. Each set will differ, however, in the pattern of loci showing gene mutation, so that the genetic properties of the offspring will reflect a blending of characteristics inherited from each parent.

The process of meiosis, reviewed in the last chapter, gives new gene combinations. Recall that, during pachytene, segments of homologous chromosomes are interchanged by crossing over. Therefore, a chromosomal segment that enters meiotic prophase in association with a centromere of paternal origin may, as a result of crossing over, separate

4.2 Genetic Variability Is Restricted in Asexually Reproducing Populations

In organisms that reproduce asexually, by whatever means, the new individual inherits exactly the same genome as its parent. An entire population that arises from a single individual by successive asexual generations is called a **clone.** All members of a clone are, therefore, genetically alike. A clonal population thus lacks the genetic reserve inherent in a sexually reproducing population and can be thought of as occupying an evolutionary "blind alley." That is, it has not the genetic reserve requisite to evolutionary adaptation. The only source of genetic variation is the infrequent gene change known as **mutation.**

4.3 Questions for Thought and Review

1. What is meant by the fitness of a species for survival in a particular environment? This question has many implications. List as many aspects of fitness as you can.

2. What are the chief ways in which meiosis acts as a mechanism for survival of a species?

3. Reference is made in the text to randomly breeding populations. Of what importance might be the matter of randomness of breeding activity in the survival of a species?

4. What is a clone? In what respects are members of a clone alike?

THREE

Production of Gametes and Initiation of Development

For the successful production of a new generation, the parental generation must first provide gametes in sufficient quantity and under such circumstances that fertilization can take place. Moreover, this must occur under conditions such that an adequate number of zygotes may survive and develop to maturity.

In temperate zones reproduction in nature usually occurs in seasonal cycles, and the young tend to emerge in the spring and summer when conditions for nourishment and growth are optimal. Organisms living in the tropics often show year-round breeding activity, but reproduction is tied to the annual cycles of rainfall in many forms. In higher plants and for most higher animals, including both invertebrates and vertebrates, reproductive cycles and breeding activities are under the immediate control of **hormones**. The hormones involved in reproduction, in turn, are usually produced or released in response to cues from the environment. In temperate zones these cues are largely provided by changing day lengths correlated with the changing seasons. Reproductive cycles and their hormonal controls are treated in Chapter 5.

Chapter 6 is concerned with gametogenesis and fertilization and early development in some plants. The physiology of gametogenesis and fertilization in plants, which is generally less well understood than that in animals, receives little attention. Instead, this chapter deals with the contrasting roles of alternating sporophyte and gametophyte generations in some representative plants, emphasizing the production of the germ cells by the gametophyte and the description of their union, and with the early events in embryo formation in higher plants.

Gametogenesis and fertilization in animals are treated at some length in Chapters 7 and 8. The origin of sperm cells, **spermatogenesis**, is considered with particular emphasis on the ultrastructural changes and gene action involved in the process. The variety of animal eggs and their membranes is noted, and special attention is given to the role of hormones and to gene action and protein synthesis during growth and maturation of the eggs and shedding of the gametes. The morphological changes at fertilization and activation of new physiological patterns and patterns of macromolecular synthesis are treated, with due concern for the insights that modern analytical methods have given us with respect to these processes.

Comparative aspects of early stages in animal development are considered in Chapter 9, and Chapter 10 is devoted to the important problem of the patterns of gene action in early development.

The reproductive organs, or **gonads**, of male and female multicellular animals are the **testes** and **ovaries**, respectively. Reproductive organs of plants are of various kinds, depending on the taxonomic group, but for higher plants, they generally consist of modified leaves. In flowering plants, which are the only ones considered in this chapter, the flower parts contain the reproductive organs. **Stamens**, each with a terminal **anther**, comprise the male organs. The female organ is the **pistil**, consisting of **stigma**, **style**, and **ovary**. The pistil originates, according to the kind of plant, by the fusion of one or more modified leaves known as **carpels**. The anthers produce the **microspores** from which the pollen grain, the **male gametophyte**, is derived, and in the ovary are found ovules in which the **female gametophyte**, or **embryo sac**, arises (Figure 5.1). The origin and significance of the male and female gametophytes are discussed further in the next chapter, which deals with fertilization and early development in plants.

Reproductive cycles involve the cyclic appearance of floral parts in plants and the cyclic maturation of germ cells in the gonads of animals. **Peren-**

5

Reproductive Cycles and Their Hormonal Control

nial plants, often requiring some years to reach sexual maturity, thereafter come into flower annually during the spring, summer, or fall. **Annual** plants develop from seed that germinates in the spring; they come into flower, produce seeds, and then die. **Biennial** plants, which are found in the temperate zones, are unable to complete their life cycle in one year, their buds requiring a more or less prolonged period of cold at 5°C, or lower, if they are to form flowers when growth resumes.

Reproductive cycles among animals, likewise, show great variety. Some animals, like the Pacific salmon, spend several years in active growth in one or more larval stages before coming to sexual maturity. They then breed once and die at the spawning grounds. Among insects it is common to find that several generations may be produced during warm months, whereupon most adults die, leaving next year's crop of adults as overwintering eggs or dormant larvae, pupae, or adults (see Chapter 17).

Most kinds of vertebrates have rather prolonged reproductive lives and, in the wild, are **seasonal breeders.** As the breeding season approaches, the gonads enlarge and produce mature gametes. At the end of the breeding period the gonads regress in size and mature gametes are absent. In the temper-

25

FIGURE 5.1 Diagrammatic representation of the complete floral parts of an angiosperm, represented as cut in a medial section and showing half of the flower parts. Within the carpel a single ovule is shown within the ovary. [Reprinted with permission of Macmillan Publishing Co., Inc., from *Development in Flowering Plants* by J. G. Torrey, p. 54. Copyright © 1967, J. G. Torrey.]

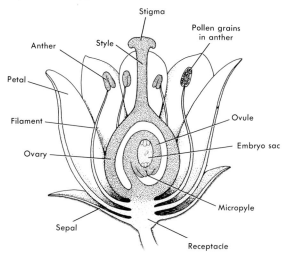

TABLE 5.1 Breeding Seasons of Several North American Mammals in the Wild

Badger	August–September
Black bear	June
Coyote	February–April
Mule deer	November–December
Red fox	December–February
Short-tailed shrew	March–October
Muskrat (Iowa)	April–August
Woodchuck	February–March
Rocky Mountain goat	November

ate zones most lower vertebrates mate only once or only during a brief period. Among some birds, if a first hatch fails, mating may be repeated and a second clutch of eggs may be laid.

Most wild mammals in temperate zones are seasonal breeders. During the breeding season the female is usually receptive to the male only during restricted periods of **heat,** or **estrus** (Gr., **estrus,** mad desire). The period of receptivity corresponds approximately with the time of **ovulation** (i.e., release of the mature egg from the ovary). In some cases the female may undergo several periods of estrus and ovulation during the breeding season. Such animals are **polyestrous.** Estrus cycles cease during pregnancy. Depending on whether or not the eggs are fertilized, on the length of the gestation period, the length of the breeding season, and the length of the estrous cycle, the female may bear only one or several litters during a breeding season. Breeding seasons for several mammals are shown in Table 5.1.

Monestrous mammals show only one period of heat during the breeding season. This period may be brief, regardless of whether or not mating occurs, as in the American black bear, or it may be prolonged, terminating only when mating occurs, as in the ferret.

Some mammals are **continuous** breeders. This is true of human beings and most nonhuman primates, of most domesticated animals, and of some tropical wild animals. Continuously breeding males produce sperm throughout the year and generally breed with any receptive female. In continuously breeding populations, females that are not impregnated undergo repetitive cycles of ovulation and periods of sexual receptivity throughout the year.

In human beings and other primates, the estrus cycle is replaced by the menstrual cycle (Figure 5.2). Ovulation occurs approximately in the middle of the cycle, and if pregnancy does not occur, the uterine lining, greatly thickened in anticipation of pregnancy, is sloughed off about 2 weeks later. This is accompanied by the rupture of blood vessels, causing the **menses,** menstrual bleeding.

Reproductive activity in human beings occurs throughout the menstrual cycle, although the period of menses tends to be avoided. There seems to be a tendency among human females, also, for greater sexual activity (Udry and Morris, 1968, 1977), or at least, greater eroticism (Adams et al., 1978), at around the time of ovulation. This may be related to other physiological changes that occur at the time of ovulation. Thus, visual and olfactory acuity are high at midcycle, and sensitivity to pain and fatigue are diminished (Diamond et al., 1972). Oral contraceptives, which prevent ovulation, but not other contraceptive measures, tend to increase human female thresholds to sexual arousal at the time ovulation would normally occur (Adams et al., 1978).

26

FIGURE 5.2 Diagram of the changes in the human endometrium and serum hormone levels during a typical human menstrual cycle.

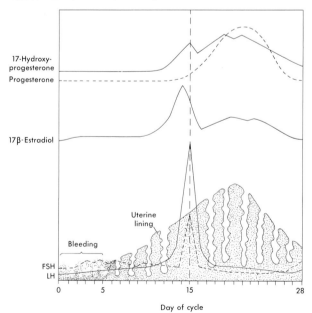

Among nonhuman primates mating tends to be restricted, although not exclusively, to the phase of the menstrual cycle during which ovulation occurs. This may be related to olfactory signals emanating from the female, as discussed later in this chapter, or to other signals, such a cyclic swelling and reddening of the skin in the region of the female genitalia correlated with the time of ovulation.

In human beings, other primates, and many laboratory animals ovulation occurs spontaneously at a relatively fixed time in the menstrual or estrus cycles regardless of mating activity. This is referred to as **spontaneous ovulation.** Among many continuously breeding mammals and among both monoestrous and polyestrous seasonal breeders ovulation occurs in response to copulation. This is called **induced ovulation,** or **reflex ovulation.** Copulation is obligatory for ovulation to occur in such animals as the black bear, domestic rabbit, cat, and ferret. Ovulation is induced by copulation under some circumstances in females that normally ovulate spontaneously. Induced ovulation is thought to occur sometimes in human females, as is treated further later on.

The remainder of this chapter is devoted to the role of hormones in the breeding cycles of plants and animals and their regulation by environmental stimuli. Much less is known about the reproductive physiology of plants than of animals, so the next section, which deals with the control of flowering, is relatively brief.

5.1 Light and the Control of Flowering

In certain species, regardless of when seeds are planted in the greenhouse in late winter or early or late spring, plants of different ages come into flower at about the same time. Some kinds are seen to flower in the spring as the days are lengthening; others in the later summer or fall, when days are shortening.

5.1.1 PHOTOPERIODISM. If a plant requires a day length, or photoperiod of longer than some critical duration, it is called a **long-day plant.** The critical

period actually may be fairly short. It is 10 hours, for example, in *Hyoscamus niger,* or henbane. Some other long-day plants are spinach, lettuce, and radish. Long-day plants do not seem to require a dark period, for they bloom under continuous illumination, and darkness can actually be somewhat inhibitory. Other species of plants require that the days be less than some critical length before flower parts start to form. These are called **short-day plants.** The common cocklebur, *Xanthium,* is an example. It will not bloom until the period of illumination diminishes to $14\frac{1}{2}$ hours or less in late summer, as the nights grow progressively longer (Figures 5.3 and 5.5).

In fact, short-day plants such as the cocklebur probably should be called **long-night plants.** Plants prevented from blooming by illumination for greater than $14\frac{1}{2}$ hours during a 24-hour day can be induced to form floral parts by exposure to a single long night before being restored to long-day conditions once more. Other short-day plants may require several long nights for floral induction.

There are also **day-neutral** or **indeterminate plants,** which flower more or less without respect to day length. Such plants include tomato, pepper, and most cultivated varieties of cotton. In such plants whether flowering occurs may depend on other factors, such as the number of nodes on a stem.

27

5.1.2 IS THERE A FLOWERING HORMONE? In flowering plants buds that would otherwise produce branches and leaves switch over to the making of flower parts in response to photoperiodic induction. The inductive stimulus is apparently generated in the leaves, however, and transported to the buds. Using the very sensitive cocklebur and appropriate opaque shielding, one may expose all of the plant except for one leaf to long-day conditions (Figure 5.3). The exceptional leaf is then submitted to the inductive short-day regime. This suffices to cause buds all over the plant to flower (Hamner and Bonner, 1938). Moreover, a leaf from an induced long-day plant can be grafted to a noninduced short-day plant of a different species and cause flowering (Hodson and Hamner, 1970).

These findings indicate that the light-receptor system is probably in the leaves of the plant, that leaves manufacture something that is transported to the buds to cause them to flower, and that what

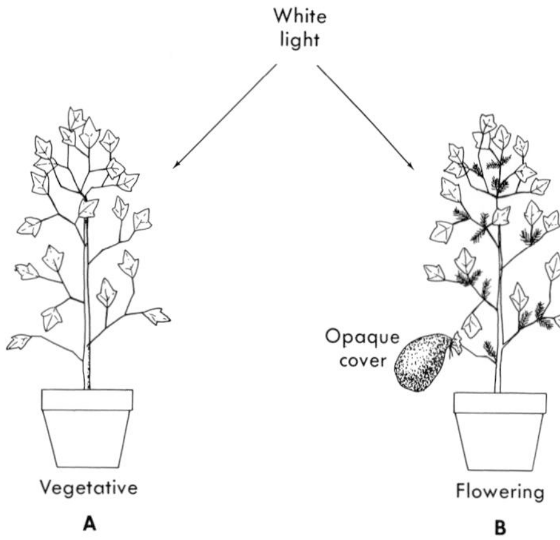

White light

Opaque cover

Vegetative

Flowering

A

B

FIGURE 5.3 Induction of flowering by exposure of a single leaf of the cocklebur, a long-night plant, to short days. A: Held at a day length of 14½ hr or more, the plant fails to flower. B: If as much as a single leaf is placed on a long-night regime by means of an opaque covering, the plant will flower. [Based on data of K. C. Hamner and J. Bonner, *Botan Gaz* 100:388–31 (1938).]

28

is transported is not species specific and apparently is the same in short- and long-day plants. These observations and deductions are compatible with the notion that there is a flowering hormone, **florigen.** As yet, however, no substance that can be identified as florigen has been isolated, nor has florigen-like activity been routinely demonstrated in extracts of leaves from induced plants.

Gibberellins are naturally occurring chemically-related plant hormones most easily extracted from immature seeds and fruits of some plants.

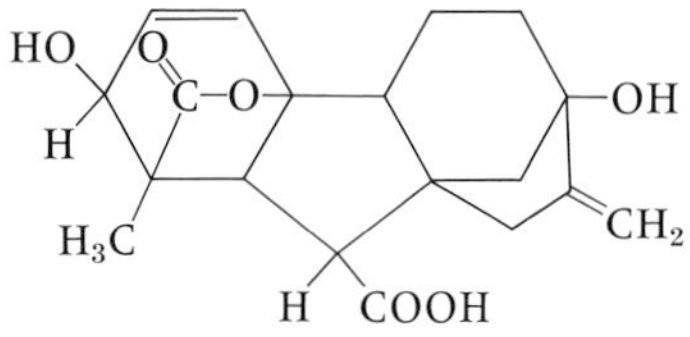

Gibberellic acid
(GA₃)

Gibberellic acid, the first of the gibberellins to be identified, is soluble in water, especially as a potassium salt, and has many remarkable effects, among which is to cause the dramatic elongation of stems of genetically dwarf plants. Lang (1957) reported that the long-day plant *Samolus* grown under short-day conditions would flower if treated daily for 3 weeks with gibberellin. Gibberellin does not affect flowering in short-day plants and thus is probably not the flowering hormone. It does cause rapid stem elongation in *Samolus* and thus may be tied up in some complex way with floral induction.

5.1.3 PHYTOCHROME, THE PHOTOPERIOD RECEPTOR. Recall that for the short-day plant cocklebur, the inducing condition is actually a long night, and that a single long night interposed in a regimen of long-day conditions is sufficient for floral induction. If, however, the dark period is interrupted at its midpoint by a brief flash of light, the inductive effect of the long night is negated, and flowering does not occur. Conversely, if a long-day plant that is being carried on an appropriate noninductive light regimen has its dark period interrupted by brief illumination, floral induction occurs (Figure 5.4).

The wavelength of light most effective in the above tests was found to be in the red region of the spectrum between approximately 620 and 680 nm (Parker et al., 1949). It would be presumed, there-

FIGURE 5.4 Reaction of short- and long-day plants under the influence of varying light regimes. Plants that require long nights for flowering, as seen on the right, do not flower if the night is of less than of some critical duration. Nor will they flower if a longer night is interrupted by a flash of red or white light. The effect of such a flash is negated, however, if it is followed by a flash of light in the far red side of the spectrum. The opposite effects are produced when long-day plants, as seen on the left, are subjected to a similar regime.

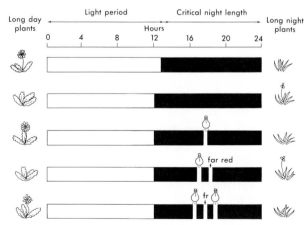

fore, that the photoperiod receptor would contain some pigment absorbing most effectively in the red region of the spectrum. This conceivably could be an open-chain tetrapyrole compound, which would absorb strongly at 620–680 nm. Chlorophyll, a closed-ring tetrapyrole, absorbs at this wavelength, but it also has a peak of absorption at 430 nm, a wavelength that is ineffective in the control of floral induction.

5.1.3.1 The Red–Far Red Effect.

A clue to the nature of the absorbing pigment **phytochrome** came through the application of earlier findings (Flint and McAlister, 1937) of the effects of red light on the germination of lettuce seeds. Some varieties of lettuce seeds germinate poorly, if at all, in the dark, but, if illuminated after they have imbibed water, they germinate rapidly. An **action spectrum** (percent germination plotted against wavelength) for the promotion of germination shows a peak in the red region of the spectrum just where maximal effects on floral induction occur. If, however, lettuce seeds are stimulated to germinate by red light and then treated with light in the far-red region, at about 720 nm, the effects of red light are reversed. This reversal, in turn, is negated by a subsequent application of red light.

The same red–far red effect was then found for floral induction, also (Figure 5.4). Flowering can be promoted by red light, subsequently reversed by far red, and the reversing effect negated by a later exposure to red light once more. This suggests that phytochrome exists in two photoreversible forms that absorb in the visible region of the spectrum: one in red region at 620–680 nm and the other in the far red at around 730 nm. The data are consistent with the idea that there is an inactive form of phytochrome, P_r, which is converted to an active form P_{fr} when it absorbs red light. The latter may be transformed back to P_r by absorption of light at 730 nm or by thermal processes proceeding in the dark (reviewed by Galston and Davies, 1970).

By means of ingenious circuitry and photocell arrangements, it has now become possible to show that exposure of leaves to light at 660 nm results in a decrease of absorbing power at that wavelength (diminution of P_r), and an increase at 730 nm. Conversely, irradiation with light at 730 nm (increase of P_{fr}) results in a decrease in optical density (i.e., decreased absorption) at this wavelength and a concurrent increase at 660 nm.

Finally a pigment, phytochrome, with properties similar to those described above, has been extracted and examined in the test tube. It is a protein with a molecular weight of about 60,000 containing a prosthetic group of the open-chain tetrapyrole type (Butler et al., 1965).

The mechanism of action of phytochrome is unknown. Besides its action in floral induction and germination, it also affects the growth of leaf and stem, rhythmic movements of leaves, and the behavior of plastids. It has been detected in the vicinity of the nuclear membrane (Galston, 1968) and may be present in the cell membrane. Conceivably, transformation of the phytochrome affects membrane permeability, and all subsequent reactions are secondary to this primary effect. If so, the mechanisms of this effect and the chain of subsequent events constitute a challenging problem.

5.2 Control of Reproductive Cycles and Mating in Animals

Animals come to reproductive maturity and engage in cyclic or continuous reproductive behavior under the influence of hormones. In the vertebrates, with which this section is chiefly concerned, these hormones originate principally in the brain, in the anterior lobe of the pituitary gland, and in the gonads. What follows concerns first the origin and nature of these hormones and, then, how extrinsic factors can affect their secretion and subsequent effects on mating behavior.

5.2.1 REPRODUCTIVE HORMONES OF THE VERTEBRATES.

Here we deal with three groups of hormones that control the cyclic activities of the vertebrate reproductive system. These groups are: the **pituitary gonadotropins,** which are produced in the **pars anterior** of the pituitary gland; the male and female **sex hormones,** which are produced in the gonads in response to the action of pituitary gonadotropins; and **hypothalamic neurohormones,** which originate in the floor of the third ventricle of the brain and are transmitted directly to blood vessels supplying the pituitary gland. These groups of hormones are found in all vertebrate animals. Male and female animals produce the same hormones, which have, to a large degree, comparable effects in the maturation and operation of reproductive cycles. The sources of these hormones and their interrelationships are diagrammed in Figure 5.5.

5.2.1.1 The Pituitary Gonadotropins. The principal pituitary gonadotropins are named for the functions first discovered for them in the mammalian ovary. They are **follicle-stimulating hormone (FSH)** and **luteinizing hormone (LH)**. They are **glycoproteins** (i.e., proteins with attached carbohydrate residues) with molecular weights of about 30,000. Each is composed of a protein core with branched carbohydrate side chains, which usually terminate in sialic acid. The protein of each hormone is made up of two nonidentical subunits designated α and β subunits, respectively. The α subunits are virtually indentical, but the β subunits are distinctive (reviewed by Vaitukaitis et al., 1976) and are responsible for differences in the specificities of the two hormones.

In the female FSH stimulates the growth of the **ovarian follicles,** the complex made up of the egg cell and surrounding follicle cells (Figure 5.6). In the male, FSH is requisite to the initiation and early progress of meiosis in the **seminiferous tubules** of the testes (see Figure 7.4). Without the hormone, sperm production ceases.

Under the influence of LH, egg cells resume their meiosis, become capable of fertilization, and

30

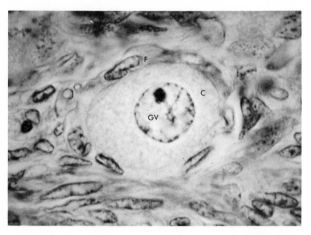

FIGURE 5.6 Mammalian oöcyte in primordial follicle. The germinal vesicle, GV, and cytoplasm, C, are small, and a layer of follicle cells, F, surrounds the oöcyte.

are **ovulated,** that is, released from the ovary. This will be described more fully in Chapter 7. In mammals the follicle cells left behind at ovulation become the **corpus luteum,** which then produces **progestins,** described later. In the male LH binds to the **interstitial cells** of the testes and is involved in the production by these cells of male sex hormones. Therefore, it is often called **interstitial cell-stimulating hormone.** The interstitial cells are frequently referred to as **Leydig cells** (Figure 5.7).

Another gonadotropic hormone of the pituitary gland is **prolactin,** a protein with a molecular weight of 25,000. It is produced by members of both sexes in all major vertebrate groups and has varied functions, affecting such matters as development of the male reproductive tract, migration, nesting, care of the young and the secretion of milk. In the rat and sheep it definitely has luteotrophic activity. That is, it promotes the formation of corpora lutea and the secretion of progestins. For this reason, it is sometimes called **luteotrophic hormone,** or LTH. In males it may act on the interstitial cells to cause them to make more male sex hormone.

5.2.1.2 The Sex Hormones—Gonadal Steroids. Sex hormones are produced in the gonads in response to the gonadotrophic hormones. The hormones of the gonad are **steroids** and the forms in the male and female are closely related chemically, being derived from **cholesterol** in relatively simple steps (Figure 5.8).

FIGURE 5.5 Sources and interrelationships of the principal hormones involved in cyclic reproduction in vertebrates.

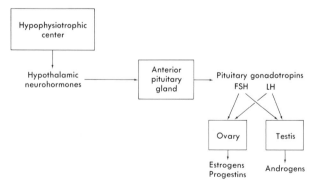

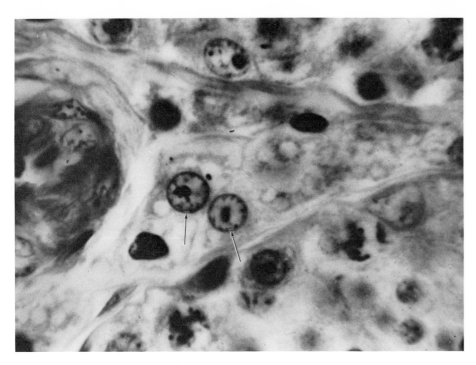

FIGURE 5.7 Leydig cells, whose nuclei are indicated by arrows, are found in connective tissue between seminiferous tubules of the testis (compare Figure 7.4).

31

Cholesterol

Ovarian steroids are classified into two groups on the basis of the time at which they are produced and act in the reproductive cycle. They are **estrogens** and **progestins.** Estrogens are responsible for the maturation of female sexual characteristics and the elicitation of female mating behavior. They are produced chiefly during the period of growth of the egg and its follicle cells. This period is called the **follicular phase** of the ovarian cycle. They also stimulate synthesis of yolk proteins by the liver in vertebrates having large-yolked eggs, as is treated in more detail in Chapter 7.

The principal estrogens are **estrone, 17β-estradiol,** and **estriol** (Figure 5.8). The term *estrogen* is often used to indicate an active female hormone preparation without specifying which molecular species or combination thereof is involved. Estrogens are produced in the ovary under the influence of FSH. In mammals they are chiefly synthesized in ovarian follicles.

Progestins are sometimes called **progestagens** or **progestational hormones.** In mammals they are essential to maintenance of the uterine lining for implantation and support of the embryos (Chapter 9). **Progesterone** is the chief progestin. It has been found in the ovaries and blood of representatives of all vertebrate groups, along with less significant amounts of related compounds having progestational activity. In female mammals **17α-OH-progesterone** is found in the blood plasma along with progesterone, but in much lower concentrations throughout estrus and menstrual cycles. In the human female and other primates it rises to a peak at the time of ovulation but quickly drops off. Thereafter, progesterone rises to a very high peak as a result of its production in great quantities by the corpus luteum (Figure 5.2). The period after ovulation is usually referred to as the **progestational phase** of the ovarian cycle.

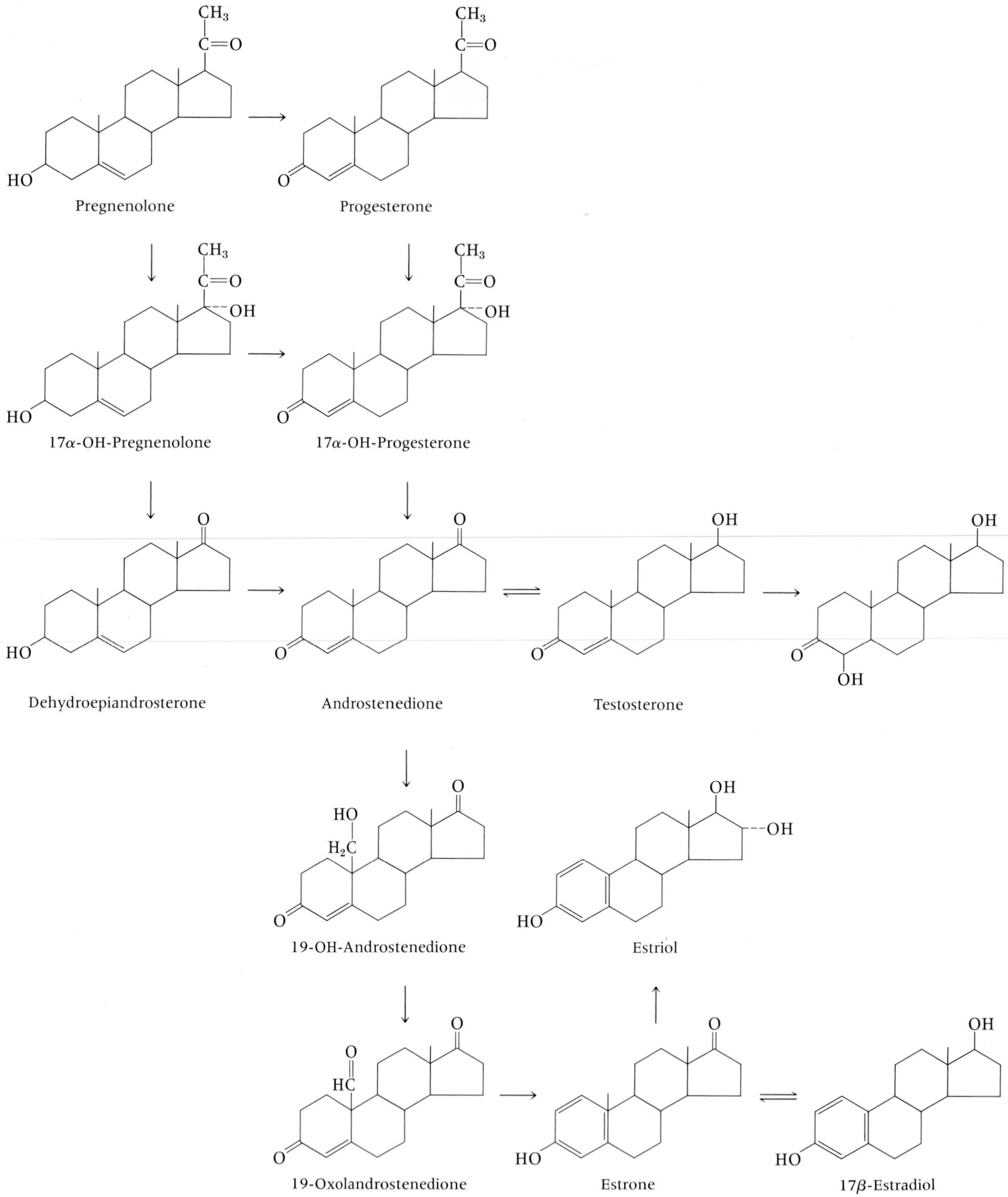

Pregnenolone

Progesterone

17α-OH-Pregnenolone

17α-OH-Progesterone

Dehydroepiandrosterone

Androstenedione

Testosterone

19-OH-Androstenedione

Estriol

19-Oxolandrostenedione

Estrone

17β-Estradiol

FIGURE 5.8 (opposite) The chief routes of synthesis of the steroid sex hormones from pregnenolone. Pregnenolone is formed by replacement of the side chain at position 17 in cholesterol with an acetate group.

The gonadal steroids of the male are called **androgens** (Gr., *andros,* man; L. *genare,* to beget). They are produced in the interstitial cells (also known as Leydig cells) of the testis (Figure 5.7) under the influence of LH. The principal hormone secreted by the testis is **testosterone,** although smaller amounts of **androstenedione** and **dehydro-epiandrosterone,** both showing weak androgenic activity, are also produced (Figure 5.8.). The ovary, too, secretes significant amounts of the latter two hormones, but very little testosterone. These weaker androgens may be converted to testosterone. Recent evidence suggests that in some target organs, for example, the prostate, testosterone is converted to **dihydrotestosterone,** which is the effective hormone (Figure 5.8). The androgens, in sum, and in whatever their effective form in each tissue, are responsible for maturation of the morphological characteristics of the male sex and the male pattern of sexual behavior.

5.2.1.3 The Hypothalamic Neurohormones.
The release of gonadotropins (and other trophic hormones of the anterior lobe of the pituitary), and probably their synthesis also, is regulated by factors that are produced in the medial basal region of the **hypothalamus** (floor and lower wall of the third ventricle of the brain). Because these factors arise in neural tissue they are called **neurohormones.** The neurohormone that brings about the release of luteinizing hormone (LH) is referred to as **luteinizing hormone–releasing hormone** (LH–RH) or **factor** (LH–RF), and that which controls release of FSH is called **follicle-stimulating hormone–releasing hormone** (FSH–RH) or **factor** (FSH–RF). Secretion of prolactin is controlled in mammals chiefly by a **prolactin-release–inhibiting hormone** (PR–IH) or factor (PR–IF). There may also be a **prolactin-releasing hormone** in mammals, but the evidence for its existence is incomplete. In birds prolactin release is definitely under the control of a prolactin-releasing hormone (P–RH) or factor (P–RF).

The releasing hormones are polypeptides of low molecular weight. Recently activities of LH–RF and FSH–RF have been identified with the same poly-

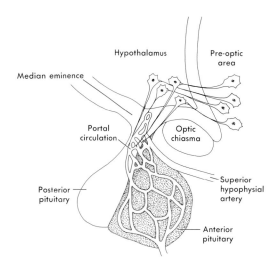

FIGURE 5.9 Neurological and hormonal relationships between the medial preoptic area, the hypophysiotropic area, and the anterior pituitary gland.

peptide, which has the amino acid sequence (pyro) Glu-His-Try-Ser-Tyr-Gly-Arg-Pro-Gly-NH₂ (Matsuo et al., 1971; Baba et al., 1971). This compound was first isolated from the hypothalamus of the pig. It also stimulates release of both LH and FSH in rats, chimpanzees, and human beings. It is now clear that there is only one releasing hormone for both LH and FSH (Schally et al., 1975). Other polypeptide neurohormones that cause release of other hormones of the anterior lobe of the pituitary have also been identified.

The hypothalamic neurohormones are produced in neurons found in the floor and lower walls of the third ventricle. As shown in Figure 5.9, these neurons surround a structure called the **median eminence** and contribute their releasing factors directly into a system of blood vessels leading into the pituitary gland. This system of neurons and the median eminence constitute the **hypophysiotrophic area.**

5.2.2 NEURAL INPUT TO THE HYPOTHALAMUS.
It is evident from the foregoing that the patterns of reproductive activity found in various kinds of animals must be correlated with patterns of release of the hypothalamic neurosecretions. But what signals the hypothalamic neurons that they should secrete their releasing factors? This question is best ap-

33

proached by looking at the results of some experiments on the laboratory rat.

The mature female rat ovulates spontaneously on a very precise schedule every 4 or 5 days, depending on conditions of caging and illumination. In the rat, as in other mammals, each ovulation is preceded by a rapid rise in the level of LH in the blood plasma (the LH "spike"; see Figure 5.2). If the hypophysiotrophic area is neurologically isolated from the rest of the brain (that is, if neural input to this region is blocked by means of microsurgery), the LH spike is abolished and ovulation fails to occur. Instead, females so treated remain constantly in estrus, and, although the ovarian follicles enlarge, ovulation does not occur. If, however, the isolation operation is incomplete anteriorly, so that certain nerves enter the hypophysiotrophic area from the **medial preoptic area** (Figure 5.9), then estrus, accompanied by ovulation, can occur (Clemens et al., 1975). Apparently, therefore, regulation of the cyclic surge in pituitary gonadotropins is exercised by neural input from the preoptic area to the hypophysiotrophic area. This is further indicated by the fact that electrical stimulation of the preoptic area brings about increased quantities of circulating LH (Clemens et al., 1971).

The male rat does not show dramatic changes in plasma LH, so he must lack the timing mechanism that is present in or acts upon the preoptic area of the female. The appropriate neural connections are present, however, for ovaries implanted in a male rat, castrated as an adult, will ovulate when the preoptic area is electrically stimulated (Quinn, 1966). The basis for the sexual differences in the neurological timing mechanisms is determined irreversibly during late prenatal and early postnatal life. Thus, the injection of androgens into a female rat perinatally (just before or after birth) causes differentiation of a male pattern of secretion of gonadotropins and consequent failure of ovulation (reviewed by Everett, 1969). Nevertheless, a female rat so treated ovulates in response to electrical stimulation of the preoptic area. (Terasawa et al., 1969). The interpretation of results such as these is not a simple one, however. For example, newborn female hamsters subsequently show masculinization when injected with estrogens (Whalen and Etgen, 1978).

5.2.3 EXTRINSIC CONTROLS OF REPRODUCTIVE ACTIVITY.

For seasonally breeding animals, signals from the cosmic environment, such as temperature, humidity, and, most important, photoperiod, are paramount factors in the maturation of gametes and the initiation of reproductive behavior. This is true both in vertebrates, which have the hypothalamo-hypophysial control mechanism, as well as for many invertebrates, whose reproductive controls remain somewhat obscure. The annual migration of many birds to their breeding grounds and, ultimately, their nesting and mating depend on the onset of lengthening days following a period of short days. A similar photoperiodic control of breeding probably operates in many mammals in the wild. In hibernating toads of the western plains, the gonads come into breeding condition as the temperature begins to rise in the spring, but breeding behavior is elicited only when the temperature and relative humidity reach appropriate levels (Bragg, 1940). The West Indian and the Pacific palolo worms release their gametes only at certain phases of the moon. For the Samoan species *Eunice viridis* this occurs twice annually, in October and November, on the sixth to eighth nights after the full moon (Smetzer, 1969).

In continuous breeders, also, ovulatory cycles and other aspects of reproductive activity are subject to considerable perturbation by environmental change. Cows show an improvement in winter fertility when the period of illumination is lengthened during the 24-hour day. In hamsters and other laboratory rodents the reproductive tract regresses notably when the animals are blinded or placed on regimens of reduced light. Among human females, girls blind from birth undergo puberty changes earlier than do controls matched for social, ethnic, and nutritional status (Zacharias and Wurtman, 1969). Nurses transferring to night duty and jet airline hostesses, who make rapid transitions between time zones, often show disturbances in their menstrual rhythm (Harris, 1969).

Once reproductive maturity is attained in various types of breeders, the stimuli that lead directly to mating and fertilization are chiefly visual, olfactory, auditory, and tactile. Presumably the sense of taste may be involved, too, as judged by the fact that many animals explore with the tongue the anogenital area and glands there and elsewhere. In many birds and insects the courting ritual of the male provides visual signals that bring the otherwise prepared female to a receptive stage (see Barfield, 1971). The potency of olfactory stimuli is readily obvious to those who have observed the behavior of a male dog when a bitch in heat comes into the neighborhood (Goodwin et al., 1979). The bull responds to olfactory stimuli in the presence of estrous cows, but blinding the bull reduces the probability that he will identify the sexual situation and respond accordingly. Auditory stimuli are important to sexual attraction and mating. Gravid females of the gobiid fish *Bathygobuis soporator* respond by making darting movements that are oriented toward the male, if he is in view (van Tienhofen, 1968). Bulls mounting model cows equipped with artificial vaginas increase the amount of semen and decrease their time for ejaculation when stimulated by the lowing (vocalization) of a real cow (de Vuyst et al., 1964).

Tactile stimuli originating from a member of the opposite sex are very important in many kinds of organisms for sexual arousal and discharge of gametes. Among human beings the tactile stimulation of erogenous zones such as the lips, nipples, genitalia, and other areas of the skin is often a part of the foreplay that leads to coupling and ejaculation. The discharge of semen occurs very quickly in animals such as the rabbit, sheep, and deer, but may require as much as 15 minutes for completion in the boar. In the case of laboratory rats and mice, several intromissions are usually required before ejaculation actually occurs, and the pattern of penile movements may affect the likelihood of impregnation.

The copulatory embrace and stimulation of the vagina by the penis are of importance in ovulation. As noted at the beginning of this chapter, some animals are reflex ovulators, that is, ovulation occurs only in response to mating. In the rabbit, the embrace of the male is often sufficient to cause reflex ovulation independently of penile intromission. The latter is required, however, for many reflex ovulators and also may advance the time of ovulation by several hours in spontaneous ovulators. Some of these considerations have been reviewed by Zarrow and Clark (1968) and Jöchle (1973). Females in a considerable number of mammalian orders are reflex ovulators. They occur especially among the Insectivora (e.g., mole shrew, hedgehog), Lagomorpha (rabbits and hares), Rodentia (e.g., 13-lined ground squirrel, muskrat, tree mouse, mongoose, nutria), and Carnivora (e.g., cat,

ferret, mink, martin, raccoon). In most of these groups, except for the Lagomorpha, the penis of the male has sharp spines, and it is believed that stimulation of the vagina by these spines induces reflex ovulation. Most members of the larger-sized mammalian groups are spontaneous ovulators, and the males lack penile spines.

Among human females, ovulation usually occurs spontaneously at about the 14th day prior to the expected onset of the menses. There are, nevertheless, convincing records that women sometimes do ovulate at other times (Clark and Zarrow, 1971). Based on the fact that sperm and unfertilized eggs seldom survive more than a day or two, there is clear evidence that conceptions have occurred before, during, and after the menses. The most convincing evidence comes from a survey of women who, caught in the path of invading armies during World War II, were raped and who conceived as a result. Of more than 1700 cases reviewed by German gynecologists, 720 were found in which the affected women had maintained careful records of their menstrual cycles prior to the occurrence of the rape, and who could give satisfactory assurances that no other cohabitations had taken place. The collected data showed that conceptions occurred on every possible day of the menstrual cycle. In cases of rape, the violence of the act itself may be responsible for the reflex surge of hormones that bring on ovulation (reviewed by Jöchle, 1973).

Experimental evidence on rats supports the conclusion that vaginal stimulation may induce ovulation. Laboratory rats are normally spontaneous ovulators, but if they are treated with certain drugs such as chlorpromazine, a central nervous system depressant, ovulation fails to occur on schedule. Such rats, however, ovulate reflexly if allowed to mate or if the vagina is stimulated by a rapidly vibrating glass rod. Presumably tactile signals from the vagina are transmitted via higher brain centers to the preoptic area and thence to the hypothalamus. In the case of the rat, this sensory input is blocked if the pelvic nerve is severed, and chlorpromazine-drugged rats do not then ovulate reflexly (Zarrow and Clark, 1968).

From the foregoing sampling, it is obvious that the control of reproductive processes by externally derived signals could be analyzed at great length. Further discussion is limited, however, to a brief account of photoperiodic induction of gonadal development and breeding behavior in sparrows and to some examples of the role of chemical stimuli (pheromones) on the sexual development and mating behavior of some other animals.

5.2.3.1 Photoperiodic Control of Reproduction in Sparrows.
In the population of white-crowned sparrows, *Zonatricha leucophrys*, that winters in the Snake River Canyon near Pullman, Washington (Farner and Wilson, 1966), the annual breeding cycle begins in April when the birds start migrating gradually to their breeding grounds in Alaska, arriving in late May. Courtship, nest building, and egg laying are complete by the end of June, followed by the hatching of the new brood. Both juvenile and adult birds molt during July and August. In late August the birds begin migrating to their wintering grounds, arriving in Washington in late September.

The gonads show remarkable changes during the breeding cycle, as depicted in Figure 5.10. In males the combined weight of the testes is less than 2 mg during November through January, and the seminiferous tubules contain only spermatogonia. The testes begin to enlarge (this is termed **recrudescence**) in late February, and they grow at an ever-increasing rate until they reach a maximum combined weight of 500–600 mg just after the birds

arrive at the breeding grounds in late May. Early in July the testes begin to regress and by mid-August are at their minimal size once more. The ovaries develop more slowly, reaching 100–200 mg upon arrival at the breeding area. They then increase to about 300 mg and spurt up to as much as 900 mg just before each ovulation. Regression begins after egg laying is finished and is complete by the end of July. A similar pattern of testicular and ovarian growth and regression is shown by nonmigrating house sparrows *(Passer domesticus)*.

The white-crowned sparrow is obligatorily phototrophic. Thus, like certain plants it requires a pattern of increasing day length in order to reproduce. Laboratory tests show that, with 8 hours or less of day length, gonadal development fails to occur, and none of the other events of the migratory-reproductive cycle ensues. Conversely, after the period of long-day illumination required to produce and hatch the brood, a period of photorefractoriness sets in. Long days will not stimulate renewed gonadal development until late October. Thereafter, birds exposed to long days show gonadal recrudescence and physical restlessness that signal the onset of migratory behavior.

What is the receptor for the photoperiodic stimulus? This was examined for the house sparrow by Menaker and his associates in 1970. They found that testicular recrudescence at the end of the refractory period occurs as readily in sparrows blinded by removal of the eyes as in controls, and thus concluded that the eyes do not participate in the perception of the photoperiod. Next, using birds with normal vision, they injected India ink under the dorsal skin of the head and then subjected the treated birds to the minimal photoperiodic stimulus required for testicular recrudescence in normal birds. At the end of the test period the testes of the control birds weighed over 300 mg, but those of the injected birds weighed only 8 mg. Thus, whereas the photoperiodic receptor is not yet known, the route by which light reaches it must lie preferentially through the dorsal surface of the skull. Eventually, whether directly or indirectly, the light affects the hypothalamus, probably through nervous connections to the posterior part of the median eminence. Destruction of this part of the median eminence prevents light-induced testicular recrudescence (Stetson, 1969).

36

FIGURE 5.10 Testicular recrudescence and regression in the house sparrow. [Based on data of M. Menaker, *Biol Reprod* 4:295–308 (1971).]

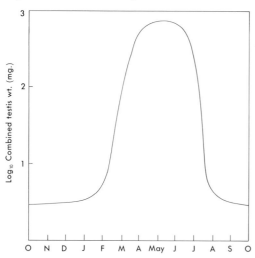

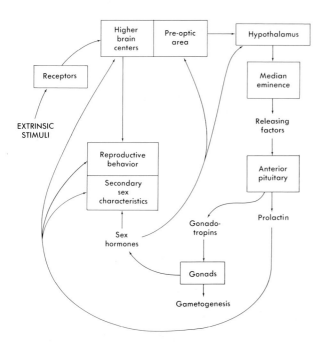

FIGURE 5.11 A generalized scheme for the control, by extrinsic factors, of the neuroendocrine pathways involved in seasonal breeding. [Redrawn with modifications from D. A. Farner and B. K. Follett, *J Anim Sci* 25:90–115 (1966).]

It is reasonable to presume that the photoperiod and other extrinsic and intrinsic signals operate in other seasonally breeding animals through the hypothalamic-hypophysial apparatus in a manner analogous to the control system in birds. A generalized scheme for the role of external and internal information in the control of reproduction in seasonal breeders is shown in Figure 5.11.

5.2.3.2 Pheromones Control Sexual Maturity and Breeding.
Pheromones are substances, produced to the exterior, that act as chemical messengers between individuals in a population (rather than between parts of the same individual, as do hormones), and that elicit a specific response in the receiver. Pheromones that are involved in the control of sexual reproduction are called *sex pheromones* and are important in reproduction in most major groups of animals. They are usually volatile substances, relatively low in molecular weight, that stimulate the olfactory organs of the recipient. Some, however, are relatively nonvolatile and pro-

duce their effects after being ingested by other members of the population. This is true for the pheromones in some social insects.

Many pheromones exercise their effects on the coming into sexual maturity of immature members of the population and on cyclic ovarian function. Volatile components of the urine of adult, noncastrated male mice are notable in this regard. Weanling female laboratory mice housed in groups of six to eight show a significant advance in the time at which they show three of the most significant signs of sexual maturity, namely, the appearance of the vaginal opening, the time of the first estrus, and the time of the first receptivity to the male when reared in the presence of a functional male or if only exposed to his urine-soiled bedding. The advance in these signs of maturity may be on the order of several weeks and is not accompanied by accelerated physical growth (Vandenbergh, 1967; Bronson and Maruniak, 1976; Vandenbergh et al., 1976). The same effects are observed in rats (Vandenbergh, 1976). In immature female mice it is found that serum LH levels, hitherto undetectable, increase to significant amounts within 30 minutes after spraying them and their bedding with male urine. This is followed by a 20- to 30-fold increase in serum estradiol. The estradiol effect is elicited independently of LH, for injections of LH alone do not lead to increased estradiol concentrations or to accelerated maturity. Estradiol injections will substitute for the presence of the male.

The puberty-accelerating pheromone of mice has been determined by Vandenbergh et al. (1976) to be a peptide of low volatility, having a molecular weight of about 860. It is present in bladder urine only of intact (noncastrated) males, and its mode of action is not yet clear. Another of its interesting effects is seen in immature mice caused to ovulate precociously by injection of gonadotropins. They produce significantly more eggs at each ovulatory cycle when placed in cages soiled by male adults. This effect does not occur if the bedding of castrated males is used or if the females are rendered **anosmic** (deprived of olfactory sense (Zarrow et al., 1970).

Mature females housed together in the absence of functional adult males tend to become pseudopregnant; that is, they cease their repetitive estrous cycles (laboratory mice are continuous breeders) and remain in the progestational stage, with active

37

corpora lutea (Ryan and Schwartz, 1977). This does not occur among anosmic females, so it is reasonable to conclude that female mice produce pheromones affecting their own reproductive cycles. Volatile materials from the urine of an adult functional male abolish pseudopregnancy in groups of females housed together (Whitten et al., 1968), but the urine of castrated or immature males is ineffective in this regard.

The urine of female rats likewise exercises an effect on the male reproductive sytem. Males exposed to the urine of females show increased plasma levels of LH (Maruniak and Bronson, 1976) and of testosterone (Purvis and Haynes, 1978).

Pheromones that affect activities of the reproductive system are well known among invertebrates, also. In the honeybee, *Apis mellifera,* a pheromone that affects reproductive maturity has been identified. The mandibular glands of the queen bee are the source of a secretion that is spread over her entire body through the grooming activities of the worker bees (Velthuis, 1972). One component of the secretion is the fatty acid 9-oxodecenoic acid, $CH_3 \cdot CO \cdot (CH_2)_5 \cdot CH{=\!=}CH \cdot COOH$. The pheromone (Callow et al., 1964) is ingested by workers, who lick it from the queen's body and then share it with their fellow workers by regurgitation. The **queen substance** is probably absorbed from the digestive tract and conceivably acts directly on the ovaries or, possibly, affects the ovaries via the **corpus allatum,** which produces a hormone required for oögenesis in insects. Injection of 9-oxodecenoic acid inhibits oögenesis in a number of kinds of arthropods.

The foregoing material provides examples of the action of **priming sex pheromones,** that is, sex pheromones that set into motion a sequence of physiological processes that eventually affect reproductive capacity or activity. Other kinds of sex pheromones, however, initiate immediate motor activity leading to mating. These are called **releasing sex pheromones.** Some of these act as attractants between members of the opposite sex, bringing them together for courtship and mating. Others act as aphrodisiacs, exciting copulatory behavior in the partner once they have come into proximity. Some releasers are both attractants and aphrodisiacs. The same queen bee substance that affects the maturation of females in a colony is also both an attractant

and an aphrodisiac to drones on the part of the unmated queen. This is a somewhat unusual case.

Pheromones involved in mating behavior are probably best known in insects, and many have been isolated and identified, particularly in those insects that damage crop plants and forest trees. Sex attractants from the females are highly species specific and are used to entice males to killing traps, thus providing a population control method much less harmful to the environment than indiscriminate chemical sprays (e.g., Beroza and Knipling, 1972).

Sex attractants and aphrodisiacs also play a significant role in mammalian reproduction. Sexually experienced male rats are particularly attracted to the odor of females in heat, even in the absence of physical, auditory, or visual contact with them. The source of the attractive odor is apparently the urine. Moreover, urine taken directly from the bladder of the female in heat is as effective in eliciting the male response as is externally voided urine. This indicates that the attractive component of the urine is probably not a component of glandular secretions along the urethra or within the vaginal orifice (Lydell and Doty, 1972). Estrous females, more so than anestrous ones, actively follow the scent of a mature intact male and will work actively to gain access to him. It has been shown for mice that normally voided male urine, but not urine taken from the bladder, provides the attractant. Voided urine is exposed to secretions of the **preputial gland** (that is, the principal gland of the foreskin), and it is apparently these secretions, possibly one or more fatty acids, that provides the attractant (Bronson and Caroom, 1971). The female attractant of male urine is thus distinct from the pheromone that evokes accelerated maturity in immature female mice.

The production of sex attractants and aphrodisiacs and the ability to react to them are clearly related to the activity of the sex hormones. This is particularly well illustrated by the reproductive activities of chimpanzees and monkeys. Chimpanzees, for example, have a 35–37 day menstrual cycle, and the first marked willingness of the female to accept the male occurs during early growth phases of the ovarian follicle after menstrual bleeding ceases. Ovulation occurs at day 18 to 20 in the cycle, and 1 or 2 days later, the period of sexual receptivity terminates in the female (see review by

Young, 1961). Among rhesus monkeys, female receptivity may or may not continue after ovulation, but, in either case, the male usually shows a declining interest in mounting her shortly thereafter (Michael and Welegalla, 1968). Effectively, therefore, mating in chimpanzees and rhesus monkeys is restricted to that time during the menstrual cycle at which ovulation and fertilization are most likely to occur. This is the height of the follicular phase of the ovarian cycle, when the production of estrogen is greatest. The decline in male interest coincides with the postovulatory, or progestational, phase of the cycle, when the concentration of progesterone is at its peak (cf. Figure 5.2).

The correlation of estrogens and progesterone with male interest is corroborated by experiments with ovariectomized adult female monkeys. Without estrogen administration such females are but little attractive to vigorous males and show little receptivity to them. When estrogen is administered, however, female attractiveness and receptivity are markedly enhanced. When ovariectomized females are treated with progesterone, the males show little interest, although the females' receptivity may be relatively high, as indicated by their invitational presentations to the males (Baum et al., 1976).

A number of reports (e.g., Michael et al., 1971) have suggested that vaginal pheromones, chiefly short-chain aliphatic acids, are produced predominantly during the follicular phase in rhesus females, and that these are a significant aspect of the elicitation of male sexual behavior. In fact, it was reported that human female vaginal washings at midcycle constitute aphrodisiac pheromones to rhesus males when used to anoint the vaginal area of ovariectomized rhesus females. These results have been disputed as artifactual. Goldfoot et al. (1976) found that vaginal washings from normal females in the follicular phase were ineffective in evoking male sexual interest when applied to the genitalia of females ovariectomized some time previously. They also found to be ineffective mixtures of aliphatic acids proposed by Curtis and his associates (1971) as the pheromone responsible for eliciting the male sexual response. Indeed, males deprived of the sense of smell by drugs do not show an altered copulatory behavior with normally cycling females (Goldfoot et al., 1978).

It has been suggested (e.g., by Comfort, 1972)

that specific naturally occurring body odors may have previously unsuspected influences on human sexual behavior. It is difficult to ascertain to what degree sex pheromones are involved in human reproduction, but the question constitutes an important area of investigation.

5.3 Summary

The primary reproductive organs of animals are the gonads—testis and ovary for male and female, respectively. Flowers bear the reproductive structures in plants, anthers and pistils being male and female parts, respectively. For organisms in temperate zones under natural conditions, the reproductive organs produce gametes only seasonally.

The photoperiod controls flowering in many kinds of plants. Long-day plants require illumination of longer than some critical period, whereas short-day plants require days of less than some critical period in order to flower. Day-neutral plants are not directly dependent on the photoperiod for flowering.

Leaves are the principal receptors of the photoperiodic stimulus, producing an as yet unidentified hormone, florigen, that is transported to the buds, switching their vegetative growth to floral growth. Gibberellins, which affect patterns of stem elongation, appear to be involved, but in obscure fashion, with floral induction.

The receptor of the photoperiodic stimulus appears to be phytochrome. It is activated by red light and reversibly inactivated by far red light. This is called the red–far red effect. How the active form of phytochrome participates in floral induction is not known.

Reproductive cycles in animals are controlled by hormones, and these, in turn, are secreted in response to external and internal stimuli. The principal groups of hormones are the releasing hormones of the hypothalamus (LH–RH and FSH–RH), the pituitary gonadotropins, LH, FSH and prolactin, and the sex hormones, estrogens and androgens. Input to the hypothalamus from the preoptic region of the brain controls the production of the releasing hormones, and these, in turn, produce the gonadotropins, which act on the gonads. They cause the gonads to mature and to produce gametes ready for fertilization. The gonads also respond by producing the sex hormones, which are responsible for morphological and physiological sexual maturity and reproductive activity.

The photoperiod is a principal factor in the reproduction of seasonal breeders, but light affects the reproductive activity of continuous breeders also. Seasonal recrudescence of the gonads and mating behavior are brought about in obligatory manner by changing day length for many animals in the wild. The photoperiodic receptor is not known, but for certain sparrows, at least, the eye is not necessary.

Pheromones, especially certain volatile compounds voided with the urine, are important in the reproductive maturation and mating behavior of many mammals. Some are priming pheromones, which set into motion long-term physiological effects. Others are releasing pheromones, which elicit immediate patterns of motor activity leading to mating. Such pheromones include sex attractants and aphrodisiacs. Sex attractants are used as important components of control procedures for many insect pests.

Sex pheromones are produced under hormonal control. There is little doubt of their important and often essential role in many animal forms, but their role in reproductive behavior in human and non-human primates remains obscure.

5.4 Questions for Thought and Review

1. What resemblances and what differences do you see in the annual reproductive patterns of plants and animals?

2. Of what advantage might induced ovulation be in the preservation of a mammalian species?

3. Documentary evidence about rape cases suggests that women can be induced to ovulate by sexual intercourse. On the basis of what we know about laboratory animals, what would likely be the sequence of neuronal and hormonal events involved?

4. What is a short-day plant? Why might it better be described as a "long-night" plant? Must the period of light be of less duration than the period of night in a short-day plant?

5. What is the evidence that a substance capable of transport from leaf to bud acts as a flowering hormone? Do long-day and short-day plants have the same or similar florigens? What is the evidence on which your answer is based?

6. How would you determine what wavelengths of light are most effective in inducing flowering in a short-day plant? In doing this, you would construct an action spectrum. What is an action spectrum?

7. Why is chlorophyll not considered to be a mediator of the photoinductive effect on flowering?

8. Describe phytochrome. What is the red–far red effect on phytochrome? What wavelengths are absorbed most effectively by phytochrome in different states? What wavelength of light is absorbed to give the form of phytochrome active in floral induction? What is the evidence on which your answer is based?

9. What are the chief sources of hormones directly involved in vertebrate reproductive cycles? What are the names of each and to what major chemical group does each belong? What is the action of each hormone?

10. What is the hypophysiotrophic area? What nervous input to this area is involved in reproductive cycles? What is the evidence on which you base your answer?

11. What is the "LH spike" in reference to ovulation? Would you expect to find this spike in an unmated cat or ferret? Why?

12. On the basis of what evidence do we conclude that neural pathways from the preoptic area to the hypophysiotrophic area are the same in male and female rats? On what a priori grounds might we expect them to be different?

13. Is the retina of the eye involved in photoperiodic induction of testicular recrudescence in house sparrows? What is the evidence on which you base your answer?

14. Define a pheromone. Are all pheromones necessarily involved in sexual activity? What other kinds of pheromones might you know about?

15. Distinguish a priming sex pheromone from a releasing sex pheromone. The urine of male laboratory rodents contains substances that act as primers and as releasers. Give examples of each of these kinds of pheromonal action.

5.5 Suggestions for Further Reading

Austin, C. H., and R. V. Short, eds. 1972. *Hormones in Reproduction.* Cambridge: Cambridge University Press.

40

BRONSON, F. H. 1971. Rodent pheromones. *Biol Reprod* 4:344–57.

BUTLER, G. G. 1976. Insect pheromones. *Biol Rev* 42:42–87.

DUNHAM, P. J. 1978. Sex hormones in Crustacea. *Biol Rev* 53:555–83.

GUILLEMIN, R. 1977. Purification, isolation and primary structure of a hypothalamic luteinizing hormone-releasing factor of ovine origin. A historical account. *Am J Obstet Gynecol* 129:214–18.

GUILLEMIN, R., and R. BURGUS. 1972. The hormones of the hypothalamus. *Sci Am* (November) 24–33.

HAFEZ, E. S. E., and T. N. EVANS, eds. 1973. *Human Reproduction*. New York: Harper & Row.

HEDDEN, P. 1978. The metabolism of the giberellins. *Annu Rev Plant Physiol* 29:149–92.

LICHT, O., H. PAPKOFF, S. W. FARMER, D. H. MULLER, H. W. TSUI, and D. CREWS. 1977. Evolution of gonadotropin structure and function. *Recent Prog Horm Res* 33:169–248.

PALEG, L. 1965. Physiological effects of gibberellins. *Annu Rev Plant Physiol* 16:291–322.

RYAN, E. L., and A. I. FRANKEL. 1978. Studies on the role of the medial preoptic area in sexual behavior and hormonal response to sexual behavior in the mature male laboratory rat. *Biol Reprod* 19:971–83.

SAWYER, C. H. 1978. History of the neurovascular concept of hypothalamo-hypophysial control. *Biol Reprod* 18:325–28.

SCHOPFER, P. 1977. Phytochrome control of enzymes. *Annu Rev Plant Physiol* 28:223–52.

SEVITT, S. 1946. Early ovulation. *Lancet* 2:448–50.

SIEGELMAN, H. W., and W. L. BUTLER. 1965. Properties of phytochrome. *Annu Rev Plant Physiol* 16:383–92.

WURTMAN, R. J. 1975. The effects of light on man and other mammals. *Annu Rev Physiol* 37:467–83.

5.6   References

ADAMS, D. B., A. R. GOLD, and A. D. BURT. 1978. Rise in female-initiated sexual activity at ovulation and its suppression by oral contraceptives. *N Engl J Med* 299:1145–50.

BABA, Y., H. MATSUO, and A. V. SCHALLY. 1971. Structure of porcine LH- and FSH-releasing hormone. II. Confirmation of the proposed structure by conventional sequential analysis. *Biochem Biophys Res Commun* 44:459–63.

BARFIELD, R. J. 1971. Gonadtrophic hormone secretion in the female ring dove in response to visual and auditory stimulation by the male. *J Endocrinol* 49:305–10.

BAUM, M. J., B. J. EVERETT, J. HERBERT, E. B. KEVERNE, and W. J. DE GREEF. 1976. Reduction of sexual interaction in rhesus monkeys by a vaginal action of progesterone. *Nature* 263:606–608.

BEROZA, M., and E. F. KNIPLING. 1972. Gypsy moth control with the sex attractant pheromone. *Science* 177:19–27.

BRAGG, A. N. 1940. Observations on the ecology and natural history of anura. I. Habits, habitat, and breeding of *Bufo cognatus* Say. *Am Nat* 74:322–49.

BRONSON, F. H., and D. CAROOM. 1971. Preputial gland of the male mouse: attractant function. *J Reprod Fertil* 25:279–82.

BRONSON, F. H., and J. A. MARUNIAK, 1976. Differential effects of male stimuli on follicle-stimulating hormone, luteinizing hormone, and prolactin secretion in prepubertal female mice. *Endocrinology* 98:1101–108.

BUTLER, W. L., S. B. HENDRICKS, and H. W. SIEGELMAN. 1965. Purification and properties of phytochrome. In T. W. Goodman, ed. *The Chemistry and Biochemistry of Plant Pigments*. New York: Academic Press, pp. 197–210.

CALLOW, R. K., J. R. CHAPMAN, and P. N. PATON. 1964. Pheromones of the honey bee: chemical studies of the mandibular gland secretion of the queen. *J Apicult Res* 3:77–89.

CLARK, J. H., and M. X. ZARROW. 1971. Influence of copulation on time of ovulation in women. *Am J Obstet Gynecol* 109:1083–85.

CLEMENS, J. A., C. J. SHAAR, J. W. KLEBER, and W. A. TANDY. 1971. Areas of the brain stimulatory to LH and FSH secretion. *Endocrinology* 88:180–84.

CLEMENS, J. A., E. B. SMALSTIG, and B. D. SAWYER, 1975. Studies on the role of the preoptic area in the control of reproductive function in the rat. *Endocrinology* 99:728–35.

COMFORT, A. 1972. *The Joys of Sex.* New York: Crown.

CURTIS, R. F., J. A. BALLANTINE, J. A. KEVERNE, R. W. BONSALL, and R. P. MICHAEL. 1971. Identification of primate sexual pheromones and the properties of synthetic attractants. *Nature* 232:396–98.

DIAMOND, M., L. DIAMOND, and M. MAST. 1972. Visual sensitivity and sexual arousal levels during the menstrual cycle. *J. Nerv Ment Dis* 155:170–76.

DE VUYST, A., G. THINÈS, and M. SOFFIÉ, 1964. Influence des stimulations auditives sur le comportement sexuel du taureau. *Experientia* 20:648–50.

EVERETT, J. W. 1969. Neuroendocrine aspects of mammalian reproduction. *Annu Rev Physiol* 31:383–416.

FARNER, D. S., and A. C. WILSON. 1966. A quantitative examination of testicular growth in the white crowned sparrow. *Biol Bull* 113:254–67.

FLINT, L. H., and E. D. MCALISTER. 1937. Wavelengths of radiation in the visible spectrum promoting the germination of light-sensitive lettuce seed. *Smithsonian Inst Misc Collec* 92(2):1–8.

GALSTON, A. W. 1968. Microspectrophotometric evidence

41

for phytochrome in plant nuclei. *Proc Natl Acad Sci USA* 61: 454–60.

GALSTON, A. W. and P. J. DAVIES. 1970. *Control Mechanisms in Plant Development.* Englewood Cliffs, NJ: Prentice-Hall.

GOLDFOOT, D. A., S. M. ESSOCK-VITALE, C. S. ASA, J. E. THORNTON, and A. I. LESHNER. 1978. Anosmia in male rhesus monkeys does not alter copulatory activity with cycling females. *Science* 199:1095–96.

GOLDFOOT, D. A., M. A. KRAVETZ, R. W. GOY, and S. K. FREEMAN. 1976. Lack of effect of vaginal lavages and aliphatic acids on ejaculatory responses in rhesus monkeys: behavioral and chemical analyses. *Horm Behav* 7:1–27.

GOODWIN, M., K. M. GOODING, and F. REGNIER. 1979. Sex pheromone in the dog. *Science* 203:559–61.

HAMNER, K. C., and J. BONNER. 1938. Photoperiodism in relation to hormones as factors in floral initiation and development. *Bot Gaz* 100:388–431.

HARRIS, G. W. 1969. Ovulation. *Am J Obstet Gynecol* 105:659–69.

HODSON, H. K., and K. C. HAMNER. 1970. Floral-inducing extract from *Xanthium. Science* 167:384–85.

JÖCHLE, W. 1973. Coitus-induced ovulation. *Contraception* 7:523–64.

LANG, A. 1957. The effect of gibberellin on flower formation. *Proc Natl Acad Sci USA* 43:709–20.

LYDELL, K., and R. L. DOTY. 1972. Male rat odor preferences for female urine as a function of sexual experience, urine age and urine source. *Horm Behav* 2:205–12.

MARUNIAK, J. A., and F. H. BRONSON. 1976. Gonadotropic responses of male mice to female urine. *Endocrinology* 99:963–69.

MATSUO, H., Y. BABA, R. M. G. NAIR, A. ARIMURA, and A. V. SCHALLY. 1971. Structure of the porcine LH- and FSH-releasing hormone. I. The proposed amino acid sequences. *Biochem Biophys Res Commun* 43:1334–39.

MENAKER, M., R. ROBERTS, J. ELLIOTT, and H. UNDERWOOD. 1970. Extraretinal light perception in the sparrow. III. The eyes do not participate in photoperiodic photoreception. *Proc Natl Acad Sci USA* 67:320–25.

MICHAEL, R. P., E. B. KEVERNE, and R. W. BONSALL. 1971. Pheromones: isolation of male sex attractants from a female primate. *Science* 172:964–66.

MICHAEL, R. P., and J. WELEGALLA. 1968. Ovarian hormones and the sexual behavior of the female rhesus monkey *(Macaca mulatta)* under laboratory conditions. *J. Endocrinol* 41:407–20.

PARKER, M. W., S. B. HENDRICKS, S. B. BORTHWICK, and F. W. WENT. 1949. Spectral sensitivities for leaf and stem growth of etiolated pea seedlings and their similarity to action spectra for photoperiodism. *Am J Bot* 36:194–204.

PURVIS, K., and N. B. HAYNES. 1978. Effect on the odour of female rat urine on plasma testosterone concentrations in male rats. *J Reprod Fertil* 53:63–65.

QUINN, D. L. 1966. Luteinizing hormone release following preoptic stimulation in the male rat. *Nature* 209: 891–92.

RYAN, K. D., and N. B. SCHWARTZ. 1977. Grouped female mice: demonstration of pseudopregnancy. *Biol Reprod* 17:578–83.

SCHALLY, A. V., A. J. KASTIN and A. ARIMURA. 1975. The hypothalamus and reproduction. *Am J Obstet Gynecol* 122:857–62.

SMETZER, B. 1969. Night of the palolo. *Natural History,* November 1969, pp. 64–71.

STETSON, M. H. 1969. The role of the median eminence in control of photoperiodically induced testicular growth in the white-crowned sparrow, *Zonotrichia leucophrys gambelii. Z. Zellforsch* 93:369–94.

TERASAWA, E., M. KAWAKAMI, and C. H. SAWYER. 1969. Induction of ovulation in androgenized and spontaneously constant-estrous rats. *Proc Soc Exp Biol Med* 132:497–501.

UDRY, J. R., and N. M. MORRIS. 1968. Distribution of coitus in the menstrual cycle. *Nature* 200:593–96.

UDRY, J. R., and N. M. MORRIS. 1977. The distribution of events in the human menstrual cycle. *J Reprod Fertil* 51:260–65.

VAITUKAITIS, J. L., G. T. ROSS, G. D. BRAUNSTEIN, and P. L. RAYFORD. 1976. Gonadotropins and their subunits: basic and clinical studies. *Recent Prog Horm Res* 32:289–331.

VANDENBERGH, J. G. 1967. Effect of the presence of a male on the sexual maturation of female mice. *Endocrinology* 81:345–49.

VANDENBERGH, J. G. 1976. Acceleration of sexual maturation in female rats by male stimulation. *J Reprod Fertil* 46:451–53.

VANDENBERGH, J. G., J. S. FINLAYSON, W. J. DOBROGOSZ, S. S. DILLS, and T. A. KOST. 1976. Chromatographic separation of puberty accelerating pheromone from male mouse urine. *Biol Reprod* 15:260–65.

VAN TIENHOVEN, A. 1968. *Reproductive Physiology of Vertebrates.* Philadelphia: Saunders.

VELTHUIS, H. H. W. 1972. Observations on the transmission of queen substances in the honey bee colony by the attendants of the queen. *Behavior* 41:105–29.

WHALEN, R. E. and A. M. ETGEN. 1978. Masculinization and defeminization induced in female hamsters by neonatal treatment with estradiol benzoate and RU-2858. *Horm Behav* 10:170–77.

WHITTEN, W. K., F. H. BRONSON, and J. A. GREENSTEIN. 1968. Estrus-inducing pheromone of male mice: transport by movement of air. *Science* 161:584–85.

YOUNG, W. C. 1961. The mammalian ovary. In W. C.

42

Young ed. *Sex and Internal Secretions,* Vol. 1. Baltimore: Williams and Wilkins, pp. 449–96.

ZACHARIAS, L., and R. J. WURTMAN. 1969. Blindness and menarche. *J Obstet Gynecol,* 33:603–608.

ZARROW, M. X., and J. H. CLARK. 1968. Ovulation follow-ing vaginal stimulation in a spontaneous ovulator and its implications. *J. Endocrinol* 40:343–52.

ZARROW, M. X., S. A. ESTES, V. H. DENEBERG, and J. H. CLARK. 1970. Pheromonal facilitation of ovulation in the immature mouse. *J Reprod Fertil* 23:357–60.

6

Gametes, Fertilization, and Early Development in Plants

leafy part of a diploid plant. The stem, called a **rhizome,** is underground, its growing tip pushing through the soil and reproducing new stem and leaf parts through vegetative growth. From the stem roots are formed at frequent intervals (Figure 6.1). This plant is called the **sporophyte.** It is the spore-producing phase of the life cycle. On the underside of the leaves there appear brownish spots called **sori.** These consist of clusters of spore cases called **sporangia,** and within these are **spore mother cells,** each of which undergoes meiosis to produce four **spores.** Spores are haploid cells usually containing a certain amount of reserve matter in the form of starch, oil, and protein and often containing numerous chloroplasts. In some ferns entire fronds are devoted exclusively to the production of sporangia.

44 In the plant kingdom there is tremendous variability in the patterns of reproduction from one major group to the next and within members of each group. From the comparative viewpoint, a striking feature of reproduction in plants is the alternation of an asexually-propagating diploid portion of the life cycle with a haploid sexual phase. In the latter gametes are produced, whose union restores the diploid phase once more. To begin this chapter, therefore, we describe briefly the reproductive cycle of ferns, in which both asexual and sexual phases of the life cycle have been fairly well studied. Then attention is turned to the flowering plants for a closer look into the origin of the gametes, fertilization, and early stages of development.

6.1 Reproduction in Ferns

The fern that we customarily observe in moist woods, the fronds of which are often used to enhance the beauty of floral displays, consists of the

FIGURE 6.1 Life cycle of a fern. The germinating spore (top) initiates the gametophyte generation, which terminates when the egg is fertilized. Fertilization marks the start of the sporophyte generation. Development of the sporophyte begins within the walls of the archegonium. The sporophyte thus begins its development anchored to the gametophyte and draws its nourishment therefrom until it develops its own root system (bottom and lower left).

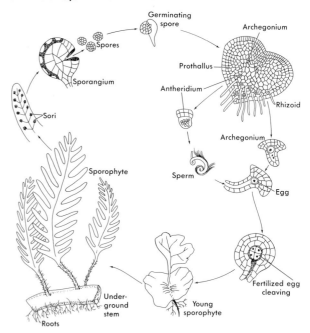

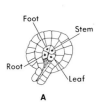

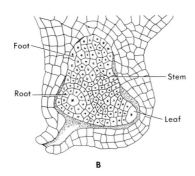

FIGURE 6.2 Cleavage pattern in a fern. Within the walls of the archegonium, the zygote undergoes successive divisions that give rise to zones of cells that may be identified as to their future morphogenetic fates.

Released from the sporangia, the spores germinate, giving rise to the haploid gametophyte generation. Germination requires moisture and, usually, red light. Germination normally leads to the formation of a heart-shaped haploid gametophyte called the **prothallium,** rich in chloroplasts and having rhizoids projecting from the under surface as nutritive organs. The achievement of this form, however, requires blue light or high intensities of white light. In darkness or in continued red light, growth of the prothallium is filamentous, and few rhizoids form.

Male and female sex organs are **antheridia** and **archegonia,** respectively. Depending on the species, they may be produced on the same or different prothallia. The sex organs are numerous, and each arises from a single cell on the under surface of the prothallium. Antheridia arise first, contain numerous sperm, and lie among the rhizoids near the base of the prothallium. Archegonia, each containing but one egg, lie near the notch at its apex. The distribution of antheridia and archegonia is probably determined by hormones. Two such hormones, called antheridogens, have been isolated (Näf, 1968; Nakanishi et al., 1971) (Figure 6.1).

When the gametes are ripe, and in the presence of adequate moisture, the sperm are released from the antheridia and swim to the archegonia, enter the neck canal, and fertilize the egg, thus initiating the sporophyte generation. The movement of the sperm toward the archegonia is not fortuitous but the result of oriented swimming movements. In the laboratory sperm show a positive response (positive chemotaxis) to sources of malic acid, which may be the chemical attractant of the archegonium.

Only one sperm fertilizes the egg, but there is essentially no research on the question as to why others do not enter it. The zygote produces a cellular wall and then begins to divide in a characteristic pattern (Figure 6.2), which is referred to as a **cleavage pattern,** in keeping with the terminology used in describing early stages in animal development (Chapter 9). The first early divisions cleave the zygote into four groups of cells whose position with respect to the axes of the prothallium and of the archegonium are such that each can usually be identified with respect to its future fate in the sporophyte. One group is the ancestral cell of the stem, one forms the root, another the first leaf, and the fourth forms a **foot,** which anchors the sporophyte to the gametophyte and serves initially to draw nutrients from it. The arrangement of the primitive tissues of the sporophyte is shown in Figure 6.2B.

From these beginnings the vegetative sporophyte phase of the life cycle begins anew.

The walls of the archegonium play an important part in the cleavage and later development of the zygote. In the fern *Phlebodium aureum* the cleavage planes are quite irregular if the walls of the archegonium are experimentally slit, and the growth of the sporophyte is slower and disorderly. If the zygote of *Todea barbara* is excised from the archegonium within 4 days of fertilization and planted in an artificial medium, it forms a prothallium rather than a leaf-bearing fern. Explanted after 4 days, however, it forms the typical sporophyte. In another fern *Pteridium aquilinum* if normal fertilization is artificially prevented by withholding moisture, and sugar is added to the medium, sporophytic plants originate from the haploid vegetative cells of the prothallium. Clearly, therefore, the morphology of the gametophyte and of the sporophyte are not determined by their diploid or haploid chromosome complement but by extrinsic factors.

45

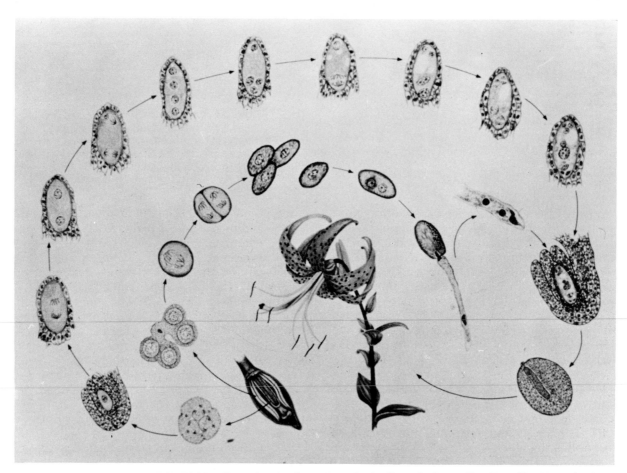

FIGURE 6.3 Life cycle of a flowering plant, *Lilium tigrinum*, the tiger lily. At center is the mature diploid plant with a perfect flower, showing male and female structures (compare Figure 5.1). At bottom right is shown an immature flower. As the flower develops, the anthers and ovaries mature. The development of the pollen grain and the embryo sac may be followed in the counterclockwise direction. The outer tier of drawings shows the development of the female gametophyte; meiosis leads to the formation of four haploid nuclei, all of which participate in formation of the embryo sac (upper left). The inner tier of figures shows the formation of the four haploid microspores; from each a male gametophyte, the pollen grain, arises. The pollen grain germinates and fertilization takes place (left center). Fertilization initiates a new diploid generation (lower left). [Courtesy of Carolina Biological Supply Company.]

6.2 Gametogenesis and Early Developmental Stages in Flowering Plants

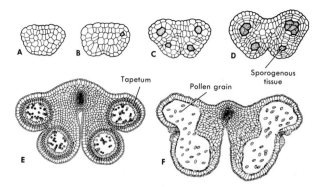

FIGURE 6.4 Stages in the development of an anther in a typical flowering plant as seen in cross section. The primordium of the anther (A) enlarges and four compartments (B–D) arise within it. Within these compartments sporogenous tissue surrounded by tapetal cells develop (E). The sporogenous tissue gives rise to microspore mother cells, which undergo meiosis, producing microspores. From each microspore a binucleate pollen grain (F) is formed. [Reprinted with permission of Macmillan Publishing Co., Inc., from Development in Flowering Plants by J. G. Torrey, p. 56. Copyright © 1967, J. G. Torrey.]

Flowering plants also show an alternation of diploid and haploid generations, but the gametophyte generation is greatly reduced. The sporophyte is the plant that we see, and the gametophytes arise and complete their differentiation within the floral parts, which bear the male and female reproductive organs (Figures 5.1 and 6.3). In "perfect" flowers both male and female reproductive organs are formed. Some species of plants have "imperfect" flowers, and in these the gametophyte of only one sex is present. "Male" and "female" flowers may arise in different plants or from different locations on the same plant.

The male gametophyte is the pollen grain, which contains a **vegetative** nucleus and **generative** nucleus. From the latter two sperm nuclei are produced when the male gametophyte matures. The female gametophyte is the embryo sac, which may have from four to 16 nuclei, one of which is the egg cell.

6.2.1 DEVELOPMENT OF THE MALE GAMETO-PHYTE — ORIGIN OF THE POLLEN.

The early development of the anther (Figure 6.4) results in the segregation of groups of cells into four compartments called **microsporangia.** Each compartment contains **sporogenous** cells surrounded by a layer of **tapetal** cells. Typically, four sporangia arise in each anther, but these subsequently fuse to form two **pollen sacs.** The sporogenous cells divide mitotically to form a population of **microspore mother cells.** These undergo meiosis, and from each arise four **microspores,** each of which then forms a pollen grain. Each microspore undergoes mitotic division shortly before the pollen is shed from the anther. This division produces the vegetative cell and the generative cell (Figures 6.5C and 6.7).

Prior to meiosis the sporogenous cells are connected with one another and with the tapetum by fine cytoplasmic connections called **plasmo-**

47

desmata. Just before meiosis begins, the spore mother cells synthesize large quantities of polysaccharide **callose** (a β-1,3-glucan polymer), surrounding themselves with a thick layer of the material. By the stage of leptotene, the plasmodesmata have disappeared, but the microspore mother cells are kept in contact by virtue of larger cytoplasmic channels through the callose. It has been calculated that as much as 24% of the surface of a spore mother cell may be involved in contact with other microspore mother cells at this time. This degree of connectivity almost certainly is involved in the fact that the spore mother cells enter into meiosis and proceed to the formation of pollen in tight synchrony (reviewed by Mascarenhas, 1975). Soon after meiosis begins the large cytoplasmic connections break down. More callose is then synthesized and disposed around each of the four microspores, which now are arranged in the form of a tetrad (Figure 6.5B).

Within each haploid microspore the complex wall of the pollen grain is synthesized and deposited. The outer layer is called the **exine.** It is composed of **sporopollenin,** a highly resistant material.

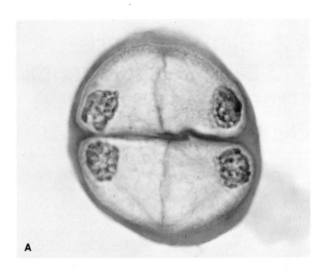

FIGURE 6.5   A: The second meiotic division of the microspore mother cell in *Lilium*. Note the heavy callose wall surrounding both cells and separating them. B: The haploid products of the second meiotic division in *Lilium* showing the tetrad configuration. C: Mature pollen grains of *Tradescantia*; the rounded nucleus of the generative cell lies within the arms of the crescentic generative cell, whose cell wall can be resolved only with the electron microscope. A and B courtesy of Carolina Biological Supply Company; C Courtesy of Dr. Joseph P. Mascarenhas.]

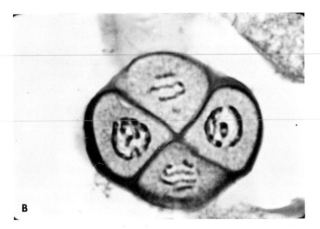

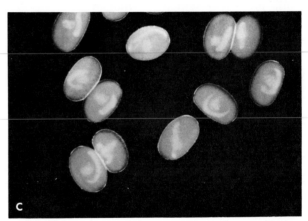

Beneath the exine an **intine** layer of cellulose is deposited (Figure 6.6). As time goes on, the tapetum secretes an enzyme **callase,** which dissolves the callose, thus freeing the individual pollen grains.

The synthetic processes that result in the production of the sculptured wall of the pollen grain go on within the individual microspores independently of concurrent gene action. This is proved by the fact that, now and again **sterile pollen grains** appear, that is, pollen grains that lack a nucleus. The resistant covering is, nevertheless, produced in the species-specific pattern (Heslop-Harrison, 1968).

The male gametophyte does not complete its development until it has been transported by wind or an animal vector to the stigma of the pistil. In most flowering plants the male gametophyte, as re-leased from the anther, consists of two cells produced by mitotic division of the microspore. These are encased within a single pollen wall, but are separate cells. These are the vegetative cell and the generative cell, as previously mentioned. The manner of their formation is illustrated for bluebell pollen in Figure 6.7. In some kinds of pollen the generative cell divides into two sperm cells before dispersal from the anther, but not until later in others, particularly the grasses. In some the generative cell is in the prophase of mitotic division as germination begins. Both sperm cells subsequently are used in fertilization.

Degeneration is normally the fate of the vegetative cell, but if microspores are isolated just before or immediately after mitosis and placed in an ap-

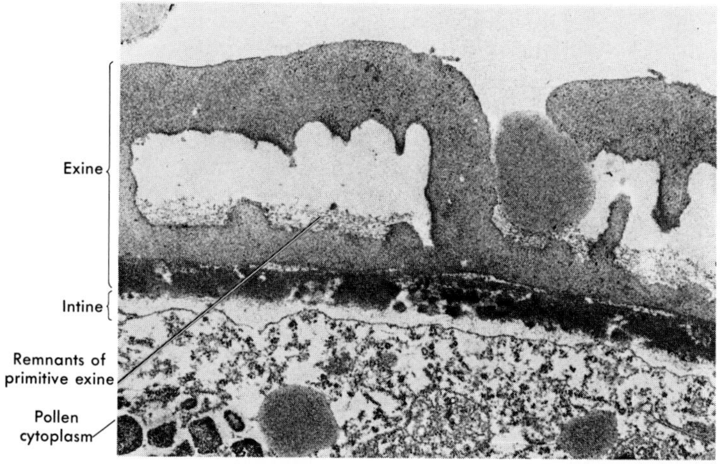

Exine

Intine

Remnants of
primitive exine

Pollen
cytoplasm

FIGURE 6.6 Electron micrograph of the wall
of a pollen grain that is almost mature. The
intine layer and the complex exine layer are
signaled, but the subdivisions of the exine
layer are not identified here. [Photograph
courtesy of Drs. Joseph Mascarenhas and
James T. Flynn.]

propriate medium, of which sucrose is a necessary component, the vegetative cell, but not the generative cell, divides mitotically to yield two identical cells, which then continue dividing and eventually form a plant embryo that can be cultured and brought to full development as a haploid plant of the sporophytic configuration (Nitsch and Nitsch, 1969).

This observation provides us with two special insights into the control of development in flowering plants. The first is that in flowering plants, even as in ferns, the pattern of development as sporophyte or gametophyte is not determined by whether the nucleus is haploid or diploid. The second is that the cytoplasmic surroundings of a nucleus determine its developmental response to environmental stimuli, for, whereas the vegetative and generative cells have sister nuclei, which are genetically identical, the cytoplasms surrounding each must be different. Cytoplasmic control of gene activity is considered at greater length later in this book.

6.2.1.1 Gene Action During Pollen Formation. Gene action consists, of course, in the transcription, or copying, of segments of the genomic DNA in the form of ribose nucleic acid (RNA) molecules. These are ribosomal RNA (rRNA), messenger RNA (mRNA), and transfer RNA (tRNA). The bulk of RNA synthesis is rRNA, and essentially that is the only aspect of gene action about which we have much information during pollen development. A good deal of rRNA is synthesized in the microspore nucleus immediately before mitosis and during a

short period in the vegetative nucleus after mitosis (Mascarenhas and Bell, 1970). Thereafter RNA synthesis ceases. The sperm nucleus apparently is not a site of RNA transcription.

Probably some mRNA and tRNA are produced

49

FIGURE 6.7 Schematic representation of the formation of the generative cell after microspore mitosis in the bluebell *Endymion non-scriptus*. A: Microspore mitosis. B–E: Formation of a wall around the nucleus of the generative cell; droplets of lipid accumulate in the vegetative cytoplasm lining the wall of the generative cell. F: In the mature pollen grain the wall of the generative cell becomes quite thin and can be distinguished only with some difficulty. [Redrawn with modifications from R. E. Angold, *J Cell Sci* 3:573–77 (1968).]

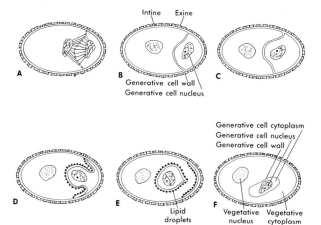

Intine   Exine

A

B

Generative cell wall
Generative cell nucleus

C

D

E

Lipid
droplets

F

Generative cell cytoplasm
Generative cell nucleus
Generative cell wall

Vegetative   Vegetative
nucleus   cytoplasm

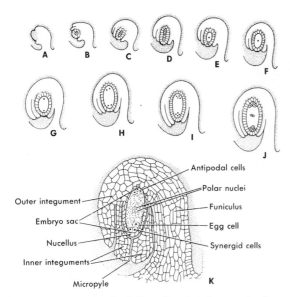

FIGURE 6.8 Scheme depicting the development of an ovule in a dicotyledonous plant. Within the nucellus a megaspore mother cell differentiates (B). The nucleus completes the first meiotic division (C) and then the second (D). Three of the resulting haploid nuclei degenerate (E), leaving the functional megaspore (F) with a single haploid nucleus. Three divisions then produce 8 haploid nuclei (G–I), which segregate, as shown in J and K. [Reprinted with permission of Macmillan Publishing Co., Inc., from *Development in Flowering Plants* by J. G. Torrey, p. 57. Copyright © 1967, J. G. Torrey.]

50

during the same period, for when radioactive uridine (which is incorporated into RNA, but not DNA) is administered, radioactively labeled RNA can be found on the chromosomes. These RNA species and the rRNA are subsequently utilized during pollen tube outgrowth, as described in Section 6.2.3.

6.2.2 ORIGIN OF THE EGG AND EMBRYO SAC.

The female reproductive organs are formed from one or more modified leaves, **carpels,** which form a **pistil** (Figures 5.1 and 6.3). The pistil is composed of ovary, stigma, and style. On the interior ovarian wall ovules appear, consisting of the **nucellus** and **megaspore mother cell** (Figure 6.8A, B). As development proceeds, the nucellus is raised on a stalk, and one or two protective layers, called **integuments,** grow from its base, completely surrounding

the developing megaspore, except for a small opening, the micropyle. As the integuments develop, the megaspore mother cell undergoes meiosis, forming a row of four haploid nuclei (Figure 6.8C, D).

6.2.2.1 Development of the Megagametophyte.

In about 70% of the species of flowering plants, only one of the haploid products of megaspore meiosis persists (Figure 6.8E, F). This survivor, the **megaspore nucleus,** gives rise to the **embryo sac,** which is the female gametophyte. Typically, the megaspore nucleus divides and its daughter cells and their daughter cells divide to give an eight-celled embryosac (Figure 6.8G–J). Three of the eight nuclei migrate to the pole opposite the micropyle forming the **antipodal cells:** two **polar nuclei** are positioned in the center of the embryo sac, and three go to the micropylar end. The latter give rise to the **egg cell** and two **synergids** (Figure 6.8K). The antipodals and synergids do not usually persist in the embryo. No proved function has been assigned to the antipodals, but the synergids and polar nuclei are important in fertilization.

6.2.3 FERTILIZATION.

Both of the sperm cells derived by division of the generative cell are involved in fertilization in flowering plants. One unites with the egg nucleus to form the zygote, from which the plant embryo develops, and the other fuses with the polar nuclei to form an **endosperm nucleus,** whose division and cellular compartmentation give rise to the **endosperm,** a nutritive tissue for the embryo. This is referred to as double fertilization (Figure 6.7F).

6.2.3.1 Formation of the Pollen Tube.

The pollen tube germinates when it lands on the stigma. Germination consists in the projection of a tube from the pollen grain and the growth of this tube through the style and into the ovary (Figure 6.9). There it enters the ovule through the micropyle. The tip of the tube penetrates between the cells of the tissues it traverses rather than through the cells. Its progress is probably aided by the enzymes, **callase, cellulase,** and **pectinase,** which are secreted by the pollen as soon as germination begins. These digest, respectively, callose, cellulose, and pectin, all complex polysaccharides, which make up plant cell walls and their strengthening framework.

Elongation of the pollen tube occurs entirely at its tip, where new cell wall material is progressively deposited. This material is formed from precursors

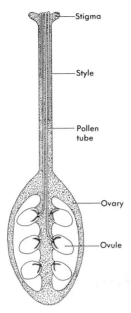

FIGURE 6.9 Germination of pollen grains that have landed on the stigma. Pollen tubes traverse the tissue of the stigma and fertilize the eggs.

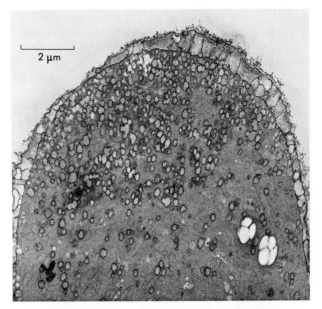

FIGURE 6.10 Electron micrograph of a longitudinal section through the tip of a growing pollen tube of *Lilium longiflorum*. A compartmented cap over the tip appears to receive contributions from subjacent cytoplasmic vesicles. [From W. G. Rosen and W. R. Gawlik, *Protoplasma* 61:181–91 (1966), by courtesy of the authors and permission of Springer-Verlag.]

51

that are packaged in Golgi-derived vesicles. These vesicles constitute the great bulk of material at the tip of the pollen tube, where their contents are discharged to the exterior through coaleacence with the plasma membrane (Figure 6.10).

Elongation of the pollen tube is characterized by intensive cytoplasmic streaming, predominantly in the direction of the tip. This streaming movement presumably carries not only Golgi vesicles and their intracellular organelles apically, but also transports the sperm cells and vegetative cell toward the tip. The vegetative cell is not involved in fertilization, but usually precedes the sperm cells during pollen tube formation.

The pollen of many plants will germinate, some with astonishing rapidity when placed in simple sugar solutions or, in some cases, in distilled water (Figure 6.11). The pollen tube of *Tradescantia*, for example, becomes visible in about 10 minutes. In about 50% of higher plants examined, pollen tubes caused to germinate in vitro will direct their growth toward bits of tissue excised from style or ovary and placed near them or toward a source of material extracted from female tissue. This is another case of

positive chemotaxis (see Section 6.2.3). Mascarenhas and Machlis (1964) suggested that calcium ion is the effective agent in snapdragon and in other plants that they studied, but Rosen (1964) found calcium to be ineffective as a chemotaxic

FIGURE 6.11 Pollen tube of the tiger lily germinated in vitro and stained to show the nuclei. The generative nucleus, as yet undivided, is indicated by arrow. [Courtesy of Carolina Biological Supply Company.]

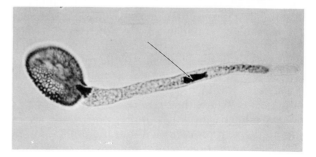

agent in *Lilium*. He reported that extracts of the lily pistil are effective, however. On the other hand, Jaffe et al. (1975) reported that *Lilium* pollen tubes, germinating in the presence of $^{45}Ca^{2+}$-labeled medium, accumulate calcium at the tip of the pollen tube in concentrations more than 100 times greater than elsewhere. This may not be related to chemotaxis, of course, but to the secretion of vesicles at the growing tip, as illustrated in Figure 6.10.

6.2.3.2 Gene Action During Pollen Germination.

Mascarenhas and Bell (1970) added radioactively labeled amino acids to the medium used for germinating *Tradescantia* pollen. They found that the radioactivity is incorporated into protein on preformed polyribosomes within 2 minutes. After 5 minutes polyribosomes become more numerous, as single ribosomes, meanwhile, diminish in number.

The pollen tube of *Tradescantia* germinated in vitro grows to a length of 6–8 mm in as many hours. One may ask whether this process requires gene activity. Mascarenhas and Bell (1970) germinated pollen on medium containing Actinomycin D, a drug that inhibits RNA synthesis, in order to answer this question. They found that the initial outgrowth of the pollen tube proceeds to a length of 250 μm, but no further, in the absence of RNA synthesis. Nevertheless, during the initial period, protein synthesis does take place, as indicated by the incorporation of radioactively labeled amino acids on new and preexisting polyribosomes.

This observation demonstrates that gene action is not required for the initiation of tube outgrowth, but is required for its continuation. The protein synthesized during the first phases of tube formation proceeds by utilizing RNA synthesized during the maturation of the gametophyte.

Mascarenhas and Bell then examined the incorporation of radioactive uridine (the distinctive nucleotide precursor of RNA) into RNA throughout germination in the absence of actinomycin D. Separation and fractionation of the newly synthesized RNA showed radioactivity only in relatively small RNA species in the size range sedimenting at about 8 Svedberg units (8 S), a size reasonable for mRNA. No activity was incorporated in rRNA, however, so one must conclude that, whereas the completion of pollen tube outgrowth requires the synthesis of some mRNA, it proceeds without the production of rRNA. In sum, pollen tube outgrowth is initiated

through the use of preformed protein-synthesizing machinery. Thereafter it proceeds only if new mRNA is transcribed, but continues in the absence of rRNA synthesis. The production of mRNA and rRNA in anticipation of their later utilization for the synthesis of protein is a phenomenon also observed in many animal eggs and embryos, as is treated later.

6.2.3.3 Entrance of the Sperm Cells into the Egg—Union of Nuclei.

In most plants the pollen tube enters the embryo sac at the micropyle. It then encounters the apex of the **filiform** apparatus, a complex thickening of the outer walls of the synergids at the micropylar end, which is penetrated by complex extensions of the synergid cell membranes. As the pollen tube approaches the synergids, one of them disintegrates. The tip of the pollen tube passes between the walls of the two synergids (Figure 6.12), turns, and enters the degenerating one, discharging both sperm cells into its cytoplasm through a hole that appears just behind the advancing tip. The sperm nuclei then pass to the

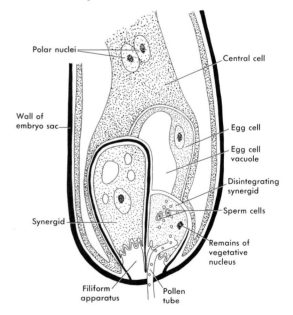

FIGURE 6.12 Discharge of the pollen tube. The pollen tube has entered the disintegrating synergid on the right and discharged both sperm. [Redrawn with modifications from W. A. Jensen, *Planta* (*Berl*) 78:158–83 (1968).]

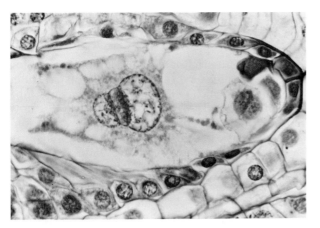

FIGURE 6.13 Double fertilization in *Lilium*. The zygote nucleus and synergids are seen at the right. At center are the fusing nuclei, a tetraploid and haploid nucleus from the embryo sac and the haploid sperm nucleus. Three triploid antipodals are not visible in this section. Note that the nuclear constitution of the embryo sac of the lily is not representative of what is found in most plants (see Figure 6.8). [Courtesy of Carolina Biological Supply Company.]

egg cell and polar nuclei, respectively, and join with them in a process of double fertilization (Figure 6.13).

The details of fertilization are poorly understood for plant gametes as compared with those of animals. Jensen and Fisher (1968) reported for cotton that each sperm nucleus in the pollen tube is surrounded by its own cytoplasm and plasma membrane, but no trace of these is found after fertilization. The fusion of egg and sperm nuclei in cotton is very similar to that of some animals (Chapter 8), in that it occurs without breakdown of either nuclear membrane (Jensen, 1964). In other plants, the male and female chromosome complements are formed out of separate nuclei, coming together eventually on a common mitotic spindle.

Except for certain rare plants in which the egg cell forms the sporophyte without fertilization, the egg sac is in physiological stasis and will not develop in the absence of sperm entry. The details of activation of the static egg have been but little studied in plants. Whatever occurs, however, must involve the reactivation of DNA synthesis in both male and female nuclei and in the polar nuclei. After fertilization both the zygote and endosperm nuclei begin to

divide, the latter more rapidly, to form the embryo and its nutritive endosperm.

6.2.4 FORMATION OF EMBRYO AND SEED.

The seed of a flowering plant consists of an embryo embedded in endosperm and encased in derivatives of the integuments and ovarian wall. Most seeds undergo a period of dormancy that ends with germination and formation of the seedling. In essentially all seeds, the embryo shows a distinct polarity, with an embryonic shoot apex, or **plumule** at one end and an embryonic root apex, or **radicle,** at the other. Embryonic leaves, as seed leaves, or **cotyledons,** attach to the embryonic axis at an intermediate portion, the **hypocotyl.** **Monocotyledonous** plants have only one seed leaf. This is often differentiated as a **scutellum** separating the embryo from the endosperm. In the grasses, which constitute the great majority of monocots, the first foliage leaves are formed within the seed and are encased in a specialized sheath, the **coleoptile** (Figure 6.14). **Dicotyledonous** plants have two seed leaves. In some dicots the rudiments of the first foliage leaves are also present in the seed.

In most plants the first cleavage of the egg cell is a transverse one, separating a **basal** cell, nearer the micropyle, from a **terminal** cell. Thereafter, cleavage planes vary in different patterns according to

53

FIGURE 6.14 Structure of the seed of wheat. In wheat seed the embryo is at a more advanced stage of development at maturity than is the embryo in the seed of *Capsella*. The wheat seed has a large starchy endosperm, which supplies energy for the early development of the plant, as described in Chapter 12. [Reprinted with permission of Macmillan Publishing Co., Inc., from *Development in Flowering Plants* by J. G. Torrey, p. 70. Copyright © 1967, J. G. Torrey.]

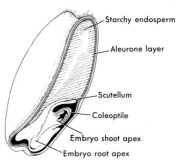

Starchy endosperm

Aleurone layer

Scutellum

Coleoptile

Embryo shoot apex

Embryo root apex

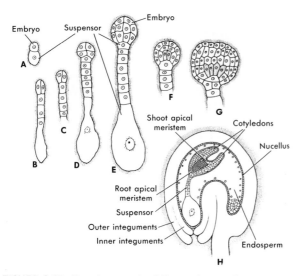

FIGURE 6.15 Development of the embryo of shepherd's purse, *Capsella bursa-pastoris*. A–G are drawn at the same magnification and show the formation of the embryo and suspensor. The older embryo (H), drawn at about half the magnification of A–G, shows the parts that are found in the mature seed. These parts should be identified by the reader in Figure 6.16B. [Reprinted with permission of Macmillan Publishing Co., Inc., from *Development in Flowering Plants* by J. G. Torrey, p. 63. Copyright © 1967, J. G. Torrey.]

the degree to which the basal cell contributes to the embryo, which arises chiefly from the terminal cell, and according to whether the first division of the terminal cell is transverse or longitudinal. Cleavage patterns in several kinds of flowering plants are shown in a book by Maheshwari (1950), a classical text in plant embryogenesis.

The development of *Capsella* is a frequently illustrated case in which there is little contribution of the basal cell to the embryo. In this plant (Figures 6.15 and 6.16) the basal cell becomes swollen and vesicular but gives rise to six to ten descendants, which form a **suspensor.** The elongation of the suspensor pushes the embryo deeper into the endosperm. Meanwhile, the terminal cell undergoes three divisions, each at right angles to the previous one, forming two tiers of four cells each. The four nearest the suspensor form the hypocotyl, and the other four give rise to the plumule and cotyledons. Only one cell of the suspensor is involved in forma-

tion of the embryo proper. It is the nearest one to the embryo, and it forms the **root cap** and **root cortex** (Figure 6.15F).

A series of **periclinal** divisions (divisions cutting off cells parallel to the surface) cuts off an outer layer of cells, which will form the outer layers of the future plant, and an inner mass. These cells now multiply, largely by transverse divisions, and elongate, especially marginally, to form the cotyledons. Divisions that separate cells along planes perpendicular to the surface are called **anticlinal.** Periclinal divisions and anticlinal divisions cooperate in shaping the embryo (Figure 6.15G).

FIGURE 6.16 Late stages in the development of the embryo of *Capsella bursa-pastoris*. A: Young embryo, showing cotyledons and suspensor. B: Mature embryo; the seed coat is forming from the integuments (the remaining components of the embryo are identified in Figure 6.15). [Courtesy of Carolina Biological Supply Company.]

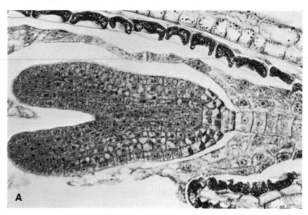

A

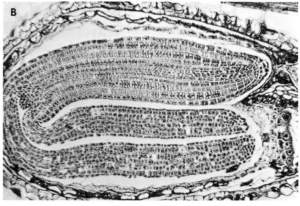

B

The endosperm is very important in formation of the seed. In some species, corn and wheat, for example (Figure 6.14), the bulk of the seed is a starchy endosperm that largely is not utilized until germination. In others, notably among legumes, the entire endosperm is transformed into storage materials in the cotyledons. This is the case for *Capsella* also (Figures 6.15 and 6.16); its cotyledons provide the energy for germination and growth of the seedling that is provided directly by the endosperm in other forms.

In all flowering plants the ovary develops into a **fruit,** which may or may not be familiarly known as such. Thus in the peach it forms a stony covering for the seed surrounded by a fleshy, edible pulp and covered with hairy skin. In grasses the ovary remains as a thin protective covering over the seed coat, which is derived from the integuments.

6.2.5 APOMIXIS AND PARTHENOGENESIS. Apomixis is the substitution for the sexual process in plant reproduction of a different process that does not involve any nuclear fusion. It takes various forms in the normal life cycle of different kinds of flowering plants, many of which are species or races found in taxonomic groups that usually show the normal fertilization process. In some cases the embryo sac and egg are formed without reduction divisions. The diploid egg is activated spontaneously and forms an embryo in an otherwise normal manner. This is called **diploid parthenogenesis** (Gr., virgin birth). Sometimes spontaneous activation also occurs in eggs that have been formed by meiosis. This is **haploid parthenogenesis.** In other forms of apomixis, the embryo arises from the nucellus or from one of the integuments. This is sometimes called **adventive embryony** or **sporophytic budding.**

Many attempts have been made to cause activation by artificial means of egg and polar nuclei that normally are activated by sperm. In contrast to results with some kinds of animal eggs (Chapter 8), in which parthenogenesis can be induced easily, little success has attended efforts to bring about artificial parthenogenesis in plants. Exposure of the flower to abnormal temperature has had limited success with oats *(Avena sativa),* and delayed pollination has produced some cases of parthenogenesis in *Petunia.*

6.3 Summary

In the life cycles of plants there is alternation of a vegetatively growing diploid form called the sporophyte generation and a haploid gametophyte generation. The sporophyte produces spore mother cells, which form spores after meiotic division. The haploid spores germinate to form the gametophyte, from which gametes arise.

In ferns the sporophyte is the more prominent generation, and the gametophyte consists of a small prothallium on which both male and female reproductive organs, antheridia and archegonia, respectively, are produced. Mature sperm are guided to the archegonia by chemotaxis.

Under altered conditions of growth, the zygote can make a new prothallium instead of a sporophyte. Likewise, sporophytes can arise directly as haploid structures from prothallial cells under certain conditions. The form of growth of each generation is thus dependent on environmental circumstances rather than on the number of gene sets.

In flowering plants the gametophyte generations are much reduced, and the male and female gametophytes are separate structures carried on the same or different flowers. The male gametophyte develops from spores that arise in sporogenous tissue born within an anther on the stamen. Sporogenous cells divide mitotically to produce microspore mother cells. They undergo meiosis synchronously to form haploid microspores. The synchrony of meiosis is probably promoted by the presence of cytoplasmic connections between the microspore mother cells.

Each microspore forms a pollen grain covered with a characteristic sculptured layer of protective material and having within a generative cell and a vegetative cell formed by microspore mitosis. The generative cell divides again by mitosis to form two sperm cells. This may occur before or after the pollen is shed from the anther.

Normally the vegetative cell degenerates, but it can divide mitotically on artificial media if the microspore is isolated at the appropriate time, and it can then give rise to a haploid sporophyte.

Before its degeneration the vegetative cell produces all of the ribosomal RNA that is to be used throughout pollen tube germination. It also pro-

55

duces the other RNA species required during the initial stages of germination. Germinating pollen may show chemotaxis toward material extracted from the female tissue.

The female gametophyte, megagametophyte, is formed in the pistil, which is composed of ovary, stigma, and style. The megaspore mother cell is surrounded by tissue of the nucellus. It undergoes meiosis to form the megaspore, which then divides mitotically to form the embryo sac containing the egg cell, and synergids near the micropyle, polar nuclei in the center, and antipodal cells at the other end.

The pollen tube passes through the micropyle, runs between the walls of the filiform apparatus at the base of the synergids, and then enters one synergid and discharges the two sperm cells. These then exit from the synergid, now degenerating, and fuse with the egg cell and polar nuclei respectively.

The endosperm arises by division of the fertilized polar nuclei. The embryo arises from the zygote by means of a species-specific pattern of cleavage divisions that segregate its materials into a spatial pattern that corresponds with the future stem, leaves, and root of the plant.

56

The expanding embryo is pushed into the endosperm by the elongation of suspensor cells derived from one daughter cell of the first cleavage. The embryo produces the primordial parts of the future plant, and then development comes to a halt. The integument derived from the nucellus contributes to special coats surrounding the embryo and endosperm. Together these constitute the seed, which is enclosed by derivatives of the ovary.

6.4 Questions for Thought and Review

1. Compare and contrast the life cycles of ferns and flowering plants.

2. What adaptive advantage might result from the reduction of the gametophyte in evolution of the land plants?

3. Malic acid has been identified as a chemical attractant for fern sperm, possibly guiding them to the archegonium. How could you distinguish whether the malic acid serves as a chemotactic substance or simply as a trap for sperm that swim into a high concentration of it?

4. What evolutionary advantages and disadvantages might accrue to plants that normally reproduce by parthenogenesis or some other form of apomixis.

5. The pattern of growth shown by the sporophyte or gametophyte is not a function of the $1n$ or the $2n$ condition. Evaluate this proposition and adduce the evidence required to defend your position.

6. Protein synthesis occurs during growth of the pollen tube. What is the evidence for this? The initial phases of pollen germination require protein synthesis but not RNA synthesis? Is this statement correct? What is the evidence on which you base your answer?

7. When is the ribosomal RNA produced that is used during germination of the pollen tube? What experiment would you undertake to answer this question?

6.5 Suggestions for Further Reading

DICKINSON, H. G., and J. LAWSON. 1975. The growth of the pollen tube wall in *Oenothera organesis*. *J Cell Sci* 18:519–32.

GALSTON, A. W. 1967. Regulatory systems in higher plants. *Am Sci* 55:144–60.

HESLOP-HARRISON, J. 1966. Cytoplasmic continuities during spore formation in flowering plants. *Endeavour* 25:65–72.

HESLOP-HARRISON, J. 1975. The Croonian Lecture, 1974. The physiology of the pollen grain surface. *Proc R Soc Lond* [B] 190:275–99.

JENSEN, W. A. 1965. The ultrastructure and histochemistry of the synergids of cotton. *Am J Bot* 52:238–56.

MACKENZIE, J. M., Jr., R. A. COLEMAN, W. R. BRIGGS, and L. H. PRATT. 1975. Reversible redistribution of phytochrome within the cell upon conversion to its physiologically active form. *Proc Natl Acad Sci USA* 72:799–803.

MASCARENHAS, J. P. 1975. The biochemistry of angiosperm pollen development. *Bot Rev* 41:259–314.

MULLER, W. H. 1979. *Botany: A Functional Approach.* New York: Macmillan.

MYLES, D. G. 1978. The fine structure of fertilization in the fern *Marsilea vestita*. *J Cell Sci* 30:265–82.

NÄF, U., J. SULLIVAN, and M. CUMMINS. 1974. Fern antheridiogen: cancellation of a light-dependent block to antheridium formation. *Dev Biol* 40:355–65.

RAGHAVEN, V. 1964. Interaction of growth substances on growth and organ initiation in the embryos of *Capsella*. *Plant Physiol* 39:816–21.

RIOPEL, J. L. 1973. *Experiments in Developmental Botany.* Dubuque, Iowa: Brown.

SCHULZ, SISTER R., and W. JENSEN. 1968a. *Capsella* embryogenesis: the early embryo. *J Ultrastruct Res* 22: 376–92.

SCHULZ, SISTER R., and W. A. JENSEN. 1968b. *Capsella* embryogenesis: the egg, zygote, and young embryo. *Am J Bot* 55:807–19.

6.6 References

HESLOP-HARRISON, J. 1968. The emergence of pattern in the cell walls of higher plants. In M. Locke, ed. *The Emergence of Order in Developing Systems*. New York: Academic Press, pp. 118–50.

JAFFE, L. A., M. H. WEISENSEEL, and L. F. JAFFE. 1975. Calcium accumulations within the growing tips of pollen tubes. *J Cell Biol* 67:488–92.

JENSEN, W. A. 1964. Observations on the fusion of nuclei in plants. *J Cell Biol* 23:669–72.

JENSEN, W. A., and D. B. FISHER. 1968. Cotton embryogenesis: the entrance and discharge of the pollen tube in the embryo sac. *Planta (Berl)* 78:158–83.

MAHESHWARI, P. 1950. *An Introduction to the Embryology of Angiosperms*. New York: McGraw-Hill.

MASCARENHAS, J. P. 1975. "The Biology of Pollen," *W. A. Benjamin Module in Biology N. 14*. Menlo Park, CA: W. A. Benjamin, 30 pages.

MASCARENHAS, J. P., and E. BELL. 1970. RNA synthesis during development of the male gametophyte of *Tradescantia*. *Dev Biol* 21:475–90.

MASCARENHAS, J. P., and L. MACHLIS. 1964. Chemotropic response of the pollen of *Antirrhinum majus* to calcium. *Plant Physiol* 39:70–77.

NÄF, V. 1968. On the separation and identity of fern antheridogens. *Plant and Cell Physiol* 9:27–33.

NAKANISHI, K., M. ENDO, V. NÄF, and L. F. JOHNSON, 1971. Structure of the antheridium-inducing factor of the fern *Anemia phyllides*. *J Am Chem Soc* 93:5579–81.

NITSCH, J. P. and C. NITSCH. 1969. Haploid plants from pollen grains. *Science* 168:85–87.

ROSEN, W. G. 1964. Chemotropism and fine structure of pollen tubes. In H. F. Linskens, ed. *Pollen Physiology and Fertilization*. Amsterdam: North-Holland, pp. 159–66.

7

Gametogenesis
in Animals

58 Animal gametes are produced in the **gonads** (Gr., *gonos*, referring to procreation), eggs in the female gonad or **ovary** and sperm in the **testis.** In each kind of gonad the gametes take their origin from a primitive line of cells that was set apart from the general body cells at some early stage of embryonic development. These are the **primordial germ cells.** Arising outside of the gonad, they migrate into it as it is formed, later differentiating into male or female gametes according to whether the gonad is a testis or an ovary (see Chapter 11).

After entering the gonads the primordial germ cells may continue to increase in number, but usually cease multiplying sometime before birth or hatching. Thereafter, they acquire different cytological characteristics, becoming **gonial** cells (i.e., **spermatogonia** in the male and **oögonia** in the female). The transformation of primordial germ cells into gonial cells has been but little studied, and the process remains obscure. Gonial cells also divide. In the testis, spermatogonia are present throughout reproductive life providing stem cells from whose progeny **spermatozoa** are produced seasonally or in wave-like cycles (reviewed by Roosen-Runge, 1962, and Clermont, 1972). In ovaries of most higher vertebrates such as the human being (Baker, 1963), the rat (Beaumont and Mandl, 1962), and in the chick (Hughes, 1963), however, the gonial divisions usually cease before birth, or shortly after, and thereafter no more oögonia are produced. In the rat, which is born on about the 22nd day of gestation, the peak of oögonial divisions occurs at about 15 days (Franchi and Mandl, 1962). Lower vertebrates, such as many teleosts and most or all amphibians, retain oögonia in the peripheral region of the ovary. At the start of each reproductive season, these give rise to additional oögonia which then begin meiosis, grow, and become mature.

The process whereby sperm are formed from spermatogonia is called **spermatogenesis.** The formation of the ovum or egg cell is called **oögenesis.** Both oögenesis and spermatogenesis involve drastic changes in the cytoplasm and in the nucleus, such that the end-product little resembles the gonial cell. The cytoplasmic changes that occur in the male germ cell result in a sperm cell that is small and motile; those that take place in the female result in a large nonmotile egg that, in many species, contains a considerable amount of stored energy in the form of protein and lipid.

This chapter deals with the structure of sperm and of eggs and the method of their production. It emphasizes the action of the genes during gametogenesis, hormonal factors controlling ovulation in higher animals, and some aspects of spawning in lower forms.

7.1 Spermatogenesis

The male gamete called a **spermatozoon,** or simply, a sperm, arises from the gonial cells through meiosis and a subsequent sequence of complex cytological changes. These changes constitute spermatogenesis.

Sperm cells are terminally differentiated cells whose only function is to engage in fertilization. Those formed by different kinds of organisms are variously constructed, but all have in common that the genetic material is in a highly condensed and essentially inactive form in the nucleus. Sperm are, moreover, usually endowed with some form of comotory apparatus and with mitochondria. The

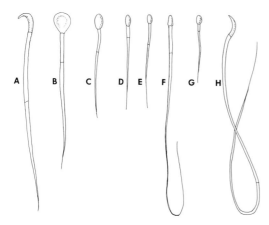

FIGURE 7.1 Examples of various mammalian sperm:
(A) rat, (B) guinea pig, (C) bull, (D) cat, (E) dog,
(F) bandicoat, (G) human, (H) hamster.

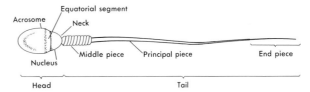

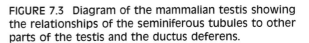

FIGURE 7.2 The parts of a typical mammalian sperm,
as reconstructed from electron microscopic studies of
sperm of bull, bat, and man. The middle piece
characteristically consists of a spiral or annular array of
mitochondria.

latter serve as energy transducers to provide the power for locomotion. Most commonly the loco-motory apparatus consists of a long tail, but it may be in the form of stiff spines, blunt pseudopodia, or undulating membranes. Some of the forms shown by animal sperms are illustrated in Figure 7.1.

7.1.1 STRUCTURE OF SPERM CELLS.
Most sperms consist of two chief parts, **head** and **tail.** The head consists of a nucleus and **acrosome.** The latter is a membrane-bounded vesicle containing **hydrolytic enzymes.** In human beings and many mammals it tends to fit as a cap over the nucleus, and in inverte-brates it is more likely to be located entirely at the apex of the nucleus. In mammals the head is joined to the tail by a **connecting piece,** and the tail con-sists of a **middle piece,** a **principal piece,** and an **end piece.** These divisions are lacking in most in-vertebrate sperm. The middle piece is characterized by an array of mitochondria (reviewed by Phillips, 1977) that extends for a variable length, terminat-ing in a ring, or **annulus.** In both vertebrate and invertebrate sperm one or more centrioles is pres-ent in association with the base of the nucleus. Two are usually present in invertebrates, the distalmost one being associated with or giving rise to the axial filament complex of the tail. In invertebrates, the axial filament complex is essentially the only com-ponent of the tail, whereas the tail of higher verte-brates shows massive columns of fibrous elements in addition. These and other details of a typical

mammalian sperm are illustrated in Figure 7.2. Other illustrations are found in Chapter 8, which deals with fertilization of animal eggs. Fawcett (1975) reviewed results of ultrastructural studies of mammalian sperm in considerable detail.

7.1.2 THE SEMINIFEROUS EPITHELIUM.
In higher animals, particularly, the sperm arise in a complex array of seminiferous tubules (Figure 7.3) in the testis. The **seminiferous epithelium** is a complex stratified epithelium that surrounds the lumen of the seminiferous tubule. This epithelium consists of

59

FIGURE 7.3 Diagram of the mammalian testis showing the relationships of the seminiferous tubules to other parts of the testis and the ductus deferens.

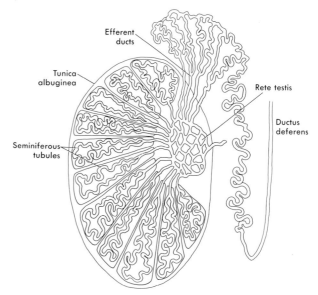

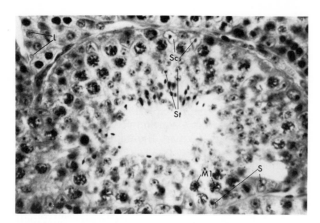

FIGURE 7.4 Cross section of a mammalian seminiferous tubule and interstitial tissue: L, cells of Leydig; M1, primary spermatocyte; Sc, Sertoli cell; St, spermatids. The dark-staining nuclei of maturing spermatids are seen adjacent to the lumen of the tubule.

the spermatogenic cells and extend through the spermatogenic layer, but their nuclei always remain adjacent to the basement membrane of the tubule (Figure 7.6). The relationships of spermatogenic cells to Sertoli cells have been analyzed in a number of vertebrates by Dym and Fawcett (1971), Vitale-Calpe (1970), Burgos and Vitale-Calpe (1967), and Gilula et al. (1976).

7.1.3 SPERMATELEOSIS.

Spermateleosis is the transformation of the spermatid into the spermatozoon. It is a complex process that proceeds through a series of cytological changes designated as **steps.**

60

spermatogenic cells and Sertoli cells (Figure 7.4). The former are cells in various stages of spermatogenesis; the latter are cells that are involved in the support and nourishment of the spermatogenic cells. Spermatogonia multiply adjacent to the basement membrane, but are moved toward the lumen as they go through meiosis and transform into sperm cells. When a cell enters the prophase of the first meiotic division it is called a **primary spermatocyte.** At the completion of the first division the two daughter cells are **secondary spermatocytes.** Secondary spermatocytes divide to form **spermatids,** and these then undergo a drastic transformation into spermatozoa, or sperm cells. Each spermatocyte thus gives rise to four spermatozoa.

In human beings, and probably in most animals, dividing spermatogonia and spermatocytes are found in clusters connected by means of cytoplasmic bridges (Dym and Fawcett, 1971). Groups of interconnected spermatogonia divide synchronously and proceed into meiosis together, retaining cytoplasmic continuity until the mature sperm are ready for release to the lumen of the seminiferous tubule (Figure 7.5).

Division of spermatogonia and spermatocytes and the subsequent differentiation of spermatids into spermatozoa occur within recesses that extend deep into the Sertoli cells. The cytoplasmic extensions of the Sertoli cells conform to the contours of

FIGURE 7.5 Schematic representation of the connection, by means of cytoplasmic channels, of cells of common descent progressing simultaneously through successive stages of spermatogenesis. [Redrawn with modifications from M. Dym and D. W. Fawcett, *Biol Reprod* 4:195–215 (1971).]

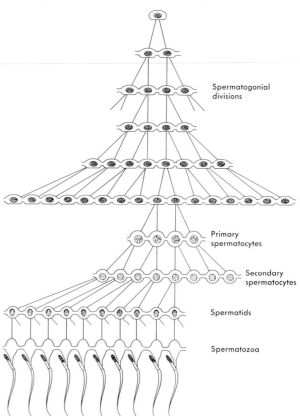

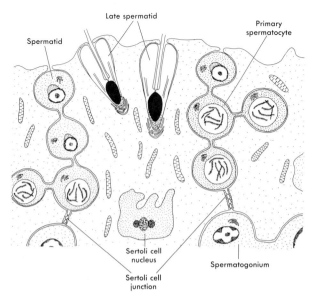

FIGURE 7.6 Schematic representation of the relationship of Sertoli cells to spermatogenic components of the seminiferous epithelium. The Sertoli cells extend the full thickness of the seminiferous epithelium, whereas the spermatogenic cells occupy deep recesses between them, moving toward the lumen as they differentiate. [Redrawn with modifications from D. W. Fawcett in C. L. Markert and J. Papaconstantinonu, eds., *The Developmental Biology of Reproduction*, Academic Press, New York, p. 29.]

Twelve steps are recognized in man (Clermont and Leblond, 1955), 19 in the rat, 13 in monkeys, and 16 in mice. These steps are grouped into **phases** whose names are descriptive of the principal activities that characterize them. Some of the steps in the phases of human spermateleosis are shown in Figure 7.7.

The first three steps constitute the **Golgi** phase (Figure 7.7A–C) during which the Golgi apparatus begins production of the **acrosomal vesicle** at the side of the nucleus facing the lumen of the seminiferous tubule. This marks the anterior end of the future head of the sperm. At the opposite side of the spermatid the centrioles begin to organize the tail fibers.

During the **cap** phase (Figure 7.7D, E), the acrosome spreads over the surface of the nucleus, now condensing, and the spermatid rotates, so the acrosomal complex faces the basement membrane of the seminiferous tubule, and the developing tail faces the lumen.

During the **acrosome** phase (Figure 7.7F, G), the nucleus elongates and the cytoplasm is displaced towards the growing flagellum, leaving the acrosome covered only by the plasma membrane anteriorly. An equatorial ring of fibrous material around the nucleus serves as the origin of an array of microtubules, the **manchette,** which extends as a cylinder into the distal cytoplasm. The manchette is possibly involved in the displacement of the cytoplasm of the spermatid. It disappears as the cytoplasm recedes from the nucleus.

The final steps of spermateleosis are the **maturation** phase (Figure 7.7H). The nucleus becomes flattened and highly condensed. The centriole and columns of the connecting piece appear, and the mitochondria become organized around the proximal part of the flagellum to form the midpiece.

FIGURE 7.7 Phases in the transformation of the spermatid to the spermatozoon: (A–C) Golgi phase; (D, E) cap phase; (F, G) acrosome phase; (H) maturation. [Adapted from a variety of sources, especially J. Clermont and C. P. Leblond, *Am J Anat* 96:229–53 (1955).]

61

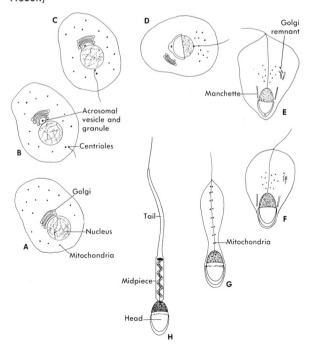

7.1.4 GENE ACTION AND PROTEIN SYNTHESIS DUR-ING SPERMATOGENESIS.

The synthesis of DNA ceases in spermatogenic cells at the last premeiotic replication of the genome, except as it is involved in formation of the synaptonemal complex and in the repair of chromosome breaks that occur in crossing over (Chapter 3). Transcription of the genome, however, continues during spermatogenesis to an extent that is variable among different kinds of organisms. Protein synthesis generally continues until the late stages of spermateleosis.

Gene action during spermatogenesis has been extensively studied in several species of *Drosophila*. In these insects, transcription ceases during the primary spermatocyte stage. Thereafter protein synthesis continues and viable sperm are produced even if most of the genome is eliminated. Thus, in certain mutant strains of *Drosophila melanogaster*, chromosomal deficiencies are produced at the end of the first meiotic division with the result that spermatids are formed lacking chromosome 2 or chromosome 3, or both, but having the very small chromosome 4 and either the X or the Y chromosome. These spermatids differentiate normally, becoming fully viable and fertile (Lindsley and Grell, 1969).

In 1968 Hennig showed that testicular tissue of males of *Drosophila hydei* contains nonribosomal RNA species that are not found in females or in nontesticular tissue of males. These are produced in first meiotic prophase and persist into the stages of spermateleosis. As females have no Y chromosomes, the male-specific RNA is presumably transcribed from the Y in males. This interpretation is supported by the fact that males having two Y chromosomes produce more of the specific RNA than do those with only one.

In some species the chromosomes of spermatocytes in first meiotic prophase show more or less elaborate loops of chromatin that are believed to be the sites of mRNA synthesis. Such extensions are frequently seen during oögenesis, and chromosomes bearing them are called lampbrush chromosomes (see Figure 7.26). Lampbrush chromosomes are usually hard to distinguish in spermatogenic cells. In *D. hydei* there are five distinct chromatin loops that project from the body of the Y chromosome (Figure 7.8). These are probably the sites at which some of the Y-specific RNA is synthesized. The RNA is significantly involved in the formation of the sperm tail, for if one or more of the loops is

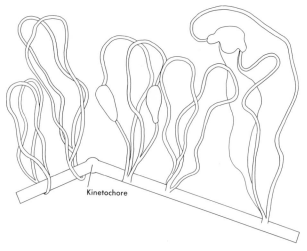

FIGURE 7.8 Diagram of the relative portions of loop-forming sites in the Y chromosome of *Drosophila hydei* at the lampbrush stage. Each loop is distinctive and occupies a particular mapped site on the long or short arm of the chromosome. [Redrawn and simplified from O. Hess, *Genetics* 56:283–95 (1967).]

deleted or suppressed by gene mutation, the resulting sperm show structural defects in the tail and are infertile. Interestingly, all of the normal components of the tail are present, mitochondria, axial filaments, and so on, but their organization is quite abnormal (Hess, 1973). Because all structural components of the tail are present, it is clear that the loops are not involved in the synthesis of mRNA for structural protein of the tail but, rather, are the templates for proteins having to do with regulating the assembly of tail proteins.

Studies of a number of other animals have shown that RNA synthesis occurs both during the first meiotic prophase and in the spermatid. In mice the peak of incorporation of ^{3}H-uridine into RNA occurs during pachytene. During the meiotic divisions the rate of incorporation of labeled uridine is low but, during the first half of spermatoleosis, it may peak at about 80% of the rate at pachytene. Synthesis of RNA ceases when the nucleus begins to elongate. Elongation begins at step 9, half way through the 16 steps of spermateleosis in mice. The synthesis of mRNA and rRNA occurs in spermatids as well as in pachytene spermatocytes, but the production of rRNA proceeds very slowly in spermatids

Kinetochore

(Geremia et al., 1978; Iatrou and Dixon, 1978; Schmid et al., 1977; Erickson et al., 1980). Kalt (1979) dissociated spermatogenic cells of *Xenopus,* labeled them with ^{3}H-uridine, and then separated them into populations enriched for various meiotic stages. Labeled mRNA, rRNA, and tRNA were produced by spermatocytes and by spermatids in all but the latest stages of spermateleosis.

The incorporation of amino acids into protein continues until the sperm is almost mature. In the mouse, labeled amino acids are incorporated into both cytoplasmic and nuclear proteins until step 15, when the final stage of maturation of the spermatid begins. Because RNA synthesis ceases at step 9 of spermateleosis in the mouse, subsequent protein synthesis must utilize previously synthesized long-lived RNA. This RNA would necessarily include mRNAs for the cytoplasmic proteins found in the acrosome and its membranes and those making up the complex elements of the tail. The long-lived RNA would also include mRNA for special proteins that accumulate in the nucleus as it elongates and condenses. The nuclear protein of somatic cells, spermatogonia, spermatocytes, and early spermatids includes **lysine-rich histones,** a set of basic proteins closely associated with the DNA double helix. In the mouse the lysine-rich histones are replaced by more basic **arginine-rich histones** during steps 11–14 (Figure 7.9). The arginine-rich histones are the nuclear proteins of the definitive mouse sperm.

FIGURE 7.9 Incorporation of labeled arginine into nuclear protein during spermatogenesis in the mouse: (A) resting stage after premeiotic S, 1.5 days; (B) the period of leptotene and zygotene, 3 days; (C) pachytene, 7 days; (D) diplotene and diakinesis, 1 day; (E) first and second meiotic divisions completed, 0.5 day; (F) spermateleosis, 13.5 days; (G) synthetic period for arginine-rich histones during steps 11–14.

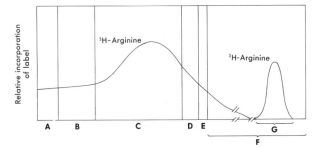

In most kinds of organisms the final type of protein found in the nucleus is likewise arginine rich and is called **protamine.** The protamines are relatively small polypeptides, being composed of as few as 31 or 32 amino acids, of which as many as 90% may be arginine. The replacement of histones by protamines has been carefully studied during spermatogenesis in the rainbow trout *Salmo gairdnerii.* Iatrou et al. (1978) prepared DNA complementary to protamine mRNA (cDNA). The cDNA was then used as a probe to determine at which stages of spermatogenesis protamine mRNA was transcribed. It was found that mRNA sequences complementary to the cDNA probe are present in both nucleus and cytoplasm as early as the primary spermatocyte stage. Protamine mRNA was not found on cytoplasmic polyribosomes, however, until protamine synthesis could be detected in spermatids, having meanwhile been stored in the cytoplasm as inactive ribonucleoprotein.

Condensation of the chromatin in conjunction with the accumulation of arginine-rich protein might reasonably be considered as a mechanism for denying access to the genome of RNA polymerase, thus preventing genetic transcription. Chromatin isolated from trout testes at progressively later stages of spermatogenesis (Marushige and Dixon, 1969) is progressively less able to show template activity when tested in vitro with RNA polymerase and the appropriate nucleotides. Pure undenatured nucleoprotamine is completely inactive in support of RNA synthesis. It may be, however, that nuclear condensation and accumulation of basic protein have their chief importance in connection with preparation for transporting the nucleus to the egg rather than the denial of genetic transcription.

7.2   Oögenesis

Egg cells are usually quite large in comparison to the size of sperm, and most kinds are spherical in shape. The human egg, for example, has a diameter of about 140 μm at the time of **ovulation** (release from the ovary). The human sperm, however, whose total length consists mostly of the tail, is only 70 μm long, and the greatest dimension of the head is only about 5 μm. The sizes of eggs found

63

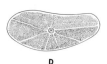

FIGURE 7.10 The distribution of yolk (stipples) and position of the nucleus in isolecithal (A), telolecithal (B, C), and centrolecithal (D) eggs.

among various kinds of animals vary over a tremendous range, up to several centimeters in birds and in sharks, whose eggs are the largest of all. The content of large eggs is chiefly a protein food reserve, **yolk.**

7.2.1 THE ORGANIZATION OF ANIMAL EGGS.

For most kinds of eggs the nucleus is relatively small as compared to the amount of cytoplasm, and its chromatin is rather diffuse. The nucleus may be centrally located, as in human eggs, which contain no yolk, but in eggs that have yolk, it is most often found eccentrically, in a zone of cytoplasm that is relatively yolk free. In such eggs the position of the yolk-free zone marks the **animal pole** of the egg. The opposite pole is referred to as the **vegetal pole,** and the imaginary line connecting the poles is the **animal–vegetal axis.** The yolk is usually arranged in a gradient of increasing concentration from animal to vegetal poles (Figure 7.10B). Cytoplasmic organelles are few in yolky areas; elsewhere, mitochondria are numerous but endoplasmic reticulum often is relatively sparse. Many kinds of eggs store considerable quantities of ribosomes and glycogen. Lipid is present in the form of small droplets interspersed with the yolk or in large, macroscopically visible spheres. Golgi material is present, often in numerous complexes, and there are often stacks of flattened cisternal plates with annular rings in register. These are called **annulate lamellae** (Figure 7.11). In oögonia of most animals, invertebrate and vertebrate, including those of the human ovary, there have been described conspicuous paranuclear masses composed of mitochondria, smooth endoplasmic reticulum, scattered Golgi complexes, annulate lamellae, and dense, vacuolated compound aggregates otherwise unidentified. These surround a mass of dense granules, closely packed vesicles, and fibers, classically termed the **cytocentrum,** or **cell center** (Figure 7.12). The whole constitutes the **Balbiani body,** first described in 1864 by E. G. Balbiani and often referred to as **Balbiani's vitelline body** (Hertig, 1968). This is a transient structure that is dissipated before meiosis begins.

In many kinds of eggs that are ready for fertilization a **cortex** or outer protoplasmic layer is recog-

FIGURE 7.11 Annulate lamellae in a primordial oöcyte in the human ovary as shown by electron microscopy. The nucleus is seen at the bottom, and the external layer of the nuclear envelope appears to give rise to the lamellae by evaginating into multiple folds. [From A. T. Hertig, *Am J Anat* 122:107–38, (1968), by permission of Alan R. Liss, Inc.]

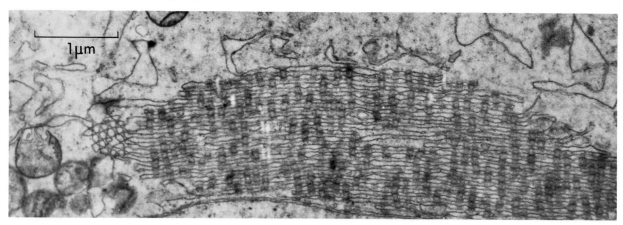

1μm

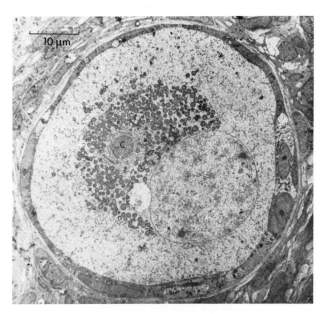

FIGURE 7.12 Electron micrograph of a human primordial oöcyte depicting the crescentic Balbiani body to the left of the nucleus. The centrosome, C, is surrounded by a halo of less electron-dense material that is, in turn, flanked by massed mitochondria. [From A. T. Hertig, *Am J Anat* 122:107–38 (1968), by permission of Alan R. Liss, Inc.]

nized. It is relatively thin and free of subcellular organelles and yolk. Pigment granules of kinds that are peculiar to the species are found in the cortex, often varying in concentration along the animal–vegetal axis. In all species there are in the cortex distinctive membrane-bounded vesicles that contain glycoproteins. These vesicles are variously called **cortical alveoli, cortical vesicles,** or **cortical granules.** They are usually about 0.5 μm in diameter and are found beneath the egg-cell membrane or in contact with it. In most animals, these vesicles rupture when the sperm makes contact with the cortex of the egg or, in some cases, when the egg is pricked with a needle. The organization of the cortical granules and their behavior during and immediately following fertilization are described in the next chapter. Their origin in the hamster was described by Selman and Anderson, 1975.

7.2.2 COVERING MEMBRANES. In addition to the oölemma, or plasma membrane, eggs are covered by one or more protective membranes. The **primary egg membrane** is produced by the egg cell alone or in cooperation with the follicle cells of the ovary as the egg undergoes growth and maturation. In some kinds of animals, secondary egg membranes are added as the egg passes through the oviduct.

The primary egg membrane is often called the **vitelline membrane.** In eggs of sea urchins it consists of only a very thin layer of acid mucopolysaccharide. In other cases it is a complex structure of fibrous proteins produced jointly by the egg and its surrounding follicle cells. It may be penetrated partially or completely by the microvilli. The vitelline membrane in man and other mammals (Figure 7.13) is called the **zona pellucida** after ovulation. It is derived from the **zona radiata** pictured in Figure 7.13. The term **chorion** is applied to the thick fibrous vitelline membrane covering eggs of fishes. In insects a chorion is added by activities of the ovarian follicle cells after the vitelline membrane is produced. Eggs with chorions usually have a **micropyle,** a funnel shaped opening through which the sperm gains access to the surface of the egg.

Familiar secondary membranes are the jelly that covers the eggs of amphibians (Figure 7.14A) and hard protective shells such as those found in birds (Figure 7.14B) and reptiles. Some shells, notably those of many parasites, are quite impermeable, even to gases, and so resistant to chemical attack as to be dissolved only in the presence of digestive enzymes or special hatching enzymes produced by the embryo. Others, such as those covering the eggs of birds and reptiles, are permeable to respiratory gases and water vapor. Eggs of birds and reptiles probably have the most complex systems of secondary membranes of all vertebrates. As the avian egg enters the oviduct, a fine fibrous layer is added to the vitelline membrane. Then a complex, highly hydrated covering of protein (mostly albumins) is added, then two layers of matted keratin fibers, which constitute the shell membrane, and finally, the shell. The latter is chiefly composed of calcium carbonate. It is pierced with fine pores within which is a collagen-like protein.

7.2.3 NATURE, ORIGIN, AND DISTRIBUTION OF YOLK. What we speak of as the "yolk" of an egg is not a chemical entity but a mixture of proteins, phospholipids, and neutral fat. Proteins constitute the principal component of the yolk, however, and

65

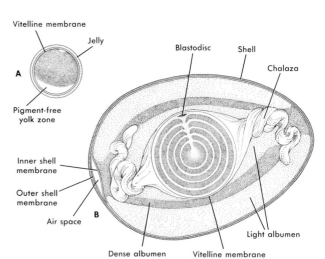

FIGURE 7.13 Relationships between oöcyte and follicle cells in the mammalian ovary. Protoplasmic projections from the follicle cells, FP, penetrate the zona radiata, ZR, and make contact with the oöcyte surface. In the zona the microvilli, MV, of the oöcyte are seen parallel to the follicular cell projections. ER, endoplasmic reticulum; N, nucleus; NU, nucleolus; M, mitochondrion. [From E. Anderson and H. W. Beams, *Ultrastruct Res* 3:432–46 (1960), by courtesy of the authors and permission of Academic Press, Inc., New York.]

FIGURE 7.14 Primary and secondary membranes of the unfertilized egg of the frog (A) and of the newly laid egg of the hen (B).

their origin and composition are treated later. In amphibians the yolk proteins are localized in **platelets** as a crystalline lattice within a membrane that is largely acid mucopolysaccharide (Figure 7.15). The yolk of the hen's egg (Bellairs, 1961, 1967) consists of an aqueous phase containing free-floating droplets that are not membrane bounded, and a variety of spheres, enclosed in membranes, within which are subdroplets enclosing lipids and droplets of soluble protein. This yolk cannot be so neatly described as that of amphibians. During fixation for electron microscopy, protein precipitates from the aqueous phase in irregular granules. The protein droplets in the yolk spheres show electron-dense granules but no crystalline structure. Wallace (1963) suggested that such dense granules are homologous with the crystalline component of the amphibian yolk platelets.

Yolky eggs show a great size increase during their differentiation. This is chiefly because of their

FIGURE 7.15 Electron micrograph of a yolk platelet of *Ambystoma maculatum*, showing the crystalline organization of the yolk and close association with a mitochondrion. [Photograph graciously supplied by Dr. R. W. Rice.]

accumulation of yolk. In vertebrates with yolky eggs, such as fishes, amphibians, reptiles and birds, and insects, the principal yolk proteins, designated **vitellogenic proteins** or **vitellogenins,** are produced exogenously (that is, outside of the ovary). They are then carried by the blood (or hemolymph in the case of the insects) to the ovary, where they are taken up selectively by the oöcytes during the **vitellogenic** phase of oöcyte growth. This is the phase during which they show their most rapid growth. During the previtellogenic phase of oögenesis in most vertebrates some yolk is synthesized endogenously, as illustrated in Figure 7.16. Among crustaceans, there is evidence from several species that much of the yolk protein is produced endogenously (Figure 7.17) (Beams and Kessel, 1963; Lui and O'Connor, 1977), but there is considerable evidence that a vitellogenin, almost entirely lipovitellin, is taken up from the hemolymph (Wolin et al., 1973).

The vitellogenic proteins are calcium-binding lipophosphoproteins of relatively high molecular weight. In vertebrates with large-yolked eggs they are produced in the liver in response to estrogens. Vitellogenins are normally produced in the ovarian follicles of the female during the breeding season, but during the nonbreeding season estrogen injections elicit from the livers of either male or female animals an almost complete switch-over to the production of vitellogenin, glycogen reserves of the liver being depleted to provide energy for this process (Wallace and Jared, 1968; Bergink et al., 1974). The dramatic change in liver cells switching to vitellogenin synthesis in the amphibian is illustrated in Figure 7.18. Released from the bloodstream, the vitellogenin is taken up by the oöcyte and broken down into distinctive moieties designated as **phosvitin** and **lipovitellin,** the former having a high content of phosphorus and the latter a major portion of lipid. The phosvitins and lipovitellins of different organisms show differing contents of phosphorus and lipid and different amino acid sequences (Wallace, et al., 1967; Jared and Wallace, 1968; Bergink and Wallace, 1974; R. C. Clark, 1972, 1976; Lui and O'Connor, 1977). The vitellogenin of *Xenopus laevis* has been intensively studied. It leaves the liver as a single polypeptide (Whali et al., 1978), complexed with lipid, having a molecular weight of 450,000; 12% of this is lipid. In the ovary the vitellogenin is selectively taken up by the oöcytes (Wallace and Jared, 1976). There it is transformed into phosvitin and lipovitellin. The physical uptake of vitellogenin has been followed using labeled vitellogenin. The vitellogenin is taken up by the formation of invaginations of the oölemma, which are then taken into the cytoplasm. This is called **endocytosis.** The vacuoles formed are called

FIGURE 7.16 Scheme depicting the synthesis of yolk in eggs of the zebrafish, *Brachydanio rerio,* during arbitrary developmental stages. Radioactive amino acids were administered to the mother by intraperitoneal injection, and the appearance of radioactivity in oöcytes and liver was monitored radioautographically. Radioactivity in yolk droplets is indicated by solid black or heavy stippling. Note that previtellogenic oöcytes (upper tier) do not incorporate radioactivity into yolk bodies as late as 48 hr after injection. Larger oöcytes (second tier; about 240 μm in diameter) incorporate radioactivity into endogenously synthesized yolk shortly after injection, as indicated by the appearance of labeled material in vesicles scattered throughout the cytoplasm. Older oöcytes (lower tier; about 750 μm in diameter) do not show incorporation of label until 3 hr after injection. Presumably 3 hr is the time required for maximal incorporation of label into vitellogenin produced in the liver. Carried to the ovary in the blood, vitellogenin is then taken into the oöcytes by pinocytosis. Continued production and export of vitellogenin depletes the level of label in the liver. [Redrawn with modifications from K. H. Korfsmeier, *Zeit Zellf* 71:283–96 (1966).]

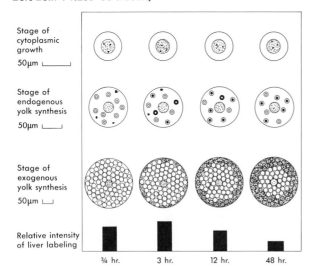

Stage of cytoplasmic growth
50μm

Stage of endogenous yolk synthesis
50μm

Stage of exogenous yolk synthesis
50μm

Relative intensity of liver labeling

¾ hr.   3 hr.   12 hr.   48 hr.

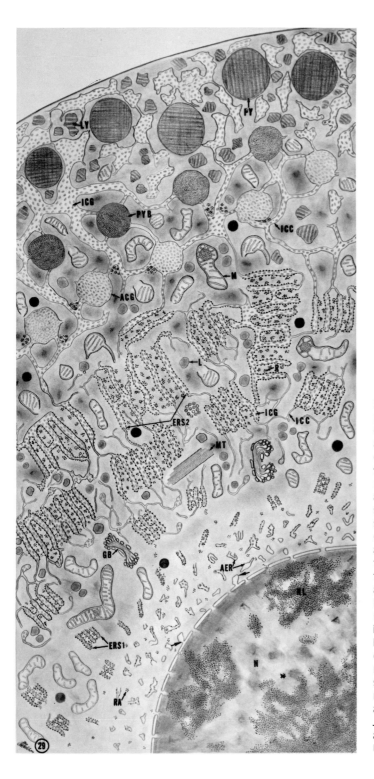

FIGURE 7.17 Scheme for the synthesis of proteinaceous yolk in the oöcyte of the crayfish. Agranular endoplasmic reticulum, AER, arises by blebbing (unlabeled arrows) from the outer lamella of the nuclear envelope and forms stacks of rough-surfaced endoplasmic reticulum, ERS1, that acquires more ribosomes and becomes ordered into compact cisternal stacks, ERS2, that are interconnected and that send branches in all directions. Intracisternal granules, ICG, thought to be yolk precursors, are presumably assembled within cisternae of the endoplasmic reticulum. They flow out into the unoriented branches where they collect and expand the cisternae as immature yolk bodies, PYB. These are eventually pinched off as mature yolk granules, PY. ACG, aggregates of intercisternal granules; GB, Golgi complexes, ICC, intercommunicating smooth-surfaced cisternae; LY, lipid yolk; R, ribosomes; RA, ribosomal aggregate; M, mitochondria; MT, striated microtubules; N, nucleus; NL, nucleolus; NP, nuclear pore; RA, ribosomal aggregates. [From H. W. Beams and R. G. Kessel, *J Cell Biol* 18:621–49 (1963), by courtesy of the authors and permission of the Rockefeller University Press.]

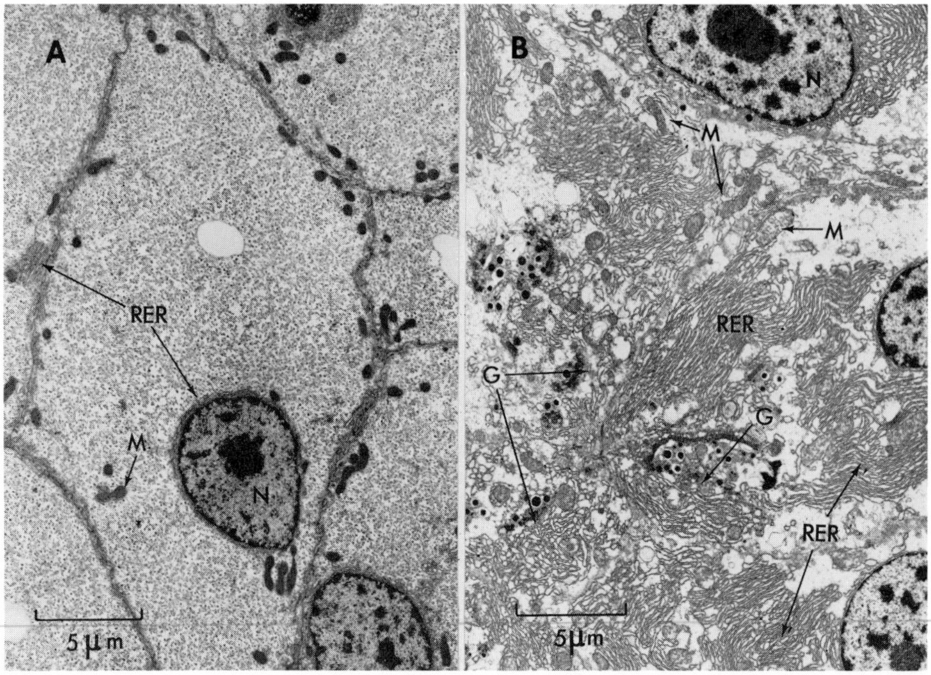

FIGURE 7.18   The effect of estradiol on the ultrastructure of hepatic cells in *Xenopus laevis* prior to hormone administration (A) and 8 days after beginning a 4-day course of estradiol injections. Note the expansion of *Golgi* complexes, G, increased number of mitochondria, M, and extensive rough endoplasmic reticulum, RER, in the cells of the treated animal. [From J. K. Skipper and T. H. Hamilton, *Proc Natl Acad Sci USA* 74:2384–88 (1977), by courtesy of the authors.]

endosomes. Fusion of endosomes produces primordial yolk platelets, which then fuse with each other and with definitive yolk platelets. It is not known how the physical transport of vitellogenin into the oöcyte and its modification to lipovitellin and phosvitin are related (Brummett and Dumont, 1977). Further remarks are directed to the processing of amphibian vitellogenin in Chapter 18. Interestingly, full-grown amphibian oöcytes (Stage 6, Figure 7.27), which have ceased the uptake of vitellogenin in vivo, will take it up in vitro from vitellogenin-laden serum (Jared and Wallace, 1969; Wallace and Bergink, 1974). Progesterone, which promotes the final stages of maturation, as described in some detail in Section 7.2.5, inhibits incorporation of vitellogenin in vitro (Schuetz, 1977; Schuetz et al., 1977).

In insects the vitellogenin is produced by the fat bodies of females under the influence of the juvenile hormone from the corpus allatum, a gland near the brain. (The corpus allatum and its secretion are defined and further considered in Chapter 17). Insect vitellogenin is a lipophosphoprotein of very low phosphorus content, and it is apparently taken up and stored without further processing in the oöcyte. Production of the vitellogenin occurs in vitro as well as in vivo in the fat bodies of females. Until recently it was thought that fat bodies of male insects cannot produce vitellogenin. Mundall et al. (1979), however, showed that male cockroaches can produce the yolk protein if they receive an implant of the female corpus allatum or if an analogue of juvenile hormone is applied. Moreover, oöcytes implanted in these males take up the vitellogenin. The basic research on insect vitellogenesis was reviewed by Telfer (1965), whose pioneering work (1954) is greatly responsible for the present state of our knowledge of insect vitellogenesis.

Eggs of various organisms differ considerably in the amount of yolk they contain, in the way it is distributed, and in the time at which it is utilized. The eggs of many marine invertebrates have little yolk and develop within a short time after fertilization into larval forms that, although composed of only a few cells, are capable of swimming and feeding. Obviously, these eggs need but little stored energy for such limited development. After a period of feeding, the larvae undergo metamorphosis into adults. Other marine eggs have considerable yolk, and many of them bypass the free larval stages, using their energy supply to develop quickly to the feeding-adult stage.

The egg of the mammal has little or no yolk. Unlike marine eggs similarly endowed, it does not develop quickly to an independent form; instead, it rapidly develops structures that enable it to relate to maternal tissues and to receive an energy supply from the bloodstream of the mother (see Chapter 9). In other organisms the egg has an abundant yolk, which is not used up during embryonic life. The tadpole of the frog, for example, retains in the cells of its alimentary tract large quantities of yolk that are used for continued growth and differentiation after hatching. Some newly hatched fish actually swim about with the yolk sac attached to the belly. For aquatic vertebrates such as these, there is probably a considerable survival advantage in a period of rapid development that leaves a free-swimming larval form able to escape readily from the egg clutch and swim about with its own energy supply.

As noted earlier, in most animal eggs there is a gradient in the concentration of yolk, the concentration being greatest at the vegetal pole. Such eggs are called **telolecithal** (yolk at one end) eggs. Most amphibians have eggs in which there is a very strong yolk gradient, the region of the animal pole showing small and scattered yolk platelets that grade progressively into the larger yolk masses occupying the vegetal pole, almost to the exclusion of the cytoplasm (Figure 7.10B). Eggs of reptiles and birds present extreme cases in this category of the egg types. The great bulk of the bird's egg consists of yolk, with only a thin disc of nucleated protoplasm at the animal pole (Figure 7.10C). In fish eggs, which are somewhat similar, some cytoplasm is present, especially peripherally, throughout the yolk zone before fertilization, but most of this flows into a protoplasmic disc at the animal pole after fertilization (see Figure 9.16A, B).

Eggs that have essentially no yolk are referred to as **alecithal** (no yolk). Human eggs are of this kind. Others have the yolk, usually in small amount, scattered evenly throughout the egg; these are called **isolecithal** (Figure 7.10A).

Eggs of many arthropods and some coelenterates are described as **centrolecithal.** They are relatively large, usually elongate, and have a very great amount of yolk (Figure 7.10D). The nucleus lies near the geometric center of the yolk mass, surrounded by a small amount of cytoplasm. A thin cytoplasmic layer covers the surface of the yolk and sends fine strands into the zone occupied by the nucleus.

7.2.4 MEIOSIS IN RELATION TO GROWTH AND MATURATION OF THE EGG.

As cited in Chapter 5, the germ cells in mammals are essentially all in meiotic prophase at the time of birth. Prior to and at the onset of meiosis, the oöcytes are found in clusters connected by cytoplasmic bridges (Figure 7.19). They remain so until prospective follicle cells insinuate themselves between the oöcytes to form primordial follicles (Figure 7.20). As this occurs, the zona pellucida begins to form and the oöcytes proceed to the diplotene stage, where they remain until their growth and further maturation are elicited by hormones (Figure 7.21). Meiosis does not progress further, however, until the egg has completed its growth, but the nucleus, in the meantime, enlarges considerably, and the staining properties of the chromatin become greatly diluted. In its enlarged stage, the oöcyte nucleus is called the **germinal vesicle.**

From each oöcyte that completes meiosis, only one definitive egg cell is formed. This is because the cytoplasm of the egg is not divided equally during the meiotic divisions. The metaphase spindle of the first division forms at the animal pole of the egg, and, at telophase, one daughter nucleus is pinched off from the egg together with a covering plasma membrane and very little cytoplasm. This is called the **first polar body,** which may or may not divide. The egg, now a secondary oöcyte, then produces a **second polar body** and has a haploid nucleus to bring to the fertilization process. The haploid nucleus of the oöcyte is called the **female pronucleus.**

The meiotic divisions of the oöcyte are often re-

71

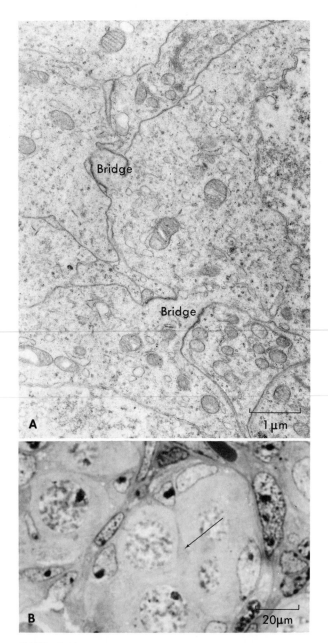

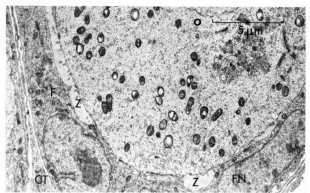

FIGURE 7.20 Origin of an individual follicle in the mouse ovary on the second postnatal day. The oöcyte has enlarged, and a partially discontinuous zona layer, Z, separates the oöcyte from surrounding follicle cells, F. CT, connective tissue; FN, nucleus of a follicle cell. [From D. L. Odor and R. J. Blandau, *Am J Anat* 124:163–86 (1969), by courtesy of the authors and permission of Alan R. Liss, Inc.]

FIGURE 7.19 A: Intercellular bridges between clustered mouse oöcytes at 17–18 days of gestation. B: The insinuation of a follicle cell process (arrow) between members of a similar cluster. The cells are in the pachytene stage of meiosis. [From J. R. Ruby, R. F. Dyer, and R. G. Skalko, *J Morphol* 127:307–340 (1969), by courtesy of the authors and permission of Alan R. Liss, Inc.]

ferred to as the **maturation divisions.** But **physiological maturation,** in the sense of ripeness for fertilization and release from the ovary, occurs at various stages of meiosis in eggs of different species (Figure 7.22). At one extreme are eggs that are mature when the germinal vesicle is intact, that is, during first meiotic prophase. At the other extreme are those that do not mature until meiosis is complete. These extremes are connected by a graded series of varying relationships between the time of cytoplasmic maturity and meiotic stage. Note that for many invertebrates maturity is achieved at first meiotic metaphase or anaphase. In most vertebrates, including human beings, the egg is fertilizable at second meiotic metaphase.

The completion of maturation and release of the egg from the ovarian follicle are usually, but not always, concurrent events. The dog and fox are distinctive among mammals in that the germinal vesicle is intact at ovulation, but then meiosis proceeds to first metaphase, at which time the egg is capable of activation. In most amphibians the germinal vesicle breaks down at the time of ovulation. The egg then proceeds to second metaphase in the body cavity and oviduct, only then becoming mature.

For eggs that reach maturity prior to the completion of meiosis, the general rule is that fertilization is required in order for meiosis to be com-

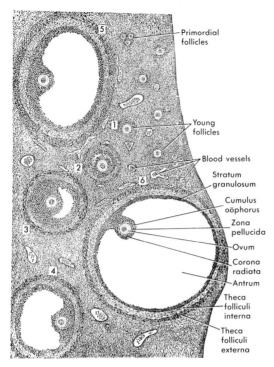

FIGURE 7.21 Portion of a mammalian ovary showing the progression of stages from young to mature ovarian follicles numbered in sequence. Follicle 6 is almost ready to ovulate. [Reprinted with permission of Macmillan Publishing Co., Inc., from *Fundamentals of Comparative Embryology of the Vertebrates*, rev. ed., by A. F. Huettner. Copyright 1949, Macmillan Publishing Co., Inc., renewed 1977 by Mary R. Huettner.]

pleted. In the absence of fertilization the egg gradually loses its fertilizability and dies. An important exception to this rule is found in the eggs of most starfishes, which are shed into the seawater with the germinal vesicle intact. They are immediately fertilizable, but, even without fertilization, the

meiotic divisions proceed to completion, the egg remaining fertilizable the while.

From the foregoing it is evident that for most eggs, inspection of the meiotic stage is sufficient to determine whether or not the egg is fertilizable. In fact, however, both the progress of meiosis and the accomplishment of physiological maturation are under the control of cytoplasmic factors that are, in turn, under hormonal control. Thus an immature amphibian oöcyte, deprived of its germinal vesicle by microsurgery, will come to maturity under proper hormone treatment.

7.2.5 HORMONAL CONTROL OF MATURATION AND OVULATION.

Each cycle of ovulation, whether in continuous or seasonal breeders, occurs in response to cyclical changes in the release of pituitary gonadotropins, as outlined in Chapter 5. In response to FSH, a number of ovarian follicles grow during each cycle, producing and releasing estrogens. They then resume meiosis, come to physiological maturity, and ovulate in response to LH.

The control of maturation and ovulation has been intensively studied in the ovaries of *Rana pipiens* and *Xenopus laevis* (Schuetz, 1977; Schorderet-Slatkine and Baulieu, 1977). In these amphibians hundreds of eggs grow to their definitive size each season. At the appropriate stage, fragments of ovary containing several immature eggs can be cut from the ovary and placed for observation in a dish containing an appropriate salt solution. Ovarian fragments can be maintained in this condition for several days without change. If, however, fragments of pituitary gland are cultured along with them, the eggs undergo **germinal vesicle breakdown (GVBD)** (that is, the nuclear membrane of the germinal vesicle disappears), and the chromosomes condense, form the first polar body, proceed to second meiotic metaphase, and are then released from the ovary.

73

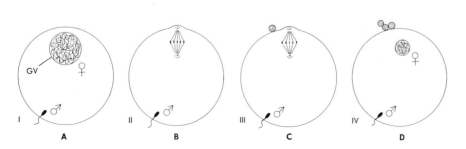

FIGURE 7.22 Ripeness of eggs for fertilization as related to stages of meiosis. A: Type I: fertilization occurs before meiosis begins. B, C: Types II and III: egg is ripe at first and second meiotic divisions, respectively. D: Type IV: the sperm enters after meiosis is completed. GV, germinal vesicle.

Pricked with a sharp needle at this stage, they show breakdown of the cortical granules, a sign that the egg has been activated, as normally occurs at fertilization. (They cannot actually be fertilized by sperm, for fertilization requires the presence of factors derived from the oviduct; see Chapter 8.)

These changes do not occur in immature eggs similarly treated after being artifically stripped from the ovary and divested of adhering follicle cells (Masui, 1967). If a few follicle cells are left attached to the otherwise naked immature oöcytes, however, they will come to maturity. Clearly, therefore, the follicle cells mediate and are essential to the action of the pituitary gonadotropins (Smith et al., 1968; Subtelny et al., 1968).

Moreover, whatever the follicle cells do in response to pituitary factors involves the synthesis of RNA and of protein. If ovarian fragments containing immature oöcytes are exposed to pituitary gland preparations in medium containing the drug actinomycin D, maturation does not occur. Likewise, maturation is prevented in similar preparations by cycloheximide or puromycin, inhibitors of protein synthesis (Wasserman and Masui, 1975b; Detlaff, 1966).

Maturation probably occurs in response to progesterone released by the follicle cells in response to gonadotrophic hormones. Naked immature oöcytes exposed to progesterone in vitro in the absence of pituitary hormones undergo breakdown of the germinal vesicle, form the first polar body, and are capable of activation. Actinomycin D does not inhibit this process, but puromycin does. It appears, therefore, that gene action (i.e., synthesis of informational RNA) is not required for the maturation of oöcytes, but synthesis of protein is. The protein must, therefore, be translated from previously synthesized mRNA (Smith and Ecker, 1969; Merriam, 1972, also cf. Snyder and Schuetz, 1973). This conclusion is further reinforced by the observation that progesterone-treated oöcytes from which the germinal vesicle has been removed (enucleated oöcytes) also undergo maturation in the sense that they are capable of activation and of abortive cleavage (Smith and Ecker, 1969).

The action of progesterone causes the production (or activation) in the oöcyte cytoplasm of a **maturation-promoting factor** (MPF). Cytoplasm from an activated intact or enucleated oöcyte, in-

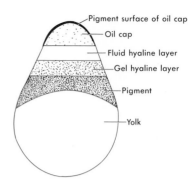

FIGURE 7.23 Centrifugal stratification of the components of a progesterone-treated oöcyte of *Rana pipiens*. Injection of material from the two hyaline layers into immature oöcytes leads to breakdown of the germinal vesicle. [Redrawn with modifications from Y. Masui, *J Exp Zool* 179:365–77 (1972).]

jected into an immature oöcyte, even of a different amphibian genus or order (Reynhout and Smith, 1974), causes maturation of the second oöcyte. Serial transfer of mature cytoplasm to successive immature oöcytes can be carried out indefinitely, thus indicating that progesterone is not the effective transferred agent (Smith and Ecker, 1971; Schorderet-Slatkine and Drury, 1973; Reynhout and Smith, 1974). Masui (1972) found that when progesterone-treated oöcytes are centrifuged for 2 hours at 4000g, the maturation factor is segregated in a clear layer of protoplasm immediately beneath a centripetal cap of oil material (Figure 7.23). MPF cannot be demonstrated in the hyaline layer of centrifuged untreated immature oöcytes or in preparations from oöcytes treated with progesterone in the presence of cycloheximide. It is sensitive to proteases and inactivated by free Ca^{2+}. It can be separated by centrifugation on a sucrose gradient (Wasserman and Masui, 1976; Masui et al., 1977). Drury (1978) tentatively characterized MPF as a phosphoprotein.

Oöcytes that mature as a result of successive serial transfers of mature cytoplasm develop a full complement of MPF by the time GVBD occurs. For *Rana pipiens* eggs, 10 serial transfers of 70 nl of cytoplasm would effect a dilution of the original injection of about 10^{12} in the final recipient oöcyte (Reynhout and Smith, 1974). Because each oöcyte in the series matures and produces MPF capable of

74

inducing GVBD in a successive oöcyte, MPF must be amplified during maturation. Masui and Markert (1971) referred to this phenomenon as **autocatalytic amplification.**

There is some question whether or to what extent amplification of MPF is dependent on the synthesis of protein. The presence of cycloheximide prevents the initiation of maturation by progesterone, but if administration of cycloheximide is delayed until approximately the time that MPF can first be assayed [after 4–6 hours in *Xenopus* (Wasserman and Masui, 1974)], GVBD occurs. This observation suggests the question as to whether inhibitors of protein synthesis could prevent the amplification of MPF in immature oöcytes injected with mature cytoplasm. Wasserman and Masui (1975b) injected mature cytoplasm from a progesterone-treated oöcyte into immature oöcytes bathed in medium with or without cycloheximide. They found that both groups of recipient oöcytes showed GVBD, even though protein synthesis was inhibited by more than 95% in the cycloheximide-treated group. Moreover, GVBD could be induced in a second cycloheximide-treated immature oöcyte and so on, serially, through a succession of immature oöcytes in which protein synthesis was inhibited. In contrast to these results, Schuetz and Samson (1979) found that cycloheximide-treated immature oöcytes of *Rana pipiens* failed to mature when injected with mature cytoplasm from a progesterone-treated donor. The basis for the differences in results obtained with *Xenopus* and *Rana* oöcytes has not been resolved. Perhaps the generation or action of MPF in *Rana* oöcytes involves a different mechanism than that found in *Xenopus*. The different results are all the more puzzling because mature cytoplasm from different species, suborders, and orders of amphibia show no specificity in reciprocal tests (Reynhout and Smith, 1974).

Whereas progesterone is probably the agent normally responsible for initiating the process of maturation, a number of natural and synthetic steroids are likewise effective. Several lines of evidence, moreover, indicate that the initial physiological action of the effective agents is exercised on the plasma membrane of the oöcyte.

1. Injection of the steroid directly into the cytoplasm is ineffective in inducing GVBD; this has been shown in several amphibians and in at least one fish (Smith and Ecker, 1969; Masui and Markert, 1971; Schorderet-Slatkine, 1972; Iwamatsu, 1978).
2. Although steroids enter the oöcyte and may be metabolized (Reynhout and Smith, 1973), it is not necessary that they do so in order to be effective. Godeau et al. (1978) covalently linked polyethylene oxide, having a molecular weight of 12,000, to an active progesterone analog, 3-oxo-4-andostenene-17β-carboxylic acid, and tested this macromolecule for its ability to induce GVBD in immature *Xenopus* oöcytes in vitro. Polyethylene oxide is very hydrophilic and is believed not to be incorporated into cells. Nevertheless, the progesterone analog combined with it induced maturation. Tests employing a radioactive label on the analog showed that probably none of it entered the oöcyte.
3. When only a portion of the egg surface is exposed to hormone, higher concentrations than normal are required to induce maturation. The surface of the animal half of the egg is more sensitive to hormone than is the surface of the vegetal half (Cloud and Schuetz, 1977).

An additional body of evidence indicates that the fundamental role of progesterone in the initiation of maturation is the release of free Ca^{2+} from a bound form in the cortex of the amphibian egg.

1. Immature oöcytes exposed to medium containing progesterone show a dramatic increase in intracellular free Ca^{2+} within 1 minute (Wasserman et al., 1980).
2. When calcium ions are introduced into the cortex of immature oöcytes by means of iontophoresis (i.e., by means of a microelectrode), they induce maturation in the absence of progesterone. Iontophoresis of Ca^{2+} into the endoplasm does not lead to GVBD, nor does the iontophoresis of other ions into the cortex or endoplasm (Moreau et al., 1976).
3. The essential role of Ca^{2+} may be inferred from the fact that a wide variety of chemicals that cause the release of bound calcium or inhibit the efflux of Ca^{2+} from the oöcyte can induce maturation in oöcytes that were not progesterone-treated. These chemicals include a variety of

75

pharmacological agents such as local anesthetics, antidepressants, and anorexiants. Also included is the lanthanum ion, which blocks Ca^{2+} flux through the plasma membrane and releases internally sequestered calcium.

4. A carboxylic acid antibiotic, ionophore A23187 (Reed and Lardy, 1972), promotes a net exchange of protons for divalent cations across biological membranes. Oöcytes of *Xenopus* exposed to high concentrations of Ca^{2+} (e.g., 10 mM) with A23187 in the medium show GVBD in the absence of progesterone (Wasserman and Masui, 1975a). With higher concentrations of calcium in the medium (e.g., 30 or 40 mM) GVBD is inhibited unless equal amounts of Mg^{2+} are present (Masui et al., 1977). [Mg^{2+} is needed for the stabilization of MPF, which is sensitive to Ca^{2+} (Wasserman and Masui, 1976).]

5. Progesterone-treated oöcytes of *Xenopus* do not mature in the absence of both Ca^{2+} and Mg^{2+}, but will do so if only Mg^{2+} is present in the medium. If, however, a calcium-specific chelating agent, ethyleneglycol-bis(β-aminoethyl ether)-N,N'-tetraacetic acid (abbreviated as EGTA), injected into the oöcyte in the presence of Mg^{2+}, maturation fails unless the injection is delayed until MPF begins to appear (Masui et al., 1977).

6. A number of substances such as theophylline or procaine, which inhibit the effect of progesterone, block an increase in free intracellular calcium, whereas all treatments that permit the initiation of maturation bring about an increase in free intracellular calcium.

How does one make the connection between an increase in free Ca^{2+}, the production of MPF, and the completion of maturation? Maturation can occur in oöcytes of *Xenopus* exposed to progesterone for 5 minutes, but production of MPF cannot be detected for several hours. The intervening steps are only partially known and some are still speculative.

Wasserman and Masui (1975b) and Masui and Clarke (1979) have suggested that MPF is present in the immature oöcyte in precursor form. The formation of active MPF involves action of a hitherto inactive initiator, which catalyzes the phosphorylation of inactive MPF, thus rendering it active. Active MPF then autocatalytically phosphorylates precursor MPF to form additional active MPF, which then

induces GVBD. Both the initiator and MPF are correctly called **protein kinases,** a term indicating that they catalyze the phosphorylation of other proteins.

The initiator protein has not been identified, but a reasonable scheme for the control of its activity can be made. Initiator is presumed to be constantly synthesized and degraded prior to the onset of maturation. Its activity as initiator, however, is blocked by a high steady-state level of phosphorylation resulting from the action of a protein kinase that, unlike initiator protein and MPF, is dependent on cyclic adenosine monophosphate (cAMP) for activity. The availability of cAMP is regulated by the activity of **adenyl cyclase,** the enzyme that promotes the formation of cAMP from ATP and by the activity of **phosphodiesterase,** which catalyzes the breakdown of cyclic AMP to 5′-AMP. The activity of these enzymes is modulated by a common Ca^{2+}-binding protein, **calmodulin,** which modulates many other Ca^{2+}-dependent reactions, also.

3′,5′ cyclic AMP

From the above we can construct a tentative sequence of events leading to production of MPF.

1. Progesterone acts on the surface of the oöcyte, causing the release of Ca^{2+} previously sequestered in the cortex.

2. Ca^{2+} then binds to calmodulin, which either activates phosphodiesterase or inhibits adenyl cyclase, resulting in a depletion of cAMP.

3. The cAMP-dependent phosphorylation of initiator protein ceases.

4. Initiator protein catalyzes the cAMP-independent phosphorylation of inactive MPF precursor.

76

The strongest point in this scheme is the progesterone-induced release of Ca^{2+}, which was well documented earlier in this section. The binding of calcium to calmodulin as a step in the sequence has not been documented, but calmodulin has been isolated from oöcytes of *Xenopus* (Wasserman et al., 1980) and thus is a possible link in the chain of events leading to maturation. Critical to this scheme is the depletion of cAMP, but there is some uncertainty whether this occurs. It is persuasive of the validity of the scheme that dibutyryl cAMP, which mimics the action of cAMP but is hydrolyzed less readily in the cell, blocks maturation. So, too, does cholera toxin, which stimulates cAMP levels in the cell (Godeau et al., 1977). O'Connor and Smith (1976), however, found no change in intracellular cAMP during progesterone-induced maturation in *Xenopus laevis* oöcytes. Neither did they find an increase in cAMP in oöcytes treated with theophylline, a drug that inhibits the activity of phosphodiesterase and blocks maturation. More recent studies suggest that progesterone affects cAMP by depressing the activity of adenyl cyclase. Mulner et al. (1979) showed that *Xenopus* oöcytes treated with cholera toxin show increased concentrations of cAMP and an enhanced rate of new synthesis of cAMP from injected ATP. Concurrent treatment with progesterone, however, greatly diminishes this effect. Moreover, in oöcytes treated with progesterone alone there is an inverse correlation between the concentration of progesterone and the new synthesis of cAMP in the oöcyte. In 1980 Kostellow et al. reported that *Rana pipiens* oöcytes show a steady-state level of about 5 picomoles of cAMP per oöcyte. By 90 minutes after the onset of progesterone treatment this value drops to approximately 2 picomoles per oöcyte. The amount of labeled ATP incorporated into cAMP during the first hour of progesterone treatment, moreover, is about 10-fold less than the amount incorporated in control oöcytes. These observations strongly suggest that, in *Rana pipiens* at least, progesterone treatment down-regulates the activity of adenyl cyclase. Because adenyl cyclase is anchored in the cell membrane and because progesterone exercises its effect at the cell membrane, it is reasonable to suggest that the progesterone receptors in the cell membrane are closely associated with regulatory sites for adenyl cyclase.

cAMP-dependent protein kinase activity is involved in the regulation of cell division and other cellular activities. In all tissues that have been studied, the enzyme consists of regulatory (R) and catalytic (C) subunits. The nature of these subunits is evolutionarily conserved, and subunits from different species can be reassociated to form recombinant molecules. The combined form of the subunits is enzymatically inactive, but the binding of cAMP causes dissociation of the subunits, freeing the catalytic protein.

$$R_2C_2 + 2\ cAMP \rightleftharpoons R_2(cAMP)_2 + 2\ C \quad (1)$$

C can then catalyze the phosphorylation of protein.

$$Protein + ATP \rightleftharpoons Protein\text{-}P + ADP \quad (2)$$

The evidence that cAMP-dependent protein kinase is involved in maintaining the block to maturation is quite persuasive. The enzyme is present in oöcytes and, if there should be a diminished availability of cAMP (through hydrolysis or sequestration), reaction (1) would be driven to the left; C would be inactivated and no longer available to catalyze the phosphorylation of a putative initiator protein. More importantly one would expect that reaction (2) would be driven to the left by an excess of the $R_2(cAMP)_2$ complex. Maller and Krebs (1977) isolated the complex from beef heart and injected it into immature *Xenopus* oöcytes. At a dose of 0.2 picomole of $R_2(cAMP)_2$ per oöcyte, GVBD occurred spontaneously in 100% of the cases.

On the basis of this result one would predict that injection of the C subunit into progesterone-treated oöcytes would block maturation. A dose of 0.6 picomole of C per oöcyte inhibited maturation in 100% of cases. This result, together with the foregoing one, strongly urges the view that phosphorylation of a protein, catalyzed by cAMP-dependent protein kinase, maintains the block to maturation in immature oöcytes (but see O'Connor and Smith, 1979).

Once the block is removed, how does MPF act to initiate GVBD? This is still unknown, but it is reasonable to assume that MPF must have direct access to the germinal vesicle. Masui (1972) displaced the germinal vesicle in *Rana pipiens* oöcytes from the region of clear cytoplasm near the animal pole into the yolky portion of the egg at the vegetal pole. The oöcytes were then exposed to progesterone. The onset of GVBD in these eggs was considerably delayed as compared to eggs with the germinal vesicle

77

in the animal half. As Cloud and Schuetz (1977) showed, the animal pole is more sensitive to progesterone than the vegetal pole is, so it seems likely that MPF is produced first in the animal hemisphere and then spreads to the vegetal hemisphere.

7.2.5.1 Ovulation in Vertebrates.
Attention is now directed more particularly to release of the oöcyte from the ovarian follicle. Ovulation depends on antecedent events in the follicle cells surrounding the oöcyte that, under normal conditions, are also involved in maturation of the oöcyte. This seems to be true both in lower vertebrates and in mammals. Progesterone apparently plays a significant role in both processes. It is normally formed from pregnenelone (Figure 5.8) in a reaction catalyzed by 3β-hydroxysteroid dehydrogenase, an enzyme that is inhibited by cyanoketone, which binds irreversibly to it. Cyanoketone blocks the effects of pituitary gonadotropin-induced oöcyte maturation and ovulation in ovarian fragments in vitro, but does not inhibit progesterone-induced maturation of oöcytes. The inhibitor also blocks ovulation in rats and rabbits (Snyder and Schuetz, 1973; M. R. Clark et al., 1978).

In preovulatory stages LH stimulates the activity of the enzyme adenyl cyclase, which results, in rabbits, in as much as a 20-fold increase in cyclic AMP. This is possibly involved in the action of a protein kinase required for steroid hormone synthesis. There is also an increase in compounds of the group known as prostaglandins prior to ovulation. These are all lipid-soluble hydroxy acids containing 20 carbon atoms. Prostaglandins are found in all tissues, but are especially concentrated in semen. In the ovary they appear only in follicles that are to be ovulated. Thus, although several actively growing follicles may be present in a rat ovary, for example, only those that ovulate show increased amounts of prostaglandins. Compounds that block prostaglandins have been shown to block ovulation in rats and rabbits (summarized by LeMaire and Marsh, 1975, and M. R. Clark et al., 1978). This blockade cannot be removed by LH or by progesterone, but can be overcome by prostaglandin administration. Moreover, prostaglandins can induce ovulation in many species, including lower vertebrates such as the trout (Jalabert and Szöllösi, 1975).

Also involved in ovulation, especially in mammals, is the activity of the enzyme collagenase that, at the time of ovulation, effects a reduction in the breaking strength of the follicular wall such that the intraovarian pressure causes its rupture, with consequent release of the ovum. According to Rondell (1970), diminished elasticity can be induced in strips of follicle wall in tissue culture by LH, cAMP, or progesterone. Effects of LH and cAMP can be blocked by cyanoketone.

The total picture of ovulatory control is still obscure. Both FSH and LH are essential. FSH induces follicle growth and estrogen production, which then set up the antecedent conditions for the production of cAMP and prostaglandins. Possibly cAMP is involved in the activity of protein kinases required for progesterone and prostaglandin production. Progesterone may be connected in some way with the activation of collagenase. These sequences seem reasonable, but not all data are in agreement. How does one reconcile, for example, the observation of Magnusson and Hillensjö (1977) that cAMP inhibits maturation of rat oöcytes? Perhaps cAMP plays its role later in preovulatory stages. This remains to be clarified.

The immediate preparations for ovulation also involve changes in the relationship of follicle cells to the oöcyte. In mice, the cumulus cells separate from the oöcytes and become sensitive to disruption by hyaluronidase, which is important for fertilization, as described in the next chapter. Furthermore, the junctions between cumulus cells and oöcyte, where they interdigitate across the zona pellucida, become interrupted. Further discussion of this topic is found in a paper by Schuetz and Swartz (1979).

7.2.5.2 Neuroendocrine Control of Maturation and Spawning in Starfish.
The release of gametes into the water in proximity so as to assure fertilization is called **spawning.** In many species of invertebrate animals living in the sea, unpaired male and female organisms release their gametes more or less simultaneously at the appropriate season. Quite unknown as yet are the mechanisms that cue the simultaneous release of gametes by members of each sex in those kinds of animals that do not have a mating behavior.

Whatever the spawning signals may be, they apparently are mediated by neuroendocrine mechanisms, at least in starfish and other echinoderms. The first and most complete information to this effect comes from investigations of starfish spawning. Along each arm of a starfish is a radial nerve that is closely associated with gonadal tissue, which also

78

occupies each arm. Some years ago Chaet and McConnaughy (1959) discovered that an aqueous extract of the radial nerve could cause spawning when injected into starfish having mature gonads. This discovery led to a long line of investigations on the control of oöcyte ovulation and maturation. The active factor from the radial nerve was later determined for the Japanese starfish *Asterias amurensis* to be a polypeptide with a molecular weight of about 2100 (Kanatani et al., 1971). It was called gonad-stimulating substance (GSS). More recently it was discovered that concanavalin A, a material isolated from castor bean seeds, will mimic the action of GSS (Kubota and Kanatani, 1975 a, b). This material binds to mannose residues of cell-surface glycoproteins. It is required to be bound to the follicle cells surface to be effective (Kubota and Kanatani, 1978).

Meanwhile, Schuetz and Biggers (1968) showed that the addition of GSS to oöcyte-free starfish ovarian tissue in vitro would cause release of a material that would promote ovulation and would induce immature oöcytes to undergo GVBD and to complete meiosis. The substance was later identified by Kanatani and Shirai (1972) as 1-methyladenine (1-MeA). It is also called mitosis-inducing subtance (MIS). It is produced by the follicle cells under the stimulus of GSS.

1-Methyladenine
(1-MeA)

MIS is active in gamete maturation in all species of starfishes tested. It can be isolated from the testes of male starfish (Cochran and Kanatani, 1975; Kubota et al., 1977) and from testes and ovaries of a number of kinds of sea urchins (Kanatani, 1974). In both males and females the amount of MIS that can be isolated from the gonads increases as the breeding season approaches (Kanatani, 1975). The gonad of the sea cucumber *Leptosynapta inhaerens* is an ovotestis. MIS-treated fragments of the gonad release sperm very rapidly and oöcytes more slowly (Ikegami et al., 1976b).

There are many parallels between the action of 1-MeA and progesterone with respect to maturation-promoting activity in echinoderms and in amphibians, respectively. MIS is produced in follicle cells, as is progesterone, and, like progesterone, is ineffective if injected directly into the egg (Kanatani and Hiramoto, 1970). After normal passage through the cell membrane, however, it induces a cytoplasmic maturation factor that is then responsible for germinal vesicle breakdown and meiotic maturation. The induction of the factor can be transferred by inoculation in serial fashion from maturing oöcytes (Kishimoto and Kanatani, 1976) and again, as in the case of amphibians, is not species specific (Kishimoto and Kanatani, 1977). When the content of the maturing oöcyte is centrifugally stratified, the factor is localized predominantly in a layer of clear protoplasm immediately beneath an oily centripetal cap (Kishimoto et al., 1977). Finally, maturation of starfish as in amphibians can be induced by ionophore A23187 or excess Ca^{2+} (Moreau et al., 1978).

The action of 1-MeA can be mimicked by chemical agents such as dithiothreitol that reduce disulfide bonds (Kishimoto and Kanatani, 1973) and blocked by substances such as *p*-chloromercurobenzoate that tend to keep them oxidized. Normal oöcytes stimulated to mature by 1-MeA show an increase in cortical proteins with reduced —SH groups (Kishimoto et al., 1976). Apparently, the initial effect of 1-MeA is on the oöcyte membrane or the immediately subjacent cortical materials. The 1-MeA receptor is not known, but it is removed from the oöcyte surface by treatment with the detergent Triton X-100. The receptor is reconstituted, however, once the oöcyte is removed to normal seawater (Morisawa and Kanatani, 1978).

A question currently under investigation is the nature of the relationship between GSS and 1-MeA. GSS increases as the breeding season approaches. Is its role perhaps the abolition or bypass of some inhibitor of 1-MeA synthesis? This is suggested by results of Ikegami (1976) and Ikegami et al. (1976a), who isolated steroid glycosides from ovaries. These compounds inhibit the production of 1-MeA in follicle cells, but this inhibition can be overcome by GSS. 1-MeA is thought to be produced by the action of a methionine donor (Shirai et al., 1972; Shirai, 1973) that produces a methylated precursor, such as 1-methyladenosine monophosphate (Shirai and Kanatani, 1972, 1973), which is successively

79

cleaved to 1-methyladenosine and then 1-methyladenine. It is not evident that the postulated inhibitors of Ikegami et al. (1976) interfere with this sequence, however.

7.2.6 GENE ACTION DURING OÖGENESIS.

Considerably more is known about gene action during oögenesis than during spermatogenesis. This is because, for most kinds of eggs, although to no great extent in those of mammals, genes for the protein synthetic machinery as well as mRNA for specific proteins are made throughout the prolonged period of oögenesis and are available for use during the period of rapid cell division that follows fertilization. Thus, for many kinds of eggs, postfertilization development can proceed to a considerable extent in the absence of further gene action. The classic case is that of the fertilized egg of the sea urchin. It divides and undergoes early stages of embryo formation (cf. Chapter 9), even though all gene action is abolished by treatment with actinomycin D (Gross and Cousineau, 1964).

7.2.6.1 Production of Transfer RNA, 5 S RNA, Ribosomal RNA.

The low molecular weight species of RNA, namely the transfer RNAs, which sediment at 4 Svedberg units (4 S RNA), and 5 S RNA, which later functions in conjunction with the ribosomes, are the principal gene transcripts made during previtellogenic stages of oögenesis in amphibians (Stages 1 and 2; Figure 9.27), whose RNA metabolism is best known (van Gansen et al., 1976). Both of these kinds of RNA are made from genes that are present in multiple copies (i.e., they are reiterated many times, or are redundant; see Chapter 18 and Appendix B).

There are 120 nucleotides in 5 S RNA. Significantly, approximately 90% of the 5 S RNA in *Xenopus* oöcytes differs in six nucleotide positions from the 5 S RNA of somatic cells. About 10% of oöcyte 5 S RNA is identical to that of somatic cells, but there is also a trace of another 5 S RNA, differing in another set of six nucleotides, which is distinctive of oöcytes. Brown et al. (1977) showed that whereas somatic cells do not have the oöcyte-specific 5 S RNAs, they do have distinctive DNA sequences corresponding to them. This is important because it indicates, first, that the different 5 S RNAs do not arise as modifications of a single transcript and, second, that the oöcytes and somatic cells differ in the control mechanisms for determin-

ing which of the 5 S DNA sequences are transcribed or exported to the cytoplasm. Other vertebrate species, but not all, also show an oöcyte-specific 5 S RNA (Denis and Wegnez, 1977a; Wegnez et al., 1978). The transcription of the genes for 4 S and 5 S RNA continues throughout oögenesis in *Xenopus*. About one half the amount of tRNA and 5 S ribosomal RNA found in the mature oöcyte is produced during the previtellogenic period, however, and is stored in the cytoplasm in combination with protein. About 70% of the 5 S is stored as a ribonucleoprotein particle that sediments at 7 S and that consists of one molecule of 5 S RNA and one molecule of protein. Most of the remaining 5 S RNA is found in a particle that sediments at 42 S and that contains one molecule of 5 S RNA, three molecules of tRNA, and one molecule each of two different proteins (Picard and Wegnez, 1979; Pelham and Brown, 1980). The somatic type of 5 S RNA is not stored (Denis and Wegnez, 1977b). Storage particles containing 5 S RNA remain in the cytoplasm throughout the previtellogenic period. During vitellogenesis the storage particles progressively disappear. Synthesis of 28 S and 18 S ribosomal RNA begins during the vitellogenic period. At the end of the vitellogenic period all 5 S RNA is found in ribosomes in association with a protein different from any of those found in storage particles (Picard and Wegnez, 1979).

Ribosomal RNA is transcribed from ribosomal genes referred to as rDNA. Ribosomal RNA is synthesized in nucleoli, and rDNA is found in one or more nucleolar-organizing regions of each haploid genome. In all organisms that have been examined, rDNA is present in multiple copies in the nucleolar-organizing region. In addition, many organisms have evolved mechanisms that provide the developing oöcyte with the opportunity to receive rapidly great quantities of transcripts of the ribosomal genes. In some cases, notably in certain insects, **nurse cells** produce the ribosomes which move to the oöcyte through cytoplasmic channels. The nurse cells have **polytenic** nuclei, that is, nuclei in which the genome is replicated many times, over 1000-fold in some cases. Thus, there is opportunity for simultaneous transcription in each nurse cell of rRNA from rDNA on many replicate chromatin strands.

Among organisms in which great quantities of rRNA are produced within the oöcyte itself, one finds that the redundant DNA of the nucleolar organizing region becomes **amplified** during oögenesis. Amplification refers to the repetitive copying of the rDNA in the form of additional nonchromosomal rDNA sequences, which then form the cores of nucleoli in which rRNA is transcribed. In the house cricket, *Acheta domestica,* for example, chromosomes designated as numbers 6 and 11 have nucleolar-organizing regions. During oögenesis their DNA is copied many times and is released during pachytene to form a large nucleolus that is the center of rRNA synthesis (Lima-di-Faria et al., 1973). The general topic of the differential amplification of rDNA was reviewed by Gall (1969), one of the pioneer investigators in the field.

Amplification of rDNA and its transcription as rRNA are probably best known for *Xenopus laevis.* In *Xenopus* there is one nucleolar-organizing region on each haploid genome. Thus there are four such regions in the $4C$ nucleus as meiosis begins. At the early pachytene stage, while the paired chromosomes are still in a bouquet formation, an intranuclear "cap" of DNA appears opposite the base of the chromatin loops. This cap stains positively with the DNA-specific Feulgen reagent. After appropriate treatment (denaturation) the contents of this cap will bind to radioactive *Xenopus* rRNA, indicating that its nucleotide sequences are complementary to rRNA (Ficq and Brachet, 1971). The cap thus consists of rDNA. As the cap reaches its maximal development, each nucleus contains a total of 43 picograms (1 pg = 10^{-12} g) of DNA, of which only 12 pg is chromosomal DNA. During the ensuing diplotene stage, the cap disperses and gradually about 1200–1600 nucleoli appear (Brown and Dawid, 1968) (Figure 7.24). Each of these consists of a core of DNA and a mass of ribosomal RNA. In these nucleoli there are, in all, 3000–5000 nonchromosomal copies of the nucleolar-organizing region.

By knowing the percent of DNA that is complementary to rRNA, and by knowing the approximate molecular weight of rDNA, one may calculate that each nucleolar-organizing region of the haploid *Xenopus* genome contains about 450 copies of rDNA. Amplification, producing 3000–5000 copies of the nucleolar-organizing region, thus provides for the possibility that, at any one time, $13.5–22.5 \times 10^5$ rDNA sequences are available for transcription in each oöcyte. This provides an immense capacity for rRNA synthesis. Brown (1966)

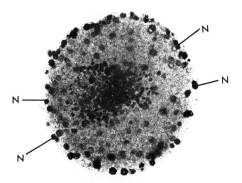

FIGURE 7.24   The germinal vesicle of *Xenopus laevis* isolated from the oöcyte. Hundreds of nucleoli, N, are present. Each of them is the site of synthesis of ribosomal RNA; each carries a copy of DNA from that segment of the genome that codes for ribosomal RNA synthesis. [From D. D. Brown and I. B. Dawid, *Science* 160:272–79 (1968), by courtesy of the authors and with permission. Copyright 1968 by the American Association for the Advancement of Science.]

calculated that an oöcyte of *Xenopus laevis* produces, at the peak of its rRNA synthetic capacity, as much ribosomal material as is produced by a mass of adult frog liver containing approximately 200,000 cells!

The transcription of rRNA from the nucleolar cores of amphibian oöcytes (Miller and Beatty, 1969) (and in some other organisms also) can be visualized by means of the electron microscope. Figure 7.25 shows a preparation made by spreading isolated nucleoli from oöcytes of a newt. The elongating chains of rRNA, which show similar polarity, are transcribed from segments of rDNA that are separated by nontranscribed spacer segments. The original transcript consists of a RNA sequence that sediments at 40 S. This is subsequently processed into 18 S and 28 S segments, which are the basis of the smaller and larger ribosomal subunits, respectively. As is normal, transcription is in the 5' to 3' direction, with the 18 S segment being transcribed first. It is separated from the 28 S segment by a transcribed spacer of which a 5.8 S segment is conserved and later hydrogen-bonded to the 28 S segment (see Figure 18.6). Reeder in 1974 reviewed much of what was known of the production and structure of rRNA in *Xenopus* at that time. More recent studies by Wellauer and Reeder (1975) and by Wellauer et al. (1976) considered in more detail the

81

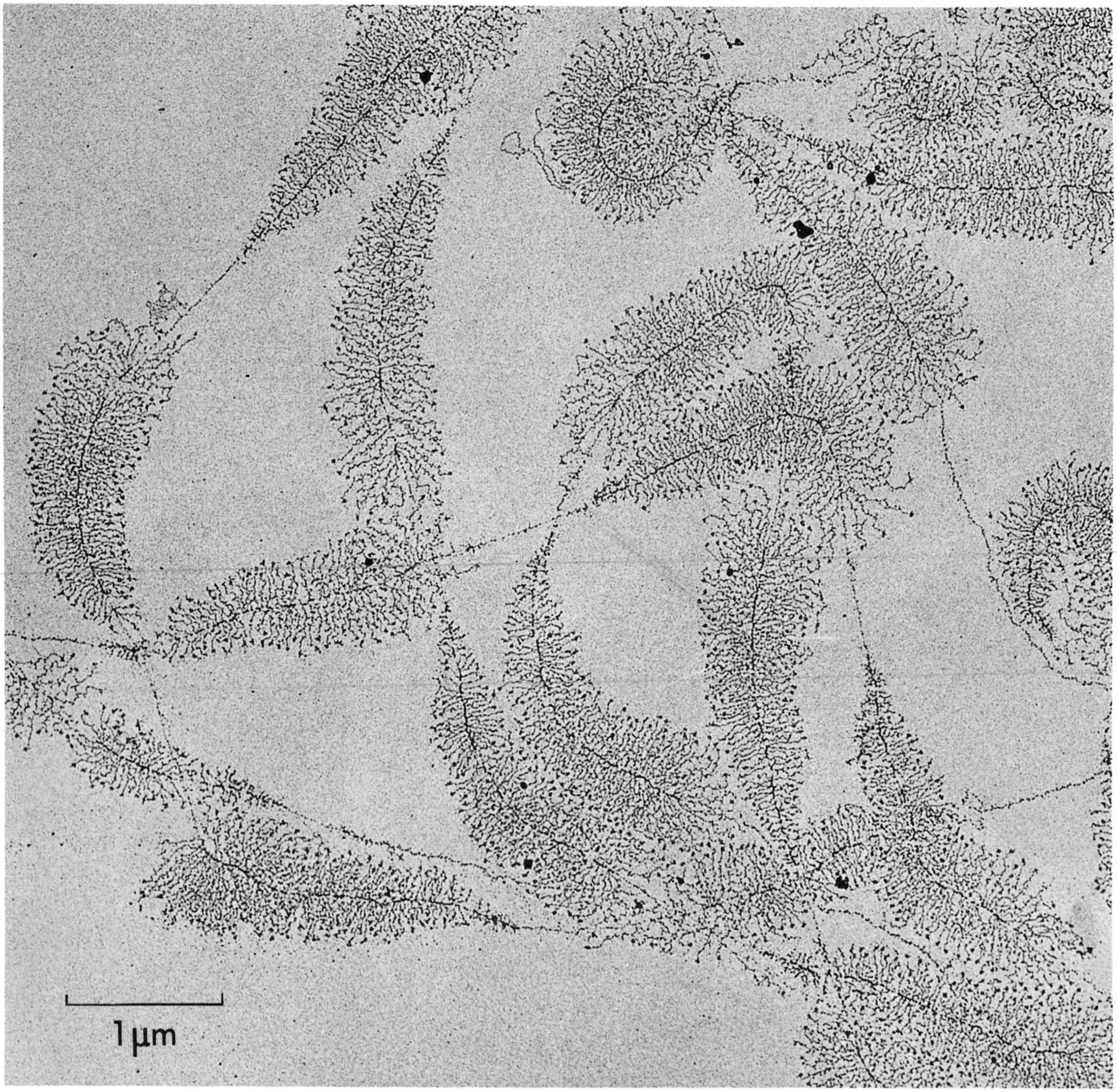

FIGURE 7.25 Electron micrograph of a portion of a nucleolar core, isolated from an oöcyte of *Notophthalamus viridescens,* showing copying of the ribosomal DNA. Transcription of the DNA core is interrupted by nontranscribed spacer DNA. Each strand extending laterally from the core is nascent rRNA. Dense particles along the matrix are thought to be molecules of RNA polymerase; note that they appear to traverse the spacer regions without transcribing them. [From O. L. Miller and B. R. Beatty, *Science* 164(3882): cover photograph (May 23, 1969), by courtesy of Dr. Miller and with permission. Copyright 1969 by the American Association for the Advancement of Science.]

relationship of transcribed and spacer regions in *Xenopus*.

Ribosomes produced in the multiple nucleoli of the amphibian oöcyte exit to the cytoplasm through the nuclear pores. Ribosome production ceases as the oöcytes mature, and the nucleoli are discharged into the cytoplasm when the germinal vesicle breaks down. Their DNA cores accumulate at the animal pole where they can be detected by means of the Feulgen reaction until the first polar body is formed (Steinert et al., 1976). Thereafter, the rDNA cannot be detected by means of the Feulgen reaction. Thomas et al. (1977) showed, however, that cytoplasmic rDNA is present throughout the first series of cell divisions after fertilization, during which time it is neither replicated nor degraded. They were able to make this determination by vir-

tue of the fact that when bulk DNA is extracted, rDNA can be separated by virtue of its greater buoyant density, as determined by centrifugation in a CsCl gradient. The amount of rDNA complementary to labeled rRNA is essentially the same at the 16-cell stage as in the oöcyte. How long it persists thereafter was not determined.

7.2.6.2 Synthesis and Storage of Messenger RNA.

Most kinds of oöcytes prior to fertilization contain large quantities of mRNA. Only a small fraction of this mRNA is associated with polyribosomes and thus is active in protein synthesis. This fraction probably amounts to less than 15% in some sea urchins (Humphreys, 1971) and to about 10% in *Xenopus* (Rosbash and Ford, 1974). The remaining mRNA is stored in the cytoplasm in ribonucleoprotein (RNP) particles and can be recruited from

FIGURE 7.26 A lampbrush chromosome from the oöcyte of *Triturus*. After meiotic pairing, the homologous chromosomes, held together only at chiasmata, Ch, have thrown out numerous side loops from the central core. [From J. D. Gall in D. Prescott, ed., *Methods in Cell Physiology*, Vol. 2, 1967, p. 39, by courtesy of Dr. Gall and permission of Academic Press, Inc., New York.]

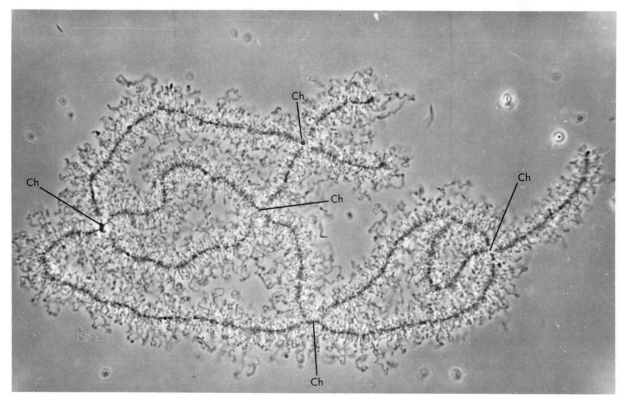

the cytoplasm to participate in protein synthesis when an egg is released from developmental stasis.

Sea urchin eggs have few polyribosomes and a low rate of protein synthesis prior to fertilization. About 5 minutes after fertilization the rate of protein synthesis increases several fold and the number of polyribosomes also increases (cf. Chapters 8 and 9). These changes likewise occur in eggs in which genetic transcription has been blocked by the drug Actinomycin D. Nevertheless, development of these eggs then proceeds normally through early embryonic stages. Quite evidently mRNA produced in the oöcyte during oögenesis suffices for protein synthetic activities required during early development. This oöcyte mRNA is referred to as **maternal mRNA** or **maternal template.** Because the greater part of maternal template is sequestered in the cytoplasm, presumably unavailable for translation, the term **masked mRNA** became a popular one. Gross (1967) reviewed the early studies that demonstrated the existence and utilization of masked mRNA.

Much remains to be learned about the production, stability and utilization of maternal mRNAs in sea urchin oögenesis. In mature oöcytes of the sea urchin *Strongylocentrotus purpuratus,* as many as 20,000 distinct mRNA species are sequestered in mRNP particles. The sequestered mRNAs are copies of genes that are present only once per haploid genome. In previtellogenic oöcytes about 14,000 distinctive mRNA species are present, 80% of which are associated with polyribosomes at any one time. Moreover, these same species are present in mature oöcytes, where, however, their number is augmented by additional species of mRNA produced during vitellogenesis. It would appear, therefore, that a large set of mRNAs is required for translation during early stages of oögenesis and that this set is also included in the mRNA that is found at maturity in the form of cytoplasmic mRNP (Galau et al., 1976; Hough-Evans et al., 1977). The sequestered mRNA in the sea urchin is apparently relatively stable for Dworkin and Infante (1978) found that mRNA that is newly synthesized and exported to the cytoplasm decays with a very short half-life. The rate of production, however, is sufficient to maintain the low level of protein synthesis that occurs prior to fertilization.

In *Xenopus,* as in sea urchins, all the mRNA that is needed for the early stages of embryogenesis ac-

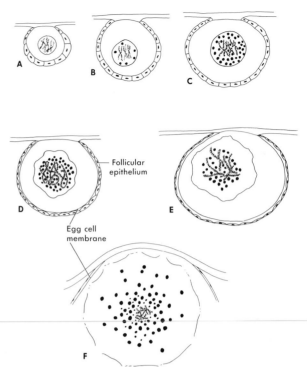

FIGURE 7.27 Successive stages of nuclear growth during development of frog eggs. [Redrawn with modifications from W. R. Duryee, *Ann NY Acad Sci* 50 (Article 8). 920–53.]

cumulates during oögenesis. Unlike the sea urchin oöcyte, however, all mRNA species present at the end of oögenesis are present prior to the onset of vitellogenesis and the amount of mRNA per oöcyte, 86 ng, does not increase during vitellogenesis (Dolecki and Smith, 1979). During the previtellogenic period (Figure 7.27A, B) 75% of the mRNA present in the oöcyte is associated with polyribosomes (Darnbrough and Ford, 1976), but at the end of oögenesis the great bulk of mRNA, 90%, is found in RNP particles (Rosbash and Ford, 1974).

The mRNA that is produced during the previtellogenic period appears to be highly stable, for radioactive precursor taken up by very young previtellogenic oöcytes diminishes very little, at least until vitellogenesis begins (Figure 7.27C) over a year later. Thereafter degradation and replacement apparently occur, for whereas the total amount of mRNA does not increase, the rate of its production triples during vitellogenesis. Dolecki and Smith (1979) presented persuasive arguments that whereas much of the newly synthesized mRNA decays rapidly, a slower-decaying portion presumably replaces damaged and decaying maternal message synthesized and sequestered earlier.

Finally, we turn to another aspect of RNA synthesis during oögenesis that has proved to be very puzzling. As noted briefly in Chapter 3, diplotene chromosomes in many kinds of oöcytes throw out loops of chromatin from a linear core such that they remind one of the appearance of an old-fashioned lampbrush or a test tube brush with bristles projecting from a twisted wire core. In these chromosomes, however, the ''bristles'' are loops of chromatin (Figure 7.26). Each of the loops is the site of synthesis of RNA to which protein is rapidly bound. Synthesis proceeds in linear polarized fashion around each loop and can be visualized by means of electron microscopy. The pattern resembles very closely that depicted in Figure 7.25 for transcription of rRNA on isolated nucleolar DNA cores. The distribution of lampbrush chromosomes among various organisms was tabulated by Davidson (1976), who also reviewed extensively the whole question of their structure and transcriptional activity.

Not only are lampbrush chromosomes found in a wide variety of animal species, but the lampbrush stage persists for a relatively long period. It may last as long as 3 months in *Xenopus* and 7 months in some urodeles. This suggests that the lampbrush configuration may play some fundamental role in the synthesis of maternal mRNA. Just what this role is remains unclear, for the qualitative pattern of RNA synthesis appears to be the same in oöcytes in the lampbrush stage and in full-grown oöcytes of *Xenopus laevis* (Anderson and Smith, 1978). The importance of the lampbrush stage may be in the response of the chromosomes to hormonal stimulation, resulting in a more rapid accumulation of cytoplasmic RNA. Thus Anderson and Smith showed that after stimulation with human chorionic gonadotropin (which has LH-like action), the fraction of mRNA that appears to be stable and that accumulates in the cytoplasm is twice as great in stimulated oöcytes as it is in unstimulated ones.

On the other hand, it has been estimated that the total amount of RNA transcribed from lampbrush chromosomes is many times that required to produce the functional mRNA of the oöcyte. Work-

ing with oöcytes of *Triton cristatus,* Varley et al. (1980) showed that much of the DNA that is transcribed from lampbrush chromosomes is satellite DNA. Satellite DNA consists of short base sequences that are repeated millions of times in the genome. Most of this DNA is located in regions near the centromeres and in chromosome 1, which shows little or no chiasma formation. The significance of RNA transcripts of satellite DNA is not known. Possibly it regulates in some way translatable sequences in the germinal vesicle.

7.2.6.3 Cytoplasmic Control of Nuclear Activity During Oögenesis.

The foregoing material has provided undisputable evidence that cytoplasmic factors, arising by virtue of outside stimuli and independently of concurrent gene action, affect the behavior of the nucleus and the structure of the chromosomes. In what follows further evidence is considered bearing both on the cytoplasmic control of nuclear and chromosomal morphology and behavior and on cytoplasmic controls of genetic transcription during oögenesis. Cytoplasmic controls of gene action are discussed in other contexts in later chapters, also.

The following experiments show that *Xenopus* egg cytoplasm at different stages of oögenesis contains factors that exercise rapid and well-defined effects on nuclear activity. Brain nuclei from adult *Xenopus,* which synthesize rRNA but seldom if ever make DNA, were injected into the cytoplasm of three types of recipient *Xenopus* eggs, which may or may not have been enucleated previously.

1. Medium to large oöcytes with an intact germinal vesicle, which synthesize RNA, especially rRNA, but no DNA.
2. Oöcytes that had just undergone rupture of the germinal vesicle and whose chromosomes were condensed and proceeding to the first meiotic spindle; these synthesize neither RNA nor DNA.
3. Ovulated oöcytes that have formed the first polar body; these would be expected to synthesize DNA after fertilization and formation of the zygote nucleus, but would not make RNA for some time thereafter.

In each instance the adult brain nuclei changed in morphology and in synthetic activity to conform to the character of the cell into which they were injected, regardless of whether or not the oöcyte nuclei were present. In immature, unstimulated oöcyte cytoplasm the nuclei enlarged, rRNA was synthesized, and enlarged nucleoli appeared; no DNA synthesis occurred. Upon injection into oöcytes in which the first meiotic spindle was forming, the brain nuclear membrane broke down, and the chromosomes condensed and became organized on spindles; neither RNA nor DNA synthesis occurred. Injected into cytoplasm of ovulated oöcytes, the brain nuclei enlarged and began synthesis of DNA but showed no nucleoli nor synthesis of RNA (Graham et al., 1966; Gurdon, 1968).

After fertilization, the amphibian egg divides and its daughter cells divide successively until hundreds of cells are produced. During this time the embryo undergoes changes in structure identified successively as blastula, gastrula, and neurula (cf. Chapter 10). It does not resume rRNA synthesis until the gastrula stage begins, and by the neurula stage, all cells are synthesizing all forms of RNA rapidly and have prominent nucleoli. If, however, the isolated nucleus of a neurula cell is introduced into the cytoplasm of a mature oöcyte, all RNA synthesis ceases in about 20 minutes (Gurdon and Woodland, 1969).

It would seem reasonable that the cytoplasmic factors that control nuclear activities in a stage-specific manner might be isolated from cytoplasmic fractions. This has proved to be the case. Shiokawa et al. (1977) reported that acid-soluble materials from full-sized *Xenopus* oöcytes, from unfertilized eggs, and early embryos incubated with isolated cells of embryos at the neurula stage, strongly and specifically inhibit their synthesis of rRNA. Similar extracts from younger oöcytes or from embryos at the gastrula stage are not inhibitory. If one examines factors from growing oöcytes, however (Crampton and Woodland, 1979), one may isolate four fractions which stimulate RNA synthesis in isolated nuclei from *Xenopus* tissue culture cells. Three of these factors stimulate the activity of the enzymes for synthesis of all classes of RNA (RNA polymerases I, II, and III; cf. Chapter 18 and Appendix B) five- to eightfold, and the other specifically stimulates the activity of RNA Polymerase I, increasing the rate of rRNA production by over 20 times. The latter factor and one of the others are essentially absent from mature eggs and from early dividing stages of fertilized eggs.

The mature amphibian oöcyte ceases meiosis

when the second metaphase spindle is formed. The chromosomes remain on the spindle until the egg is fertilized or is artificially activated. What inhibits the continuation of meiosis? In 1977 Meyerhof and Masui succeeded in extracting from mature oöcytes of *Rana pipiens*, but not from immature ones, a factor that is capable of arresting cell division when injected into a fertilized egg or into one of the daughter cells of the first division. They later isolated a similar factor from eggs of *Xenopus laevis* (Meyerhof and Masui, 1979). This factor, termed **cytostatic factor,** or CSF, is present in the clear supernatant fluid after centrifugation of the dejellied crushed oöcytes for 2 hours at $125,000 \times G$. It is inactivated by Ca^{2+} and also inactivated in the absence of Mg^{2+}. Although not further characterized, the CSF is presumably formed or activated during maturation and is responsible for meiotic arrest. Important for its activity must be the decline in free Ca^{2+} during the late phases of maturation. Release of the nuclear stasis at fertilization is presumably related to the increase in free Ca^{2+} that occurs at that time, as discussed in the next chapter.

7.3 Summary

Sperm, consisting chiefly of a compact nucleus, locomotive structures, and a supply of hydrolytic enzymes are produced in the seminiferous tubules of the testes where all stages of meiosis and spermateleosis are found in characteristic cycles of the seminiferous epithelium. All stages of spermatogenesis proceed in close association with Sertoli cells, from which mature sperm are eventually released.

Gene action in primary spermatocytes is concerned with production of the structural and regulatory proteins required in spermateleosis. During spermateleosis genetic transcription is progressively shut down, but protein synthesis continues through the use of previously transcribed templates and ribosomes. A good deal of this synthesis is involved in the production of arginine-rich proteins, which replace the lysine-rich histones that characterize the chromatin of the spermatid nucleus.

Eggs are large, compared to sperm, and are described as alecithal, isolecithal, telolecithal, or centrolecithal according to the amount and distribution of their yolk. They have primary protective membranes such as the vitelline membranes and the chorion, which are produced in the ovary, and some have secondary membranes such as jelly layers or shells that are added as the egg passes through the oviduct.

In vertebrates with large-yolked eggs, the yolk precursors are produced in the liver under the influence of estrogen and taken up from the blood by the growing oöcytes. In insects the vitellogenic proteins (vitellogenins) are taken up from the hemolymph. The fat bodies of insects are known to produce the yolk precursor in females under the influence of secretions from the corpus allatum. In crustaceans vitellogenins are taken up from the hemolymph, but a considerable amount of yolk is synthesized by the egg in situ.

During their growth phase most eggs are primary oöcytes in the prophase of the first meiotic division. Each primary oöcyte forms only one egg cell, the meiotic divisions giving rise to polar bodies at the animal pole instead of gametes. The haploid nucleus resulting from meiosis is the female pronucleus.

Eggs are fertilizable at various stages of meiosis in different organisms. The condition of fertilizability (physiological maturity) and the achievement of the characteristic meiotic stages are brought about by the action of hormones. Physiological maturity is a cytoplasmic condition, however, and is achieved independently of concurrent genetic transcription. It is dependent, however, on proteins synthesized on mRNA produced prior to hormonal stimulation leading to maturation and ovulation.

In vertebrates oöcytes begin meiosis and come to physiological maturation under the influence of pituitary gonadotropins. In frogs, at least, LH apparently stimulates the production of the steroid hormone progesterone by the ovarian follicular cells, and this in turn interacts with the cell surface before entering the oöcyte in effective form to induce maturation. In frogs, the cytoplasm of one egg, induced to mature by progesterone, can be injected into immature eggs to bring about their maturity. The action of progesterone requires Ca^{2+}, but direct injection of the ion is ineffective. Ca^{2+} introduced into the egg cortex by iontophoresis or by Ionophore A23187 will induce maturation in the absence of progesterone.

Possibly the phosphorylated proteins, maintained at a high level by the activity of cAMP-dependent protein kinase, may prevent the spontaneous onset and completion of maturation. Calcium, by activating phosphodiesterase, conceivably could promote the formation of inactive protein kinase by catalyzing the conversion of cAMP to AMP, forcing the recombination of active kinase with its regulator. This remains problematical, however.

Ovulation in vertebrates possibly involves increased concentrations of cyclic AMP and prostaglandins. Cyclic AMP presumably is involved in progesterone-mediated effects on enzyme activities involved in follicular rupture.

Spawning in starfish is induced by a polypeptide GSS, produced by the radial nerve. Part of the action of this neurohormone is to cause release of oöcytes and their expulsion to the exterior. It binds to the follicle cells of the ovary and stimulates their production of 1-methyladenine, which induces the completion of meiosis. Its action is parallel in most respects to that shown by progesterone in amphibian meiosis.

In most kinds of organisms oöcytes apparently accumulate stores of the various classes of RNA, namely, 4 S and 5 S RNA, mRNA, and rRNA, which are then used during early developmental stages after fertilization. There is little doubt that rRNA is stockpiled, but there is some question as to whether mRNA is truly stockpiled and unused (masked) or whether it is synthesized and degraded at a steady-state level. Oöcytes of a number of lower organisms have one or two 5 S RNAs that are oöcyte specific. These appear to be stockpiled for future use. They differ only slightly from somatic-type 5 S RNA, which is not stored.

Ribosomal RNA is produced in great quantities during oögenesis from ribosomal genes that are highly reiterated (i.e., exist in many copies in the genome). In several cases the rDNA is amplified, in the sense that multiple copies of the reiterated rDNA are made, released, and used as multiple nonchromosomal centers of rRNA synthesis.

Many organisms show a lampbrush stage during oögenesis. This stage may be quite prolonged and is denoted by the formation of large chromatin loops thrown out from diplotene chromosomes. These are the sites of active RNA transcription. It is presumed that transcripts from lampbrush chromo-

87

somes contribute to the supply of maternal mRNAs available to the fertilized egg, but the extent of this contribution is not clear.

At all stages of oögenesis cytoplasmic factors are present that control both the morphology of the nucleus and its chromosomes and the amount and kind of gene activity. This has been shown by experiments in which adult nuclei have been injected into oöcytes at various stages of development. From growing oöcytes, factors have been isolated that promote the synthesis of all RNAs in isolated nuclei. Maturing oöcytes contain factors that promote chromosome condensation and the formation of meiotic figures. Finally, mature oöcytes contain a cytostatic factor that, in amphibians at least, prevents the continuation of meiosis in mature unfertilized eggs.

7.4 Questions for Thought and Review

1. Recall the definitions of spermatogonia, spermatocytes, and spermatids. What are comparable stages in the differentiation of the female gametes in animals?

2. Of what significance might be the fact that cytoplasmic connections are retained between dividing spermatogonia and the subsequently arising spermatocytes and differentiating sperm?

3. What are the principal subcellular events that occur during spermateleosis? What relationships to the Sertoli cells change during spermateleosis?

4. What is meant by the steps in spermateleosis?

5. At what stage in spermatogenesis are the genes transcribed that determine the synthesis and assembly of the proteins in the sperm? Cite two lines of evidence on which you base your answer.

6. Of what significance is the incorporation of amino acids into protein during late stages of spermateleosis? How can this incorporation be accounted for, considering that RNA synthesis ceased in advance of it?

7. What is the likely role of the arginine-rich proteins produced during spermateleosis?

8. What are the chief components of yolk in animal eggs? What are the principal patterns in which yolk is distributed in eggs?

9. Distinguish between primary and secondary egg membranes. What are they and where are they produced?

10. What are vitellogenic proteins? Where are they produced? Under what control? Can males synthesize these proteins?

11. What is meant by the term physiological maturation as applied to animal oöcytes? What is the relationship between physiological maturation and meiosis in animal eggs? What experiments have revealed the nature of this relationship?

12. What sequence of biochemical events occurs in the ovary in response to the ovulatory surge of LH, resulting in ovulation?

13. Is gene action required in the oöcyte in response to LH in order that the egg may be activated? What is the evidence on which you base your answer?

14. What substances have been shown to be involved in spawning in starfish? What is the relationship between them?

15. What is redundant DNA? What is amplified DNA? In what ways have redundant and amplified DNA been shown to be involved in oögenesis?

16. What are lampbrush chromosomes? Of what significance might they be in oögenesis? What experiments bear on their presumed significance?

17. What evidence adduced from the study of oögenesis indicates that nuclear activity can be determined by cytoplasmic factors?

7.5 Suggestions for Further Reading

BALAKIER, H. 1978. Induction of maturation in small oöcytes from sexually immature mice by fusion with meiotic or mitotic cells. *Exp Cell Res* 112:137–41.

BELLAIRS, R. 1964. Biological aspects of the yolk of hen's egg. *Adv Morphogenesis* 4:217–72.

BIGGERS, J. D., and A. W. SCHUETZ, eds. 1972. *Oögenesis.* Baltimore: University Park Press.

FAWCETT, D. W. 1975. Gametogenesis in the male: prospects for its control. In C. L. Markert and J. Papaconstantinou, eds., *The Developmental Biology of Reproduction.* New York: Academic Press, pp. 25–53.

GURDON, J. B. 1974. *The Control of Gene Expression in Animal Development.* Cambridge, Mass.: Harvard University Press.

GURDON, J. B. 1976. Egg cytoplasm and gene control in development. *Proc R Soc Lond* [B] 198:211–47.

GURDON, J. B., and H. R. WOODLAND. 1968. The cytoplasmic control of nuclear activity in animal development. *Biol Rev* 43:233–67.

HENNIG, W., G. F. MEYER, I. HENNIG, and O. LEOCINI. 1973. Structure and function of the Y chromosome of *Drosophila hydei*. *CSH Symp Quant Biol* 38:673–83.

KANATANI, H. 1973. Maturation-inducing substance in starfishes. *Int Rev Cyt* 35:253–98.

KOTITE, N. J., S. N. NAYFEH, and F. S. FRENCH. 1978. FSH and androgen regulation of sertoli cell function in the immature rat. *Biol Reprod* 18:65–73.

LIFSCHYTZ, E., and D. HAREVEN. 1977. Gene expression and the control of spermatid morphogenesis in *Drosophila melanogaster*. *Dev Biol* 58:276–94.

MILLER, O. L., and B. R. BEATTY. 1969. Portrait of a gene. *J Cell Physiol* Symposium on Protein–Nucleic Acid Interaction. Gatlinburg, Tennessee, March 31–April 3, 1969.

ROSENBERG, M. P., R. HOESCH, and H. H. LEE. 1977. The relationship between 1-methyladenine-induced surface changes and fertilization in starfish oöcytes. *Exp Cell Res* 107:239–45.

SMITH, L. D. 1975. Molecular events during oöcyte maturation. In R. Weber, ed. *The Biochemistry of Animal Development*, Vol. III. New York: Academic Press, pp. 1–46.

SOMMERVILLE, J., D. B. MALCOLM, and H. G. CALLAN. 1978. The organization of transcription on lampbrush chromosomes. *Phil Trans R Soc Lond* [*Biol*] 283:359–66.

WASSARMAN, P. M., R. M. SCHULTZ, and G. E. LETOURNEAU. 1979. Protein synthesis during meiotic maturation of mouse oöcytes in vitro. Synthesis and phosphorylation of a protein localized in the germinal vesicle. *Dev Biol* 69:94–107.

7.6 References

ANDERSON, D. M., and L. D. SMITH. 1978. Patterns of synthesis and accumulation of heterogeneous RNA in lampbrush stage oöcytes of *Xenopus laevis* (Daudin). *Dev Biol* 67:274–85.

BAKER, T. G. 1963. A quantitative and cytological study of germ cells in human ovaries. *Proc R Soc* [B]158:417–33.

BEAMS, H. W., and R. G. KESSEL. 1963. Electron microscope studies on developing crayfish oöcytes with special reference to the origin of yolk. *J Cell Biol* 18:621–49.

BEAUMONT, H. M., and A. M. MANDL. 1962. A quantitative and cytological study of oögonia and oöcytes in the foetal and neonatal rat. *Proc R Soc* [B]155:557–59.

BELLAIRS, R. 1961. The structure of the yolk of the hen's egg as studied by electron microscopy. I. The yolk of the unincubated egg. *J Biophys Biochem Cytol* 11:207–25.

BELLAIRS, R. 1967. Aspects of the development of yolk spheres in the hen's oöcyte, studied by electron microscopy. *J Embryol Exp Morphol* 17:267–81.

BERGINK, E. W., and R. A. WALLACE. 1974. Precursor-products relationships between amphibian vitellogenin and the yolk proteins, lipovitellin and phosvitin. *J Biol Chem* 249:2897–903.

BERGINK, E. W., and R. A. WALLACE. 1974. Precursor-BOS, M. GRUBER, and A. B. GEERT. 1974. Estrogen-induced synthesis of yolk proteins in roosters. *Am Zool* 14:1177–93.

BROWN, D. D. 1966. The nucleolus and synthesis of ribosomal RNA during oögenesis and embryogenesis of *Xenopus laevis*. *Natl Cancer Inst Monogr* 23:297–309.

BROWN, D. D., D. CARROLL, and R. D. BROWN. 1977. The isolation and characterization of a second oöcyte 5 S DNA from *Xenopus laevis*. *Cell* 12:1045–56.

BROWN, D. D., and I. B. DAWID. 1968. Specific gene amplification in oöcytes. *Science* 160:272–80.

BROWN, D. D., P. C. WENSINK, and E. JORDAN. 1971. Purification and some characteristics of 5 S DNA from *Xenopus laevis*. *Proc Natl Acad Sci USA* 68:3175–79.

BRUMMETT, A. R., and J. N. DUMONT. 1977. Intracellular transport of vitellogenin in *Xenopus* oöcytes: An autoradiographic study. *Dev Biol* 60:482–86.

BURGOS, M. H., and R. VITALE-CALPE. 1967. The mechanism of spermiation in the toad. *Am J Anat* 120:227–52.

CHAET, A. B., and A. R. MCCONNAUGHY. 1959. Physiologic activity of nerve extracts. *Biol Bull* 117:407–408.

CLARK, M. R., W. F. TRIEBWASSER, J. M. MARSH, and W. J. LEMAIRE. 1978. Prostaglandins in ovulation. *Ann Biol Anim Biochim Biophys* 18:427–34.

CLARK, R. C. 1972. Sephadex fractionation of phosvitins from duck, turkey and ostrich egg yolk. *Comp Biochem Physiol* 41B:891–903.

CLARK, R. C. 1976. The composition and distribution of carbohydrate in phosvitin from hen, duck, turkey, ostrich and crocodile egg yolk. *Int J Biochem* 7:569–72.

CLERMONT, Y. 1972. Kinetics of spermatogenesis in mammals: seminiferous epithelium cycle and spermatogonial renewal. *Physiol Rev* 52:198–236.

CLERMONT, Y., and C. P. LEBLOND. 1955. Spermiogenesis in man, monkey, ram and other mammals as shown by the periodic acid-Schiff technique. *Am J Anat* 96:229–53.

CLOUD, J. C., and A. W. SCHUETZ. 1977. Interaction of progesterone with all or isolated portions of the amphibian (*Rana pipiens*) oöcyte surface. *Dev Biol* 60:359–70.

COCHRAN, R. C., and H. KANATANI. 1975. Site of production of maturation-inducing substance in the starfish testis. *Biol Bull* 149:424.

CRAMPTON, J. M., and H. R. WOODLAND. 1979. Isolation from *Xenopus laevis* embryonic cells of a factor which stimulates ribosomal RNA synthesis by isolated nuclei. *Dev Biol* 70:467–78.

89

DARNBOROUGH, C., and C. J. FORD. 1976. Cell-free translation of messenger RNA from oöcytes of *Xenopus laevis*. *Dev Biol* 50:285–301.

DAVIDSON, E. H. 1976. *Gene Activity in Early Development*, 2nd ed. New York: Academic Press.

DELAUNAY, J., M. WEGNEZ, and H. DENIS. 1975. Biochemical research on oögenesis. Comparison between the proteins of the ribosomes and the proteins of the 42 S particles from small oöcytes of *Xenopus laevis*. *Dev Biol* 42:379–87.

DENIS, H., and M. WEGNEZ. 1977a. Biochemical research on oögenesis. Oöcytes and liver cells of the teleost fish *Tinca tinca*. Different kinds of 5 S RNA. *Dev Biol* 59:228–36.

DENIS, H., and M. WEGNEZ. 1977b. Biochemical research on oögenesis. Oöcytes of *Xenopus laevis* synthesize but do not accumulate 5 S RNA of somatic type. *Dev Biol* 58:212–17.

DETLAFF, T. A. 1966. Action of actinomycin and puromycin upon frog oöcyte maturation. *J Embryol Exp Morphol* 16:183–95.

DOLECKI, G. J., and D. L. SMITH. 1979. Poly(A)$^+$ RNA metabolism during oögenesis in *Xenopus laevis*. *Dev Biol* 69:217–36.

DRURY, K. 1978. Method for the preparation of active maturation-promoting factor (MPF) from in vitro matured oöcytes of *Xenopus laevis*. *Differentiation* 10:181–86.

DWORKIN, M. B., and A. A. INFANTE. 1978. RNA synthesis in unfertilized sea urchin eggs. *Dev Biol* 62:247–57.

DYM, M., and D. W. FAWCETT, 1971. Further observations on the numbers of spermatogonia, spermatocytes, and spermatids connected by intercellular bridges in the mammalian testis. *Biol Reprod* 4:195–215.

ERICKSON, R. P., J. M. ERICKSON, C. J. BETLACH, and M. L. MEISTRICH. 1980. Further evidence for haploid gene expression during spermatogenesis: heterogeneous, poly(A)-containing RNA is synthesized postmeiotically. *J Exp Zool* 214:13–19.

FAWCETT, D. N. 1975. The mammalian spermatozoon. *Dev Biol* 44:394–436.

FICQ, A., and J. BRACHET. 1971. RNA-dependent DNA polymerase: possible role in the amplification of ribosomal DNA in *Xenopus* oöcytes. *Proc Natl Acad Sci USA* 68:2774–76.

FORD, P. J., T. MATHIESON, and M. ROSBASH. 1977. Very long-lived messenger RNA in ovaries of *Xenopus laevis*. *Dev Biol* 57:417–26.

FRANCHI, L. L., and A. M. MANDL. 1962. The ultrastructure of oögonia and oöcytes in the foetal and neonatal rat. *Proc R Soc* [B]157:99–114.

GALL, J. G. 1969. The genes for ribosomal RNA during oögenesis. *Genetics* 61:*Suppl. 1*:121–32.

GEREMIA, R., A. D'AGOSTINO, and V. MONESI. 1978. Bio-

chemical evidence of haploid gene activity in spermatogenesis of the mouse. *Exp Cell Res* 111:23–30.

GILULA, N. B., D. W. FAWCELL, and A. AOKI. 1976. The Sertoli cell occluding junctions and gap junction in mature and developing mammalian testis. *Dev Biol* 50:142–68.

GODEAU, J. F., S. SCHORDERET-SLATKINE, P. HUBERT, and E.-E. BAULIER. 1978. Induction of maturation in *Xenopus laevis* oöcytes by a steroid linked to a polymer. *Proc Natl Acad Sci USA* 75:2353–57.

GRAHAM, C. F., K. ARMS, and J. B. GURDON. 1966. The induction of DNA synthesis by frog egg cytoplasm. *Dev Biol* 14:349–81.

GROSS, P. 1967. The control of protein synthesis in embryonic development and differentiation. *Cur Top Dev Biol* 1:1–46.

GROSS, P., and G. COUSINEAU. 1964. Macromolecule synthesis and the influence of actinomycin on early development. *Exp Cell Res* 33:368–95.

GURDON, J. B. 1968. Changes in somatic cell nuclei inserted into growing and maturing amphibian oöcytes. *J Embryol Exp Morphol* 20:401–14.

GURDON, J. B., and H. R. WOODLAND. 1969. The influence of the cytoplasm on the nucleus during cell differentiation with special reference to RNA synthesis during amphibian cleavage. *Proc R Soc* [B]173:99–111.

HENNIG, W. 1968. Ribonucleic acid synthesis of the Y-chromosome of *Drosophila hydei*. *J Mol Biol* 38:227–39.

HERTIG, A. T. 1968. The primary human oöcyte: some observations on the fine structure of Balbiani's vitelline body and the origin of the annulate lamellae. *Am J Anat* 122:107–38.

HESS, O. 1973. Local structural variations in the Y-chromosome of *Drosophila hydei* and their correlation to genetic activity. *CSH Symp Quant Biol* 38:663–83.

HIRAI, S., J. KUBOTA, and H. KANATANI. 1971. Induction of cytoplasmic maturation by 1-methyladenine in starfish oöcytes after removal of the germinal vesicle. *Exp Cell Res* 68:137–43.

HOUGH-EVANS, B. R., B. J. WOLD, S. G. ERNST, R. J. BRITTEN, and E. H. DAVIDSON. 1977. Appearance and persistence of maternal RNA sequences in sea urchin development. *Dev Biol* 60:258–77.

HUGHES, G. C. 1963. The population of germ cells in the developing female chick. *J Embryol Exp Morphol* 11:513–36.

HUMPHREYS, T. 1971. Measurements of messenger RNA entering polysomes upon fertilization of sea urchin eggs. *Dev Biol* 26:201–208.

IATROU, K., A. W. SPIRA, and G. H. DIXON. 1978. Protamine messenger RNA: evidence for early synthesis and accumulation during spermatogenesis in rainbow trout. *Dev Biol* 64:82–98.

IKEGAMI, S. 1976. Role of asterosaponin A in starfish

spawning induced by gonad-stimulating substance and 1-methyladenine. *J Exp Zool* 198:359–66.

IKEGAMI, S., Y. KAMIYA, and H. SHIRAI. 1976a. Characterization and action of meiotic maturation inhibitors in starfish ovary. *Exp Cell Res* 103:233–39.

IKEGAMI, S., H. KANATANI, and S. S. KOIDE. 1976b. Gamete-release by 1-methyladenine in vitro in the sea cucumber, *Leptosynapta inhaerens*. *Biol Bull* 150:402–10.

IWAMATSU, T. 1978. Studies on oöcytes maturation of the medaka, *Oryzias latipes*. V. On the structure of steroids that induce maturation in vitro. *J Exp Zool* 204:401–408.

JALABERT, B., and D. SZÖLLÖSI. 1975. In vitro ovulation of trout oöcytes: effects of prostaglandins on smooth muscle-like cells of the theca. *Prostaglandins* 9:765–78.

JARED, D. W., and R. A. WALLACE. 1968. Comparative chromatography of the yolk proteins of teleosts. *Comp Biochem Physiol* 24:437–43.

JARED, D. W., and R. A. WALLACE. 1969. Protein uptake in vitro by amphibian oöcytes. *Exp Cell Res* 57:454–57.

KALT, M. R. 1979. *In vitro* synthesis of RNA by *Xenopus* spermatogenic cells. I. Evidence for polyadenylated and non-polyadenylated RNA synthesis in different cell populations. *J Exp Zool* 208:77–95.

KANATANI, H. 1974. Presence of 1-methyladenine in sea urchin gonad and its relation to oöcyte maturation. *Dev Growth Differ* 16:159–70.

KANATANI, H. 1975. Maturation-inducing substances in asteroid and echinoid oöcytes. *Am Zool* 15:493–505.

KANATANI, H., and Y. HIRAMOTO. 1970. Site of action of 1-methyladenine in inducing oöcyte maturation in starfish. *Exp Cell Res* 61:280–84.

KANATANI, H., S. IKEGAMI, H. SHIRAI, H. OIDE, and S. TAMURA. 1971. Purification of gonad-stimulating substance obtained from radial nerves of the starfish, *Asterias amurensis*. *Dev Growth Differ* 13:151–64.

KANATANI, H., and H. SHIRAI. 1972. On the maturation-inducing substance produced in starfish gonad by neuronal substance. *Gen Comp Endocrinol Suppl* 3:571–79.

KISHIMOTO, T., M. L. CAYER, and H. KANATANI. 1976. Starfish oöcyte maturation and reduction of disulfide-bond on oöcyte surface. *Exp Cell Res* 101:104–10.

KISHIMOTO, T., S. HIRAI, and H. KANATANI. 1981. Role of germinal vesicle material in producing maturation-promoting factor in starfish oöcyte. *Dev Biol* 81:177–81.

KISHIMOTO, T., and H. KANATANI. 1973. Induction of starfish oöcyte maturation by disulfide-reducing agents. *Exp Cell Res* 82:296–302.

KISHIMOTO, T., and H. KANATANI. 1976. Cytoplasmic factor responsible for germinal vesicle breakdown and meiotic maturation in starfish oöcyte. *Nature* 260:321–22.

KISHIMOTO, T., and H. KANATANI. 1977. Lack of species specificity of starfish maturation-promoting factor. *Gen Comp Endocrinol* 33:41–44.

KISHIMOTO, T., J. KUBOTA, and H. KANATANI. 1977. Distribution of maturation-promoting factor in starfish oöcyte stratified by centrifugation. *Dev Growth Differ* 19:283–88.

KOSTELLOW, A. B., D. ZIEGLER, and G. A. MORRILL. 1980. Regulation of Ca^{2+} and cyclic AMP during the first meiotic division in amphibian oöcytes. *J Cyclic Nucleotide Res* 6:347–58.

KUBOTA, J., and H. KANATANI. 1975a. Concanavalin A: its action in inducing oöcyte maturation-inducing substance in starfish follicle cells. *Science* 187:654–55.

KUBOTA, J., and H. KANATANI. 1975b. Production of 1-methyladenine induced by concanavalin A in starfish follicle cells. *Dev Growth Differ* 17:177–85.

KUBOTA, J., and H. KANATANI. 1978. Binding of concanavalin A to the surface of starfish follicle cells and production of 1-methyladenine. *Dev Growth Differ* 20:349–51.

KUBOTA, J., K. NAKAO, H. SHIRAI, and H. KANATANI. 1977. 1-methyladenine-producing cell in starfish testis. *Exp Cell Res* 106:63–70.

LEMAIRE, W. J., and J. M. MARSH. 1975. Interrelationships between prostaglandins, cyclic AMP and steroids in ovulation. *J Reprod Fertil* 22:53–74.

LIMA-DE-FARIA, A., H. JAWORSKA, and T. GUSTAFSSON. 1973. Release of amplified ribosomal DNA from the chromomeres of *Acheta*. *Proc Natl Acad Sci USA* 70:80–83.

LINDSLEY, D. L., and E. H. GRELL. 1969. Spermiogenesis without chromosomes in *Drosophila melanogaster*. *Genetics* 61:Suppl 1:69–78.

LUI, C. W., and J. D. O'CONNOR. 1977. Biosynthesis of crustacean lipovitellin. III. The incorporation of labeled amino acids into the purified lipovitellin of the crab, *Pachygrapsus crassipes*. *J Exp Zool* 199:105–08.

MAGNUSSON, C. and T. HILLENSJÖ. 1977. Inhibition of maturation and metabolism in rat oöcytes by cyclic AMP. *J Exp Zool* 201:139–47.

MALLER, J. L., and E. G. KREBS. 1977. Progesterone-stimulated meiotic cell division in *Xenopus* oöcytes. *J Biol Chem* 252:1712–18.

MARUSHIGE, J., and G. H. DIXON. 1969. Developmental changes in chromosomal composition and template activity during spermatogenesis in trout testis. *Dev Biol* 19:397–414.

MASUI, Y. 1967. Relative roles of the pituitary, follicle cells and progesterone in the induction of oöcyte maturation in *Rana pipiens*. *J Exp Zool* 166:365–76.

MASUI, Y. 1972. Distribution of the cytoplasmic activity inducing germinal vesicle breakdown in frog oöcytes. *J Exp Zool* 179:365–78.

91

MASUI, Y., and H. J. CLARKE. 1979. Oöcyte maturation. *Int Rev Cytol* 57:185–282.

MASUI, Y., and C. L. MARKERT. 1971. Cytoplasmic control of nuclear behavior during meiotic maturation of frog oöcytes. *J Exp Zool* 177:129–46.

MASUI, Y., P. G. MEYERHOF, M. A. MILLER, and W. J. WASSERMAN. 1977. Roles of divalent cations in maturation and activation of vertebrate oöcytes. *Differentiation* 9:49–57.

MERRIAM, R. W. 1972. On the mechanism of action in gonadotropic stimulation of oöcyte maturation in *Xenopus laevis. J Exp Zool* 180:421–26.

MEYERHOF, P. G., and Y. MASUI. 1977. Ca and Mg control of cytostatic factors from *Rana pipiens* oöcytes which cause metaphase and cleavage arrest. *Dev Biol* 61:214–29.

MEYERHOF, P. G., and Y. MASUI. 1979. Properties of cytostatic factor from *Xenopus laevis* eggs. *Dev Biol* 72:182–87.

MILLER, O. L., JR., and B. R. BEATTY. 1969. Visualization of nucleolar genes. *Science* 164:955–57.

MOREAU, M., M. DOREE, and P. GUERRIER. 1976. Electrophoretic introduction of calcium ions into the cortex of *Xenopus laevis* oöcytes triggers meiosis reinitiation. *J Exp Zool* 197:443–49.

MOREAU, M., P. GUERRIER, M. DOREE, and C. C. ASHLEY. 1978. Hormone-induced release of intracellular Ca^{2+} triggers meiosis in starfish oöcytes. *Nature* 272:251–53.

MORISAWA, M., and H. KANATANI. 1978. Oöcyte-surface factor responsible for 1-methyladenine-induced oöcyte maturation in starfish. *Gamete Res* 1:157–64.

MULNER, O., D. HUCHON, C. THIBER, and R. OZON. 1979. Cyclic AMP synthesis in *Xenopus laevis* oöcytes. Inhibition by progesterone. *Biochim Biophys Acta* 582:179–84.

MUNDALL, E. C., S. S. TOBE, and B. STAY. 1979. Induction of vitellogenin and growth of implanted oöcytes in male cockroaches. *Nature* 282:97–98.

O'CONNOR, C. M., K. R. ROBINSON, and L. D. SMITH. 1977. Calcium, potassium, and sodium exchange by full-grown and maturing *Xenopus laevis* oöcytes. *Dev Biol* 61:28–40.

O'CONNOR, C. M., and L. D. SMITH. 1976. Inhibition of oöcyte maturation by theophylline: possible mechanism of action. *Dev Biol* 52:318–22.

O'CONNOR, C. M., and L. D. SMITH. 1979. *Xenopus* oöcyte cAMP-dependent protein kinases before and during progesterone-induced maturation. *J Exp Zool* 207:367–74.

PELHAM, H., and D. D. BROWN. 1980. The 5 S DNA transcription factor is the same protein that complexes 5 S RNA in oöcytes. Annual Report of the Director, Department of Embryology, Carnegie Institution of Washington, pp. 82–85.

PHILLIPS, D. M. 1977. Mitochondrial disposition in mammalian spermatozoa. *J Ultrastruct Res* 58:144–54.

PICARD, B., and M. WEGNEZ. 1979. Isolation of a 7 S particle from *Xenopus laevis* oöcytes: a 5 S RNA–protein complex. *Proc Natl Acad Sci USA* 76:241–45.

REED, P. W., and H. A. LARDY. 1972. A23187: a divalent cation ionophore. *J Biol Chem* 247:6890–97.

REEDER, R. H. 1974. Ribosomes from eukaryotes: genetics. In M. Nomura, A. Tissieres, and P. Lengyel, eds. *Ribosomes.* Cold Spring Harbor, New York: Cold Spring Harbor Laboratory, pp. 489–518.

REYNHOUT, J. K., and L. D. SMITH. 1973. Evidence for steroid metabolism during the in vitro induction of maturation in oöcytes of *Rana pipiens. Dev Biol* 30:392–402.

REYNHOUT, J. K., and L. D. SMITH. 1974. Studies on the appearance and nature of a maturation-inducing factor in the cytoplasm of amphibian oöcytes exposed to progesterone. *Dev Biol* 38:394–400.

RONDELL, P. 1970. Follicular processes in ovulation. *Fed Proc* 29:1875–79.

ROOSEN-RUNGE, E. C. 1962. The process of spermatogenesis in mammals. *Biol Rev* 37:343–77.

ROSBASH, M., and P. J. FORD. 1974. Polyadenylic acid containing RNA in *Xenopus laevis* oöcytes. *J Mol Biol* 85:87–101.

SCHMID, M., F. J. HOFGÄRTNER, M. T. ZENZES, and W. ENGEL. 1977. Evidence for postmeiotic expression of ribosomal RNA genes during male gametogenesis. *Hum Genet* 38:279–84.

SCHORDERET-SLATKINE, S. 1972. Action of progesterone and related steroids on oöcyte maturation in *Xenopus laevis.* An in vitro study. *Cell Differ* 1:179–89.

SCHORDERET-SLATKINE, S. and E.-E. BAULIEU. 1977. Induction by progesterone and a "maturation-promoting factor" of soluble proteins in *Xenopus laevis* oöcytes in vitro. *Ann NY Acad Sci* 286:421–33.

SCHORDERET-SLATKINE, S., and K. C. DRURY. 1973. Progesterone induced maturation in oöcytes of *Xenopus laevis.* Appearance of a "maturation promoting factor" in enucleated oöcytes. *Cell Differ* 2:247–54.

SCHORDERET-SLATKINE, S., M. SCHORDERET, P. BOQUET, F. GODEAU, and E.-E. BAULIEU. 1978. Progesterone-induced meiosis in *Xenopus laevis* oöcytes: a role for cAMP at the "maturation-promoting factor" level. *Cell* 15:1269–76.

SCHUETZ, A. W. 1977. Induction of oöcyte maturation and differentiation: mode of progesterone action. *Ann NY Acad Sci* 286:408–20.

SCHUETZ, A. W., and J. D. BIGGERS. 1968. Effect of calcium on the structure and functional response of the starfish ovary to radial nerve factor. *J Exp Zool* 168:1–10.

SCHUETZ, A. W., T. G. HOLLINGER, R. A. WALLACE, and D. A. SAMSON. 1977. Inhibition of [^{3}H]-vitellogenin uptake by

92

isolated amphibian oöcytes injected with cytoplasm from progesterone-treated mature oöcytes. *Dev Biol* 58:428–33.

SCHUETZ, A. W., and D. SAMSON. 1979. Protein synthesis requirement for maturation promoting factor (MPF) initiation of meiotic maturation in *Rana* oöcytes. *Dev Biol* 68:636–42.

SCHUETZ, A. W., and W. J. SWARTZ. 1979. Intrafollicular cumulus cell transformations associated with oöcyte maturation following gonadotropic hormone stimulation of adult mice. *J Exp Zool* 207:399–406.

SELMAN, K., and E. ANDERSON. 1975. The formation and cytochemical characterization of cortical granules in ovarian oöcytes of the golden hamster (*Mesocricetus auratus*). *J Morphol* 147:251–72.

SHIOKAWA, K., A. KAWAHARA, Y. MISUMI, Y. YASUDA, and K. YAMANA. 1977. Inhibitor of ribosomal RNA synthesis in *Xenopus laevis* embryos. VI. Occurrence in matured oöcytes and unfertilized eggs. *Dev Biol* 57:210–14.

SHIRAI, H. 1973. Effects of methionine and S-adenosylmethionine on production of 1-methyladenine in starfish follicle cells. *Growth Differ* 15:307–313.

SHIRAI, H., and H. KANATANI. 1972. 1-methyladenosine ribohydrolase in the starfish ovary and its relation to oöcyte maturation. *Exp Cell Res* 75:79–88.

SHIRAI, H., and H. KANATANI. 1973. Induction of spawning and oöcyte maturation in starfish by 1-methyladenosine monophosphate. *Growth Differ* 15:217–24.

SHIRAI, H., H. KANATANI, and S. TAGUCHI. 1972. 1-methyladenine biosynthesis in starfish ovary: Action of gonad-stimulating hormone in methylation. *Science* 175:1366–68.

SLATER, J., D. GILLESPIE, and D. W. SLATER. 1973. Cytoplasmic adenylation and processing of maternal RNA. *Proc Natl Acad Sci USA* 70:406–11.

SMITH, L. D., and R. E. ECKER. 1969. Role of the oöcyte nucleus in physiological maturation in *Rana pipiens*. *Dev Biol* 19:281–309.

SMITH, L. D., and R. E. ECKER. 1971. The interaction of steroids with *Rana pipiens* oöcytes in the induction of maturation. *Dev Biol* 25:232–47.

SMITH, L. D., R. E. ECKER, and S. SUBTELNY. 1968. In vitro induction of physiological maturation in *Rana pipiens* oöcytes removed from their ovarian follicles. *Dev Biol* 17:627–43.

SNYDER, B. W., and A. W. SCHUETZ. 1973. In vitro evidence of steroidogenesis in the amphibian (*Rana pipiens*) ovarian follicle and its relationship to meiotic maturation and ovulation. *J Exp Zool* 183:333–42.

STEINERT, G., C. THOMAS, and J. BRACHET. 1976. Localization by in situ hybridization of amplified ribosomal DNA during *Xenopus laevis* oöcyte maturation. (a light and electron microscopy study). *Proc Natl Acad Sci USA* 73:833–36.

SUBTELNY, S., L. D. SMITH, and R. E. ECKER. 1968. Maturation of ovarian frog eggs without ovulation. *J Exp Zool* 168:39–48.

TELFER, W. H. 1954. Immunological studies of insect metamorphosis. II. The role of a sex-limited blood protein in egg formation by cecropia silkworm. *J Gen Physiol* 37:539–58.

TELFER, W. H. 1965. The mechanism and control of yolk formation. *Annu Rev Entomol* 10:161–84.

THOMAS, C., F. HANOCQ, and V. HEILPORN. 1977. Persistence of oöcyte amplified rDNA during early development of *Xenopus laevis* eggs. *Dev Biol* 57:226–29.

VAN GANSEN, P., C. THOMAS, and A. SCHRAM. 1976. Nucleolar activity and RNA metabolism in previtellogenic and vitellogenic oöcytes of *Xenopus laevis*. A biochemical and autoradiographical, light and EM study. *Exp Cell Res* 98:111–19.

VARLEY, J. M., H. C. MACGREGOR, and H. P. ERBA. 1980. Satellite DNA is transcribed on lampbrush chromosomes. *Nature* 283:686–88.

VITALE-CALPE, R. 1970. Ultrastructural studies of spontaneous spermiation in the guinea pig. *Z Zellforsch* 105:222–33.

WALLACE, R. A. 1963. Studies on amphibian yolk. III. A resolution of yolk platelet components. *Biochim Biophys Acta* 74:495–504.

WALLACE, R. A., and E. W. BERGINK. 1974. Amphibian vitellogenin: properties, hormonal regulation of hepatic synthesis and ovarian uptake, and conversion of yolk proteins. *Am Zool* 14:1159–75.

WALLACE, R. A., and D. W. JARED. 1968. Estrogen induces lipophosphoprotein in serum of male *Xenopus laevis*. *Science* 160:91–92.

WALLACE, R. A. and D. W. JARED. 1976. Protein incorporation by isolated amphibian oöcytes. V. Specificity for vitellogenin incorporation. *J Cell Biol* 69:345–51.

WALLACE, R. A., S. L. WALKER, and P. V. HAUSCHKA. 1967. Crustacean lipovitellin. Isolation and characterization of the major high-density lipoprotein from the eggs of decapods. *Biochem* 6:1582–90.

WASSERMAN, W. J., and Y. MASUI. 1974. A study on gonadotropin action in the induction of oöcyte maturation in *Xenopus laevis*. *Biol Reprod* 11:133–44.

WASSERMAN, W. J., and Y. MASUI. 1975a. Initiation of meiotic maturation in *Xenopus laevis* oöcytes by the combination of divalent cations and ionophore A23187. *J Exp Zool* 193:369–75.

WASSERMAN, W. J., and Y. MASUI. 1975b. Effects of cycloheximide on a cytoplasmic factor initiating meiotic maturation in *Xenopus* oöcytes. *Exp Cell Res* 91:381–86.

WASSERMAN, W. J., and Y. MASUI. 1976. A cytoplasmic factor promoting oöcyte maturation: Its extraction and preliminary characterization. *Science* 191:1266–68.

WASSERMAN, W. J., L. H. PINTO, C. M. O'CONNOR, and L. D.

93

SMITH. 1980. Progesterone induces a rapid increase in [Ca²⁺] in *Xenopus laevis* oöcytes. *Proc Natl Acad Sci USA* 77:1534–36.

WEGNEZ, M., H. DENIS, A. MAZABRAUD, and J.-C. CLEROT. 1978. Biochemical research on oögenesis. RNA accumulation during oögenesis of the dogfish *Scyliorhinus caniculus*. *Dev Biol* 62:99–111.

WELLAUER, P. K., I. B. DAWID, D. D. BROWN, and R. H. REEDER. 1976. Molecular basis for length heterogeneity in ribosomal DNA from *Xenopus laevis*. *J Mol Biol* 105:461–86.

WELLAUER, P. K., and R. H. REEDER. 1975. A comparison of the structural organization of amplified ribosomal DNA from *Xenopus mulleri* and *Xenopus laevis*. *J Mol Biol* 94:151–61.

WHALI, W., G. U. RYFFEL, T. WYLER, R. B. JAGGI, R. WEBER, and I. B. DAWID. 1978. Cloning and characterization of synthetic sequences for the *Xenopus laevis* vitellogenin structural gene. *Dev Biol* 67:371–83.

WOLIN, E. M., H. LAUFER, and D. F. ALBERTINI. 1973. Uptake of the yolk protein, lipovitellin, by developing crustacean oöcytes. *Dev Biol* 35:160–70.

Fertilization, although a complex process, can be defined simply as the union of sperm and egg to form a zygote. The chief components of fertilization are the approach of the sperm to the egg, contact between them, penetration of the sperm through the membranes surrounding the egg, and entrance of the sperm contents into the egg cytoplasm. The process is completed when the gametes initiate the synthesis of DNA and the mitotic apparatus for the first cell division of the new generation is formed. The analysis of this series of changes presents a number of challenging problems.

These problems are basically the same in whatever organism one studies, and some of them may be presented by way of introduction. How does the motile sperm find the egg? Is it attracted by the egg or does it encounter it by chance? By what means does the sperm penetrate the membranes that cover the egg? Does more than one sperm enter the egg? What is the role of the egg in effecting union with the sperm? Does the egg take an active part in the entrance of the sperm, or is it penetrated passively? Eggs that are not fertilized die, usually a short time after ovulation. What properties of the egg are altered when the sperm enters, such that it may live and develop? Are repressed segments of the genome derepressed? Are new or masked RNA templates made available for protein synthesis? Does the egg show changes in permeability? In respiratory rate? In protoplasmic viscosity? In surface tension? Which of these changes are significant for further development of the egg?

Questions such as the foregoing, and many related ones, have been raised by numerous investigators, and there is abundant literature on the subject of fertilization that dates back to the middle of the nineteenth century. The more recent literature is principally concerned with ultrastructural changes in egg and sperm at the time of fertilization, as revealed by use of the electron microscope, and with effects of fertilization on gene activity and protein synthesis. The techniques of electron microscopy and the sophisticated analytical methods of molecular biology have brought many new insights into the fertilization process, and these emerge as this chapter proceeds.

A very large proportion of research on fertilization has been carried out on the eggs of marine organisms, both invertebrates and vertebrates, that shed their gametes directly into the seawater. Gametes from these organisms can usually be obtained in considerable quantities and can be brought together by the investigator under conditions in which they can most conveniently be studied. Insights into fertilization processes that have resulted from the study of marine organisms have guided investigations into many aspects of fertilization in domestic animals and in man. It is now quite clear that fundamentals of the process are the same in all kinds of animals. This is an important generalization. An intimate knowledge of fertilization is important in animal breeding, and hence in the provision of a large part of the world's food supply. It is also important in the development of methods to control our expanding human population.

8

Fertilization and Parthenogenesis in Animal Development

8.1 Dispersal of the Gametes

In general, eggs are nonmotile, and sperm swim to them by lashing movements of their flagella. The medium in which the sperm approaches the egg is an aqueous one, whether it be the water of pond, stream, or ocean, or the fluids provided by secretions of cells lining the female reproductive tract. The most primitive conditions for fertilization are provided by sedentary marine animals such as sponges, bivalve mollusks, barnacles, and other encrusting organisms that cast their sperm and eggs widely into the water in the same general area. The joining of gametes is facilitated by the swimming movements of the sperm, but it is compromised by currents, wave action, and other chance events. Under these circumstances, and with little or no provision for protection of the fertilized eggs, gametes must be produced in prodigious numbers so that a sufficient number of zygotes and surviving embryos will be provided for the next generation.

Many kinds of organisms that shed gametes into water have evolved mechanisms to assure a high concentration of gametes from both sexes in the same spot, thus promoting a larger number of successful fertilizations. The male of the squid *Loligo pealii*, for example, uses a specialized tentacle to pick up a packet of sperm (spermatophore) from his mantle cavity and to transfer it to the body of the female by way of her funnel. The eggs are thus fertilized in seawater in the mantle cavity and then they emerge through the funnel as a jelly-covered string. Or, if the female is not yet ready to produce her egg strings, the male may deposit the spermatophore in a special niche in the buccal membrane, just below the mouth. Sperm slowly released during the next day or two are then available for fertilizing the eggs as they appear (Arnold, 1962). More familiar to most people is the behavior of frogs and toads in **amplexus** and egg laying; the male, grasping the female with his forelimbs from behind, sheds sperm over her eggs as they emerge from the cloaca.

Another kind of restriction on the dispersal of gametes before fertilization is found among organisms that have **internal fertilization.** In these the male deposits a dense suspension of sperm **(semen)** directly into the genital tract of the female. Many kinds of anatomical and physiological mechanisms have been evolved for the transfer of sperm into the genital tract of the female. In insects the male and female unite their genital openings by means of elaborate interlocking chitinous plates. Elasmobranchs and some fishes use modified anal fins as intromittent organs. Some reptiles and birds have rather intricate penis-like structures with an external groove that serves as a sperm channel. The mammalian penis contains erectile tissue that becomes engorged with blood at the time of sexual arousal. This permits intromission, which is followed by ejaculation of semen, sometimes immediately or, depending on the kind of organism, after back and forth movements of the penis in the vagina.

8.2   Do Eggs Attract Sperm?

In order for a substance to be identified as attracting sperm, it must be shown that it causes directed swimming movements **(chemotaxis)** of the sperm toward its source. Until recently there was no proof that any animal eggs produced sperm attractants. True, cases have been known in which egg-associated materials bring about a high concentration of sperm in the vicinity of the egg. The spermatozoa of herring are inactive when set free into sea water, but become intensely active in the vicinity of the micropyle of the mature egg (Figure 8.1). In a number of hydroids, the funnel tissue of the female gonangium, but not the eggs or other structures, is the source of a species-specific sperm attractant (Figure 8.2), probably a polypeptide of low molecular weight (Miller and Tseng, 1974). Miller (1975, 1979a, b) found that extracts of eggs of a number of species of urochordates and of a number of kinds of hydroids show species-specific effects on the behavior of sperm. In the presence of the homologous extracts, inactive sperm are activated and tend to turn toward the source of the extract more frequently than away from it.

A clear and unequivocal case of the timed release of a sperm attractant has been found in the hydroid *Orthopyxis caliculata*. The eggs are spawned after the first polar body is formed. If sperm are present at that time, they are not attracted to the eggs, but their random motions may carry them into the jelly layer surrounding the egg where they remain relatively inactive. The second polar body is

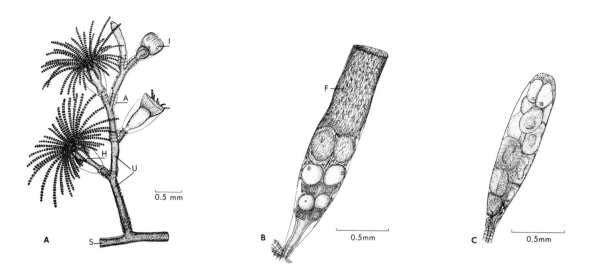

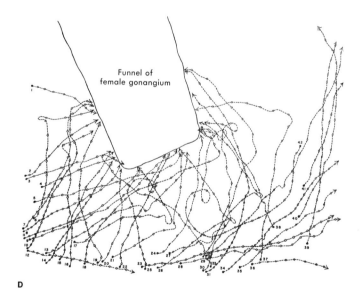

FIGURE 8.1 Chemotaxis in sperm of *Campanularia flexuosa.* A: Small portion of a colony showing stolon, D, upright, U, and mature, H, and immature, I, hydranths, both with annulations, A. B, C: Reproductive hydranths, or gonangia. The female gonangium (B) shows a large funnel, F, leading to ova within; C shows a male gonangium containing sperm sacs. D: Plot of 42 sperm trails in the vicinity of the female gonangium. The solid circles indicate the start of each trail, and open circles give the position of each sperm at 0.45-sec intervals. Note that most trails lead to the funnel tissue, the distances between open circles increasing the while, indicating a speeding up of sperm movement with proximity to the funnel. [A from R. L. Miller, *J Exp Zool* 161:23–44, Fig 1, by courtesy of Dr. Miller and permission of Alan R. Liss, Inc.]

formed about 10 minutes after spawning and, within 1 or 2 minutes, the sperm trapped on the jelly become active once more and move towards the egg surface. At the point on the egg surface where the second polar body was released motile sperm attach to the egg and their flagella become oriented perpendicularly to the surface, beating synchronously. More sperm attach to those already adhering, and a clot of sperm heads builds up. After 10 minutes the clot of sperm breaks up, sperm activity decreases, and the attractant power of the egg is lost. Meanwhile, a sperm, presumably one of those of the clot, will have successfully fertilized the egg (Miller, 1978).

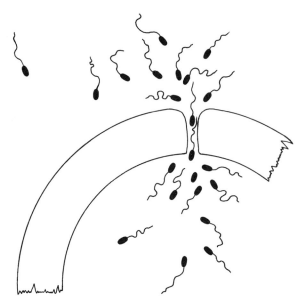

FIGURE 8.2 Aggregation of sperm about the micropyle of an isolated section of the chorion of the herring egg. This property of the micropylar region is retained during 30 min of vigorous washing in Ringer's solution. [Adapted from R. Yanagimachi, *Annot Zool Japon* 30:114–19 (1957).]

98

8.3 Activation of the Sperm

Prior to being shed, mature sperm are physically inactive. In most instances their locomotor activity is triggered by dilution with marine or fresh water, as appropriate for organisms with external fertilization or, as in mammals, by a mixture with the secretions of the bulbourethral and prostate glands of the male reproductive tract. For some invertebrates it has been reported that sperm, already active by virtue of dilution with seawater, increase their rate of movement and their oxygen consumption in the presence of water in which eggs have been standing. Likewise, it appears that mammalian sperm show increased motility after more or less prolonged exposure to the fluids of the female reproductive tract. These matters are treated further later. What now follows is a discussion of two as-

pects of sperm activation that render the sperm capable of fertilizing the egg. The first of these is the **acrosome reaction,** a complex change in the relationship of the acrosome to the sperm head, found in both vertebrates and invertebrates. The second aspect of activation is found in all mammals and is

FIGURE 8.3 The acrosome reaction in the sea urchin sperm. A: The head of the unreacted sperm, showing the acrosomal complex, including acrosomal granule, A, surrounded by the acrosomal membrane, in which one distinguishes inner, IAM and outer, OAM, parts. The acrosome is covered by the plasma membrane, PM, that extends around the entire sperm. The acrosomal granule caps an indentation in the nucleus, N, that contains unoriented fibrous material, FM. At the opposite end, the nucleus is indented by the attachment of the tail, T; mitochondria, M, surround the tail at its insertion. B–E: Successive stages in the acrosome reaction. Note the dissolution of the plasma membrane and outer acrosomal membrane at the tip; they fuse at X. The acrosomal tubule, AT, protrudes as the fibrous material of the nuclear indentation swells and pushes out the inner acrosomal membrane that now becomes its covering layer. The covering of the acrosomal tubule is now continuous at X with the original plasma membrane of the remainder of the sperm. [Adapted from various sources, chiefly J. C. Dan in C. B. Metz and A. Monroy, eds., *Fertilization*, Vol. 1, New York, Academic Press, 1967, pp. 237–93.]

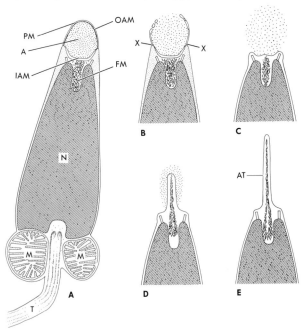

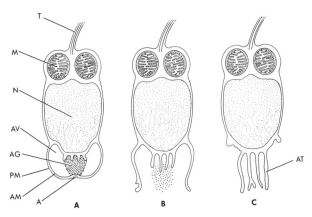

FIGURE 8.4 The acrosome reaction in *Hydroides*. A: Head of the unreacted sperm. B: Rupture of the acrosomal vesicle and fusion of the acrosomal membrane and overlying plasma membrane; dissolution of acrosomal granule. C: Protrusion of acrosomal tubules. A, apical vesicle; AG, acrosomal granule; AM, acrosomal membrane; AT, acrosomal tubule; AV, acrosomal vesicle; M, mitochondrion; N, nucleus; PM, plasma membrane; T, tail, or flagellum. [Adapted from various drawings and photographs by L. H. Colwin and A. L. Colwin, *J Biophys Biochem Cytol* 161:231–54 (1961).]

required before the acrosome reaction can occur. It is called **capacitation**. Because the acrosome reaction has very similar elements in most forms, it is described first.

8.3.1 THE ACROSOME REACTION. In most kinds of invertebrate sperm the acrosome reaction occurs when the sperm makes contact with the egg or its surrounding membranes or, in some cases, when sperm are introduced into seawater in which eggs have been standing **(egg water)**. Calcium is required for the reaction, and, in sea urchins and some other forms, it can be induced by elevated Ca^{2+} concentrations (Levine et al., 1978).

The components of the acrosome reaction are illustrated for sea urchin sperm in Figure 8.3. They consist in

1. Vesiculation and dissolution of the plasma membrane and of the distal portion of the membrane surrounding the acrosomal vesicle
2. The union of the remaining portion of the acrosomal membrane with the plasma membrane so that the two become confluent

3. Release of the contents of the acrosomal vesicle
4. Projection of the original inner portion of the acrosomal vesicle as the acrosomal tubule
5. Projection and fibrification of the content of the subacrosomal space to form the core of the acrosomal tubule (Dan, 1970).

The core of the tubule is polymerized actin. With myosin it forms a contractile system that may be involved in the entrance of the sperm into the egg (Tilney et al., 1978).

There are a number of variations of the invertebrate acrosomal reaction. In some species, particularly among those with eggs surrounded by a thick jelly coat, the acrosomal tubule may be long and thick, penetrating through the jelly to the vitelline membrane of the egg. In other forms the tubule may be short, or it may comprise multiple projections instead of a single one (Figure 8.4). In forms such as *Nereis* (Figure 8.5) an electron-dense **acrosomal rod**, originally housed in a nuclear indentation, fills the core of the tubule. The composition of this rod has not been analyzed.

In mammals the acrosomal reaction also in-

99

FIGURE 8.5 The acrosome reaction in *Nereis*. A: Head of the unreacted spermatozoon. B: Onset of the acrosome reaction. The content of the acrosome has been released and the sperm plasma membrane has fused with the acrosomal membrane; the acrosomal tubule and its axial rod are beginning to project. C: Reaction essentially complete, A, acrosome; AR, axial rod; AT, acrosomal tubule; D, distal centriole; F, fold of composite acrosome-plasma membrane; M, mitochondrion; N, nucleus; P, proximal centriole. [From J. F. Fallon and C. R. Austin, *J Exp Zool* 166:225–42 (1967), by courtesy of the authors and permission of Alan R. Liss, Inc.]

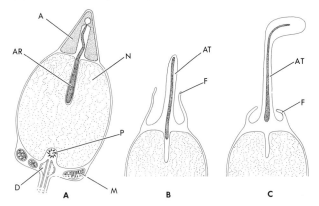

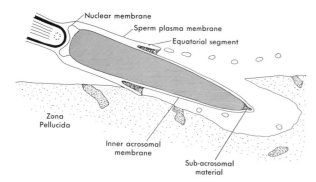

FIGURE 8.6 Penetration of the zona pellucida by the activated sperm of the rabbit. The acrosomal reaction has occurred, and loose vesicles formed by the fusion of the sperm plasma membrane and outer acrosomal membrane are being dispersed. The inner acrosomal membrane covers the subacrosomal material and is continuous with the sperm plasma membrane at the equatorial segment. [An interpretation of data from J. M. Bedford, *Am J Anat* 133:213–54 (1972).]

year, Heller and Raftery isolated and purified three similar lysins from *Megathura* sperm. ATP-ase activity has been localized in the acrosomal granule in both starfish and bivalve mollusks by Mabuchi and Mabuchi (1973). Others have noted jelly-dispersing activities in sea urchin sperm (reviewed by Levine et al., 1978). The enzyme phospholipase A, which is possibly involved in the fusion of egg and sperm as

FIGURE 8.7 A sequence of events in the acrosome reaction in the golden hamster. The sperm plasma membrane and outer acrosomal membrane undergo fusion and vesiculation and then detach from the sperm head, beginning at the side opposite the curved tip of the acrosomal region (arrow in A). The detached acrosomal structure, somewhat swollen, retains its cap-like shape, as seen in D. Compare with Figure 8.8. [From L. E. Franklin et al., *Biol Reprod* 3:180–200 (1970), by courtesy of Dr. Franklin and permission of Academic Press, Inc., New York.]

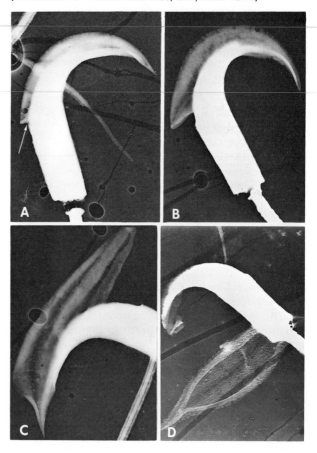

100

volves progressive breakdown and fusion of the plasma membrane and outer acrosomal membrane. In the rabbit the combined membranes form a cloud of loose vesicles about the sperm head (Figure 8.6), the inner acrosomal membrane remaining as the limiting membrane at the anterior end of the sperm, covering the subacrosomal material. The loose vesicles are lost as the sperm begins to penetrate the zona pellucida. In the hamster the fusion of the outer acrosomal membrane and plasma membrane produces a vesiculated shroud that retains the basic shape of the acrosome and is detached as a unit from the sperm head (Figures 8.7 and 8.8).

The acrosomal content that is released or exposed during the acrosomal reaction consists of hydrolytic enzymes, or **egg membrane lysins.** The latter term is chiefly used with respect to acrosomal enzymes in invertebrates. Best known are the egg membrane lysins of *Mytilus edulis,* the common edible mussel of the Atlantic shore and of the giant keyhole limpet *Megathura crenulata.* Seawater from suspensions of *Mytilus* sperm that have been induced to undergo the acrosome reaction dissolve *Mytilus* egg membranes, whereas supernatants from unreacted sperm do not (Wada et al., 1956). Haushka isolated and characterized an egg membrane lysin from *Mytilus* sperm in 1973. In the same

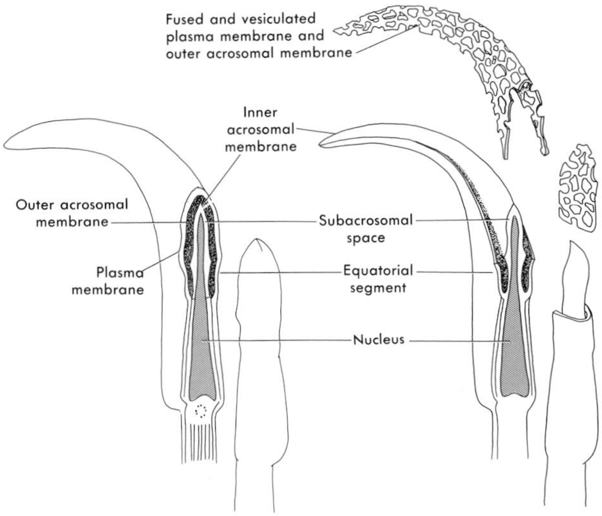

Figure 8.8 [From L. E. Franklin et al., *Biol Reprod* 3:180–200 (1970), by courtesy of Dr. Franklin and permission of Academic Press, Inc., New York.]

described later in this chapter, was demonstrated in suspensions of sea urchin sperm activated by egg water (Conway and Metz, 1976).

Very importantly, Levine et al. (1978) have demonstrated in sperm of the West Coast sea urchin *Strongylocentrotus purpuratus* the presence of a trypsin-like protease quite similar in properties to a protease, **acrosin,** that is well known in mammalian sperm. These sperm show little or no protease activity before activation but do so after treatment with egg water. Most of the activity remains bound to the sperm. Homogenates of unreacted sperm show approximately the same activity, suggesting that it is present in the acrosome in an active form. The protease is apparently important in fertilization for, whereas, preactivated sperm in this species can fertilize eggs, they will not do so if treated with a trypsin inhibitor.

The acrosomal enzymes have been much better characterized in mammalian sperm. In human beings and most other mammals, the egg enters the oviduct, surrounded by a cloud of cumulus cells and the corona radiata (Figure 7.21). These are cellular layers derived from the ovarian follicle as described in Chapter 7. The egg is fertilized high in the oviduct before these layers are lost. Penetrating sperm must pass between cumulus cells and then trans-

verse the more tightly packed corona radiata and, finally, the zona pellucida, before reaching the plasma membrane of the egg. Enzymes carried by the sperm show activities that are related rather specifically to these barriers. The first of these enzymes to be recognized for its significance in mammalian fertilization was **hyaluronidase** (Austin, 1948). It hydrolyzes hyaluronic acid, a major component of most of the intercellular matrix. Preparations of hyaluronidase or of dense sperm suspensions disperse the cumulus layer, but do not affect the corona radiata or underlying zona pellucida. Sperm that are treated with univalent, nonagglutinating antibodies against their species-specific hyaluronidase do not fertilize (Metz, 1972).

A material that can be isolated from the acrosomal content attacks the intercellular material of the corona radiata, leaving the eggs dispersed but intact and apparently viable (Zaneveld and Williams, 1970; Bradford et al., 1976). This substance is heat labile, nondialyzable, and apparently enzymatic in nature. It apparently facilitates passage of the sperm through the corona radiata and is called **corona penetrating enzyme** (CPE). It has no effect on the zona pellucida.

Acrosomes also contain a trypsin-like enzyme that dissolves the zona pellucida (Stambaugh and Buckley, 1969). It has been purified from rabbit sperm (Stambaugh and Smith, 1974), has been determined to have a molecular weight of about 22,000, and is very like human pancreatic trypsin in amino acid content and reaction to inhibitors. Called acrosin, it has been found in sperm of all mammalian species that have been examined. As already noted, it has recently been found in invertebrate sperm also. Acrosin is required for the penetration of the sperm through the zona pellucida, although other proteinases may be involved also (Srivastava et al., 1979). Inhibitors of acrosin interfere with fertilization both in vitro and in vivo and are being tested as contraceptive agents added to vaginal creams (Zaneveld et al., 1979). Mammalian acrosin is present in the sperm in an inactive form, **proacrosin,** whose conversion to acrosin is stimulated by oviductal fluid (Stambaugh and Mastroianni, 1980). In human sperm an inhibitor **acrostatin** has been found in association with proacrosin but the precise relationship is not known (Bhattacharyya and Zaneveld, 1978).

In sum, the acrosomal content of enzymes ap-

101

pears to be highly adapted to facilitating the access of sperm to egg. Other enzymes are also present, including β-N-acetyglucosamidase, acid phosphatase, aryl sulfatase phosphatase, and neuraminidase. Because these enzymes are characteristically present in **lysosomes,** which are found in all kinds of cells, it has been suggested (Allison and Hartree, 1970) that the acrosome is a specialized lysosome evolved to facilitate fertilization in multicellular organisms. The enzyme activities associated with mammalian sperm were summarized by Stambaugh in 1978.

8.3.2 CAPACITATION.

As already noted, mammalian sperm require a period of residence in the female reproductive tract before they undergo the acrosome reaction and become capable of fertilizing the egg. This period may vary from less than 1 hour in the mouse to as much as 5 or 6 hours in human beings. The requirement for **capacitation** (a term coined by Austin in 1952) was first noted by Austin and Chang independently in 1951. Since that time capacitation has been the object of an enormous amount of research, the most important of which was reviewed by Gwatkin in 1976. His paper can be consulted for documentation of most of what follows.

Capacitation potential is variable as between uterus and fallopian tubes in different kinds of mammals, but is generally best completed in the fallopian tubes. It occurs rapidly after ovulation and can be induced by cumulus cells in vitro. Oviductal and uterine contents show diminished capacitation potential during the progestational phase of the reproductive cycle as compared to the follicular and ovulatory phases.

Capacitation may be evoked in vitro by oviductal fluids and by fluid withdrawn from large ovarian follicles. It can also be elicited by incubation in blood serum from animals of the same or different species. What common chemical properties are held by these fluids that renders them capable of capacitating sperm is unknown.

A number of observers have reported that the beating of the flagellum in capacitated sperm is considerably greater than in noncapacitated sperm. This would presumably facilitate their penetration of the materials surrounding the egg. Capacitation is also accompanied by increased respiration and enhanced glycolysis. This might well be expected in correlation with increased motility. The availability of glucose and the reactions of glycolysis are obligatory, in the mouse at least, for initiation of the acrosome reaction and increased motility of the sperm (Fraser and Quinn, 1981).

Sperm that have been capacitated can be decapacitated once more by exposure to sperm-free ejaculated semen. It is known that there are several sperm-coating proteins, derived from the epididymal and seminal fluids, which are removed from the surface of the sperm during capacitation. Presumably decapacitation involves the rebinding of certain proteins to the sperm surface. From semen of bulls and of rabbits a protein has been isolated that inhibits the fertilizing ability of previously capacitated sperm. Called **decapacitation factor,** it is presumed to block the ability of the corona penetration enzyme, but does not affect the activity of acrosin or of trypsin.

The activity of acrosin is higher in sperm taken from the epididymus than in ejaculated sperm. This has been proposed as resulting from the contribution to semen of an inhibitory protein secreted by the seminal vesicles. Such a protein has been described. It inhibits the action of acrosin and of trypsin and has been called **seminal plasma trypsin inhibitor.** It is doubtful that removal of this inhibitor has any effect on the capacitation reaction, however, for acrosin is apparently involved only in the penetration of the zona pellucida, which occurs after capacitation and the acrosome reaction have occurred.

The simplest view of capacitation is that it involves the removal by enzymes of the female reproductive tract of sperm-coating proteins or components thereof. A number of enzymes known to be present in the female tract have such capabilities. There is a high level of proteolytic activity, for example, in the estrous uterus and a high protease-inhibitor content in the progestational uterus.

A requirement for capacitation is rare among sperm of nonmammalian species. O'Rand (1972a, b) found that sperm of *Campanularia flexuosa* (Figure 8.1) require contact with the epithelial cells that surround egg packets within the gonophore in order to be able to fertilize. From contact with the surfaces of these cells they acquire a coating of trypsin-sensitive material. Eggs removed from their packets cannot be fertilized by sperm unless the latter have been exposed to nontrypsinized female epithelial cells.

The possibility that egg jelly, added in the oviduct, provides capacitating factors for amphibian sperm has been raised many times. Ovulated eggs, taken from the body cavity are not readily fertilizable, but can be fertilized in the presence of egg water or jelly fragments. It has been considered that the jelly contains oviductal factors that capacitate the sperm. It now appears, however, that a portion of the oviduct, the **pars recta,** is the source of factors that make the vitelline membrane sensitive to the acrosomal proteinase. This is presumably an acrosin-like enzyme, for fertilization of body-cavity eggs in the presence of an extract of the pars recta is prevented by trypsin inhibitors (Cabada et al., 1978; see also Elinson, 1973, and Grey et al., 1977).

8.4   Fertilizin

Some years ago, Lillie (1913) reported that supernatant sea water from heavy suspensions of the eggs of the sea urchin *Arbacia pundulata* or of the clam worm *Nereis limbata* each has the capacity to cause the sperm of its own species to form huge aggregates in the form of spherical masses visible to the unaided eye. Sperm heads were seen to be directed inwardly in these masses, the tails meanwhile projecting radially and lashing vigorously. Lillie called the active material of the egg water **fertilizin.** He believed that the head-to-head aggregation of sperm was an agglutination reaction resulting from an antigen–antibody-like combination of the fertilizin with a complementary sperm substance, **antifertilizin.**

A great many kinds of eggs release substances that bind to sperm and might be called fertilizins. In only a few cases, however, have egg substances caused the reaction that Lillie called agglutination. Moreover, it now seems that the reaction of sperm to egg water is not comparable to an immunological isoagglutination reaction. If it were, sperm would remain aggregated when immobilized by means of inhibitors such as azide or cyanide. Collins (1976), however, showed that the aggregates promptly

come apart when sperm are immobilized. What Lillie perceived as agglutination in the presence of egg water is instead the swarming of freely moving sperm to a common focus. The sperm aggregations break up spontaneously after 5–10 minutes.

Since Lillie's time a great deal of research, chiefly employing sea urchin gametes, has been carried out on fertilizin. The earlier work was summarized by Lillie (1919), Tyler (1965), and Metz (1967, 1978). Sea urchin fertilizin has been identified as the chief component of the egg jelly coat, which slowly dissolves in sea water. It is a sialopolysaccharide–protein complex of high molecular weight. It is highly acidic because of the high concentration of carboxyl residues in the bound sialic acids and because of the numerous sulfate groups linked to polyfucose (Hotta et al., 1970).

Fertilizin was once thought to be involved in binding the sperm to the egg, but, as demonstrated below, other factors have now been shown to have this primary responsibility. So, what role does the egg jelly serve? Presumably it exercises a protective function. Because sperm that have stood in egg water have diminished fertilizing power it has also been suggested that it is one of the mechanisms involved in preventing the entry of supernumerary sperm.

8.5   Entrance of the Sperm into the Egg

Details of the incorporation of the sperm into the egg are best known for several invertebrates and for mammals. The following account will be relatively brief, and readers are referred to reviews by Epel (1978, 1979), Gwatkin (1976) and Metz (1978) for more information.

The entry process in invertebrates is initiated when the tip of the acrosomal tubule makes contact with the vitelline membrane or the egg plasma membrane. In the sea urchin *Strongylocentrotus purpuratus* acrosomal reaction exposes at the tip of the acrosomal tubule a species-specific protein with a molecular weight of about 30,500, which has been termed "**bindin**" (Vacquier and Moy, 1977; Bellet et al., 1977). This binds to a high molecular weight (5×10^6) glycoprotein aggregate in the vitelline membrane (Aketa, 1973, 1975; Glabe and Vacquier, 1978) in a reaction that requires the presence of Ca^{2+}. Bindin is identified with the electron-dense, insoluble component of the acrosomal vesicle. Isolated bindin causes the agglutination of jelly-free eggs and of isolated vitelline membranes, but this reaction is blocked by pretreatment of the bindin with vitelline membrane glycoproteins. Another substance, a species-specific carbohydrate with few or no peptide residues, was isolated by Aketa et al. (1978) from sperm of a Japanese sea urchin, *Hemicentrotus pulcherrimus.* Jelly-free eggs treated with this carbohydrate are not fertilizable. If, however, the sperm substance is first reacted with vitelline membrane glycoproteins, then eggs exposed to sperm in the reaction mixture are readily fertilized. Unlike bindin, the sperm carbohydrate does not cause the agglutination of jelly-free eggs, but, like bindin, it is Ca^{2+} dependent. Its localization on the sperm apparently is not known, nor is its relationship to bindin in the fertilization process.

Motion picture analysis of sea urchin fertilization reveals that many sperm bind to the egg, continually rotating about the axis of the binding point

for 15 to 20 seconds after the initial attachment. During this period one may presume that the sperm protease described by Levine et al. (1978), which was mentioned above, is dissolving components of the vitelline membrane, possibly the glycoprotein receptor, which is trypsin sensitive. The first sperm to penetrate becomes immobile and begins to enter the egg. The remaining sperm continue movement until detached, as described below.

When the tip of the acrosomal tubule comes into contact with the egg cell membrane the apposed membranes break down into vesicular remnants that soon disappear. This process, possibly facilitated by sperm phospholipase, as suggested by Conway and Metz (1976), opens a channel of continuity between the egg cytoplasm and the sperm contents (Figure 8.9). At the edge of this channel the plasma membranes of sperm and egg are in contact, and they fuse to become a continuous membrane. Meanwhile, there is a surge of egg cytoplasm to the point of sperm entry so that the sperm is situated in a small elevation, the fertilization cone (Figures 8.10 and 8.11), which is adorned, in the sea urchin at least, with elongate microvilli (Schatten and Mazia, 1976). As the cone continues to elevate, its cytoplasm gradually flows under the sperm plasma membrane and around the nucleus, mitochondria, and tail filaments. Once within, the sperm takes a position parallel to the surface and then rotates so that the tip of the nucleus is directed away from the center of the egg. As the fertilization cone subsides, the sperm plasma membrane is incorporated into the egg membrane (Colwin and Colwin, 1963).

Presumably the entrance of the sperm is chiefly the result of the activity of the egg. The presence of actin in the acrosomal tubule, however (Tilney et al., 1978) suggests that contractile activity of the tubule is involved. It does shorten during early phases of the entry process.

In mammalian fertilization capacitated sperm are generally thought to undergo the acrosome reaction in contact with the cumulus cells, releasing enzymes that facilitate movement of the sperm between the cells investing the ovum, as described earlier. Reaching the zona pellucida the sperm first attach loosely to it and are easily dislodged. After 30 to 40 minutes, as described by Hartmann and Hutchison (1974) for hamster gametes, the loose attachment is followed by a tight binding. The

104

FIGURE 8.9 Union of sperm and egg membranes in the sea urchin. A: The acrosome reaction has taken place, and the tip of the acrosomal tubule has made contact with the vitelline membrane of the egg. B: The tip of the acrosomal membrane has fused with the egg plasma membrane, and sperm and egg contents have become confluent. Presumably contraction of the fibrous actin filaments in the core of the acrosomal tubule promotes the initial phases of sperm entry into the egg.

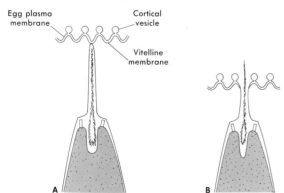

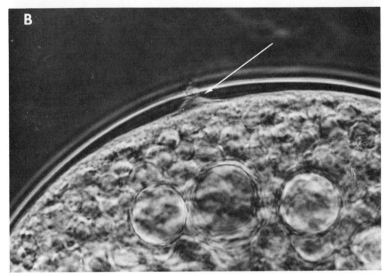

FIGURE 8.10 Approach of the sperm and formation of the fertilization cone in *Nereis*. A: The head of the sperm has made contact with the vitelline membrane. B: 12 min later, the egg has elevated a fertilization cone (arrow), which is beginning to engulf the sperm. [Photograph graciously supplied by Dr. John F. Fallon.]

sperm cannot now be dislodged readily and, after approximately 20 minutes, will have penetrated the zona pellucida.

This view of the sequential relationships between capacitation, the acrosome reaction and sperm binding may require modification in view of some recent observations. Overstreet and Cooper (1979) reported that sperm showing modified acrosomes may be found in any region of the oviduct in mated rabbits 6–8 hours post coitum. Possibly, however, the reaction is not completed until

the sperm are in the vicinity of the eggs in the ampulla of the oviduct. On the other hand, Saling et al. (1979) reported that capacitated sperm bind to the zona pellucida of mouse eggs freed of cumulus cells and then undergo the acrosome reaction only after a relatively long period of tight binding to the zona. In Meizel's laboratory (Cornett and Meizel, 1978; Cornett et al., 1979; Meizel and Working, 1980) it was found that epinephrine and norepinephrine (catecholamines), added to simple, otherwise non-capacitating media, evoke a normal acrosome reac-

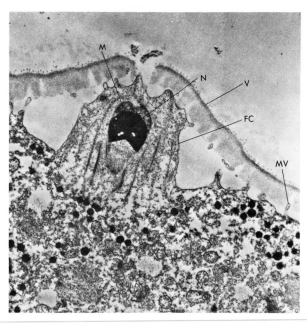

FIGURE 8.11 Electron micrograph of the sperm head of the marine worm *Urechis caupo* within the fertilization cone formed by the egg cortex, 5 min after fertilization: FC, fertilizaton cone; M, mitochondrion; MV, microvillus; N, nucleus; V, complex vitelline membrane. [Reprinted from A. Tyler, *Am Nat* 99:309–44 (1965), by courtesy of the late Dr. Tyler and permission of The University of Chicago Press.]

the zona and which were presumably derived from acrosin that entered with the sperm.

In most mammals the sperm, having penetrated the zona, makes almost immediate contact with the egg plasma membrane and becomes immobile. The mode of initial attachment between sperm and egg is somewhat different in mammals from what it is in invertebrates. In the latter it is the tip of the inner acrosomal membrane, the covering membrane of the acrosomal tubule, that first fuses with the egg plasma membrane. Not so in mammals. The inner acrosomal membrane apparently plays little role in the binding of sperm to egg. Instead, the initial fusion is between the plasma membrane of the post-acrosomal region of the sperm head and the egg plasma membrane. This has been studied in some detail during fertilization of the egg of the golden hamster. As soon as the sperm makes contact, the surface of the hamster egg produces numerous microvilli that surround it (Figure 8.12). In the postacrosomal region the microvilli fuse broadly with the sperm plasma membrane, and the double membrane so formed then apparently vesiculates, creating a broad channel of access of the egg cytoplasm to the sperm nucleus. At the margin of the channel, egg and sperm membranes become continuous. This was first described for the rat (Figure 8.13) and confirmed later for the hamster (Figure 8.14). Thus, a single continuous membrane, a mo-

tion in hamster sperm. Hamster ova, freed of follicular cells and inseminated in the same media in the presence of the hormones, were fertilized. This suggests that the catecholamines induced both capacitation and the acrosome reaction. Catecholamines are present in uterine and oviductal tissues, but there seems to be no information about their concentration in fluids of the reproductive tract.

There is considerable, although not universal, agreement that penetration of the zona is accomplished by acrosin that is bound either to the inner acrosomal membrane, exposed during the acrosomal reaction or, as described for rabbits and rhesus monkeys by Stambaugh (1978), to microtubular arrays in the acrosomal matrix. Such structures had hitherto escaped detection. In traversing the zona, the sperm usually takes a diagonal or slightly curving path, effecting a tunnel (Figure 8.6) which, according to Stambaugh and Smith (1976), bears zones of proteolytic activity, not found elsewhere in

FIGURE 8.12 An interpretation of the binding of the hamster sperm head to the egg surface by means of extensive microvilli. [Adapted from R. Yanagimachi and Y. D. Noda, *Experientia* 28:69–72 (1972).]

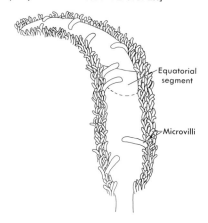

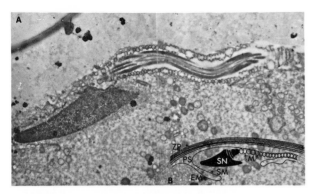

FIGURE 8.13 Rat sperm penetrating the egg. A: Electron micrograph of a longitudinal section of a sperm near the egg surface. B: Relationship of egg and sperm membranes based on the analysis of the micrograph. EM, egg plasma membrane; M, mitochondria of sperm; SM, sperm plasma membrane; SN, sperm nucleus; PS, perivitelline space; ZP, zona pellucida. [From D. Szollosi and H. Ris, *J Biochem. and Biophys. Cytol.*, 10:275–83, (1961), by courtesy of the authors and Rockefeller University Press.]

saic of egg and sperm membranes, covers the entire egg-sperm complex.

It was the Colwins (Colwin and Colwin, 1963), studying fertilization in the enteropneust *Saccoglossus kowlevskii*, who first proposed that the sperm plasma membrane becomes incorporated as a mosaic patch in the plasma membrane of the zygote. There were some who doubted this interpretation and some who suggested that, because of the fluid nature of plasma membranes, the sperm membrane could not remain as an intact patch in the surface of the zygote. Experiments of Gabel et al. (1979), however, provide ample evidence of the validity

FIGURE 8.14 Continuity of sperm content and egg cytoplasm during entry of the golden hamster sperm into the egg. [Adapted from R. Yanagimachi in S. Segal, ed., *The Regulation of Mammalian Reproduction*, Springfield, IL, C. C Thomas, 1973, pp. 215–27.]

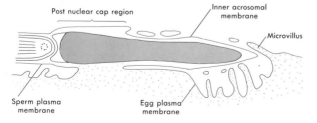

Post nuclear cap region

Inner acrosomal membrane

Microvillus

Sperm plasma membrane

Egg plasma membrane

and generality of the Colwins' observations. They labeled the plasma membranes of both sea urchin and mouse sperm with fluorescein isothiocyanate, a dye that gives a fluorescent glow when exposed to ultraviolet light. Sperm so treated are viable and capable of fertilizing eggs of their respective species. Following the development of fertilized eggs by means of light and ultraviolet microscopy, they were able to trace the fluorescent patch of sperm plasma membrane into advanced cleavage stages and, in the case of the sea urchin, through gastrulation. The fluorescent patch became smaller as development proceeded. It usually was restricted to one blastomere, although it was sometimes subdivided. That the mosaic patch is, indeed, an integral and functional part of the surface of the zygote is further indicated by an earlier observation of O'Rand (1977). He showed that antibodies against rabbit sperm cause cytolysis when applied to fertilized rabbit eggs, whereas they do not affect the integrity of unfertilized eggs.

8.6 Why Does Only One Sperm Fertilize an Egg?

107

The answer to this question is now fairly well known for sea urchins and starfish, and the same answer may apply to other marine invertebrates. For mammals, the answer is less completely known. For sea urchins and their relatives, many hundreds of sperm may bind to the egg, but only one enters. Should more than one do so (**polyspermy**), development would be abnormal. A rapid block to polyspermy would seem to be necessary to assure that normal development could occur. In the case of mammals, relatively few sperm reach the egg. A rapid block to polyspermy would thus be less important, and, in fact, there does not seem to be such a block. Mammalian eggs fertilized in vitro with heavy suspensions of sperm are frequently polyspermic. In echinoderms there is a rapid block to polyspermy, which is followed by a slower block. The latter is important, for the fast block is incomplete.

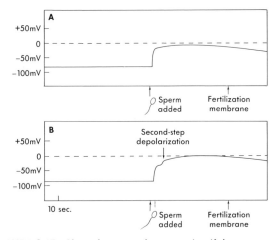

FIGURE 8.15 Changing membrane potentials associated with fertilization in the sea urchin *Strongylocentrotus purpuratus*. A: The solid line shows the membrane potential prior to and subsequent the presumed contact of a successful sperm. The change measured subsequent to insemination is referred to as the activation potential. Eggs that reach an activation potential above −10 millivolts seldom become polyspermic. B: a case in which the membrane potential showed a second-step change, presumably indicating the successful contact of a second sperm. The potentials were measured by means of a microelectrode inserted into eggs from which the jelly had been removed. [Adapted from L. A. Jaffe, *Nature* 261:68–71 (1976).]

108

The fast block consists in a change in the voltage between the inside and the outside of the cell that is induced within 1 second of the contact of the fertilizing sperm. L. A. Jaffe (1976) inserted a microelectrode into the egg of *Strongylocentrotus purpuratus* and recorded a negative potential (membrane potential) of about −70 mV. Immediately at fertilization, however, the voltage underwent a positive change, rising to as much as +20 mV in some cases. This change results from a flow of sodium ions into the cell, similar to what occurs when a nerve fiber is activated. In the egg, the elevated membrane potential persists for about 60 seconds. The voltage change is apparently responsible for preventing a second sperm from entering the egg. If this change were not sufficiently great, a second sperm could enter (Figure 8.15). Dr. Jaffe found that if she artificially raised the voltage across the membrane of unfertilized eggs above about +5 mV, fertilization

would not occur. Similar results to those of Jaffe have been reported for a Japanese starfish *Asterina pectinifera* by Miyazaki and Hirai (1979) and by Miyazaki (1979).

The block imposed by the voltage change is incomplete, for if eggs are inseminated with such dense sperm suspensions that all are fertilized instantly, an increasing proportion of them show polyspermy for a minute or more after insemination (Figure 8.16). This is the time required for the egg cortex to undergo a massive structural change called the **cortical reaction**. This results in the discharge of the cortical alveoli into the space beneath the vitelline membrane and the transformation of the latter into a **fertilization membrane.** It is the formation of this membrane that constitutes the final block to polyspermy. Electrically mediated fast blocks to polyspermy have also been reported for the marine worm *Urechis caupo* (Gould-Somero et al., 1979; Holland and Gould-Somero, 1981) and for the anurans *Rana pipiens* and *Bufo americanus* (Cross and Elinson, 1980).

8.6.1 THE CORTICAL REACTION. The cortical reaction is the first visible sign of egg **activation.** As noted in an earlier section frog eggs that are hormonally matured may be activated by pricking with a

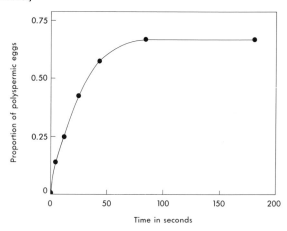

FIGURE 8.16 The proportion of polyspermic eggs as a function of time after heavy insemination in the sea urchin *Psammechinus miliaris.* In this experiment the sperm density was 9.11×10^7 per ml. [Replotted from L. Rothschild and M. M. Swann, *J Exp Biol* 29:469–83 (1952).]

needle. Activation by chemical agents can also occur, as is discussed in different contexts in subsequent sections. Here we concentrate on the cortical reaction as normally elicited by sperm in the sea urchin egg and in mammals.

In sea urchins the contact of the fertilizing sperm is followed after about 20 seconds by the rupture of the cortical alveoli, which fuse with the egg plasma membrane and discharge their contents into the space between it and the overlying vitelline membrane, creating a **perivitelline** space (Figures 8.17 and 8.18). This reaction begins at the point of sperm entry and spreads through the cortex in all directions to the opposite pole of the egg. During the lag period before rupture of the alveoli, sperm continue to bind to the egg, but as the alveoli rupture, these sperm begin to detach in a wave that follows the progress of this breakdown. Detachment of all but the successful sperm is complete about 60 seconds after insemination (Figure 8.19).

Rupture of the cortical granules releases a mixture of enzymes, structural proteins, and mucopolysaccharides. One of the enzymes specifically modifies the sperm receptor glycoproteins so that the supernumerary sperm are detached. Another breaks down the proteins that bind the vitelline membrane to the plasma membrane (Carroll and Epel, 1975). The mucopolysaccharides especially cause the osmotic movement of water into the perivitelline space, elevating the vitelline membrane. The latter receives contributions of structural proteins from the discharged alveoli, transforming it into the fertilization membrane or **activation calyx** (Anderson, 1968). Another enzyme, a peroxidase, further hardens the fertilization membrane by catalyzing the cross-linking of proteins through their tyrosyl residues (Foerder and Shapiro, 1977). Another protein covers the surface of the egg to form a transparent layer, the hyaline layer (Figure 8.20).

Mammalian eggs also show a cortical reaction. Once the sperm has attached to the plasma membrane of the egg, a wave of breakdown of the cortical granules begins, and a perivitelline space is created beneath the zona pellucida. The block to polyspermy is tied to this reaction in mammalian eggs, but in a variable manner (Gwatkin, 1976). In some cases, such as the dog, pig, sheep, and hamster, the discharge of the cortical granules causes the zona to be refractory to the binding of sperm.

This probably results from the activity of proteases discharged by the granules. This is called the **zona reaction.** The rabbit egg does not show a zona reaction, but the plasma membrane does not readily bind to a second sperm. Sato (1979) inseminated mouse eggs in vitro after removing cumulus cells. He observed that binding of spermatozoa to the zona occurs readily during the first minute, but less readily thereafter. Also, when a second sperm penetrated the zona within 1 minute after a first sperm contacted the egg plasma membrane, polyspermy could occur. The likelihood of polyspermy was diminished thereafter, however. The zona block and vitelline block to polyspermy, under the circumstances of Sato's experiments, apparently occur much more rapidly than they do in vivo in the mouse and in several other organisms, as Sato himself noted.

In many organisms the relationships of the cortical granules to the vitelline membrane are much different from what one sees in sea urchins and mammals. In *Nereis,* as described by Fallon and Austin (1967), the vitelline membrane is about 0.7 μm thick and is traversed at intervals by microvilli that project slightly beyond it (Figure 8.21). A thick cortical region with multiple layers of cortical alveoli lies beneath the egg membrane. The alveoli rupture, beginning at the point of sperm contact, spreading around the egg. Their breakdown effectively destroys the oölemma and releases a great deal of jelly-like substance through the ''pores'' of the vitelline membrane once occupied by the microvilli. This produces a protective layer of jelly around the egg. The vitelline membrane is now the fertilization membrane, and a new limiting membrane for the egg forms beneath the perivitelline space, possibly incorporating fragments of the membranes that once surrounded the cortical granules. The block to polyspermy is not known for *Nereis.*

In the case of some fish eggs, which have a chorion with micropyle, the latter is plugged as soon as the first sperm enters, thus preventing polyspermy (Brummett and Dumont, 1979). Cortical alveoli rupture, and a perivitelline space appears because of the shrinkage of the egg volume and osmotic entrance of water. In a number of organisms, such as the annelid *Hydroides* and *Mytilus* the mussel, cortical granules do not rupture upon fertilization and no perivitelline space appears. The block to poly-

109

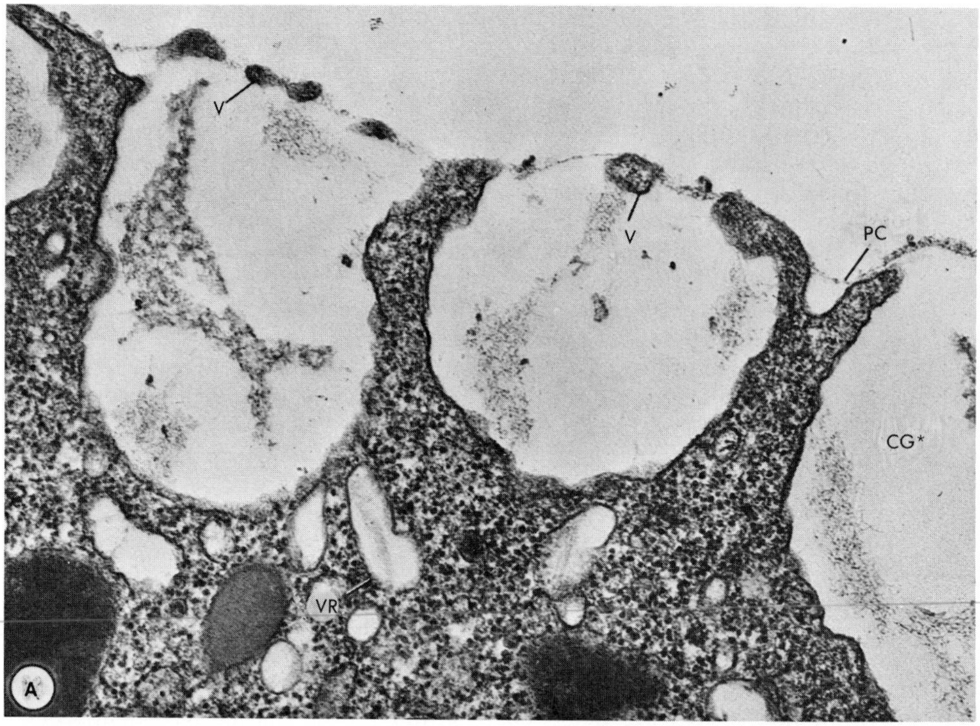

FIGURE 8.17 (Above and on facing page) Separation of the primary egg coat and formation of the activation calyx in the sea urchin egg. A: Section through three partly released cortical granules, CG, showing the primary coat, PC, membrane-bounded vesicle, V, and a vesicle, VR, containing rod-like structures. B: Elevation and thickening of the activation calyx, AC, and formation of the perivitelline space. An unreleased cortical granule is at the right near two microvilli, MV. Insets (a) and (b): The egg before and after elevation of the fertilization membrane, or activation calyx. Inset (c): A tangential section showing the primary coat, PC, on the egg surface and covering microvilli. [From E. Anderson, *J. Cell Biol* 37:514–39, (1968), by courtesy of the author and permission of the Rockefeller University Press.]

FIGURE 8.18 Cortical reaction in the sea urchin egg. Successive stages showing contact of the sperm, the wave of breakdown of the cortical granules, elevation of the fertilization membrane, and formation of the hyaline layer. CG, cortical granule; FM, fertilization membrane; HL, hyaline layer; PM, egg plasma membrane; VM, vitelline membrane. See also Figure 8.17.

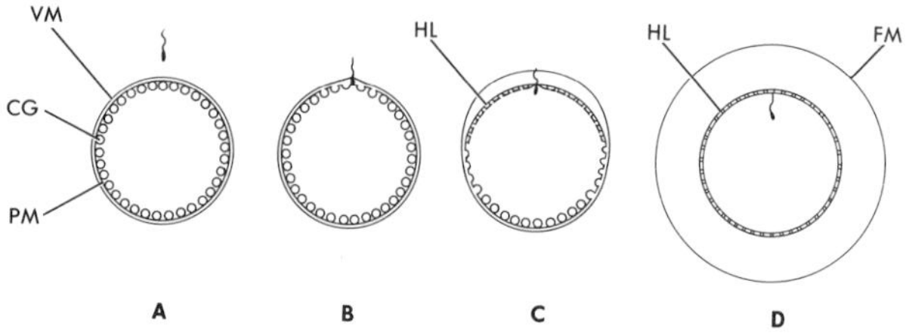

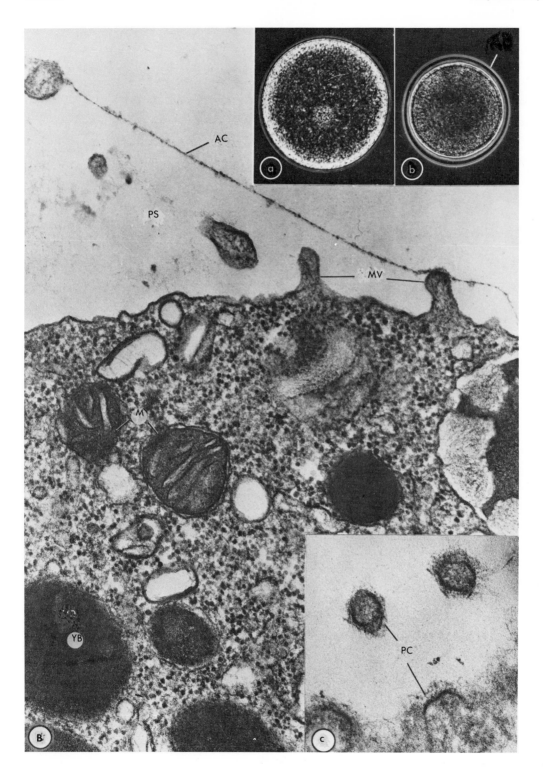

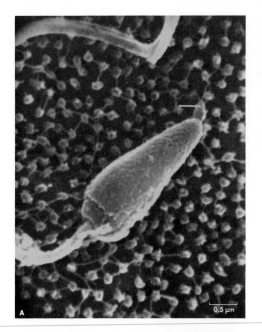

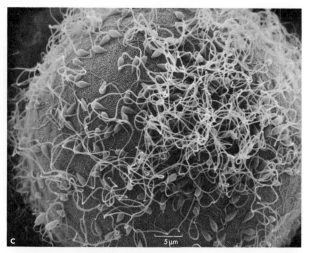

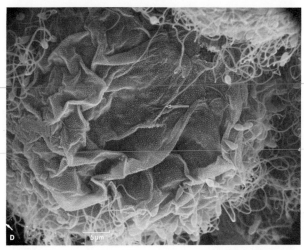

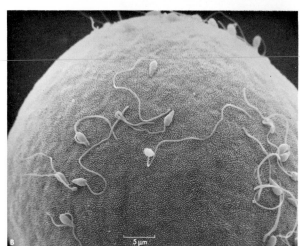

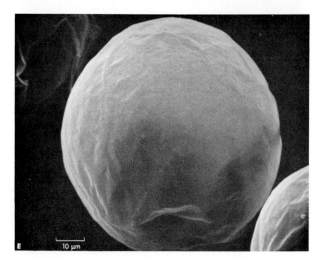

FIGURE 8.19 (Opposite page) Binding and detachment of sperm from the egg surface in the sea urchin *Strongylocentrotus purpuratus* as seen in scanning electromicrographs. A: Initial binding of the acrosomal tubule (arrow) to the vitelline surface of the dejellied egg. The dense array of projections seen at the egg surface are microvilli and the covering vitelline layer, which follows their contours. At 22°C, and using a sperm concentration of 3×10^8 per ml of seawater, the number of sperm bound increases progressively for 25 seconds when the cortical reaction and sperm detachment begin. B: 1 sec after insemination. C: 15 sec after insemination. D: 30 sec after insemination; the arrow points to the tail of the fertilizing spermatozoan and the hole it has made in the vitelline membrane. The extent of the cortical reaction is signalled by the zone of sperm detachment surrounding the successful sperm; the wrinkled surface of the fertilization membrane is an artifact of fixation. E: The surface of the fertilization membrane as seen at lower magnification 3 min after insemination; it is now hardened and retains its rounded configuration during fixation. [From M. J. Tegner and D. Epel, *Science,* 179:685–88 (1973), by courtesy of the authors and with permission. Copyright 1973 by The American Association for the Advancement of Science.]

113

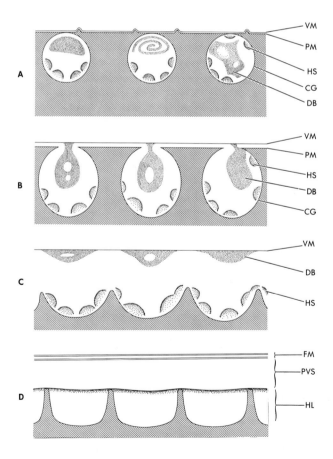

FIGURE 8.20 Formation of the fertilization membrane and hyaline layer in the sea urchin. An interpretation based on electron micrographs. A: Surface of the unfertilized egg. B: Beginning of rupture of cortical granules and extrusion of dense material into the perivitelline space. C: Dense material adheres to the vitelline membrane and hemispheric globules originally lining the granules spread over the new egg surface. D: The definitive fertilization membrane formed by fusion of the dense material and vitelline membrane; the hyaline layer arises from the material of the hemispheric globules. VM, vitelline membrane; PM, plasma membrane; HS, hemispheric globules; CG, cortical granule; DB, dense body; FM, fertilization membrane; PVS, perivitelline space; HL, hyaline layer. [From Y. Endo, *Exp Cell Res* 25:383–97 (1961), by permission of the author and Academic Press, Inc., New York.]

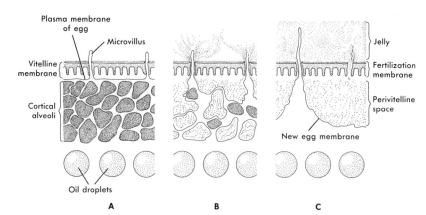

FIGURE 8.21 The fertilization reaction in the cortex of the egg of *Nereis*, showing fusion and rupture of the cortical alveoli, formation of a new cytoplasmic surface, the release of jelly precursor, and its passage through the vitelline membrane. A: Unreacted egg cortex. B: 10 min after fertilization. C: 20 min after fertilization. [Original drawings supplied by Dr. John F. Fallon and Dr. C. R. Austin.]

spermy is not known, but it must involve changes in the vitelline membrane, and what these changes are seems not to have been investigated.

For large, yolky eggs of birds, reptiles, some amphibians, and some insects there is no block to polyspermy. Several sperm may enter the egg, but only one contributes the genetic material of the male parent. The others disintegrate when the first of their number reaches the female pronucleus (Fankhauser, 1945).

8.6.2 CAN AN EGG BE REFERTILIZED? A number of experiments point to the fact that fertilization, once accomplished, does not so alter the egg surface that it cannot be fertilized again. Eggs of some sea urchins can be refertilized as late as the two-cell stage after removal of the fertilization membrane and exposure to calcium- and magnesium-free seawater (Sugiyama, 1951) or other solutions, such as isotonic urea, that dissolve the hyaline layer (Nakano, 1956). Removing the fertilization membrane from *Saccoglossus* eggs at the eight- to sixteen-cell stage permits refertilization in the absence of other treatment (Colwin and Colwin, 1967). Eggs of the sea urchins *Lytechinus pictus* and *L. variegatus* can be refertilized when the fertilization membrane is removed in the absence of other treatment up to one half hour after the initial fertilization (Tyler et al., 1956).

Refertilized eggs receive the sperm by forming a fertilization cone in the normal manner, although a new fertilization membrane does not appear. Moreover, refertilized eggs become highly polyspermic, even though they are inseminated with sperm suspensions at concentrations that normally give monospermic fertilization. Evidently, therefore, the passage of the fertilization wave per se does not modify the reactivity of the egg cortex to sperm in the case of many marine eggs or of some mammals, too (Barros et al., 1972).

8.7 Formation and Union of the Pronuclei

When the egg completes meiosis, the reduced chromosome complement that remains in the egg at the animal pole after extrusion of the polar bodies acquires a nuclear envelope and becomes the **female pronucleus.** The sperm nucleus loses its enveloping membrane after it enters the egg and then, as its compact content of chromatin loosens up, it acquires a new envelope and becomes the **male pronucleus** (Longo and Kunkle, 1978). The female pronucleus then moves deeper into the egg and the male pronucleus moves along an intersecting path (Figure 8.22). The mechanisms of pronuclear movements are not well understood, but microtubules seem to play an important role. In *Arbacia* (Longo and Anderson, 1968; Longo, 1976) the proximal centriole of the sperm becomes the center of origin of a radiating array of microtubules that constitutes the **sperm aster.** The sperm aster then leads the way in migration to the female pronucleus. Pronuclear migration ceases, however, if the

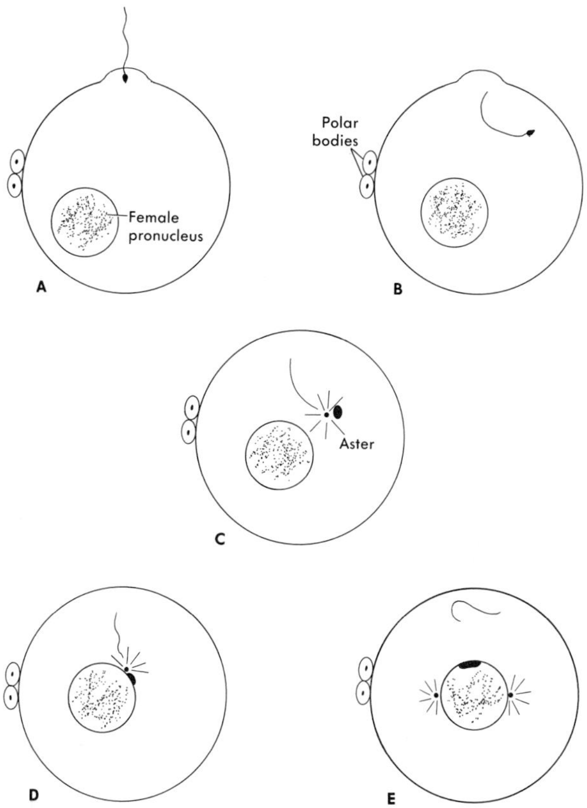

FIGURE 8.22 The sequence of events leading to the fusion of pronuclei in the sea urchin *Arbacia punctulata*. The fertilization membrane is not shown, nor is there any effort to depict the loss of the sperm nuclear membrane and the formation of the membrane of the male pronucleus.

egg is transferred to seawater containing colcemid, a drug that disperses microtubules into their macromolecular subunits. When the treated eggs are returned to normal seawater microtubules re-form, and pronuclear migration resumes (Zimmerman and Zimmerman, 1967). Eventually contents of both pronuclei blend indistinguishably into a single spheroid body. The sperm aster and another aster, arising either from the distal centriole of the sperm or from the egg, then constitute centers for the poles of the microtubular array that constitutes the first division spindle. Chromosomes condense, the nuclear membrane breaks down, and mitotic cell division occurs.

Actual fusion of the pronuclei within a single nuclear membrane occurs chiefly in those kinds of eggs in which meiosis is completed prior to fertilization. In eggs that are fertilized in the germinal vesicle stage, such as those of *Ascaris,* the pronuclei not only fail to fuse, but give rise to the groups of chromosomes that form separately on the equatorial plate of the first division spindle. In *Mytilus,* which is inseminated at first meiotic metaphase, the pronuclei lie side by side without extensive interdigitation. Their envelopes break down separately before the chromosomes become arrayed on the equatorial plate. In the rabbit egg, which is fertilized at second meiotic metaphase, the pronuclei likewise fail to fuse, but they do develop extensive interdigitation and appear to be tightly attached. Subsequently the nuclear membranes break down and the chromosomes condense and move together onto a single mitotic spindle. (Reviewed by Wilson, 1925, and by Longo and Kunkle, 1978).

The timing of DNA synthesis in preparation for the first mitosis of the zygote occurs variably in different kinds of organisms. In the sand dollar DNA synthesis begins in the pronuclei about 20 minutes after fertilization and is completed prior to their union (Simmel and Karnofsky, 1961). In sea urchins, the pronuclei unite before DNA synthesis begins (Longo and Plunkett, 1973), but when pronuclear fusion is inhibited by colcemid, DNA synthesis proceeds in the separated pronuclei (Zimmerman and Zimmerman, 1967).

115

8.8 Biochemistry of Fertilization

This section deals with the sequence of biochemical changes that is stimulated by union of the sperm with the egg, especially as is known for sea urchins. This sequence is broken down into two phases; (1) early changes that occur during the first 60 seconds after the egg and sperm make contact and (2) a series of late events that begin about 5 minutes after contact and lead to the first cell division. These changes are summarized in Table 8.1. Information contained in this table is largely based on excellent reviews by Epel (1978, 1979).

TABLE 8.1 The Program of Fertilization in Sea Urchin Eggs

Time, sec	Sequence of Events
1	Sperm binds to egg
2	Minor influx of Na$^+$ Fast block to polyspermy
8	Release of Ca^{2+} from intracellular depots begins
20	Rupture of cortical granules begins Major influx of Na$^+$ begins Release of protons begins pH begins to rise from 6.6
30	Activation of NAD kinase Formation of NADP from NAD Increased activity of G-6-p dehydrogenase
40	Respiratory rate increases
60	Fertilization membrane complete Block to polyspermy complete
200	Intracellular [Na$^+$] reaches plateau pH reaches plateau at 7.2 Intracellular [Ca^{2+}] returns to prefertilization level
350	Rate of protein synthesis rises
400	Membrane potential becomes K$^+$-dependent Permeability to amino acids increases
600	Intracellular [Na$^+$] begins to decrease pH begins to decrease
1200	Intracellular Na$^+$ and pH return to prefertilization levels
1600	Initiation of DNA synthesis
5500	First mitosis completed

116

The influx of Na$^+$ that constitutes the fast block to polyspermy (Figure 8.15) triggers a wave of Ca^{2+} release (Figure 8.23), which is directly responsible for rupture of the cortical granules. A major influx of Na$^+$ then raises the intracellular pH. This pH change is believed to bring about the activation of enzymes that are responsible for an increase in the respiratory rate of the egg. This increase occurs rapidly, preceding completion of the fertilization membrane. The fertilization membrane is completely elevated in about 60 seconds.

The early changes, noted above, pave the way for a continued rise in pH, which leads, eventually, to the sequence of late changes. The late changes are marked by the mobilization of previously sequestered maternal mRNA, a 30% increase in the number of polysomes (Humphreys, 1971), the onset of a rapid increase in the rate of protein synthesis (Figure 8.24), DNA synthesis in the pronuclei, and the first cell division. During the period of late changes there is notable increase in the extent to which adenosine residues are added sequentially and turned over at the 3' terminal of certain maternal mRNAs (Wilt, 1973). The significance of polyadenylated mRNA is treated further in Chapter 18.

The release of Ca^{2+} is alone sufficient to initiate the entire fertilization program. The program can be activated even in Ca^{2+}-free sea water in the presence of ionophore A23187, which releases bound intracellular Ca^{2+}. Thereupon both early and late changes proceed normally. Most of the late changes are not, however, dependent on the occurrence of the early changes, for in sea urchin eggs the early changes may be bypassed by addition of NH$_4$Cl or NH$_4$OH to the sea water. Activation by the ammonium ion completely bypasses the early changes, but sets in motion the late changes, including acid release, the activation of transport systems, K$^+$ conductance, release of sequestered mRNA, additional polyadenylation (Wilt and Mazia, 1974), and chromosome condensation.

The elevation of pH is delayed after ammonium activation, and so is the onset of protein synthesis, which is dependent on pH increase (Grainger et al., 1979). A normal increase in polyribosomes takes place, however, indicating that mRNAs are recruited from the maternal supply as in normal fertilization. The increased availability of mRNAs promotes a rate of protein synthesis considerably above that of the unfertilized egg, but the higher rate is, nevertheless, considerably below the level found in normally fertilized or Ca^{2+}-activated eggs. The reason for this is found in the fact that, whereas in normally activated eggs there occurs a two- to three-fold increase in the efficiency of translation of mRNAs, in the ammonium-activated egg the increase in translational efficiency fails to occur (Brandis and Raff, 1979; Hille and Albers, 1979). It must be concluded, therefore, that early changes induced by fertilization or Ca^{2+} activation are required in order for an increased translational efficiency to occur later on.

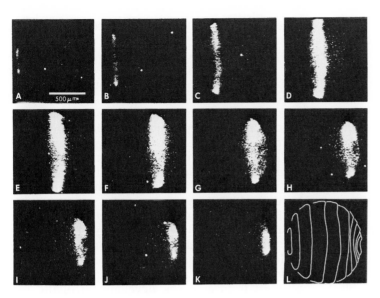

FIGURE 8.23 A cortical wave of calcium release and resequestration is visualized in a sperm-activated medaka (rice-paddy fish) egg that was injected prior to fertilization with aequorin, a protein that reacts specifically to free Ca^{2+} by emitting light. The passage of the light wave is shown in successive photographs (A–K) taken 10 seconds apart, and displayed from left to right in successive rows from top to bottom. The egg axis is horizontal, with the sperm-entrance point to the left. The lower right frame is a tracing showing the leading edge of the wave fronts in each of the eleven pictures. [From J. C. Gilkey et al., *J Cell Biol* 76:448–66 (1978), by courtesy of Dr. L. F. Jaffe, who kindly supplied the illustration, and permission of the Rockefeller University Press.]

There are several points at which control of translation might be exercised. These are discussed more fully in Chapter 18. Thus far there is no clear evidence, for differences in components of the translational machinery in unfertilized sea urchin eggs or for the presence of inhibitors that are removed after activation of the egg (but see Monroy et al., 1965; Metafora et al., 1971; Mano and Nagano, 1970). Ilan and Ilan (1978) have, in fact, demonstrated that in homogenates of unfertilized sea urchin eggs, maternal messages in the form of mRNPs are readily translated in vitro in the presence of 12 μM Mg^{2+}. This is in the physiological range of free Mg^{2+} for sea urchin eggs (Steinhardt and Epel, 1974). So, what does activation do in order to facilitate the translation of maternal mRNPs in ovo?

8.8.1 PHYSIOLOGICAL DEREPRESSION IN OTHER ORGANISMS. Sea urchin eggs are quite evidently in a state of physiological repression prior to fertilization. The sperm, Ca^{2+}, or other activators are responsible for physiological derepression in sea urchin eggs, and this change is rather dramatic, as the foregoing material demonstrates. We may ask whether the pattern of derepression found in sea urchins occurs in other organisms, also. This seems to be so for in a great number of animals from different phyla, the introduction of Ca^{2+} by ionophore A23187 or microinjection of Ca^{2+} or high Ca^{2+} in the medium elicits an activation response. There is a difference, however, in the timing of this response with respect to the stage at which the egg is normally fertilized. Sea urchins have completed meio-

117

FIGURE 8.24 The effect of fertilization on protein synthesis in eggs of the sea urchin *Lytechinus pictus*. The abscissa shows the cumulative incorporation of ^{14}C-leucine into protein expressed as disintegrations per minute per standard sample of eggs. [Redrawn from D. Epel, *Proc Natl Acad Sci USA*, 57:899–906 (1967).]

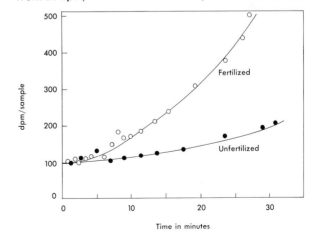

sis and are strongly repressed physiologically at fertilization. In most other forms, however, eggs are repressed before meiosis is complete. Their derepression occurs under hormonal stimulation or in response to Ca^{2+} and involves the resumption of meiosis in addition to the other activation events. Thus, in most starfishes responding to 1-methyladenine or to ionophore A23187 meiosis resumes and respiration and protein synthesis are enhanced. There is no further increase in protein synthesis at fertilization. Likewise, Ca^{2+} is required for the initiation of meiotic maturation in *Rana pipiens* and *Xenopus laevis* (Ecker and Smith, 1971; Merriam, 1971). In *R. pipiens* the rate of protein synthesis increases after 20 hours of pituitary stimulation and shows no further increase at fertilization (Ecker and Smith, 1968). Finally, it may be noted that injection of Ca^{2+} into mature mouse eggs induces cortical granule breakdown, release of the meiotic block and cell division to the stage when implantation into the uterine wall normally occurs (Fulton and Whittingham, 1978).

118

8.9 Parthenogenesis and Gynogenesis

In nature and under a variety of artificial conditions, embryonic development can be initiated and carried to completion in eggs that receive no contribution from the male genome. The term **parthenogenesis** (Gr., *parthenos,* maiden) is applied in these instances as it is in plants. For cases in which the sperm activates the egg but contributes no genetic material, the term **gynogenesis** (Gr., *gyne,* woman) is used.

8.9.1 NATURAL PARTHENOGENESIS. Parthenogenesis is a common mode of reproduction among many organisms. It was first described by Bonnet, who observed in 1762 that aphids (plant lice) reproduce parthenogenetically during the summer. The occurrence of parthenogenesis has since been analyzed in a number of parasitic and free-living forms. It occurs in many species of rotifers, nema-

todes, helminths, annelids, and insects, among the invertebrates, and is found, though rarely, in lizards and birds. A flock of turkeys showing a high incidence of parthenogenetically developing eggs is maintained at the Experiment Station of the U.S. Department of Agriculture, at Beltsville, Maryland (see Olsen, 1969), but parthenogenesis is not a significant mode of reproduction in birds. There are a number of reports (reviewed by Beatty, 1967, 1972) of the spontaneous activation of eggs of virgin laboratory mammals, but there are apparently no cases in which development proceeded beyond early stages.

In many species parthenogenetically produced offspring are constituted exclusively of males bearing the haploid chromosome complement. The male-producing eggs will have undergone a normal meiosis, but, in the absence of a chromosome complement from a sperm, development of the new individual proceeds with only the reduced chromosome number. This occurs in a number of rotifers and arachnids, many insects of the order Hymenoptera (including familiar honeybees), two families of beetles (order Coleoptera), and two families of coccids (order Homoptera). Exceptionally diploid males are produced. This is the case for the coccid *Lecanium putmani* and for the Beltsville turkeys noted above. Diploidy is restored in these cases when the haploid nucleus divides without cytokinesis. Both daughter nuclei then form a common metaphase plate for the second division of the zygote.

In numerous other species, notably among insects of the orders Hymenoptera, Homoptera, Diptera, and Coleoptera, parthenogenetic development of the egg issues only in the production of diploid females. In some such species there is not true meiosis during oögenesis and the egg cells are left with a single diploid nucleus. In other cases the diploid number is restored by the fusion of two meiotic nuclei, or by the fusion of the haploid daughter nuclei produced by mitosis. This also occurs in parthenogenetic species of lizards. Obviously, parthenogenesis is obligatory as the mode of reproduction in populations of animals in which no males occur or in which their number is so few that they can play no significant reproductive role. Obligatory parthenogenesis is found among all major groups in which parthenogenesis occurs at all.

Other organisms may reproduce parthenogenetically or not, according to circumstances. This pattern of reproduction is referred to as **facultative parthenogenesis.** Facultative parthenogenesis is shown, for example, by the queen honeybee who lays an unfertilized egg in comb cells destined for the development of haploid males (drones) and fertilized eggs in those destined for workers.

The question of whether a given egg will be fertilized or not depends on a combination of genetic and environmental factors. Thus, the female alfalfa leaf-cutter bee, *Megachile rotunda,* in the act of depositing eggs to be fertilized, ceases contracting the abdomen when about two thirds of the egg is forced out. A pause then follows during which the stored sperm are pumped into the micropyle of the egg, whereupon oviposition is completed and a female develops from the fertilized egg. If a male-forming egg (haploid) is to be produced, there is no pause during oviposition. In this case, the stimulus inducing fertilization appears to be associated with the depth of the nesting tunnels. Female-producing eggs are laid in the inner deeper tunnels of the nest and male eggs in the outer shallower tunnels (Gerber and Klostermeyer, 1970). Among the Hymenoptera *Microplecton fuscipennis*, which parasitizes the spruce sawfly, lays fertilized eggs on large hosts and unfertilized eggs on small ones. In other instances, the female has no obvious behavioral control over fertilization. In the wasp *Dahlbominus fulginosus*, the percentage of female-producing eggs is determined by a factor that is of genetic origin and is sex limited, being transmitted by the females to their sons (Wilkes, 1964).

The diet of the mother is a controlling factor for some rotifers in determining whether their offspring will be capable of producing fertilized eggs or not. On a diet low in vitamin E, the population consists essentially of **amictic** females, that is, females whose oöcytes do not undergo meiosis. These oöcytes develop parthenogenetically within the body of the mother and develop into females that are, likewise, amictic. If, however, the amictic female receives a diet high in vitamin E during early developmental stages of her offspring, the latter will differentiate as **mictic** females, that is, females whose oöcytes undergo meiosis. Oöcytes of mictic females, if fertilized, form eggs that are released as resting forms that later develop into amictic females. If not fertilized, however, these oöcytes develop parthenogenetically as males! (Birky and Gilbert, 1971.)

8.9.2 ARTIFICIAL PARTHENOGENESIS. A large variety of physical and chemical agents have been used to activate the eggs of various animals. Hypertonic or hypotonic seawater, acids, bases, alkaloids, heat, cold, pricking with a needle, electrical currents, and other treatments have been used with some degree of success, usually very modest. After activation many kinds of eggs develop to larval stages but very few to adult life. Adults have been obtained from artificially stimulated eggs of sea urchins, starfish, silkmoths, fishes, and frogs (see reviews by Tyler, 1965, and Beatty, 1967, and 1972).

Some organisms that develop following parthenogenetic activation do so with the haploid chromosome number. It appears that, for greatest success in development past the stage of the cortical reaction, however, some mechanism for restoration of the diploid chromosome number must take place. Known methods include fusion of haploid nuclei after their first division, fusion of the female pronucleus with a polar body nucleus, and utilization of a polar meiotic spindle as a mitotic spindle (i.e., suppression of the second meiotic division).

Frog eggs, activated by pricking with a clean needle, show the cortical reaction and complete the second meiotic division but do not divide. If, however, the activating needle is dipped in blood or any of a variety of tissues and fractions thereof, cell division is initiated and parthenogenetic tadpoles are formed with some frequency. The great majority of these are dwarfed, weak, and unviable. These prove generally to be haploid. If, however, the second meiotic division is suppressed by cold treatment after activation, vigorous tadpoles of diploid constitution are formed (see review of Fankhauser, 1945).

Shaver (1953) investigated the nature of tissue factors required to initiate division in artificially activated frog eggs. He homogenized various tissues and tested fractions thereof for ability to promote parthenogenetic development when injected into pricked eggs. He found that an active large-granule fraction sedimenting centrifugally at $10,000 \times G$, could be obtained from various adult and embryonic frog tissues and from embryonic tissues of mouse and chick. Later Fraser (1971) and Masui et al. (1978), as well as several other investigators, found that the cleavage-initiating activity (CIA) of the $10,000 \times G$ fraction binds to colchicine, which is a property of microtubular protein. Brain, a rich

119

source of microtubules, is a rich source of CIA. Injected into nucleated unfertilized eggs of *Rana pipiens,* it can induce parthenogenetic development of haploid swimming tadpoles (Fraser, 1971). It will not, however, induce cleavage in enucleated eggs, although isolated mitotic spindles of sea urchin eggs will do so (Masui et al., 1978). Frog eggs contain microtubular protein but possibly require microtubular segments provided by CIA for the initiation of division spindles. In contrast to frog eggs, many invertebrate eggs do not require an exogenous nucleating agent to promote microtubular assembly, for upon parthenogenetic stimulation they sometimes form numerous "cytasters," which are radial arrays of microtubules. These can become arrayed on mitotic spindles (Figure 8.25) with or without chromosomes.

The ovulated eggs of a number of virgin female mammals tend to show signs of activation as they age, and some can form transplantable teratomas. Freshly ovulated eggs also respond to a number of treatments by activation and cell division. Treatment with hyaluronidase (Kaufman and Sachs,

1976), heat shock and electric shock elicit activation and development of mouse eggs to the **blastocyst** stage, which is the stage at which implantation into the uterine wall begins (cf. Chapter 9). Such embryos, introduced into the uterus of a hormonally prepared female, sometimes develop to fairly advanced stages (reviewed by Graham, 1974, and by Tarkowski, 1975). Earlier reports of the development to term of parthenogenetically activated rabbit eggs (Pincus, 1939) have not been confirmed.

This brief discussion of artificial parthenogenesis may be terminated by pointing out that the recognition of the role of Ca^{2+} in activation of eggs may lead the way to the successful production of many kinds of parthenogenic organisms. Hamster eggs have been successfully activated by pricking with an appropriately sized needle in the presence of Ca^{2+} (Uehara and Yanagimachi, 1977) and, as noted above, Fulton and Wittingham (1978) found that mouse eggs can be activated by Ca^{2+} in culture medium. Such eggs show the normal series of activation reactions and have been cultured in vitro to the blastocyst stage. Very promising is the high level of success that has been achieved in rearing to adulthood diploid parthenogenic female sea urchins that have been activated by means of ionophore A23187 (Brandriff et al., 1978). These results offer encouragement for progress in studying the genetics, especially of marine organisms, a relatively unexplored area of research.

120

FIGURE 8.25 Cytasters formed in eggs of *Toxapneustes* after treatment with hypertonic sea water. [Redrawn from E. B. Wilson, *The Cell*, 3rd ed., New York: Macmillan, 1928.]

8.9.3 GYNOGENESIS.
There are natural populations of fish and salamanders that consist only of female individuals. They do not reproduce parthenogenetically, but their eggs, diploid or sometimes triploid, are activated by sperm from males of related bisexual species. Paternal chromosomes are not incorporated into the zygote. Therefore, the offspring are replicates of the mother in morphology and genetic constitution. Reproduction is thus **gynogenetic.**

Poecilia formosa (the familiar Amazon molly) is a species, consisting of only diploid females, which inhabits freshwater streams and coastal lagoons of northeastern Mexico and southern Texas. Males of either of the related bisexual species *P. mexicana* and *P. latispinna* provide the sperm that stimulate ova of *P. formosa* to develop. In the same family are gynogenetic triploid females that probably arose from hybrids involving an unreduced (i.e., diploid) egg from one species and the haploid sperm of another. One of these is *Peociliopsis 2 monacha-lucida* (i.e., two sets of *P. monacha* genes and one set of *P. lucida* genes). It uses sperm of the bisexually reproducing *P. monacha*. These and similar cases were reviewed by Schultz (1971).

A similar triploid *Peociliopsis monacha 2-lucida* uses sperm of either *P. monacha* or *P. lucida*. Cimino (cited by Schultz, 1971) examined the behavior of chromosomes during oögenesis of the triploid and found that meiosis is preceded by an **endomitotic** division (mitosis without cell division) that raises the chromosome number to hexaploid. This is followed by two meiotic divisions that reduce the chromosome number to triploid once more.

Two all-female triploid species of *Ambystoma* are known. These are *A. tremblayi* and *A. platineum*, believed to have originated about 10,000 years ago from hybrids of the bisexual species *A. laterale* and *A. jeffersonianum*. Apparently, normal meiosis in ova of these triploid species is also preceded by an endomitotic increase of the chromosome number to $6N$ (see review by Asher and Nace, 1971).

In the laboratory one can produce gynogenesis artificially by treating unfertilized eggs with sperm that have been treated with X-rays, ultraviolet irradiation, or various dyes. Such sperm, although capable of stimulating the onset of development, degenerate within the egg cytoplasm and contribute no genetic material. The eggs complete meiosis normally and begin development as haploids unless the diploid number is restored by endomitosis.

8.10   Summary

In animal fertilization the egg and sperm come together in a process that releases the gametes from developmental stasis and unites the haploid genome from male and female parents in a diploid zygote. Union of the gametes occurs in an aqueous medium and is facilitated by the production of large numbers of them and by anatomical and behavioral attributes of the parents that cause them to be deposited in proximity.

Very few animal eggs attract sperm, but substances emanating from female tissue or egg membranes may activate or bind to them. For sperm to fertilize an egg, they must first undergo the acrosome reaction. In the case of mammalian eggs, and perhaps some other forms, the acrosome reaction requires the prior occurrence of a process called capacitation. The acrosome reaction is a change in the acrosomal complex involving the fusion and vesicular breakdown of the sperm plasma membrane and the outer membrane of the acrosomal vesicle. This allows the release of acrosomal enzymes, or egg membrane lysins, including proteases and glucanases. In mammals acrosomal hyaluronidase assists in dispersing the cumulus layer around the ovum. Another acrosomal enzyme, corona penetrating enzyme (CPE), assists in passage of the sperm through the corona radiata. Yet another acts on the zona pellucida. It is trypsin-like in its biochemical properties and is called acrosin.

Mammalian sperm undergo capacitation during a period of residence in the female reproductive tract. Enzymes or other components of the female tract apparently remove or inactivate fertility-inhibiting factors that are found in the seminal plasma. One of these inhibits CPE, and another inhibits the action of acrosin. Removal of the latter may not be involved in capacitation. A form of capacitation is required for fertilization in some hydroids, but except for these, mammals are the only forms in which a requirement for capacitation has been clearly demonstrated.

Sperm-binding substances have been found in eggs or their jelly envelopes in most species. They are called fertilizins. Despite over a half-century of research on fertilizin, its significance in the sperm-egg interaction is not clear. The principal component of jelly-covered eggs of echinoderms, it may have a role in reducing polyspermy, the entry of excess sperm into the egg.

In the case of fertilization in invertebrates, the first contact of egg and sperm usually involves the egg plasma membrane and the inner membrane of the acrosomal vesicle, now covering one or more prolonged acrosomal tubules. This membrane is continuous with the sperm plasma membrane. In sea urchins, the binding of sperm to egg is mediated by a glycoprotein receptor in the vitelline membrane, which shows species specificity for a matching protein, bindin, at the tip of the acrosomal tubule. Although many actively gyrating sperm bind to the egg, only one usually penetrates and makes contact with the egg surface at the tip of the acro-

121

somal tubule. In mammalian fertilization first contact is between the postacrosomal region of the sperm head and the egg membrane. In each case, however, the contacting membranes of egg and sperm fuse and undergo vesicular disintegration, creating an open passage that permits confluence of their contents. A prominent surge of egg cytoplasm through this passage envelopes the sperm nucleus, forming a fertilization cone. The cone is less prominent during mammalian fertilization. The sperm plasma membrane becomes incorporated in the zygote membrane as the fertilization cone subsides.

Most eggs admit only one sperm. Polyspermy is prevented in sea urchin eggs by an incomplete fast block, which consists in depolarization of the plasma membrane immediately upon successful sperm-egg contact. It is also inhibited by a slow block, which is constituted by the formation of the fertilization membrane as a consequence of the cortical reaction. The cortical reaction consists in a wave of rupture of the cortical alveoli, starting at the point of sperm entry. The alveolar contents consist of enzymes and other materials that detach supernumerary sperm, elevate and harden the fertilization membrane, and create, by osmotic effects, a perivitelline space. Mammalian eggs do not show the fast block, but breakdown of the cortical alveoli in most mammals modifies the zona pellucida in such a way that supernumerary sperm cannot bind to and penetrate it. This is called the zona reaction.

After entrance of the sperm, the egg nucleus completes meiosis, if it has not already done so, and forms the female pronucleus. The nuclear membrane of the sperm disintegrates, and the sperm chromatin loses its compact configuration. A new nuclear membrane then forms around the male pronucleus. The pronuclei move together and chromosomes condense and become arranged on the first mitotic spindle of the new generation.

Fertilization of the egg or, in many cases, the hormonal events that prepare the egg for fertilization initiate a series of events that constitute a release from physiological stasis. In sea urchins, which have completed meiosis prior to fertilization, activation of the egg by the sperm or by artificial means initiates a series of early Ca^{2+}-dependent events that result in the cortical reaction. Calcium influx and temporary release of intracellularly bound calcium then provides conditions that lead to a massive efflux of protons with consequent in-

crease in intracellular pH. This in turn brings about an increase in K^+ conductance and permeability to amino acids, the release of sequestered maternal mRNA, an increase in the rate of protein synthesis, and increased polyadenylation of mRNA. These are termed late changes. These events are followed by fusion of the pronuclei, DNA synthesis, and the first division of the zygote. Ammonium ion can activate the late changes in the absence of the early ones.

In eggs that go into physiological stasis prior to the completion of meiosis, the break of physiological stasis is accomplished by hormonal stimulation or by increased intracellular Ca^{2+}. The fertilization-induced changes in these eggs are less dramatic than those seen in sea urchins, for the fertilization program is already under way prior to the availability of sperm. Thus such eggs usually show little or no increase in protein synthesis after fertilization, having already done so in response to hormonal stimulation.

Activation of the egg without the participation of sperm occurs naturally in many kinds of organisms, leading, in many cases, to all-female populations and in other cases to populations with haploid males. This is called natural parthenogenesis. Many kinds of eggs can be activated artificially by means of chemical and physical agents. The characteristic physical and chemical changes of fertilization follow artificial activation, but the completion of normal embryonic development occurs in only a few kinds of animals whose eggs are stimulated parthenogenetically. The discovery that increased intracellular Ca^{2+} is the critical event for bringing about the early changes in fertilization and that it sets up conditions for the late ones leads to the hope that artificial parthenogenesis can constitute a powerful tool for genetic studies in a variety of organisms that normally reproduce sexually.

In some natural populations eggs are fertilized by males of a different species, but the male genome does not contribute to development of the embryo. This is called gynogenesis. Gynogenetic populations consist only of females. In the laboratory gynogenetic development is brought about by fertilization with sperm that have been irradiated or otherwise treated so as to be incapable of forming a functional pronucleus. Eggs so treated develop as haploids unless the diploid chromosome number is restored by endomitosis.

8.11 Questions for Thought and Review

1. Are animal sperm attracted to eggs? How would one distinguish an attractive action of the egg or its surrounding tissues from a trapping action or simply an induction of motility?

2. What is fertilizin? Could it possibly play a role in preventing polyspermy? How?

3. What are the principal components of the acrosomal reaction in invertebrate sperm? How does the acrosomal reaction differ in sperm of marine invertebrates and mammals?

4. What evidence suggests that acrosomes of invertebrate sperm contain lytic enzymes? What is the presumed role of these enzymes?

5. What are the chief components of acrosomes in mammalian sperm? What specific actions have been identified in these components?

6. What is the phenomenon of capacitation with reference to mammalian fertilization? How might a knowledge of capacitation factors be put to use in fertility control?

7. Compare and contrast the mechanism of penetration of the egg by the sperm in sea urchins and in mammals?

8. What are the components of the cortical reaction in marine eggs. Is the breakdown of cortical granules an essential component of fertilization? What is the evidence?

9. How can one reconcile the observed facts that, on the one hand, sea urchin eggs usually are monospermic and the unsuccessful sperm are detached as the cortical reaction proceeds around the egg and, on the other hand, eggs can be made polyspermic by dense sperm suspension during the entire interval required for the fertilization membrane to be formed?

10. Compare and contrast the events involved in the formation of male and female pronuclei in sea urchin eggs.

11. What is the zona reaction? What factors are thought to be involved in the occurrence of the reaction?

12. Compare and contrast the release of physiological stasis in eggs that are activated before and after the completion of meiosis.

13. What are the "early" changes that occur in fertilization of sea urchin eggs? What are the "late" changes? What is the relationship between the early and late changes?

14. Parthenogenetic development of sea urchins through metamorphosis has been achieved after parthenogenetic stimulation of eggs either with ionophore A23187 or with NH_4Cl or NH_4OH. Better results are achieved with the ionophore. Why might this be expected? The answer to this question might be reconsidered after reading the next chapter.

15. Certain parthenogenetically reproducing *Ambystoma* species appear to be triploid and to have arisen from hybrids. How might the occurrence and survival of such triploid species be explained?

8.12 Suggestions for Further Reading

Austin, C. R. 1974. Principles of Fertilization. *Proc R Soc Med* 67:925–27.

Austin, C. R. 1978. Patterns in metazoan fertilization. *Curr Top Dev Biol* 12:1–9.

Austin, C. R., and R. V. Short, eds. 1972. *Reproduction in Mammals.* I. *Germ Cells and Fertilization.* Cambridge: Cambridge University Press.

Barber, M. L. 1979. Changes in enzyme activities and lipid content of echinoderm egg membranes, at maturation and fertilization. *Am Zool* 19:821–37.

Bedford, J. M. 1970. Sperm capacitation and fertilization in mammals. *Biol Reprod Suppl* 2:128–158.

Franklin, L. E. 1970. Fertilization and the role of the acrosomal region in non-mammals. *Biol Reprod Suppl* 2:159–76.

Jaenike, J., and Robert K. Selander. 1979. Evolution and ecology of parthenogenesis in earthworms. *Am Zool* 19:729–37.

Lallier, R. 1977. The problem of sea urchin egg fertilization and its implications for biological studies. *Experientia* 33:1263–67.

Longo, F. J. 1976. Ultrastructural aspects of fertilization in spiralian eggs. *Am Zool* 16:375–94.

Marx, J. L. 1978. The mating game: what happens when sperm meets egg. *Science* 200:1256–59.

Maslin, T. P. 1971. Parthenogenesis in reptiles. *Am Zool* 11:361–80.

Paul, M., and R. N. Johnston. 1978. Uptake of Ca^{2+} is one of the earliest responses to fertilization of sea urchin eggs. *J Exp Zool* 203:143–49.

Schatten, G., and D. Mazia. 1976. The penetration of the spermatozoon through the sea urchin egg surface at fertilization. Observations from the outside on whole eggs and from the inside on isolated surfaces. *Exp Cell Res* 98:325–37.

Schmell, E., B. J. Earles, C. Breaux, and W. J. Lennarz. 1977. Identification of a sperm receptor on the surface

123

of the eggs of the sea urchin *Arbacia punctulata. J Cell Biol* 72:35–46.

SHAPIRO, B. M., R. W. SCHACKMANN, C. A. GABEL, C. A. FOERDER, M. L. FARANCE, E. M. EDDY, and S. J. KLEBANOFF. 1979. Molecular alterations in gamete surfaces during fertilization and early development. In S. Subtelny and N. K. Wessels, eds. *The Cell Surface: Mediator of Developmental Processes.* New York: Academic Press, pp. 127–49.

TALBOT, P., and L. E. FRANKLIN. 1976. Morphology and kinetics of the hamster sperm acrosome reaction. *J Exp Zool* 198:163–76.

TALBOT, P., and L. E. FRANKLIN. 1978. Surface modification of guinea pig sperm during in vitro capacitation: an assessment using lectin-induced agglutination of living sperm. *J Exp Zool* 203:1–14.

TILNEY, L. G. 1975. The role of actin in nonmuscle cell motility. In S. Inoue and R. E. Stephens, eds. *Molecules and Cell Movement.* New York: Raven, pp. 339–85.

TYLER, A., and B. S. TYLER. 1966. Physiology of fertilization and early development. In R. A. Boolootian, ed. *Physiology of Echinodermata.* New York: Wiley–Interscience, pp. 683–741.

VACQUIER, V. D. 1979. The adhesion of sperm to sea urchin eggs. In S. Subtelny and N. K. Wessels, eds. *The Cell Surface: Mediator of Developmental Processes.* New York: Academic Press, pp. 151–168.

VACQUIER, V. D. 1979. The interactions of sea urchin gametes during fertilization. *Am Zool* 19:839–49.

YANAGIMACHI, R. 1978. Sperm-egg association in mammals. *Curr Top Dev Biol* 12:83–105.

8.13 References

AKETA, K. 1973. Physiological studies on the sperm surface component responsible for sperm-egg binding in sea urchin fertilization. I. Effect of sperm-binding protein on the fertilizing capacity of sperm. *Exp Cell Res* 80:439–41.

AKETA, K. 1975. Physiological studies on the sperm surface component responsible for sperm-egg binding in sea urchin fertilization, II. Effect of concanavalin A on the fertilizing capacity of sperm. *Exp Cell Res* 90:56–62.

AKETA, K., S. MIYAZAKI, M. YOSHIDA, and H. TSUZUKI. 1978. A sperm factor as the counterpart to the sperm-binding factor of the homologous eggs. *Biochem Biophys Res Commun* 80:917–22.

AKETA, K., and T. OHTA. 1977. When do sperm of the sea urchin, *Pseudocentrotus depressus,* undergo the acrosome reaction at fertilization? *Dev Biol* 61:366–72.

ALLISON, A. C., and E. F. HARTREE. 1970. Lysosomal enzymes in the acrosome and their possible role in fertilization. *J Reprod Fertil* 21:501–15.

ANDERSON, E. 1968. Oöcyte differentiation in the sea urchin, *Arbacia punctulata,* with particular reference to the origin of cortical granules and their participation in the cortical reaction. *J Cell Biol* 37:514–39.

ARNOLD, J. M. 1962. Mating behavior and social structure in *Loligo pealii. Biol Bull* 123:53–57.

ASHER, J. H., and G. W. NACE. 1971. The genetic structure and evolutionary fate of parthenogenetic amphibian populations. *Am Zool* 11:381–98.

AUSTIN, C. R. 1948. Function of hyaluronidase in fertilization. *Nature* 162:63–64.

AUSTIN, C. R. 1951. Observations on the penetration of sperm into the mammalian egg. *Aust J Sci Res [Biol]* 40:581–96.

AUSTIN, C. R. 1952. The capacitation of mammalian sperm. *Nature* 170:326.

BARROS, C., A. M. VLIEGENTHART, and L. E. FRANKLIN. 1972. Polyspermic fertilization of hamster eggs in vitro. *J Reprod Fertil* 28:117–20.

BEATTY, R. A. 1967. Parthenogenesis in vertebrates. In C. B. Metz and A. Monroy, eds. *Fertilization.* Vol. 1. New York: Academic Press, pp. 413–40.

BEATTY, R. A. 1972. Parthenogenesis and heteroploidy in the mammalian egg. In J. D. Biggers and A. W. Schuetz, eds. *Oögenesis.* Baltimore: University Park Press, pp. 277–99.

BELLET, N. F., J. P. VACQUIER, and V. D. VACQUIER. 1977. Characterization and comparison of "bindin" isolated from sperm of two species of sea urchins. *Biochem Biophys Res Commun* 79:159–65.

BHATTACHARYYA, A. K., and L. J. D. ZANEVELD. 1978. Release of acrosin and acrosin inhibitor from human spermatozoa. *Fertil Steril* 30:70–78.

BIRKEY, C. W., and J. J. GILBERT. 1971. Parthenogenesis in rotifers: the control of sexual and asexual reproduction. *Am Zool* 11:245–66.

BRADFORD, M. M., R. A. MCRORIE, and W. L. WILLIAMS. 1976. Involvement of esterases in sperm penetration of the corona radiata of the ovum. *Biol Reprod* 15:102–106.

BRANDIS, J. W., and R. A. RAFF. 1978. Translation of oogenic mRNA in sea urchin eggs and early embryos. Demonstration of a change in translational efficiency following fertilization. *Dev Biol* 67:99–113.

BRANDIS, J. W., and R. A. RAFF. 1979. Elevation of protein synthesis is a complex response to fertilization. *Nature* 278:467–69.

BRANDRIFF, B., R. T. HINEGARDNER, and R. STEINHARDT. 1978. Development and life cycle of the parthenogenetically activated sea urchin embryo. *J Exp Zool* 192:13–24.

BRUMMETT, A. R., and J. N. DUMONT. 1979. Initial stages of sperm penetration into the egg of *Fundulus heteroclitus*. *J Exp Zool* 210:417–34.

CABADA, M. O., M. I. MARIANO, and J. S. RAISMAN. 1978. Effect of trypsin inhibitors and concanavalin A on the fertilization of *Bufo arenarum* coelomic oöcytes. *J Exp Zool* 204:409–16.

CARROLL, E. J., Jr., and D. EPEL. 1975. Isolation and biological activity of the proteases released by sea urchin eggs following fertilization. *Dev Biol* 44:22–32.

CHANG, M. C. 1951. The fertilizing capacity of spermatozoa deposited into the fallopian tubes. *Nature* 168:697–98.

COLLINS, F. 1976. A reevaluation of the fertilizin hypothesis of sperm agglutination and the description of a novel form of sperm adhesion. *Dev Biol* 49:381–94.

COLWIN, L. H., and A. L. COLWIN. 1963. Role of the gamete membranes in fertilization in *Sacroglossus kowlevskii* (enteropneusta). II. Zygote formation by membrane fusion. *J Cell Biol* 19:501–18.

COLWIN, A. L., and L. H. COLWIN. 1967. Behavior of the spermatozoon during sperm-blastomere fusion and its significance for fertilization. *Z Zellforsch* 78:208–220.

CONWAY, A. F., and C. B. METZ. 1976. Phospholipase activity of sea urchin sperm: its possible involvement in membrane fusion. *J Exp Zool* 198:39–48.

CORNETT, L. E., B. D. BAVISTER, and S. MEIZEL. 1979. Adrenergic stimulation of fertilizing ability in hamster spermatozoa. *Biol Reprod* 20:925–29.

CORNETT, L. E., and S. MEIZEL. 1978. Stimulation of *in vitro* activation and the acrosome reaction of hamster spermatozoa by catecholamines. *Proc Natl Acad Sci USA* 75:4954–58.

CROSS, N., and R. P. ELINSON. 1980. A fast block to polyspermy in frogs mediated by changes in the membrane potential. *Dev Biol* 75:187–98.

DAN, J. C. 1970. Morphogenetic aspects of acrosome formation and reaction. *Adv Morph* 8:1–39.

ECKER, R. E., and L. D. SMITH. 1968. Protein synthesis in amphibian oöcytes and early embryos. *Dev Biol* 18:232–49.

ECKER, R. E., and L. D. SMITH. 1971. Influence of exogenous ions on the events of maturation in *Rana pipiens* oöcytes. *J Cell Physiol* 77:61–70.

ELINSON, R. P. 1973. Fertilization of frog body cavity eggs enhanced by treatments affecting the vitelline coat. *J Exp Zool* 183:291–302.

EPEL, D. 1978. Mechanisms of activation of sperm and egg during fertilization of sea urchin gametes. *Curr Top Dev Biol* 12:186–246.

EPEL, C. 1979. Experimental analysis of the role of intracellular calcium in the activation of the sea urchin egg at fertilization. In S. Subtelny and N. K. Wessels, eds.

The Cell Surface: Mediator of Developmental Processes. New York: Academic Press, pp. 169–85.

FALLON, J. F., and C. R. AUSTIN. 1967. Fine structure of the gametes of *Nereis limbata* before and after interaction. *J Exp Zool* 166:225–42.

FANKHAUSER, G. 1945. The effects of changes in chromosome number on amphibian development. *Quart Rev Biol* 20:20–78.

FOERDER, C. A., and B. M. SHAPIRO. 1977. Release of ovoperoxidase from sea urchin eggs hardens the fertilization membrane with tyrosine crosslinks. *Proc Natl Acad Sci USA* 74:4214–18.

FRASER, L. R. 1971. Physico-chemical properties of an agent that induces parthenogenesis in *Rana pipiens* eggs. *J Exp Zool* 177:153–72.

FRASER, L. R., and P. J. QUINN. 1981. A glycolytic product is obligatory for initiation of the sperm acrosome reaction and whiplash motility required for fertilization in the mouse. *J Reprod Fert* 61:25–35.

FULTON, B. P., and D. G. WHTTINGHAM. 1978. Activation of mammalian oöcytes by intracellular injection of calcium. *Nature* 273:149–51.

GABEL, C. A., E. M. EDDY, and B. M. SHAPIRO. 1979. After fertilization sperm surface components remain as a patch in sea urchin and mouse embryos. *Cell* 18:207–15.

GERBER, H. S., and E. C. KLOSTERMEYER. 1970. Sex control by bees: a voluntary act of egg fertilization during oviposition. *Science* 167:82–84.

GLABE, C. G., and V. D. VACQUIER. 1978. Egg surface glycoprotein receptor for sea urchin sperm bindin. *Proc Natl Acad Sci USA* 75:881–85.

GOULD-SOMERO, M., L. A. JAFFE, and L. Z. HOLLAND. 1979. Electrically mediated polyspermy block in eggs of the marine worm *Urechis caupo*. *J Cell Biol* 82:426–40.

GRAHAM, C. F. 1974. The production of parthenogenetic mammalian embryos and their use in biological research. *Biol Rev* 49:399–422.

GRAINGER, J. L., M. M. WINKLER, S. S. SHEN, and R. A. STEINHARDT. 1979. Intracellular pH controls protein synthesis rate in the sea urchin egg and early embryo. *Dev Biol* 68:396–406.

GREY, R. D., P. K. WORKING, and J. L. HEDRICK. 1977. Alteration of structure and penetrability of the vitelline envelope after passage of eggs from coelom to oviduct in *Xenopus laevis*. *J Exp Zool* 201:73–84.

GWATKIN, R. B. L. 1976. Fertilization. In G. Poste and G. L. Nicolson, eds. *The Cell Surface in Animal Embryogenesis and Development.* Amsterdam: Elsevier/North-Holland Biomedical Press, pp. 1–54.

HARTMANN, J. F., and C. F. HUTCHISON. 1974. Mammalian fertilization in vitro: sperm-induced preparation of the zona pellucida of golden hamster ova for final binding. *J Reprod Fertil* 37:443–45.

125

HAUSHKA, S. D. 1973. Purification and characterization of *Mytilus* egg membrane lysin from sperm. *Biol Bull* 125:363 (Abstr.).

HELLER, E., and M. A. RAFTERY. 1973. Isolation and purification of three egg-membrane lysins from sperm of the marine invertebrate *Megathura crenulata* (giant keyhole limpet). *Biochemistry* 12:4106–13.

HILLE, M. B., and A. A. ALBERS. 1979. Efficiency of protein synthesis after fertilization of sea urchin eggs. *Nature* 278:469–71.

HOLLAND, L., and M. GOULD-SOMERO. 1981. Electrophysiological response to insemination in oöcytes of *Urechis caupo. Dev Biol* 83:90–100.

HOTTA, K., M. KUROKAWA, and S. ISAKA. 1970. Isolation and identification of two sialic acids from the jelly coat of sea urchin eggs. *J Biol Chem* 245:6307–11.

HUMPHREYS, T. 1971. Measurement of messenger RNA entering polysomes upon fertilization of sea urchin eggs. *Dev Biol* 26:201–208.

ILAN, Judith, and J. ILAN. 1978. Translation of maternal messenger ribonucleoprotein particles from sea urchin in a cell-free system from unfertilized eggs. *Dev Biol* 66:375–85.

JAFFE, L. A. 1976. Fast block to polyspermy in sea urchin eggs is electrically mediated. *Nature* 261:68–71.

KAUFMAN, M. H., and L. SACHS. 1976. Complete preimplantation development in culture of parthenogenetic mouse embryos. *J Embryol Exp Morphol* 35:179–90.

LEVINE, A. E., K. A. WALSH, and E. J. B. FODOR. 1978. Evidence for an acrosin-like enzyme in sea urchin sperm. *Dev Biol* 63:299–306.

LILLIE, F. R. 1913. Studies on fertilization. V. The behavior of the spermatozoa of *Nereis* and *Arbacia* with special reference to egg extractives. *J Exp Zool* 14:515–74.

LILLIE, F. R. 1919. *Problems of Fertilization.* Chicago: University of Chicago Press.

LONGO, F. J. 1976. Derivation of the membrane comprising the male pronuclear envelope in inseminated sea urchin eggs. *Dev Biol* 49:347–68.

LONGO, F. J., and E. ANDERSON. 1968. The fine structure of pronuclear development and fusion in the sea urchin, *Arbacia punctulata. J Cell Biol* 39:339–68.

LONGO, F. J., and M. KUNKLE. 1978. Transformations of sperm nuclei upon insemination. *Curr Top Dev Biol* 12:149–84.

LONGO, F. J., and W. PLUNKETT. 1973. The onset of DNA synthesis and its relation to morphogenetic events of the pronuclei in activated eggs of the sea urchin, *Arbacia punctulata. Dev Biol* 30:56–67.

MABUCHI, Y., and I. MABUCHI. 1973. Acrosomal ATPase in starfish and bivalve mollusk spermatozoa. *Exp Cell Res* 82:271–79.

MANO, Y., and H. NAGANO. 1970. Mechanism of release of maternal messenger RNA induced by fertilization in sea urchin eggs. *J Biochem (Tokyo)* 67:611–28.

MASUI, Y., A. FORER, and A. M. ZIMMERMAN. 1978. Induction of cleavage in nucleated and enucleated frog eggs by injection of isolated sea-urchin mitotic apparatus. *J Cell Sci* 31:117–35.

MEIZEL, S., and P. K. WORKING. 1980. Further evidence suggesting the hormonal stimulation of hamster sperm acrosome reactions by catecholamines in vitro. *Biol Reprod* 22:211–16.

MERRIAM, R. W. 1971. Progesterone induced maturational events in oöcytes of *Xenopus laevis*. I. Continuous necessity for diffusible calcium and magnesium. *Exp Cell Res* 69:75–80.

METAFORA, S., L. FELICETTI, and R. GAMBINO. 1971. The mechanism of protein synthesis activation after fertilization of sea urchin eggs. *Proc Natl Acad Sci USA* 71:690–93.

METZ, C. B. 1967. Gamete surface components and their role in fertilization. In C. B. Metz and A. Monroy, eds. *Fertilization*, Vol. 1. New York: Academic Press, pp. 163–236.

METZ, C. B. 1972. Effects of antibodies on gametes and fertilization. *Biol Reprod* 6:358–83.

METZ, C. B. 1978. Sperm and egg receptors involved in fertilization. *Curr Top Dev Biol* 12:107–47.

MILLER, R. L. 1966. Chemotaxis during fertilization in the hydroid campanularia. *J Exp Zool* 161:23–44.

MILLER, R. L. 1975. Chemotaxis of the spermatozoa of *Ciona intestinalis. Nature* 254:244–45.

MILLER, R. L. 1978. Site-specific sperm agglutination and the timed release of a sperm chemo-attractant by the egg of the leptomedusan, *Orthopyxis caliculata. J Exp Zool* 205:385–92.

MILLER, R. L. 1979a. Sperm chemotaxis in the hydromedusae. I. Species specificity and sperm behavior. *Marine Biol* 53:99–113.

MILLER, R. L. 1979b. Sperm chemotaxis in the hydromedusae. II. Some chemical properties of the sperm attractants. *Marine Biol* 53:115–24.

MILLER, R. L., and C. Y. TSENG. 1974. Properties and partial purification of the sperm attractant of *Tubularia. Am Zool* 14:467–86.

MIYAZAKI, S.-I. 1979. Fast polyspermy block and activation potential. Electrophysical basis for their changes during oöcyte maturation of a starfish. *Dev Biol* 70:341–54.

MIYAZAKI, S.-I., and S. HIRAI. 1979. Fast polyspermy block and activation potential. Correlated changes during oöcyte maturation of a starfish. *Dev Biol* 70:327–40.

MONROY, A., R. MAGGIO, and A. M. RINALDI. 1965. Experimentally induced activation of the ribosomes of the unfertilized sea urchin egg. *Proc Natl Acad Sci USA* 54:107–111.

NAKANO, E. 1956. Physiological studies on refertilization of the sea urchin egg. *Embryologia* 139–65.

OLSEN, M. V. 1969. Potential uses of parthenogenetic development in turkeys. *J. Hered.* 60:346–48.

O'RAND, M. G. 1972a. In vitro fertilization and capacitation-like interaction in the hydroid *Campanularia flexuosa. J Exp Zool* 182:299–306.

O'RAND, M. G. 1972b. A soluble cell surface material required for spermatozoan-epithelial cell interaction during fertilization in *Campanularia flexuosa. Biol Bull* 143:472.

O'RAND, M. G. 1977. The presence of sperm-specific surface isoantigens on the egg following fertilization. *J Exp Zool* 202:267–73.

OVERSTREET, J. W., and G. W. COOPER. 1979. The time and location of the acrosome reaction during sperm transport in the female rabbit. *J Exp Zool* 209:97–104.

PINCUS, G. 1939. The comparative behavior of mammalian eggs in vivo and in vitro. IV. The development of fertilized and artificially activated rabbit eggs. *J Exp Zool* 82:85–129.

SALING, P. M., J. SOWINSKI, and B. T. STOREY. 1979. An ultrastructural study of epididymal mouse spermatozoa binding to zona pellucidae in vitro: sequential relationship to the acrosome reaction. *J Exp Zool* 209: 229–38.

SATO, K. 1979. Polyspermy-preventing mechanisms in mouse eggs fertilized in vitro. *J Exp Zool* 210:353–59.

SCHATTEN, G., and D. MAZIA. 1976. The surface events at fertilization: the movements of the spermatozoon through the sea urchin egg surface and the roles of the surface layers. *J Supramol Struct* 5:343–69.

SCHULTZ, R. J. 1971. Special adaptive problems associated with unisexual fishes. *Am Zool* 11:351–60.

SHAVER, J. 1953. Studies on the initiation of cleavage in the frog egg. *J Exp Zool* 122:169–92.

SIMMEL, E. B., and D. KARNOFSKY. 1961. Observations on the uptake of tritiated thymidine in the pronuclei of fertilized sand dollar embryos. *J Biophys Biochem Cytol* 10:59–65.

SRIVASTAVA, P. N., S. R. AKRUK, and W. L. WILLIAMS. 1979. Dissolution of rabbit zona by sperm acrosomal extract: effect of calcium. *J Exp Zool* 207:521–29.

STAMBAUGH, R. 1978. Enzymatic and morphological events in mammalian fertilization. *Gamete* 1:65–85.

STAMBAUGH, R., and J. BUCKLEY. 1969. Identification and subcellular localization of the enzymes effecting penetration of the zona pellucida by rabbit spermatozoa. *J Reprod Fertil* 19:423–32.

STAMBAUGH, R., and L. MASTROIANNI. 1980. Stimulation of rhesus monkey *(Macaca mulatta)* proacrosin activation by oviduct fluid. *J Reprod Fertil* 59:479–84.

STAMBAUGH, R., and M. SMITH. 1974. Amino acid content of rabbit acrosomal proteinase and its similarity to human trypsin. *Science* 186:745–46.

STAMBAUGH, R. and M. SMITH. 1976. Sperm proteinase release during fertilization of rabbit ova. *J Exp Zool* 197:121–25.

STEINHARDT, R. A., and D. EPEL. 1974. Activation of sea-urchin eggs by a calcium ionophore. *Proc Natl Acad Sci USA* 71:1915–19.

SUGIYAMA, M. 1951. Re-fertilization of the fertilized eggs of the sea urchin. *Biol Bull* 101:335–44.

SUZUKI, R. 1958. Sperm activation and aggregation during fertilization in some fishes. I. Behavior of spermatozoa around the micropyle. *Embryologia* 4:93–102.

TARKOWSKI, A. K. 1975. Induced parthenogenesis in the mouse. In C. L. Markert and J. Papconstantinou, eds. *The Developmental Biology of Reproduction.* New York: Academic Press, pp. 107–29.

TILNEY, L., D. P. KIEHART, S. SARDET, and M. TILNEY. 1978. Polymerization of actin. IV. The role of Ca^{++} and H^+ in the assembly of actin and in membrane fusion in the acrosomal reaction of echinoderm sperm. *J Cell Biol* 77:536–50.

TYLER, A. 1965. The biology and chemistry of fertilization. *Am Naturalist* 99:309–34.

TYLER, A., A. MONROY, and C. B. METZ. 1956. Fertilization of fertilized sea urchin eggs. *Biol Bull* 110:184–95.

UEHARA, T., and R. YANAGIMACHI. 1977. Activation of hamster eggs by pricking. *J Exp Zool* 199:269–74.

VACQUIER, V. D., and G. W. MOY. 1977. Isolation of bindin: the protein responsible for adhesion of sperm to sea-urchin eggs. *Proc Natl Acad Sci USA* 74:2456–60.

WADA, S. K., J. R. COLLIER, and J. C. DAN. 1956. Studies of the acrosome. V. An egg-membrane lysin from the acrosomes of *Mytilus edulis* spermatozoa. *Exp Cell Res* 10:168–80.

WILKES, A. 1964. Inherited male-producing factor in an insect that produces its males from unfertilized eggs. *Science* 144:305–307.

WILSON, E. B. 1928. *The Cell.* New York: Macmillan.

WILT, F. H. 1973. Polyadenylation of maternal mRNA of sea urchin eggs after fertilization. *Proc Natl Acad Sci USA* 70:2345–49.

WILT, F. H., and D. MAZIA. 1974. The stimulation of cytoplasmic polyadenylation in sea urchin eggs by ammonia. *Dev Biol* 37:422–24.

YANAGIMACHI, R. 1957. Studies of fertilization in *Clupea pallasii.* III. Manner of sperm entrance into the egg. *Dobutugaku Zasshi* 32:226–33.

ZANEVELD, L. J. D., S. A. BEYLER, D. S. KIM, and A. K. BATTACHARYYA. 1979. Acrosin inhibitors as vaginal contraceptives in the primate and their acute toxicity. *Biol Reprod* 20:1045–54.

ZANEVELD, L. J. D., and W. L. WILLIAMS. 1970. A sperm enzyme that disperses the corona radiata and its inhibition by decapacitation factor. *Biol. Reprod* 2:363–68.

ZIMMERMAN, A. M., and S. ZIMMERMAN. 1967. Action of colcemid in sea urchin eggs. *J Cell Biol* 34:483–88.

127

9

Sorting the Materials of the Animal Egg— Formation of the Germ Layers

9.1 The Animal Body Plan—Basically a Simple One

Most familiar animals are **triploblastic,** that is, they have body plans in which the components are arranged in three basic layers called **germ layers.** To analyze the triploblastic body plan we first recognize that a part of each organism's body makes contact with the external environment. For simplicity, let us call that part of the body the **outer layer** and schematize it this way.

The early stages in the development of the embryo are marked by the partitioning of the substance of the egg among successive generations of daughter cells produced by mitotic divisions. These cells continue to divide until the new individual, often referred to as the **germ,** comprises a great many cells. But cell division alone does not suffice to transform the substance of the germ into a functioning organism. Rather, the cells must be positioned according to the body plan appropriate to the species and must differentiate the functional characteristics required of each part of the body plan. In this chapter we first consider the patterns in which the materials of various kinds of eggs are segregated by mitotic divisions after fertilization. Next, we describe the gross morphological changes that occur as the cells of the embryo are shifted into positions corresponding to the future body plan. Thereafter, we consider a number of questions related to the mechanisms that determine the patterns in which the materials of the egg are segregated. Finally we turn to the question of the cellular mechanisms underlying the morphogenetic movements that are involved in the realization of the definitive body plan in various animal groups.

The outer layer comprises the skin, central nervous system, and sense organs. It is the covering of the animal, the part that communicates directly with the exterior and that senses the external environment and mediates the reactions of the organism to it.

In the basic body plan we recognize another part whose function it is to receive food and to process it mechanically and chemically so that its high-energy compounds and protoplasmic building blocks are available to the cellular machinery throughout the organism. These functons are carried out by a part of the animal that we shall call the **inner layer.** We recognize the inner layer as the alimentary tract—pharynx, esophagus, stomach, liver, pancreas, and small and large intestines. It originates essentially as a tube that connects the anterior and posterior openings of the outer layer. It may be diagrammed this way.

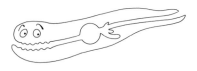

The final aspect of the plan must represent the parts that nourish, support, and move the structures formed by the outer and inner layers. These parts are the vascular system, bone, cartilage, tendon, skeletal and smooth muscle, and so on; they compose what we shall call the **middle layer.** The middle layer can be schematized as occupying the space between the inner and outer layers as shown here.

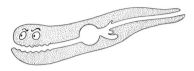

9.1.1 THREE LAYERS ARE FORMED FROM ONE.
Simply stated, it is the task of the egg to make three layers of tissue and to arrange them into inner, middle, and outer regions. It does this according to a construction plan that can be represented with great simplicity, if we forget details for a moment. First, by mitotic cell division the fertilized egg becomes many cells. These then arrange themselves in a fluid-filled ball. If we regard the walls of this ball as a layer, then we have at this time simulated a one-layered animal.

By a relatively simple (simple to draw, at any rate) maneuver, the ball becomes two layered. Imagine that you push in the wall of a dead tennis ball with your index finger; do the same to a fluid-filled ball of cells. The one-layered structure transforms into a two-layered one as shown here. Many animals, such as sponges and coelenterates, have a two-layered, or **diploblastic** body plan.

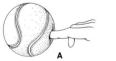

There are several ways in which the three-layered form can arise from the two-layered configuration, but the simplest case to illustrate is the one in which some of the cells from the inner and outer layers escape from their neighbors and invade the

space between the first two layers. The three germ layers have been given names from Greek roots that can be translated into terms similar to those we have already applied to them. The outer layer is **ectoderm** (*ekto* means outside and *derma* means skin); the inner layer is **endoderm** (*endos* means within); the middle layer is **mesoderm** (*meso* means middle).

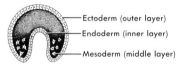

— Ectoderm (outer layer)
— Endoderm (inner layer)
— Mesoderm (middle layer)

We speak of skin, sense organs, and nervous system as arising from ectoderm; of the muscles, blood, and bone as mesodermal derivatives; and of the alimentary tract as originating from endoderm.

The foregoing discussion should suggest that understanding the sequence of morphologic changes that take place in development is not an impossible task—in outline at least. In the following sections we examine the development of representatives from different groups of organisms to learn their patterns of morphological change in achieving the three-layered condition. Thereafter, we analyze the cellular mechanisms underlying some of these morphological changes.

129

9.2 Cleavage and Gastrulation

Mitotic divisions of the embryo partition the substance of the egg among some dozens, hundreds, or thousands of cells, depending on the kind of egg. The partitioning process is called **cleavage.** The cleavage divisions separate daughter cells along planes, **cleavage planes,** that are marked by **cleavage furrows.** The cleavage furrows are the sites of the initial constriction of the cytoplasm during anaphase. The products of cleavage are called **blastomeres.** Blastomeres may be very similar in appearance, or they may be distinctive by virtue of differences in size and in their content of yolk, mitochondria, lipid droplets, and so on.

In most kinds of organisms the period of cleavage comes to an end when the blastomeres are arranged about a central cavity. The embryo is then called a **blastula,** and the cavity is the **blastocoele.** The blastula undergoes mechanical changes that lead to the three-layered condition; during this process, which is called **gastrulation;** the germ is then referred to as a **gastrula.**

The various ways in which protoplasm and yolk are arranged in eggs of different kinds dictate differences in the processes of cleavage and gastrulation. Such differences notwithstanding, the basic plan of cleavage and gastrulation may be learned from certain kinds of eggs that show the processes in relatively uncomplicated fashion; the developmental patterns of other forms can then be analyzed as variations of a basic pattern.

9.2.1 THE SIMPLEST PATTERNS OCCUR IN ECHINODERM EGGS.
Cleavage is described in two members of this group: the first, because its divisions occur with such a precise geometry that it can serve as a model or type from which other patterns of cleavage can be derived; the second, because it has been a favorite organism for the experimental analysis of some phases of development, and because repeated references to it are made later in the book.

9.2.1.1 Cleavage in *Synapta Digitata*.
A degenerate relative of starfishes, brittle stars, sea urchins, and other echinoderms, *Synapta digitata* has an egg that contains very little yolk. Its cleavage was described in the previous century by Selenka, and it

provides the basic pattern that is our starting point in the analysis of cleavage.

After union of the pronuclei, the first mitotic spindle forms with its polar axis at right angles to the animal-vegetal axis of the egg. At telophase the cytoplasm constricts in a furrow that passes directly through the animal and vegetal poles, dividing the egg into two daughter cells of equal size. The first cleavage furrow is said to be meridianal, for it passes through opposite poles of the egg, just as a meridian of longitude passes through north and south poles of our planet (Figure 9.1A).

The daughter cells lie side by side, somewhat flattened against each other. Soon their nuclei prepare for the second cleavage; the mitotic spindles form anew, again with their axes at right angles to the animal-vegetal axis of the egg. In this division, however their axes are also at right angles to the axis of the first cleavage spindle. Cleavage furrows appear simultaneously in each cell and pass through the animal and vegetal poles at right angles to the plane of the first division (Figure 9.1B–D). Two successive meridianal divisions thus produce four blastomeres.

The four blastomeres now divide, but this time the mitotic spindles are aligned parallel to the animal-vegetal axis and, simultaneously in each cell, an equatorial cleavage furrow appears. There are now eight blastomeres (Figure 9.1E). Successive simultaneous divisions form 16, 32, 64, and 128 cells, with division furrows alternating between meridianal and latitudinal (Figure 9.1F–H). This

130

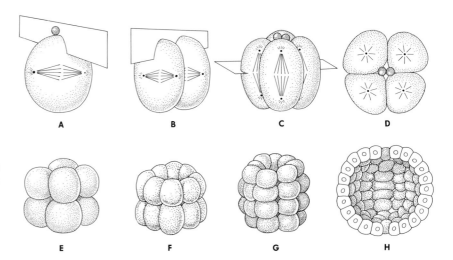

FIGURE 9.1 Cleavage and blastula formation in *Synapta*. Holoblastic cleavage forms blastomeres of equal size, and a symmetric blastula results. Details in text. [Redrawn from various sources after Selenka.]

A   B   C   D

E   F   G   H

results in an arrangement of equal blastomeres in horizontal and vertical rows about a central cavity. The blastomeres shift toward each other at the poles so that there results by the 256-cell stage a hollow sphere composed of a single layer of cells; this is the blastula. The central cavity is the **blastocoele.**

Throughout the cleavage period in *Synapta,* the egg may be bisected through any meridian to give mirror-image halves. The egg is, therefore, radially symmetrical, and the cleavage pattern is often termed **radial cleavage.** Because the entire egg is cut by each cleavage furrow, the term **holoblastic** (entire germ) cleavage is also used.

9.2.1.2 Cleavage in Sea Urchins: Prospective Fates of Blastomeres.
Numerous species of sea urchins (Figure 9.2) are found in coastal waters throughout the world, from just below the tidal zone down to considerable depths. Their eggs are favorite ones for embryologic study, for they can be obtained in great numbers from one species or another during most of the year, and the plan of cleavage and gastrulation shown by many of them is particularly suitable for the experimental analysis of cleavage and of morphogenetic movements. They also constitute especially favorable material for biochemical studies.

Cleavage in most frequently studied sea urchins (Figure 9.3) is radial and holoblastic, as it is in *Synapta.* The pattern starts in the same way, but the third cleavage furrow is usually not at the equator, forming just above it instead, so that, at the eight-cell stage, the four cells at the vegetal pole are slightly larger than those at the animal pole. Then, at the fourth cleavage, there is a radical departure from the classic pattern shown by *Synapta:* the

FIGURE 9.2 The sea urchin *Arbacia punctulata* is common to the coastal waters of southern New England. It is a favorite organism for embryologic research.

upper tier of cells divides meridianally to form a ring of eight cells, but in the lower tier the spindles arrange themselves obliquely so that four small cells are pinched off from the lower inner sides of the larger cells. The small cells are called **micromeres,** and the larger cells are **macromeres.** The cells of the upper tier, intermediate in size, are called **mesomeres** (Figure 9.3E).

Subsequent cleavages proceed more or less synchronously for some time in all cells. As the 32-cell stage is formed, the eight mesomeres are cut by horizontal furrows to give two rings of eight cells each, An_1 and An_2, one above the other; the four macromeres are cut meridianally, however, so that they form a ring of eight cells below An_2 (Figure 9.3E, F). At the sixth cleavage all division planes are

131

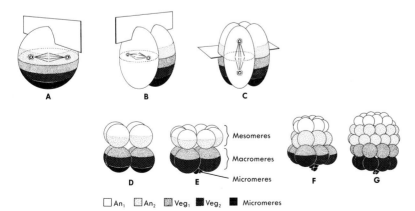

An₁ ☐   An₂ ☐   Veg₁ ▨   Veg₂ ■   Micromeres ■

FIGURE 9.3 Cleavage in the egg of the sea urchin. A, B, C: Positions of the cleavage spindles in the 1-, 2-, and 4-cell stages, respectively. D–G: 8-, 16-, 32-, and 64-cell stages resulting from the third, fourth, fifth, and sixth cleavages, respectively. [Adapted from C. H. Waddington, *Principles of Embryology,* Macmillan, New York, 1956, p. 80.]

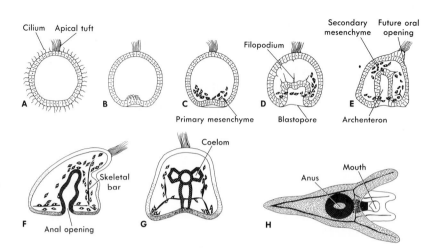

FIGURE 9.4 Gastrulation and formation of the pluteus larva in the sea urchin: (A) blastula; (B) ingression of the primary mesenchyme; (C) the mesenchyme blastula; (D) invagination of the endoderm; (E) continued invagination and formation of the secondary mesenchyme; (F) prism larva viewed from the right side; (G) later prism larva viewed from the oral side, showing coelomic sacs; (H) pluteus viewed from the aboral side.

132

horizontal, so there are formed two rings of eight cells in An_1 and An_2, and two rings of daughter cells from each of the macromeres—the upper one Veg_1, the lower one Veg_2. At the vegetal pole, by this time, are 16 micromeres (Figure 9.3G). In the next division, a meridianal one, the stage of 128 cells is reached. By this time the cells are arranged in the typical configuration of the blastula, that is, about a fluid-filled center (Figure 9.4A).

Division continues, the blastula wall meanwhile remaining one cell thick. The cells then produce motile cilia on their outer surfaces. At the animal pole the cilia are longer and coarser, forming an **apical tuft** (Figure 9.4A). When the cilia are formed, the blastula begins to rotate within the fertilization membrane. Shortly thereafter, as the fertilization membrane is weakened through the action of a hatching enzyme, the blastula emerges from it and swims freely.

Gastrulation then ensues, resulting in a **prism-shaped larva** (Figure 9.4F, G) that rapidly changes form to become a **pluteus** (Figure 9.4H). At the start of gastrulation, the blastula wall opposite the apical tuft flattens, forming a **basal plate.** Cells detach from the center of the basal plate (Figure 9.4C) and move singly into the blastocoele. This process is called **ingression,** and the cells that participate in this process are the descendants of the cleavage micromeres. Once in the blastocoele they are called **primary mesenchyme.** Mesenchyme is a term applied to loosely arranged cells, usually of mesodermal origin.

As ingression proceeds, the remainder of the basal plate, consisting of cells derived from Veg_2 invaginates (L., *in*, in; *vagina*, sheath), that is, it turns inward as a continuous sheet of cells into the blastocoele, pushing toward the animal pole (Figure 9.4D). Invagination of the basal plate creates a new cavity, the **archenteron,** or **primitive gut.** The opening of the archenteron to the exterior is the **blastopore,** which becomes the anal opening of the pluteus. The primary mesenchyme cells form a ring around the base of the primitive gut, accumulating especially on right and left sides in the region of the future ventral portion of the embryo. There they begin the production of skeletal bars, which first appear as triradiate spicules (Figure 9.4E).

Cells at the tip of the invaginating archenteron now begin to spin out **filopodia** (Figure 9.4D) into the blastocoele, making contact with the body wall in the region of the animal pole. Contraction of these filopodia is thought to bring about, in part at least, the final phases in the invagination of the archenteron and its orientation toward the future oral region. Gustafson and Wolpert (1967) indicated that the coelomic sacs (Figure 9.4G) are pulled out as extensions of the archenteron wall by means of similar contractile processes. Eventually, however, and in orderly manner, the archenteron tip cells detach from it as **secondary mesenchyme.** Some of the secondary mesenchyme cells are pigmented and may enter the ectoderm. Others attach to the walls of the esophagus and eventually anchor it to the dorsal side of the larva.

The dorsal ectoderm of the gastrulating embryo grows faster than the ventral (Figure 9.4E, G), giving the larva a prism shape. From the ventral ectoderm an oral invagination meets and fuses with the tip of the archenteron forming the mouth and esophagus and completing the digestive tract. Animals whose embryos form the mouth as a secondary opening into the digestive tract are classified among the **Deuterostomia** (Gr. *deuteros,* second; *stoma,* mouth). Echinodermata and Chordata are the most important deuterostome phyla; others are Chaetognatha, Pogonophora, and Hemichordata.

The prospective fates of the various regions of the echinoderm blastula can be followed by staining with vital dyes (Figures 9.3 and 9.4) or by following the disposition of pigment granules that occur naturally in the subequatorial region of the cortex. The pigment is distributed essentially to cells of Veg_1 and Veg_2. As already noted, Veg_2 contributes the endoderm and secondary mesenchyme. Veg_1 provides the ectoderm that covers the anal side of the pluteus, and the rest of the ectoderm is provided by An_1 and An_2. Thus, at the 64-cell stage the prospective fates of cells in the various tiers can be assigned in a generally way. Without resort to artificial markings, however, right and left, dorsal and ventral, cannot be distinguished by simple inspection until gastrulation, when the primary mesenchyme accumulates in a bilaterally symmetrical pattern near the future ventral side of the embryo (Figure 9.4D, E).

9.2.2 SPIRAL CLEAVAGE IN ANNELIDS AND MOL-LUSKS, AND ORIGIN OF THE TROCHOSPHERE LARVA.
A variation of the radial cleavage pattern is called **spiral cleavage.** It is holoblastic and unequal and is found in many annelids, mollusks, nemerteans, and some flatworms. Spirally cleaving eggs are useful for embryological study because, in many of them, each blastomere can be recognized by its size and position relative to other blastomeres, and its descendants can be followed to their unvarying developmental fates. The lineage of each cell thus can be fully known. This property is found in some other kinds of eggs, also. Those that have it are said to have **determinate** cleavage. Those that lack this property have **indeterminate** cleavage.

An important study of cell lineage was published for the egg of the eastern shoreline mollusk *Crepidula fornicata* by E. G. Conklin, whose system of numbers and letters for designating the descent and fate of each blastomere is still in use in studies of

spirally cleaving eggs. In *Crepidula* (Figure 9.5) the first two cleavages pass through the animal vegetal axis at right angles to each other and produce four large blastomeres called **macromeres.** These are designated A, B, C, and D. Macromere D is larger than the others, which are about equal in size. The third cleavage is quite unequal, for the cleavage planes are such that each of the four macromeres pinches off a small cell, a micromere, at the animal pole. In the formation of this **quartet** of micromeres (designated 1a, 1b, 1c, and 1d), as viewed from the animal pole, each cell is shifted to the right

FIGURE 9.5 Cleavage and gastrulation *Crepidula*. (A–F) views from the animal pole; (G, H) views from the side. The first meridianal division is completed in A, and the first cleavage furrow bends to the right in B. In C, the spindles rotate with their left ends (in the counter-clockwise sense) higher toward the animal pole than their right ends, so that cells A and C are cut off in D at a higher level of focus than B and D. In E, the spindles for the third cleavage form with their upper ends closer to the animal-vegetal axis than their lower ends; the upper ends swing to the right (arrows). In F, small daughter cells, micromeres, are cut off to the right of the lower macromeres. The 29-cell stage is shown in G. In H is shown the late gastrula, with archenteron, A, stomodeum, S, and mesoderm, M, the latter derived from the *4d* cell. [Adapted from E. G. Conklin, *J Morphol* 13:1–226, 1897.]

133

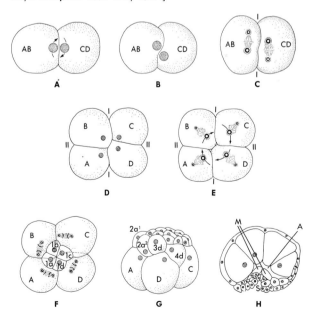

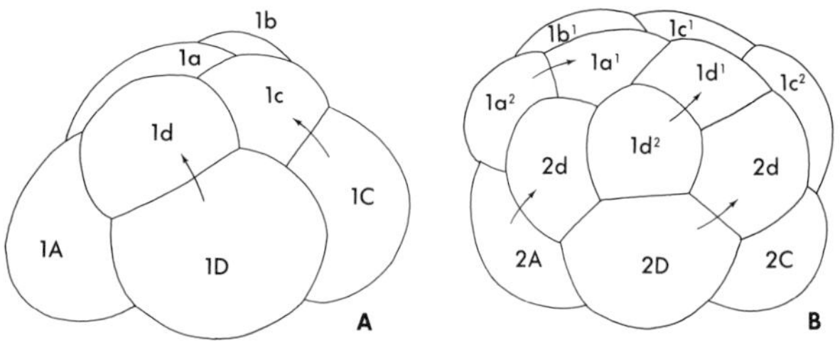

FIGURE 9.6 Cleavage of *Trochus* viewed from the side, the animal pole being directed upward. The D quadrant marks the future dorsal side. A: 8-cell stage. B: 16-cell stage. [Redrawn from various sources, including A. Richards, *Outline of Comparative Embryology*, Wiley, New York, 1931.]

134

(clockwise) of its macromere of origin. Such a view is shown for *Crepidula* in Figure 9.5E and F. At the next division each macromere pinches off a second quartet of micromeres (2a, 2b, 2c, 2d). This time, however, the micromeres are displaced to the left. Shortly thereafter the first quartet divides, producing daughter cells displaced to the left (counter clockwise). The third and fourth cleavages of *Trochus*, another mollusk, are illustrated in side view in Figure 9.6.

At subsequent cleavages the macromeres pinch off successive quartets of micromeres that are displaced alternately to the right and the left. The successively spiraling divisions leave the macromeres with a cap of micromeres at the animal pole (Figure 9.5G). This is essentially what constitutes the blastula stage in *Crepidula* and several other spirally cleaving forms. There is little or no blastocoele, except in embryos with small A, B, C, and D cells. Among annelids, mollusks, and other forms with spiral cleavage there is a considerable variation in the size ratio between micromeres and macromeres (Figures 9.5 and 9.6), but comparative studies of cell lineage show that the successive quartets are formed in the same way and, with but minor variations, corresponding blastomeres have the same developmental value in different kinds of organisms.

Gastrulation in annelids and mollusks leads to the formation of a free-swimming larva called a **trochophore** or, sometimes, a **trochosphere.** The trochophore of *Patella,* a gastropod mollusk, is shown in Figure 9.7A and B. That of an annelid

Arenicola is illustrated in Figure 10.4. In most mollusks one recognizes a more mature larval form, the **veliger,** which is derived from the trochophore. An early veliger stage of *Crepidula* is illustrated in Figure 9.7C. In both kinds of organisms, gastrulation is accomplished by the overgrowth of the macromeres by the micromeres. This process is called **epiboly.** The macromeres shift to fashion a small archenteron (Figure 9.5H). The endoderm is formed from the macromeres with contributions from three of the fourth quartet of micromeres, 4a, 4b, and 4c. The first three quartets contribute the ectoderm and the third also contributes some larval mesoderm, but the major part of the mesoderm arises from the 4d cell (Figure 9.5G). This cell is budded off precociously beneath overlying micromeres of the third quartet and shifts inward dividing to form right and left stem cells that come to lie posterior to the stomodeum, which forms in the region of the blastopore (Figure 9.7). Later, its descendants produce the mesoderm of the body cavity, gonads, and musculature. The fate of the 4d cell is considered further in Chapter 10 in another context.

Let us note that, in contrast to the echinoderms, which are forms in which the blastopore becomes the larval anus, the blastoporal region of the annelids, mollusks, and similar forms becomes the mouth. The anus develops as a secondary opening posteriorly from the archenteron. Animal phyla in which the mouth is closely associated with the primary opening of the archenteron are grouped together as **Prostomia** (Gr., *protos,* first; *stoma,* mouth). They include most of the major invertebrate groups.

9.2.3 ORIGIN OF THE PRIMARY ORGAN RUDIMENTS IN AMPHIBIA.
Amphibian eggs are usually somewhat more than a millimeter in diameter and are encased in a multilayered jelly coat that is exterior to the vitelline membrane. The cortical cytoplasm is densely populated with dark pigment granules in the animal half of the egg. The vegetal half is lightly pigmented and contains dense masses of yolk (Figures 9.8 and 9.9).

At fertilization the cortical alveoli break down, the vitelline membrane is lifted off to form the fertilization membrane, and a perivitelline space is formed. Within this space the egg is free to rotate, and it does so with the heavier vegetal pole assuming the lower position. The second polar body then

FIGURE 9.7 Molluskan larvae: (A): an early trochophore of *Patella*, as seen in saggital section; (B): a later stage in the development of the trochophore of *Patella* as viewed from the oral side; (C): early veliger larva of *Crepidula*, as seen in saggital section. The mesoderm cell seen in A is one of the daughter cells of 4*d*, one going to the right side of the larva and one to the left. This pair of cells gives rise to the mesodermal bands seen on right and left sides of the oral opening in B. The comparable cells in *Crepidula* give rise to the suboral mesodermal cells of the foot region shown in C. [A and B redrawn from various sources, including E. Korschelt and K. Heider, *Textbook of Embryology of the Invertebrates IV*, English translation, Macmillan, New York, 1900. C adapted from E. G. Conklin, *J Morphol* 13:1–226 (1897).]

135

FIGURE 9.8 Rotation-induced localization of the gray crescent. An interpretative scheme depicting the "rotation of symmetrization" as described by Ancel and Vintemberger [*Bull Biol France Belg* 31:1–182 (1948)]. A: The unfertilized egg is restrained within its jelly in such a way that the animal–vegetal axis is rotated 135° with respect to the vertical. B: Following artificial activation and formation of the perivitelline space, the egg is free to rotate under the influence of gravity, with the unpigmented heavier yolky region swinging downward. C: The gray crescent arises around that portion of the border between light and dark regions of the egg cortex that was highest on the inclined egg. The plane in which the animal–vegetal axis moves coincides with the median plane of the future embryo. A, animal pole; V, vegetal pole.

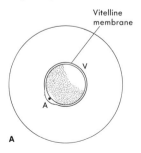
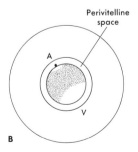
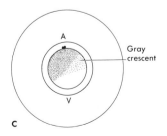

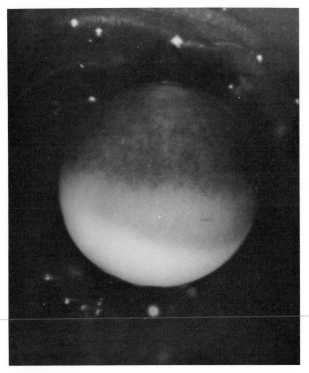

FIGURE 9.9 The gray crescent in the freshly spawned egg of *Ambystoma mexicanum*. The photograph shows an unfertilized egg in which the gray crescent was induced to appear by heating at 35–36°C for 10 minutes. [From H. H. Benford and M. Namenwirth, *Dev Biol* 39:172–76 (1974), by courtesy of the authors and permission of Academic Press, Inc., New York.]

136

appears as a minute black dot at the upper, or animal pole, denoting completion of the second meiotic division (Figure 9.8).

9.2.3.1 Appearance of the Gray Crescent.

About 1½ to 2 hours after insemination in frog eggs a cortical change appears on one side of the egg. This change results in a diminution in intensity of the pigmentation in a crescentic zone at the margin between the densely pigmented animal half of the egg and the relatively unpigmented vegetal half. This zone is referred to as the **gray crescent** (Figures 9.8, 9.9, and 9.10). In eggs of the axolotl *Ambystoma mexicanum* the crescent appears after 4 hours or more (Malacinski et al., 1975). The presence of a gray crescent can be discerned only with difficulty in some batches of eggs and is variable with respect to its appearance or nonappearance in eggs of dif-

ferent species. The presence of the gray crescent makes the egg a bilaterally symmetrical structure before cleavage begins. In eggs of anurans the first cleavage usually passes through the center of the gray crescent or within a few degrees of the center.

The gray crescent is an important landmark. It defines the future dorsal side of the embryo, the point at which the blastopore forms as gastrulation begins and the direction of the longitudinal axis of the embryo. The origin of the crescent is thus of some importance, but the factors that determine its position are not clear, despite the fact that a great deal of research has been devoted to this topic (reviewed by Brachet, 1977). If sperm are applied locally to one side of a frog egg, the gray crescent appears opposite the point of sperm entry (Ancel and Vintemberger, 1948). As Elinson (1975) showed, activation of the egg by pricking or by contact with the sperm causes a symmetrical contraction of the cortex over the entire pigmented portion of the egg in *Rana pipiens*. This contraction gradually relaxes during the next hour. Meanwhile the contraction has pulled the sperm closer to the animal pole, and it moves to meet the female pronucleus, which migrates toward the center of the egg. Conceivably these combined movements could issue in an asymmetric shift of yolky material opposite the point of sperm entry, as suggested by some (see Brachet's 1977 review). Also, the position of the gray crescent can be determined by gravity.

FIGURE 9.10 Cleavage of the amphibian egg. The uncleaved zygote is seen from the left side in A and from the future posterior side in B. In C, the first cleavage is shown to bisect the gray crescent. D: 4-cell stage. E: 8 cells. F: 16 cells. The early blastula is shown in exterior view in G and with one side cut away in H.

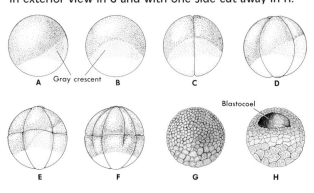

Ancel and Vintemberger (1948) mounted unfertilized frog eggs in their jelly membranes with the animal pole tilted 135° from the vertical and then stimulated them parthenogenetically by means of electric shock. Activation was followed by rotation of the egg within the perivitelline space, the heavier vegetal pole swinging downward (Figure 9.8). In each case the gray crescent formed around that part of the marginal zone between dark and light pigment that was highest on the inclined egg before rotation began (see also Kirchner et al., 1980; Chung and Malacinski, 1980).

On the other hand, there is evidence suggestive of a prelocalization of factors controlling the position of the gray crescent. Urodele eggs are normally polyspermic and all sperm move within the egg cytoplasm. Yet, a single gray crescent forms. Moreover, unfertilized eggs of *Ambystoma mexicanum* show a distinct gray crescent (Figure 9.9) after being placed in a water bath at 35–37° C for 10 minutes. (Benford and Namenwirth, 1974).

The gray crescent has long been considered to play a significant role in the organization of the embryonic body. This concept is discussed in Chapter 13.

9.2.3.2 Formation of the Blastula.

Amphibian eggs cleave holoblastically beginning with two equal meridional divisions followed by an unequal latitudinal division near the animal pole, as shown as Figure 9.10. The fourth cleavage, a meridional one is completed more rapidly in the upper tier of cells than in the yolk-filled cells of the lower tier.

Subsequent divisions occur rapidly and with a high degree of synchrony in cells of the animal half of the egg, but in the vegetal half they occur more slowly, so that, at any stage of cleavage, vegetal cells are larger than those in the animal half.

By the time cleavage has produced several hundred cells, the blastocoele, or segmentation cavity, is a relatively large cavity located nearer to the animal pole (Figure 9.10H). Because it is nearer the animal pole, it is roofed by small, rapidly dividing cells, but its floor is made up of larger yolk-filled cells. In accommodating to this distribution of cells, in the amphibian a pattern of gastrular movements different from that of other forms has evolved.

9.2.3.3 What Does Gastrulation Lead to?

In amphibians, gastrulation shifts the cells that will form the future larval organs into positions corresponding to their prospective fates, but this occurs without the functional differentiation of these cells. This is in contrast to many marine invertebrates, in which gastrulation essentially accomplishes the morphogenesis of a free-swimming organism with recognizable and functional larval organs. In order to understand better what is involved in gastrulation in the amphibian, therefore, it is perhaps better to examine first its consequences as expressed in the appearance of the next developmental stage after gastrulation, namely the **neurula** stage.

The body plan of an early neurula stage is shown in Figure 9.11. After gastrulation the ectoderm of the dorsal side thickens, forming the **neural plate.** The sides of the neural plate are then ele-

137

FIGURE 9.11 Neurula stage of the anuran amphibian. A: As seen in dorsal view, with planes of section for B and C shown by dashed lines a and b, respectively.

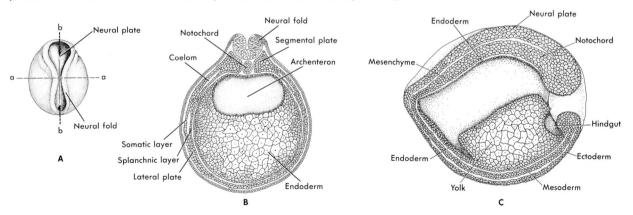

vated and turn in, meeting at the midline to form a tube. From this tube the central nervous system later forms. Ventral to the neural tube, mesodermal cells are arranged in a rod, the **notochord,** which is a primitive supporting rod running the length of the embryo. Lateral to the notochord, the mesoderm forms a band of tissue, the **segmental plate.** This is continuous with the **lateral** plate, which surrounds the inner mass of endoderm. The segmental plate subsequently becomes organized into blocks of cells, **somites,** arranged in a longitudinal row on each side of the neural tube. These give rise to most of the skeleton and to the skeletal musculature of the body. The lateral plate separates into the somatic mesoderm and the splanchnic mesoderm. The former, together with the overlying ectoderm, is termed the **somatopleure.** Somatic mesoderm contributes to the appendicular skeleton and connective tissues of the appendages and body wall and, together with ectoderm, forms the major part of the integument and its derivatives. The splanchnic mesoderm, together with the endoderm, constitutes the splanchnopleure, perhaps better visualized in forms with less intraembryonic yolk than the amphibian (see Figure 11.46). The splanchnic mesoderm invests the visceral organs, supplying their connective tissues and vascular elements. The bulk of the neurula internally is made up of yolky endoderm cells which form the lining of the digestive tract and give rise to the major digestive glands. At the anterior end of the neurula the neural folds are farther apart and larger. Posteriorly, they are lower and eventually merge with material around the closing blastopore. From this zone of confluence there will develop essentially all the postanal portion of the body and the tail. These matters are reviewed and extended in Chapter 11.

9.2.3.4 Dye Marks Are Used to Follow Gastrular Movements. To understand the pattern of cellular movement during gastrulation, it is helpful to know the position in the blastula of the presumptive germ layers. **Fate maps** for blastulas of a number of kinds of amphibians have been made by following the disposition of marked cells during gastrulation and neurulation. Groups of cells on the surface of the blastula are marked, after removal of the jelly, by applying small films of agar or cellulose acetate impregnated with any of several basic vital dyes, such as nile blue, neutral red, or bismark brown. The dyes bind strongly to the pigment and

138

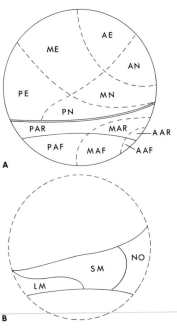

FIGURE 9.12 Fate maps for the beginning gastrula of *Xenopus laevis* as viewed from the prospective left side. A: The fates of cells in the superficial layers are demarcated by dashed lines, the site of the future blastopore is denoted by a solid line, and the limit of endodermal involution is indicated by a double line. Note that cells for forming only ectodermal and endodermal structures are in the superficial layers as gastrulation begins. AE, ME, and PE: prospective anterior, middle, and posterior epidermal regions. AN, MN, and PN: prospective anterior, medial and posterior neural areas. AAR, MAR, and PAR: anterior, middle, and posterior regions of the archenteron roof. AAF, MAF, and PAF: corresponding levels of the floor of the archenteron. B: Fate map for the mesoderm; the dashed line indicates that the cells in question are located below the ectodermal surface. The central yolk mass is not shown. The approximate distribution of cells forming head mesoderm and notochord is denoted by NO, that of future somite cells by SM, and that of prospective lateral mesoderm by LM. [Adapted from R. Keller, *Dev Biol:* A, 42:222–41 (1975); B, 51:118–37 (1976).]

lipid granules in the cells, creating long-lasting marks. If dyes of different colors are used, the subsequent fates of several regions of the blastula may be followed simultaneously.

Recent studies of gastrulation in *Xenopus laevis* reveal that the prospective ectodermal organs—

body epidermis and nervous system—are derived from the darkly pigmented cells of the animal half of the embryo plus the superficial cells of the **marginal zone,** a more lightly pigmented circumferential band that extends below the equator. Cells of the remaining portion of the embryo give rise to the roof and floor of the archenteron, the primitive gut (Figure 9.12A). Cells of the prospective mesoderm lie deep to the future ectodermal cells of the marginal zone, sandwiched between them and the central yolk mass (Figure 9.12B).

The first external sign of gastrulation in most amphibians is the appearance of a darkly pigmented crescentic groove in the marginal zone at the future dorsal side of the embryo. Through this groove endoderm **involutes** (turns in as a coherent cellular sheet) from both above and below. The groove deepens, and its lateral reaches are prolonged around the marginal zone, eventually meeting on the opposite side of the embryo (Figure 9.13). The groove thus encircles the lightly colored vegetal region, forming a **yolk plug,** which is progressively internalized and disappears. As Figure 9.13 shows, there is essentially no open blastopore in amphibian embryos. The blastopore is, in fact, constituted by the circular groove just described. The site at which the groove first appears is referred to as the **dorsal lip of the blastopore.** As the blastopore extends around the marginal zone, one refers to its **lateral lips** and then to its **ventral lip.**

The archenteron is formed by invagination at the blastoporal groove. The groove is at first a slit-like cavity lined with future endodermal cells and having prospective mesoderm and ectoderm, first dorsal and lateral to it, and eventually, ventral to it also. As gastrulation proceeds, the archenteron enlarges and eventually displaces the blastocoele, or **segmentation cavity,** as it is sometimes called (Figure 9.13).

As the prospective endoderm is internalized and forms the archenteron, the ectoderm expands, spreads toward the vegetal pole, and eventually covers the entire embryo. The expansion of the ectodermal layer over the internal mass is called **epiboly.** Epiboly requires a considerable expansion and rearrangement of the cells of the animal half of the egg and of the superficial ectoderm of the marginal zone. Even as the ectoderm expands, however, in the region of the marginal zone, it must likewise undergo shrinkage in the region of the

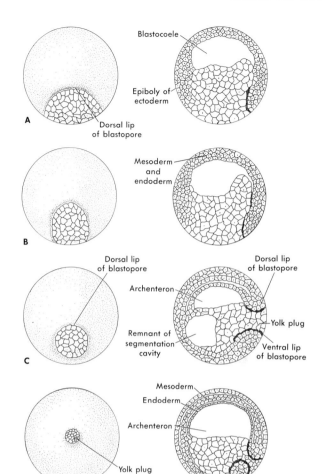

FIGURE 9.13 Stages of gastrulation in the embryo of an anuran. A–D (left): The sequence of events in surface view from the future postero-dorsal angle. On the right the internal changes are shown for sagittal sections of the corresponding embryos viewed from the left side. [Adapted and modified from A. F. Heuttner, *Fundamentals of Comparative Embryology of the Vertebrates*, Macmillan, New York, 1949, pp. 78, 79.]

139

blastopore as the yolk plug is internalized and the lips of the blastopore converge.

Epiboly likewise carries the mesoderm of the marginal zone toward the vegetal pole. This direction of movement is contrary to that required for the interposition of a mesodermal layer between endoderm and mesoderm anteriorly. Counteracting this rearward movement, the mesoderm undergoes involution, turning inward and forward between endoderm and ectoderm first dorsally, then laterally and ventrally. Having undergone involution the forward-streaming mesodermal cells tend to converge dorsally and laterally in anticipation of the formation of the notochord and somites (Figure 9.14); see also Figure 9.11B).

In anuran amphibians, as shown in Figure 9.13 and as treated further in Section 9.4.2.2 (Figure 9.67), the pattern of gastrulation movements is such that the entire archenteron is lined with endoderm. In urodeles, however, less endoderm is carried forward on the dorsal side during gastrulation so that the roof of the archenteron is initially mesodermal. The mesodermal roof is, of course, continuous with the endodermal sidewalls and floor, but these elements gradually disjoin, and the freed endodermal edges converge to the dorsal midline and fuse (Figure 9.15), thus completing the triploblastic body plan.

9.2.4 EARLY DEVELOPMENT IN FISHES AND BIRDS.

These vertebrates have eggs with a great deal of yolk and proportionately very little protoplasm. The nucleus and cytoplasm of the fertilized egg are confined essentially to a disc at the animal pole, the **blastodisc,** and the cleavage divisions are restricted to the nucleated zone without involving the yolk. The cleavage pattern is often referred to as **dis-**

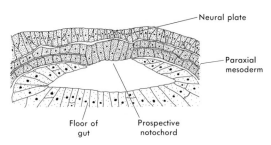

FIGURE 9.15 Semidiagrammatic illustration of a cross section through the dorsal region of a beginning urodele neurula showing initiation of the neural plate and convergence of lateral endoderm toward the dorsal midline to form the roof of the gut.

coidal. It is also called **meroblastic** (Gr., *meros,* part), meaning that cleavage cuts through only a part of the germ. Within these three vertebrate groups, the early development of birds, especially of the chicken, has been most studied, and thus will receive a larger proportion of attention in what follows. The development of bony fishes, especially of the brackish water minnow *Fundulus heteroclitus* and several species of trout (*Salmo*), has been the experimental subject of many studies, notably by Oppenheimer, Trinkaus, Devillers, Ballard, and others.

9.2.4.1 Cleavage and Gastrulation in Teleost Fishes.

In eggs of teleosts mitotic division of the blastodisc produces a layer of cells, the **blastoderm,** which becomes elevated in the form of a hemispherical mass rising above the yolk at the animal pole (Figure 9.16A). During gastrulation the epibolic spread of the blastoderm over the spherical yolk occurs at essentially a uniform rate along all meridians. As epiboly progresses, the embryonic body appears along one meridian (Figure 9.16G). By the time the advancing edges of the blastoderm meet at the vegetal pole, the principal organ rudiments of the embryo are well defined: regional differentiation of the nervous system has occurred; the sense organs have begun development; and the mesoderm is organized in somites corresponding to the segmentation of the future axial musculature.

Before fertilization the egg cytoplasm is distributed over the yolk in a thin cortical layer. When the egg is activated, most of the cytoplasm streams to the animal pole forming the blastodisc (Figures 9.16B and 9.17A). Elsewhere the yolk mass re-

140

FIGURE 9.14 Direction of mesodermal movements in an anuran gastrula shown with ectoderm removed. A: Gastrula viewed facing the blastopore. B: Viewed from the future left side of the embryo. Heavy lines indicate the involution of mesoderm, and the corresponding thin lines show the convergence of mesodermal cells toward the future dorsal side of the embryo. [From A. F. Heuttner, *Fundamentals of Comparative Embryology of the Vertebrates*, Macmillan, New York, 1949, as adapted and modified from papers of Vogt.]

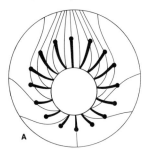 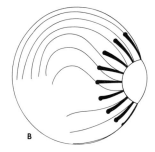

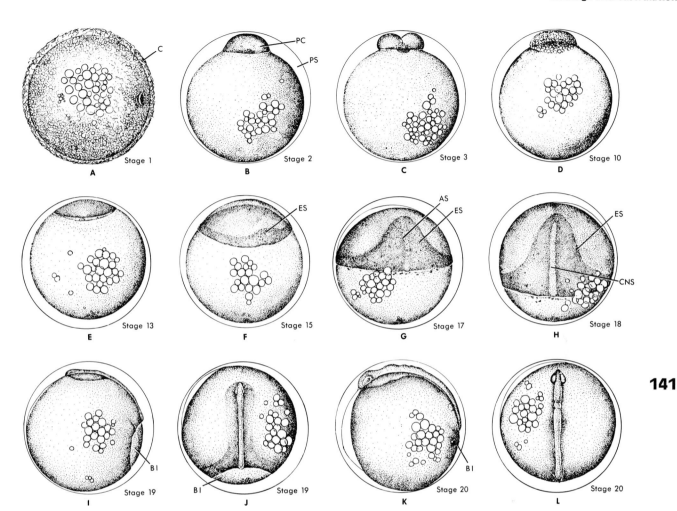

FIGURE 9.16 Cleavage, gastrulation, and embryo formation in *Fundulus heteroclitus*. Stages are those arbitrarily designated by Armstrong and Child (1966). A: Unfertilized egg showing fibrous chorion, C, and oil droplets within the yolk mass. B: The fertilized egg, showing the protoplasmic cap, PC, and perivitelline space, PS. C: First cleavage. D: High blastula. E: Beginning gastrulation. F–H: Embryonic shield, ES, in successive stages of growth; the future central nervous system, CNS, derived from the axial strand, AS, seen in G, becomes evident in H. The emergence of the body form and closure of the blastopore, Bl, occur progressively and are completed by stage 20. I–L: These events in profile (I, K) and in dorsal view (J, L). [From P. B. Armstrong and J. S. Swope, *Biol Bull* 128:143–68 (1966), by permission of the authors and the *Biological Bulletin*.]

mains covered with a thin layer of cytoplasm, the **yolk cytoplasmic layer.** Electron micrographs (Lentz and Trinkaus, 1967) show that this layer is separated from the yolk by a thin membrane continuous with the membrane that underlies the blastodisc. In the latter location it is highly convoluted, thus presenting a greater surface area that presumably facilitates exchange of materials between yolk and cytoplasm (Figure 9.17A).

The furrows of the first several cleavage divi-

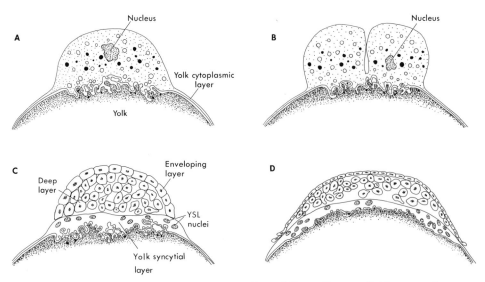

FIGURE 9.17 Origin of the blastoderm, formation of the yolk syncytial layer, and relationships between these components through the stage of beginning gastrulation in *Fundulus heteroclitis*. A: The protoplasmic cap before the onset of cleavage (cf. Fig. 9.16A, B). Note that the protoplasmic cap is continuous with the yolk cytoplasmic layer, which surrounds the egg and has its own external and internal limiting membranes; and that the yolk compartment is isolated within its own limiting membrane; this compartment interdigitates extensively with the protoplasmic compartment above. B: Section through two cells of a 4-cell blastoderm. As the cleavage furrow does not extend through the protoplasmic cap, the blastomeres are still joined by a protoplasmic bridge; the yolk cytoplasmic layer remains continuous with the margin of the blastomeres. C: Section through a blastoderm at stage 8. Tangential cleavages have segregated the blastoderm proper from the yolk cytoplasmic layer, leaving it as the yolk syncytial layer, which separates the blastomeres from the yolk. The blastoderm is now organized into a coherent outer layer, the enveloping layer, which terminates at the margin of the periblast. Beneath this layer the deep blastomeres remain loosely organized and relatively mobile within the segmentation cavity. The latter is bounded by the enveloping layer and the periblast membrane. D: Diagram of a beginning gastrula, stage 12. The blastoderm retains the same basic organization, but is flattened, and cells of the deeper layers are more closely united. Epiboly of the periblast has begun as nuclei migrate into the yolk cytoplasmic layer, and this is followed by epiboly of the blastoderm, which is attached to the periblast only by marginal cells of the enveloping layer [A, an interpretation based on observations of T. Lentz and J. P. Trinkaus; B–D redrawn with modifications from T. Lentz and J. P. Trinkaus, *J Cell Biol* 32:121–38 (1967).]

142

sions occur synchronously and pass through the blastodisc in a direction perpendicular to the surface of the yolk. They give rise to a fairly regular and predictable pattern (Figure 9.18). The vertical furrows do not, however, extend all the way to the yolk (Figure 9.17B), thus leaving an uncleaved basal layer at the zone of contact with the yolk. When cleavages parallel to the surface of the yolk begin to occur, uncleaved basal cytoplasm is cut off

to form a multinucleated syncytial layer that is continuous with the yolk cytoplasmic layer (Figure 9.17C, D). This syncytium is referred to as the **yolk syncytial layer,** or **YSL.** Nuclei in the YSL continue to divide, and some of them migrate into the yolk cytoplasmic layer. During epiboly these nuclei, many of them polyploid (Figure 9.19), proceed into the yolk cytoplasmic layer in advance of the edge of the blastoderm, thus creating the **external yolk**

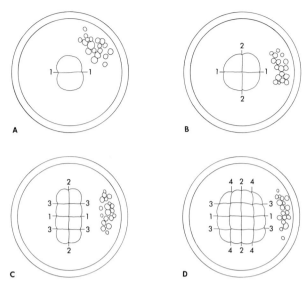

FIGURE 9.18 An analysis of the cleavage pattern in *Fundulus heteroclitus*. A–D: The products of, respectively, the first, second, third, and fourth cleavages as seen from the animal pole. [Based on studies of J. Oppenheimer, *J Exp Zool* 73:405–44 (1936).]

syncytial layer, or E-YSL. The portion of the YSL beneath the blastoderm is the **internal yolk syncytial layer,** or **I-YSL.**

The blastoderm, as distinguished from the YSL, now becomes organized as a blastula. The outermost layer of the blastula is a tightly coherent sheet, one cell layer in thickness, attached to the YSL around its entire margin. This layer remains throughout blastula and gastrula stages. Because of its position relative to deeper cells of the blastoderm, it is called the **enveloping layer** [also referred to in many English-language accounts as the **Deckschicht** (German) or as the **couche enveloppant** (French)]. Beneath the enveloping layer are rounded or lobular deep blastomeres separated from each other and from the enveloping layer by spaces of varying dimensions. These spaces are the remnants of the cleavage furrows that have merged and form collectively the segmentation cavity (Figure 9.17C, D). The blastula reaches its climax at the stage of high blastula (Figure 9.17C). The high blastula has its contours well elevated above the yolk. It flattens as the morphogenetic movements of gastrulation begin (Figure 9.17D).

As the blastula flattens, it begins to follow the

E-YSL in epiboly over the surface of the yolk. Regional differentiations in the blastoderm now become evident. These are the **germ ring** and the **embryonic shield.** The germ ring rims the blastoderm and consists of the margin of the enveloping layer, deeper cells called **epiblast,** and a diffuse **hypoblast.** The embryonic shield is a shield-shaped

FIGURE 9.19 Blastoderm margin and nuclei of the yolk syncytial layer in the early embryo of the muskellunge. A: The coherent cells at the edge of the enveloping layer are seen above in the relationship to the advancing YSL. A gigantic, presumably polyploid, nucleus (arrow) is seen in the YSL along with other YSL nuclei that are usually of smaller diameter than those seen in cells of the blastoderm. B: Gigantic and smaller YSL nuclei seen at higher magnification. (A from W. Bachop and J. W. Price, *J Morphol* 135:239–46 (1971), by courtesy of Dr. Bachop and permission of Alan R. Liss, Inc., B, an unpublished photograph kindly supplied by Dr. Bachop.]

143

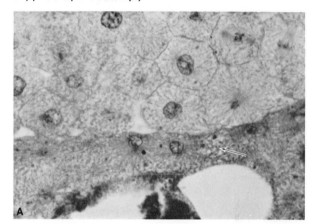

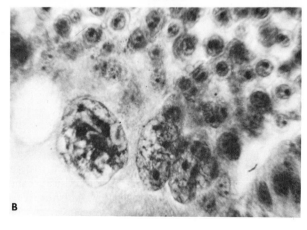

zone that is continuous with the germ ring at the future posterior side of the embryo. As the blastoderm advances over the yolk, the embryonic axis becomes defined within the embryonic shield, its parts appearing in anteroposterior sequence. The first embryonic structure to emerge is the **axial strand** (Figure 9.16G, H), which is constituted of cells destined to form the neural tube and notochord. The axial strand arises from materials that converge upon the midline beneath the enveloping layer to form a solid **keel**. This later separates into a ventral notochord and overlying neural strand. The lumen of the neural tube arises secondarily by virtue of an internal split that begins anteriorly and extends posteriorly. Elongation of the neural tube and notochord occurs entirely from material within the shield. Materials for the future gut are also localized within the shield and are incorporated into the body as it elongates. Meanwhile, the somites and paraxial mesoderm arise by convergence of the deeper cells of the germ ring, as epiboly continues.

The germ layers do not arise by invagination but by rearrangements of the cells beneath the enveloping layer. At the high blastula stage, the cells beneath the enveloping layer are separate from each other and migrate somewhat randomly, forming and retracting blunt protrusions called **lobopodia**. As the high dome of the blastodisc flattens and becomes more uniform in thickness in preparation for gastrulation, the cells become more coherent. They cease their migration and form a consolidated blastodisc arched over a shallow subgerminal cavity, whose bottom is formed by the I-YSL. Maps depicting the prospective fates of different regions of the blastoderm at this stage have been made (Figure 9.20) by Ballard (1973) for the trout *Salmo gairdneri*.

The onset of epiboly is marked by the onset of a process of disengagement of the deep cells from the under surface of the blastodisc. These disengaged cells accumulate in increasing numbers in a crescentic area at the future posterior end, where the embryonic shield appears, and peripherally to form the germ ring. This movement leaves the central area of the blastoderm to consist only of the enveloping layer and a thin layer of epiblast. In the shield and germ ring the disengaged cells constitute the hypoblast. The cells that form the endoderm disengage principally from the posterior quadrant of the blastodisc and form initially a single layer of cells

144

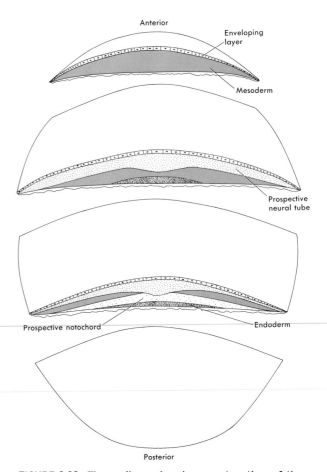

FIGURE 9.20 Three-dimensional reconstruction of the blastodisc of *Salmo gairdneri* shown as sliced transversely. [Redrawn with modifications from W. W. Ballard, *J Exp Zool* 184:49–74 (1973).]

beneath the embryonic shield. These cells are incorporated into the emerging body as epiboly progresses.

Epiboly is marked by a distinctly different behavior of the epiblast and hypoblast. The former spreads in epiboly, drawn along by the attachment of the enveloping layer which itself attaches to the YSL. The epiblast contributes no cells to the hypoblast during epiboly. The hypoblast of the germ ring, however, shows two component movements. On the one hand, the cells spread epibolically over the yolk, but at the same time they move posteriorly, converging upon the embryonic shield (Figures 9.21 and 9.22) and adding successively more posterior portions of the body.

9.2.4.2 Cleavage and Gastrulation in Hen's Egg. Discoidal cleavage is shown by avian eggs, as in teleosts, but details of cleavage, epiboly, and embryo formation differ considerably. When ready for fertilization, the egg lies high in the oviduct unencumbered by secondary membranes (cf. Figure 7.14). It consists of a spherical yolk mass, which may be as much as several centimeters in diameter in the case of large birds. At the animal pole of the sphere is an island of protoplasm, 3.5–4 mm in diameter in the case of the chick, containing the female pronucleus. This protoplasmic region is called the **blastodisc** or the **germ.** It overlies a zone of liquefied yolk and, after proper fixation, may be removed intact from the yolk sphere during the early cleavage stages (Eyal-Giladi and Kochav, 1976). Yolk and germ are encased in a relatively tough and

rather complex fibrous vitelline membrane whose structure has been analyzed by Bellairs et al. (1963) and others. No circumferential perivitelline space is found after fertilization. As the germ divides and forms a **blastoderm,** the attachment of the protoplasmic zone to the vitelline membrane is progressively lost except where the leading edge of the blastoderm binds tightly to its under surface during epiboly.

Fertilization occurs in the upper region of the oviduct, and cleavage begins as the egg moves towards the cloaca, successively acquiring its coverings of albumin, shell membranes, and shell on the way. In laying hens, the first cleavage is initiated about $5\frac{1}{2}$ hours after the previous egg was laid.

The early cleavages are in planes perpendicular to the yolk surface (Figure 9.23). The position of the first furrow is often eccentric, and its direction usually has no bearing on the direction of the future

FIGURE 9.21 Movements of deep cells of the germ ring during early stages of gastrulation in *Salmo*. A: The margin of the blastoderm of an embryo at arbitrary stage 7 is represented within the margin of a blastoderm at stage 8. B: The contour of the blastoderm at stage 8 is represented within the contours of blastoderm at stage 9. In embryos at stages 7 and 8, clumps of finely divided particles of colored chalk were deposited among and on the surfaces of deep cells at the margin of the germ ring at various positions designated by small circles; the disposition of the chalk particles was then examined at stages 8 and 9, respectively. Arrows show the predominant direction of initial shifting of the chalk particles, and stippling indicates the approximate extent to which the particles comprising each inserted clump were displaced. Note that marks placed in the region of the developing embryonic shield are shifted posteriorly, whereas those placed in other regions of the germ ring shift circumferentially toward the region of the shield. [Redrawn with modifications from W. W. Ballard, *J Exp Zool* 184:27–48 (1973).]

145

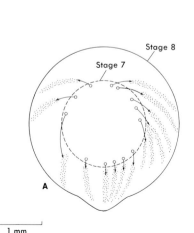
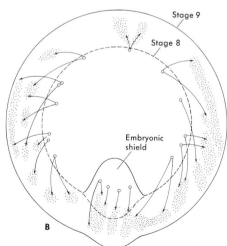

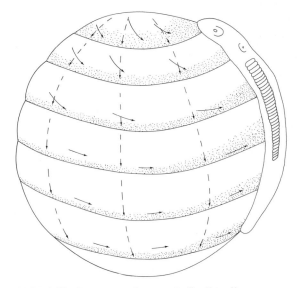

FIGURE 9.22 Summary diagram indicating the functional distribution of epiblast and deep cells of the germ ring during epiboly and embryo formation in the trout. Epiblast cells (dashed arrows) undergo epiboly, descending toward the blastopore while attached to the periblast. Deeper cells descend with the germ ring but simultaneously converge toward the body axis, contributing to it as they do so. [Redrawn with modifications from W. W. Ballard, *J Exp Zool* 184:27–48 (1973).]

146

embryonic axis. Succeeding divisions have been reported as synchronous (Emanuelsson, 1965), but most observers have noted little regularity in their timing and pattern. They occur rapidly, at first, passing only partially through the germ but subsequently becoming delimited basally as well. Tangential cleavages soon begin, and, after about 5 hours, 250–300 closed blastomeres are visible on the upper surface of the germ. At the time, however, only 80–90 ventrally closed cells can be seen. These are centrally located in the germ. By 10–11 hours after the start of cleavage, the entire cytoplasmic mass of the germinal disc is cleaved on both upper and lower surfaces and thus constitutes a true blastoderm. As the cells close ventrally, a **subgerminal space** progressively extends to the periphery of the blastoderm (Figure 9.24A). The marginal cells are completely closed, and no yolk syncytial layer is formed in avian eggs, as in the case of eggs of teleosts.

After cellulation is complete, the cells of the upper surface divide intensively, becoming much smaller and thus quite distinctive from large slower-dividing yolk-filled cells of the lower surface. The latter cells then begin to detach from the upper tier of cells (Figure 9.24B) first near the future posterior side of the blastoderm, and progressing anteriorly. This leaves a somewhat translucent central region of the blastoderm, the **area pellucida,** surrounded marginally by a more opaque zone, the **area opaca** (Figures 9.24C, D, and 9.25A). The detached cells, individually and in clusters, come to rest on top of the yolk at the bottom of the subgerminal activity. The fate of these large cells is not clear. At the time of egg laying the lowermost cells of the blastoderm are detached and the upper layer has organized a coherent epithelium (Figure 9.24E–G). From its lower side, beginning posteriorly, the epithelium proliferates a meshwork of hypoblast cells. The overlying cells become organized as an epiblast of columnar epithelium (Figure 9.24F–H). The hypoblast becomes a continuous layer, thicker posteriorly than anteriorly, as can be seen through the transparent epiblast. This creates the appearance of the embryonic shield, as illustrated in Figure 9.25A. As first formed, the hypoblast does not extend to the margin of the area pellucida (Figure 9.24C), but shortly after egg laying, a bridge of cells forms between the posterior margin of the hypoblast and the area opaca. In this bridge a new structure arises. It is the **primitive streak** (Figures 9.25B and 9.26).

The primitive streak is significant for the formation of the embryonic endoderm and the mesoderm, as is described here. After it appears (Figure

FIGURE 9.23 Sequence of early cleavages in the bird's egg as seen surface view. Early cleavages are in planes perpendicular to the surface of the yolk, forming a single layer of cells. The blastoderm becomes multilayered later as a consequence of cell divisions in planes tangential to the surface. [Interpretative drawings based on photographs of living and fixed eggs of the hen by M. W. Olsen, *J Morphol* 70:513–33 (1942), and of the pigeon by M. Blount, *Biol Bull* 13:231–50, (1907).]

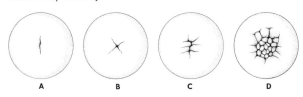

A   B   C   D

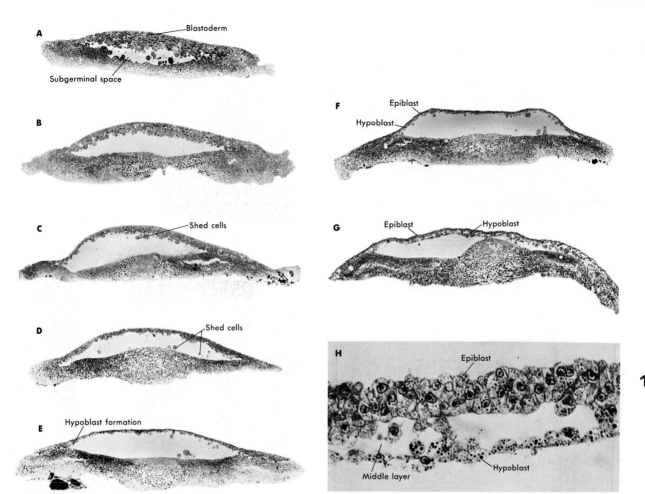

FIGURE 9.24 origin of the hypoblast in the embryo of the chick as seen in saggital sections at various stages of development of egg in the uterus and shortly after laying. In each figure the future posterior side of the embryo is at the left. A: Stage VI; a true blastoderm, 4–5 cells thick, surmounts a fluid-filled subgerminal space; this is the condition of the embryo 10–11 hr after the previous egg was laid. B–D: Progressive shedding of cells from the lower side of the blastoderm into the subgerminal space during stages VII, VIII, and IX, respectively; the events shown occur between about 12 and 19 hr after the laying of the previous egg. The fate of the detached cells is not known. E: Stage XI; the blastoderm at the time of egg-laying. Cells of the lower layers of the blastoderm have detached and some are seen on the floor of the subgerminal cavity. The uppermost layer of the germ is now an organized epithelium. It gives rise on its lower tide to a new generation of small cells; these constitute the hypoblast. Hypoblast cells first appear in a more coherent pattern posteriorly, but occur in patches anteriorly. F, G: The hypoblast becomes organized as a coherent layer progressing from posterior to anterior shortly after egg laying. H: The epiblast is a columnar epithelium; between it and the hypoblast a middle layer of cells appears. These could constitute part of the embryonic endoderm, as opposed to extraembryonic endoderm, which possibly comes entirely from the primitive hypoblast. (From S. Kochav et al., *Dev Biol* 79:296–308 (1980).]

147

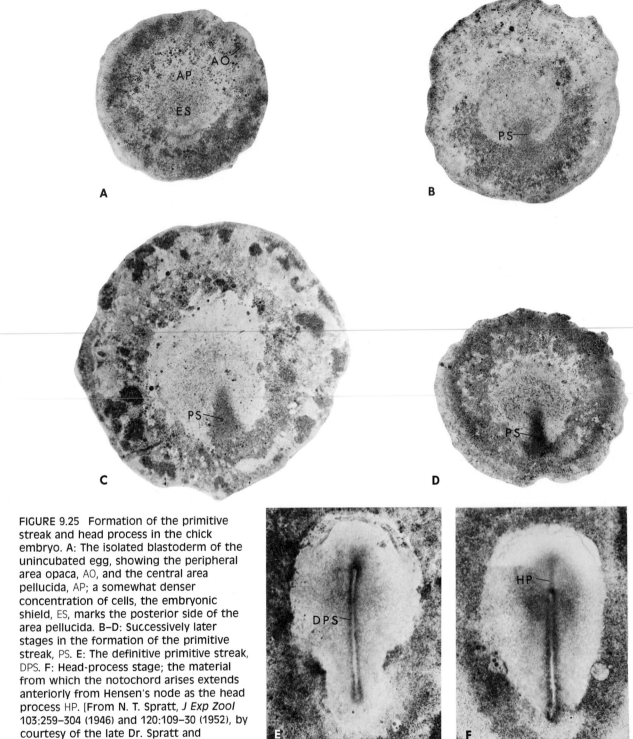

FIGURE 9.25 Formation of the primitive streak and head process in the chick embryo. A: The isolated blastoderm of the unincubated egg, showing the peripheral area opaca, AO, and the central area pellucida, AP; a somewhat denser concentration of cells, the embryonic shield, ES, marks the posterior side of the area pellucida. B–D: Successively later stages in the formation of the primitive streak, PS. E: The definitive primitive streak, DPS. F: Head-process stage; the material from which the notochord arises extends anteriorly from Hensen's node as the head process HP. [From N. T. Spratt, *J Exp Zool* 103:259–304 (1946) and 120:109–30 (1952), by courtesy of the late Dr. Spratt and permission of Alan R. Liss, Inc.]

FIGURE 9.26 The ventral surface of the chick blastoderm, isolated shortly after egg laying, as viewed by means of scanning electron microscopy. The posterior margin of the blastoderm is at the upper left. The beginning primitive streak forms a shelf of cells that is seen extending downward and to the right. Large round cells are in process of separating from the streak and adjacent epiblast. [From L. L. Litke, *J Submicr Cytol* 10:1–13 (1978), by courtesy of Dr. Litke and Karger Publishing Co., Basel.]

149

9.25B, C), the area pellucida assumes a pear-shaped outline (Figure 9.25D, E). The primitive streak itself soon takes the form of a groove that terminates anteriorly in a slightly deeper pit, the **primitive pit.** The walls of the pit are slightly elevated as **Hensen's node** (Figure 9.25E). In this condition the primitive streak is referred to as the **definitive primitive streak.** Subsequently there extends anteriorly from the node, beneath the epiblast, a strand of tissue called the **head process,** which is the beginning of the future notochord (Figure 9.25F).

Whereas in teleosts the emergence of the body form is generally correlated with the epibolic encasement of the yolk sphere, in birds the gastrulatory movements that issue in the triploblastic body form occur within the area pellucida and can be considered separately from the epibolic movements of the blastoderm over the yolk. The latter give rise to a vascular **yolk sac,** which is the site of movements of nutrients from yolk to the embryonic circulation. Epibolic movements of both chick and teleost are considered further later. For now, attention is directed to considerations of the origin of the germ layers in birds as seen in the area pellucida.

Rosenquist (1972) excised chick blastoderms prior to and approximately at the stage shown in Figure 9.25 and cultured them in vitro. He excised small fragments of hypoblast from various marked positions and substituted for them pieces of hypoblast from identical positions in donor embryos whose nuclei were labeled with ^{3}H-thymidine. He was thus able to follow the fates of the various grafts (Figure 9.27). He found that marks placed at the level of the anterior end of the primitive streak at the stage illustrated contributed to the endoderm of the embryonic body. Other marked areas contributed, instead, to the endoderm of the future yolk sac. It is clear, therefore, that the formation of the embryonic endoderm involves the primitive streak.

In earlier experiments Spratt (1946) isolated chick blastoderms at various early stages, including that of the embryonic shield (Figure 9.25A), and cultured them in vitro with the epiblast facing upward. By following the movements of marked cell groups, he showed that the primitive streak apparently arises by convergent movements of the epiblast, as illustrated in Figure 9.28. By inserting blocks of thymidine-labeled cells into the epiblast and following their subsequent movements, Rosenquist (1966, 1972) and Nicolet (1970) showed that the cells contributing the entire embryonic endoderm are derived from epiblast that invaginates through the primitive streak. By the head process stage (Figure 9.24F) all of the embryonic endoderm has entered the hypoblast. The position of the vari-

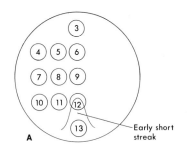

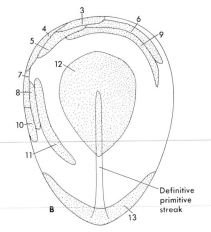

FIGURE 9.27 Ventral views illustrating the disposition of thymidine-labeled hypoblast cells placed in the corresponding position of an unlabeled host chick embryo. A: Numbered positions at which labeled blocks of cells were inserted at the stage of the short primitive streak. B: The distribution of the labeled cells at the stage of the definitive primitive streak. Note that only the labeled material at position 12 spreads into the central area of the blastoderm to give rise to the embryonic endoderm. Cells of the posterior part of the streak and from other areas of the hypoblast are all destined to form extraembryonic endoderm. [Adapted from G. C. Rosenquist, *J Exp Zool* 180:95–104 (1972).]

the endoderm and overlying layer (Figure 9.30). At Hensen's node, the entering cells contribute chiefly to notochord. Posteriorly, the first-entering mesoderm spreads widely to form mesoderm of the yolk sac, whereas the later-entering material gives rise to the body mesoderm (Nicolet, 1970).

As the primitive streak proceeds to achieve its definitive length, the area pellucida becomes some-

FIGURE 9.28 Movements of the epiblast during early stages of gastrulation in the chick embryo. Carbon particles (dots) placed in patterns on the epiblast of the prestreak blastoderm, as shown in the left-hand column, are distributed at successively later stages as shown in the middle and right-hand columns. (From N. T. Spratt, *J Exp Zool* 103:259–304 (1946), by permission of Alan R. Liss, Inc.]

TYPE OF MARKING	GENERALIZED RESULTS	
	SHORT STREAK	MEDIUM-BROAD STREAK

ous regions of the future gut can then be plotted by analysis of the disposition of marked cells placed in the hypoblast (Figure 9.29).

Invagination of the endoderm occurs throughout the length of the streak, according to Nicolet (1970), the last apparently entering anteriorly Meanwhile, continued convergence of the epiblast toward the primitive streak brings in prospective body mesoderm, which spreads laterally between

150

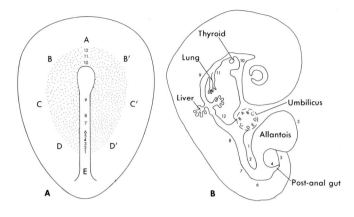

FIGURE 9.29 A: Diagram illustrating the position of pregut endoderm cells after their migration from the primitive streak. The stippled area indicates the extent of pregut endoderm. The numbered regions and the lettered boundary correspond to the regions of the definitive gut shown in B. [Adapted with modifications from G. C. Rosenquist, *Dev Biol* 26:323–35 (1971).]

what pear shaped, and then elongates further as the head process forms. Further elongation of the area pellucida is marked by a shifting posteriorly and shortening of the streak (Figure 9.31). The regression of the streak leaves ahead of its anterior end progressively more posterior levels of mesoblast interposed between the underlying hypoblast and overlying epiblast (Figure 9.30). The latter is now entirely ectodermal ahead of the node and begins the differentiation of anterior neural structures and future epidermis. Beneath the forming forebrain, folding of the blastoderm gives rise to the cavity of the foregut (Figures 9.31C and 9.32).

The formation of neural folds and their union to form a neural tube proceeds from anterior to posterior following the direction of the regressing primitive streak. As the neural folds close over, the subjacent mesoderm (the segmental plate) on either side is blocked into somites (Figure 9.32) in anteroposterior order, beginning at the level of the hindbrain, and continuing in the wake of the closing neural tube.

The primitive streak continues to regress until about 21 somites have been formed. Thereafter its remaining part becomes incorporated into the end bud (see Figure 11.10B) whose growth provides material for axial organs of the postanal level.

9.2.5 INITIATION OF DEVELOPMENT IN INSECTS.

Insects are more numerous and of more different species than almost any other major group of organisms, and they are of tremendous economic and medical importance. Their genetics, life cycles, modes of reproduction, and development have been the objects of intensive and sophisticated analysis for many years, and have contributed greatly to our understanding of genetic and developmental mechanisms. The initiation of development in insects is, therefore, of great concern to students of biomedical science. But, this fact notwithstanding, very few modern textbooks dealing with animal development devote space to the topic. Perhaps this is because developmental patterns among insects are so different from what they are in other kinds of animals, and furthermore, so different among various kinds of insects, as to make it difficult to handle the topic in a concise and economical way.

151

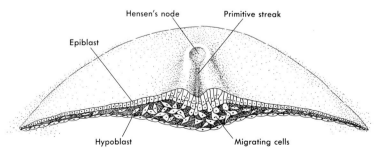

FIGURE 9.30 Three-dimensional representation of the migration of mesoblast cells through the primitive streak and their spreading between epiblast and hypoblast. For details of the relationships of the migrating cells to each other and to the epiblast see Figure 9.73.

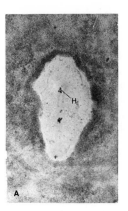
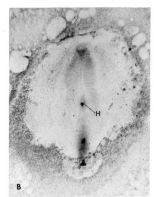
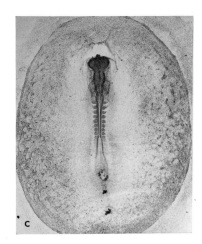

FIGURE 9.31 Regression and shortening of the primitive streak in the chick embryo. A: The blastoderm at the stage of the definitive primitive streak; clumps of finely divided carbon particles mark Hensen's node, *H*, and the middle and posterior end of streak. B: After several hours, the head process has formed anterior to the node, and the streak has moved posteriorly. C: The streak continues to shorten and regress as the longitudinal axis of the embryo appears anterior to it. [From N. T. Spratt, *J Exp Zool* 104:77 (1947), by courtesy of the late Dr. Spratt and permission of Alan R. Liss, Inc.]

152   FIGURE 9.32 Chick embryo incubated about 24 hr and in process of forming the sixth pair of somites.

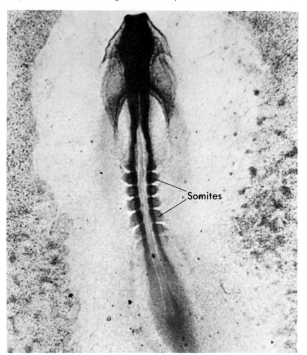

Somites

The account of early development that follows is necessarily sketchy, but it should provide a background for certain analytical studies that are presented further on; and it is hoped that it will provide the student with sufficient background to approach confidently the reading of research papers and review articles in the field of insect development.

9.2.5.1 The Insect Blastoderm. Eggs of insects are elongate (Figures 7.10D and 9.33) and filled with yolk, except for a thin peripheral rim of cytoplasm. This rim is connected by slender strands to an island of cytoplasm near the geometric center of the yolk, and therein lies the oöcyte nucleus. When the sperm enters the egg (Figure 9.33), the nucleus migrates toward the periphery and produces the polar bodies. The female pronucleus then shifts once more toward the center of the egg to meet the sperm pronucleus.

The cleavage of insect eggs is meroblastic and results in the formation of a single layer of cells surrounding the central yolk. The zygote divides mitotically (Figure 9.34A) and the daughter nuclei of successive divisions separate, eventually populating the yolk mass. These are called **cleavage nuclei,** even though no cleavage furrows are interposed

Each division brings about not only separation of daughter nuclei but also a division of the cytoplasmic island around each. These islands of cytoplasm remain connected with each other and with the peripheral cytoplasm, however, by means of a cytoplasmic reticulum that branches throughout the yolk.

After the period of synchronous division is completed, the peripheral cytoplasm becomes transformed into a cellular blastoderm (Figure 9.34B). In *Drosophila montana* (Fullilove and Jacobson, 1971) the cleavage nuclei, having fully populated the

FIGURE 9.33 Fertilization and formation of the cleavage nucleus in *Drosophila*. A: Sperm enters through the micropyle, and the oöcyte goes into anaphase of the first meiotic division. B: The first meiotic division is completed C: The second division begins in both daughter nuclei, and the sperm begins to form the male pronucleus. D–F: The second meiotic division is completed, forming three polar body nuclei, P, and one female pronucleus; the male pronucleus is formed. G, H: Both pronuclei form condensed chromosomes that align as separate sets in the first cleavage spindle. The polar bodies likewise form chromosomes, a set of four from each so that twelve chromosomes appear. Each of these is double, apparently having replicated at the same time as the chromosomes of the pronuclei prepared for division. The polar body chromosomes soon disintegrate and do not partake in development of the egg. [From A. F. Huettner, *J Morphol* 38:249–65 (1924), by permission of Alan R. Liss, Inc.]

between them. The cleavage nuclei divide synchronously, and often with considerable rapidity, for a period that is variable in different groups of insects. In the damselfly *Platycnemis* the first nine divisions are synchronous, and in the fruit fly *Drosophila* the first 12 occur in synchrony. In the former, the first four or five cleavages require about 2 hours to complete each mitotic cycle, but 9 hours is required before the products of the ninth division double their number. In *Drosophila,* in contrast, the 12 synchronous divisions proceed with great rapidity, only 10 minutes being required for the completion of each mitotic cycle.

FIGURE 9.34 Cleavage and blastoderm formation in the egg of the honey bee. A: Two cleavage nuclei, N, each with its own island of cytoplasm, separated by yolk globules. B: Cellular blastoderm at the periphery of the egg; yolk nuclei, YN, remaining within, will subsequently be cellulated and become vitellophages.

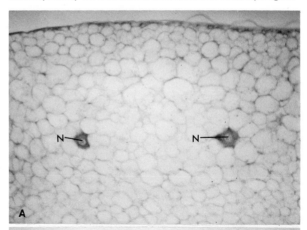

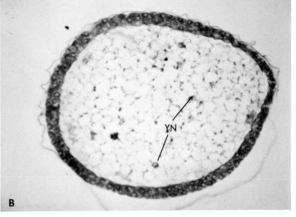

153

yolk, begin to populate the peripheral cytoplasm during the eighth and ninth divisions. Three more synchronous divisions then occur, giving rise to a syncytial blastoderm constituted of about 3500 cells, others remaining in the yolk mass as **vitello-phages** ("yolk eaters").

At this time the peripheral nuclei undergo a dramatic change in shape and volume. Within a period of 35 to 40 minutes the spherical nuclei (Figure 9.35A), with a diameter of about 4.9 μm, elongate to 12.7 μm in length (perpendicular to the surface), meanwhile suffering a reduction of only 0.2 μm in diameter (Figure 9.35B). Their volume thus increases 2.4 times in less than 1 hour! Simultaneously with the increase in length of the nuclei is the appearance of microtubules that are oriented per-

pendicularly to the surface of the egg, extending from points near the surface of the egg to the base of the elongating nucleus. These conceivably play a role in the formation of the cleavage furrows that transform the peripheral syncytium into a cellular blastoderm.

As seen in surface view, the elongating nuclei are packed in hexagonal array (Figure 9.36). Cleavage furrows then appear simultaneously in hexagonal patterns around each nucleus. The advancing tips of the furrows are expanded into a "canal" (Figure 9.37A, B) that surrounds each cytoplasmic unit in a plane transverse to the long axis of the nucleus. The canal around each unit is connected at each corner of the hexagon with the furrow canal around each of the six surrounding cells. As the fur-

FIGURE 9.35 A: Electron micrograph of the periphery of the egg of *Drosophila montana* after the migration of the nuclei but prior to cellulation. The surface of the egg is elevated and bears numerous microvilli over each nucleus. In the valleys between elevations the egg surface is flattened. In the valley the cellular furrows will form. B: Periphery of an egg fixed at a stage about 15 min later than the one shown in A. Nuclear elongation has occurred and furrows, F, ending in furrow canals, FC, surround each nucleus peripherally. [From S. L. Fullilove and A. G. Jacobson, *Dev Biol* 26:560–77 (1971), by courtesy of the authors and permission of Academic Press, Inc., New York.]

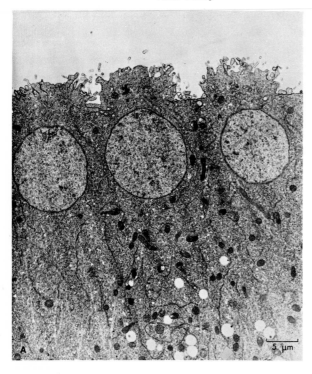
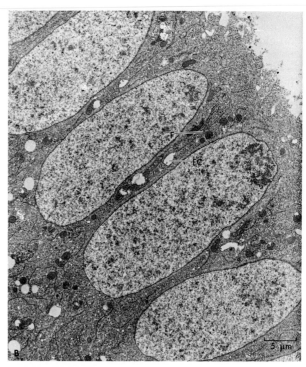

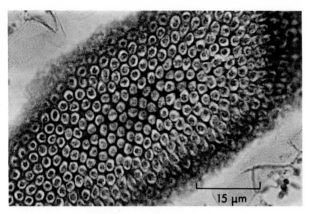

FIGURE 9.36 The hexagonal packing of cells in the blastoderm of the egg of *Drosophila montana* as revealed by a tangential section through the egg. [From S. L. Fullilove and A. G. Jacobson, *Dev Biol* 26:560–577 (1971), by courtesy of the authors and permission of Academic Press, Inc., New York.]

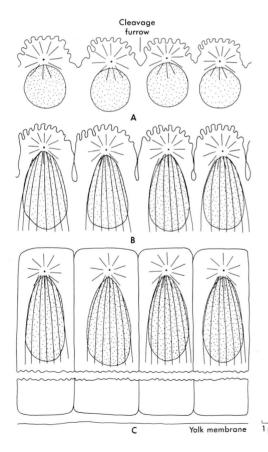

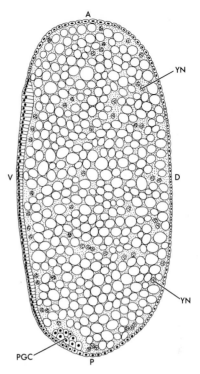

FIGURE 9.38 Germ band in *Clytia*. Saggital section showing a thickening of the blastoderm on the future ventral side of the embryo of a chrysomelid beetle. A, anterior; D, dorsal; P, posterior; PGC, primordial germ cells. V, ventral; YN, yolk nucleus. [From A. Lecallion, *Arch d'Anat Micr* 2:89–176 (1898).]

FIGURE 9.37 Successive stages in the elongation of blastoderm nuclei and formation of the cellulated blastoderm in *Drosophila montana*. A: Nuclei that have just completed their final division; cleavage furrows are beginning to form. B: The extension of microtubules from the centriolar regions and the formation of cleavage furrows. Note that the advancing tip of the furrow is expanded; a section tangential to the surface of the egg passing through the expanded region of the furrow would show it as a circular canal (ring canal) surrounding the cytoplasm of each cell. C: The completion of cellulation. The ring canals surrounding each cell have expanded and merged, thus forming the basal plasma membrane of each cell and a membrane delimiting the yolk. The portion omitted from each cell is about 10 mm, a distance slightly less than the length of the nucleus. [Redrawn with modifications from S. L. Fullilove and A. G. Jacobson, *Dev Biol* 26:560–77 (1971).]

rows proceed past the basal ends of the nuclei, their canals expand and fuse, cutting off cellular units from one another and from the yolk beneath (Figure 9.37C). Thus the cellular blastoderm is formed. Presumably this is comparable to the blastula stage in the development of other forms.

9.2.5.2 Formation of the Germ Band and Origin of the Germ Layers.

The first visible sign of differentiation in the blastoderm occurs when cells along one side of the posterior end of the egg thicken in an oval-shaped region called the **germ band.** From this region the embryo develops, and the position of the band determines its future ventral side (Figure 9.38). The way in which morphogenesis of the embryonic body is accomplished from the germ band, however, is quite variable in different kinds of insects. In beetles a median groove appears as cells of the germ band pass inward toward the yolk and spread out interiorly (Figure 9.39). The inner layer is the source of mesoderm and endoderm, and the outer layer forms the ectoderm plus a membrane called an **amnion.** Orthopterans, as exemplified by the dragonfly, form the amnion and reverse the polarity of the germ band by invagination and elongation, as shown in Figure 9.40. These evolutions are comparable to gastrula movements in other kinds of animals. Regardless of how the amnion is formed, the germ band is very early divided by transverse furrows into consecutive metameres corresponding to the segments of the definitive body (Figure 9.41).

9.2.5.3 Factors Determining Early Embryonic Organization.

In the case of a number of insects, its has been shown that the portion of the egg marking the future posterior side is the site of an **activation center** that is required in order for germ-band formation to occur. This center has been especially studied in the case of the dragonfly *Platycnemius pennipes* (reviewed by Bodenstein, 1953). The egg of this insect is about 370 μm in length. If it is constricted midway along its length before the nuclei have moved to the periphery, a blastoderm and germ band form only in the posterior part of the egg. If a constriction is placed about the egg less than 90 μm from its posterior end (Figure 9.42A), a complete embryo forms anteriorly to the constriction (Figure 9.42B). But, if the constriction cuts off a slightly larger portion posteriorly, a cellular blastoderm without germ band forms anteriorly (Figure 9.42C). Once the cellular blastoderm has formed, however, ligation does not inhibit formation of a germ band and embryo anteriorly (Figure 9.42C).

The passage of the activating principle is not inhibited by a partial constriction. Moreover, its anterior movement is marked by a notable clearing of the yolk and a contraction of the yolk system that results in a localized aggregation of the blastoderm cells for formation of the germ band.

The first sign of morphological differentiation is the aggregation of blastoderm cells for that portion of the future germ band that has its midpoint at

156

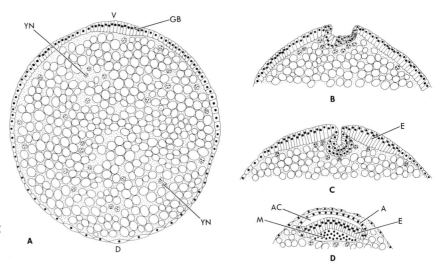

FIGURE 9.39 Successive stages in the formation of the germ layers and amnion in *Clytia*. A, amnion; AC, amniotic cavity; D, dorsal; E, ectoderm; GB, germ band; M, mesendoderm; V, ventral; YN, yolk nucleus. [From A. Lecallion, *Arch d'Anat Micro* 2:89–176 (1898).]

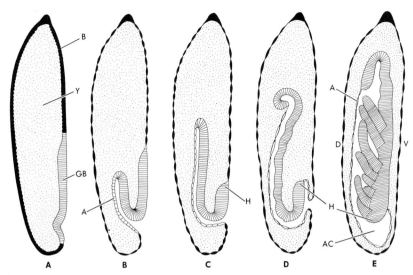

FIGURE 9.40 Invagination and amnion formation in the embryo of the dragonfly. Successive stages in the inturning of the germ band, its segmentation, and formation of the amnion. The rudiments of the appendages appear in E. Note that the posterior end of the embryo invaginates and faces the anterior end of the egg. At a later stage of development, the embryo reverses to its original orientation with respect to the egg axis. A, amnion; AC, amniotic cavity; B, blastoderm; D, dorsal; H, head; GB, germ band; V, ventral; Y, yolk. [From E. Korschelt and K. Heider, *Textbook of the Embryology of Invertebrates*, vol. 3, English translation, Macmillan, New York, 1899, p. 275.]

approximately the level of the presumptive thorax. This is about 125 μm anterior to the posterior end of the egg. This is an important center. If a constriction is placed loosley about the egg at this level, a germ band forms on both sides of the ligature. If, however, the ligature is placed in front or in back of this center, a germ band develops only in that portion containing the center. This important region is referred to as the **differentiation center.** From the earlier-noted experiments, furthermore, it is evident that for the differentiation center to form, the

FIGURE 9.42 Demonstration of the activation center in the egg of the dragon fly *Platycnemius penniper*. A: Ligation close to the posterior pole does not interfere with production of the germ band if positioned prior to formation of the cellular blastoderm. B: Resulting embryo. C: Ligation further anteriorly prior to formation of the cellular blastoderm permits the blastoderm to form, but no germ band appears. D: Embryo formed subsequent to ligation applied after the formation of the cellular blastoderm. [Adapted from A. Kühn, *Lectures on Developmental Physiology*, 2nd ed., Roger Milkman (trans.), Springer-Verlag, Berlin, 1971.]

157

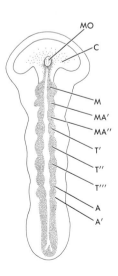

FIGURE 9.41 Segmentation of the germ band in a beetle embryo. Ventral view of the germ band of *Lina*, a chrysomelid, as seen through the covering amnion. A,A', first and second abdominal segments; C, cerebral lobe; T', T'', T''', first, second, and third thoracic segments, respectively; M, mandibular segment; MA', MA'', first and second maxillary segments, respectively; MO, mouth. [From E. Korschelt and K. Heider, *Textbook of the Embryology of Invertebrates*, Vol. 3, English translation, Macmillan, New York, 1899, p. 291.]

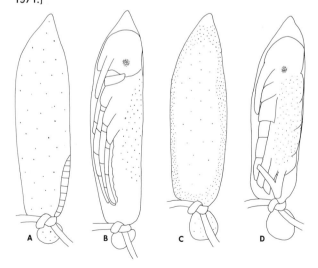

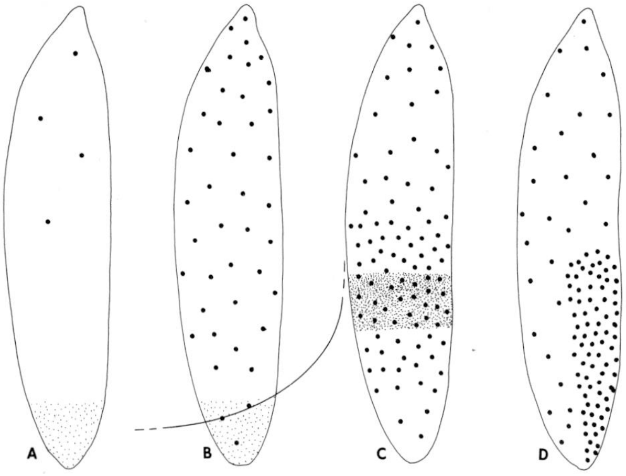

FIGURE 9.43 Activation center and differentiation center in the egg of the dragon fly. A: Cleavage has begun (0 hr). B: After 19 hr, the activation center has become nucleated, a requisite for subsequent formation of a germ band (cf. Figure 9.40) and for the forward spreading of a product of the activation center (curved line). C: By 46 hr, the differentiation center is established in the region where the future thorax will appear. This region is indicated by heavy stippling. If a ligature is placed around the egg at this point, a germ band forms on both sides. D: 64 hr after cleavage began, the cleavage nuclei begin to aggregate preferentially in the region of the future germ band. [Adapted from D. Bodenstein, "Embryonic Development" in K. D. Roeder, ed., *Insect Physiology*, Wiley, New York, 1953.]

prior action of the activation center is required (Figure 9.43).

9.3 Early Development of Mammals

Mammalian embryos (except those of the egg-laying monotremes) develop in the uterus of the mother. They are endowed with little energy reserve, and their survival depends on their importation of nutrients from maternal tissue. Thus, early events must involve the origin of mechanisms for establishing access to the maternal circulation, which serves as a source of nutrients and provides for the exchange of gases.

Both maternal and embryonic tissue contribute to the origin of these mechanisms. The cleaving egg becomes a **blastocyst** which is specialized for attaching to and implanting in the uterine wall. The endometrium (lining of the uterus) responds to the presence of the blastocyst by increasing its blood supply and capillary permeability. Subsequent intimacy of association is achieved in the formation of the **placenta,** a highly specialized structure for maternal–fetal exchange involving extraembryonic membranes of the embryo and the endometrium of the uterus.

The mechanisms of implantation and placentation vary among different kinds of mammals and so do the degrees of intimacy that are established between the maternal circulation and that of the embryo (later called the **fetus**). Next we describe cleavage, implantation, and placentation with emphasis on human and other anthropoid material.

9.3.1 CLEAVAGE AND PREIMPLANTATION EVENTS.
Fertilization of mammalian eggs occurs high in the oviduct, cleavage divisions (Figure 9.44) proceeding slowly as the ovum moves toward the uterus. Cleavages are holoblastic and slightly unequal as a rule. They are not synchronous and produce a loose ball of cells, the **morula** (Figure 9.45) within the intact zona pellucida. The morula comprises 16 or more cells by the time it reaches the uterus, which occurs in 3 to 5 days in the human being. Once in the uterus, its cells begin to divide rapidly and spaces appear between them. By the time about 100 cells are present, the embryo is in the form of a fluid-filled vesicle, the **blastocyst** (Figure 9.46) with a small mass of cells, the **inner cell mass** confined to one section of its interior. At first the wall of the vesicle is only a single layer of cells. It is called the **trophoblast.** The inner cell mass will form the embryonic body and the trophoblast provides the attachment of the embryo to the wall of the uterus.

In the uterus the blastocyst expands rapidly by cell division and by the accumulation of fluid in the vesicle. As the time for implantation approaches, other changes occur in the blastocyst and in the uterine fluids. Pregnancy-specific proteins appear, including proteases that aid in removing the zona pellucida. The surface glycoproteins of the blastocyst change, too, bringing about regional differences in the adhesivity of the trophoblastic cells for the uterine lining. In primates, the ability of the

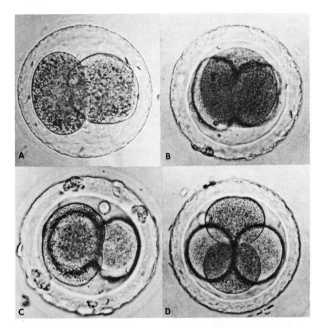

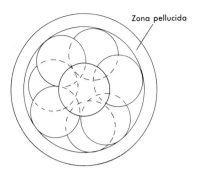

FIGURE 9.45 Eight-celled morula of the rhesus monkey *Macacus rhesus*. [Redrawn with modifications from W. H. Lewis and C. G. Hartman, *Carnegie Contribs Embryol* 24:187–201 (1941).]

FIGURE 9.44 Second cleavage in the egg of the rabbit. A: The product of the first cleavage as viewed from the animal pole. The blastomere on the left, designated the "first blastomere," is elongating in preparation for the second cleavage, which is meridianal. B: The first blastomere, now seen in equatorial view, has completed its division, having formed two daughter cells of somewhat unequal size. The other blastomere, designated the "second blastomere," has not yet divided, so that a transient three-cell stage occurs. C: Equatorial view of the three-celled embryo from a point 90° of arc from the view shown in b. The second blastomere (right) has begun to elongate in a direction parallel to the original animal–vegetal axis and the direction of its cleavage furrow is transverse to the animal–vegetal axis. D: Equatorial view of the four-cell stage as seen from a point 180° from the view shown in B. The daughter cells of the first blastomere are in sharp focus, whereas the daughter cells of the second blastomere are slightly out of focus. Note the crosswise configuration of the blastomere pairs. Where the second blastomere cleaves with less delay than depicted here, the furrow begins in the meridianal plane, but the cell rotates 90° by the time cytokinesis is completed, thus giving an identical crosswise configuration. [From B. J. Gulyas, *J Exp Zool* 193:235–48 (1975, by courtesy of Dr. Gulyas and permission of Alan R. Liss, Inc.]

blastocyst to attach and invade is restricted to the embryonal pole of the trophoblast (Figures 9.46 and 9.48), that is, the region overlying the inner cell mass. In some other forms, the guinea pig, for example, the site of attachment is the opposite, or abembryonal, pole.

In human beings, other primates, and many

159

FIGURE 9.46 Blastocyst of a baboon recovered on day 7 after fertilization. The inner cell mass is easily recognizd in the upper right quadrant. [From A. G. Hendrickx and D. C. Kraemer, *Anat Rec* 162:11–120 (1968), by courtesy of the authors and permission of Alan R. Liss, Inc.]

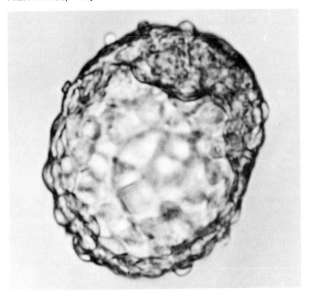

familiar mammals, implantation begins 1 or 2 days after the blastocyst enters the uterus. But for numerous kinds of animals there is normally a prolonged delay in implantation, the blastocyst meanwhile remaining viable and free in the uterine fluids, with or without its zona pellucida (this varies with the species), and growing very slowly, if at all. Delayed implantation is probably of considerable adaptive value to many once-a-year breeders, to animals whose young are born in an immature condition or that suckle for prolonged periods, and to many polyestrous animals whose production of successive litters without delay would cause the later ones to be born under adverse climatic conditions.

Northern fur seals provide an interesting example. They haul out in their breeding grounds in Northern Alaska annually each spring. The females produce their pups in June and early July and then become impregnated promptly during a short period of sexual receptivity after parturition. Having weaned their pups, the pregnant females put out to sea not to return until the following May or June. Although the gestation period is thus almost a full year, the embryo actually remains dormant in the blastocyst stage for almost 4 months, growing only slowly and exhibiting little or no mitotic activity, but consuming glucose and synthesizing small quantities of protein. The blastocyst becomes "activated" in early November, increasing its mitotic rate and enlarging in size (Daniel, 1971). Concurrently, the corpus luteum enlarges, presumably releasing increasing quantities of progesterone, and the uterine wall thickens and produces quantities of protein. Implantation occurs in mid-November, presumably when estrogen concentrations have risen sufficiently to cause the progesterone-sensitized endometrium to respond to the enlarging blastocyst.

Among other animals showing delayed implantation are the American black bear, weasel, martin, and mink. In the latter three cases, a period of increasing day lengths after December normally creates the hormonal conditions necessary for implantation. In rats and mice delayed implantation occurs in the presence of suckling young. The delay is influenced by the duration, timing, and intensity of the suckling stimulus (Bindon, 1969). Delayed implantation is probably not a significant feature in human reproduction, but ovulation is inhibited to

160

some degree in nursing mothers, and this notably affects the fecundity of some human populations.

9.3.2 IMPLANTATION AND FORMATION OF THE GERM LAYERS.

At the time implantation begins, the exposed surface of the uterine lining, the endo-

FIGURE 9.47 Human endometrium at the time of implantation, about 11 days after fertilization. [From A. T. Hertig and J. Rock, *Carnegie Contribs Embryol* 29:127–56, (1941), by permission of the Carnegie Institution of Washington, Department of Embryology, Davis Branch.]

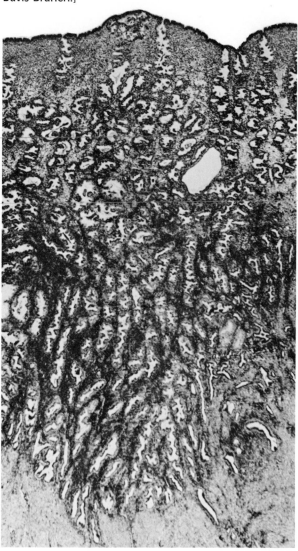

metrium, is a single-layered epithelial sheet, which forms numerous slender-necked tubular uterine glands. These pass inward through the superficial **compact layer** of the endometrium and are greatly dilated in the deeper lying **spongy** layer (Figure 9.47).

Having adhered to the epithelium, the trophoblastic cells penetrate it (Figure 9.48A) and erode it away, meanwhile multiplying and spreading into the underlying compact layer. The endometrium responds by a dramatic increase in vascularity and capillary permeability, by marked enlargement of the uterine glands, and by the production of characteristic large, rounded cells, rich in glycogen and lipid and often with more than one nucleus. The latter are diagnostic of pregnancy and are called decidual cells. These changes, which constitute the **decidual reaction,** are elicited in the human endometrium only after erosion of the surface epithelium, but erosion is not required in the mouse and many other mammals. As the invading trophoblastic cells multiply they differentiate into two layers, an inner cellular **cytotrophoblast** and an outer layer, the **syntrophoblast,** in which the nuclei are not separated by cell membranes. This differentiation proceeds at the invading front of the blastocyst and proceeds around the periphery as **nidation** continues (*nidus* = nest, referring to the nesting of the conceptus in the uterine wall).

The decidual reaction and subsequent implantation of the blastocyst are apparently mediated by the interaction of estrogen, produced by the blastocyst itself (George and Wilson, 1978), and estrogen-binding sites in the wall of the uterus. Preimplantation blastocysts of rats, rabbits, and mice produce estrogens and do so more actively on the day of implantation. When estrogen antagonists are introduced experimentally, however, they displace estrogen from the uterine receptors and block implantation (Gupta et al., 1977). Moreover, the introduction of a number of rat blastocysts into the uterus diminishes the ability of the uterus to bind exogenously administered labeled estrogen. This effect is exercised as much as 20 hours before the decidual reaction can normally be detected, and is presumably due to the saturation of uterine estrogen-binding sites by estrogen released from the blastocysts (Ward et al., 1978). Finally, it may be noted that whereas four- to eight-cell mouse embryos do not induce a decidual reaction when intro-

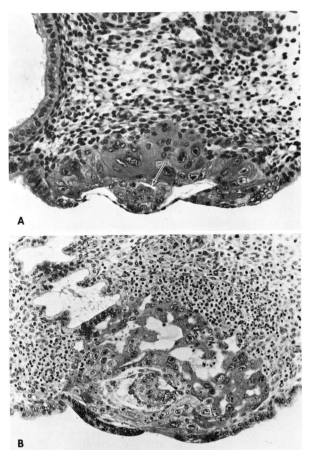

FIGURE 9.48 A: Human conceptus at the initiation of implantation, presumably at 7 days. The transition from the thin trophoblast at the abembryonic pole to the thick solid layer at the embryonic pole is dramatically evident. Large masses of syntrophoblast project into the stroma of the endometrium. The amniotic cavity is now evident (arrow), and the inner cell mass is seen as a double-layered disc. B: Fully implanted blastocyst at about 9 days. [From A. T. Hertig and J. Rock, *Carnegie Contribs Embryol* 31:65–84 (1945) and 33:169–86, (1949), by permission of the Carnegie Institution of Washington, Department of Embryology, Davis Branch.]

duced into the uterus, they will do so if previously treated with estrogen (Dickmann et al., 1977).

The human ovum is fully implanted at 9 days, the syntrophoblast progressively and rapidly enlarging the scope of its invasion of the endometrium. The spreading of the syntrophoblast creates a maze of strands lined with cytotrophoblast. These are called **primary villi.** They branch and anastomose to enclose **lacunae,** fluid-filled spaces in the eroded endometrium. Blood from maternal vessels invaded by the advancing syntrophoblast fills these spaces (Figure 9.48B).

Very early in its penetration of the endometrium, the trophoblast initiates secretion of a hormone, **chorionic gonadotropin.** This hormone is a glycoprotein whose action is similar to that of LH and, in fact, its α-protein subunit is almost identical to that of LH. The function of human chorionic gonadotropin (HCG) is to maintain the functional status of the corpus luteum during the first few weeks of pregnancy, providing the secretion of progesterone essential to the maintenance of pregnancy. After about the 40th day of pregnancy the corpus luteum of pregnancy can be dispensed with, for the placenta then produces adequate quantities of progesterone.

HCG is detectable in the blood and urine a few days after conception and thus this can serve as the basis of pregnancy tests.

As nidation begins, the cells of the inner mass become arranged as a bilaminar disc, the **embryonic disc.** The lower layer of the disc is the primary endoderm. The **epiblast** lies above this and will form ectoderm and mesoderm. Between the epiblast and cytotrophoblast a fluid-filled cleft appears (Figure 9.48A), which expands into a cavity, the **amniotic cavity.** The roof of the cavity at this stage is derived from the cytotrophoblast and is called the **amniogenic layer.** This gives rise to the **amnion,** a vesicular membrane within whose fluid-filled interior the embryo undergoes most of its development (Figures 9.48 and 9.50).

By 9 days the endoderm is continuous with a layer of cells that lines the interior of the blastocyst. This is the **primary yolk sac** (Figure 9.49), or **exocoelomic membrane.** The edges of the epiblast, meanwhile, remain continuous with the amniogenic layer. With further development, the amnion enlarges and the primary yolk sac constricts (Figure 9.50) so that the embryonic disc forms a shelf between the amniotic cavity on one side and the cavity of the reduced **secondary yolk sac** on the other. At later stages the embryonic disc is somewhat comparable in its organization to the area pellucida of the unincubated chick blastoderm (Figures 9.51 and 9.52). In it a primitive streak will form and

162

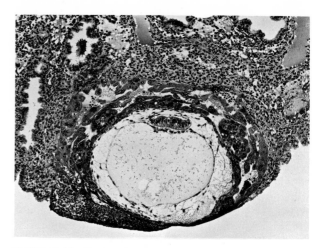

FIGURE 9.49 Human blastocyst fully implanted at about 12 days showing extensive invasion of the endometrium by the trophoblast. The amnion has enlarged, and the endoderm is continuous with the exocoelomic membrane. [From A. J. Hertig and J. Rock, *Carnegie Contribs Embryol* 29:126–46 (1941), by permission of the Carnegie Institution of Washington, Department of Embryology, Davis Branch.]

FIGURE 9.50 Human conceptus of 16 days. The secondary yolk sac has formed and chorionic villi, now highly branched, surround the entire vesicle. A loose mesenchyme fills the spaces between the embryo and the wall of the conceptus. This view shows the relationships of the chorionic villi and the extraembryonic coelom. [From C. H. Heuser, J. Rock, and A. T. Hertig, *Carnegie Contribs Embryol* 31:85–99 (1945), by permission of the Carnegie Institution of Washington, Department of Embryology, Davis Branch.]

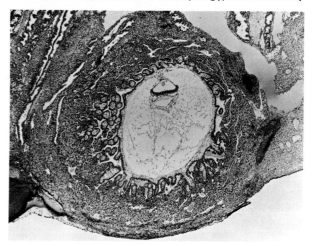

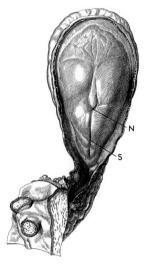

FIGURE 9.51 Dorsal view of human embryo at about 8 days. The amnion has been cut away to reveal the surface of the embryonic disc. The anterior end of the embryo faces the top of the page. The primitive node, N, and primitive streak, S, are readily recognized. [From C. H. Heuser, *Carnegie Contribs Embryol* 23:251–67 (1932), by permission of the Carnegie Institution of Washington, Department of Embryology, Davis Branch.]

FIGURE 9.52 Schematic saggital section of a human embryo at 19 days showing the relationship of the allantois and body stalk and the beginning of mesoderm formation. [Adapted with modifications from H. O. Jones and J. I. Brewer, *Carnegie Contribs Embryol* 29:157–65 (1941).]

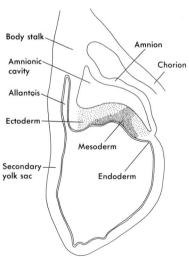

through the streak epiblast will invaginate to form mesoderm.

By 11 days these structures are covered with a loose mesenchyme that is designated as **extra-embryonic mesoderm.** It fills the spaces between the embryo and the wall of the blastocyst, and pushes out into the primary villi providing them each with a mesodermal core. The villi are now called **secondary villi** or **chorionic villi** (Figure 9.50).

The source of the extraembryonic mesoderm is not clear. Some or all of it may come from the cytotrophoblast lining the blastocyst, but Luckett (1971) has suggested that it may be derived from the first cells that invaginate through the primitive streak, which appears very precociously at the caudal margin of the embryonic disc during the 12th day. Cells that invaginate through the streak at this time move posteriorly and dorsally along the caudal margin of the amnion and into the cores of the adjacent primary villi and over the inner surface of the blastocyst. By the 14th day chorionic villi are established over the entire surface of the blastocyst, which is now referred to as the **chorionic vesicle.**

Spaces within the extraembryonic mesoderm coalesce to form the **extraembryonic coelom.** The coelom intervenes as a fluid-filled cavity between the embryo and the wall of the chorionic vesicle except at the future posteriodorsal region of the embryo. Here the chorionic mesoderm is continuous with the amniogenic layer and the posterior end of the embryo. This mesoderm remains as a permanent connection between the embryo and the wall of the chorionic vesicle, and is called the **body stalk** (Figure 9.52). An endodermal diverticulum from the posterior end of the future embryo pushes into the body stalk. This is the **allantois.** The allantois and body stalk are the forerunners of the **umbilical cord,** which carries the embryonic blood vessels involved in maternal–fetal exchange.

9.3.3 DECIDUAE AND PLACENTAE. In human pregnancy, the fusion that occurs between the expanding chorionic vesicle and the endometrium is so intimate as to lead to the shedding of essentially all of the uterine lining at birth. The endometrium of pregnancy is therefore named the **decidua** (L., *decidere,* to fall off). One distinguishes three regions of the decidua (Figure 9.53): (1) the **decidua parietalis,** which is the general lining of the uterus not

163

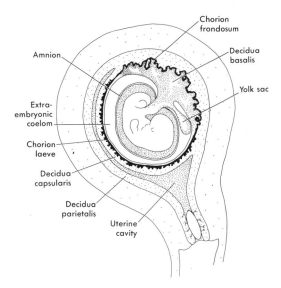

FIGURE 9.53 Relationships of the young human embryo and its membranes to the uterus. The chorionic villi on the side facing the decidua basalis continue to enlarge and branch whereas those facing the decidua capsularis regress.

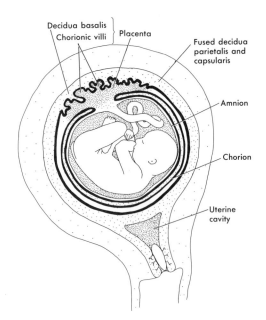

FIGURE 9.54 Relationships of the advanced human fetus and its membranes to the uterus. The decidua capsularis has fused with the uterine wall, and chorionic villi persist only in the placenta.

164

occupied by the embryo; (2) the **decidua capsularis,** the compact layer of the endometrium covering the chorionic vesicle and interposed between it and the uterine cavity; (3) the **decidua basalis,** the region beneath the chorionic sac between it and the muscular wall of the uterus.

During early pregnancy the amnion expands, surrounding the embryo with a watery cushion, and by the end of the 2nd month it has pressed against the chorionic wall, completely obliterating the extraembryonic coelom. The chorionic vesicle expands, too, pushing the decidua capsularis into the lumen, ultimately obliterating the uterine cavity. By the beginning of the 4th month, the decidua capsularis will have fused with the decidua parietalis of the remaining uterine wall. Essentially, therefore, the only cavity in the uterus thereafter is that of the amnion.

Until about 9 weeks after conception, the chorionic villi cover the entire surface of the vesicle. They then progressively disappear beginning on the abembryonal side as the decidual capsularis is compressed and its vascularity reduced. They are then progressively restricted to a discoidal region around

the end of the umbilical cord (Figure 9.54). Here they persist, establishing together with the underlying maternal tissue, the complex anatomical and physiological unit that is called the placenta. For the rest of development the placenta regulates the interchanges between mother and fetus. It also serves as a major source of progesterone in women beginning during the third month of pregnancy.

Various kinds of mammals, even closely related ones, differ greatly in the way maternal–fetal relationships are established in the uterus. In human beings implantation is **interstitial,** that is, the conceptus becomes embedded completely in the endometrium. But in another primate, the baboon, implantation is **superficial,** the chorionic villi forming only at the embryonal pole and the rest of the chorionic vesicle projecting into the uterine lumen without a decidua capsularis. In human beings, some apes, the bear, and some other animals the chorionic villi persist only in a disc-shaped patch, and the fully developed placenta is referred to as **discoid.** The types of tissue/fetal–maternal associations found in a number of kinds of mammals are described by Torrey and Feduccia (1979).

9.4 Problems of Cleavage and Gastrulation

The foregoing discussion outlines patterns whereby the germ layers are formed in different kinds of eggs. The next step is the analysis of the mechanisms that bring about these patterns. In what follows, a few of the numerous problems that this analysis imposes are discussed, and the experiments that suggest solutions to some of them are cited. However, many more questions are raised than can be answered.

9.4.1 WHAT DETERMINES THE CLEAVAGE PATTERN?

The pattern of cleavage emerges from the positioning of the cleavage furrows and the timing and order of their appearance. An outstanding aspect of cleavage in most eggs is the synchrony with which the blastomeres divide. Does synchrony result simply from an essential identity of metabolic processes in cleavage cells, or are there intercellular controls that keep the cells in synchrony? Synchrony is eventually lost, presumably because of the differential partitioning of cytoplasmic components by the cleavage furrows and because of other local differences attributable to the position of the cell in the cleaving mass: availability of oxygen, diffusion of carbon dioxide and hydrogen ions, and so on.

One might conceive that synchrony of division might be mediated by virtue of the flow of ions or small molecules between blastomeres. Cells permitting this flow would be **electrically coupled**, a property that could be revealed by passing ions by means of an electrode into one blastomere and then measuring the current flow as picked up by another electrode in another blastomere. In fact, however, there seems to be no relationship between electrical coupling and synchrony of cleavage divisions. Daughter cells of the first division in amphibian eggs are electrically coupled (Ito and Loewenstein, 1969), but synchronously dividing blastomeres of the starfish do not show electrical communication until the 32-cell stage (Tupper and Saunders, 1972).

9.4.1.1 Coordination of Nuclear and Cytoplasmic events.

Cell division involves both nuclear division (mitosis) and partitioning of the cytoplasm (cytokinesis). In the cleavages that are most studied these events are usually coordinated so that cytokinesis occurs during mitotic anaphase and telophase. But this coordination does not always occur; recall that cytoplasmic partitioning comes only after many nuclear divisions in the formation of the insect blastoderm. In the typical case, however, the formation of daughter nuclei and cytokinesis are coordinated in the production of the cleavage pattern. Because the cleavage furrow passes through the cell in a plane perpendicular to the long axis of the mitotic spindle, it is apparent that the factors determining the position of the spindle will affect the pattern of cells that appears during the cleavage period.

9.4.1.2 Cytoplasmic Factors Position The Spindle.

In almost all eggs and blastomeres the mitotic spindle forms approximately in the geometric center of the yolk-free area of the cytoplasm. In most cases this is nearer the animal pole, and the more yolk there is, the more eccentrically the spindle is placed. But factors other than yolk distribution affect the location of the spindle. Lillie (1909) subjected unfertilized eggs of *Chaetopterus* (an annelid) to centrifugal force applied at various angles with respect to the original animal-vegetal axis. This treatment caused the yolk to stratify at the centrifugal pole, but after fertilization the first cleavage spindle and furrow formed quite independently of the location of the yolk!

What does position the spindle? How is its orientation changed in preparation for successive cleavages? Apparently the controlling factors are cytoplasmic, and they proceed independently regardless of the number of mitotic divisions that occur. This independence is shown by an experiment of Hörstadius (1928) (Figure 9.55). He was able to delay the rate of cell division in newly fertilized sea urchin eggs for varying periods of time by the use of dilute seawater or by shaking. If, for example, the first cleavage was delayed until the time the third cleavage of control eggs occurred, the first spindle formed parallel to the animal-vegetal axis instead of perpendicular to it, as is normal for first cleavage. This resulted in the formation of one cell at the animal pole and one at the vegetal. At the

165

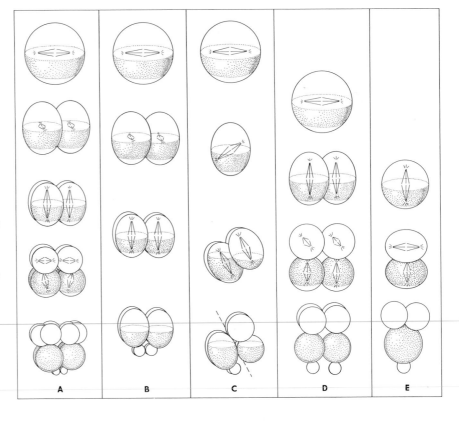

FIGURE 9.55 Effects of delayed mitosis on the cleavage pattern of a sea urchin egg. Column A, top to bottom. Successive positions of cleavage spindles and distribution of blastomeres in normally developing eggs. Columns B–E: Spindle positions and patterns of blastomeres in eggs in which the initiation of cleavage was delayed for successively longer intervals. [Adapted with modifications from S. Hörstadius, *Acta Zool* 9:1–191 (1928).]

166

next cleavage the spindles formed in the same planes as did those of the control eggs, giving the T-shaped arrangement of blastomeres shown in Figure 9.55E.

9.4.1.3 The Cleavage Pattern Is Fixed Soon After Fertilization.
If one separates the blastomeres formed by the first cleavage of a sea urchin egg, each blastomere continues to cleave as though it were part of a whole (Figure 9.56): at the next division it divides meridianally; then equatorially, producing two cells at the animal pole and two at

FIGURE 9.56 Half-cleavage pattern formed after three successive divisions of one blastomere of the sea urchin egg isolated after the first cleavage. [Adapted from S. Hörstadius, *Acta Zool* 9:1–191 (1928).]

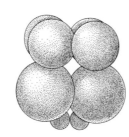

the vegetal; in the next division, the upper cells divide meridianally and the lower ones cut off micromeres. The total effect is, therefore, a half-cleavage pattern. But, if the egg is cut meridianally before fertilization and the nucleated half is fertilized, that half then cleaves in a whole pattern. The cortical control factors are, therefore, not patterned prior to fertilization. Hörstadius (1928) showed that the pattern becomes fixed progressively during the first few minutes after fertilization.

9.4.1.4 The Control Factors May Be Located in the Egg Cortex.
In the sea urchin egg, as in most eggs, the outer layer of cytoplasm, the cortex, is somewhat more rigid than the interior cytoplasm or endoplasm. The cortex remains relatively undisturbed by light centrifugal forces that easily displace the pigment granules, mitochondria, and other formed elements of the endoplasm. Hörstadius, Lorch, and Danielli (1950) found normal cleavage and development of sea urchin eggs after removing 50% of the endoplasm by means of a micropipet.

9.4.1.5 What Is the Mechanism for Forming the Cleavage Furrow?

The furrowing mechanism seems to operate chiefly in that region of the cortex where the furrow forms. Thus, if the furrow region of the cortex is physically restrained by means of suction with micropipettes, cleavage is inhibited (Rappaport and Ratner, 1967). Such restraint applied elsewhere, however, does not inhibit progress of the furrow.

There is strong evidence that the furrowing force is exercised by contractile material that is assembled in the cortex at the site of the impending cleavage plane. Microneedles inserted in the path of the furrow in holoblastically cleaving eggs impede its progress (Figure 9.57), presumably by creating a physical barrier to the closing of a ring of contractile

FIGURE 9.57 First cleavage of the egg of the sand dollar *Echinarachnius parma* with two microneedles inserted through the division plane. A: Appearance of the cleaving egg shortly after insertion of the needles. B: Later the furrow has deepened and pushed the needles together. If the needles are allowed to remain, the furrow does not proceed further. [From R. Rappaport, *J Exp Zool* 161:1–8 (1966), by courtesy of Dr. Rappaport and permission of Alan R. Liss, Inc.]

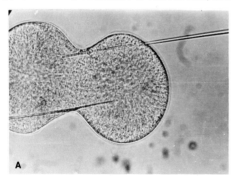

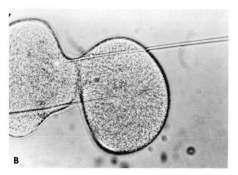

material. A smaller band of contractile material is found at the base of the cleavage furrow in the superficially cleaving egg of the squid. When this band is severed, the furrow disappears (Arnold, 1971, 1976).

Studies carried out by means of the electron microscope reveal that the contractile ring is made up of *microfilaments*. These are polymers of an actin-like protein (akin to muscle actin, which, with myosin, makes up the contractile element of muscle fibers). Microfilaments are approximately 5 nm in diameter, quite long, but of undetermined ultimate length. In cleaving eggs, microfibrillar material is found in randomly ordered array beneath the cell surface. In the region of the furrow, however, the microfibrils become aligned parallel to the plane of cleavage, forming distinctive linear arrays (Figure 9.58). Such arrays have been described for eggs of sea urchins (Schroeder, 1972), squid (Arnold, 1969), hydromedusae and polychaete worms (Szollosi, 1970), newts (Selman and Perry, 1970), and *Xenopus* (Luchtel et al., 1976). In the sea urchin *Arbacia* (Schroeder, 1972) the microfilaments become aligned in a circumferential ring in the cleavage plane a few seconds after the anaphasic separation of chromosomes begins, and persist at the base of the advancing cleavage furrow as a contractile ring until the furrow is completed.

Certain compounds known as cytochalasins have the property of causing microfilaments to become disassembled into their component protein subunits. In this condition, contractility is lost. Not surprisingly, therefore, cytochalasins block cytokinesis in cleaving eggs and cultured cells, but they do this without interfering with mitotic division of the nucleus. Schroeder has shown for cleaving eggs of the sea urchin *Arbacia* that application of cytochalasin B arrests cleavage in the matter of 60 seconds, by

167

Cytochalasin B
(a drug that causes microfilament disassembly)

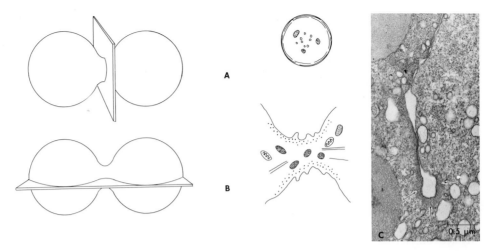

FIGURE 9.58 Distribution of microfilaments in the contractile band of the egg cortex during the process of furrow formation. A: As schematized for an echinoderm cell sectioned through the furrow in the direction perpendicular to the spindle axis (i.e., parallel to the direction of the furrow). B: As diagrammed for a similar cell in the region of the furrow in a section cut parallel to the spindle axis (i.e., perpendicular to the direction of furrowing). C: As seen by transmission electron microscopy in a cross section through a cleavage furrow in the egg of the squid. The electron-dense material at and around (arrows) the base of the furrow consists of dense aggregations of microfilaments seen in cross section. [Photograph courtesy of Dr. John M. Arnold.]

which time contractile ring filaments are no longer visible. Within this time the furrow will have begun to recede, although mitosis continues. The simultaneous disruption of furrowing and of the contractile ring seems to confirm a significant role for the latter in cytokinesis.

9.4.1.6 Interaction of Astral Rays and Cortex. The nonchromosomal elements of the mitotic figure consist of the **spindle fibers** and the **asters.** The latter comprise an array of fibers originating from the centrioles and, in cleaving eggs, extending radially to considerable lengths. Modern ultrastructural research suggests that both the spindle fibers and astral rays are microtubules, subcellular structures commonly noted in actively moving, elongating, or cleaving cells. Astral rays are formed by the polymerization of a protein, tubulin, about an organizing center, possibly a centriole.

The initiation of furrowing is possibly dependent on interaction between the astral rays and the egg cortex (Rappaport, 1971). When a spherical echinoderm egg is converted into a torus (the shape of a doughnut) by compressing it between two glass surfaces, with a small glass bead in the center (Fig-

ure 9.59A), the first cleavage produces a horseshoe-shaped binucleate cell (Figure 9.60B). In preparation for the second cleavage, mitotic spindles form in each arm of the horseshoe in such a way that two asters without an intervening mitotic spindle confront each other across the bend of the horseshoe. At cleavage, a furrow appears in the horseshoe bend between these asters shortly after the furrows appear in other parts of the cell (Figure 9.59C). From this experiment one may conclude that formation of a furrow does not require participation of a mitotic spindle and that the presence of appropriately positioned asters can bring about the surface differentiation required for normal furrow formation. Presumably this differentiation consists in the ordering of microfilaments below the cell membrane to form a contractile ring in the cleavage plane.

It requires the presence of two asters in order for a cleavage furrow to form. Moreover, whether the furrow appears or not is affected by the distance between the asters and by their distance from the cell surface (Figure 9.60C). In addition, the astral influence must be exercised for a given period in

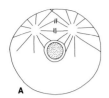
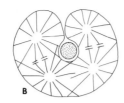
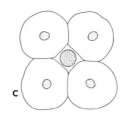

FIGURE 9.59 Cleavage in a torus-shaped egg of the sand dollar *Echinarachnius parma*. A: A small glass bead is pushed through a cell by compression under a cover slip: the mitotic apparatus is displaced to one side of the "hole in the doughnut." B: The first cleavage occurs as a result of furrowing on one side of the bead. The daughter nuclei have formed spindles for the next division. C: Formation of four blastomeres as a result of furrow formation between all astral centers. [An interpretative drawing based on photographs in R. Rappaport, *Dev Growth Differentiation* 12:31–40 (1970).]

order that a furrow may be initiated. In eggs of sand dollars under appropriate conditions (Figure 9.61) the required period of stimulation is 1 minute at 18° C (Rappaport and Ebstein, 1965).

Although the correlation between the appearance of the contractile band of microfilaments and the impingement of the astral rays on the cell cortex is highly suggestive of the functional connection between these elements in determining cleavage, the mechanism that positions the furrow remains unclear. In later experiments Rappaport (1978) showed that rapid stirring of the cytoplasm intervening between the spindle and egg cortex fails to inhibit furrowing. On the other hand, when the cytoplasm is agitated before the appearance of the astral structures, the formation of the mitotic apparatus is retarded and furrowing is inhibited. Therefore, early events in the development of the astral apparatus are presumably of greater importance for the assembly of components of the furrowing mechanisms than are later ones.

9.4.2 SPECIAL PROBLEMS OF BLASTULA AND GASTRULA.
The problems posed by the configuration of the blastula in many animal forms and by the behavior of cells and cell sheets are considered here. Blastulae are of several kinds. The typical echinoderm blastula, called an **equal coeloblastula**, is hollow and the cells are rather uniform in size. Amphibians have an **unequal coeloblastula**. The **discoblastula** of bird and fish comprises a layer of cells capping a very large yolk mass; insects and a number of crustaceans have a **superficial blastula**, the blastoderm consisting of a layer of cells sur-

rounding a large yolk mass; in primates one finds the **morula**, a solid mass of cells from which a blastocyst later differentiates.

The rearrangements that occur when these various kinds of blastulas undergo gastrulation are described earlier in this chapter. These rearrangements, as Trinkhaus (1969) has noted, "combine cell movements which in their aggregate are the most sweeping to occur in all of embryogenesis."

FIGURE 9.60 Critical distances for furrow formation in eggs of the sea urchin *Hemicentrotus pulcherrimus*. A: Displaced mitotic spindle in a flattened egg. B: Furrow formation is restricted to the region of the former mitotic spindle; daughter nuclei are on opposite sides of the furrow. C: Diagram of distances measured in determining conditions for furrow formation. In normal eggs at first cleavage, the asters are about 29.5 μm apart. In flattened eggs, furrowing can occur if the asters are no more than approximately 32.5 μm apart (axis *a*) provided that the spindle-to-surface distance (axis *b*) is around 35 μm. [An interpretative drawing of data from R. Rappaport, *J Exp Zool* 171:59–68 (1969).]

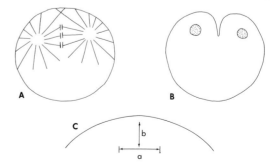

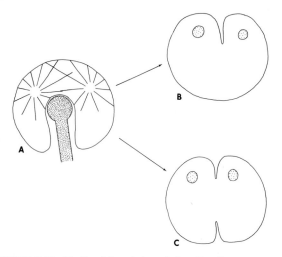

FIGURE 9.61 Method for determining the time required for exposure of the egg cortex to astral rays in order for furrow formation to occur in *Echinarachnius parma*. The mitotic apparatus, pushed to an eccentric position, as in Figure 9.60A, would give a unilateral furrow, as shown in Figure 9.60B. In this experiment, when the more distant surface is pushed into the zone between the asters and held for less than 1 min at 18°C, the same result occurs, as shown in B. But, if the distant surface is held in this zone for more than 1 min, but less than 3½ mins, the cell returns to its original shape, but a furrow then appears in the more distant surface as shown in C. [Interpretative drawing based on R. Rappaport and R. P. Ebstein, *J Exp Zool* 158:373–82 (1965).]

They involve the movements of sheets of cells, as in invagination and epiboly, and the detachment and ingression of individual cells from cell sheets. But it is important to recognize that gastrulation is a process in which the whole of the germ is totally and simultaneously involved in such a way that the movement of each of its components is integrated with that of every other one in a unified pattern that is characteristic for each species. This means that it is almost impossible to find common ground for analyzing gastrulatory patterns in different kinds of organisms. Accordingly, salient features of cellular mechanisms underlying gastrulatory movements are considered in some of the material that follows.

9.4.2.1 Sea Urchin Gastrulation.

At the time gastrulation is to begin, the sea urchin blastula consists of a single layer of cells, spherically arranged, whose exterior sides are firmly anchored in the hyaline layer (Figures 8.20 and 9.62). The cells adhere to each other rather uniformly around the circumference of the sphere. At the onset of gastrulation, a thickening (Figure 9.63) of the cellular wall of the vegetal region arises in a zone forming a ring around the vegetal pole. This thickening results from an increase in the length of contact between adjacent cells, which thus assume a columnar configuration. The increased length of contact is interpreted as resulting from increased adhesiveness of the lateral membranes of the cells comprising the ring.

Cells in the sheet circumscribed by the ring then become less cohesive with their neighbors, thus diminishing height, so that the sheet tends to expand and flatten, forming the vegetal plate. Because the cells of the vegetal plate are anchored in the

FIGURE 9.62 Diagram of the relationship between the cells of the blastula wall and the hyaline membrane in sea urchin eggs. Cellular processes are anchored in the layers of the hyaline membrane externally. Bands of septate desmosomes run entirely around each cell near its external border, linking adjoining cells firmly together. [Adapted from T. Gustafson and L. Wolpert, *Biol Rev* 42:442–98 (1967).]

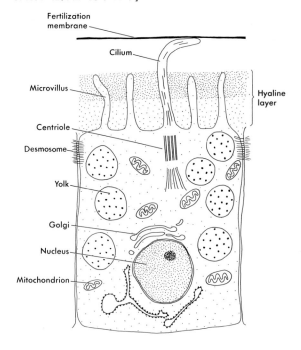

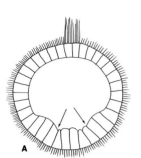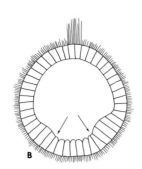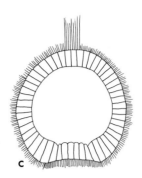

FIGURE 9.63 Postulated changes in cellular relationships as gastrulation begins in the sea urchin *Psaminechinus miliaris*. Increased adhesion between cells leads to enhanced cell length in a circumferential ring around the vegetal zone, as designated by arrows in A and B. Expansion of the intervening cells of the vegetal plate in the presence of the surrounding ring and relatively inelastic hyaline layer (see Figure 9.62) causes invagination to begin as shown in C. [Adapted from T. Gustafson and L. Wolpert, *Biol Rev* 42:442–98 (1967).]

relatively inelastic hyaline layer (Figure 9.62), however, their tendency to expand is restricted on the outer side and thus continued expansion forces the vegetal plate to curve inward (Figures 9.4B and 9.62C). If the hyaline layer is removed by digestion with proteolytic enzymes or weakened by use of calcium- and magnesium-free seawater, the outer surfaces of cells at the vegetal plate show pulsatory movements similar to those of the inner surface, and primary invagination of the archenteron fails to occur, or is highly abnormal (Citkowitz, 1971). These observations are consistent with the notion that initial phases of invagination are caused by diminished adhesivity of vegetal plate cells whose expansion is restricted at their outer surfaces.

The flattening of the vegetal plate is followed by enhanced pulsatory activity on the part of the presumptive primary mesenchyme cells (Figure 9.4B). These then leave the blastula wall as rounded cells, singly or in groups, apparently by virtue of a decrease in their adhesivity both for each other and for the hyaline layer. First piling up at the vegetal pole, they begin to send out long filipodia, apparently randomly, into the blastocoele. Where the filaments make stable contact with the blastocoele wall, their contraction furnishes the basis for migration of the cells. Contacts are made and broken, and the mesenchyme cells may become considerably dispersed around the walls of the blastocoele, extending as much as halfway to the animal pole.

Eventually, however, they assume an ordered arrangement in a ring around the lower half of the blastula wall (Figure 9.4C). The distribution of cells in the ring is not uniform, for they tend to collect in greater numbers at two points, one on each side of the invaginating archenteron at the ventrolateral sides (Figure 9.4D). From these clusters cells then extend toward the animal pole in cable-like strands. Where these strands and the vegetal ring intersect, the mesenchyme cells begin secreting the skeletal rods of the larva.

Motion picture analysis of the movements of the primary mesenchyme cells suggests that their typical pattern is the result of the random exploration of the blastocoele wall by their filopodia, which attach to the wall with varying degrees of stability. The final pattern, therefore, reflects those regions of the wall where the contact is most stable. This interpretation requires, of course, the further postulate that the inner wall of the blastula will have developed a pattern of selective adhesive properties in relation to the primary mesenchyme.

Invagination of the ventral plate occurs in isolated vegetal halves, extending as mush as one third of the normal distance to the animal pole before ceasing its inward movement (Moore and Burt, 1939). In the intact animal this is about the point at which there occurs a plateau in the invagination time curve (Figure 9.64). Invagination thus occurs in two phases. Clearly, the primary invagination

171

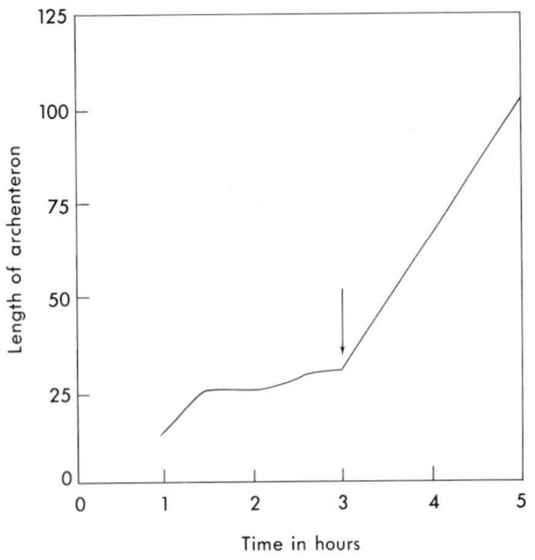

FIGURE 9.64 Primary and secondary invagination of the archenteron in *Psammechinus miliaris* are illustrated in this plot of archenteron length against time. The primary phase is characterized by pulsatile movements of the cells of the archenteron tip and is relatively slow. The secondary phase, beginning at the point designated by an arrow, is much more rapid. Filopodia now form, attach to the blastocoele wall, and presumably accelerate invagination by contracting. [Adapted from T. Gustafson and L. Wolpert, *Exp Cell Res* 22:437–49 (1961).]

occurs by virtue of forces intrinsic to the vegetal plate, presumably those outlined above. The secondary advance of the archenteron then is correlated with the spinning out of long filopodia from pulsating cells at its tip. These are the secondary mesenchyme cells, which will eventually detach from the archenteron. First, however, their filopodia vigorously explore the inner wall of the blastula near the animal pole, making and breaking attachments, but apparently finding their most stable contacts at the inner junctions of the ectodermal cells.

Secondary invagination is apparently brought about largely by the contraction of the filopodia of the secondary mesenchyme. The filopodia exert sufficient contractile tension to pull the blastocoele inward at their attachment points, and they can be observed to shorten as the secondary phase of invagination begins. If the filopodia are cut, the archenteron immediately retracts, and does not resume its advance until new filopodal connections with the animal pole are made (Trinkhaus, 1969).

What is the structural basis of the probing activi-

ties and contractile power of the filopodia? Studies with the electron microscope (Tilney and Gibbins, 1969) reveal that the base of the filopodium and adjacent cell body are endowed with a cytoskeleton of abundant microtubules. These presumably act as stiffeners for the exploring filopodia, but probably are not a part of the contractile apparatus. The slender distal part of the filopodium, however, is filled with elongate filaments 50 Å in diameter. These very much resemble the actin-like microfilaments found at the base of the cleavage furrow in various kinds of eggs, and which are thought to function by contraction.

9.4.2.2 Cellular Basis for Amphibian Gastrulation. Although the cell movements of gastrulation in amphibian embryos have probably been studied longer and more intensively than in embryos of any other vertebrate group, their underlying basis is only now beginning to be understood, albeit still imperfectly. Important to the achievement of this understanding is the recognition, emphasized by Trinkaus (1976), that the superficial cells of the blastula, comprising both ectoderm and endoderm, constitute a continuous sheet of cells, confluent and firmly adherent to one another. This condition provides a basis for the precise integration of the complex interplay of the component morphogenetic processes of gastrulation—ingression, involution, invagination, and epiboly.

The ingression of superficial cells is associated with the first signs of gastrulation. These cells, called **bottle** or **flask** cells, project the bulk of their bodies deep among the yolky cells of the interior of the embryo, meanwhile retaining their connection to the surface by their long, attenuated necks (Figure 9.65). At the surface the bottle cells show extensive interlocking connections with the neighboring cells (Figure 9.66). As more bottle cells appear, there occurs a surface indentation, which is the beginning of the blastoporal groove. The elongated neck of the bottle cell gives every indication of being under tension, for if the neck is detached from the surface, it retracts into the cell body. Holtfreter (1943) thought of the bottle cells as towing the surface of the blastoporal cells inward, communicating their pull to the adjacent cells and thus initiating invagination.

Bottle cells are associated with the apex of the archenteron for a while after invagination begins (Figure 9.67B, C), but it is doubtful that any loco-

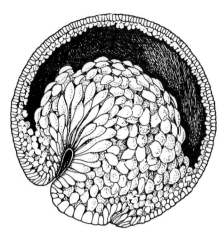

FIGURE 9.65 "Bottle cells" in amphibian gastrulation. Schematized saggital section through an early gastrula showing bottle cells attached to the surface of the blastoporal groove. [From J. Holtfreter, *J Exp Zool* 94:261–318 (1943), by courtesy of Dr. Holtfreter and permission of Alan R. Liss, Inc.]

motor action on their part is responsible for deepening the blastoporal groove. In fact, when the bottle cells are removed at the site of the nascent dorasl lip of the blastopore, the archenteron forms normally without them (Cooke, 1975). Furthermore, the bottle cells lose their connection with the surface relatively early in gastrulation (Figure 9.67C, D) and merge with other endoderm cells in the floor of the archenteron. It would appear, therefore, an active inward movement of bottle cells can account for only the beginning phase of invagination and not for the involution through the blastopore of the endodermal mass.

It seems reasonable to propose, rather, that the mesoderm is responsible for the involution of endoderm and the enlargement of the archenteron (Trinkaus, 1976). Mesoderm from the isolated dorsal lip of the blastopore shows an autonomous capacity to expand, and thus its anterior extension during gastrulation can be attributed to intrinsic rather than to extrinsic forces. Moreover, as Schechtman (1942) showed, prospective head mesoderm will migrate across the inner surface of prospective ectoderm in explants, carrying with it adherent prospective endoderm. Drawing heavily on observations of Keller and Schoenwolf (1977), Trinkaus (1976) suggested that at the late blastula stage, pro-

spective mesodermal cells become locomotory, moving first deeper and then forward (Figure 9.67B) on the substratum provided by the overlying ectodermal cells. Prospective endodermal cells adhere to the forward-moving mesodermal cells and are carried along with them. Because the superficial endodermal cells are confluent and firmly adherent to each other, they move as a sheet, creating both the roof and floor of the archenteron and accounting for the involution of endoderm on both sides of the blastoporal groove.

Because the more superficial prospective mesodermal cells are carried toward the blastopore even as the deeper cells are moving in the opposite direction during gastrulation, the dorsal lip of the blastopore is the site of a massive involution of mesoderm. The direction of involution is denoted in Figure 9.67B– C by means of curved arrows. In anurans it now seems to be clear that no surface cells of the blastula undergo involution and thus they do not participate in formation of the notochord and other mesodermal derivatives. Among urodeles, however, it has been reported that material for forming the notochord is located in the surface layer prior to gastrulation and involutes through the dorsal lip of the blastopore. Other mesodermal derivatives are not represented in the surface layer (Løvtrup, 1966).

Finally we may turn to the cellular processes that account for the epibolic spread of ectoderm during gastrulation. It is not yet clear to what extent epiboly represents the active spreading of ectoderm as opposed to its passive stretching. Keller (1980) studied the changing shapes of ectodermal cells during gastrulation in *Xenopus.* At the beginning of gastrulation the prospective ectoderm of the dorsal and ventral marginal zones consists of a superficial single layer of contiguous, tightly adhering cells, about 50 μm thick, underlaid by a so-called interdeep layer of more loosely adhering cells of about equal thickness and a somewhat thicker, deeper layer of cells referred to as the inner-deep layer. As gastrulation proceeds, cells from the two deep layers move between one another (interdigitation), forming a single layer of slightly columnar cells. Concomitantly the superficial cells flatten, increasing their surface area. With continued progress of gastrulation, the inner layer of cells becomes shorter, and the superficial cells flatten and spread. A similar pattern of changes occurs in the ectoderm

173

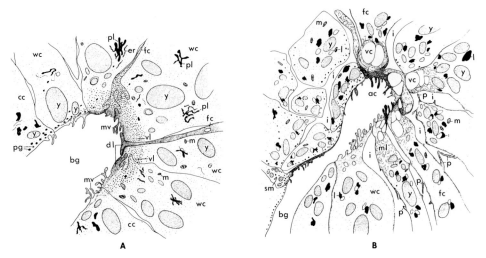

FIGURE 9.66 Interpretive drawings of electron microscopic views of parasaggital sections through the blastopore of the frog *Hyla regilla* at the stage of the beginning (A) and crescent-shaped (B) blastopore. In A the blastopore groove, bg, is U-shaped, with the base of the U constituting the innermost section of the groove. The groove is lined with long-necked flask or bottle cells, fc; wedge-shaped cells, wc; and cuboidal cells, cc. The cortical zone of the invaginating cells contains electron-dense material, dl, presumably microfilaments, and vesicular material. B: The archenteric cavity, ac, is now enlarging, and electron-lucent intercellular spaces, i, bridged by projections, p, are evident. Lipid droplets, l, are present but not so frequently in polymorphic form, pl, as in A. Membranous layers, ml, have appeared in many invaginating cells. Electron-dense cortical material lines the archenteric cavity, being particularly prominent in certain specialized mesodermal cells, sm. Of particular note in both A and B are the interdigitating and overlapping cellular margins in the blastoporal groove. Abbreviations not otherwise noted are er, endoplasmic reticulum; m, mitochondria; vc, vacuoles; y, yolk platelets. [From P. C. Baker, *J Cell Biol* 24:95–116 (1965), by courtesy of Dr. Baker and permission of the Rockefeller University Press.]

of the blastocoele roof, but the flattening of superficial and interdigitated deep cells occurs more slowly and is not so extreme. In the prospective neural ectoderm, interdigitation of the cells of the deep layers is incomplete, the cells remaining in a pseudostratified columnar configuration. The superficial cells, however, are somewhat flattened.

Conceivably, the epibolic spread of the ectoderm could be brought about autonomously by the active force generating interdigitation and flattening of deep cells, particularly in the marginal zones. On the other hand, it is conceivable that during epiboly the superficial layer of ectoderm is stretched passively by tension generated at the margin of the blastopore. As the superficial cells stretch, the deep cells would rearrange themselves to occupy the increased area available to them, but might not generate a spreading force by their interdigitation. Further studies of cellular behavior and of mechanical stresses in the ectoderm are required to understand the basis of epiboly in amphibian gastrulation.

9.4.2.3 Epiboly in Teleosts. The formation of the germ layers in teleosts is described earlier in this chapter. Here we turn our attention to the cellular basis for the dramatic epibolic movements that occur in teleosts.

There are two chief components of epiboly in teleost eggs: epiboly of the yolk syncytial layer and epiboly of the blastoderm (cf., Figure 9.17 and 9.19). The epiboly of the YSL involves a controlled flow of cytoplasm and nuclei from the thicker YSL at the animal pole progressively into the thinner yolk cytoplasmic layer with which it is continuous. As noted in Section 9.2.4.1, the E-YSL precedes the

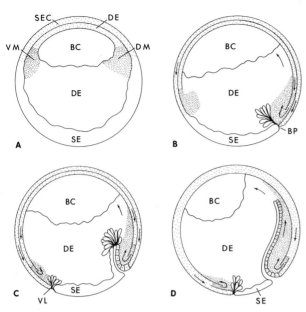

FIGURE 9.67 Schematic sagittal sections of the embryo of *Xenopus laevis* at the late blastula stage (A) and at progressively later stages (B–D) of gastrulation showing the distribution of materials for forming each of the germ layers. Prospective ectoderm is designated by means of light stippling and prospective mesoderm by means of dark stippling. Endoderm is unstippled. Cellular outlines are shown only for bottle cells and for endodermal cells of the archenteron roof. Arrows show the direction of cellular movements in the ectoderm and endoderm and the involuting mesoderm. BC, blastocoele; BP, blastopore; DE, deep endoderm; DEC, deep ectoderm; DM, dorsal mesoderm; SE, superficial endoderm; SEC, superficial ectoderm; VM, ventral mesoderm. [Adapted from R. E. Keller and G. C. Schoenwolf, *Wilhelm Roux' Arch* 182:165–86 (1977).]

gether with accompanying growth could create an outward push, driving the blastoderm over the yolk. The fact is, however, that for the teleost *Fundulus heteroclitus,* at least, epiboly of the blastoderm proceeds to completion in the presence of a concentration of the drug colchicine sufficient to block all mitoses. Thus, epiboly of the blastoderm is not dependent on division of its constituent cells. Another possibility is that marginal cells of the enveloping layer move actively in epiboly, towing and stretching the remainder of the enveloping layer by migrating vegetally, using the YSL as a substratum. That this may be so is suggested by the fact that a blastoderm detached microsurgically from the YSL and then allowed to reattach will first contract and then expand and eventually catch up with the rapidly advancing margin of the E-YSL (Figure 9.68). Thereafter the blastoderm completes its epiboly behind the advancing front of the E-YSL. Yet, it must be recognized that any epithelium, artificially provided with a free edge by cutting, will expand if provided with an appropriate substratum. The I-YSL presumably provides such a substratum, but the fact that the vegetalward progress of blastoderm epiboly is tied to the progress of epiboly of the E-YSL suggests that one must look for clues as to the mechanism for epiboly of the blastoderm in the relationship between the marginal zone of the enveloping layer and the E-YSL.

These relationships have been analyzed in *Fundulus* embryos by Betchaku and Trinkaus (1978) by means of electron microscopy. These investigators found that throughout the course of epiboly the vegetalward edge of each marginal cell of the enveloping layer is firmly attached to the E-YSL by means of an extensive junctional complex involving interdigitating folds of apposed membranes and a mixture of tight and close junctions. When the enveloping layer is forcibly detached from the YSL, the marginal cells tear where joined to the YSL and leave parts of themselves behind. The fact that the marginal cells of the enveloping layer are firmly embedded in the YSL, which is capable of undergoing epiboly independently and more rapidly in the absence of the blastoderm, suggests that the blastoderm moves passively in epiboly rather than actively.

This possibility is reinforced by the observation that, prior to the onset of epiboly of the blastoderm, the E-YSL extends about 400 μm beyond the

175

blastoderm in epiboly to the vegetal pole but is in no way dependent on epiboly of the blastoderm for its progress. In fact, if the marginal cells of the enveloping layer of the blastoderm are detached from the E-YSL, the vegetalward progress of the E-YSL accelerates considerably, as though it was freed from some constraint. The E-YSL then goes on to complete its epiboly at the vegetal pole. Obviously, the blastoderm, is not required in order for epiboly of the YSL to occur, but, thus far, the motive force for YSL epiboly is obscure.

What, then, is responsible for epiboly of the blastoderm? One possiblity is that cell division to-

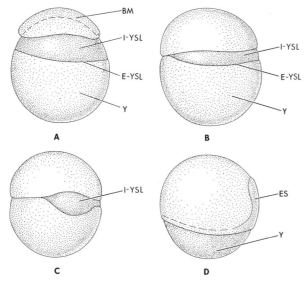

FIGURE 9.68 Reattachment and epiboly of the blastoderm of *Fundulus heteroclitus* after detachment from the yolk syncytial layer. A: 30 min after detachment; the blastoderm, BM, has contracted and thus exposed the internal yolk syncytial layer, I-YSL. The margin of the external yolk syncytial layer, E-YSL, is advancing in epiboly at an accelerated rate. B: After 105 min the blastoderm has expanded, migrating over the internal yolk syncytial layer and partially reattaching to the external yolk syncytial layer; the latter, meanwhile, continues to envelop the yolk, Y. C: After 330 min the blastoderm has overtaken the advancing E-YSL around most of its margin. D: After 480 min the E-YSL and blastoderm continue in epiboly over the remaining yolk; the embryonic shield, ES, seen in profile, is laying down the axial organs of the embryo. [From J. P. Trinkaus, *J Exp Zool* 118:269–320 (1951), by courtesy of Dr. Trinkaus and permission of Alan R. Liss, Inc.]

margin of the blastoderm and shows a smooth surface. But, as epiboly begins, the E-YSL progressively narrows to about 35–40 μm in width and its surface becomes highly folded and convoluted. This convolution of the E-YSL is accompanied by a thickening of the cortical network of microfilaments that fill the folds and suggests that a contractile force resides in the E-YSL. This force narrows the width of the E-YSL and throws its surface into folds, meanwhile exerting tension on the attached margin of the enveloping layer and stretching it under tension toward the vegetal pole. Importantly, narrowing and folding of the E-YSL occurs even after

detachment of the blastoderm early in epiboly, thus increasing the likelihood that active contractile narrowng of the E-YSL is the motive force for expansion of the blastoderm, at least during epiboly past the equator of the egg, when folding is no longer seen.

9.4.2.4 Epiboly and Gastrulation in the Chick Embryo.

In the chick embryo the margin of the blastoderm spreads over the yolk beneath the vitelline membrane. Eventually the yolk is completely enveloped, and the enclosing membrane becomes the vascular yolk sac (Figure 9.69). The embryo proper is not involved in this expansion. It is necessary, therefore, to discuss separately the cellular basis for the epibolic movements involved in yolk sac formation and the mechanisms whereby the cells of the bilaminar blastoderm form the germ layers. Epiboly is discussed first.

FIGURE 9.69 Regional differentiation of the blastoderm and formation of the yolk sac in the chick embryo. A: The embryo, area pellucida, and area vasculosa viewed from the animal pole at about 40 hours of incubation. The yolk sac (stippled) has spread almost to the equator of the yolk. B: After about three days of incubation the area vasculosa has greatly expanded, and the yolk sac has almost completely enveloped the yolk.

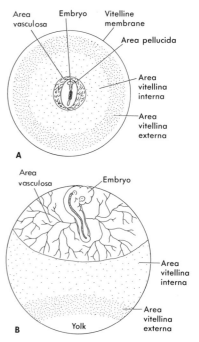

The leading edge of the expanding blastoderm consists of a thin layer of ectodermal cells, called the **margin of overgrowth.** Further back, the ectoderm is underlaid with endoderm constituting the **area vitellina.** Mesoderm invades the area vitellina, bringing in blood vessels and constituting the **area vasculosa** (Figure 9.69). The ectodermal cells that make up the margin of overgrowth appear to be the chief locomotor organs of the expanding blastoderm. These are the only cells that are strongly adherent to the underside of the vitelline membrane; when they are detached from the vitelline membrane in blastoderms isolated in vitro, the blastoderm separates completely from the membrane.

The attached cells are greatly flattened in a zone about 80 μm wide. Here and there can be seen villous protrusions of their contact surfaces into spaces in the fibrous vitelline membrane, and, with proper fixation, electron dense placques are found in zones of intimate contact with the substratum (cf., Chapter 11). The most distal cell of the attached zone extends its leading lamella over the vitelline surface and is itself partially underlapped by the lamella of the next most proximal cell (Figure 9.70). Scanning electron micrographs of the advancing marginal zone reveal that the marginal lamellae extend filopodia as much as 30 μm in length in advance of their pathway of expansion (Figure 9.71). The ends of these filopodia may be rounded, knob-like, or flattened, and appear to be anchored firmly to the vitelline membrane. Presumably, contraction of these filipodia is an important mechanism for advance of the blastoderm over the yolk.

The marginal region of a blastoderm will spread

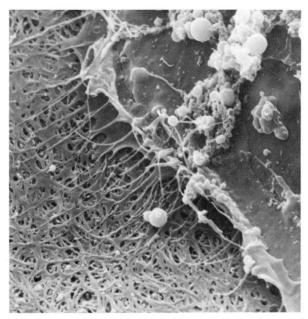

FIGURE 9.71 Scanning electron micrograph of the edge of the epiblast extending over the inner face of the vitelline membrane in the chick. Numerous filopodia extend across the vitelline membrane. A portion of the advancing edge is curled back, revealing the presence of more than one layer of filopodia. [From E. A. G. Chernoff and J. Overton, *Dev Biol* 57:33–46 (1977), by courtesy of the authors and permission of Alan R. Liss, Inc.]

177

even if isolated from the rest of the blastoderm, but, when the blastoderm is present, this spreading tendency tends to pull apart the nonmarginal cells. As Trinkaus (1969) pointed out, it is significant that the ectodermal cells of the blastoderm are held together only loosely during the first 18 hours of incubation, when the tension is not yet severe. But, as the stretching becomes more intense, specialized junctions appear, such as desmosomes, which bind cells more tightly together.

As noted, formation of the embryonic body in the chick occurs within the area pellucida. The germ layers arise and undergo their morphogenesis independently of yolk-sac epiboly. Earlier in the chapter, the origin of the hypoblast, formation of the primitive streak, and ingression of endoderm and mesoderm through the primitive streak are discussed. Attention is now directed to some aspects of

FIGURE 9.70 Diagrammatic section through the marginal zone of the chick blastoderm. [Adapted from J. R. Downie and S. M. Pegrum, *J Embryol Exp Morphol* 26:623–35 (1971).]

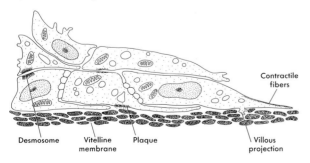

Contractile fibers

Desmosome   Vitelline   Plaque   Villous
               membrane   projection

the cellular basis for these morphogenetic movements.

The question has been raised many times whether differential mitotic rates in the epiblast are responsible for the formation of the primitive streak and the movements of the epiblast during avian gastrulation. Most investigators, however, have found no local differences in mitotic rate in the epiblast and primitive streak (reviewed by Bellairs, 1971). Spratt analyzed the pattern of cellular movements in the epiblast by marking the cells with powdered blood charcoal (Figure 9.28). The movements of the charcoal marks reflect with some accuracy the movements of the epiblast cells in relation to the forming primitive streak. These movements are presumably due to autonomous tendencies for the cellular sheet to deform according to a particular pattern. Possibly, however, this pattern is imposed in some measure by the underlying endoderm. If the endoderm is removed from a blastoderm at an early primitive streak stage and replaced at right angles to its original axial orientation, the primitive streak continues to form, but it bends in the direction of the original anterior end of the endoderm (Waddington, 1933; Vakaet, 1967; also Bellairs, 1971).

Studies with the electron microscope have provided some new insights into the formation of the mesoderm from epiblast (Figures 9.30 and 9.72). Lateral to the streak, the epiblast cells consist of a low columnar or almost cuboidal epithelium. The boundaries of adjacent cells are essentially parallel to each other and perpendicular to the surface of the epithelium. The space between the membranes of adjoining cells is around 113 Å (Balinsky and Walther, 1961), and no specialized attachments are found between them. A discontinuous basement lamina is found under the epiblast.

In the region of the primitive streak, Trelstad et al. (1967) have observed that adjacent cells are in contact by means of minute **tight junctions.** These are focal points at which no space can be resolved between apposed cell membranes. There are also focal **close junctions,** small regions where the space between neighboring cell membranes is less than 100 Å.

As the epiblast cells approach the streak, they assume a tall columnar shape, and the epithelium appears to be stratified. Toward the upper surface of the streak the cell membranes show irregular fold-

178

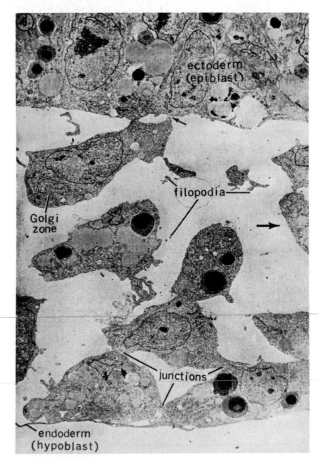

FIGURE 9.72 Mesodermal cells migrating laterally between epiblast and endoderm (hypoblast) after leaving the primitive streak. The direction of migration is to the right, as seen in this illustration. The mesoderm migrates as a continuous layer, its cells in contact with each other either broadly or by means of filopodia and also in contact with epiblast above and hypoblast below. [From R. L. Trelstad et al., *Dev Biol* 16:78–106 (1967), by courtesy of the authors and permission of Academic Press, Inc., New York.]

ings and interdigitation. Now one begins to see among cells of the central part of the primitive streak some that are extremely elongated, their deeper ends broadened, but with attenuated necks extending to the surface of the mesoblast. These are similar to the bottle cells of the amphibian gastrula. Balinsky and Walther (1960) reported that the contents of these cells gradually move deeper, attenuating the neck progressively, until it loosens its hold to the surface. The neck is withdrawn and the cell becomes a mesenchyme cell, a part of the mesodermal layer.

The cells that leave the primitive streak are polarized with respect to the distribution of their contents. The original basal (inner) end of the cell containing the nucleus leads the way, and the apical end containing the Golgi trails behind (Figure 9.73). The leading end, chiefly, develops filopodia that contain filaments about 50 Å in diameter, which are possibly contractile and serve in cellular locomotion and lateral movement of the mesodermal sheet. Microtubules are present in the cell body, perhaps serving as the cytoskeleton (Trelstad et al., 1967).

Until fairly recently students of the chick embryo have suffered from the illusion that the mesoderm is formed by the migration of single cells away from the primitive streak. The work of Trelstad et al. (1967), however, reveals that the expansion of the mesoblast should be regarded as the spreading of a cellular sheet. Extensive areas of membrane apposition occur between cells of the mesenchyme, and there are numerous focal junctions of both the close and tight variety. Many of the junctions may be newly formed by exploring filopodia, but, conceivably, some of the early connections developed in the epiblast may be retained in the expanding mesoblast.

Near the primitive streak, the mesoblast is many layered, but near its margin single cells may bridge the gap between epiblast and hypoblast. Filopodia from the mesenchyme seem to probe and feel their way along the incomplete basement lamina of epiblast and hypoblast, often contacting cells in lamina-free regions and forming a tight junction between them (Figure 9.72).

The observations recorded here are marks of progress in our understanding of gastrulation in avian embryos. Clearly it involves chiefly movements of cohesive cellular sheets, but what are the forces that direct these movements? The presumption has been that all of the cells of the epiblast that converge upon the primitive streak pass through it, but do they? Probably, but this is not absolutely certain. What is the significance of close junctions as contrasted with tight junctions between cells in gastrulation? Do either or both of these lead to formation of permanent junctions? These are all ques-

tions related to "the most difficult and most neglected of all basic problems in morphogenesis . . . that of supracellular integration" (Weiss, 1951).

9.5 Summary

The animal body plan is fundamentally simple: an outer layer, ectoderm, covers a middle layer, mesoderm, which surrounds an inner, tubular layer, the endoderm. To begin constructing this body, the zygote divides mitotically and the daughter cells repeatedly divide until many cells are formed.

The early divisions of the embryo are called cleavage divisions. The pattern in which these divisions occur varies according to the relative proportions and the distribution of yolk in the egg. The cleavages may be holoblastic or meroblastic, depending on whether they divide the entire germ or not, and they may be radial or bilateral, depending on whether several or only one meridianal plane divides the egg into mirror-twin halves. They are discoidal if confined to a nucleated protoplasmic disc at one pole of the egg, and superficial if they lead to a blastoderm surrounding a central yolk. Spiral cleavage is a variant pattern of radial cleavage.

At the end of cleavage, the embryo is a blastula or, in mammals, a morula. By a variety of marking experiments fate maps of the blastula can be constructed. These show the prospective fates realized by the parts of the blastula in subsequent development.

In various ways, according to the cleavage pattern, cells of the blastula are transformed into a three-layered configuration. The transformation is termed gastrulation, and it involves chiefly processes of invagination, ingression, and epiboly, these processes participating to different degrees in gastrulation in different kinds of embryos.

In mammalian embryos the morula forms a vesicular blastocyst containing an inner cell mass, and an outer layer, the trophoblast. The blastocyst implants in the wall of the uterus, and the trophoblast, in conjunction with maternal tissue, forms the placenta.

The cleavage pattern is fixed soon after fertilization, and the factors that determine it lie in the cortex of the egg. These then determine the position of the mitotic spindles of the cleavage divisions. The position of the cleavage furrow is determined by the asters whose fibers, microtubules, interact with the cortex to bring about the assembly of actin microfibrils into a contractile ring or band at the site of the impending cleavage furrow.

Gastrulation combines sweeping movements of sheets of cells in invagination and epiboly and the detachment, ingression, and reconnection of individual cells. Because eggs cleave in such varied patterns and form blastulas of widely diverging kinds, it is impossible to find common ground for the comparative analysis of gastrular patterns. The analysis, instead, must be made at the cellular level. Of particular note in sea urchin embryos are changing cellular adhesivities involved in the ingression of the primary mesoderm, and "searching" and contractile movements of filopodia, and their responsibility for invagination and positioning of the archenteron. In amphibian embryos the salient features are the sinking in of the flask cells, initiating blastopore formation, and the coordinated expansion of prospective ectoderm over the involuting mesoderm and endoderm. In embryos of the teleosts, epiboly of the blastoderm is dependent entirely on attachment of the envelope layer to the yolk cytoplasmic layer which precedes the blastoderm over the surface of the yolk. Teleosts, in contrast to amphibians, reptiles, and birds, seem to form their germ layers entirely by rearrangement of the cells of the deeper layers of the blastoderm, without involving movements of invagination or ingression. Birds are distinctive in that they form their embryonic germ layers by invagination through a primitive streak, a process that occurs in modified form in mammals. Epiboly in birds is distinctive from that of teleosts in that it does not involve a periblast and epiboly plays no role in the formation of the embryo.

180

9.6　Questions for Thought and Review

1. What are the names of the germ layers? From what germ layers are the sensory cells of the eyes derived? The cells that line the digestive tract? The blood vessels?

2. Describe the sequence of cleavage planes in the egg of *Synapta*. How does cleavage differ in the sea urchin, frog, and chick? What is spiral cleavage?

3. What is meant by the expression prospective fate? What is the prospective fate of cells near the vegetal pole of an amphibian blastula? Of the descendants of the micromeres in a sea urchin embryo?

4. What is the cavity of the blastula called? What new cavity is formed during gastrulation?

5. What is the gray crescent of the amphibian embryo? How is its material distributed during cleavage?

6. Describe and name the major mass movements of cells that occur during amphibian gastrulation.

7. In amphibian gastrulation what is the dorsal lip of the blastopore? What is the fate of materials moving through it?

8. Describe the basic body plan of a vertebrate as exemplified by the amphibian neurula seen in cross section.

9. What is the roof of the archenteron? How does it differ in origin in frogs and urodeles?

10. What is meant by the term blastoderm as applied to the bird embryo? From where does it arise? How does it differ in the insect?

11. In what kinds of embryos is a primitive streak found? What are its parts? What is its significance in gastrulation?

12. How is mesoderm formed in the eggs of insects?

13. Can an organism that shows a half-cleavage pattern form a whole embryo? Think about this again after reading Chapter 10.

14. When sea urchin eggs are stimulated parthenogenetically, extra asters sometimes form. Might these possibly become involved in furrow formation? If so, under what circumstances?

15. What is the evidence for involvement of the cortex in positioning of the cleavage furrows?

16. Distinguish between microfilaments and microtubules. How do they differ structurally, and what distinctive roles do they play in cleaving cells?

17. What is the significance of the hyaline layer in cleavage and gastrulation in sea urchin eggs?

18. What is the experimental evidence for the involvement of filopodia in gastrulation in the sea urchin?

19. Describe the presumed role of the flask cells in amphibian gastrulation? What coordinates the expansion tendencies of the prospective ectoderm?

20. Distinguish between epiboly of the periblast and epiboly of the blastoderm in teleost eggs. What do the terms mean? How does each component accomplish its epiboly?

21. What is the role of the vitelline membrane in epiboly of the yolk sac in the chick? What is the margin of overgrowth?

22. Describe movements of the epiblast during formation of the primitive streak in the chick embryo.

23. What is regression of the primitive streak in the chick embryo.

24. What experiments show that the pattern of epiblastic movement in the chick epiblast is affected by the hypoblast?

25. What is the mechanism of migration of mesoderm through the primitive streak in the chick embryo, as revealed by electron microscopy?

9.7 Suggestions for Further Reading

COSTELLO, D. P., and C. HENLEY. 1976. Spiralian development: a perspective. *Am Zool* 16:277–91.

ENDERS, A. C., ed. 1963. *Delayed Implantation*. Chicago: University of Chicago Press.

FINN, C. A., and L. MARTIN. 1974. The control of implantation. *J Reprod Fertil* 39:195–206.

HAFEZ, E. S. E., and T. N. EVANS, eds. 1973. *Human Reproduction*. New York: Harper & Row.

KUCKETT, W. P. 1978. Origin and differentiation of the yolk sac and extraembryonic mesoderm in presomite human and rhesus monkey embryos. *Am J Anat* 152:59–98.

LØVTRUP, S, U. LANDSTROM, and H. LØVTRUP-REIN. 1978. Polarities, cell differentiation and primary induction in the amphibian embryo. *Biol Rev* 53:1–42.

RAVEN, C. P. 1976. Morphogenetic analysis of spiralian development. *Amer Zool* 16:395–403.

SHUR, B. D. 1977. Cell-surface glycosyltransferases in gastrulating chick embryos. I. Temporally and spatially specific patterns of four endogenous glycosyltransferase activities. *Dev Biol* 58:23–39.

SNOW, M. H. L., J. AITKEN, and J. D. ANSELL. 1976. Role of the inner cell mass in controlling implantation in the mouse. *J Reprod Fertil* 48:403–404.

TILNEY, L. G., and J. R. GIBBINS. 1969. Microtubules in the formation and development of the primary mesenchyme in *Arbacia punctulata*. II. An experimental analysis of their role in development and maintenance of cell shape. *J Cell Biol* 41:227–50.

TURNER, F. R., and A. P. MAHOWALD. 1976. Scanning electron microscopy of *Drosophila* embryogenesis. I. The structure of the egg envelopes and the formation of the cellular blastoderm. *Dev Biol* 50:95–108.

TURNER, F. R., and A. P. MAHOWALD. 1977. Scanning electron microscopy of *Drosophila melanogaster* embryogenesis. II. Gastrulation and segmentation. *Dev Biol* 57:403–16.

9.8 References

ANCEL, P., and P. VINTEMBERGER. 1948. Recherches sur le détérminieme de la symétrie bilaterale dans l'oeuf des amphibiens. *Bull Biol France et Belg Suppl* 31:1–182.

ARNOLD, J. M. 1969. Cleavage furrow formation in a telolecithal egg. I. Filaments in early furrow formation. *J Cell Biol* 41:894–904.

ARNOLD, J. M. 1971. Cleavage furrow formation in a telolecithal egg *(Loligo pealii)*. II. Direct evidence for the contraction of the cleavage furrow base. *J Exp Zool* 176:73–86.

ARNOLD, J. M. 1976. Cytokinesis in animal cells; new answers to old questions. In G. Poste and G. L. Nicolson, eds. *The Cell Surface in Animal Embryogenesis and Development*. Amsterdam: North-Holland, pp. 55–80.

BALINSKY, B. J., and H. WALTHER. 1961. The immigration of presumptive mesoblast from the primitive streak in the chick as studied with the electron microscope. *Acta Embryol Morphol Exp* 4:261–83.

BALLARD, W. W. 1973. A new fate map for *Salmo gairdneri*. *J Exp Zool* 184:49–74.

BELLAIRS, R. 1971. *Developmental Processes in Higher Vertebrates*. Coral Gables, FL: University of Miami Press.

BELLAIRS, R., M. HARKNESS, and R. D. HARKNESS. 1963. The vitelline membrane of the hen's egg: a chemical and electron microscopical study. *J Ultrastr Res* 8:339–59.

BETCHAKU, T., and J. P. TRINKAUS. 1978. Contact relations, surface activity, and cortical microfilaments of marginal cells of the enveloping layer and of the yolk syncytial and yolk cytoplasmic layers of Fundulus before and during epiboly. *J Exp Zool* 206:381–426.

BENFORD, H. H., and M. NAMENWIRTH. 1974. Precocious appearance of the gray crescent in heat-shocked axolotl eggs. *Dev Biol* 39:172–76.

BINDON, B. M. 1969. Mechanism of inhibition of implantation in suckling mice. *J Endocrinol* 44:357–62.

BODENSTEIN, D. 1953. Embryonic development. In K. D. Roeder, ed. *Insect Physiology*. New York: Wiley, pp. 780–821.

BRACHET, J. 1977. An old enigma: the gray crescent of amphibian eggs. *Curr Top Dev Biol* 11:133–86.

CHUNG, H.-M., and G. M. MALACINSKI. 1980. Establishment of the dorsal/ventral polarity of the amphibian embryo: use of ultraviolet irradiation and egg rotation as probes. *Dev Biol* 80:120–33.

CITKOWITZ, H. 1971. The hyaline layer: its isolation and role in echinoderm development. *Dev Biol* 24:384–62.

COOKE, J. 1975. Local autonomy of gastrulation movements after dorsal lip removal in two anuran amphibians. *J Embryol Exp Morphol* 33:147–57.

181

DANIEL, J. C. 1971. Growth of the preimplantation embryo of the northern fur seal and its correlation with changes in uterine protein. *Dev Biol* 26:316–2.

DICKMANN, Z., J. S. GUPTA, and S. K. DEY. 1977. Does "Blastocyst estrogen" initiate implantation? *Science* 195:687–88.

ELINSON, R. P. 1975. Site of sperm entry and a cortical contraction associated with egg activation in the frog *Rana pipiens. Dev Biol* 47:257–68.

EMMANUELSSON, H. 1965. Cell multiplication in the chick blastoderm up to the time of laying. *Exp Cell Res* 39:386–99.

EYAL-GILADI, H., and S. KOCHAV. 1976. From cleavage to primitive streak formation: a complementary normal table and a new look at the first stages of development of the chick. I. General Morphology. *Dev Biol* 49:321–37.

FULLILOVE, S. L., and A. G. JACOBSON. 1971. Nuclear elongation and cytokinesis in *Drosophila montana. Dev Biol* 26:560–77.

GEORGE, F. W., and J. D. WILSON. 1978. Estrogen formation in the early rabbit embryo. *Science* 199:200–201.

GUPTA, J. S., K. S. DEY, and Z. DICKMANN. 1977. Evidence that "embryonic estrogen" is a factor which controls the development of the mouse preimplantation embryo. *Steroids* 29:363–69.

GUSTAFSON, T., and L. WOLPERT. 1967. Cellular movement and contact in sea urchin morphogenesis. *Biol Rev* 42:442–98.

HOLTFRETER, J. 1943. A study of the mechanics of gastrulation. *J Exp Zool* 94:261–318.

HÖRSTADIUS, S. 1928. Über die Determination des Keimes bei Echinodermen. *Acta Zool* 9:1–91.

HÖRSTADIUS, S., L. J. LORCH, and J. F. DANIELLI. 1950. Differentiation of the sea urchin egg following reduction of the interior cytoplasm in relation to the cortex. *Exp Cell Res* 1:188–93.

ITO, S., and W. R. LOEWENSTEIN. 1969. Ionic communication between early embryonic cells. *Dev Biol* 19:228–43.

KELLER, R. E. 1980. The cellular basis of epiboly: an SEM study of deep-cell rearrangement during gastrulation in *Xenopus lavis. J Embryol Exp Morphol* 60:201–34.

KELLER, R. E., and G. C. SCHOENWOLF. 1977. An SEM study of cellular morphology, contact, and arrangement, as related to gastrulation in *Xenopus laevis. Wilhelm Roux' Arch* 182:165–86.

KIRCHNER, M., J. C. GERHART, K. HARA, and G. A. UBBELS. 1980. Initiation of the cell cycle and establishment of bilateral symmetry in *Xenopus* eggs. In S. Subtelny and N. K. Wessels, ed. *The Cell Surface: Mediator of Developmental Processes.* New York: Academic Press, pp. 187–215.

LENTZ, T. L., and J. P. TRINKAUS. 1967. A fine structural study of cytodifferentiation during cleavage, blastula, and gastrula stages of *Fundulus heteroclitus. J Cell Biol* 32:121–38.

LENTZ, T. L., and J. P. TRINKAUS. 1971. Differentiation of the junctional complex of surface cells in the developing *Fundulus* blastoderm. *J Cell Biol* 48:455–72.

LILLIE, F. R. 1909. Polarity and bilaterality of the annelid egg. Experiments with centrifugal force. *Biol Bull* 16:54–79.

LØVTRUP, S., 1966. Morphogenesis in the amphibian embryo. Cell type distribution, germ layers, and fate maps. *Acta Zool* (Stockholm) 47:209–76.

LUCHTEL, D., J. G. BLUEMINK, and S. W. DELATT. 1976. The effect of injected cytochalasin B on filament organization in the cleaving egg of *Xenopus laevis. J Ultrastruct Res* 54:406–16.

LUCKETT, W. P. 1971. The origin of extraembryonic mesoderm in the early human and rhesus monkey embryos. *Anat Rec* 169:369–70.

LUTHER, W. 1935. Entwicklungsphysiologische Untersuchungers am Forellenkeim: die Rolle des Organizationszentrums bei der Enstehung der Embryonalanlage." *Biologisches Zentralteblatt* 55:114–37.

MALACINSKI, G. M., H. BENFORD, and H.-M. CHUNG. 1975. Association of an ultraviolet irradiation sensitive cytoplasmic localization with the future dorsal side of the amphibian egg. *J Exp Zool* 191:97–110.

MALACINSKI, G. M., H.-M. CHUNG, and M. ASASHIMA. 1980. The association of primary embryonic organizer activity with the future dorsal side of amphibian eggs and early embryos. *Dev Biol* 77:449–62.

MOORE, A. R., and A. S. BURT. 1939. On the locus and nature of the forces causing gastrulation in the embryos of *Dendraster excentricus. J Exp Zool* 82:159–71.

NICOLET. G. 1970. Analyse autoradiographique de la localization des différéntes ébauches présomptovies dans la ligne primitive de l'embryon de poulet. *J Embryol Exp Morphol* 23:79–108.

RAPPAPORT, R. 1971. Cytokinesis in animal cells. *Int Rev Cytol* 31:169–213.

RAPPAPORT. R. 1978. Effects of continual mechanical agitation prior to cleavage in echinoderm eggs. *J Exp Zool* 206:1–12.

RAPPAPORT, R., and R. P. EBSTEIN. 1965. Duration of stimulus and latent periods preceding furrow formation in sand dollar eggs. *J Exp Zool* 158:373–82.

RAPPAPORT, R., and J. H. RATNER. 1967. Cleavage of sand dollar eggs with altered patterns of new surface formation. *J Exp Zool* 165:89–100.

ROSENQUIST, G. C. 1966. A radioautographic study of labeled grafts in the chick blastoderm. *Carnegie Contribs Embryol* 38:71–110.

ROSENQUIST, G. C. 1972. Endoderm movements in the chick embryo between the early short streak and head process stages. *J Exp Zool* 180:95–104.

SCHECHTMAN, A. M. 1942. The mechanism of amphibian

gastrulation. I. Gastrulation-promoting interactions between various regions of an anuran egg. *Univ Calif Publ Zool* 51:1–40.

SCHROEDER, T. E. 1972. The contractile ring. II. Determining its brief existence, volumetric changes, and vital role in cleaving *Arbacia* eggs. *J Cell Biol* 53:419–34.

SELMAN, G. G., and M. M. PERRY. 1970. Ultrastructural changes in the surface layers of the newt's egg in relation to the mechanism of its cleavage. *J Cell Sci* 6:207–27.

SPRATT, N. T., Jr. 1946. Formation of the primitive streak in the explanted chick blastoderm marked with carbon particles. *J Exp Zool* 103:259–304.

SZÖLLÖSI, D. 1970. Cortical cytoplasmic filaments of cleaving eggs: a structural element corresponding to the contractile ring. *J Cell Biol* 44:192–209.

TILNEY, L. G., and J. R. GIBBINS. 1969. Microtubules and filaments in the filopodia of the secondary mesenchyme cells of *Arbacia punctulata* and *Echinarachnius parma*. *J Cell Sci* 5:195–210.

TORREY, T. W., and A. FEDUCCIA. 1979. *Morphogenesis of the Vertebrates*, 4th ed. New York: Wiley.

TRELSTAD, R. L., E. D. HAY, and J.-P. REVEL. 1967. Cell contact during early morphogenesis in the chick embryo. *Dev Biol* 16:78–106.

TRINKAUS, J. P. 1951. A study of the mechanism of epiboly in the egg of *Fundulus heteroclitus*. *J Exp Zool* 118:269–320.

TRINKAUS, J. P. 1969. *Cells into Organs*. Englewood Cliffs, NJ: Prentice-Hall.

TRINKAUS, J. P. 1976. On the mechanism of metazoan cell movements. In G. Poste and G. O. Nicolson, eds. *The Cell Surface in Animal Embryogenesis and Development*. Amsterdam: North-Holland, pp. 225–329.

TRINKAUS, J. P., and T. L. LENTZ. 1967. Surface specializations of *Fundulus* cells and their relation to cell movements during gastrulation. *J Cell Biol* 32:139–54.

TUPPER, J. T. and J. W. SAUNDERS, Jr. 1972. Intercellular permeability in the early *Asterias* embryo. *Dev Biol* 27:546–54.

VAKAET, L. 1967. Contribution a l'étude de la prégastrulation et de la gastrulation de l'embryon de poulet en culture in vitro. *Mem Acad R Med Biol* 5:235–37.

WADDINGTON, C. H. 1933. Induction by the endoderm in birds. *Wilhelm Roux' Arch* 128:502–21.

WARD, W. F., A. G. FROST, and M. W. ORSINI. 1978. Estrogen binding by embryonic and interembryonic segments of the rat uterus prior to implantation. *Biol Reprod* 18:598–601.

WEISS, P. 1951. The outlook in morphogenesis. *Coll Intern Centre National Rech Sci Paris* 28:563–82.

183

10

Maternal and Embryonic Controls of Animal Development

tute organ-forming materials, as such, but, rather, they differentially affect patterns of gene activity in the various blastomeres.

Are maternal RNAs used by the developing embryo? Surely they are involved in the enhanced protein synthesis that immediately follows fertilization (Chapter 9). But do they serve a further function? Are they involved in the formation of macromolecules important in morphogenesis? When does the embryonic genome start to be transcribed? Does the early embryo synthesize macromolecules distinctive from those of the oöcyte? Of what significance in early development are the first products of the embryonic genome? What determines the nature and sequence of products of the embryonic genome? These are some of the questions to be examined after we have investigated the effects of the oöcyte cytoplasm on postfertilization development. Chapter 18 also treats some of these matters in the context of the genetic control of development.

The thrust of this chapter is the elucidation of the relative roles of maternal and embryonic genomes in determining the course of early development after fertilization. Of major importance in this context is the fact that the entire content of the mature oöcyte is produced under the exclusive control of the maternal genome. This content consists of direct gene products stored during oögenesis—rRNA, mRNA, and tRNA—and of a varying content of structural and enzymatic proteins, yolk, mitochondria, and other formed elements of the cytoplasm. The distribution of cytoplasmic components may or may not change at fertilization and varies greatly from one kind of egg to another. Cleavage in many kinds of eggs partitions zones of cytoplasm whose distinctive content marks them as destined for a particular and invariable fate. The cleavage planes themselves change direction in a spatiotemporal pattern that proceeds independently of the timing of the nuclear divisions.

Is the zygote, in fact, a mosaic work of materials, each part of which is predestined by the maternal genome to form a particular part of the organism? Seeking the answer to this question we shall find that almost all eggs show some prelocalization of morphogenetic determiners. These do not consti-

10.1 Mosaic and Regulative Development

In the majority of invertebrate groups, eggs are predominately of the type in which cytoplasmic zones of distinctive appearance are segregated by cleavage, and these zones can be traced to particular organs or systems. Such eggs have **determinate** cleavage, as explained in Chapter 9. But, as cleavage begins they are also, in most cases, rigid mosaics of developmental potentialities, for, if an organ-forming zone is delected experimentally, the structure it normally would have formed is absent in the resulting organism. Such eggs are called **mosaic eggs** and are said to have **mosaic development.**

In contrast, eggs of most vertebrates and many invertebrates show **regulative development.** These have the ability to form a normal embryo or larva despite the experimental deletion of blastomeres. **Regulative eggs** are, with few exceptions, those in which individual blastomeres are not readily identifiable as to their developmental fate. They are, therefore, principally eggs that have **indeterminate cleavage.**

The concepts of determinate and indeterminate

cleavage and of mosaic and regulative development have arisen largely from the experimental study of animal eggs. Accordingly, the following discussion does not include plant material. In fact, however, the concepts are possibly quite applicable in cleavage stages in plant embryos, but these are less accessible to experimental analysis. Cleavage in plants segregates cells whose subsequent fates are readily traced and which can be identified at each stage by virtue of their positions and shapes. Nevertheless, the prospective fate of a cell may very well be determined in a plant embryo by its position rather than by a distinctive cytoplasmic content.

10.1.1 SOME EXAMPLES OF THE DEVELOPMENT OF MOSAIC EGGS.

Ascidians are primitive chordates, classified in the subphylum Urochordata, whose eggs form free-swimming tadpole-shaped larvae (Figure 10.1). After fertilization the eggs of many ascidians show a bilaterally arranged pattern of distinctively tinted cytoplasmic regions (Figure 10.2). The cleavage planes then segregate the different cytoplasmic regions according to a very precise geometric plan so that individual cells can be identified after each cleavage as to ancestry and prospective fate. If the blastomeres are separated after the first cleavage and reared separately, each continues to cleave as though it were part of a whole, and each forms a right or a left half-larva according to the plan of cell lineage. Effects of various other manipulations of ascidian blastomeres were reviewed by Kuhn (1965).

Eggs of annelids and mollusks have almost identical patterns of spiral cleavage, and their blastomeres may be homologized on a one-to-one basis (Figure 10.3; see also Figure 9.6). Likewise they form **trochophore** larvae with the same basic body plan (Figures 9.7 and 10.4). Each blastomere is specified as to its future fate by the particular cytoplas-

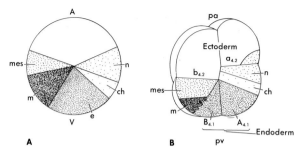

FIGURE 10.2 A: The segregation of organ-forming plasms in an ascidian egg. The clear area at the animal pole will be segregated into cells that form ectoderm. B: The 8-cell stage as seen from the right side of the bilaterally symmetrical embryo. Conventionally the blastomeres of the right side are designated with block letters and numerical subscripts; those of the left side, with italic lettering (not shown). Organ-forming plasms are designated as follows: ch, notochord; e, endoderm; m, muscle; mes, mesenchyme; n, notochord. [Adapted from G. Reverberi in G. Reverberi, ed., *Experimental Embryology of Marine and Fresh Water Invertebrates*, North Holland, Amsterdam, 1971, pp. 507–50.]

185

mic region segregated within it, and rather complete cell lineage charts have been made for a number of annelids and mollusks (Kolalewski, 1883; Wilson, 1892; Conklin, 1897). The development of isolated blastomeres from such eggs (Figure 10.5) and of embryos deprived of specific blastomeres supports the general rule that these eggs are of a highly mosaic type.

FIGURE 10.3 Views from the animal pole of 16-celled embryos of *Trochus*, a mollusk (A), and *Arenicola*, an annelid (B). Blastomeres are conventionally numbered in the same manner in both forms, denoting their homologies. The cells of the progressive prototroch (trochoblasts) are stippled. See also Figures 10.4, 10.5, and 9.7.

FIGURE 10.1 Stylized sketch of an ascidian tadpole-shaped larva.

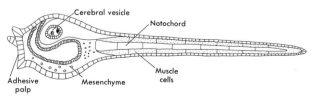

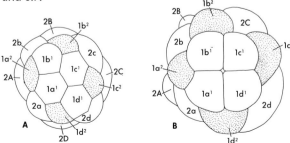

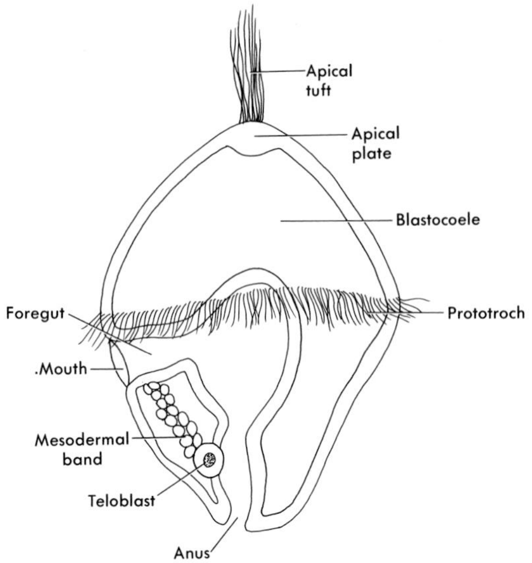

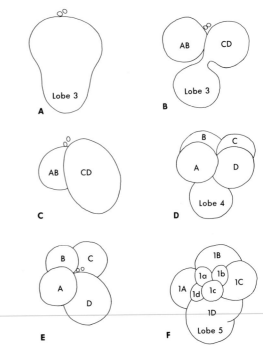

FIGURE 10.6 Polar lobe formation and early cleavage stages in *Ilyanassa obsoleta*. A protrusion of the vegetal pole forms at the time of the formation of each of the polar bodies. A third lobe appears during the first cleavage (A, B) and is resorbed into the CD cell at the end of that cleavage. A fourth lobe appears during the second cleavage and is resorbed into the D cell (D, E). When the first quartet of micromeres appears a small fifth lobe is seen (F). [Adapted from A. C. Clement, *J Exp Zool* 121:593–626 (1952).]

FIGURE 10.4 The trochopore of *Arenicola* as seen from the left side. Mesodermal bands of left and right sides will be extended posteriorly chiefly by proliferation of the teloblast cells, and will give rise to the segmented coelom of the adult worm.

186

Of particular interest among spirally cleaving eggs are those that form a polar lobe. A prominent example of such an egg is that of *Ilyanassa obsoleta,* the common mud snail of the eastern coast of the United States. The polar lobe is a protrusion of the

FIGURE 10.5 Development of the isolated primary trochoblast in *Patella*: (A) the embryo at the 16-cell stage; (B) an isolated trochoblast; (C, D) division of the trochoblast; (E) differentiation of the division products as ciliated prototroch cells. [Adapted from E. B. Wilson, *J Exp Zool* 1:197–268 (1904).]

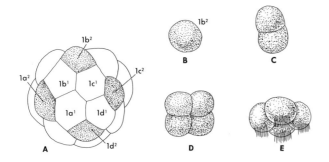

vegetal side of the egg that appears at the time each polar body is formed and that appears again prior to the first division (Figure 10.6A). As division occurs the lobe constricts off from the rest of the egg giving the embryo a **trefoil** configuration (Figure 10.6B).

If the polar lobe is removed at the trefoil stage, a number of larval organs fail to appear and others are highly disorganized. Eyes, foot, heart, and shell gland are absent, the intestine fails to appear, and the organization of the rest of the digestive tract is imperfect (Clement, 1952, 1976; Atkinson, 1971).

Clement (1952) made a very meticulous analysis of the cleavage pattern in normal and lobeless eggs of *Ilyanassa*. In the normal egg (Figure 10.6F, left column), the D macromere is the largest of the first four cells, but the 1d cell, which separates from it, is smaller than the other cells of the first quartet (1a–1c), and it subdivides at a slower tempo than they do. Cell 2d, however, is the same size as 2a–2c (Figure 10.7A), but its descendant $2d^{11}$ is larger than the corresponding other descendants of the second quartet (Figure 10.8C). Important is the fact that the 4d cell is produced precociously from the 4D macromere, well ahead of the other members of the fourth quartet. The latter are quite yolky, whereas 4d, although having some yolk, has a much clearer cytoplasm. When 4d divides, it forms ME^1 and ME^2 (Figure 10.8B, C). When the ME cells divide, the yolky component goes to the endoblasts E_1 and E_2, and the **primary mesoblasts**, Me^1 and Me^2, receive the clear cyto-

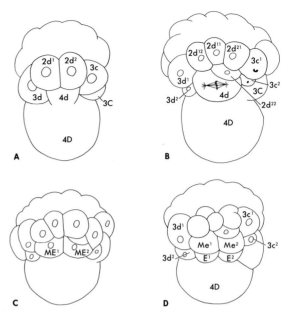

FIGURE 10.8 Normal stages showing for *Ilyanassa obsoleta* the origin of the 4d cell (A), its division to form the mesentoblast cells ME^1 and ME^2 (B, C) and the origin of the primary mesoblast cells ME^1 and ME^2 (D). [Adapted from A. C. Clement, *J Exp Zool* 121:593–626 (1952).]

187

FIGURE 10.7 Contrasting patterns of cleavage in lobed (left column) and lobeless (right column) embryos of *Ilyanassa obsoleta* with emphasis on differences in the performance of the 1d cell of the first quartet and its descendants at successive stages designated A–C. See explanation in the text. [Adapted from A. C. Clement, *J Exp Zool* 121:593–626 (1952).]

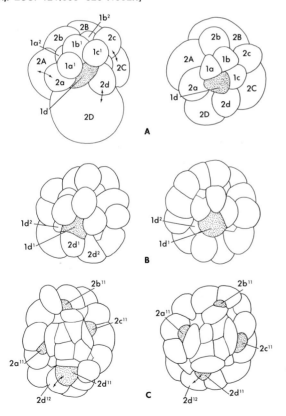

plasm (Figure 10.8D). Because of the distinctive size and position of the D macromere and its contributions to the first four quartets, the embryo is bilaterally symmetrical (Figure 10.8).

After removal of the polar lobe at the trefoil stage, all of the distinctive features of the D quadrant are lacking, and the embryo develops with radial symmetry (Figure 10.7, right column) rather than with bilateral symmetry, as is normal. The D macromere and its descendants are no longer distinguishable from the components of the other quadrants, and the 4d cell forms only the yolky endoblast cells and no primary mesoblast cells. The resulting monstrous larva, therefore, completely lacks mesodermal bands (cf. Figure 9.7 for *Patella*) that would have come from Me^1 and Me^2 (Clement, 1952).

In later experiments Clement (1962) studied the development of the ABC quadrant isolated by destruction of the D macromere after the formation of each of the first four quartet generations. Of chief importance is his observation that the removal of the 4D macromere (i.e., after formation of 4d) permits the formation of a structurally complete larva, differing from controls only by virtue of its smaller size. Removal of the D macromere either before or after formation of the first quartet has essentially the same developmental effect as removal of the polar lobe at the trefoil stage (Figure 10.6B). Deletion of the D macromere after formation of the third quartet leads to formation of larvae of intermediate quality. After production of 4d, however, the 4D macromere apparently serves no essential morphogenetic function.

Distinctive cytoplasmic structures have been described in polar lobes of several mollusks (reviewed by Dohmen and Verdonk, 1979). *Bithynia tentaculata* provides a very interesting case, for a distinctive vegetal body can be seen in the polar lobe prior to the second cleavage. This body is composed of electron-dense vesicles, probably containing RNA. These vesicles are dispersed and shunted into the D cell at the completion of the second cleavage. If the vegetal body in the uncleaved egg is shifted centrifugally so that it does not enter the polar lobe, removal of the lobe permits the formation of relatively normal larvae. When the operation is performed on uncentrifuged eggs such adult structures as eyes, foot, shell, and so on do not form. Thus, in *Bithynia* morphogenetic determinants are initially localized in the vegetal body.

10.1.1.1 Cephalopods Show a Remarkable Form of Mosaic Development.

Although cephalopods are mollusks, their cleavage is entirely different from that of other molluskan forms. As described for *Loligo peali* by Arnold (1971), the unfertilized egg is ovate and consists of a central mass of membrane-bounded yolk platelets covered by a thin layer of cytoplasm rich in mitochondria, Golgi bodies, and small vesicles. A protoplasmic disc forms at the animal pole after fertilization, somewhat as in the case of the teleost egg, and cleavage is meroblastic. The blastoderm is at first a single layer of cells, the marginal ones of which are called **blastocones** and the central ones blastomeres. Expanding over the yolk, the blastocones give rise to a marginal syncytium that eventually spreads apically and underlies the entire blastoderm. This syncytium is called the **yolk epithelium.** It serves to digest yolk platelets and also is significant in organ formation, as is described shortly. Mitotic figures at the margin of the blastoderm also produce a layer of

cells intermediate between the yolk epithelium and the overlying layer.

This multilayered structure expands by marginal division until the equator of the egg is reached. Meanwhile, the intermediate layer and outer layer begin to form the definitive body parts, the remainder of the egg being involved only in forming an external yolk sac. Shortly after this stage, the anlagen of the various organ primordia can be seen as thickenings of the outer layer of the blastoderm arranged in a pattern of bilateral symmetry. One can pick out, for example, the primordia of the eyes, tentacles, shell gland, otocysts, and so on. If one now deletes one or more of these primordia by gentle rubbing with a steel needle, but without disturbing the underlying yolk epithelium, cells from the margin of the wound grow over the yolk epithelium and a completely normal organ develops. If the underlying yolk epithelium is removed, however, the organ fails to form. Correspondingly, if the eye primordium, for example, is excised and cultured in vitro along with the corresponding underlying yolk epithelium it will form an eye, but it will not do so in the absence of the yolk epithelium.

The yolk epithelium would seem, therefore, to be composed of a mosaic of potentialities for inducing the formation of organs from overlying tissues in their appropriate pattern. But whence did the yolk epithelium acquire its pattern of morphogenetic information? Presumably, this information is present in the egg cortex even before the yolk epithelium is formed. Arnold (1968) made local defects to the egg cortex at the time of first cleavage, at some distance from the protoplasmic disc, by means of a microbeam of ultraviolet irradiation. Cellulation of the irradiated area proceeded normally, but if, for example, a future arm region were irradiated, no arm would form. This result suggests that the informational pattern in the egg cortex may have been derived from follicle cells in the gonad. This pattern, transferred to the yolk epithelium, becomes a morphogenetic map that regulates the pattern of embryogenesis.

10.1.2 REGULATIVE DEVELOPMENT.

The ability of fertilized eggs to show regulative development is highly correlated with the extent to which they show cytoplasmic homogeneity. Mammalian eggs, which have a remarkably homogeneous cytoplasm, are highly regulative, for example, whereas am-

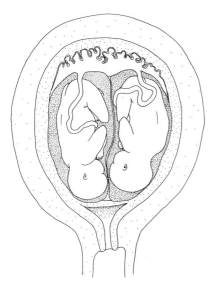

FIGURE 10.9 Intrauterine relationships of human monozygotic twins sharing a single placenta.

phibian eggs, which show a strong animal-vegetal gradient in the distribution of yolk and which have an asymmetrically located gray crescent, have more limited regulative powers.

Regulative powers of mammalian eggs actually extend to postimplantation stages. Human identical twins are referred to as **monozygotic,** for they develop from a single zygote. In about 75% of the cases they develop within a common chorion and share a common placenta (Figure 10.9). Two individuals presumably form in these cases by subdivision of the inner cell mass after implantation. The armadillo always bears quadruplets that are produced by the separation of the inner cell mass of the blastocyst into four parts.

In about 25% of human identical twins, development occurs within individual chorionic sacs. This means that division of the original embryonic mass occurred prior to blastocyst formation, so that each fragment formed its own trophoblast and inner cell mass. In mice, individual blastomeres from the eight-cell stage will form normal blastocysts in vitro (Tarkowski, 1959), so it is conceivable that some identical human twins or higher mutliples might arise from single blastomeres or from other fragments of the morula.

The great regulative capacity of mammalian

eggs is further signaled by the fact that blastomeres or blastocysts from one or more embryos may be fused. These give rise to a single giant blastocyst which, when placed in the uterus of a hormonally prepared foster mother, will develop to term in a high percentage of cases. Beatrice Mintz, of the Institute for Cancer Research, regularly creates **tetraparental** mice to study interactions between cells of different genotype during development. A "tetraparental" mouse is made by combining a cleaving egg obtained from one inseminated female with one obtained from another female inseminated by a different male. In 1970 Dr. Mintz reported that her laboratory had produced 1014 mice of dual embryonic origin, comprising 47 different paired genotypic combinations. Subsequently, Markert and Petters (1979) successfully reared hexaparental mice; that is, mice originating from three morulas.

When fusion technique is carried out on blastomeres of differing genotypes or with blastomeres marked by injection of visible material, one may show that the fate of each blastomere is determined by the position that it occupies with respect to other blastomeres: outer cells of the morula tend to become trophoblast cells, whereas those on the inside contribute to the inner cell mass, and thus to the embryo (Hillman et al., 1972; Gardner and Johnson, 1972; Kelly, 1977, 1979). Stern and Wilson (1972) have shown that presumptive trophoblast derived from embryos at the early blastocyst stage can become incorporated into the inner cell mass of chimaeric embryos. When the inner cell mass and trophoblast are separated from later blastocysts, however, the two are seen to have distinctly different properties (Gardner and Johnson, 1972). Trophoblast will form vesicles and induce the decidual reaction in uterine tissue that is properly prepared hormonally, but it will not form an inner cell mass and soon ceases proliferation. The isolated inner cell mass does not re-form a blastocyst, nor will it induce a decidual reaction. Thus the regulative property of the mammalian egg is restricted at the blastocyst stage, at least so far as the differentiation of embryonic and extraembryonic (trophoblastic) tissue is concerned. The restriction of regulative capacity is correlated with the loss of communication by means of gap junctions between cells of the inner cell mass and the trophoblast (Lo and Gilula, 1979) and by changing patterns of protein synthesis in the two tissues (Handyside and Johnson, 1978).

189

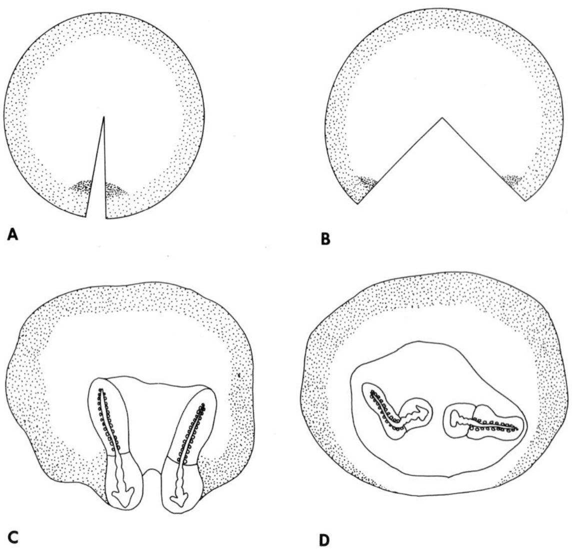

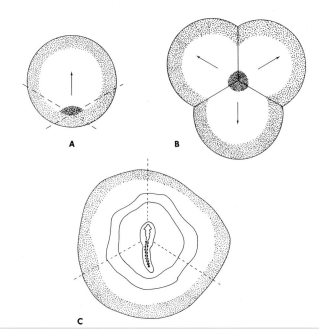

FIGURE 10.10 Formation of double embryos from a single chick blastoderm cultured *in vitro*. The blastoderm is incised (A) posteriorly through the region of the embryonic shield, and the cut edges are shifted apart (B). Twin embryos frequently result, sometimes showing reversed polarity, (C) and sometimes polarity corresponding to the positions occupied by the separated parts of the shield (D). [Adapted from N. T. Spratt and H. Haas, *J Exp Zool* 147:57–94 (1961).]

190

FIGURE 10.11 Formation of a single chick embryo from three fused blastoderms. Blastoderms containing the major portion of the embryonic shield were isolated (A) and fused *in vitro* (B). Arrows indicate the original anteroposterior axis. C: Resulting single embryo. [Adapted from N. T. Spratt and H. Haas, *J Exp Zool* 147:271–93 (1961).]

Developing eggs of birds and teleosts are also highly regulative until relatively late stages. Spratt and Haas (1961a, b) obtained double embryos from single unincubated chick blastoderms cultured in vitro (Figure 10.10) and single ones from several fused embryos (Figure 10.11). Lutz and his colleagues (1963) regularly produce twin duck embryos by making a single transverse cut of the early blastoderm in ovo. In the brook trout Luther (1937) excised the half of the blastoderm containing the embryonic shield from the yolk and replaced it with a half blastoderm of the opposite side from a donor egg. A complete embryo then formed from the donor tissue, which normally would have given rise mostly to the yolk sac.

The sea urchin provided the first example of the experimental production of a whole embryo from isolated blastomeres; therefore, it is often cited as an outstanding case of regulative development. As will be documented below, however, its regulative capacities are far less than those of a mammal.

Driesch (1891, 1892) removed the fertilization membrane by gently shaking newly fertilized sea urchin eggs and then separated the blastomeres by more vigorous shaking at the two- and four-cell stages. Each blastomere divides as though part of a whole. The one-half embryo forms two micromeres after three more divisions, and the one-fourth embryo produces a single micromere after two additional divisions (Figure 9.56). Eventually, however, a small blastula is formed, gastrulation occurs, and a pluteus larva develops (less frequently when one-fourth embryos are used). This was studied more extensively by Hörstadius (1928). He also showed that two entire blastulae can be fused, forming a larger but perfectly organized pluteus, so long as the animal-vegetal axes of both embryos coincide.

Eggs of *Cerebratulus*, a nemertean worm, have a very symmetrical pattern of spiral cleavage (Figure 10.12). But, in contrast to most spirally cleaving eggs, those of *Cerebratulus* have considerable regulative power. Isolated blastomeres of the two-cell

FIGURE 10.12 The 8-cell stage in the cleavage of *Cerebratulus*. This spirally cleaving egg is unusual in that it produces larger blastomeres at the animal pole, uppermost in this figure, than at the vegetal pole. Moreover, it is highly regulative, as explained in the text. [Adapted from T. H. Morgan, *Experimental Embryology*, Columbia University Press, New York, 1927, p. 266.]

stage produce normal larvae. Each blastomere cleaves as though it were still part of a whole, however, and the blastocoele is wide open before closing over prior to gastrulation. Blastomeres isolated at the four-cell stage likewise undergo partial cleavage and sometimes gastrulate, forming structures resembling the normal larva.

Numerous other examples of regulative development are found among both invertebrates and vertebrates. Some of these are noted in the next section.

10.1.3 UBIQUITY OF PRECOCIOUS CYTOPLASMIC LOCALIZATION.
With the possible exception of mammalian eggs, essentially all kinds of fertilized eggs show some degree of cytoplasmic prelocalization. Thus, the sea urchin egg shows an animal-

FIGURE 10.13 The morphogenesis of isolated animal and vegetal halves of the sea urchin blastula. [Based on S. Hörstadius, *Biol Revs* 14:132–79 (1939).]

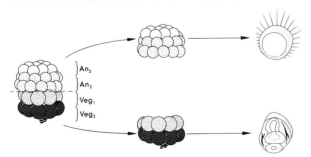

vegetal gradient in the distribution of yolk and pigment. Correspondingly, meridional halves of cleaving sea urchin eggs or blastulae form normal plutei, but isolated animal and vegetal halves do not (Figure 10.13). Instead, the animal half usually lacks a digestive tract and skeleton, and the vegetal half **exogastrulates** (i.e., the endoderm pushes outward instead of inward) and forms skeletal parts, but it is deficient in ectoderm and lacks an apical tuft. Animal and vegetal halves of fertilized and cleaving *Cerebratulus* eggs also are unable to form complete larvae in isolation. Amphibian blastomeres separated at the two-cell stage can form twin embryos but will usually do so only if the first cleavage passes through at least part of the gray crescent (Figure 9.10). In newts the first cleavage is usually at right angles to the plane of bilateral symmetry (i.e., the plane bisecting the gray crescent), and, therefore, only the cell mass that contains the gray crescent gastrulates.

10.1.4 EPIGENESIS OF PRECOCIOUS SPECIFICATION.
With Wilson (1928, p. 1093), we ask "whether cytoplasmic germinal prelocalization exists from the beginning or is itself a product of antecedent development." If localization occurs progressively as a consequence of prior events, it would be said to arise by **epigenesis** (Gr., *epi*, upon or after; *genesis*, from the root of *gignesthai*, to be born). **Epigenesis** is contrasted with **preformation**.

The preponderance of evidence supports the epigenetic origin of cytoplasmic localization. We have already seen that in the regulative eggs of sea urchins and of *Cerebratulus*, isolated blastomeres cleave as though they were still parts of a whole egg. What pattern would be shown by fragments of uncleaved eggs? Eggs of *Cerebratulus* are shed with the germinal vesicle intact, and the animal-vegetal axis may be recognized, first, by the shape of the egg and later by the position of the first polar body. After discharge of the egg into the water, the germinal vesicle breaks down and the first meiotic metaphase spindle is formed. There meiosis halts until fertilization occurs. At any point after its discharge, the egg may be cut in any desired plane, and both fragments may be isolated and fertilized. If the operation is performed before the germinal vesicle breaks down, only the nucleated part develops upon fertilization. But, regardless of the plane of section, the nucleated fragment cleaves in every

191

detail like a whole egg (Figure 10.12) of diminished size. When fragments are produced after breakdown of the germinal vesicle, but before fertilization, both cleave normally upon subsequent fertilization and produce dwarf larvae of normal configuration. Apparently, the contents of the germinal vesicle are required for cleavage. Fragments produced by cutting after fertilization, however, tend to cleave in a partial pattern and often form abnormal larvae. But, as already noted, either of the first blastomeres, when separated from the other, cleaves in a half-pattern but forms a normal, though dwarfed larva (reviewed by Wilson, 1928).

In rigidly mosaic eggs, such as those of ascidians, cytoplasmic prelocalization of organ-forming regions is likewise the product of progressive change based upon antecedent development. When the ascidian egg is shed, the large germinal vesicle is intact and the pigment-laden cortex surrounds a central yolk mass. The germinal vesicle promptly moves to the animal pole, and its nuclear membrane breaks down, releasing a volume of nucleoplasm, which mixes with egg cytoplasm to produce a clear zone at the animal pole. Within this zone the small meiotic spindle forms. At this stage fertilization occurs, the sperm entering near the vegetal pole. Entrance of the sperm triggers the completion of meiosis and initiates a sequence of events leading to the redistribution of egg content into the bilaterally symmetrical pattern of organ-forming areas previously described (Figure 10.2) As already mentioned, the separated blastomeres of the two-cell stage each cleave in a half-pattern and form incomplete embryos. At later stages the removal of specific blastomeres leads to the formation of larvae with regional defects. Reverberi and Ortolani (1962) showed, however, that if one divides the unfertilized egg of the tunicate *Ascidia malaca* by means of equatorial, meridional, or oblique sections, each half may be fertilized and then will cleave in a normal pattern, and form a complete larva, normally proportioned, but smaller than the controls. Nucleated meridianal halves will likewise form normal tadpoles after fertilization and after extrusion of one or both polar bodies. Vegetal halves will also cleave in a normal pattern and form complete larvae if isolated before or just after the extrusion of the second polar body, for the sperm enters at the vegetal pole and cytoplasmic redistribution has not yet occurred. Animal halves isolated

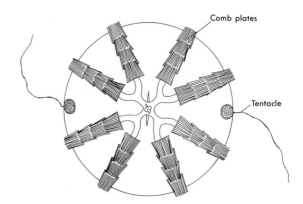

FIGURE 10.14 Young *Mnemiopsis leidyi* viewed from the aboral pole showing eight rows of comb plates symmetrically arranged. In this species, light-emitting organs (not shown here) are associated with each row of comb plates. [Adapted with modifications from G. Freeman and G. T. Reynolds, *Dev Biol* 31:61–100 (1973).]

after formation of the first polar body will neither cleave normally nor form complete larvae, however (Ortolani, 1958).

Ctenophore eggs also show a progression of early changes leading to what appears to be a very precocious localization of cytoplasmic elements destined for particular organelles. The more generalized members of the phylum are free-swimming marine invertebrates, walnut-shaped and of jellylike consistency. They propel themselves through the water by the action of eight rows of ciliated **comb plates** (Figure 10.14). In each plate the cilia are arranged like the teeth of a comb, hence the name **comb jellies** is applied to these animals. It has been known for many years (Fischel, 1898) that during cleavage (Figure 10.15) material localized in the egg cortex enters micromeres which form the swim plates. Deletion of micromeres affects the number of swim plates. Yatsu (1912) found that this localization occurs only after fertilization and polar body formation in the Mediterranean ctenophore *Beröe ovata*. Nucleated egg fragments isolated prior to this time develop normally. More recently the progressing localization of swim-plate material has been reexamined by Freeman (1976) using the familiar *Mnemiopsis* of eastern American waters (Figures 10.14 and 10.15).

These examples make it clear that the cytoplasmic localization of morphogenetic potential occurs

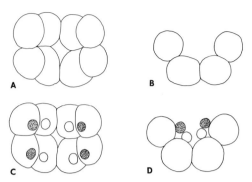

FIGURE 10.15 A: The 8-cell stage in the cleavage of *Mnemiopsis leidyi* as seen from the animal pole, which corresponds to the future aboral side. B: The same, as seen from the side. C and D illustrate the positions of the micromeres formed at the fourth cleavage. In the stippled micromeres resides the potential for forming comb plates. The unstippled ones are destined to become luminescent organs. [Adapted from G. Freemen, *Dev Biol* 49:143–66 (1976).]

progressively as a consequence of antecedent development. The differences between regulative and mosaic eggs simply reflect differences in the tempo of events that partition cytoplasm of different morphogenetic significance among the blastomere nuclei. What is of greater importance, however, is that the cytoplasmic components that progressively achieve a regionally specific distribution in the egg were produced through the operation of the maternal genome.

10.1.5 EFFECTS OF REGIONAL DIFFERENCES IN THE EGG CYTOPLASM ON THE ACTIVITIES OF BLASTO-MERE NUCLEI.

From what has preceded, it follows that blastomere nuclei and their descendants must differ in the expression of their genetic repertory of developmental possibilities according to the regional character of the egg cytoplasm that they inherit. This fact is most clearly and unequivocally demonstrated by cases in which the zygote shows a region of specialized cytoplasm called **germ plasm.** Blastomere nuclei relegated to the germ plasm in normal cleavage or that are exposed to it experimentally become primordial germ cells, producing the gametes of the next generation.

The classic case invoked to illustrate the determinative role of the germ plasm is provided by the zygote of the nematode parasite of the horse, *Ascaris*

megalocephala univalens. In this organism, the germ line stem cell retains the chromosomal complement of the zygote, namely two long chromosomes, but all somatic (body) cells have a diminished amount

FIGURE 10.16 Segregation of the germ plasm and diminution of chromatin in the parasitic, nematode *Ascaris megalocephala*. The germ is viewed from the future right side. The first cleavage (A) produces a somatic cell, S_1, at the animal pole whose daughter nuclei will undergo chromatin diminution during the second cleavage as shown in B. The lower cell, P, cleaves in a furrow perpendicular to the first, producing blastomeres P_2 and S_2, as shown in B and C. S_2 undergoes chromatin diminution when it next divides as shown in E and F, but P_2 does not. P_2 gives rise to P_3 and S_3, as seen in F. At the next division, not shown, S_3 undergoes chromatin diminution and P_3 produces P_4 and S_4. S_4 loses chromatin at its next division. P_4 remains quiescent, but later resumes division to produce the germ cell line. [Redrawn with modifications from A. Richards, *Outline of comparative Embryology*, Wiley, New York, 1931, pp. 47–48.]

193

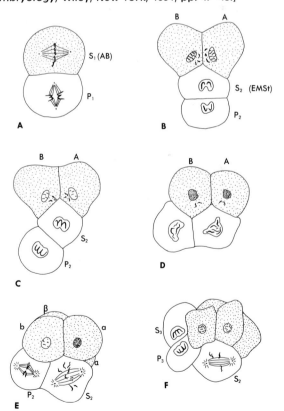

of chromatin that is reorganized into numerous small chromosomes. At the two-cell stage (Figure 10.16) the germ plasm is segregated in one of the blastomeres. The other blastomere undergoes **chromatin diminution** at its next division. A prospective somatic cell nucleus segregates from the germinal cell line at each successive cleavage, until, at the 16-cell stage, there are only two cells left with undiminished chromatin. One of these becomes the ancestor of all the germ cells, undergoing no further diminution of chromatin and forming no more somatic cells. Its sister cell loses chromatin at the next division, and her daughter cells and their descendants are somatic cells.

Boveri (1910) showed that whether a nucleus retains its chromatin undiminished depends on its association with a special region of the cytoplasm near the vegetal pole. In eggs that cleave abnormally as a consequence of dispermy or after centrifugation (Figure 10.17) this zone may be partitioned among more than one blastomere and thus two or more germ line stem cells may develop. The effective component of the cytoplasm is probably a nucleic acid, for it is inactivated by ultraviolet light, which is strongly absorbed by DNA and RNA.

Tobler, Smith, and Ursprung (1972) analyzed the DNA of diminished and undiminished genomes of *Ascaris* and found that about 27% of germ line DNA is lost in somatic cells. The eliminated DNA contains both unique and repeated nucleotide sequences, but it probably does not contain any ribosomal DNA cistrons.

Many other organisms, notably insects, show chromatin diminution in the blastomeres destined to form somatic cells. The implication, of course, is that the lost chromatin contains genetic information relating only to some phase of the production of gametes. Experiments carried out by Geyer-Duszyńska (1959, 1966) on eggs of the midge *Wachtliella persicariae* support this idea. In this midge, a distinctive germinal plasm, rich in RNA-protein granules, is present at the posterior pole of the egg (opposite the micropyle). Nuclei that enter the polar zone during cleavage retain their chromatin complement of 40 and become the germ cell nuclei. Other nuclei continue to divide but proceed to reduce the chromosome number to eight. If cleavage nuclei are prevented from entering the polar plasm by means of a ligature about the posterior end, all lose chromosomes and the resulting insect develops with sterile gonads, gonads lacking gametes (Figure 10.18).

If the ligature is removed after chromatin diminution has taken place, a nucleus with diminished chromatin may wander into the polar plasm, become surrounded by a cell membrane when the blastoderm forms, and then give rise to a line of germ cells. Females developing from such ligatured eggs appear normal and they have primordial germ cells, but are unable to carry out oögenesis. Males from similarly treated eggs may actually form sperm, but these die during metamorphosis (cf. Chapter 17). Kunz and his collaborators (1970) showed that the chromosomes that are eliminated in cleavage are distinctive during oögenesis. Throughout most of the growth of the oöcyte, they are diffusely organized (despiralized) and are active sites of mRNA synthesis, whereas the eight somatic chromosomes remain compact and relatively inactive. Thus the cleavage-eliminated chromosomes presumably contain genes that are active only during oögenesis.

A distinctive germinal plasm described for *Drosophila* and other dipterans by Hegner is illustrated in Figure 10.19. This plasm accumulates in a thin disc at the posterior end of the egg during oögenesis (Figure 10.19A) and becomes associated with nuclei that enter this region of the egg during cleavage (Figure 10.19B, C). As seen in the electron microscope, the polar plasm contains distinctive polar

194

FIGURE 10.17 Abnormal cleavage and distribution of germ cell determinant cytoplasm in a dispermic (fertilized by two sperm) *Ascaris* egg. The atypical cleavage pattern shown in A results in the formation of three stem cells (stippled) that divide with undiminished chromatin as shown in B. [Adapted from E. B. Wilson, *The Cell*, 3rd ed., Macmillan, New York, 1928, pp. 1090–91.]

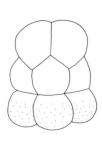

A   B

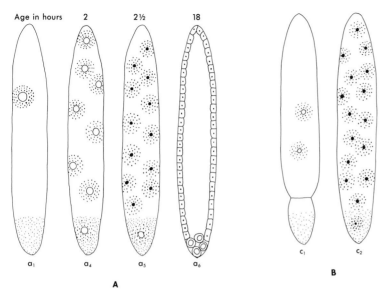

A

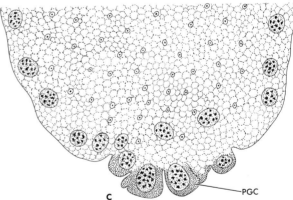

B

FIGURE 10.18 Elimination of chromosomes and segregation of the germ line in *Wachtiella persicariae*. A: At stage a_1, 1 hr after fertilization, the zygote is formed and the first three divisions are normal, with one nucleus entering the pole plasm at stage a_4, 2 hr after fertilization. At the next division all nuclei except the one in the pole plasm lose 32 out of 40 chromosomes. B: Effects of a ligature experiment. If the ligature is imposed such that no nucleus can enter the pole plasm at the 8-cell stage, c_1, all nuclei undergo chromatin diminution at the 16-cell stage, c_2, even if the ligature is removed. [Adapted from I. Geyer-Duszynska, *J Exp Zool* 141:391–441 (1959).]

195

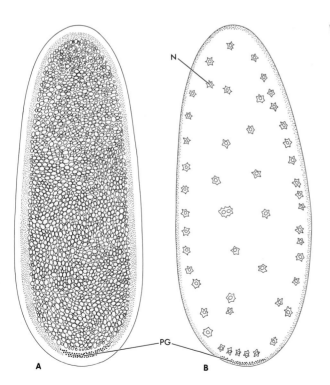

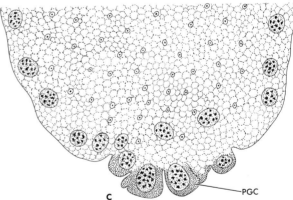

C

FIGURE 10.19 Pole plasm and primordial germ cells in dipterans. A: Egg, showing location of polar granules in *Drosophila*. B: Cleavage in the egg of *Drosophila* showing migration of nuclei to the pole plasm. C: Pinching off of the pole cells prior to blastoderm formation in another fly *Campsilura*. N, nucleus; PG, polar granules; PGC, primordial germ cell. [A and B from R. W. Hegner, *The Germ Cell Cycle in Animals*, Macmillan, New York, 1914, p. 114; C from R. W. Hegner, *J Morphol* 25:375–509 (1914), by permission of Alan R. Liss, Inc.]

granules, each appearing as a mesh of fibrous elements less than 1 μm in diameter. During vitellogenesis they are associated with mitochondria, but become detached from them after fertilization. They then break up into smaller rod-shaped or spherical structures and clusters of ribosomes become associated with them. The cleavage nuclei that enter the region of these granules are precociously cut off by cytoplasmic furrows (Figure 10.19C), and thereafter RNA is no longer detected in association with them. The pole cells later enter the gonads and become the primordial germ cells, meanwhile retaining a distinctive content of polar plasm. Mahowald, who has studied the polar granules of *Drosophila* and extensively reviewed the literature on them (Mahowald et al., 1979), suggested in 1968 that the polar granule functions as a storage element for mRNA used in the synthesis of proteins that determines the pole cells as future germ cells.

Illmensee and Mahowald (1973, 1974, Illmensee et al., 1976) later showed that pole plasm can bring about differentiation of germ cells after experimental transfer to another region of the egg. They transferred polar plasm from a genetically wild type *Drosophila melanogaster* egg to the anterior portion of a mutant embryo of the same stage. Mutant nuclei in contact with the polar plasm took on the cytological characteristics of germ cell nuclei and, when removed and placed in the polar region of another embryo of known different genotype, formed viable gametes that bred true for the mutant strain of their origins. In 1976 Mahowald et al. carried out transplantation of pole plasm between *D. melanogaster* and *D. immigrans* and obtained comparable results.

In newly formed pole cells unique nucleolar organelles appear. These are composed of protein, are highly electron dense, and acquire a hollow spherical configuration. The walls of the sphere are distinctively different in pole cell nuclei of *D. melanogaster* and *D. immigrans* in that the latter show electron lucent regions of the spherical wall. Very important when nuclei of *D. melanogaster* are exposed to pole plasm of *D. immigrans,* they become pole cells whose nuclear organelle is identical to that of *D. immigrans* pole cells. This suggests that the protein of the nucleolar body is possibly determined by species-specific pole plasm RNA.

A germ line determinant containing RNA has also been found in some species of frogs. Certain

experiments involving this material will be treated in Chapter 11 in connection with considerations of the origin of the urogenital system in vertebrates.

10.2 Gene Action During Early Embryogenesis

In Chapter 7 it is pointed out that, in many kinds of animals, oögenesis is marked by the accumulation of mRNA in a form that is made available for translation during maturation or after fertilization (Chapter 8). They also accumulate the ribosomal and transfer RNAs required for protein synthesis. It is now clear that, for eggs in which maternal gene products are stockpiled, the initial events of morphogenesis and underlying patterns of protein synthesis can proceed independently of the embryonic genome. This was suggested some years ago by the fact that enucleated fragments of sea urchin eggs form chromosome-free spindles and asters (**cleavage amphiasters**) and undergo cleavage (Harvey, 1936) but do not form normal blastulas. Likewise, enucleated frog oöcytes "fertilized" with lethally irradiated sperm cleave and form abnormal blastulas with a small cavity. Development during early stages proceeds more normally, however, in embryos treated with dosages of actinomycin D that are sufficient to prevent transcription of the maternal genome but not so great as to prevent DNA synthesis. Gross and Cousineau (1963, 1964) found that, whereas RNA synthesis can be completely blocked in sea urchin eggs, cleavage proceeds, nevertheless, to the blastula stage and the rate of protein synthesis continues unimpaired (Figure 10.20). Gastrulation fails to occur, however. There is an absolute requirement for new protein synthesis during sea urchin cleavage for when eggs are treated with drugs such as cycloheximide and puromycin, which inhibit the assembly of polypeptide chains on the ribosomes, cleavage stops (Hultin, 1961; Karnofsky and Simmel, 1963). Similar results have been obtained with eggs of a number of amphibians, mollusks, ascidians, and teleosts. Eggs that fail to stockpile genetic information and translational machinery undergo only limited cleavage in the absence of new transcription after fertiliza-

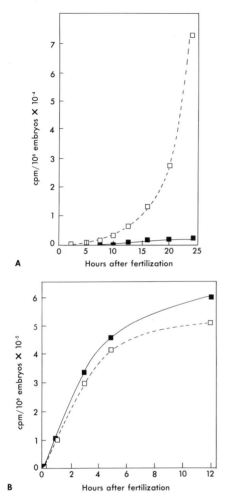

FIGURE 10.20 Effect of actinomycin D on the synthesis of RNA and protein in fertilized eggs of the sea urchin *Arbacia punctulata*. A: Uptake of ¹⁴C-uracil into acid-insoluble material in the absence (□——□) and in the presence (■——■) of 115 mg/ml actinomycin D in seawater. Eggs at a concentration of 3.7 × 10⁴/ml. B: Incorporation of ¹⁴C-valine into protein in control (■——■) embryos and in embryos at a concentration of 3.6 × 10⁴/ml in seawater with actinomycin D at a concentration of 115 mg/ml. [Redrawn with modifications from P. R. Gross and G. H. Cousineau, *Exp Cell Res* 33:268–395 (1964).]

stores accumulated during oögenesis or that can be produced rapidly after activation of the egg. Especially important would be histone proteins, which are involved in the organization of chromatin, and enzymes such as DNA and RNA polymerases, proteins that form structural elements such as membranes, spindle microtubules, cilia, contractile proteins involved in cleavage, and various enzymatic proteins essential to the assembly of these elements.

Histones are relatively small proteins whose synthesis is dependent on mRNAs that have a sedimentation coefficient of about 9 S. In fact cytoplasmic 9 S mRNA consists almost entirely of templates for histone synthesis. Histones are not stockpiled in sea urchin oöcytes, and histone synthesis is activated along with total protein synthesis immediately after fertilization. Thereafter, the absolute rate of histone synthesis is tightly linked to the rate of cell division throughout cleavage. Histone mRNA is present in cytoplasmic RNPs of sea urchin oöcytes, but additional messages are made during cleavage. Skoultchi and Gross (1973) isolated radioactively labeled 9 S RNA from polysomes of sea urchin blastulas. They then tested the ability of unlabeled unfertilized egg RNA and blastula polysomal RNA to compete with the labeled RNA for binding to denatured (i.e., single-stranded) sea urchin total DNA (Figure 10.21). On the basis of their results they calculated that the amount of histone mRNA in the egg is about one fourth of that on polysomes at the blastula stage.

In contrast to the case for sea urchins, *Xenopus* oöcytes stockpile histones. This is important, for, despite the fact that they also stockpile histone mRNA and transcribe new histone mRNA during cleavage, the DNA synthesis rate far outstrips the rate of histone synthesis past the blastula stage. *Xenopus* has only about 30 copies of the various histone genes per haploid genome, whereas sea urchin embryos may have as many as 1000. Moreover, *Xenopus* makes about 1000 times as much DNA per embryo each minute during cleavage as does the sea urchin. The advantage to *Xenopus* of stockpiling both histones and their mRNAs during oögenesis is thus obvious (evidence reviewed by Davidson, 1976; see also Adamson and Woodland, 1977).

In the organisms in which they have been studied, chiefly sea urchins, *Xenopus* and *Drosophila*, the

197

tion. This is the case for mammals (see Davidson, 1976, for an extensive review of these matters).

Cleaving eggs obviously require also that various structural proteins be available either from

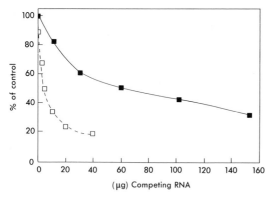

FIGURE 10.21 Binding of 9 S RNA from the early blastula of the sea urchin to denatured total sea urchin DNA in competition with total polysomal RNA of the early blastula (□——□) and with unfertilized egg RNA (■——■). 9 S RNA is essentially all histone mRNA. These results identify histone mRNA as a major component of material RNA and also suggest that a considerable portion of the histone mRNA transcribed in the blastula is the same kind as that stored in the oöcyte. [Redrawn with modifications from A. Skoultchi and P. R. Gross, *Proc Nat Acad Sci USA* 70:2840–44 (1973).]

198

various nucleotide polymerases are accumulated during oögenesis in such amounts as to carry the embryo to the postgastrular period. Thereafter, new gene expression occurs, for polymerase activity increases and new forms of the enzymes appear.

10.2.1 CONTROLLED TRANSLATION OF MATERNAL TEMPLATES. Are all species of maternal mRNAs translated immediately upon fertilization, or only some? Are they translated at the same rate? Are maternal templates differentially localized in the egg cytoplasm? Some observations that relate to these questions are presented here.

Sea urchin eggs and embryos contain large quantities of tubulin not yet assembled into macromolecular structures. After fertilization the synthesis of new tubulin molecules begins before the first cleavage. Tubulin is present in the unfertilized egg, but the pool of unassembled tubulin subunits remains constant throughout cleavage. Thus there is a continuing breakdown and resynthesis of tubulin during cleavage (Raff et al., 1971). As the blastula forms its cilia preparatory to hatching out of the fertilization membrane, tubulin shows a marked increase in its rate of synthesis (Stephens, 1972). This

increase is not affected by treating eggs with actinomycin D. Hence the synthesis of tubulin is largely, if not exclusively, on maternal templates during cleavage, and the enhanced rate of its production prior to hatching must result from a more efficient translation of these templates. Points of translational control of protein synthesis and their developmental significance are discussed at some length in Chapter 18.

There are also mechanisms that determine the stage at which translation products of maternal templates become detectable. An example of apparently delayed translation of maternal templates in sea urchins is provided by the case of the enzyme that catalyzes the reduction of the ribonucleotide diphosphates to deoxyribonucleotide diphosphates. This enzyme is essential to the synthesis of DNA in the absence of an adequate pool of deoxynucleotide diphosphates. It cannot be detected in unfertilized eggs and can first be assayed 1 hour after insemination, even in the presence of actinomycin D. Its appearance is prevented, however, by treatment with inhibitors of protein synthesis. Evidently, therefore, the enzyme system is absent in unfertilized eggs and is synthesized, using preexisting mRNA, in response to fertilization (Noronha et al., 1972).

An important landmark in early development of the sea urchin is the formation of cilia. These appear in embryos treated with actinomycin D, so their components must either be present at fertilization or be made on maternal templates later. A number of proteins can be extracted from isolated cilia and, most of them, including tubulin, as noted above, are present at fertilization and are also synthesized during cleavage. The ciliary **ATPase** (enzyme that hydrolyzes ATP to ADP with release of energy), which is important in ciliary movement, has two components, one of which is present before fertilization and one of which is synthesized after. Moreover, several proteins of yet unknown function, which are present in minor amounts in cilia, are synthesized only as the cilia appear (Stephens, 1972). Their appearance, too, is apparently under translational control.

The existence of highly mosaic eggs suggests the possibility that regionally distinctive egg plasms are possibly endowed with species of mRNA that might be unique to those plasms or that might be selectively translated there. Newrock and Raff (1975) examined the question in embryos of *Ilyanassa*. They found that proteins synthesized in the presence of actinomycin D by normal embryos and those deprived of the polar lobe give different patterns when separated by means of electrophoresis on polyacrylamide gels. These differences could arise either because factors segregated in the polar lobe determine differences in the maternal templates to be translated or because different messages are segregated there.

In the ascidian *Ciona* the cells of the future endoderm begin to show alkaline phosphatase activity at the late gastrula stage which occurs at about 6 hours after fertilization at 18°C. Its activity increases markedly thereafter as the gut is formed (Whittaker, 1977, 1979). This phosphatase is distinctive from other alkaline phosphatases of the embryo in that it is inhibited by L-phenylalanine in histochemical reactions. In embryos reared in the presence of actinomycin D, phenylalanine-sensitive alkaline phosphatase activity appears in cells of the prospective gut, indicating that maternal mRNA is involved directly or indirectly in its synthesis. The enzyme does not appear, however, in the presence of puromycin. Whittaker (1977) inhibited cleavage by means of cytochalasin B at various stages and observed the subsequent appearance of an active alkaline phosphatase. At the 16-cell stage, there are six blastomeres in the endodermal cell lineage. When cytochalasin is applied at this stage, which occurs $2\frac{3}{4}$ hours after fertilization, gut-specific alkaline phosphatase activity appears in the endodermal cells some hours later. Puromycin prevents the appearance of the phosphatase in cytochalasin-treated cells, but actinomycin D does not. Conceivably puromycin sensitivity could indicate the localized synthesis of an activator protein of a preformed enzyme, but the quantitatitve increase in enzyme activity during development is so great as to suggest that this is not so. Unfortunately, as in the case of the polar lobe of *Ilyanassa,* the evidence in this case does not rule out the possibility that the phenylalanine-sensitive alkaline phosphatase mRNA is uniformly distributed and that what is segregated in the endodermal cell lineage is a selective translational control factor (cf. Chapter 18).

10.2.2 PERSISTENCE OF MATERNALLY TRANSCRIBED TEMPLATES. To what extent do maternally transcribed templates persist and undergo

translation in the embryo? This question will be answered briefly, as it applies to the sea urchin. There is a voluminous literature related to the problem, but no attempt will be made here to supply detailed documentation concerning it. For key references one may consult the paper of Hough-Evans et al. (1977) and Davidson (1976).

During the first hours after fertilization all or almost all of the maternally synthesized species of mRNA begin to be translated on polysomes. This includes histone mRNAs, which are transcribed from genes present in multiple copies in the genome, and a large number of mRNAs that are copied from genes present in only one copy per haploid genome. These are called **single-copy** genes. It is the latter group whose disposition during early development is best known. It has been estimated that there are about 6.1×10^8 nucleotide pairs present in single copy sequences on the sea urchin genome. This number is referred to as the **complexity** of the single-copy genome. Some of the single-copy genes may be transcribed and represented as cytoplasmic mRNAs and others may not. In the mature oöcyte of the sea urchin the complexity of cytoplasmic mRNA is about 37×10^6. The mean length of sea urchin mRNA is about 1800 nucleotides, not counting the untranslated terminal poly(A) sequences. The median length is about 1200 nucleotides. This suggests that as many as 20,000–30,000 different proteins could be synthesized in the oöcyte using maternal mRNAs.

One may now ask to what extent do the mRNAs of maternal origin continue to be used during embryonic development. One may further inquire whether the embryonic genome produces and utilizes messages for the same proteins as those transcribed from maternal templates. In answer to the first question it is probable that during early cleavage the single-copy maternal messages begin to break down in a random manner. Their half-life (the time for one half of existing messages to break down) is about 5.7 hours, at least from the blastula stage on. In fact, by the blastula stage, almost all of the maternally synthesized mRNAs are no longer on polysomes; that is, they are broken down, unavailable for translation, or are poor competitors with embryonic mRNA for the translational machinery (cf. Chapter 18). Their competitive effectiveness may be diminished by the fact that the maternal messages have much less protein associated with them than do the embryonic mRNAs, which begin to be transcribed at the 8–16-cell stage (Wilt, 1970; Young and Raff, 1979).

Despite the fact that maternally synthesized messages rapidly disappear from polysomes after cleavage begins, the embryo continues to transcribe many mRNAs that are identical to the maternal mRNAs. Thus, at the gastrula stage almost all mRNA nucleotide sequences on polysomes are identical to maternally synthesized templates found in the oöcyte. The same is true at the pluteus stage. Even in adult tissues of the sea urchin a major segment of polysomal mRNA is identical in nucleotide sequences to those found in oöcytes.

10.2.3 NARROWING PATTERN OF GENE TRANSCRIPTION IN THE EMBRYO.

Does the complexity of gene transcription change as development proceeds? The answer to this question is in the affirmative, and, moreover, the change is in the direction of diminishing complexity. As noted above, the complexity of mRNA in the mature sea urchin oöcyte is 37×10^8 nucleotides. This represents about 6% of the total number of base pairs involved in single-copy DNA. At the blastula stage the complexity of RNA on polyribosomes diminishes to about 27×10^6 nucleotides and at gastrulation to 17×10^6 nucleotides, which is slightly over 2% of the total number of nucleotide pairs representing single-copy genes in the haploid genome. Polysomal mRNAs from adult sea urchin tissues have a complexity of less than 1% of the complexity of single-copy genes in DNA.

10.2.4 ARE NOVEL MESSAGES SYNTHESIZED DURING EARLY EMBRYOGENESIS?

Although at any developmental stage genes being transcribed and translated are genes that were also transcribed during oögenesis and whose products were present in the oöcyte, hitherto untranscribed genes do become transcribed and translated during development. The majority of kinds of mRNAs, however, are present in relatively few copies per cell, and the proteins for which they code are not readily resolved by methods presently available. Some structural genes, however, are represented far more extensively on polysomal mRNA than are others. These constitute a minority of the structural genes, but their transcripts may represent as much as 90% of polysomal mRNA. These structural genes are termed the prevalent class, and it is possible to detect some changes from stage to stage in proteins translated from newly produced transcripts of the prevalent class. One may also detect the appearance of new proteins translated from mRNA of multiple-copy genes in certain cases, as is illustrated further along.

Brandhorst (1976) used the method of two-dimensional gel electrophoresis to detect radioactively labeled proteins synthesized in oöcytes, fertilized eggs, and early embryos of sea urchins. This method is not explained here, but readers may turn to the paper of O'Farrell (1975), who developed it, for details. The proteins are separated in two dimensions of space on sheets of polyacrylamide gel. Their presence is revealed as separate spots of radioactivity when x-ray film is applied to the gels and then developed. By this means over 400 separate proteins, mostly of the prevalent class, can be detected during the stages studied. Brandhorst found a few spots of radioactivity restricted to oöcyte preparations, but, for the most part, the pattern of spots was identical. The patterns of spots displayed by proteins of early and late blastulas are, also, much like those of the oöcyte, although some spots not common to the earlier stages are found. By the early gastrula stage, however, numerous new spots appear and the relative intensities of other spots diminish (indicating diminished rates of synthesis of proteins in these spots). The pluteus also shows a few new spots and changing relative intensities of others.

These results give positive assurance that quantitative changes in the rates of production of different proteins occur as morphogenesis begins in the sea urchin. Because of difficulties in detecting proteins present in only a few copies per cell, it is not certain that major qualitative changes occur. Westin (1972) found that antigenically distinctive proteins appear at the end of the blastula stage in sea urchins. Their appearance is prevented by allowing the embryos to develop in the presence of sufficient actinomycin D to block transcription. Gastrulation is also blocked.

Giudice et al. (1968) examined the question as to when the mRNAs requisite to gastrulation must be present in the pregastrular sea urchin embryo. They exposed sea urchin eggs to actinomycin D for various periods at different times after fertilization up to the time the prism larva would normally arise (Figure 10.22). In the sea urchin *Paracentrotus*

200

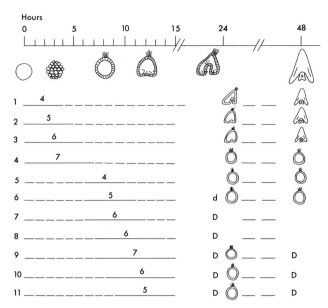

FIGURE 10.22 Effects of short treatments with actinomycin D applied to the sea urchin egg during the first 24 hours of development. At the top are shown the times and staging of normal development. Eleven different experimental treatments are indicated by the numbers at the left and by the lengths and positions of solid lines. Note that treatment with the drug during the first 6 hours of development does not affect the formation of a normal pluteus. Applied at later periods, the drug slows development and, as the time for gastrulation approaches, it causes death, D, prior to or at the blastula stage. [Adapted from G. Giudice, V. Mutolo and G. Donanuti, *Arch Entw Organ* 161:118–28 (1968).]

lividus the hatching blastula stage is attained in 11 hours at a temperature of 18°C. Treatment with actinomycin D during the first 6 hours after fertilization causes a delay in development, but the embryos eventually reach the pluteus stage. Treatments of varying duration beginning after 6 hours prevent gastrulation and the embryos usually degenerate. Examination of Figure 10.22 suggests that during the period between appearance of the blastula cavity and hatching, the continuous synthesis of certain novel species of mRNA is required for the synthesis of proteins involved in gastrulation.

One of the novel genes that is presumably in this group is a gene for one of the classes of histones that is designated H1. The histone genes, of which there are five major classes, are readily separable from other proteins and from one another. The H1 class is particularly rich in the amino acid lysine. A few years ago it was noted that the H1 histone that is isolated from chromatin of the cleavage-stage sea urchin is gradually replaced in the blastula and early gastrula by another lysine-rich histone of different electrophoretic mobility. The earlier-appearing H1 protein was designated $H1_m$ (the subscript m indicating morula, a term applied to the cleaving ovum before a blastula cavity appears; historically this term was used chiefly for the preblastocyst stage of mammalian embryos). The later-appearing H1 protein was designated $H1_g$ (gastrula). It was first thought that the mRNAs for both histones were present as maternal mRNAs. Presumably $H1_m$ was translated during cleavage stages and $H1_g$ was translated as the time for gastrulation approached. This presumption was tested by Arceci et al. (1976). They used a cell-free translating system made from wheat embryos (cf., Chapter 18), which readily translates exogenously supplied mRNA. They found that, whereas this system would readily translate $H1_m$ from mRNAs of oöcytes and embryos of early cleavage stages, $H1_g$ did not appear among the translation products. If, however, mRNAs were isolated from postgastrular stages and tested in the same system, $H1_g$ appeared along with $H1_m$ at a ratio of approximately 1:1. The evidence in this case strongly suggests that a novel H_l histone gene is transcribed in preparation for gastrulation in sea urchins.

10.2.4.1 When Are Paternal Genes Expressed?

There is abundant evidence that the synthesis of RNA, as indicated by its uptake of radioactive uridine, begins in the male pronucleus shortly after fertilization. This has been reported for a number of species, including nematodes, sea urchins, sand dollars, and mammals (reviewed by Longo and Kunkle, 1977). But it is not known what classes of mRNA are produced in the pronuclei or whether they are exported to the cytoplasm.

The approach generally taken to determine when the paternal genome is activated has been that of detecting the presence in the embryo of proteins, usually enzymes, that differ from maternal enzymes that catalyze the same reaction by different specific activities or, especially, by different electrophoretic mobilities. In general, it appears that when a particular gene is due to be activated, alleles

201

for both maternal and paternal enzymes are transcribed simultaneously (reviewed by Etkin, 1977), except in cases in which viable hybrids between different species are examined. In hybrids the expression of at least some paternal genes may be delayed. For example, in the cross of *Ambystoma mexicanum* and *A. texanum,* whose alcohol dehydrogenases show different electrophoretic mobilities, the enzyme appears in the liver at 4 weeks, which is normal. But it is of the maternal type, regardless of which species provided the eggs. The alcohol dehydrogenase of the paternal type does not appear until 3–5 weeks later.

The general impression is that, in the case of hybrids, the preferential expression of the maternal allele is due to the presence of regulatory factors synthesized during oögenesis. These factors act when genes coding for specific proteins are to be transcribed. If the paternal genome is contributed by another species, the regulatory site on the paternal gene may not be able to respond as efficiently to the regulatory substance as does the similar site on the maternal genome. In fact, there are cases in which certain paternal genes in hybrid embryos are never transcribed or, if transcribed, are not translated (Castro-Sierra and Ohno, 1968).

Suppose, however, that one is not dealing with hybrids, but rather with crosses between members of the same species that differ only in having detectably different forms of the same enzyme. In these instances it is usually seen that proteins encoded in the male and female genomes appear simultaneously. Thus in intraspecific *Ambystoma texanum* crosses between individuals differing only in having variant forms of alcohol dehydrogenase, both maternal and paternal genes are expressed simultaneously (Etkin, 1977). Ohno et al. (1968) showed for the Japanese quail that both paternal alleles at the gene locus for 6-phosphoglucose dehydrogenase are activated simultaneously after about 36 hours of incubation. Prior to that time the embryo utilizes a steadily declining supply of maternally stored enzyme. The quail flock used by Ohno and his associates showed four different alleles for the enzyme, designated A, B, C, and D. The enzyme, a dimer, is active in any combination of the translation products of these genes. Interestingly, in matings between males homozygous for A and females heterozygous for B and D, the dimer was of the AB combination twice as frequently as it

was in the AD combination. Thus the maternal genome in some way affects the proportion in which the B gene product is utilized, but whether in transcription, translation, or in posttranslational events is not evident.

10.2.5 CYTOPLASMIC INHIBITORS OF EMBRYONIC GENE ACTION. Embryos from different kinds of organisms differ somewhat in the stage of development at which different classes of RNA begin to be transcribed. These differences may be correlated to a great extent with differences in the extent to which RNA classes and proteins are stockpiled during development. In general, radioactive uridine administered to cleaving eggs is found first in the nuclei in high molecular weight **heterogenous RNA** (HnRNA, cf. Chapter 18), some of which may appear as mRNA associated with ribosomes of maternal origin. Transfer RNAs and ribosomal RNAs are usually transcribed considerably later. In *Xenopus* new mRNAs appear early in cleavage, but appreciable amounts of tRNA are not synthesized until the blastula stage. Production of ribosomal RNA only becomes prominent during gastrulation. In mammals, which show little stockpiling of the protein-synthetic machinery, new RNA synthesis occurs much more rapidly. The four-cell mouse embryo is already synthesizing all classes of RNA, for example. The developmental stages at which RNA classes are first transcribed from the embryonic genome in embryos of several kinds of animals were tabulated by Davidson (1976), whose book may be consulted for appropriate references.

Cytoplasmic inhibitors may play a determining role in patterns of RNA synthesis during early embryogenesis (reviewed by Gurdon and Woodland, 1968). Endoderm nuclei from the neurula stage of *Xenopus* are active in the synthesis of rRNA, but this synthesis ceases when such nuclei are injected into recipient activated eggs whose nucleus has been killed by ultraviolet irradiation. Such eggs frequently develop normally, however. Ribosomal RNA synthesis resumes at the gastrula stage of such nuclear-transplant embryos, precisely at the time it begins in control embryos raised directly from fertilized eggs (Figure 10.23). Similarly, the synthesis of tRNA proceeds rapidly in the cells of early gastrula endoderm in *Xenopus,* but when the nucleus of an endodermal cell is transplanted to an activated egg that lacks a functional nucleus, tRNA synthesis is

202

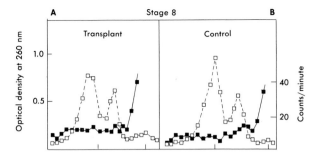

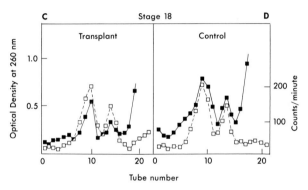

FIGURE 10.23 Sedimentation patterns on a sucrose gradient of total ³²P-labeled RNA isolated from normal fertilized eggs (B and D) and from fertilized eggs that received a transplanted nucleus after the zygote nucleus was inactivated (A and C). Optical density (□——□) and radioactive disintegration in counts per minute (■——■) are plotted against decreasing (from left to right) sedimentation velocity. Note that at stage 8 neither nuclear transplant (A) nor control (B) embryos show appreciable incorporation of radioactivity under the two principal peaks of optical density, which represent 28 S and 18 S ribosomal RNA. In contrast, at stage 18, peaks of radioactivity incorporated in RNA correspond to the peaks of optical density of rRNA. [Adapted from J. B. Gurdon and H. R. Woodland, *Biol Rev* 43:233–67 (1968).]

10.3 Stable Nuclear Activation by a Maternal Gene Product

In the Mexican axolotl, *Ambystoma mexicanum,* there occurs a mutant gene *o* that is inherited as a simple recessive that has a **maternal effect;** that is, females that inherit the o^-/o^- condition as a result of the mating of heterozygotes have offspring that are abnormal even though they may have a wild type allele. When heterozygotes are mated with each other, their offspring are of the expected genotypes, as shown in Figure 10.24. All offspring are completely normal, except that in larval life those of o^-/o^- constitution grow somewhat more slowly and have a diminished ability to regenerate amputated limbs (cf. Chapter 16). In these same animals likewise, the males are sterile, with poorly differentiated testes, but the females have normal ovaries and can be successfully mated. Regardless of whether eggs from such females are fertilized by sperm from o^+/o^+ or o^+/o^- males, cleavage proceeds normally, but the embryos are arrested in development at the time gastrulation begins. This block is preceded by a 75% reduction in the mitotic rate in the blastula and a failure of the enhanced RNA synthesis that normally precedes gastrulation.

These results suggest that the o^+ gene is required during oögenesis for the production of one or more

203

FIGURE 10.24 Inheritance and phenotypic effects of the *o* gene in *Ambystoma mexicanum.*

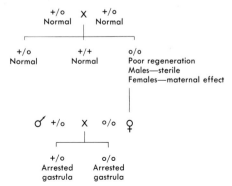

inhibited until the early gastrula stage is attained.

The factors that inhibit rRNA synthesis in the fertilized egg are carried into the blastula stage. Shiokawa and Yamana (1967) found that neurula cells from *Xenopus laevis* cultured in vitro synthesize rRNA, but this synthesis is reversibly depressed by the presence in the culture of blastula cells or of a heat-stable, dialyzable factor that blastula cells release to their surroundings. The factor does not affect the synthesis of other RNA species or of DNA.

regulatory factors that affect the activation of metabolic patterns required for gastrulation.

That this is so was convincingly demonstrated by Briggs and his collaborators (Briggs, 1973; Brothers, 1976, 1979), who showed that the gastrular block and antecedent metabolic defects can be removed by injection of undiluted whole cytoplasm from normal mature eggs into fertilized eggs of o^-/o^- mothers or by injecting the content of the germinal vesicle from oöcytes at various stages of their maturation. The effective o^+ substance is thus something that is produced in the germinal vesicle of the oöcyte and released to the cytoplasm when the germinal vesicle breaks down during the final stages of maturation. In fertilized eggs from o^+/o^- or o^+/o^+ mothers, the corrective factor may be extracted until the blastula stage, but not later. Briggs and Justus (1968) showed that the o^+ substance is inactivated by heating to 50°–55° C for 1 hour and by treatment with the enzyme trypsin. Thus it is probably a protein.

Briggs (1972) raised the question of whether activity of the o^+ gene is required at other than the time of oögenesis. When eggs of o^-/o^- females are fertilized with sperm bearing either the o^+ or the o^- gene, the resulting eggs are of the o^+/o^- and the o^-/o^- genotypes, respectively. Briggs found that they respond identically to injections of the o^+ gene product (protein) throughout gastrulation and proceed to the stage of hatching. At this time, however, half of the embryos die, the remaining half going on to reproductive maturity. Successful breeding of members of the latter group showed that they carried the o^+ gene. The dying group presumably were o^-/o^- individuals. Apparently, therefore, the o^+ substance cannot program the genome for the completion of development in the absence of the o^+ gene.

Brothers (1976, 1979) sought to determine the time during which the corrective factor coded by the o^+ gene is requred in order to program the embryonic genome past the gastrular block. In order to do so she transplanted nuclei from normal early blastulas into enucleated eggs of o^-/o^- mothers. Such diploid nuclei would have been exposed to the o^+ substance during early cleavages. These eggs developed only to the early gastrula stage, and showed the diminished mitotic rate and reduced RNA synthesis characteristic of eggs from o^-/o^- mothers, regardless of the genotype of the donor nucleus. If, however, she used normal donor nuclei

from cells of late blastulas, the recipient eggs showed normal metabolic patterns and developed to hatching. It would appear, therefore, that nuclei, even normal ones, require exposure to the o^+ substance until late blastula stages, in order to be programmed for the metabolic changes required for gastrulation.

Even more important, these changes are permanent! Early blastulas that develop from enucleated eggs of o^-/o^- mothers that had received a normal nucleus from a late blastula can then serve as successful nuclear transplant donors to similar eggs. This process can be repeated serially and some transplant recipients attain sexually mature adulthood. These experiments show that descendants of a normal nucleus that has been activated by o^+ substance retain their capacity to support normal development without futher exposure to the o^+ substance.

10.4   Summary

Animal eggs are classified as of the mosaic or regulative type. These usually show determinate or indeterminate cleavage, respectively. At or soon after fertilization, mosaic eggs show distinctive cytoplasmic regions corresponding to the particular organs in the larva or adult. Deletion of blastomeres results in corresponding defects in development. Regulative eggs, on the other hand, have to a great degree the ability to restore normal form and function after experimental defect.

Outstanding examples of mosaic eggs are found especially in ctenophores, annelids, mollusks, and ascidians. Regulative eggs are found in sea urchins and most vertebrates. Human eggs, as well as those of other mammals and those of birds and fishes, are highly regulative. *Cerebratulus*, a nemertean worm that shows spiral cleavage, is unusual in having great powers of regulation.

Actually there is no clear line of demarcation between regulative and mosaic eggs. All eggs, with the possible exception of those of mammals, show a degree of visible cytoplasmic prelocalization before cleavage begins. Moreover, if one examines so-called mosaic eggs at stages prior to fertilization, one finds that cytoplasmic prelocalization of organ-forming areas is the product of progressive change during antecedent development. At sufficiently early stages the so-called mosaic eggs are actually highly regulative, forming complete organisms despite defects wrought upon them. It is the case, nevertheless, that blastomere nuclei and their descendants differ in the expression of their genetic repertory of developmental possibilities according to the regional character of the egg cytoplasm they inherit. This is particularly illustrated by the cases in which cytoplasmic determiners of the stem cells for the line of primoridal germ cells have been found. These are regionally localized and constitute a germ plasm. In some cases blastomeres into which no germ plasm is segregated undergo chromatin diminution. In one case, at least, the lost genes are apparently required only in gametogenesis. The effective component of the germ plasm probably is RNA, for ribonucleoprotein particles characterize it, and its effect is abolished by irradiation with ultraviolet light, which is strongly absorbed by nucleic acids.

In some organism, notably sea urchins and amphibians, development can proceed after fertilization to gastrulation in the absence of a nucleus or in the presence of actinomycin D, which stops genetic transcription from the nuclear genes. Protein synthesis is required, however, and, in the absence of gene activity, must proceed using maternal mRNAs. These are not transcribed simultaneously but in temporal patterns that are apparently determined by factors of the egg cytoplasm. The disposition of maternal templates prior to their translation is not certain. Some are attached to polyribosomes at the time of fertilization, and others appear to be in the soluble phase of the cytoplasm in the form of ribonucleoprotein particles. Messenger RNAs translating for histones have been found in the soluble phase. Other messages that become polyadenylated after fertilization are subsequently translated. Some may be in polyribosomes and others in the same soluble phase of the cell sap.

Even though the embryonic genome can be dispensed with in early development, mRNAs begin to be translated early in cleavage, followed by tRNA and rRNA as gastrulation approaches. This sequence is determined by factors in the egg cytoplasm, specific inhibitors that are inactivated at the appropriate time in development.

During early development the embryo tran-

scribes and exports to the cytoplasm many mRNAs that are identical to maternal mRNAs of the oöcyte. As development proceeds, however, the complexity of single gene transcripts found on polyribosomes of the embryo diminishes to only a small fraction of the complexity of single gene transcripts in the oöcyte. Meanwhile, some novel sequences of nucleotides appear in the embryonic mRNAs. Some of these sequences are required for the synthesis of proteins required during gastrulation. In the sea urchin the most critical period for this synthesis is just prior to the stage of the hatching blastula.

There is sometimes a considerable delay before parts of the embryonic genome exercise their effects on morphogenesis. This is especially noteworthy in interspecific hybrids in which paternal alleles are either not expressed until some time after products of the corresponding maternal genes can be detected or are not expressed at all. In normal embryos cytoplasmic inhibitors are known to affect the temporal pattern of genetic transcription during early embryonic life.

An unusual mechanism for control of the genetic program is found in axolotls. A protein, o^+ substance, produced during oögenesis, is required during cleavage and blastula formation to program the genome for normal development beyond the gastrula stage.

205

10.5 Questions for Thought and Review

1. What is the basis for classifying eggs as having determinative cleavage?

2. Define mosaic development and regulative development.

3. Describe the patterns of cytoplasmic prelocalization in eggs of ascidians and ctenophores.

4. What is the significance of the polar lobe, which is found during cleavage in some mollusks and annelids? Give the experimental basis for your answer.

5. On the basis of what evidence do we consider mammalian eggs to be highly regulative?

6. What determines whether a given blastomere in the mammalian morula will become trophoblastic as opposed to embryonic tissue? What is the evidence bearing on this question?

7. In what way is the development of the *Cerebratulus* egg different from that of most other spirally cleaving eggs?

8. Explain what is meant by the statement that cytoplasmic localization arises epigenetically.

9. Regional differences in the egg cytoplasm determine the pattern of gene action in blastomere nuclei. Defend this position.

10. What is possibly the developmental significance, if any, of chromatin lost from somatic cell lines during cleavage in some organisms? Cite an experiment in support of your answer.

11. Discuss the significance of pole plasm in the development of insects?

12. Development of animal eggs until the time of gastrulation proceeds essentially independently of the action of the maternal genome. Explain this proposition critically, citing appropriate experimental evidence.

13. The egg cytoplasm contains maternal templates, mRNA transcribed during oögenesis on the maternal genome. In what form are these messages held in the zygote? What is the evidence relating to this question?

14. Maternal messages are active in protein synthesis after fertilization and during cleavage. Are all messages translated at the same time? What is meant by translational control of protein synthesis? What is the evidence that such controls are exercised during cleavage?

15. How may one determine whether the embryonic genome is transcribed during cleavage? Are all species of RNA produced simultaneously? How does the pattern of RNA synthesis differ during cleavage in sea urchins and amphibians as compared with mammals?

16. What visible change in the blastomere nucleus is usually associated with the onset of rRNA synthesis?

17. Describe an experiment designed to determine at what developmental stages the genes required for gastrulation are transcribed in sea urchin development.

18. What is the o^+ gene in *Amblystoma*? What is the experimental evidence relating to the time at which this gene is active?

10.6 Suggestions For Further Reading

ANDERSON, D. T. 1973. *Embryology and Phylogeny in Annelids and Arthropods.* Oxford: Fergamon, 495 pages.

ARNOLD, J. M., and L. D. WILLIAMS-ARNOLD. 1976. The egg cortex problem as seen through the squid eye. *Am Zool* 16:421–46.

BROWN, D. D. 1967. The genes for ribosomal RNA and their transcription during amphibian development. *Curr Top Dev Biol* 2:48–73.

COLLIER, J. R. 1966. The transcription of genetic information in the spiralian embryo. *Curr Top Dev Biol* 1:39–59.

COLLIER, J. R. 1976. Nucleic acid chemistry of the *Ilyanassa* embryo. *Am Zool* 16:483–500.

DENIS, H. 1977. Role of messenger ribonucleic acid in embryonic development. *Adv Morphol* 7:115–50.

FREEMAN, G. 1977. The multiple roles which cell division can play in the localization of developmental potential. In S. Subtelny and I. R. Konigsberg, eds. *Determinants of Spatial Organization.* New York: Academic Press, pp. 53–76.

GIUDICE, G. 1973. *Developmental Biology of the Sea Urchin Embryo.* New York: Academic Press, 469 pages.

GROSS, P. R. 1967. The control of protein synthesis in embryonic development and differentiation. *Curr Top Dev Biol* 2:1–46.

GURDON, J. B. 1969. Intracellular communication in early animal development. In A. Lang, ed. *Communication in Development.* New York: Academic Press, pp. 59–82.

GURDON, J. B. 1974. *The Control of Gene Expression in Animal Development.* Cambridge: Harvard University Press.

KIDDER, G. M. 1976. RNA synthesis and the ribosomal cistrons in early molluscan development. *Am Zool* 16:501–20.

NEYFAKH, A. A. 1971. Steps of realization of genetic information in early development. *Curr Top Dev Biol* 6:45–77.

RAFF, R. A., K. M. NEWROCK, D. D. SECRIST, and F. R. TURNER. 1976. Regulation of protein synthesis in embryos of *Ilyanassa obsoleta. Am Zool* 16:529–45.

REVERBERI, G. 1971. *Experimental Embryology of Marine and Fresh-Water Invertebrates.* Amsterdam: North-Holland, 587 pages.

ROSSANT, J. 1976 Investigation of inner cell mass determination by aggregation of isolated rat inner cell masses with mouse morulae. *J Embryol Exp Morphol* 36:163–74.

SPINDLE, A. I. 1978. Trophoblast regeneration by inner cell masses isolated from cultured mouse embryos. *J Exp Zool* 203:483–89.

WATTERSON, R. L. 1955. Selected invertebrates. In B. H. Willier, P. Weiss, and V. Hamburger, eds. *Analysis of Development.* Philadelphia: Saunders, pp. 315–36.

10.7 References

ADAMSON, E. D., and H. R. WOODLAND. 1977. Changes in the rate of histone synthesis during oöcyte maturation and very early development in *Xenopus laevis*. *Dev Biol* 57:136–49.

ARCECI, R. J., D. R. SENGER, and P. R. GROSS. 1976. The programmed switch in lysine-rich histone synthesis at gastrulation. *Cell* 9:171–78.

ARNOLD, J. 1968. The role of the egg cortex in cephalopod development. *Dev Biol* 18:180–97.

ARNOLD, J. M. 1971. Cephalopods. In G. Reverberi, ed. *Experimental Embryology of Marine and Fresh Water Invertebrates*. Amsterdam: North-Holland, pp. 265–311.

ATKINSON, J. W. 1971. Organogenesis in normal and lobeless embryos of the marine prosobranch gastropod *Ilyanassa obsoleta*. *J Morphol* 133:339–52.

BOVERI, T. 1910. Die Potenzen der *Ascaris*-Blastomeren bei abeänderter Furchung, zugleich en Beitrag zur Frage qualitativungleicher Chromosomen-Theilung. *Festschrift R Hertwig* 3:131–214. (Jena: G Fischer).

BRANDHORST, B. P. 1976. The two-dimensional gel patterns of protein synthesis before and after fertilization of sea urchin eggs. *Dev Biol* 52:310–17.

BRIGGS, R. 1972. Further studies on the maternal effect of the *o* gene in the mexican axolotl. *J Exp Zool* 181:271–80.

BRIGGS, R. 1973. Developmental genetics of the axolotl. In F. D. Ruddle, ed. *Genetic Mechanisms of Development*. New York: Academic Press, pp. 169–99.

BRIGGS, R., and J. T. JUSTUS. 1968. Partial characterization of the component from normal eggs which corrects the maternal effect of gene *o* in the Mexican axolotl (*Ambystoma mexicanum*). *J Exp Zool* 167:105–16.

BROTHERS, A. J., 1976. Stable nuclear activation dependent on a protein synthesized during oögenesis. *Nature* 260:112–15.

BROTHERS, A. J. 1979. A specific case of genetic control of early development: the *o* maternal effect mutation of the mexican axolotl. In S. Subtelny and I. R. Konigsbers, eds. *Determinants of Spatial Organization*. New York: Academic Press, pp. 167–83.

CASTRO-SIERRA, E., and S. OHNO. 1968. Allelic inhibition at the autosomally inherited gene locus for liver ADH in chicken-quail hybrids. *Biochem Genet* 1:323–39.

CLEMENT, A. C. 1952. Experimental studies on germinal localization in *Ilyanassa*. I. The role of the polar lobe in determination of the cleavage pattern and its influence in later development. *J Exp Zool* 121:593–626.

CLEMENT, A. C. 1962. Development of *Ilyanassa* following removal of the D macromere at successive cleavage stages. *J Exp Zool* 149:193–216.

CLEMENT, A. C. 1976. Cell determination and organogenesis in molluscan development: a reappraisal based on deletion experiments in *Ilyanassa*. *Am Zool* 16:447–53.

CONKLIN, E. G. 1897. The embryology of *Crepidula*. *J Morphol* 13:1–216.

DAVIDSON, E. H. 1976. *Gene Activity in Early Development*, 2nd ed. New York: Academic Press.

DOHMEN, M. R., and N. H. VERDONK. 1979. The ultrastructure and role of the polar lobe in development of molluscs. In S. Subtelny and I. K. Konigsberg, eds. *Determinants of Spatial Organization*. New York: Academic Press, pp. 3–27.

DRIESCH, H. 1891. Entwicklungsmechanische Studien I–II. *Zeit Wiss Zool* 53:160–84.

DRIESCH, H. 1892. Entwicklungsmechanische Studien III–VI. *Zeit Wiss Zool* 55:1–62.

ETKIN, L. D. 1977. Preferential expression of the maternal allele for alcohol dehydrogenase (ADH) in the amphibian hybrid *Ambystoma mexicanum* (axolotl) × *Ambystoma texanum*. *Dev Biol* 60:93–100.

FISCHEL, A. 1897. Experimentelle Untersuchungen am Ctenophorenei. I. Von der Entwickelung isolirter Eitheile. *Wilhelm Roux' Arch* 6:109–30.

FISCHEL, A. 1898. Experimentelle Untersuchungen am Ctenophorenei. II. Von der küstlichen Erseugung (Halber-) Doppel- und Missbildungen. III. Über Regulationen der Entwickelung. IV. Über den Entwickelungsgang und die Organisationsstufe des Ctenophoreneies. *Wilhelm Roux' Arch* 7:557–630.

FREEMAN, G. 1976. The role of cleavage in the localization of developmental potential in the ctenophore *Mnemiopsis leidyi*. *Dev Biol* 49:143–77.

GARDNER, R. L., and M. H. JOHNSON. 1972. An investigation of inner cell mass and trophoblast tissues following their isolation from the mouse blastocyst. *J Embryol Exp Morphol* 28:279–312.

GEYER-DUSZYŃSKA, I. 1959. Experimental research on chromosome diminution in *Cecidomiidae* (Diptera). *J Exp Zool* 141:391–441.

GEYER-DUSZYŃSKA, I, 1966. Genetic factors in oögenesis and spermatogenesis in *Cecidomyiidae*. *Chromosomes Today* 1:174–78.

GIUDICE, G., V. MUTOLO, and G. DONATUTI. 1968. Gene expression in sea urchin development. *Wilhelm Roux' Arch* 161:118–28.

GROSS, P. R., and G. H. COUSINEAU. 1963. Effects of actinomycin D on macromolecule synthesis and early development in sea urchin eggs. *Biochem Biophys Res Comm* 10:321–26.

GROSS, P. R., and G. H. COUSINEAU. 1964. Macromolecule synthesis and the influence of actinomycin on early development. *Exp Cell Res* 33:368–95.

GURDON, J. B., and H. R. WOODLAND. 1968. The cytoplasmic control of nuclear activity in animal development. *Biol Rev* 43:233–67.

HANDYSIDE, A. H., and M. H. JOHNSON. 1978. Temporal and spatial patterns of the synthesis of tissue-specific polypeptides in the reimplantation mouse embryo. *J Embryol Exp Morphol* 44:191–99.

HARVEY, E. B. 1936. Parthenogenetic merogony or cleavage without nuclei in *Arbacia punctulata*. *Biol Bull* 71:101–21.

207

HILLMAN, N. M. I. SHERMAN, and C. GRAHAM. 1972. The effect of spatial arrangement on cell determination during mouse development. *J Embryol Exp Morphol* 28:263–78.

HÖRSTADIUS, S. 1928. Über die Determination des Keimes bei Echinodermen. *Acta Zoologica* 9:1–191.

HOUGH-EVAN B. R., B. J. WOLD, S. G. ERNST, R. J. BRITTEN, and E. H. DAVIDSON. 1977. Appearance and persistence of maternal RNA sequences in sea urchin development. *Dev Biol* 60:258–77.

HULTIN, T. 1961. The effect of puromycin on protein metabolism and cell division in fertilized sea urchin eggs *Experientia* 17:410–11.

ILLMENSEE, K., and A. P. MAHOWALD, 1973. Induction of germ cells at the anterior pole of the *Drosophila* embryo. *J Cell Biol* 59:154a (Abstr.).

ILLMENSEE, K., and A. P. MAHOWALD. 1974. Transplantation of posterior polar plasm in *Drosophila.* Induction of germ cells in the anterior pole of the egg. *Proc Natl Acad Sci USA* 71:1016–20.

ILLMENSEE, K., A. P. MAHOWALD, and M. R. LOMIS. 1976. The ontogeny of germ plasm during oögenesis in *Drosophila. Dev Biol* 49:40–65.

KARNOFSKY. D. A., and E. B. SIMMEL. 1963. Effects of growth-inhibiting chemicals on the sand dollar embryo, *Echinarachnius parma. Prog Exp Tumor Res* 3:254.

KEDES, L. H., and P. R. GROSS. 1969. Synthesis and function of messenger RNA during early embryonic development. *J Mol Biol* 42:559–75.

KELLY, S. J. 1977. Studies of the developmental potential of the 4- and 8-cell stage mouse blastomeres. *J Exp Zool* 200:365–76.

KELLY, S. J. 1979. Investigations into the degree of mixing that occurs between the 8-cell stage and the blastocyst stage of mouse development. *J Exp Zool* 207:121–30.

KOWALEWSKI, A. 1883. Etude sur l'embryologie du *Dentale. Ann Musée Hist Natl Marseille* I(7):1–54.

KUNZ, W., H.-H. TREPTE, and K. BIER. 1970. On the function of the germ line chromosomes in the oögenesis of *Wachtliella persicariae (Cecidomyiidae). Chromosoma* 30:180–92.

LO, C. W., and N. B. GILULA. 1979. Gap-junctional communication in the preimplantation mouse embryo. *Cell* 18:399–409.

LONGO, F. J., and M. KUNKLE. 1977. Synthesis of RNA by male pronuclei of fertilized sea urchin eggs. *J Exp Zool* 201:431–38.

LUTHER, W. 1937. Transplantations—und Defektversuche am Organisationszentrum de Forellenkeim-scheibe. *Arch Entw-mech* 137:404–24.

LUTZ, H., M. DEPARTOUT, J. HUBERT, and C. PIEAU. 1963. Contribution à l'étude de la potentialité du blastoderm non incube chex les oiseaux. *Dev Biol* 6:23–44.

MAHOWALD, A. P. 1968. Polar granules of *Drosophila.* II. Ultrastructural changes during early embryogenesis. *J Exp Zool* 167:237–62.

MAHOWALD, A. P., C. D. ALLIS, K. M. KARRER, E. M. UNDERWOOD, and G. L. WARING, 1979. Germ plasm and pole cells of *Drosophila.* In S. Subtelny and I. R. Konigsberg, eds. *Determinants of Spatial Organization.* New York: Academic Press, pp. 127–46.

MAHOWALD, A. P., K. ILLMENSEE, and F. R. TURNER. 1976. Interspecific transplantation of polar plasm between *Drosophila* embryos. *J Cell Biol* 70:358–73.

MARKERT, C. L., and R. M. PETTERS. 1979. Manufactured hexaparental mice show that adults are derived from three embryonic cells. *Science* 202:56–60.

MINTZ, B. 1970. Gene expression in allophenic mice. In H. A. Padykula, ed. *Control Mechanism in the Expression of Cellular Phenotypes.* New York: Academic Press, pp. 15–42.

NEWROCK, K. M., and R. A. RAFF. 1975. Polar lobe specific regulation of translation in embryos of *Ilyanassa obsoleta. Dev Biol* 42:242–61.

NORONHA, J. M., G. H. SHEYS, and J. M. BUCHANAN. 1972. Induction of a reductive pathway for deoxyribonucleotide synthesis during early embryogenesis of the sea urchin. *Proc Natl Acad Sci USA* 69:2006–10.

O'FARRELL, P. H. 1975. High resolution two-dimensional electrophoresis of proteins. *J Biol Chem* 250:4007–21.

OHNO, S., C. STENIUS, L. C. CHRISTIAN, and C. HARRIS. 1968. Synchronous activation of both parental alleles at the 6-PGD locus of japanese quail embryos. *Biochem Genet* 2:197–204.

ORTOLANI, G. 1958. Cleavage and development of egg fragments in ascidians. *Acta Embryol Morphol Exp* 1:247–72.

RAFF, R. A., G. GREENHOUSE, K. W. GROSS, and P. R. GROSS. 1971. Synthesis and storage of microtubule proteins by sea urchin embryos. *J Cell Biol* 50:516–27.

REVERBERI, G., and G. ORTOLANI. 1962. Twin larvae from halves of the same egg in ascidians. *Dev Biol* 5:84–100.

SHIOKAWA, K., and K. YAMANA. 1967. Inhibitor of ribosomal RNA synthesis in *Xenopus laevis* embryos. *Dev Biol* 16:389–406.

SKOULTCHI, A., and P. R. GROSS. 1973. Maternal histone messenger RNA; detection by molecular hybridization. *Proc Natl Acad Sci USA* 70:2840–44.

SPRATT, N. T., and H. HAAS. 1961a. Integrative mechanisms in development of the early chick blastoderm. II. Role of morphogenetic movements and regenerative growth in synthetic and topographically disarranged blastoderms. *J Exp Zool* 147:57–94.

SPRATT, N. T., and H. HAAS. 1961b. Integrative mechanisms in development of the early chick blastoderm. III. Role of cell population size and growth potentiality in synthetic systems larger than normal. *J Exp Zool* 147:271–93.

208

STEPHENS, R. E. 1972. Studies on the development of the sea urchin *Stronglylocentrotus droebachiensis*. III. Embryonic synthesis of ciliary proteins. *Biol Bull* 142: 489–504.

STERN, M. S., and I. B. WILSON, 1972. Experimental studies on the organization of the preimplantation mouse embryo. I. Fusion of asynchronously cleaving eggs. *J Embryol Exp Morphol* 28:247–54.

TARKOWSKI, A. K. 1959. Experiments on the development of isolated blastomeres of mouse eggs. *Nature* 184:1286–87.

TOBLER, H., K. D. SMITH, and H. URSPRUNG. 1972. Molecular aspects of chromatin elimination in *Ascaris lumbricoides*. *Dev Biol* 27:190–203.

WESTIN, M. 1972. The occurrence of stage specific antigens during early sea urchin development, *J Exp Zool* 179:207–14.

WHITTAKER, J. R. 1977. Segregation during cleavage of a factor determining endodermal alkaline phosphatase development in ascidian embryos. *J Exp Zool* 202:139–53.

WHITTAKER, J. R. 1979. Cytoplasmic determinants of tissue differentiation in the ascidian egg. In S. Subtelny and I. R. Kongisberg, eds. *Determinant of Spatial Organization*. New York: Academic Press, pp. 29–51.

WILSON, E. B. 1892. The cell lineage of *Nereis. J Morphol* 6:361–470.

WILSON, E. B. 1928. *The Cell in Development and Heredity.* New York: Macmillan.

WILT, J. H. 1970. The acceleration of ribonucleic acid synthesis in cleaving sea urchin eggs. *Dev Biol* 23:444–55.

YATSU, N. 1912. Observations and experiments on the ctenophore egg. III. Experiments on germinal localization. *Annot Zool (Japan)* 8:5–13.

YOUNG, E. M. and R. A. RAFF. 1979. Messenger ribonucleoprotein particles in developing sea urchin embryos. *Dev Biol* 72:24–40.

209

11

Molding the Form
of the Animal Body

When gastrulation is complete, the embryo has taken a major step toward becoming a functionally independent organism. The germ layers are now arranged in a manner roughly corresponding to their definitive relationships, and the parts of each layer will normally exhibit only a limited range of developmental possibilities. This chapter is concerned with the molding of the germ layers into the form of organs and organ systems and the origin of the definitive body shape in vertebrate embryos. It therefore deals with **morphogenesis**, the origin of form.

11.1 An Inventory of Morphogenetic Processes

Understanding the derivation of the definitive body form from the three-layered gastrula is simplified if we first recognize a number of elementary morphogenetic processes that occur in the germ layers, both individually and jointly. These processes are relatively easy to outline, but their underlying mechanisms are often difficult to resolve.

11.1.1 EPITHELIAL FOLDING. Epithelia are layers of cells that cover external surfaces and line cavities. A number of processes are particularly prominent in the morphogenesis of epithelial layers. **Folding** is such a process. The appearance of a fold is usually preceded by a localized elongation of the cells in the direction perpendicular to their exposed surfaces (recall the initial stages in invagination in gastrulation of the sea urchin embryo, Chapter 9). This change is referred to as **palisading** (Figure 11.1). Palisading is accompanied by a reorganization of the microtubules of elongating cells such that they are predominantly oriented parallel to the direction of cellular elongation (Figure 11.2). Is the orientation of the microtubules a cause or an effect of cell elongation? (See discussion by Burnside, 1971.) This question is not yet answered.

When the subsequent folding is inward from the free surface, it is referred to as **invagination;** when outward from a surface or a lumen, as **evagination.** At the concave side of the fold the outer ends of the cells diminish their free surfaces, possibly through the action of a "purse string" of contractile microfilaments (Figures 11.2 and 11.3). Longitudinal foldings give rise to the neural tube in vertebrate embryos, as is described later in this chapter. Also, invaginations and evaginations of palisaded ectodermal cells form pits and hollow vesicles that give rise to important components of the sense organs. Both invaginations and evaginations of epithelial layers are involved in the origins of some kinds of glands.

11.1.2 FORMATION AND HOLLOWING OF SOLID CORDS. From the tips of invaginating epithelia solid cords of cells may grow into surrounding mesenchymal tissues and then hollow out. In this process cells from the interior of the cord detach from one another and rearrange themselves into tubes. This sequence of changes occurs prominently in the formation of hairs (Figure 11.4), teeth, and various digestive and endocrine glands. Solid cords of cells also arise by condensation from embryonic mesenchyme in connection with the origin of kidney tubules and other ducts. Here, too, the internal cells release their attachments and give rise to hollow tubes.

11.1.3 CELLULAR CONDENSATIONS. A very important process in the morphogenesis of many struc-

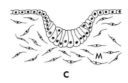

FIGURE 11.1 Palisading and invagination, or folding, of an epithelial layer. A simple epithelium in A, underlaid by mesenchyme, M, shows palisading in B and invagination in C.

tures is the localized condensation of mesenchyme cells. Such condensations occur in the origin of structures such as hairs and feathers (Figure 11.4),

FIGURE 11.2 Scheme depicting the orientation of microtubules and microfilaments in elongating cells of the neural plate of the newt *Taricha torosa* (see Figure 11.12). Numerous microtubules are aligned parallel to the long axis of the cell (paraxial microtubules, pmt). They are seen in cross section when the cell is cut transversely. Other microtubules are found parallel to the free surface of the cell at its apex amt. Also near the apex and in the same plane are microfilaments, mt, 50–70 Å in diameter, arranged in a circumferential bundle that encircle the apex in the manner of a purse string. Desmosomes, d, bind adjacent cells tightly and also serve as attachment sites for the microfilaments of either side. The net effect is that there is an interconnected network of filaments over the entire neural plate. Lipid droplets, ld, and yolk platelets, yp, are recognized in the neural plate cells. [From B. Burnside, *Dev Biol* 26:416–31 (1971), by permission of the author and Academic Press, Inc., New York.]

in the formation of capsules around the rudimentary sense organs (Figure 11.27), and the origin of the skeleton. Condensations arise by the movement of mesodermal cells toward morphogenetic centers and by differential rates of cell division.

11.1.4 DETACHMENTS AND FUSIONS. During morphogenesis cell layers often detach from one another, as, for example, in the formation of the embryonic **coelom,** or body cavity (Figures 11.15 and

FIGURE 11.3 Scheme illustrating invagination (A) and evagination (B) of a columnar epithelium in contact with mesoderm. In invagination, microfilaments are more densely ordered at the apex of the cell, a, and their contraction, as depicted on the right, diminishes the diameter of the apical end of the cell, throwing the cell membrane into folds. Conversely, in evagination, this process occurs at the basal, b, end of the cell. [Adapted from N. K. Wessels et al., *Science,* 171:135–43 (1971).]

211

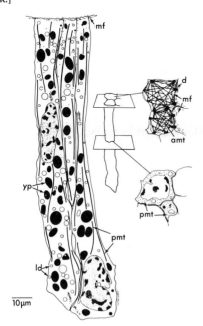

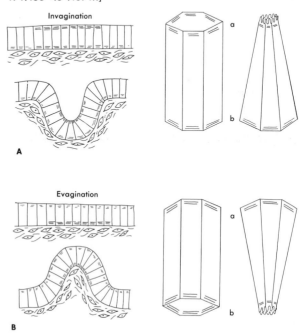

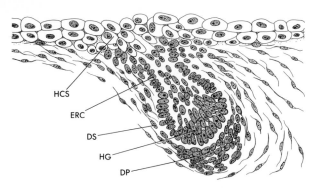

FIGURE 11.4 Origin of hair primordium by ingrowth of a cellular cord. The hair germ, HG, forms from cells at the base of the ingrowth, and the other cells become arranged as hair canal cells, HCS, and as primordia of the hair sheaths. ERC, external root sheath; DS, dermal sheath; DP; dermal papilla.

TABLE 11.1 The Distribution of Cell Deaths in Spinal Ganglia of the Chick Embryo*

Ganglia Number		4½ Days	5 Days	6 Days	7 Days
1		+	++	++	+
2		+	++	++	+
3		+	++	++	+
4		++	+++	+++	+
5		++	+++	+++	+
6		++	+++	+++	++
7		++	+++	+++	++
8		++	+++	+++	++
9		++	+++	+++	+
10		+	+++	+++	+
11		+	+++	+++	+
12		+	+++	+++	+
13	Wing	+	++	++	+
14		0	0	+	+
15		0	0	0	0
16		+	++	++	+
17		+	+++	+++	+
18		+	+++	+++	+
19		+	+++	+++	+
20		+	+++	+++	+
21		0	+++	+++	+
22		0	+++	+++	+
23	Leg	0	+++	+++	+
24		0	+++	+++	+
25		0	0	+	+
26		0	0	+	+
27		0	0	++	+
28		0	0	++	+
29		0	0	++	+
30		0	0	++	+
31		0	0	++	+
32		0	0	++	+
33		0	0	+	+
34		0	0	+	+
35		0	0	+	+

*Embryos were sacrificed on days indicated and sectioned for histology. Each spinal ganglion was examined for the presence of dead cells. Estimates of the number of dead cells in each ganglion were grouped in classes designated by a number of + signs, as shown. The absence of dead cells in a ganglion is indicated by 0. [From data of V. Hamburger and R. Lev-Montalcini, 1949, *J Exp Zool* 111:457–502. Later studies (Hamburger et al., 1981) of cell death at the wing level showed more cell deaths than previously estimated, but the general pattern of differential death shown here holds true.]

11.41). Crevices appear at intervals in the mesodermal mass parallel to the neural tube (segmental plate), blocking out and separating the somites (Figures 11.10, and 11.15) in anteroposterior sequence. There are many instances in which individual cells and groups of cells detach from major cell layers and migrate to new locations, where they form sheath cells for peripheral nerves, visceral ganglia, pigment cells of the skin, and so on. Later in this chapter examples are given of a number of morphogenetic events that involve the detachment of cells, groups of cells, and cell layers.

Fusions between cellular layers and migrating masses of cells are important morphogenetic processes. Ectodermal folds (Figures 11.10, 11.11, and 11.15) fuse to form the neural tube. The edges of small invaginations meet and fuse to form vesicles: two examples of this fusion occur, in the origin of the lens of the eye and the inner ear (Figure 11.27). The **sternum,** or breast bone, in higher vertebrates arises by the fusion of mesenchymal condensations that originally arose separately on either side of the ventral midline.

11.1.5 MORPHOGENETIC CELL DEATH.
Dead cells are found with some frequency in early embryos. Often their distribution appears to be random, and no particular morphogenetic significance is attached to their presence. In many cases, however, localized centers of **necrosis** (death) are so distrib-

uted as to suggest that cell death may play an important morphogenetic role. When cellular layers or groups of cells meet, when detachments or separations occur, and as tissues form sharp folds, death of numerous cells occurs, usually in such a position as to suggest that death facilitates the ongoing morphogenetic process. Massive necrosis plays a major role in the separation of the tissues that form the digits of the extremities in higher vertebrates (Figure 11.42) and in determining the number of cells in the motor columns of the spinal cord and the sensory ganglia (Figure 11.19 and Table 11.1). Death and resorption of entire organs and organ systems occur during the metamorphosis of many insects, in amphibians, in many fishes, in cyclostomes, and in marine invertebrates (reviewed by Saunders, 1966).

11.1.6 MOVEMENT AND CHANGING SHAPE OF CELLS.

The movement of cells from one place to another is involved in the morphogenesis of many tissues and organ systems. Recall, for example, that primordial germ cells originate outside the gonad and subsequently migrate into it (Chapter 7; see also Section 11.2.2.8). This example indicates that cells are motile and that there are factors that guide them to their new location. A brief description follows of the movement of tissue cells (as distinguished from amoeboid, ciliated, and flagellated cells) and of the effects of the cell's substratum on its shape and rate of locomotion.

Cells of metazoan animals usually lack rigidity, and, suspended in a liquid medium out of contact with other cells or surfaces, they tend to take a spherical shape, but are abundantly endowed with microvilli and blebs that protrude and retract. When they settle on a planar surface, the cell margins begin to spread and the blebs and microvilli diminish or disappear, as though serving as reserves of cell membrane utilized in spreading. The marginal spreading gives rise to a broad leading lamella about 0.4 μm in thickness. As one zone of spreading becomes larger, locomotion takes place in that direction. In rapidly moving fibroblasts the **leading lamella** spreads out in fan shape as the rest of the cell is drawn out as a triangle (Figure 11.5A). If, however, several leading lamellae spread in different directions, the cell as a whole may not move until one becomes larger. The leading lamella is extended as cytoplasm flows into it (Figures 11.6 and 11.7A). As it advances it forms broad regions of uniformly **close contact**, the membrane of the lower surface of the cell having a separation distance of about 30 nm from the substratum (Figures 11.6 and

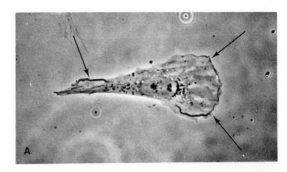

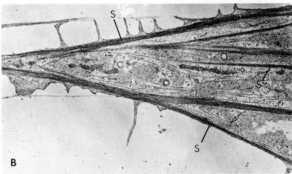

FIGURE 11.5 A: Light microscopic photograph of a living glial cell from a spinal ganglion of an 8-day chick embryo. The leading lamella is at the right, which is the direction of locomotion. Wave-like ruffles, indicated by arrows, are seen at the anterior edge as well as in a small lateral lamella near the tail of the cell. B: Low power electron micrograph of a section through the tail of a cell similar to that shown in A. The section is parallel to the surface of the substrate and illustrates prominent microfilament bundles, S, also called stress filaments or sheath bundles. Compare with Figures 11.6 and 11.8. [From N. K. Wessels, B. S. Spooner, and M. A. Ludueña, in *Locomotion of Tissue Cells*, Ciba Foundation Symposium 14, North-Holland, Amsterdam, 1973, pp. 53–82. By courtesy of Dr. Wessels and permission of North-Holland Publishers.]

213

11.7). The formation and persistence of the zones of close contact are associated specifically with the spreading process. When spreading ceases, close contacts are lost. Variously, under the main body of the cell, and here and there under the leading lamella, are interspersed areas of separation of 100–140 nm (Figures 11.6 and 11.7).

In many cultured cells the forward spreading of the leading lamella is accompanied by **ruffling.**

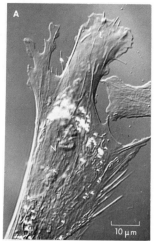

FIGURE 11.6 An embryonic chick heart fibroblast moving over a glass surface as viewed by means of differential interference optics (A) and 1.5 min later, by means of interference reflection optics (B). The direction of net movement is toward the top of the page. The leading edge of another fibroblast is approaching from the right. A: A leading lamella, L, shows at its tip a broad lamellipodium bearing several microspikes. A prominent stress fiber is at the left edge of the lamella and other stress fibers extend back into the cell body. To the far left is another spreading process showing several small lamellipodia and stress fibers. To the right of the nucleus, N, is a zone of retraction in which the cytoplasm is flowing back into the cell body around the stress fibers, which remain attached distally. B gives a great deal of information about the relationship of the lower cell surface to the substratum and the sites of insertion of the stress fibers. The very lightest areas at the fringes of the leading lamella are the lamellipodia, which are lifted some distance above the substrate. The very darkest patches are areas of focal contact, where separation of the lower surface of the cell membrane from the substratum is only 10–15 nm. The gray areas are regions of close contact, representing a separation distance of about 30 nm. The light areas within the cell are zones of greater separation, in the range of 100–140 nm. It is most instructive to examine each of these photographs in relation to the other. Note that the prominent stress fiber seen at the left edge of the leading lamella in A inserts in the cell membrane at a plaque near the distal end as seen in B. Other stress fibers in the leading lamella also can readily be associated with particular plaques. At the base of the lamellipodium microspikes and small bundles of fibers are seen in A, and these can be related to small plaques in B. In the area of retraction numerous attachment plaques can be found in B, corresponding to the insertions of stress fibers seen in A. Also worthy of note are the numerous points of focal contact in the cell approaching from the left. Because photograph B was taken 1.5 min later than A, some positional changes have occurred. Note that the lamellipodia on the left portion of the Y-shaped process to the left of the field at the top are less prominent, whereas those of the right portion have expanded. In the broader leading lamella, two closely associated microspikes have moved closer together. These tips show as dark regions, which means that they have either reached down to the substratum or are elevated considerably above the lamellipodium. [Photographs, hitherto unpublished, kindly supplied by Dr. C. Izzard, whose courtesy is gratefully acknowledged.]

214

Ruffles are local protrusions of the leading lamella that lift some distance, even vertically, from the substratum over a front of varying width (Figures 11.5, 11.6, and 11.7). They are often propagated back toward the nucleus in a wave-like front. Ranks of such ruffles are often seen in regions of lamellar spreading (Figures 11.5A and 11.7B). The term **lamellipodium** is used by some (e.g., Abercrombie et al., 1970a–c, 1971; Izzard and Lochner, 1976, 1980) to describe the uplifted leading edge, whereas others (e.g., Trinkaus, 1976) use the term synonymously with what we have termed the leading lamella.

When the leading edge of a chick heart fibroblast has advanced some distance and the cell has elongated in the direction of movement, the cell becomes quite taut and is under considerable tension. As tension builds up, the trailing edge of the cell is pulled loose from the substratum and retracts toward the next point of adhesion in the direction of the leading edge. Continued advance of the leading lamella, followed by detachment at the trailing edge, results in a net forward displacement of the cell.

When cells are allowed to spread on a deformable substratum of very thin silicone rubber (Harris et al., 1980), the tension that they develop is transmitted to the substratum and causes it to form wrinkles perpendicular to the direction of cellular movement.

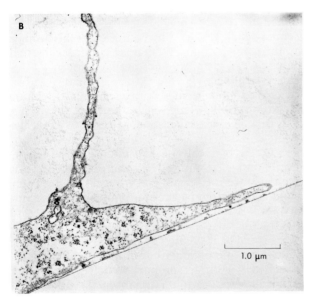

The tension developed as the leading lamella advances is transmitted to the cell body by microfilament bundles, or **stress fibers** (Figures 11.5B, 11.6A, 11.7A, and 11.8), that appear to insert in discrete structures called **plaques** in the leading lamella (Figures 11.6B and 11.7A). In electron micrographs the plaques are seen as condensations of

FIGURE 11.8 A: Stress fibers, S, and microtubules, T, as seen at higher magnification of the right hand portion of the cell shown in Figure 11.5B. The section is cut in a plane very close to the substratum, such that the empty space, E, lies in a zone of greater separation of cell from substratum. B: Edge of another glial cell sectioned parallel to and very close to the substrate illustrating the complex web of microfilaments that makes up the cortex of the cell and their relationships to the stress fibers. [A from N. K. Wessels, B. K. Spooner, and M. A. Ludueña, in *Locomotion of Tissue Cells*, Ciba Foundation Symposium 14, North Holland, Amsterdam, 1973, pp. 53–82. By courtesy of Dr. Wessels and permission of North-Holland Publishers; B a hitherto unpublished photograph kindly supplied by Dr. Wessels, whose courtesy is gratefully acknowledged.]

215

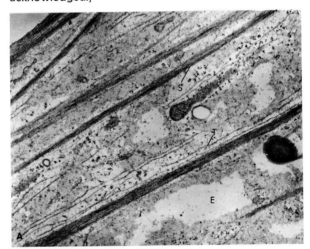

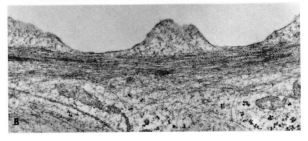

FIGURE 11.7 Leading lamellae and lamellipodia of chick heart fibroblasts grown on plastic and sectioned perpendicular to the substrate and parallel to the direction of advance. A: A lamellipodium is raised from the edge of the leading lamella. Beneath the forward edge of the leading lamella is a zone of close contact, and immediately behind is a placque, or focal contact. An area of greater separation between cell and substrate is to the left. Microfilaments can be seen extending back into the leading lamella from the placque. B: A lamellipodium, reared upward, is moving toward the cell body (to the left) as the leading edge of the cell (right) begins to lift from the substratum, presumably as an incipient lamellipodium. [From M. Abercrombie, J. E. M. Heaysman, and S. M. Peguem, *Exp Cell Res* (67:359–67 (1971), courtesy of Dr. Heaysman and permission of Academic Press, Inc., New York.]

microfilaments and amorphous material subjacent to the cell membrane at the points of closest cell-to-substrate contact. These points of very close contact show a separation distance of 10–15 nm, as measured by Izzard and Lochner (1976) by means of the interference reflection microscope. The development of a plaque is usually preceded by the formation of a short linear projection of the lamellipodium having a core of microfilament bundles or by the appearance of a similar fibrous assembly within the lamellipodium. As the leading lamella flows into the base of the lamellipodium, the points of focal contact develop amid the region of close contact, and the microfilament bundles progressively extend back into the cell body as stress fibers, which presumably originate by the progressive ordering of a preexisting cytoplasmic network of filaments. In some cases the lamellipodium or a microspike may dip to the surface and form focal contacts in advance of the zone of close contact (Figure 11.6), which, however, overtakes them as the cell continues to advance (Izzard and Lochner, 1980).

Many plaques appear at the periphery of the leading lamella. From these the stress fibers extend as elongate strands into the body of the cell parallel to the direction of the advance of the leading lamella (Figure 11.6). The individual plaques do not move. Instead, earlier-arising plaques fade and are replaced by new focal adhesion points formed ahead of the existing ones. Focal adhesions at the trailing edge of the cell are residual plaques from earlier zones of cell spreading opposite to the overall direction of movement. These adhesions hold the cell taut as the leading lamella advances prior to detachment of the trailing edge and retraction of the cell in the direction of the leading lamella.

It is not certain whether the forward retraction of the trailing edge is the result of passive elastic recoil or active contraction, or both. During recent years evidence has accumulated, for many kinds of cells, that the cell body is pervaded by a complex network of microfibrils and microfibrillar bundles that contain elements of the contractile proteins of muscle. Fluorescent antibodies of myosin, tropomyosin, actin, and α-actinin react with components of the network (Lazarides and Weber, 1974; Lazarides, 1975, 1976; Lazarides and Burridge, 1975; Weber and Groeschel-Stewart, 1974). In the stress fibers tropomyosin and α-actinin show an alternat-

216

ing repeat pattern and myosin shows an interrupted pattern. Actin appears to be uniformly distributed along the stress fibers and attaches to the points of close contact by means of a 130,000 molecular weight protein that seems not to be represented in the contractile system of muscle (Geiger, 1979). In view of the composition of the stress fibers and their points of firm attachment, it seems reasonable to consider that the microfilament bundles are actively involved in the generation of tension under isometric conditions. When the trailing edge of a cell is released, the cell is then presumably shortened isotonically with translocation of the cell body in the direction of the leading lamella. Indeed, microfilament bundles are capable of isotonic contraction. Isenberg et al. (1976) showed this capability using a strain of rat mammary adenocarcinoma cells that show a very distinct microfibrillar pattern in culture. When these cells are extracted with glycerol in the cold and then treated with a solution containing adenosine triphosphate (ATP), they contract, just as glycerinated muscle cells do. Moreover, microfilament bundles dissected from such glycerinated muscle cells by means of high-intensity laser microbeam contract isotonically in the presence of ATP. The advance of tissue culture cells conceivably could be mediated by the shortening of microfibrillar bundles released from isometric conditions as the trailing edge parts from the substratum. Advance could be facilitated by further shortening, for, in fact, the microfilament bundles in whole glycerinated cells contract proportionately to a greater extent in response to ATP than do the cell bodies.

One may now ask whether cells in vivo migrate in the same manner as they do on artificial surfaces. Bellairs et al. (1969) filmed the edge of the chick blastoderm migrating over the inner surface of the vitelline membrane in vitro and observed ruffled membranes similar to those described earlier in this section. Conceivably, ruffling occurs as a response to conditions in the filming chamber. Chernoff and Overton, whose work was cited in Chapter 9, observed ruffles in similar cells following incubation in the absence of yolk. Recall, however, that their study emphasized the role of long contractile filopodia in extending the margin of the blastoderm. Bard and Hay (1975) did not observe ruffles during an electron microscopic study of mesenchyme cells advancing into the stroma of the embryonic chick cornea. Also, Bard et al. (1975) showed by means of electron microscopic observations that the endothelium that forms on the inner face of the lens stroma arises from mesenchyme that migrates over the stroma, extending filopodia and lamellae without showing ruffling. Karfunkel (1977) observed the migration ventrally of the mesodermal sheet between ectoderm and endoderm during gastrulation in urodeles and anurans. He reported that cells at the leading edge of the mesoderm, as seen by means of the scanning electron microscope, are usually round with short and narrow flattened projections in the forward direction. No evidence of ruffling was seen. Poole and Steinberg (1977), likewise employing scanning electron microscopy, observed a similar locomotory mode in cells of the caudally growing tip of the pronephric duct (discussed in Section 11.2.2). In a detailed analysis of the movement of living deep cells of the *Fundulus* embryo during blastula and gastrula stages, Trinkaus (1973) observed that the translocation of cells occurs chiefly by the flowing of cell substance into blunt extensions called **lobopodia** or by extending nonruffling flattened lamellipodia. In sum, it does not appear that ruffling plays a significant role in the movement of cells in vivo.

It is probable, nevertheless, that many aspects of cellular migratory behavior in tissue culture reflect properties also shown in vivo. Tissue culture cells show a variety of different shapes depending on the physical characteristics of the substratum upon which they are grown. It is evident that cells in vivo also encounter different substrata (collagen fibers, nerve axons, blood vessels, and so on). The effect of the substratum is illustrated by the behavior of cells emigrating from a bit of living tissue cultured in a clot of blood plasma. The molecules of the protein, **fibrin,** that make up the framework of the clot may be likened to a brush-heap, or three-dimensional, network. Liquid fills the spaces in the meshwork. The fibrin molecules form a fine or coarse meshwork depending on the concentration of plasma used in making the clot or on the pH at which clotting occurs. Thus, when a high pH or low plasma concentration is used in making the clot, the fibrin meshwork is fine, and explanted cells tend to form numerous projections around their periphery and to migrate slowly, maintaining a stellate appearance (Figure 11.9A). When the clot is made at lower pH or with higher concentrations of plasma, the fibrin molecules aggregate in coarse bundles, apparently

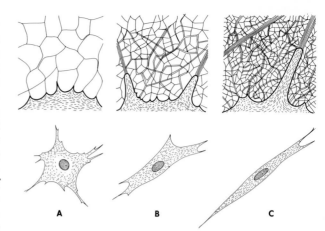

FIGURE 11.9 Cell shape and movement as affected by the substratum. The upper row of figures schematizes the configuration of fibrin micelles in clots of blood plasma made with concentrations of plasma increasing in the direction A to C. In the lower row are corresponding shapes of heart fibroblasts cultured by these clots. [From P. Weiss and B. Garber, *Proc Natl Acad Sci USA* 38:264–80 (1951), by courtesy of the authors.]

217

forming progressively broader highways (Figure 11.9B, C), which the advancing edges of cells tend to follow in preference to the finer fibrin meshes. Thus, the migrating cells tend to adopt a bipolar configuration and to migrate rapidly. Evidence of this kind, much of it assembled by Weiss and his co-workers, suggests that the migrating cell is guided by the configuration of the surface over which it moves. This is called **contact guidance** (cf. Weiss, 1958). Contact guidance certainly seems to play a role in the advancement of cells into and over the corneal stroma and in advancing edges of cell sheets generally.

Because of its significance for the understanding of morphogenesis, the matter of cellular motility deserves much more attention than can be devoted to it in this work. Readers are urged particularly to turn to the comprehensive review of Trinkaus (1976) and the meticulous studies of Izzard and Lochner (1980) and Harris et al. (1980) for further insights into the locomotory activity of cells and for more extensive references than can be provided here.

11.1.7 LOCAL PROLIFERATION. In contrast to what is seen in the development of plants (Chapter 12),

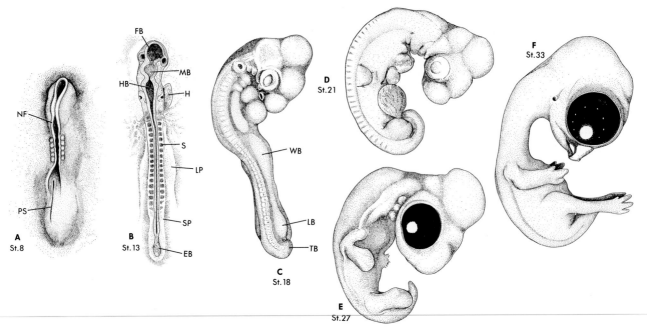

FIGURE 11.10 Stages in the development of the chick embryo. Numbers indicate developmental stages in the Hamburger–Hamilton series. EB, end bud; FB, forebrain; H, heart; HB, hindbrain; LB, leg bud; LP, lateral plate; MB; midbrain; NF, neural fold; PS, primitive streak; S, somite; SP, segmental plate; TB, tail bud; WB, wing bud. [Stages 8 and 14 after Keibel and Abraham, as adapted and modified from B. I. Balinsky, *An Introduction to Embryology*, Saunders, Philadelphia, 1965, p. 213. Stages 18, 21, 27 and 33 are interpretative drawings of corresponding stages in V. Hamburger and H. L. Hamilton, *J Morphol* 88:49–92 (1951).]

localized zones of more rapid cell division have little to do with the origin of the form of the animal body. It is only after organ rudiments are established that local centers of higher mitotic activity play a role in morphogenesis. Thus the walls of the early embryonic brain show localized bilateral thickenings, separated by constrictions. The thickenings, **neuromeres,** show regions of higher mitotic activity than is found in the valleys between them (Källen, 1962). Another example is found in the origin of the embryonic limbs in the chick embryo. The forelimb and hindlimb buds (Figure 11.10C) appear as thickenings of the body wall separated by the future flank. Prior to the appearance of the limb buds the mesoderm, along all the length of the mesodermal body wall, where the future limbs and flank will form, shows the same rate of cell division. When the limb buds appear, they do so, not because of a localized increase in cellular proliferation, but

by virtue of a decrease in the rate of mitosis in the flank (Searls and Janners, 1971).

11.2 Morphogenesis of the Germ Layers

Section 11.1 outlines a number of important components of morphogenesis. In sum, these are: tissue foldings, including invaginations and evaginations of cellular sheets; the fusion, detachment, or separation of groups of cells and cell layers; palisading, local aggregations, and condensations; the migration of cells from their sites of origin; and the localized death of cells. To a large degree these processes account for the changing external form of embryos

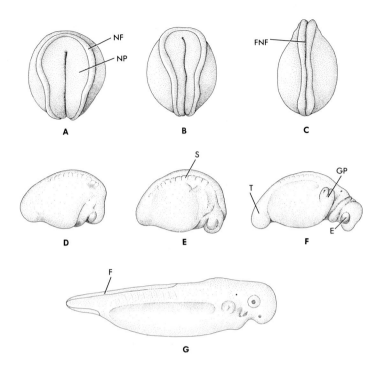

FIGURE 11.11 A series of stages in the development of *Ambystoma maculatum:* dorsal views (A–C); views from the right (D–G). E, eye primordium; F, dorsal fin; FNF, fusing neural folds; G, gill plate; NF, neural fold; NP, neural plate; S, somite; T, tail bud. [Adapted with modifications from V. Hamburger, *A Manual of Experimental Embryology*, The University of Chicago Press, Chicago, 1942, pp. 202–204.]

(Figures 11.10 and 11.11) and for the origin and shaping of internal organs. In the following sections discussions of the embryos of the higher vertebrates illustrate some of the ways in which these processes contribute to the development of the major organs that arise in each germ layer.

11.2.1 ECTODERM.
It is appropriate to begin with the ectoderm, for it is the layer that shows the first major morphogenetic change following gastrulation, namely, the origin of the central nervous system. This has been most thoroughly studied, both descriptively and experimentally, in amphibian and chick embryos.

11.2.1.1 Origin of the Neural Tube.
The most precise analyses of the origin of the neural tube have been carried out on urodeles, particularly the western newt *Taricha torosa*. At the end of gastrulation the ectoderm consists of a single layer of cells. Shortly thereafter the entire dorsal hemisphere of the embryo transforms into a flattened, compact plate of palisaded cells. This plate then assumes the shape of a "keyhole" (Figures 11.11B and 11.12G) with raised edges and a midline groove. The plate is called the **neural plate,** its raised edges are **neural folds** and the groove in the midline is the **neural groove.** Shortly after the keyhole configuration is achieved, the neural plate elongates rapidly and the neural folds converge upon each other in the dorsal midline (Figure 11.11C) and their fusion forms the neural tube, which is the primordium of most of the central nervous system. These early changes are accomplished in the absence of appreciable growth of the prospective neural tube.

The mechanism of these changes has been analyzed in the embryo of *Taricha torosa* by A. G. Jacobson and his collaborators. The keyhole pattern emerges between arbitrary stages 13 and 15 (stages of Twitty and Bodenstein, as reproduced by Rugh, 1962). Stage 13 occurs at the end of gastrulation, when only a small yolk plug remains. Individual cells of the future neural plate at stage 13 were identified at the intersections of a superimposed grid (Figure 11.12A) and thereafter followed cinematographically (Figures 11.12B–D) to their new positions at stage 15, the "keyhole" stage. Analysis of the cinematographs showed the pathways of movement of the cells at the grid intersections and changes in shape and area of the original squares of the grid. It was observed that cells generally move towards the midline and anteriorly along the midline (Burnside and A. G. Jacobson, 1968).

219

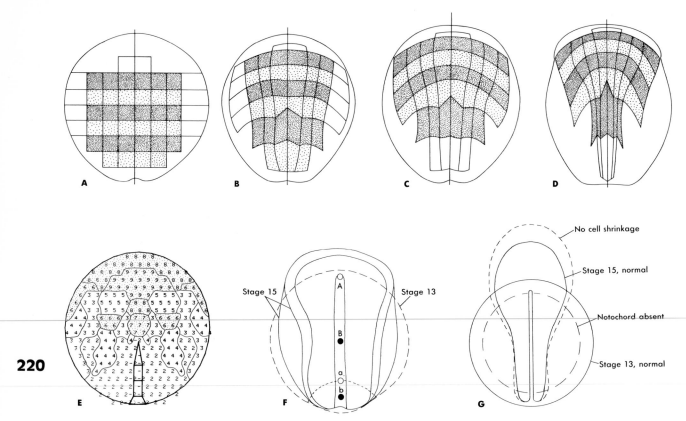

FIGURE 11.12 Analysis of neural plate formation in the embryo of the newt *Taricha torosa*. A: Dorsal view of the outline of the dorsal hemisphere at stage 13, with a superimposed grid of orthogonal coordinates. Cells located at the intersections of the grid were followed through successive stages thereafter. B–D: Positions occupied by cells at intersections of the original grid located at stages 14, $14\frac{1}{2}$, and 15, respectively. E: Shrinkage patterns of the apical surfaces of neural plate cells retroprojected from the neural plate of an embryo at stage 15 to the outline of the dorsal hemisphere at stage 13; large numbers indicate proportionately greater shrinkage. F: Superimposed outlines of the embryo at stages 13 (dashed lines) and 15 (solid lines); at stage 13 cells marked in regions a and b occupy, respectively, points at the anterior end of the notochord-notoplate and halfway between the anterior end and the blastopore. These cells retain their same relative positions in stage 15. G: The approximate shape of the newt embryo as seen from the future dorsal side in the late gastrula (stage 13) and midneurula (stage 15) is shown in solid lines. The outline labeled notochord absent is the computerized prediction of the outline that would be seen at stage 15 if the apices of the neural plate cells underwent the normal shrinkage of neurulation in the absence of the notochord. The outline labeled No cell shrinkage is the computer-predicted outline that would be found at stage 15 if the notochord underwent elongation as it usually does, but in the absence of shrinkage of the prospective neural ectoderm. [Adapted from A. G. Jacobson and R. Gordon, *J Exp Zool* 197:191–246 (1976).]

Next, changes in cell shape were measured in the mapped system between stages 13 and 15. Cells of the neural plate undergo elongation without changing their volume, contracting their apical and basal ends and extending microtubules in the direction of the elongating axis (Figure 11.2). This results in a progressive shrinkage of the surface of the neural plate. This shrinkage is, however best described as a **patterned shrinkage,** for some cells diminish their surfaces more than others. The greater the surface shrinkage, the greater the cellular elongation. Figure 11.12E shows, for an embryo of stage 13, the shrinkage pattern to be undergone by cells in various regions of the prospective neural plate as the embryo develops to stage 15. Areas of cells are identified by numbers; higher numbers signifying regions of greater shrinkage of the apical and basal cellular surfaces and, hence, regions of greater cellular elongation.

The regional pattern of shrinkage and elongation of the neural plate cells is preestablished in embryos of stage 13. If, for example, cells from the anterior portion of the neural hemisphere at stage 13, destined to undergo the greatest elongation, are transplanted to the posterolateral portion of the future neural plate, where little change in cellular elongation is expected to occur, the transplanted cells shrink their apical and basal surfaces and elongate to the same degree they would have if left in situ.

The programmed pattern of differential shrinkage naturally entrains shifts in position of adjacent cells, but differential shrinkage alone does not suffice to bring about the keyhole configuration. As the keyhole shape emerges, the neural plate elongates anteroposteriorly. This elongation is dependent on the concomitant elongation of the notochord and the portion of the neural plate directly overlying it. At stage 13, the notochord occupies a crescentic region anterior to the closing blastopore (Figure 11.12F) and is closely adherent to the overlying ectoderm, especially at its anterior end. The portion of the neural plate associated with the notochord is referred to as the **notoplate.** In normal development the notochord and notoplate extend anteriorly, not by addition from one end, but by uniform elongation along the whole length of the neural plate (Figure 11.12F). If the dorsal hemisphere of the egg is isolated at stage 13 along with the notochord, the notochord and notoplate elongate normally, and a well-formed keyhole-shaped neural plate develops. If, however, the notochord is excised from such a preparation at stage 13, the notoplate fails to elongate and the entire neural plate merely shrinks in size (Figure 11.12G). The isolated notochord, likewise, fails to elongate. Apparently only neural ectoderm and notochord are essential to the formation of the keyhole-shaped neural plate. Mechanical or other factors from ectoderm adjacent to the plate or from other adjoining tissues are not required.

When only the notoplate and underlying notochord are isolated at stage 13, and cultured on an agar surface, the whole complex elongates at a rate and to a degree comparable to the elongation of these components in the intact embryo. Apparently, therefore, in the intact embryo the forward thrust of the notoplate-notochord complex is combined with the differential shrinkage pattern to give rise to the keyhole configuration. The forward extension of this complex is coordinated with the extensive shrinkage of cells anterior to it, but is necessarily combined with a degree of shearing and consequent displacement of cells from their original neighbors along the pathway of elongation.

The closure of the neural tube occurs rapidly once the keyhole outline is achieved. The apical cell surfaces of the neural plate shrink greatly, but the basal cell surfaces do not, and the cells become wedge shaped (Figure 11.13). Wedge formation, as an active process, would be expected to help drive the formation of a tube from the keyhole plate. Wedging could, however be a relatively passive response to tube formation driven in some other way. Thus, A. G. Jacobson and Gorden (1976) noted that during the process of tube formation the elongation of the notoplate and underlying notochord in the midline proceeds at a rate approximately ten times as rapidly as it does before or after closure of the neural tube. A. G. Jacobson (1978) raised the possibility that if the neural plate were to behave as a viscoelastic sheet, stretching it along the midline could roll it up into a tube. A flat sheet of rubber pulled in the direction of a line between two points on opposite edges would roll into such a tube.

There is some suggestion that the meeting of the neural folds and the precise positioning of their juncture is assisted by the secretion of carbohydrate-rich extracellular material (Figure 11.14). Similar material is found in the neural groove of the chick

221

FIGURE 11.13 Changing cell shapes as they might be seen in a cross section of a newt embryo during the formation and curvature of the neural plate: at midgastrula (A), late gastrula (B), and midneurula (C) stages. [Adapted from B. Burnside, *Dev Biol* 26:416–41 (1971).]

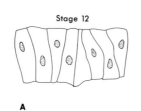

Stage 12

A

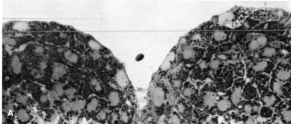

B

C

FIGURE 11.14 Neurulation in *Ambystoma maculatum*. A: Low power view of a cross section of the embryo during neurulation. Extracellular material can be seen at the bottom of the neural groove. B: Scanning electron micrograph at a later stage in neurulation, showing cell-surface material on approaching cell surfaces. [A from D. Moran and R. W. Rice, *J Cell Biol* 64:172–81 (1975); B from color photograph kindly supplied by Dr. Moran, whose courtesy is gratefully acknowledged.]

222

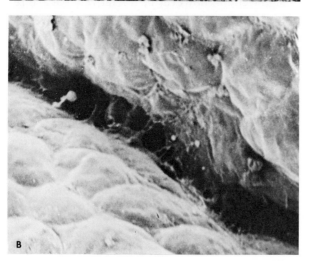

embryo (Lee et al., 1976, 1978), but its distribution seems to be limited to the tips of the approaching folds. Moran and Rice (1975) and Rice and Moran (1977) suggested that the extracellular material might provide a structural link that assures accurate spatial alignment between converging folds. As the folds meet, the edges of the adjacent ectoderm of each side fuse and become continuous over the neural tissue, which now forms a longitudinal tube. In this tube the cells appear as a pseudostratified columnar epithelium arranged around a lumen. The neural tube detaches from the ectoderm and ultimately develops into the brain and most of the spinal cord.

The closure of the neural folds and detachment of the neural tube are illustrated in schematic cross section of anuran embryos in Figure 11.15. In anurans the neural plate is several cell layers in thickness, and the formation of a neural tube involves rather complex cellular shiftings before the single-layered neural epithelium is formed (Schroeder, 1970). Karfunkel's (1974) review should be consulted for more details of the neurulation process in anuran embryos.

The major part of the neural plate and neural folds in amphibians contributes the central nervous system as far as the beginning of the tail. The extension of the neural tube into the tail arises from the **tail bud** (Figure 11.11F), which is formed from the posterior end of the neural plate and the material intervening between it and the blastopore. Within the relatively homogeneous tail bud, as it grows out, the neural tube differentiates by hollowing out and separating from the other tissues, which segregate into notochord and somites.

In the chick embryo the posterior portion of the trunk and tail likewise forms from a tail bud, often called an **end bud** (Figure 11.10B). The end bud

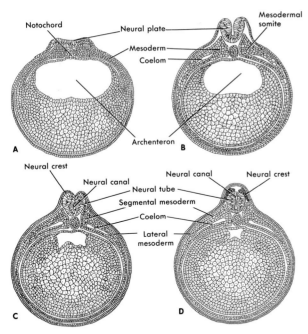

FIGURE 11.15 Formation of the neural tube. A–D: Cross sections of an anuran neurula at successive stages of neurulation, showing the meeting of the neural folds and origin of the neural crest cells from the edges of each fold. [Redrawn from A. F. Huettner, *Comparative Embryology of the Vertebrates*, Macmillan, New York, 1949, p. 88.]

forms when regression of the primitive streak is completed, namely, at the stage when about 20 somites are formed. It arises by condensation of the remnants of the primitive streak with Hensen's node. This condensation consists of a mass of homogenous-appearing cells, out of which form the neural tube, somites, connective tissue, and caudal arteries of the tail and possibly of the posterior lumbosacral region. It does not, however, contribute cells to the notochord of the tail region or to the hindgut or transient tail gut (Criley, 1969; Schoenwolf, 1977).

The morphogenesis and histogenesis of the brain and spinal cord are extremely complex, and details should be sought in textbooks on developmental anatomy and neuroanatomy. A few remarks may be addressed, nevertheless, to the differentiation of the spinal cord and to the regionalization of its own neuroblasts and the neuroblasts of its associated spinal ganglia.

The closure of the neural tube and its organization as a pseudostratified columnar epithelium are followed by cellular proliferation and the regionalization of the walls of the tube. Careful study of the early neural tube reveals that all cells of the epithelium are firmly attached to one another at the lumen of the tube, with their nuclei displaced to varying distances toward the external limiting membrane of epithelium. Initially all cells of the neural tube are capable of dividing. As one of the elongate cells enters prophase, its nucleus migrates toward the neural canal and the cell body rounds up, its distal processes being withdrawn. The rounded cells remain as such during metaphase, anaphase, and early telophase. Thereafter, the daughter cells begin to elongate peripherally once more, and their nuclei migrate distally and enter mitotic interphase. By providing radioactively labeled thymidine to early chick and mouse embryos, it has been possible to demonstrate that the DNA synthetic phase of the mitotic cycle in cells of the wall of the neural tube takes place in nuclei that are distal from the lumen (Sauer and Walker, 1959). The relationships of nuclear position to stages in the cell cycle are diagrammed for the 10-day mouse embryo in Figure 11.16.

Mitoses greatly increase the number of cells of

223

FIGURE 11.16 The pattern of nuclear migration and division in the neural tube of the 10-day mouse embryo. The lumen of the neural tube is on the right. DNA synthesis, S, occurs in the mantle zone, and then the post synthetic nuclei, G₂, migrate to the edge of the neural lumen, where mitosis, M, occurs. After cell division the daughter cells extend their processes to the margin of the tube, and the nuclei migrate to the mantle zone for DNA synthesis once more. [From S. L. Kaufman, *Dev Biol* 20:146, 157 (1969), by courtesy of the author and permission of Academic Press, Inc., New York.]

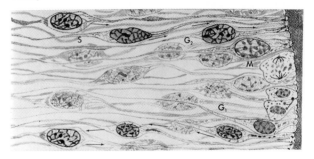

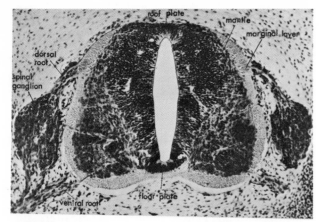

FIGURE 11.17 Spinal cord of a pig embryo of 7 mm crown–rump length sectioned transversely to show the differentiation of the mantle and marginal layers and the relationships of the dorsal and ventral roots and spinal ganglia. [From B. I. Balinsky, *An Introduction to Embryology*, 3rd ed, Saunders, Philadelphia, 1970, p. 372, by permission of the author and the W. B. Saunders Co.]

level. At the cervical level it is depleted by cellular degeneration (Figure 11.19). At the level of the limbs, however, large lateral columns of motor neuroblasts remain, supplying neurons to the musculature of the extremities. Axons from the motor neuroblasts of both the visceral and somatic columns exit from the neural tube in segmentally arranged bundles, referred to as the **ventral roots** (Figure 11.17).

The spinal ganglia arise from cells that migrate from a region of the neuroepithelium called the

FIGURE 11.18 The pattern of migration of cells in the spinal cord of the chick embryo. A: At 2 days of incubation some cells are still in the division cycle, and their nuclei migrate as in Figure 11.16, whereas others are beginning to shift ventrolaterally. B: At 4 days a uniform lateral motor column has formed. C: At 6 days migration of the neuroblasts of the preganglionic cells of the visceral motor system begins. D: The migratory cells have formed the visceral motor center of Terni and the remaining cells form the somatic motor column. [Adapted with modifications from V. Hamburger and R. Levi-Montalcini in P. Weiss, ed., *Some Aspects of Genetic Neurology*, The University of Chicago Press, Chicago, 1950, pp. 128–60.]

224

the neuroepithelium. Eventually some of the daughter cells migrate away from the lumen and take up positions immediately beneath the exterior limiting membrane of the neural tube. These cells are the **neuroblasts.** They multiply and form a densely packed layer of cells called the **mantle layer.** Axons and dendrites then grow out from the neuroblasts, forming a new outer layer, the **marginal layer.** It contains very few nuclei other than those of supporting cells of various types. The relationships among these layers are illustrated in Figure 11.17.

In the chick embryo, beginning anteriorly after the second day and progressing posteriorly, neuroblasts accumulate as a ventrolateral column along the periphery of the neural tube, meanwhile continuing to multiply. By $4\frac{1}{2}$ days of incubation these constitute a compact column of cells essentially uniformly distributed along the neural tube. By this time, however, the most medial neuroblasts start migrating dorsally (Figures 11.18 and 11.19), giving rise to the preganglionic motor neurons of the viscera. This column of neurons is referred to as the **column of Terni.** Its origin was analyzed by Levi-Montalcini in 1950.

Migration thus depletes the population of neuroblasts in the motor columns at the thoracolumbar

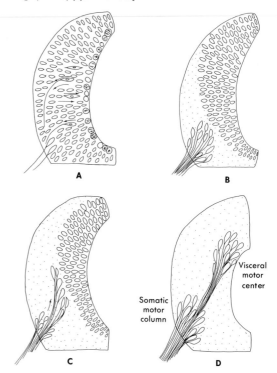

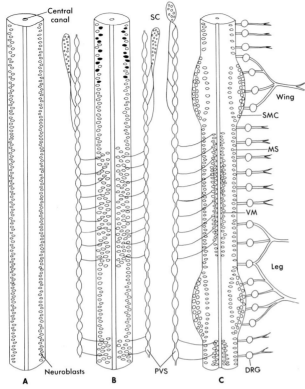

FIGURE 11.19 Diagram showing the origin, migration, and regional distribution of motor neurons along the length of the spinal cord of the chick embryo. A: Neuroblasts are distributed relatively uniformly along the cord. B: Visceral motor cells have migrated centrally to constitute the preganglionic motor supply to the paravertebral sympathetic chain, PVS. Necrosis is prominent at anterior levels of the motor columns. C: The definitive distribution pattern of motor neuroblasts is essentially complete. The spinal nerves and dorsal root ganglion, DRG, are shown on the right and the paravertebral sympathetic chain on the left of the cord. The lateral somatic motor columns, SMC, are well developed at wing and leg levels. The medial somatic motor column, MS, is less prominent. [Adapted with modifications from R. Levi-Montalcini, *J Morphol* 86:253–83 (1950).]

neural crest (Figure 11.15D), as described in the next section. These ganglia consist of sensory neuroblasts, the axons of which enter the dorsal side of the neural tube at the level of each somite. They are often referred to as **dorsal root ganglia** (Figure 11.17). The dorsal root ganglia, like the lateral motor columns, show regional differences in their

population of neuroblasts. Larger ganglia containing more neuroblasts supply sensory innervation to the limbs.

The observed correlation between populations of neuroblasts in the lateral motor columns and spinal ganglia with the tissue mass to be innervated (the **peripheral load**) suggests that the populations of neuroblasts are determined by the quantity of the peripheral load. Hamburger (1939) tested this by grafting an extra leg primordium to the prospective flank region of the chick embryo (Figure 11.20). This resulted in increased populations of neuroblasts in the spinal cord at the flank level and in the sensory ganglia supplying the limb.

Effects of the periphery on populations of central neuroblasts are apparently exercised through effects on regional patterns of cellular proliferation as well as cellular death. As Table 11.1 clearly

FIGURE 11.20 Effects on the nervous system of increasing the peripheral load in the chick embryo. A supernumerary leg bud was grafted to the right flank of the embryo prior to the ingrowth of sensory and motor fibers. As this graphic reconstruction clearly shows, the spinal ganglia supplying the supernumerary limb are considerably larger than those of the contralateral side at the same level. In sections this specimen also showed hyperplasia of the lateral motor column on the right side. [Redrawn with modifications from V. Hamburger, *Physiol Zool* 12:268–84 (1939).]

225

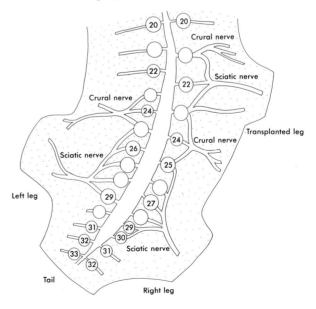

shows, sensory ganglia of the thoracolumbar level are normally depleted by cell death in the chick embryo, and these ganglia are relatively small. Cell death likewise is a normal factor in determining the population of cells in the lateral motor columns (Hamburger, 1975). Moreover, if one extirpates a limb primordium in the early chick embryo, the population of sensory neuroblasts is greatly depleted by death at the limb level, and the lateral motor columns are essentially eliminated by cataclysmic necrosis (Hamburger and Levi-Montalcini, 1949; Hamburger, 1958). On the other hand, the naturally occurring loss of cells from the lateral motor columns of the flank level is reduced by grafting an extra leg bud (Hollyday and Hamburger, 1976; Hollyday et al., 1977) to the flank of a host chick. As already noted, sensory ganglia of the flank level that supply such grafts show increased populations of neuroblasts. This results, in part at least, from an increase in the mitotic rate in these ganglia (Hamburger, 1939), but measurements of cell death in the hyperplastic ganglia have apparently not been made. A general review of the effects of deletions and additions to the peripheral load of spinal

226 ganglia and motor columns was compiled by M. Jacobson (1970).

11.2.1.2 Neural Crest. The neural crest arises in association with the fusion of the neural folds, as shown in Figure 11.15. Typically it forms a band of cells interposed between the dorsal side of the neural tube and the overlying epidermis. From this region crest cells migrate in characteristic spatiotemporal patterns to various parts of the body, beginning anteriorly and progressing posteriorly. Derivatives of the crest cells of the head region include the cranial ganglia, all cartilages and bones of the facial and oral region, most connective tissues, including the two innermost coverings of the brain **(pia mater** and **arachnoid),** the **odontoblasts** (dentine-forming cells) of the teeth, and the **calcitonin** cells of the **ultimobranchial bodies.** (See Section 11.2.3 on major endodermal derivatives.) The migratory pathway taken by marked neural crest cells in the anterior region of the brain is illustrated in Figure 11.21. Excess vitamin A inhibits the migration of neural crest cells in the head region, an effect that presumably explains the facial malformations in rat embryos whose mothers receive injections of the vitamin at the time neural crest mi-

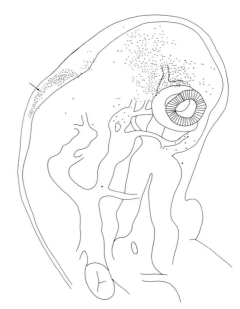

FIGURE 11.21 Distribution of neural crest cells in the head region of an unlabeled host embryo of about 48 hr of incubation that had received an implant of radioactively labeled head neural crest 22 hr previously. The site of the transplant is indicated by an arrow. Each of the black stipples in the drawing represents a labeled cell. Cells have migrated rostrally and ventrally from the implantation site. [Adapted from a photograph from D. M. Noden, *Dev Biol* 42:106–30 (1975).]

gration occurs in the head region (Hassell et al., 1977). It is possible, too, that the teratogen interferes with the normal interaction between crest cells and the pharyngeal endoderm. In the newt, at least, neural crest from head neural folds will not form cartilage in vitro unless cultured in direct contact with pharyngeal endoderm (Epperlein, 1978).

Neural crest cells originating at the trunk level migrate in two principal streams on each side, one laterally between the somites and epidermis and the other ventrally between the somites and the neural tube (Figure 11.22). The laterally migrating cells give rise primarily to pigment cells, the chief type of which is the **melanocyte,** which is responsible for depositing melanin pigments of the skin, its derivatives, and other parts of the body. The ventrally migrating cells give rise to the sensory ganglia; all ganglia of the autonomic nervous system

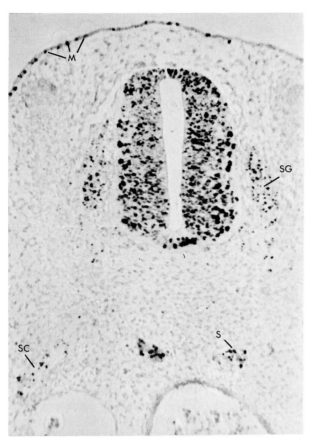

and their supporting cells (glia); **Schwann cells,** which ensheath the nerve fibers; and cells of the adrenal medulla, which is an important component of the sympathetic division of the autonomic nervous system.

Crest cells leaving the neural tube emigrate through a cell-free matrix consisting of hyaluronate and a delicate fabric of collagen fibrils. Hyaluronic acid is one of the class of **glycosaminoglycans** (GAGs),* polyanionic components of intercellular matrices of large molecular weight. Hyaluronic acid is not sulfated and can exist in relatively pure form in intercellular spaces as highly hydrated salt.

In material fixed for scanning electron microscopy in such a manner as to preserve GAGs, the intercellular matrix appears as a dense network of fibers beaded with precipitated GAGs (Tosney, 1978). As the cells move away from their origin and the amount of hyaluronate diminishes, sulfated GAGs, chiefly chondroitin sulfate, are detected in greater quantity. Neural crest cells aggregating in the spinal ganglia have very few associated GAGs (Derby, 1978). This discussion is a much abbreviated and oversimplified version of the involvement of glycosaminoglycans in neural crest migration. Readers may wish to consult papers relating to this topic, especially those of Pratt et al. (1975), Pintar

227

FIGURE 11.22 Radioautograph showing some derivatives of labeled neural crest at the trunk level in the chick embryo. The neural tube, with crest, was excised from the trunk region of a chick embryo with nuclei labeled with tritiated thymidine and grafted to a nonlabeled host in place of a portion of its own neural tube and crest. In the case illustrated here, the host embryo at 2 days of incubation received its graft from a donor of similar age and was then fixed 2 days later. Neuroblasts in spinal ganglia, SG, and sympathetic ganglia, S, are seen to be labeled, as are sheath cells, SC, of a motor nerve. Labeled cells in the epidermis, M, are presumably melanoblasts. Notably Teillet, who used quail neural tube and crest in similar grafts to chick embryos [M.-A. Teillet, *Ann. d'Embryol et de Morphogenese,* 4:95–109 (1971)], did not observe entrance of quail neural crest cells into chick epidermis until slightly later stages. [From J. A. Weston, *Dev Biol* 6:279–310 (1963), by courtesy of Dr. Weston and permission of Academic Press, Inc., New York.]

*The glycosaminoglycans are linear polymers of polysaccharides consisting of a hexosamine (either N-acetylglucosamine or N-acetylgalactosamine), alternating with a hexuronate residue (glucuronate or ioduronate). Hyaluronic acid is made up of alternating residues of N-acetylglucosamine and glucuronate. The other GAGs are sulfated and covalently bonded in mixed populations along the length of a protein core. In this condition they are called **proteoglycans.** The chief proteoglycan of cartilage matrix, for example, consists of some dozens of molecules of **chondroitin sulfate** (molecular weight 30,000–40,000) and of **keratin sulfate** (molecular weight 3000–6000) attached to a protein core about 300–400 nm in length. Proteoglycans form aggregates in which they are bound to a long chain of hyaluronate. In cartilage the length of the hyaluronate chain can be as great as 1200 nm.

The importance of the roles played by the various forms of GAGs in embryonic development is becoming increasingly understood. Reviews by Manasek (1975) and Muir (1977) are recommended reading for those who wish to improve their knowledge of the biology of these substances.

(1978), Bronner-Fraser and Cohen (1980), and Erickson et al. (1980).

Are the neural crest cells specified in advance of their migration, as to what their developmental fate is to be? LeDouarin and Teillet (1974) sought to determine whether cells from which the sympathetic nervous system and parasympathetic systems originate are predetermined for their eventual fates in advance of their migration from the neural crest. To do this they first located the region of the neural crest from which each of these systems arises, by replacing selected segments of neural tube and crest from chick embryos with homologous segments from quail embryos. Because quail cell nuclei are distinctive from those of the chick, the distribution of the grafted neural crest derivatives could be followed.

They found (Figure 11.23) that the sympathetic chain of ganglia arises from the portion of the neural crest that is located between the levels of the eighth and the 24th somites and that, within this zone, the cells for the adrenal medulla arise from crest cells that originate between the levels of the 18th and 24th somites (the **adrenomedullary area**). The parasympathetic ganglia, which are chiefly located in the walls of the digestive tract, arise from different regions of the neural crest. One of these regions, called the **vagal area**, is at the level of the first seven somites. The other, the **sacral area**, is posterior to the 24th somite.

Next, they exchanged segments of neural tube and crest between "adrenomedullary" and "vagal" levels of quail and chick embryos (Figure 11.24A) and followed the distribution of quail crest cells in the chick hosts. They found crest cells from the adrenomedullary area, grafted in place of vagal crest cells, form the various ganglionic plexi of the digestive tract as is normal for cells of the vagal area. Cells of the adrenomedullary region of the crest are not, therefore, predetermined for the adrenal medulla or sympathetic ganglia. Quail adrenomedullary crest cells grafted in place of chick adrenomedullary crest cells do not contribute to the enteric plexuses of the host. Because they do so if grafted in place of vagal crest, there must be a preferential pathway along which vagal cells normally travel to their destinations.

Vagal crest cells, on the other hand, possibly have a degree of predetermination. Grafted to the adrenomedullary area of a chick host (Figure

228

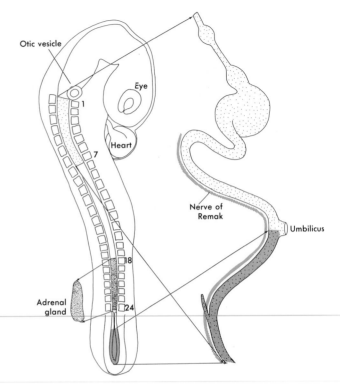

FIGURE 11.23 Diagram showing the regional origin of neural crest cells from which various components of the autonomic nervous system originate as determined by the transplantation of quail neural tube into the chick embryo. Crest cells from the level between somites 1 and 7 contribute the neuroblasts of ganglia found in the walls of the alimentary tract along its entire length. Those from the level behind somite 28 contribute only to the formation of ganglia of the postumbilical gut. Neural crest cells originating at the levels of somites 8–28 form neuroblasts of the orthosympathetic nervous system (not shown) and to the cells of the adrenal medulla, the latter arising from the crest at the levels of somites 18–24. [Adapted with modifications from N. LeDouarin and M.-A. Teillet, *Dev Biol* 41:162–84 (1974).]

11.24B), quail vagal crest cells will, indeed, populate the adrenal medulla and sympathetic ganglia, but they also contribute a few cells to plexuses of the gut wall, something that quail adrenomedullary crest cells never do normally. This fact suggests that at least some vagal crest cells are predetermined to respond to signals, of whatever kind, directing them to the intestinal wall.

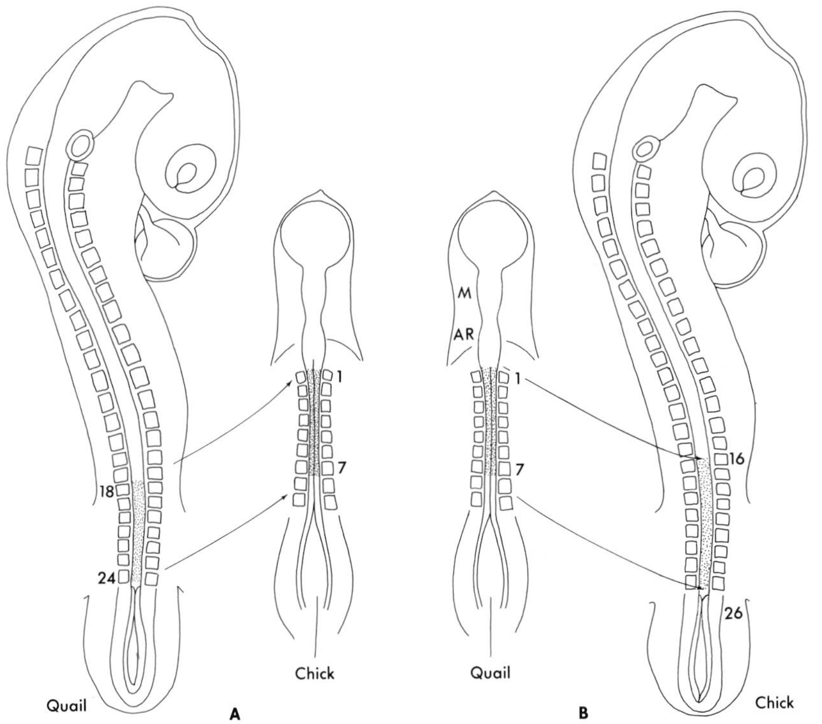

FIGURE 11.24 Heterotopic and heterochronic transplantations of neural tube and neural crest between chick and quail embryos. A: The source of cells for the adrenal medullary cells of the quail is substituted for the neural tube and crest of a younger quail embryo at the level from which cells would normally contribute neuroblasts to ganglia of the gut wall. B: Quail neural tube and crest from the levels of somites 1–7 was substituted for trunk neural tube and crest in the adrenomedullary area. [Adapted with modifications from N. LeDouarin and M.-A. Teillet, *Dev Biol* 41:162–84 (1974).]

229

There seems, however, to be little premigratory biochemical specification of neural crest cells as prospective sympathetic or parasympathetic neuroblasts. The cells of the autonomic nervous system are characterized by their content of enzymes for the production of their specific neurotransmitter substances. Dopamine β-hydroxylase catalyzes the conversion of dopamine to norepinephrine in sympathetic neurons, and choline acetyl transferase catalyzes the synthesis of acetylcholine from choline in parasympathetic neurons. Nevertheless, crest cells from the vagal area or from the trunk area can form either enzyme in the appropriate tissue environment (Le Douarin et al., 1977; Le Lievre et al., 1980) or even in the absence of specific tissue stimuli; namely, in cell cultures in vitro (Kahn et al., 1980; Sieber-Blum and Cohn, 1980).

Two types of nerve cells, both derived from the neural crest, show a remarkable developmental response to a special protein called **nerve growth factor** (NGF). These cells are the sympathetic neuroblasts and the ventrolateral neuroblasts of the spinal ganglia. The discovery of NGF resulted from the observations by Levi-Montalcini and Hamburger

(1951) that chick embryos receiving grafts of certain mouse tumors develop greatly enlarged sensory ganglia and show an even more dramatic growth of sympathetic ganglia, accompanied by massive invasion of all parts of the viscera by sympathetic fibers. Such tumors and their extracts likewise stimulate the outgrowth of nerve fibers **(neurites)** from spinal and sympathetic ganglia cultured in vitro (Levi-Montalcini et al., 1954; S. Cohen et al., 1954). Following upon this discovery, it was found that the NGF is a protein (S. Cohen, 1958) and that similar proteins can be isolated from the venoms of snakes and Gila monsters and, in especially large quantities, from the submaxillary glands of male mice. It is secreted by the convoluted tubular portion of the male mouse gland, which becomes greatly enlarged under the influence of testosterone at puberty. It is produced in lesser quantities in females and castrated males. NGF has also been shown to be produced in the prostate gland in several other mammals (Harper et al., 1980).

NGF isolated from the submaxillary gland of the mouse is a dimer whose identical subunits each consist of a chain of 118 amino acids having a molecular weight of 13,259. The dimer has a sedimentation coefficient of 2.5 S and is often referred to as 2.5 S NGF. This distinguishes it from 7 S NGF, which is a complex of the dimer with two other pairs of identical subunits. The latter may provide a storage or protective vehicle for the 2.5 S NGF.

In the chick embryo the response of the sensory ganglia to NGF is limited to days 8 through 15. During this period nerve growth factor labeled with ^{125}I binds to plasma membrane receptors which are then internalized and transported to the nucleus where they bind strongly to chromatin (Andres et al., 1977). In contrast, binding of NGF by cells of the sympathetic ganglia remains approximately constant, even postnatally, paralleling their continuing ability to respond to NGF and their presumed continuing requirement for it. That they do continue requiring it is indicated by the fact that when newly hatched chicks and young mammals are injected with antibodies produced against NGF (anti-NGF), cells of their sympathetic ganglia are cataclysmically exterminated.

NGF apparently plays two linked but experimentally separate roles in the growth and differentiation of neuroblasts. These are, respectively, neurite outgrowth and metabolic maturation. When chick sensory neuroblasts are cultured in vitro in the presence of Sepharose beads to which ^{125}I-labeled NGF is covalently bound, membrane receptors on the neuroblasts interact with the NGF, but cannot internalize it. Nevertheless, outgrowth of neurites proceeds. The cells do not however, produce tyrosine hydroxylase and other enzymes involved in the production of the neurotransmitter substance of these nerves, namely, the catecholamine norepinephrine. On the other hand, free ^{125}I-labeled NGF injected into the anterior chamber of the eye is internalized by nerves supplying the ciliary ganglion and transported to the superior cervical ganglion. There it induces the synthesis of the enzymes for production of noradrenalin (reviewed by Andres et al., 1977, and by Thoenen et al., 1978; see also Levi et al., 1980).

Observations such as these suggest that tissues receiving sympathetic innervation produce or store NGF. Uptake of NGF then occurs at the tip of the outgrowing neurite, and retrograde transport of the factor relays to the cell body the condition of the periphery. Unfortunately, except for the salivary

gland in some animals, organs that receive heavy sympathetic innervation cannot be shown to synthesize NGF, and the levels of NGF in these tissues is too low to be detected (Harper et al., 1980).

During the past few years it has been found that a variety of cell types cultured in vitro can produce a NGF that is indistinguishable from that of mouse submaxillary gland by immunological and biological criteria. These include neuroblastoma (a tumor line) cells, a variety of mouse fibroblast cell lines, and primary cultures of chick and human fibroblasts (reviewed by Murphy et al., 1977). Rat muscle cells in vitro also produce NGF, both as single cells and syncytial muscle fibers (Murphy et al., 1977). The rat muscle NGF was isolated in a high molecular weight form and is more stable than the 7 S complex described above. It is presumed to contain the same 2.5 S active core as the mouse NGF, for their activities are identical.

These observations, although resulting from studies carried out in vitro, suggest that various tissues in vivo produce the NGF essential to normal outgrowth and maintenance of nervous tissue. Does embryonic muscle tissue in vivo possibly exercise a neurotrophic effect, guiding nerves to the limbs, for example? Gunderson and Barrett (1979) found that the tips of neurites from neurons of chick dorsal root ganglia change their direction of growth toward a source of NGF diffusing from the tip of a micropipet. A related observation of some importance is that injections of NGF diminish the frequency of cell deaths in both ventrolateral and dorsomedial neuroblasts of chick spinal ganglia (Hamburger et al., 1981).

The role of NGF as related to the normal outgrowth and differentiation of neurites is far from settled. Coughlin et al. (1978), for example, explanted the superior cervical sympathetic ganglion of the 14-day mouse embryo in organ culture, using a supporting raft of cellulose filter in liquid medium. They found that abundant neurites were elaborated in the absence of added NGF and in a medium containing anti-NGF antibodies. The activity of tyrosine hydroxylase increased eightfold during the first 48 hours of culture but usually began to decline after 3 days. In the presence of NGF or of the salivary primordium from donors of the same age, the activity of the enzyme rose much more rapidly. Anti-NGF in the medium had no effect on the changes in tyrosine hydroxylase activity. These

230

results could be interpreted to indicate that 14-day superior cervical ganglia have been exposed to and carry some NGF along in organ culture in a form inaccessible to anti-NGF. Possibly, also, the submaxillary gland primordium produces and delivers NGF to neurons in an antibody-resistant form. Nevertheless, the results also suggest that growth and differentiation of the superior cervical neuroblasts are to some degree independent of NGF and, likewise, that the effect of the target organ may be independent of the factor. These matters remain unresolved.

For additional insights into the problems associated with nerve growth factor, readers are referred to excellent reviews by Levi-Montalcini (1966), Server and Shooter (1977), and Bradshaw (1978).

11.2.1.3 Origin of the Eye.   Shortly after the neural tube is formed, the future forebrain region evaginates on left and right sides, forming the bulging **optic vesicles** (Figure 11.25). These push through the loose mesoderm of the head and make contact with the ectoderm. In response to this contact (referred to as an **inductive response;** cf. Chap-

FIGURE 11.25   Early stages in morphogenesis of the vertebrate eye as seen in stylized cross section through the embryonic forebrain.

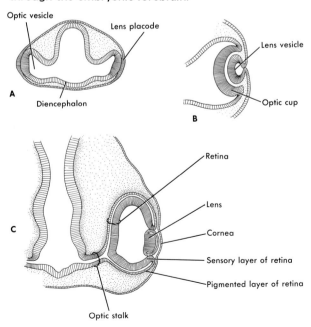

ters 13 and 14), the ectoderm cells elongate perpendicularly to the zone of contact, forming a thickening on each side, the **lens placode.** The placode then invaginates and detaches from the overlying ectoderm to form the lens of the eye. As the lens primordium invaginates, the optic vesicle reverses its outward bulge and invaginates, forming the **optic cup** (eyecup). The lens placode and optic cup mutually accommodate to each other. The lining of the eyecup subsequently forms the retina. The formation of the optic vesicles requires a prior inductive action of the prospective head mesoderm on the anterior portion of the neural plate. In the absence of this induction or in the event that the neural ectoderm is unable to respond, an eyeless condition results.

A recessive mutant gene that brings about an eyeless condition in the embryo in the axolotl *Ambystoma mexicanum* was described by Humphrey in 1969. Subsequently Van Deusen (1973) determined experimentally that the gene affects specifically the part of the prospective forebrain ectoderm from which the optic vesicles arise (Figure 11.26A). If the prospective forebrain ectoderm of the mutant is replaced by a comparable piece of ectoderm from a normal donor, normal eyes develop. When the reciprocal experiment is carried out, the genetically normal host is eyeless. These results show that the genetic lesion in the eyeless mutant affects the ability of the mutant ectoderm to respond to the inductive action of the head mesoderm (Figure 11.26B–C).

The morphogenesis of the eye is controlled by a complex series of interactions in the anterior region of the early embryo. These interactions were reviewed by Coulombre (1965) and Reyer (1977). Some of them are considered in Chapter 13. A scholarly monograph by Hay and Revel (1969) describes the origin of the cornea of the eye, as revealed by the electron microscope.

11.2.1.4 Placodes of Other Sense Organs. Nasal placodes arise anteriorly on each side of the anterior aspect of the head. These invaginate to form the **nasal grooves,** from which the sensory epithelium of the olfactory organ develops. Ear placodes form on each side of the hindbrain and then invaginate to form the **otic vesicles,** which ultimately contribute the inner ear. The distribution of these sensory placodes in vertebrate embryos is diagrammed in Figure 11.27. Factors that affect the

231

FIGURE 11.26 A: The location of the eye-forming region at the early neural fold stage of the embryo of *Ambystoma mexicanum*; the single eye field is outlined in white. B: Scheme of exchange of prospective mutant head ectoderm and prospective head mesoderm between normal (stippled) and mutant early gastrulae. C: The ectodermal exchange schematized for the late gastrula stage. The normal host fails to form eyes, whereas the mutant host forms normal eyes. D: The mesodermal exchange schematized for the late gastrula stage. The mutant host fails to form eyes, whereas, in the normal host, normal eyes appear. [Adapted with considerable modification from E. Van Deusen, *Dev Biol* 34:135–58 (1973); A is similar to Dr. Van Deusen's Figure 12, but was constructed from a photograph, of unknown origin, found in the author's laboratory.]

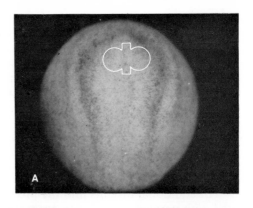

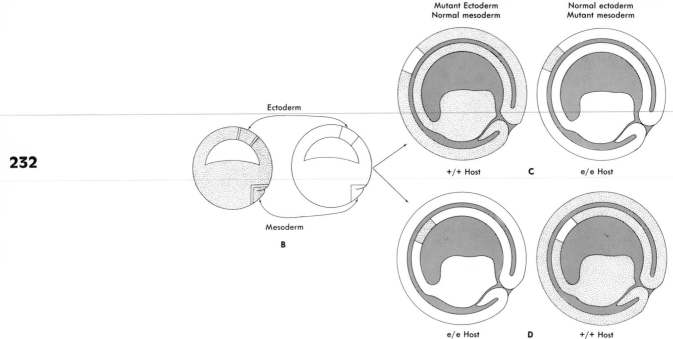

232

origin and positioning of these placodes were analyzed in an interesting paper by A. Jacobson (1966).

Other ectodermal thickenings appear in the nearby head region and contribute cells to ganglia of the cranial nerves. The ganglionic placodes do not invaginate as vesicular structures, but, rather, their cells detach and migrate centrally, coming to associate with clusters of neural crest cells. From these associations the cranial ganglia arise. The origin of one of these ganglia, the fifth, or trigeminal,

was analyzed experimentally by Hamburger (1962). Separate populations of large and small neuroblasts occur in this ganglion, and Hamburger's experiment, schematized in Figure 11.28, suggests that the larger ones are contributed by the cells arriving from the placode and that the smaller ones arise from the neural crest of the hindbrain region. When Noden (1978) grafted quail trunk neural crest to the hindbrain of a chick host, he found that it did not join with the placodal cells to form a tri-

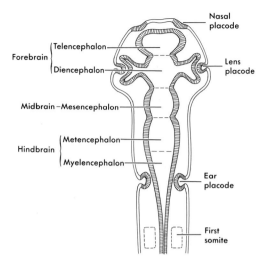

FIGURE 11.27 Scheme of the distribution of placodes of the sense organs in the head region of a vertebrate embryo. [From C. H. Waddington, *Principles of Embryology*, Macmillan, New York, 1956, p. 254.]

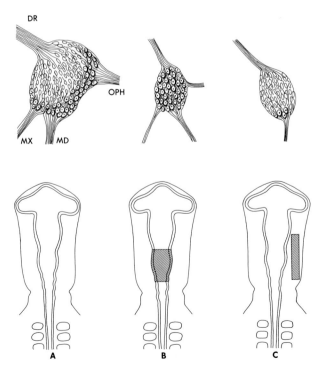

geminal ganglion. In the absence of the appropriate crest cells, placodal cells formed a ganglion containing large, but abnormally located, neuroblasts.

11.2.1.5 Guidance of Outgrowing Nerve Fibers. Once in position, neuroblasts, whether differentiating within the spinal cord or in any of the various ganglia, must make **synaptic connections** with other nerve cells and with various end organs. They do this by spinning out long tubular extensions of their own substance. The nerve fibers are called **axons** if they carry nerve impulses away from the cell body, and **dendrites** if they transmit the signal towards the cell body. The collective word for these fibers is **neurite,** a term introduced earlier in the chapter. Cells of the sensory placode produce outgrowths that connect with other nerve cells in the appropriate brain centers; motor neurons of the spinal cord send their axons to the peripheral musculature and sympathetic ganglia; from neuroblasts of the spinal ganglia, neurites connect with proprioceptors in muscle and other sensory receptors of the periphery.

The outgrowing nerve fiber protrudes as a cylindrical outgrowth from the cell body of the neuron from which is generated a stream of flowing **axoplasm** (protoplasm of the axon, or neurite). The axoplasm erupts at the tip in numerous branching pseudopodial processes that seem to compete with one another for the axial current. The principal branch eventually drains away the flow, obliterat-

233

FIGURE 11.28 Dual origin of sensory neuroblasts of the trigeminal ganglion. A: Outline of the head region of a normal 16-somite chick embryo (below) and enlarged section of the right trigeminal ganglion (above). B: Deletion of the neural crest and rostral part of the hindbrain with resulting trigeminal ganglion of the right side. Note that the ganglion comprises exclusively large neuroblasts of placodal origin, but that all rami from the ganglion are present. C: Extirpation of the trigeminal placode region of head ectoderm of the right side (below); the right trigeminal ganglion made up almost entirely of small neuroblasts of neural crest origin (above). Only the mandibular ramus of the ganglion is present. DR, dorsal root; OPH, ophthalmic ramus; MD, mandibular ramus; MX, maxillary ramus. [From V. Hamburger, *J Cell Comp Physiol* 60(Suppl. 1):84 (1962), by permission of Dr. Hamburger and Alan R. Liss, Inc.]

FIGURE 11.29 The outgrowing tip of a neurite, showing progressive definition of the channel of axoplasmic flow. The tip of the neurite is termed a growth cone. It typically shows numerous projections (more numerous and much finer than shown here) along its leading edge and upper surface. These are microspikes. Ruffling may also occur at the margin, as depicted for moving fibroblasts in Figures 11.5A and 11.6A.

ing the smaller ones, and, by its continued growth, defines the path of the nerve (Figure 11.29).

An important problem is the guidance of the filopodial tip of the neurite to its end organ. During the early stages of morphogenesis, at first only a few cells in a particular motor or sensory area send out fibers, which are called **pioneering fibers,** and they proceed quite separately from one another. But once the pioneering fibers establish contact with their peripheral goals, the fibers that are subsequently spun out by neighboring neuroblasts seem to apply themselves to the first-arriving fibers rather than to go pioneering on their own. Apparently the achievement of a peripheral goal leads to alterations of the surface of the pioneering nerve fiber, making it a preferential pathway that is followed by successively later-emerging fibers (Figure 11.30). A kind of **selective adherence** to successful pioneers would thus seem to account for the pathways taken by secondarily arising neurites. The factors that determine which neuroblasts will be pioneers have not been recognized, however, and it is not yet known what guides the pioneer to its target.

The guidance of nerve fibers has been the object of intensive research for many years. An area of particular interest is the guidance of fibers of the regenerating optic nerve in lower vertebrates. An optic nerve is made up of bundles of axons whose cell bodies are in the inner layers of the retina. During embryonic development, the optic nerves from each retina pass beneath the forebrain and cross to the opposite side of the midbrain, entering the optic lobes by way of the optic tract (Figure 11.31). There they make synaptic connections with cells of the

234

optic tectum. The axons transmit visual signals in the form of nerve impulses to specific cellular regions of the optic tectum.

In fish and urodeles one may remove the eye, sever the optic nerve, and then replace the eye in its socket with great assurance that, barring infection, the cut nerves will sprout new tips that will reconnect with the optic tectum. After sufficient time is allowed for reconnection, the animal will show a normal response to visual stimuli (i.e., it will display normal **visuomotor behavior**). If, however, one severs the optic nerve and then replaces the eye in reversed anteroposterior and dorsoventral orientation and allows time for regeneration, the animal's visuomotor behavior with respect to stimuli

FIGURE 11.30 Selective adherence of later-emerging nerve fibers to successful pioneer. A: Pioneering fiber probes the periphery. B, C: Later-emerging tips apply themselves to the successful pioneer as path to their end organs. [Adapted from P. Weiss in B. H. Willier, P. Weiss, and V. Hamburger, eds., *Analysis of Development*, Saunders, Philadelphia, 1955, p. 352.]

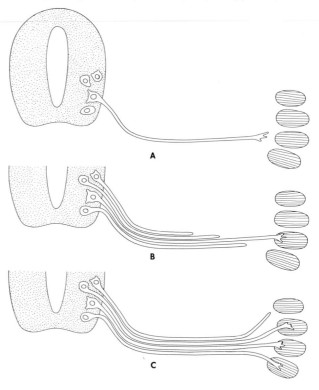

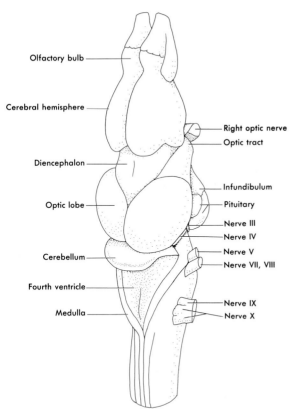

FIGURE 11.31 Right dorsolateral view of an amphibian brain. The optic tract, visible on the right, carries axons from the left eye to the right optic lobe and tectum. The optic nerve from the right eye is shown as severed. [Adapted from several sources, including T. W. Torrey and A. Feduccia, *Morphogenesis of the Vertebrates*, 4th ed., Wiley, New York, 1979, p. 480.]

corresponding regions of the tectum (see discussions by M. Jacobson, 1966; Sperry, 1965; Gaze, 1970).

The regional specificity of the relationship between retina and optic tectum is referred to as **retinotectal specificity.** When does the regional specificity of the retina arise? This question can be answered by rotating the eye primordium at various embryonic stages prior to the outgrowth of the nerve fibers and observing the visuomotor behavior of the larva or adult (Figure 11.34), or by recording electrically the input of nerve impulses to various regions of the tectum when different parts of the retina are stimulated by light. Results of such studies show that the retinal primordium initially has no positional specificity with respect to the tectum. Later, however, the positional character of retinal cells with respect to the anteroposterior axis is specified. Then, shortly afterward, the cells' specificity with respect to the dorsoventral axis is fixed.

The question now arises: Do tectal markers cor-

FIGURE 11.32 Visuomotor responses in the urodele. A: The normal target-directed response of the animal after removal of the eye and replacement in normal orientation. B: Reversed direction of the response in an animal in which the eye was excised and replaced after 180° rotation of the anteroposterior and dorsoventral axes.

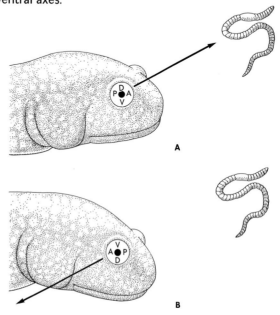

presented to that eye is reversed (Figure 11.32). Because each quadrant of the retina sends fibers to a specific part of the optic tectum (Figure 11.33), this can only mean that the severed optic nerves reestablish their synaptic connections with the same cells of the tectum with which they previously had been in contact or, at least, they connect with cells in the same general region of the tectum. The high degree of specificity with which synaptic connections are made, despite the physical handicaps imposed by eye reorientation, would seem to argue for chemical guidance of the ingrowing neurites or for a special affinity, perhaps adhesiveness, between neurites from localized regions in the retina and

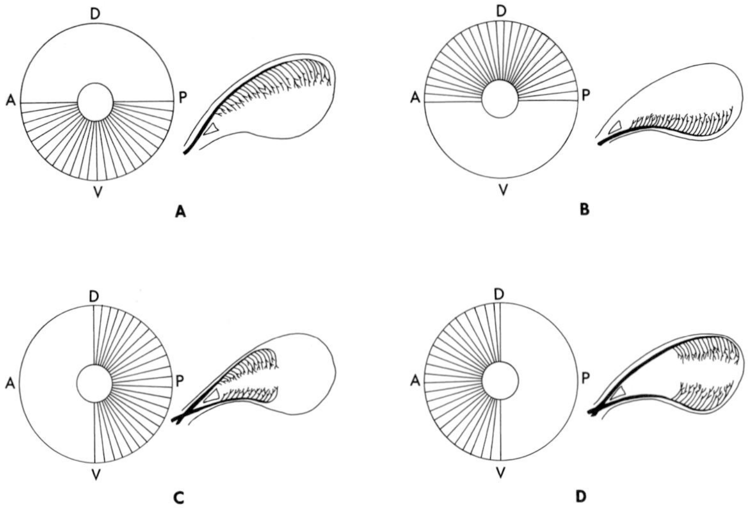

FIGURE 11.33 The regional pattern of the regenerated retinotectal projection in the goldfish after severing of the optic nerve and deletion of a retinal half. In each scheme, the retina is shown in circular outline, and to its right are shown the course and distribution of regenerated axons to the contralateral tectum. Axons enter the tectum either by a medial or lateral branch of the optic nerve and make synaptic connections respectively to the dorsomedial and ventrolateral portions of the tectum. The axial orientation of the retina is designated by letters: A, anterior; P, posterior; D, dorsal; V, ventral. A: After deletion of the dorsal half-retina, the regenerating axons from the ventral half enter the contralateral tectum by the medial route and make synaptic connections in the dorsomedial portion of the tectum, which is their normal site for synapsis. B: Similarly regenerating dorsal half-retinas enter the tectum by the lateral branch and innervate the ventrolateral region of the tectum. C, D: Posterior and anterior half-retinas utilize both medial and lateral pathways to the tectum, innervating, respectively, anterior and posterior portions of the retina. [Adapted with modifications from R. W. Sperry in R. L. DeHaan and H. Ursprung, eds., *Organogenesis*, Williams and Wilkins, Baltimore, 1965, pp. 161–86.]

responding to localized regions of the retina emerge independently of the ingrowing retinal neurites during development, or do the neurites impose their specificities on the tectum at their arrival? Several lines of evidence suggest that the latter may be the case. First, the fibers from the embryonic retina grow to the tectum in a topographically ordered pattern; those from the central retina form the core of the optic nerve, and those from progressively more peripheral parts of the retina are arranged radially about this core. This orderly arrangement is maintained. Next, we note that Chung and Cooke (1975) rotated the tectal region of the neural tube in *Xenopus* embryos before the arrival of optic fibers

and found no inversion of the resulting projection of the retina on the tectum. After the arrival of optic fibers, however, rotations of portions of the tectum result in locally rotated projections (Levine and Jacobson, 1974). Finally, it may be shown that a regenerating half-retina in the adult goldfish, although initially reinnervating only its specific half-tectal goal (Figure 11.33), will, after some months expand over the entire tectum. The specific markers previously occupying this territory or expansion can be shown, meanwhile, to have disappeared. If the nerve from the half-retina is crushed once more, the regenerating fibers return to the expanded projection (Schmidt, 1977). Although

Specification of Eyes Axes in Embryo   **Visual Behavior in Adult**

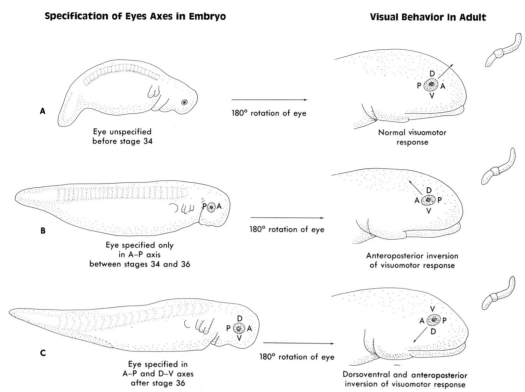

FIGURE 11.34 Ontogeny of axial order in the retinatectal projection of *Ambystoma punctatum*. Reorientation of the optic primordium prior to stage 34 does not affect subsequent visuomotor behavior. Polarization of the anteroposterior axes occurs between stages 34 and 36, and reorientation at this time affects visual behavior only with respect to the anteroposterior axis. Thereafter rotation of the eye brings about reversed visual behavior with respect to both axes. [Schematized from data of L. S. Stone, *J Exp Zool* 145:85–96 (1960).]

237

this final case is based on studies of adult material, it provides further suggestive evidence that ingrowing embryonic neurites confer specific tectal markers.

11.2.1.6 Other Ectodermal Derivatives. The foregoing discussion deals rather heavily with the nervous system, as is appropriate in view of the increasing importance of studies of neurogenesis and the origin of behavior. Before turning to the analysis of morphogenetic processes in the other germ layers, however, it should be noted that the other major derivatives of the ectodermal layer are the integument and the structures intimately associated with it. Ectoderm contributes the epidermis of the skin and a number of derivatives, such as hairs, feathers, sweat and sebaceous glands, and the hard

parts of scales. In the mouth ectoderm forms the enamel organ of the teeth and the branching ducts and secretory acini of the salivary glands (these structures are described among derivatives of the oropharynx).

The chief protein produced by the integumentary ectoderm is **keratin.** Keratins from various sources differ somewhat in their physiochemical properties, but all are fibrous proteins with a high content of the amino acid cysteine. The outer layer of the skin consists of keratinized dead cells that were originally produced by the germinal layer of the skin (Figure 11.35). These dead cells are progressively flaked off and replaced by new cells displaced upward by further proliferation from the

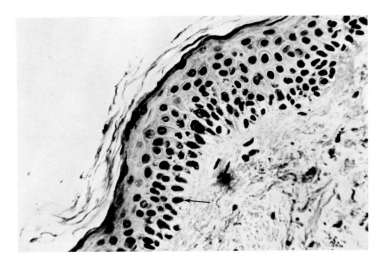

FIGURE 11.35 Section of human skin. The germinal layer is designated by an arrow, with dermis lying below. The heavily keratinized outer layer of cells and sloughing dead cells are seen at the surface of the skin.

238

germinal layer. As the new cells progress to the exterior the cells synthesize great quantities of keratin and then die.

Hairs and feathers are also made of keratin. In the feather the keratin is arranged in intricate patterns that change along the length of the feather vane. These patterns are distinctive for various body regions. Likewise, the structure and distribution of hair are regionally distinctive on the mammalian body. The origin of the regional specificity of integumentary derivatives is treated in Chapter 19.

As analyzed chiefly in mammals, the differentiation of the epidermis is greatly affected by a polypeptide that, like nerve growth factor, was first found in the submaxillary salivary glands of male mice. This factor is called **epidermal growth factor** (EGF), and it is produced in the submaxillary glands of both males and females. Like nerve growth factor, it is produced in the coiled tubular portion of the submaxillary gland and is present there in much greater quantity in males than in females.

Epidermal growth factor, injected into newborn mice, induces precocious opening of the eyelids, early eruption of the teeth and hair, and accelerated differentiation of the skin. It promotes the healing of corneal wounds in rabbits in vivo and hyperplasia of rabbit cornea in vitro. In organ cultures it stimulates the growth of mammary epithelium and chick and mammalian skin, including that of human beings. It is a potent **mitogen** (inducer of mitosis) when added to cultures of fibroblasts.

Mouse EGF is a protein composed of 53 amino

acid residues and has a molecular weight of 6045. It is present in the plasma of both male and female mice at a concentration of about 1.5 ng/ml. Of the various mouse organs studied, other than the submaxillary gland, only kidney, stomach, parotid glands, and pancreas contain appreciable concentrations of the factor. It is readily localized and measured in tissues and serum by use of antisera against EGF (anti-EGF). EGF labeled with radioactive iodine supplied to fibroblasts in vitro is found to be bound to cell surface receptors, which are then internalized. In the cytoplasm the EGF is rapidly degraded and returned to the medium. It is not clear whether internalization is required for the mitogenic effect of EGF or is simply a way of degrading the factor. EGF must be in the medium for a period of time much longer than that required to saturate the surface-binding sites before an effect on DNA synthesis is observed. Any molecular interpretation of the action of EGF (and none is obvious) must take this observation into account (Carpenter and Cohen, 1976; Das et al., 1977; Carpenter and Cohen, 1978).

The foregoing studies of internalization were carried out on human EGF. Like mouse EGF, it contains 53 amino acids, 37 of which are sequenced similarly to those of mouse EGF. It shows the same biological activity as mouse EGF whether tested on human or mouse material. Serum and urine of adult human males and females both contain EGF, but it is found in significantly higher quantities in the serum and urine of pregnant women. Pregnancy urine is thus a good source of the factor.

Epidermal growth factor may be involved in the origin of cleft palate, a congenital defect that results from the failure of the bilaterally arising rudiments of the roof of the mouth (the **palatal shelves**) and upper jaw (**maxillary arch**) to fuse. Normally the epithelial covering of the advancing edges of the palatal shelves degenerates as the elements approach the midline of the roof of the mouth. Hassell (1975) showed that the addition of epidermal growth factor to organ cultures of mouse palatal rudiments inhibits the degeneration of epithelium and prevents fusion of the two halves of the palate. Bedrick and Ladda later showed (1978) that the administration of epidermal growth factor to pregnant mice produces proliferation and thickening of the epithelium of the palatal processes in the fetuses, but does not significantly increase the incidence of cleft palate. Cortisone, a steroid hormone from the adrenal cortex, has long been known to induce cleft palate with fairly high frequency. Bedrick and Ladda (1978) found that combined treatment of pregnant mice with both EGF and cortisone results in cleft palate in all fetuses. It appears that the effect of EGF may be potentiated by cortisone. This example should serve as a further warning to prospective human mothers of possible teratogenic effects of certain hormones administered in amounts higher than normal during pregnancy.

11.2.2 MESODERM.

As already noted, for embryos such as those of the amphibian, bird, and mammal, a main component of gastrulation is the involution or ingression, of mesoderm through a blastopore or primitive streak. Once inside, the mesoderm constitutes a massive middle layer that eventually contributes the great bulk of the body, thus determining its form to a major extent.

11.2.2.1 The Notochord.

Upon the completion of gastrulation in amphibians, the mesoderm in the dorsal midline of the neurula detaches itself from the lateral mesoderm to form the rodlike notochord, which runs almost the full length of the embryo (Figures 9.11C and 11.15D). In the chick the notochord arises from the anterior end of Hensen's node as the primitive streak regresses (Figure 9.25). As the node shifts posteriorly, it creates lines of shearing between itself and adjacent tissues, separating the mesodermal layer into bilateral masses and laying down the notochord in its

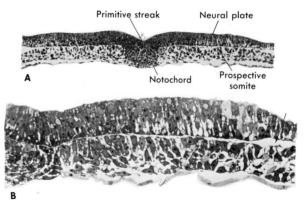

FIGURE 11.36 A: Transverse section through a chick blastoderm at the head process stage (cf. Figure 4.25 F) at the level of the future third somite. B: Similar section at higher magnification through the future third somite of an embryo at the one-somite stage, showing only one side of the future neural plate, notochord, and segmental plate (prospective somite). Note that the close adherence of somitic cells to the overlying ectoderm extends only beneath the thickened neural plate. The separation of the layers is emphasized by an arrow in B. Note that in B, the cells of the segmental plate have become more tightly packed as they become organized into a somitomere. In both A and B, the hypoblast is seen as a single layer of cells underlying the notochord and segmental plate. [From B. H. Lipton and A. G. Jacobson, *Dev Biol* 38: 73–90 (1974), by courtesy of Dr. Lipton and permission of Academic Press, Inc., New York.]

239

wake (Figures 11.36 and 11.37). The notochord lies directly beneath the neural tube and provides a rigid support along the embryonic axis. The notochord persists as the sole skeleton-like support in the Cephalochordata, but in vertebrates it mostly degenerates and contributes only to the intervertebral discs of the vertebral column.

11.2.2.2 Somites.

The mesoderm immediately adjacent to the notochord on each side is the segmental plate (Figures 11.10B and 11.39A). The plate is segmented into rows of tissue blocks, the **somites.** These tissue blocks subsequently form dermis of the skin dorsally, the vertebral column, the ribs and probably a part of the appendicular girdles, the extrinsic and intrinsic musculature of the appendages and girdles, the pectoral muscles, intercostal muscles, abdominal muscles, and dorsal and intervertebral muscles (Chevallier, 1979). The segmental arrangement of the somites is easily seen in

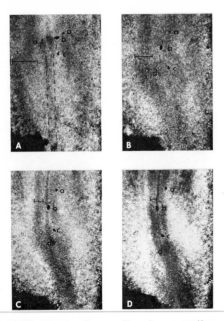

FIGURE 11.37 Frames from a time-lapse motion picture film of a chick blastoderm marked with carbon particles at sites designated a, b, c, d. The transverse bar in each frame spans the width of the denser cells of the future neural plate. In A a, b, and c were placed on the presumptive neural plate lateral to the primitive streak and anterior (a), lateral (b), or posterior (c), to the level of Hensen's node. A: The embryo at the start of the film; dashes outline the primitive streak. B: After 4 hr the node has regressed, bypassing particle c. The entire embryo is elongating as indicated by the increasing distances between a, b, and c. These marks, moreover, have moved closer to the midline. C, D: Frames taken 6.5 and 8.5 hr, respectively, after the start of filming. The node continues to regress, and lateral marks move toward the midline as the presumptive neural plate condenses. Marks b and c now lie over the notochord. [From B. H. Lipton and A. G. Jacobson, *Dev Biol* 38:73–90 (1974), by courtesy of Dr. Lipton and permission of Academic Press, Inc., New York.]

tributed by cells from the posterior half of one somite and the anterior half of the succeeding one.

The origin of the anterior sets of somites in the chick was analyzed by Lipton and Jacobson (1974 a, b) and Meier (1979). They showed that the neural plate, which is a pseudostratified columnar epithelium at the head process stage, is underlaid by mesoblast cells that are attached to it by their basal ends in a pattern extending laterally as far as the neural plate does. These are future somite cells (Figure 11.38A–C). Before the regression of Hensen's node these cells, as viewed from above, become parceled into circular domains tandemly arranged on either side of the primitive groove. These domains were recognized by Meier (1979) who called them **somitomeres.** Although consisting of distinct cellular patterns, the somitomeres are at first continuous with the axial mesoderm of the primitive groove. As Hensen's node regresses (Figure 11.37), it shears the sheet of mesoblast into right and left halves, leaving the notochord between. Meanwhile, the neural plate becomes compacted as its cells elongate, and it begins to curl upward. This movement draws the cells of its adherent somitomeres closer together (Figure 11.38A), and new attachments form between them, transforming their epithelial configuration into a rosette pattern as seen in cross section (Figure 11.38B). Both Hay (1968) and Lipton and Jacobson (1974a) observed that numerous filaments bind the medial face of the rosette to the neural tube and notochord. These connections probably relate to the role of neural tube and notochord in the stabilization and further differentiation of the somites, for Lipton and Jacobson (1974b) showed that, whereas anterior somites begin to form when mesoderm and medullary plates are separated from the prospective notochord by means of a cut lateral to the midline (Figure 11.39), these somites soon disintegrate.

Normally the rosettes first form a small lumen, surrounded, as seen in cross section, by a triangular wall. The lumen soon fills with cells that immigrate chiefly from the ventrolateral corner of the triangle where it is continuous with the lateral plate (Figure 11.38C). The ventral and ventromedial sides of the triangle then lose their epithelial character, merging with the cells of the lumen and becoming an ill-defined aggregation that extends around the ventral sides of the neural tube to the notochord. This aggregation is the **sclerotome.** Its cells, now mesen-

early embryos. This arrangement is thought by some to reflect the primitive metameric arrangement of body parts in presumed invertebrate ancestors of the vertebrates. In the vertebrates, of course, metamerism is largely lost, but aspects of it are retained in the segmental arrangement of the body-wall musculature in the lower vertebrates and in the serially ordered vertebrae, each of which is con-

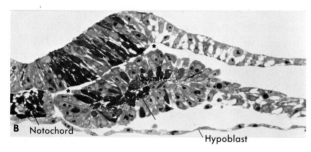

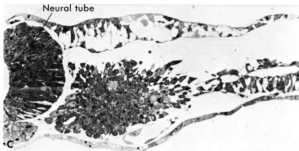

FIGURE 11.38 Somitogenesis as seen in transverse section through the level of the third somite at different developmental stages of the chick embryo. A: Transverse section at the stage of 2 or 3 somites. The section passes through the third somitomere at the time its anterior and posterior borders are becoming clearly defined. The cells have formed closer attachments at their inward-facing apices. The notochord is becoming better defined, and the neural folds are slightly elevated. B: At the 4-somite stage, the neural tube has begun to close, and the notochord becomes clearly separated from it and the somitic mesoderm. The bases of the somite cells lose their adhesivity for the notochord and neural tube, and their apices face a small central cavity (arrow), the somitocoele. C: The neural tube has closed and the somitocoele is now filled with cells proliferating from the epithelial wall of the somite. At the right, the lateral mesoderm is splitting into an upper layer, the somatic mesoderm, and a lower layer, the splanchnic mesoderm. [From B. H. Lipton and A. G. Jacobson, *Dev Biol* 38:73–90 (1974), by courtesy of Dr. Lipton and permission of Academic Press, Inc., New York.]

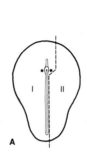

FIGURE 11.39 Effects of separating a portion of the chick blastoderm from the primitive streak. A: The right side of the blastoderm is separated from the primitive streak and Hensen's node. Regression of the streak occurring in the left side of the blastoderm results in the formation of notochord on that side only. B: Results of the operation schematized in A as seen in the living blastoderm halves. The left fragment, I, formed brain, notochord, and 15 somites (not all visible in the photograph). In the right fragment, II, 5 somites and segmental plate formed in association with a part of the neural plate (to the left of the somites). The first two somites are breaking down. No more than 7 somites formed in cases such as the one illustrated here. Left in culture for an additional 14–20 hr, all somites of the right fragment break down and the entire mesodermal layer disperses laterally. [From B. H. Lipton and A. G. Jacobson, *Dev Biol* 38:91–103 (1974), by courtesy of Dr. Lipton and permission of Academic Press, Inc., New York.]

241

chymal in character, converge around the notochord and later form the vertebral column. The dorsal part of the triangle, meanwhile, retains its definitive boundaries and epithelial characteristics. It becomes the **dermatome,** from which much of the dermis of the skin arises. Medially it is continuous with the **myotome,** which forms from the dorsomedial wall of the triangle as it folds laterally, coming to underlie the dermatome (Figure 11.40). Factors that influence the changes in shape of the somites were analyzed by Packard and Jacobson (1979). The reader is referred to their paper for details.

11.2.2.3 Roles of Intermediate and Lateral Mesoderm. Lateral to the somites there is a narrow intermediate zone of mesoderm; beyond this zone, the mesoderm separates into two layers, an

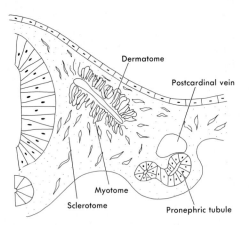

FIGURE 11.40 The differentiation of the somite into dermatome, myotome, and sclerotome as seen on one side of the neural tube and notochord.

outer somatic layer and inner splanchnic layer (Figures 11.38C and 11.41). The intermediate mass is the chief source of the urogenital organs and the adrenal cortex. The somatic layer contributes the peritoneum, the connective tissues of the limbs, the sternum, and the appendicular skeleton. The splanchnic layer provides the material of the heart, invests the digestive tract, and forms the mesenteries. The major visceral organs, the liver and pancreas, are formed cooperatively by endoderm and splanchnic mesoderm, and the splanchnic mesoderm, likewise, forms the muscular walls of the digestive tract which is lined with endoderm.

11.2.2.4 Origin of the Limbs. The primordium of a limb is called a **limb bud**. It arises as cells of the somatic mesoderm proliferate beneath the ecto-

FIGURE 11.41 Cross section of an early chick embryo, illustrating the relationships of the germ layers.

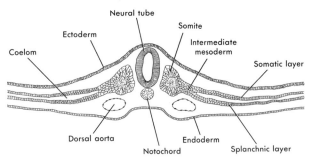

derm to form a conical (in urodeles, for example) or paddle-shaped (humans, chicks, fishes) projection (Figures 11.10C and 11.42). In fishes, *Xenopus*, reptiles, birds, and mammals this projection is rimmed by a thickening of ectoderm (Figures 14.4, 14.5, and 19.36) called the **apical ectodermal ridge (AER)** (Saunders, 1948); in urodeles the apical ectoderm forms a slightly thickened cap over the tip of the bud. There is a relationship of dependency between the apical ectodermal ridge and limb-bud mesoderm. This relationship is treated more fully in Chapter 14. Suffice it to say for now that in the absence of limb-forming mesoderm, an AER will not form. Reciprocally, in the absence of the AER (as brought about surgically or by genetic mutation, for example), the mesoderm will not grow. These relationships have been most completely analyzed in the chick embryo (reviewed by Saunders, 1977), but apply to the development of reptiles and mammals, too.

The mesoderm of the limb-forming regions of the body wall is of dual origin. Its initial component is the somatic mesoderm, which comes to lie against the body-wall ectoderm as gastrulation proceeds. The second component is contributed by cells that migrate into the limb-forming region from the ventrolateral ends of the somites. In the chick embryo the wing bud arises at the anteroposterior level defined by somites 15–20. At the time the chick embryo has 20 pairs of somites (about 50 hours after incubation begins), cells begin to migrate from the ventrolateral end of somite 15 into the mesoderm of that part of the body wall from which the wing bud will later arise. Successively, somites 16–20 contribute their cells to the prospective wing, the last ones leaving somite 20 about the time a total of 36 pairs of somites have been formed. Somewhat later the mesoderm of the future leg region receives an influx of cells from somites 26–32 (Chevallier, 1978, 1979; see also Jacob et al., 1978).

The somatopleure at the level of the prospective wing or leg may be excised before the invasion of somite cells and grafted to the body cavity of a host embryo. There it will form an appendage with skeletal parts, connective tissue, and integument, but completely lacking in muscle (Christ et al., 1977). Or, if the somites at the level of the future limb are destroyed by x-irradiation, a limb lacking its intrinsic musculature will form (Chevallier et al., 1978;

Chevallier, 1979). A very important note is that in the chick embryo the somitic mesoderm at the level of the wing may be replaced with somitic mesoderm of the quail taken from the level of neck, wing flank, or leg. In each instance the wing develops normally, but all the muscles are made up of quail cells. All other components of the wing are of chick origin, including notably the entire connective tissue network of the muscular tissue, tendons, vascularization, and skeleton (Christ et al., 1977; Chevallier et al., 1977, 1978; Kieny and Chevallier, 1979). These experiments demonstrate not only that the limb musculature is derived from cells contributed by the somites but also that the somite cells are not specified in advance as to the muscles that they will form. The ultimate spatial disposition of the muscle cells in limbs and body wall must, therefore, be dependent on positional information preexisting in the somatopleure (cf. Chapter 19).

The foregoing evidence notwithstanding, there are indications that under some circumstances, somatopleural mesoderm at the limb level may be able to form muscle cells. Chevallier et al. (1977) noted that when chick somites are used to replace those of quail at the prospective limb level prior to the normal invasion of the somatopleure by quail somitic cells, muscles of mixed chick and quail origin are usually found in the appendage that results. Moreover, in contrast to the findings of Christ et al. (1977), McLachlan and Hornbruch (1979) found that when quail somatopleure is grafted to the coelom of the chick embryo prior to the invasion of somitic cells into the quail tissue, some muscle cells of quail origin are usually found. Does the presence of differentiating muscle cells of somitic origin normally suppress an option that somatopleural mesoderm may have to form muscle? This question arises once more in Chapter 14.

The limb buds become evident as thickenings of the somatopleure during the period of invasion by somitic cells. Once in the limb bud, however, they cannot be distinguished morphologically from other mesodermal elements of the bud. As the limb bud elongates, however, the myogenic (muscle-forming) cells take positions predominantly on the dorsal and ventral sides of the outgrowth, and prospective cartilage cells, which give rise to skeletal parts, occupy the central region. This was determined by Haushka and Haney (1978), who analyzed the differentiation in organ culture of cell groups isolated from the respective limb-bud regions.

As the limb bud elongates, it progressively undergoes histological differentiation, the cartilaginous elements and muscle masses becoming visibly differentiated in proximodistal order. Likewise, it changes from a paddle-shaped or conical configuration into the definitive form of a limb. These changes, too, have been analyzed most extensively in the chick. The shaping of the limb bud into the form of a forelimb or hind limb is accomplished chiefly by differential growth, aided by differential cellular death. Growth contributes the bulk of the limb and blocks out its general form, but as the limb buds elongate, waves of cellular death sweep along the mesoderm of the anterior and posterior edges of the appendages, and their refined contours emerge as the waves pass. This process reaches a climax in the erosion of the interdigital tissues; such erosion apparently aids in the sculpturing of the toes (Figure 11.42). The role of cell death in morphogenesis of the limb was treated in some detail by Saunders et al. (1962) and by Saunders and Fallon (1967). Notably, if death in the interdigital area fails to occur, as happens after administration of the dye janus green, the toes remain joined by soft tissue. In the duck, which is normally web footed, interdigital necrosis is restricted chiefly to the distal margin of the future web. Cell death is also a major factor in

243

FIGURE 11.42 The changing shape of the leg bud in the chick embryo. Stipples show the distribution of cellular deaths. Numbers refer to developmental stages of the chick in the Hamburger–Hamilton series [V. Hamburger and H. L. Hamilton, *J Morphol* 88:49–92 (1951)].

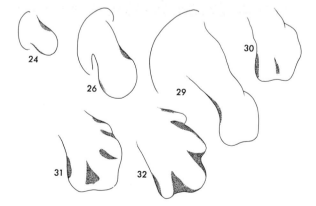

the separation of the digits in mammals, including human beings (Menkes et al., 1965) and in reptiles (Fallon and Cameron, 1977).

11.2.2.5 Origin of the Heart. Many experimental studies have been carried out on the origin of the heart. In the chick embryo the future heart-forming cells are located in the epiblast at the primitive streak stage. Rosenquist and DeHaan (1966) systematically replaced, one at a time, small areas of the epiblast from one embryo with corresponding areas from another embryo whose nuclei were radioactively labeled. By determining which of the grafted areas contributed radioactive cells to the heart of the recipient embryo, they could retro-project the heart-forming area on the primitive streak blastoderm (Figure 11.43).

Once invaginated, the heart-forming mesoderm is organized in two separate regions, one on either side of the embryonic axis (Figure 11.44), and each becomes that part of the splanchnic mesoderm, which is in contact ventrally with the endodermal region that will form the foregut of the embryo. The subsequent development of the heart now depends on the undercutting of the anterior end of the embryo from the surface of the yolk. This gives rise to the **head fold,** which involves all three germ layers and creates the **anterior intestinal portal** (Figure 11.45), the opening through which the cavity of the foregut is continuous with the cavity of the yolk sac. On either side of the foregut wall some cells of the heart-forming portion of the splanchnopleure now detach and form delicate tubes, the future lining cells of the heart, or **endocardium.** With the lengthening of the foregut, the tubes from each side

FIGURE 11.43 Migration of radioactively labeled precardiac cells of the chick embryo into the primitive streak and their subsequent appearance in the heart primordium. A: Living embryo at the primitive streak stage bearing a graft of epiblast cells from a donor embryo of the same age whose nuclei were labeled with ^{3}H-thymidine. B: Cross section of such an embryo at a later stage, showing the entrance of the labeled cells (dark spots) into the primitive streak; original position of graft shown by arrows. C: Cross section through the same embryo pictured in A, fixed after heart formation has begun, and sectioned for radioautography. Labeled cells are seen in parts of the heart and other nearby mesoderm. D: Distribution of prospective heart-forming cells in the mid-primitive-streak blastoderm of the chick. The approximate location of heart-forming cells is the stippled rectangle. Radioactively labeled epiblast from this area, grafted to the identical region of a nonlabeled host, is subsequently located in the heart. (A–C, from G. C. Rosenquist and R. L. DeHaan, *Carnegie Contribs Embryol* 38:111–21 (1966), by courtesy of the authors and permission of the Carnegie Institution of Washington, Department of Embryology, Davis Division D, an interpretation of data from the same article.]

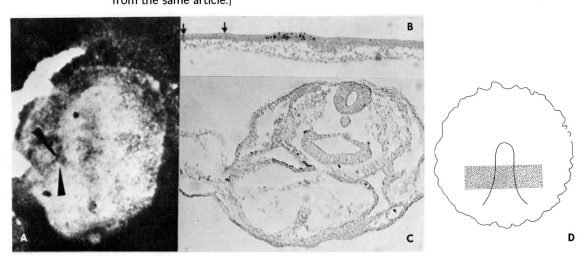

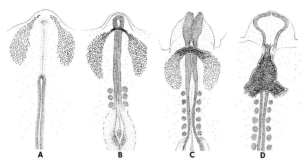

FIGURE 11.44 Bilateral origin of the heart during early stages of the chick embryo from the time the head fold originates (A) until shortly after closure of the neural folds of the head region (D). The laterally disposed precardiac cells move to the midline from each side and join (B, C) as the anterior intestinal portal shifts posteriorly and as lateral folds undercut the head region (cf. Figures 11.45 and 11.46). [Adapted from R. L. DeHaan, *Acta Embryol Morphol Exptl* 6:26–38 (1963); modifications partially based on data of H. Stalsberg and R. L. DeHaan, *Dev Biol* 19:128–59 (1969).]

fuse ahead of the anterior wall of the intestinal portal to form a single tube. The remaining sections of the heart-forming splanchnopleure are likewise brought together from each side to form the **epimycardium,** which gives rise to the muscular wall of the heart. These changes are diagrammed in

FIGURE 11.45 Saggital sections of early chick embryos. A: At the head process stage after about 18 hr of incubation. B: At the early head-fold stage, wherein special attention is directed to the anterior intestinal portal. [Adapted with modifications from B. F. Patten and B. M. Carlson, *Foundations of Embryology*, 3rd ed., McGraw-Hill, New York, 1974, pp. 149, 168.]

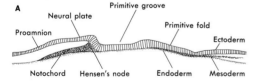

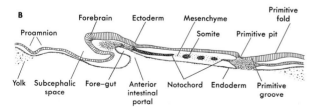

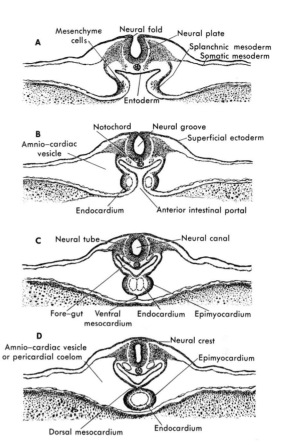

FIGURE 11.46 Formation of the chick heart as seen in transverse sections through comparable anterior regions of embryos at approximately 25, $26\frac{1}{2}$, $27\frac{1}{2}$, and 29 hr of incubation in A, B, C, and D, respectively. The appearance of heart-forming cells in the midline follows the posterior progress of the anterior intestinal portal. This sequence should be examined in conjunction with Figures 11.44 and 11.45. [From A. F. Huettner, *Comparative Embryology of the Vertebrates*, 2nd ed., Macmillan, New York, 1949, p. 171.]

245

Figures 11.44 and 11.46. The short, relatively straight heart tube becomes highly asymmetric and is subsequently thrown into loops. Parts of these loops become the different heart chambers (Figure 11.47). Formation of the heart loops has been correlated in a causal way with patterns of changing cellular shape (Manasek et al., 1973) and with differential cellular death (Pexieder, 1975; Manasek, 1976).

Two separate hearts are formed if the bilateral heart rudiments are prevented from fusing. Fusion

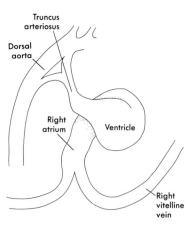

FIGURE 11.47 Primitive loops of the heart of the chick embryo and their future fate as seen from the right side.

may be prevented by nicking the wall of the anterior intestinal portal in the ventral midline by means of a microneedle, at about the stage illustrated in Figure 11.44B (DeHaan, 1959). This experiment demonstrates the essential mechanical role of the anterior intestinal portal in heart formation and also illustrates the fact that each of the bilateral heart rudiments has the property, acting alone, of making a complete heart.

11.2.2.6 Blood. The blood is constituted of the blood plasma and cellular elements. The cellular elements consist predominantly of red blood cells, or erythrocytes; in addition there are several kinds of white blood cells, or leucocytes, and blood platelets. The blood is transported through the vascular system, consisting of the arteries, veins, and capillaries. It is pumped away from the heart by the contraction of the ventricles and its return to the heart is largely brought about by the compressing action of the musculature on the veins.

The origin of the heart was described briefly in Section 11.2.2.5. The development of the accompanying vascular system proceeds according to a fascinating and complex pattern of ever-changing routes, correlated with major morphogenetic events. The morphogenesis of the vascular system has been described in considerable detail in a number of textbooks on developmental anatomy, such as those of Arey (1965) and Patten and Carlson (1974), and is not considered in this work.

In all vertebrates both the vascular and cellular

elements of the blood first differentiate in **blood islands** (cf. Chapter 10.) These islands appear initially as groups of primitive mesodermal cells. Within these groups is first segregated the endothelium, flattened cells that arrange themselves about a lumen. These cells secrete the blood plasma and constitute the future lining of the capillaries and other vessels. Surrounded by the endothelium are the primitive blood cells, which are called **hemocytoblasts.**

In higher animals, such as the bird, mouse, and man, the blood islands arise chiefly in the region of the yolk sac (recall the description of the **area vasculosa** in the chick embryo in Chapter 10). The blood vessels originate by differentiation of the endothelial layer within these islands and the subsequent extension of this endothelium to join up with the endothelium of other blood islands or with endothelium that arises in situ in mesenchyme elsewhere. To a minor extent, intraembryonic blood islands may proliferate from endothelium of the aortic arches and heart and may appear de novo in the head mesenchyme (reviewed by Romanoff, 1960).

The **hemopoietic** (blood-forming) tissues of the yolk sac chiefly engage in the formation of a primitive line of red blood cells **(erythroid cells),** which remain nucleated. When the yolk sac of the mouse is isolated in vitro, however, it also forms **granulocytes** (white blood cells of different kinds, which are distinguished by their granule content), **lymphocytes,** and **megakaryocytes** (from which blood platelets arise). Thus it would appear that the primitive hemocytoblasts of the blood islands are capable of proliferating and differentiating along several lines.

The hemopoietic function of the yolk sac is only temporary, however, being taken over by the liver and spleen in the embryo and by bone marrow, spleen, and other lymphoid organs postnatally. The schedule of changing sites of hemopoietic activity in the chick is shown in Figure 11.48.

A good deal of modern research has centered on the altered pattern of **erythropoiesis** (formation of red blood cells) that accompanies the shift of red blood cell formation from the yolk sac to the liver, for it is in the liver that red blood cells of the adult morphology and biochemical specificity first arise. These characteristics persist unchanged when erythropoiesis subsequently shifts to the bone mar-

246

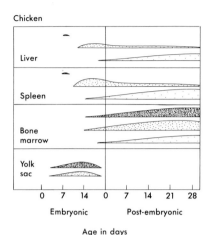

FIGURE 11.48 Scheme to illustrate the chronology of hemopoiesis in the yolk sac, liver, spleen, and bone marrow in the chick during embryonic and early postembryonic stages. Erythropoiesis is represented by dark stippling, granulocytopoiesis by intermediate stippling, and lymphopoiesis by very light stippling. [Adapted from A. L. Romanoff, *The Avian Embryo*, Macmillan, New York, 1960, p. 581.]

row. Cells with adult characteristics are said to arise from the **definitive erythroid line,** as opposed to the **primitive erythroid line** of yolk sac origin. Conceivably, the definitive erythroid line might arise by seeding of the liver by primitive hemocytoblasts originating from the yolk sac. This is probably not so, however, for if a quail embryo is substituted for the embryo of a chick blastoderm prior to the onset of hemopoiesis, the definitive erythroid cells, as well as other components of the blood, eventually are chiefly of quail type. The yolk sac produces most of the erythrocytes at the beginning of development, but from the sixth day red blood cells of embryonic origin are increasingly present in the circulation (Dieterlen-Lièvre, 1975; Martin et al., 1978).

In the chick embryo primitive erythroid cells begin to form during the second day of incubation, but, beginning during the sixth day, definitive erythrocytes are seen, and by the 16th day only 1% of the circulating red cells are of the primitive type. In the mouse the first erythropoiesis occurs in the blood islands on the eighth day, and nucleated erythrocytes enter the circulation on the ninth day. On the 10th day hemopoiesis begins in the liver

giving rise to erythroid cells that lose their nuclei and are indistinguishable from those of the adult.

In all forms that have been studied, the embryonic, or fetal, hemoglobin (Hb_F) is distinguishable from adult hemoglobin (Hb_A). In the chick embryo the primitive erythrocytes are visibly different from those that characterize the adult and that begin to be seen on day 6 (Mahoney et al., 1977). Moreover, the polypeptides (hemoglobin is made up of four polypeptides, each with a heme prosthetic group) found in the primitive line are distinctive from those in the adult form (reviewed by Wilt, 1967). Similar changes in the composition of fetal and adult hemoglobin are found in amphibians, mice, and men, and these changes are likewise correlated with the onset of erythropoiesis in the liver (see reviews by Marks and Rifkind, 1972, and by Rifkind, 1974).

In some organisms, notably adult human beings and metamorphosing amphibians, a significant proportion of fetal hemoglobin is present in the circulating blood. It is reasonable, therefore, to raise the question of whether the genes for the different polypeptide chains of adult and fetal hemoglobin are active in the same cell or in different ones. This matter was reviewed by Maniatis and Ingram (1971). Sensitive immunological methods permit it to be said that among adult human erythrocytes, as many as 5% contain some fetal hemoglobin along with adult hemoglobin. Moreover, in metamorphosing *Xenopus* tadpoles, as many as 25% of red blood cells contain both adult and fetal hemoglobins. Maniatis and Ingram investigated the same question in metamorphosing bullfrog (*Rana catesbeiana*) tadpoles, however, and found that the two hemoglobins occur in separate cells. Apparently, therefore, there are species differences in the degree to which the genes for polypeptides of fetal hemoglobin are repressed in cells of the definitive erythroid series.

It may be noted, also, that the mRNA templates for synthesis of hemoglobin polypeptide chains in both fetuses and adults are quite stable. Wilt (1965) showed that synthesis of fetal hemoglobin begins at approximately the seventh somite stage of the chick embryo. When RNA synthesis is inhibited by the drug actinomycin D, beginning at the head process stage, hemoglobin synthesis nevertheless begins on schedule, indicating that the process utilizes templates synthesized some 10 hours earlier. In the

247

mouse embryo treatment with actinomycin D on days 10 and 11 does not inhibit synthesis of hemoglobin by cells of the yolk-sac erythroid series, but does inhibit the formation of the adult hemoglobin by cells of the definitive series. By day 13, however, the cells of the definitve line are capable of producing hemoglobin of the adult type in the absence of RNA synthesis. This timing corresponds to the appearance of **polychromatophilic** and **orthochromic erythroblasts** (Figure 11.49), in which production of mRNA has essentially ceased, and which are in the process of synthesizing hemoglobin of the adult type.

FIGURE 11.49 Stages in the differentiation of the definitive line of red blood cells in mammals. The hemocytoblast (A) is the stem cell not only for cells in the erythrocytic line, shown here, but also for granulocytes and megakaryocytes. The orthochromic erythroblast (E) extrudes its nucleus and becomes a reticulocyte (F), which then becomes a typical erythrocyte. Reticulocytes are but little represented in circulating blood in normal animals, but are released from the bone marrow rapidly in animals made anemic. They then constitute the major element of the circulating blood. [Adapted from various sources, including T. L. Lenz, *Cell Fine Structure*, Philadelphia, Saunders, 1971, pp. 15–43.)

248

11.2.2.7 Urogenital System.

The organs that serve the reproductive and the urinary functions arise entirely from the intermediate mesoderm, except for a portion of the urinary bladder, which is of endodermal origin. The developmental anatomy of these systems is treated in fascinating detail in a number of standard textbooks on embryology, but it is beyond the scope of this work. This section, therefore, deals only briefly with the origin of the organs of reproduction and excretion and with the control of sex differentiation.

In vertebrates the functional embryonic kidney is a system of tubules called the **mesonephros.** The mesonephros remains as the definitive kidney in fishes and amphibians but is replaced by a later-developing kidney, the **metanephros,** in reptiles, birds, and mammals. The relationship of these kidneys to their antecedent nonfunctional **pronephros** is diagrammed in Figure 11.50.

The mesonephros arises from the intermediate mesoderm, projecting as a longitudinal ridge almost the entire length of the dorsal side of the body cavity on each side of the midline (Figure 11.51). This ridge is composed of the mesonephric tubules and their **glomeruli** (filtration units). The tubules empty into the mesonephric duct, which originates as a continuation of the duct of the pronephros and grows posteriorly to join the embryonic cloaca.

The gonads arise as localized thickenings of the ventromedial face of the mesonephros. These thickenings consist of a mesenchymal core covered by a layer of epithelium continuous with that covering the mesonephros (Figure 11.51). This epithelium grows thicker and is called the **germinal epithelium.** It is in the germinal epithelium that the primordial germ cells, arriving from outside the gonad, first take up their residence. The germinal epithelium forms an anteroposterior ridge, the **germinal ridge,** on the inner face of the mesonephros.

The initial condition of male and female gonads is the same (Figure 11.52A, B). The mesenchyme beneath the germinal epithelium condenses into cords of cells, the **primary sex cords.** These cords constitute the **medulla** of the gonad, and the germinal epithelium is the **cortex.** In this condition the gonad is in the stage of sexual indifference. In genetic male embryos (Figure 11.52C), the primordial germ cells next migrate into the primary sex cords, which later hollow out to form the seminiferous tubules in which the primordial germ cells differen-

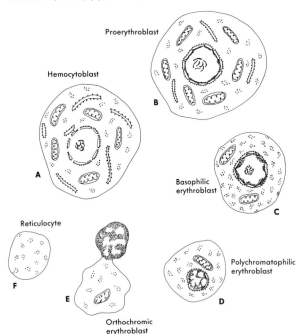

Proerythroblast

Hemocytoblast

B

A

Basophilic
erythroblast

C

Reticulocyte

Polychromatophilic
erythroblast

F

E

D

Orthochromic
erythroblast

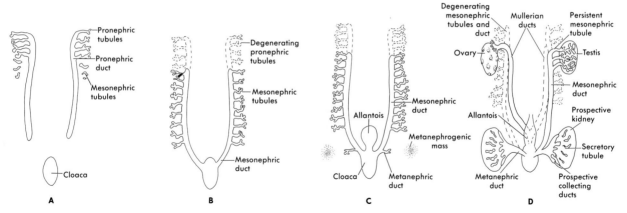

FIGURE 11.50 Schematized outline of the differentiation of the urogenital system in higher animals. A: First to form are pronephric tubules, which arise anteriorly, opening into the body cavity and connecting to the caudally growing pronephric duct. B: The pronephric tubules degenerate as mesonephric tubules form in sequence caudally. They form functional glomeruli (see Figure 11.51) and drain into the continuation of the pronephric duct, now the mesonephric duct. The latter opens into the cloaca, derived from the hindgut (cf. Figure 11.60). C: Near the cloaca the metanephric duct arises as a bud, often called the ureteric bud, from the mesonephric duct and grows dorsally toward a mesenchymal condensation, the metanephrogenic mass. D: Müllerian ducts arise in both female (left) and male (right) urogenital systems, but persist only in the female. In the male, some of the mesonephric tubules persist and become the vasa efferentia of the testis; the mesonephric duct becomes the vas deferens (cf. Figure 17.3). Degenerating ducts and tubules are indicated by dashed lines. The metanephric duct forms the ureter and collecting tubules of the definitive kidney, the metanephros, and the metanephrogenic mass forms the secretory tubules and glomeruli. [Adapted from B. F. Patten and B. M. Carlson, *Foundations of Embryology*, 3rd ed., McGraw-Hill, New York, 1974, p. 499.]

249

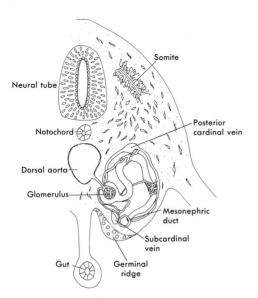

tiate into spermatogonia (cf. Chapter 7). In the genetic female (Figure 11.52D) the primary sex cords are resorbed and the interior of the gonads remains in the form of loose mesenchyme permeated with blood vessels. The cortex increases greatly in thickness as the medulla regresses. Now masses of cortical cells on the inner face of the cortex become split up into groups surrounding clusters of primordial germ cells. Subsequently, these cells are surrounded by follicle cells to give rise to primary ovarian follicles.

During the stage of sexual indifference embryos of both male and female genetic sex are equipped

FIGURE 11.51 The mesonephric tubule and its vascular supply are illustrated in simplified three-dimensional view as related to the gonadal ridge and other elements of the trunk, which are represented in cross section.

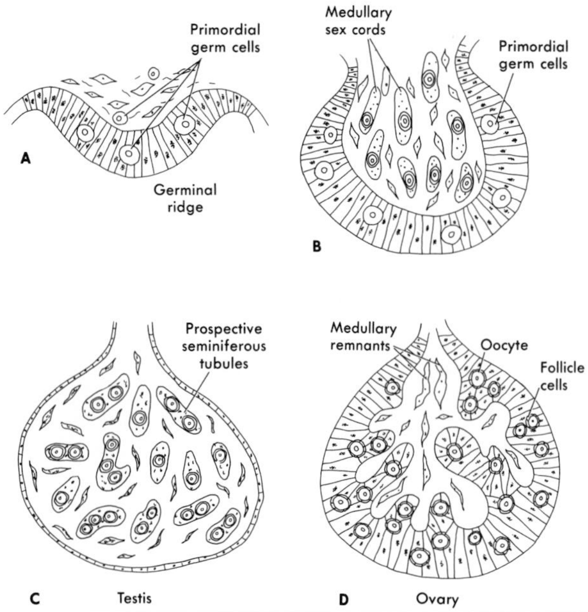

FIGURE 11.52 Schematic representation of the development of the gonads in higher vertebrates. A: The germinal ridge contains primordial germ cells and bulges into the body cavity as mesenchyme moves ventrally. B: The invading mesenchyme forms the medulla of the embryonic gonad differentiating into primary sex cords into which primordial germ cells migrate from the cortically located germinal epithelium. C: In the male the medullary element persists and forms the prospective seminiferous tubules, which are populated with primordial germ cells. D: In the female the primary medullary cords regress, and the germinal epithelium proliferates secondary sex cords populated with primordial germ cells, which then differentiate into primary oöcytes. [Adapted from B. I. Balinsky, *An Introduction to Embryology*, 3rd ed., Saunders, Philadelphia, 1970, p. 491.]

250

with a double set of sex ducts. These are the paired mesonephric ducts, which become the vasa deferentia in the male, and the paired **Müllerian** ducts, which form the oviducts, or fallopian tubes, anteriorly, and which fuse posteriorly (Figure 11.50C) to form, in the female mammal, the uterus and vagina. Both sexes likewise show, at this stage, a swelling in the ventral midline at the base of the tail, which is called the **genital tubercle.** In genetically male embryos the differentiation of the gonad as a testis is shortly followed by degeneration of the Müllerian ducts. The genital tubercle then forms

the penis and scrotal sac. In female embryos the mesonephric duct degenerates, and the genital tubercle forms the clitoris and labia.

What determines the differentiation of primary (gonadal) and secondary sex characteristics (the accessory characteristics are the sex ducts and external genitalia)? The possibility that diffusible hormones control sex differentiation was strongly suggested long ago by the fact that among cattle the female co-twin to a bull calf shows varying degrees of masculinization, apparently correlated with the occurrence of crossed circulation brought about by fusion of the chorionic vesicles in the uterus. The female is usually born sterile and is known among stockmen as a freemartin. The freemartin heifer usually has external genitalia of female type, but the clitoris is sometimes enlarged. The uterus and fallopian tubes may be reduced or completely absent, and the gonads, in extreme cases, assume the appearance of testis, and there may be epididymides (plural of epididymus), vasa deferentia, and seminal vesicles. The transformed gonads show seminiferous tubules which are devoid of germ cells (Willier, 1921).

On the basis of detailed studies of heterosexual cattle twins, Lillie (1917, 1923) concluded that the freemartin is, indeed, zygotically a female that is modified in the male direction by the action of sex hormones from the male twin. Lillie's papers stimulated an immense amount of research on the role of hormones in sex differentiation, but even so, no one has yet succeeded in producing sex reversal of a mammalian gonad by any form of hormone treatment. When male hormones are injected into cows carrying female fetuses, no modifications of the fetal gonads or Müllerian ducts occur, although, in contrast, the external genitalia are strongly modified in the male direction (reviewed by Short et al., 1969; see also Burns, 1961). This suggests that in the case of heterosexual cattle twins, something other than testosterone is transmitted to the female fetus.

There is now a considerable body of evidence that in mammals the presence of the Y chromosome determines the male sex by causing the indifferent gonad to differentiate as a testis. In the absence of the Y chromosome, the gonad differentiates as an ovary. The Y chromosomal male-determining gene product is believed to be the so-called **H-Y antigen.** This is a cell surface protein that is responsible for

causing female mice of certain inbred strains to re-
ject skin grafts from male mice of the same inbred
strains. A similar antigen is found in normal males
of all mammalian species that have been examined.

One view (Ohno, 1977) would have gonadal
cells endowed with special receptors for the H-Y
antigen, causing them to be organized in the form
of a testis. The testis would then produce androgens, which would determine male secondary sex
characters. In the case of the freemartin heifer, H-Y
bearing cells from the male twin entering the indifferent gonad would cause it to form a testis, secreting male hormones and causing the appearance of
male secondary sex characteristics. The paper of
Ohno cited above and another by Silvers and
Wachtel (1977) give thorough documentation of
the probable role of the H-Y antigen in sex determination. Erickson (1977) has expressed some reser-
vations about the role assigned to the H-Y antigen
in development, and his paper should be consulted
for a documentation of some possible difficulties
with the view presented here.

In higher animals, such as birds and mammals,
differentiation of female structures appears to be
essentially hormone independent. For example, the
Müllerian ducts of the male bird continue to develop in vitro in the absence of male hormone and
regardless of whether or not female hormone is
supplied. With male hormone they degenerate. On
the other hand, in a number of amphibians functional sex reversal has been achieved in the male or
female direction, according to the hormone administered, suggesting that both male and female hormones may play a role in the realization of the phenotypic sex (Burns, 1961).

11.2.2.8 Origin of the Primordial Germ Cells.
The origin and migration of the primordial germ
cells have been the object of a good deal of clever
research during recent years. In fact, that the prospective germ cells originate outside the gonads in
many kinds of animals is noted in Chapter 10,
where the case of the insects receives special attention. In most vertebrates the primordial germ cells
arise from the endoderm or in connection with special plasms located in yolky regions of the egg. Because the gonad itself is mesodermal, however, it is
as reasonable to discuss the origin and behavior of
the primordial germ cells in this section as it is to
treat them in Section 11.2.3, which deals with endodermal derivatives.

The primordial germ cells of some amphibians
are of particular interest, for they arise in conjunction with a visibly differentiated germinal plasm,
even as in insects. In eggs of most anurans that have
been studied there is a zone of distinctive cytoplasm, rich in ribose nucleic acid (RNA), near the
egg membrane at the vegetal pole. The material in
this zone is granular, each granule consisting of a
dense core about 10–20 nm in diameter associated
with finer components 2–8 nm in diameter (reviewed by Smith and Williams, 1975, 1979). The
granules of the germinal plasm are thus similar to
those described for *Drosophila* pole plasm, as discussed earlier. During cleavage this material is carried upward into the yolk (Figure 11.53), where it
becomes associated with the nuclei of endodermal
cells as the yolk is progressively cellulated. The
RNA-rich material is segregated during subsequent
cell divisions, setting up a special population of
daughter cells, the primordial germ cells. These are
found in the yolky floor of the blastocoele and,
later, the floor of the archenteron. After the tadpole
hatches, the primordial germ cells migrate to the
roof of the gut, then through the dorsal mesentery
to the gonads (Figure 11.54A). The events in the
germ cell lineage of *Xenopus laevis* are similar to
those in species of the genus *Rana,* and were described in some detail by Zust and Dixon in 1977.

In 1966 L. D. Smith undertook experiments designed to test whether the RNA-rich cytoplasm of
the vegetal pole in *Rana pipiens* is the determiner of
primordial germ cells or whether it is simply inci-
dental to their differentiation. He removed the dis-

FIGURE 11.53 Distribution of RNA-rich germinal plasm
in the egg of the frog. This plasm is located at the
vegetal pole of the uncleaved egg (A), but is carried
toward the animal pole along the cleavage furrows (B),
some of it reaching the floor of the future blastocoele
by the stage of 16–24 cells (C). [Adapted from L.
Bounoure, *Ann Sci Natur Zool* 10(Ser. 17):69–248, 1934.]

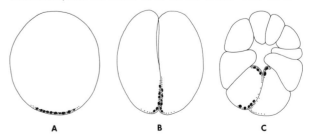

A   B   C

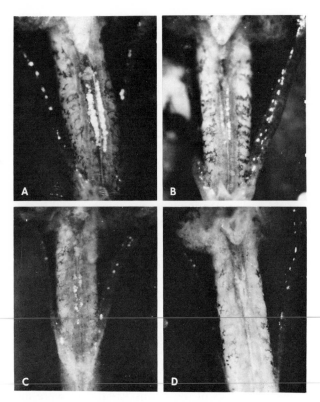

252

veloped into tadpoles with sterile gonads or, at most, with gonads greatly deficient in germ cells. Next, he injected irradiated eggs with RNA-rich material from the vegetal pole of nonirradiated eggs. The recipients developed into fertile frogs. Whether it is the RNA of the germ plasm, or some other cytoplasmic component associated with it, that determines that cells containing it shall become germ cells remains to be clarified. RNA is likely to be the responsible agent, for it strongly absorbs ultraviolet light, which sterilizes the germ plasm. Wakahara (1978) found that vegetal pole cytoplasm or a $20,000 \times G$ fraction of vegetal material from eggs of *Xenopus laevis* would reverse the sterilizing effect of ultraviolet irradiation in that species also. Moreover, he found that injection of the active fraction into the subcortical cytoplasm at the vegetal pole of nonirradiated eggs caused them to produce significantly more primordial germ cells than normally occur.

In urodeles the cells from which the primordial germ cells subsequently arise were first located in mesoderm that involutes around the ventrolateral lips of the blastopore (Figure 11.55). A distinc-

FIGURE 11.54 Effects of ultraviolet irradiation of the RNA-rich germinal plasm at the vegetal pole in *Rana pipiens*. Dejellied fertilized eggs were lightly compressed under a quartz slide with the vegetal pole uppermost, irradiated with varying dosages of ultraviolet light, released, and allowed to develop. After reaching the swimming tadpole stage, the animals were anesthetized, ventrally dissected and eviscerated, and examined for the presence of primordial germ cells in the gonadal ridges.
A: Unirradiated control; the primordial germ cells are visible as prominent bands in the germinal ridges along the dorsal body wall. B, C: Two larvae that developed from eggs that were irradiated with a dose of 5300 ergs/mm². Note the limited number of germ cells. D: Larva that developed from an egg irradiated with a dose of 15,000 ergs/mm². Germ cells are completely absent from the gonads, as confirmed by histological section. [From L. D. Smith, *Dev Biol* 14:330–47 (1966), by courtesy of Dr. Smith and permission of Academic Press, Inc., New York.]

tinctive cytoplasm microsurgically from some eggs and, in others, subjected the vegetal pole to ultraviolet irradiation. After either treatment the eggs de-

FIGURE 11.55 Localization of primordial germ cells during gastrulation in urodeles. A: Localization of cells with capacity to form primordial germ cells as determined by P. D. Nieuwkoop [*Arch neerl Zool* 8:1–205 (1947)]. B: Scheme of replacing prospective germ cells of mutant (*dd*) white axolotls (right) with those of black (*DD or Dd*) axolotls (left) as carried out by L. D. Smith (1964).

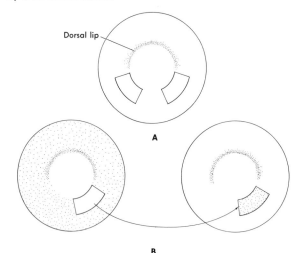

tive germinal plasm is not recognized in cells of this region, but the presence there of prospective germ cells was definitely proved by L. D. Smith (1964), who exchanged the prospective ventrolateral mesoderm between gastrulae of the black form of the Mexican axolotl (*DD* or *Dd*) and its recessive mutant (*dd*) (Figure 11.55B). When the recipient embryos developed to adulthood, it could be shown by means of test matings that their germ cells were of donor type. Thus a gastrula of mutant type (*dd*) receiving a graft of wild-type (*DD* or *Dd*) prospective mesoderm would, at sexual maturity, produce some or all black offspring when mated with the recessive mutant form (*dd*).

When the grafts failed to heal, material from the wound edges would move to the interior in place of the graft, but the gonad on the operated side of the resulting animal would be sterile. Moreover, prospective ectoderm of the gastrula, grafted in place of the ventrolateral mesoderm, although it would invaginate, did not form germ cells. In later experiments Sutasurja and Nieuwkoop (1974) combined pieces of the endoderm cell mass systematically with different portions of the animal hemisphere of the urodele blastula and showed that primordial germ cells can form from any region of the animal half, including regions that would normally never give rise to primordial germ cells. The capacity for germ cell formation increases in two gradients, one toward the ventral peripheral region of the animal hemisphere and another in the direction away from the future dorsal lip of the blastopore. The dorsal portion of the endodermal mass was most effective in eliciting the differentiation of primordial germ cells. Presumably, therefore, in urodeles the interaction of endoderm and cells of the animal hemisphere causes the bilateral localization of prospective germ cells that are then carried by the movements of gastrulation to the ventrolateral blastoporal lips and thence to the interior (see also Ikenishi and Nieuwkoop, 1978). Boterenbrood and Nieuwkoop (1973) showed that prospective ectoderm of the early blastula can form histologically distinctive primordial germ cells when it is used to enrobe a fragment of lateral (preferably) or dorsal endoderm and then cultured in vitro. Possibly, therefore, the power to differentiate as primordial germ cells can be acquired by ectodermal or mesodermal cells in particular association with endoderm at very young stages. But, by the time gastru-

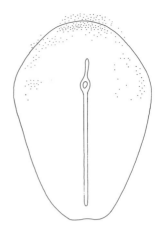

FIGURE 11.56 The distribution of primordial germ cells (stipples) in the germinal crescent in the chick embryo at the head process stage. [Adapted from B. H. Willier, *Anat Rec* 70:89–112.]

lation is well advanced, as in Smith's (1964) experiments, the ability to form primordial germ cells is restricted solely to mesodermal cells that involute around the ventrolateral blastoporal lips, as first determined by Nieuwkoop.

In embryos of birds distinctive large cells are found at the head process stage in a crescent-like zone in the endodermal layer at the anterior margin of the blastoderm. This zone is the **germinal crescent** (Figure 11.56), and the distinctive cells are primordial germ cells. A similar crescent is found in some reptiles. In birds and reptiles the primordial germ cells enter blood vessels (Figure 11.57) and make their way via the bloodstream to the gonads. If the primordial germ cells are destroyed in the crescent by ionizing radiations or by cautery, sterile gonads develop.

In the unincubated blastoderm of a chick or duck the primordial germ cells are apparently widely scattered in the hypoblast, particularly posteriorly. They are not recognizable histologically at that time, but become visibly differentiated after their arrival in the region of the germinal crescent. When fragments of the unincubated chick blastoderm are cultured on artificial media, however, the primordial germ cells become recognizable in a matter of 48 hours (Dubois, 1969). Thus the environment of the germinal crescent is not requisite to the cytological differentiation of the primordial germ cells.

In the region of the germinal crescent the pri-

253

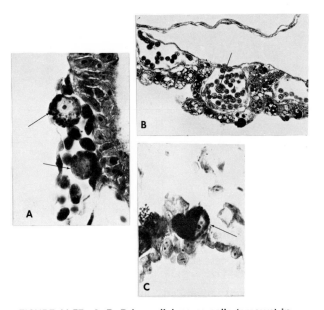

FIGURE 11.57 A, B: Primordial germ cells (arrows) in, respectively, embryonic and extraembryonic blood vessels of the snake-like lizard *Anguis fragilis.* C: The arrow points to a primordial germ cell entering the genital ridge of the snake *Vipera aspis.* [From J. Hubert, *C R Acad Sci* Paris, 266:231–33 (1968), by courtesy of Dr. Hubert and permission of Gauthier-Villars et cie.]

mordial germ cells multiply. Counts made in the duck embryo by Fargeix (1972) reveal that they number 50–70 at the primitive streak stage, increasing to 180 or more by the stage of six somites, and exceeding 300 by the time thirteen pairs of somites are formed. Apparently, the number of primordial germ cells to be produced is determined in the region of the germinal crescent. When a single unincubated duck blastoderm is transected, it often gives rise to twin embryos, and thus two germinal crescents. Rogulska (1968) found that the total number of germ cells produced by the twins is approximately double the number formed when only one embryo arises from a single blastoderm.

In the chick the primordial germ cells enter the blood vessels at the time the area vasculosa forms, leaving the endoderm by amoeboid motion. They are then carried by the blood to the germinal epithelium. The spatiotemporal change in the distribution of the primordial germ cells in the chick embryo as determined by Dubois (1969) is diagrammed in Figure 11.58.

Primordial germ cells apparently are selectively attached to the germinal epithelium. When grafts of chick germinal crescent material are placed in the coelom of another chick embryo previously deprived of its own germinal crescent, primordial

FIGURE 11.58 Scheme depicting the distribution of primordial germ cells in the chick embryo prior to and after laying. Germ cells that are being carried passively by cell movements or that are localized in the gonads are shown in black. Those entering the vascular system or being transported therein are indicated by open circles. The exclusive posterior localization of germ cells in the blastoderm in utero is tentative. Patterns of distribution at later stages are supported by positive evidence. [Adapted from R. Dubois, *J Embryol Exp Morphol* 21:255–70 (1969).]

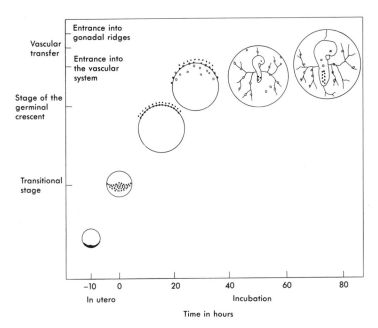

germ cells leave the graft and enter the germinal epithelium of the host. If their entrance is prevented by a barrier, such as a fragment of vitelline membrane, they will dispose themselves along the barrier adjacent to the germinal ridge (Dubois, 1968). Reynaud (1969, 1973) asked whether the attraction of primordial germ cells to the germinal epithelium is species specific. To answer this question he prepared suspensions of primordial germ cells from the germinal crescent of turkey embryos and injected them into the bloodstream of host chick embryos previously sterilized by irradiation of the germinal crescent with ultraviolet light. He reported that germ cells of the turkey colonized the gonads of the host and restored a normal germ cell population. These germ cells formed morphologically mature gametes, but successful breeding of the host birds was not achieved.

The primordial germ cells of mammals are recognizable both by their cytological characteristics, which are similar to those of other germ cells, and by their intense reaction to the histochemical assay for the enzyme alkaline phosphatase. In human beings (Figure 11.59) and other mammals they are first found in the endodermal layer in the region of the base of the allantois. It has been shown in mice that they next enter the wall of the gut, then migrate through the mesentery to the gonads. Fewer than 100 primordial germ cells are formed initially in the normal mouse, but their number increases by mitotic division to about 5000 during migration to the gonads.

Whereas in the case of birds and many reptiles, the germ cells reach the gonads via the circulatory system, mammalian germ cells are like those of amphibians and lower reptiles in that they migrate through the tissues to the germinal ridges.

Will primordial germ cells developing in gonads of the opposite genetic sex form gametes appropriate to the sex of the host? This is certainly true for many fishes, amphibians, and birds (reviewed by Witschi, 1967). In *Rana temporaria*, for example, genetic male (XY) and genetic female (XX) tadpoles develop according to their genetic sex when reared at 20°C. A temperature of 10°C, however, favors the development of the cortex of the gonad, so that all primordial germ cells, whether of XX or XY constitution, form oöcytes, and the tadpoles develop as functional females. When XY females are bred to genetic (XY) males, and the larvae are reared at

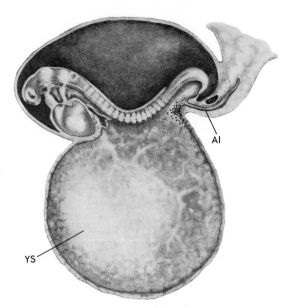

FIGURE 11.59 Origin of primordial germ cells in the human embryo. The location of the germ cells is indicated by black dots in endoderm around the base of the allantois, Al, and in adjoining regions of the yolk sac, YS. [From E. Witschi, *Carnegie Contribs Embryol* 32:67–80 (1948) by courtesy of the late Dr. Witschi and permission of the Carnegie Institution of Washington, Department of Embryology, Davis Division.]

255

20°C, a sex ratio of two females to one male is found in the offspring, for YY individuals are inviable. Thus primordial germ cells of the male genotype form female gametes in the environment of the ovary. Conversely, tadpoles reared at 30°C develop as phenotypic males regardless of genetic sex, and breeding experiments reveal that in this case functional sperm differentiate from primordial germ cells of XX individuals

Eggs of the common snapping turtle *Chelhydra serpentina*, incubated at temperatures above 31°C develop as females. Those held at temperatures between 24 and 27°C become males, and those incubated at temperatures below 24°C differentiate as females. Bull and Vogt (1979) reported these findings and also reviewed cases of temperature-controlled sexual differentiation in a number of species of turtles and of one lizard.

In genetically female birds both gonads arise, but only the left one normally becomes a functional ovary. The right one regresses to a small structure with a predominantly medullary component. In chicks, the primordial germ cells will have disappeared from the rudimentary right ovary by 3 months after hatching. If the functional ovary is removed, however, the right gonad begins to grow, but its medullary component predominates and forms a testis. When left ovariectomy is carried out before the primordial germ cells in the right gonad have degenerated, the right gonad will develop as a testis with mature sperm.

The phenotype of the gonad may not be the determinant of the gametic sex in mammals, however. Recall (Chapter 10) that tetraparental mice may be made by combining early cleavage stages from embryos of two different sets of parents. Obviously when preimplantation mouse blastocysts are united to produce tetraparental mice, 50% of the pairs will consist of male/female chimeras. When both the male and female cells survive, many of the sex chimeras, designated XY/XX, develop as phenotypically normal fertile males whose offspring show the normal sex ratios when bred to normal females. Very few develop as phenotypic females, and a few show signs of hermaphroditism (McLaren, 1975). In 1968 Mintz reported that she had examined almost 30,000 offspring of tetraparental mice mated to normals, and found no deviation from the 1:1 expected sex ratio.

If it is true, as already suggested, that the presence of cells bearing the H-Y antigen organizes the indifferent gonad into a testis, then one would expect that sex chimeras would be predominantly of male type, for XY cells would likely be represented to some degree in the gonad. It would also be likely, however, that some primordial germ cells would be of the XX genotype and would enter the testis. McLaren et al. (1972) studied the fetal gonads in a number of sex chimeras of mice and found prophase meiotic cells in specimens examined at $15\frac{1}{2}$–$17\frac{1}{2}$ days after mating. These cells were almost certainly of XX constitution for they lacked the so-called **sex vesicle**, a configuration of the male sex chromosomes, which is seen in spermatogenesis in normal males. In gonads of embryos at later stages, such cells were found degenerating. In the female mouse, as in the rat and human female (cf. Chapter 7), essentially all oögonia are in meiotic prophase prior to birth, whereas in the male, meiosis is not seen until puberty. McLaren et al. (1972) and McLaren (1980) found that in sex chimeras oöcytes

begin differentiating in gonads at the same time as they do in embryonic gonads of normal mice. Then they die. In sum, therefore, it appears that functional sex reversal of mammalian gametes does not occur. The only germ cells that differentiate are those whose genotypes are appropriate to the phenotypic sex of the gonad.

11.2.3   MAJOR ENDODERMAL DERIVATIVES. The endoderm provides the lining of the digestive tract, the secretory cells of the digestive glands that empty into it, the lungs, a number of endocrine glands, and other important structures that are noted here. The morphogenesis of the digestive tract, however, involves intimate association and interaction of the endoderm with the mesoderm. The mesoderm closely invests the endoderm and provides conditions absolutely essential to its development. Some aspects of this dependency are treated in this chapter. Others are considered in another context in Chapter 14.

We have earlier recognized the endoderm as the innermost tubular component of a three-layered embryo. In teleost fishes, birds, and mammals, the embryonic body is undercut by anterior, posterior, and lateral folds (Figures 11.45 and 11.60) such that the endoderm is contained within rather than beneath the ectoderm and mesoderm, as it was initially. It then remains in greater (birds) or lesser (mammals, such as human beings) communication with the yolk and yolk sac. In frog embryos, as shown in Chapter 9, the endodermal tube is created by the processes of invagination that give rise to the archenteron, or primitive gut. In urodeles, however, recall that the endoderm rises from the floor of the archenteron laterally (Figure 9.15) and then fuses dorsally to form the roof of the gut. In the past it was thought by some that the amphibian archenteron is obliterated in later development and that the definitive gut is created by the opening of a new passage through the yolk mass. This idea was proved to be false by Ballard (1970), who demonstrated that suspensions of carmine particles injected into the archenteron remain to occupy the definitive gut.

11.2.3.1 Origin of the Major Divisions of the Gut in the Chick. The first step in the formation of the gut occurs in the chick embryo after about 24 hours of incubation. At that time no mesoderm intervenes between the ectoderm and endoderm an-

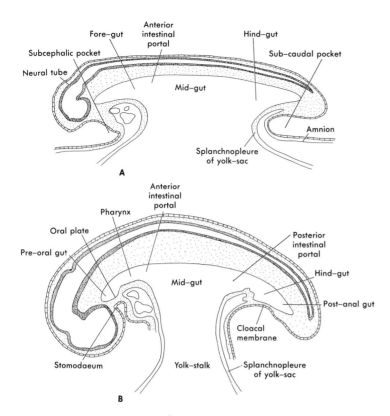

FIGURE 11.60 Schematized longitudinal sections of chick embryos of about 3 (A) and 4 (B) days of incubation, showing regionalization of the gut. The endoderm is represented as a solid line that lines the interior of the gut and combines with mesoderm to form the splanchnopleure.

terior to the apex of the notochord. This zone is thrown into a fold ventrally along a crescentic line, creating a shallow endoderm-lined pocket beneath the notochord and newly arising neural folds of the head region (Figure 11.45). The unfolded pouch is the beginning of the **foregut,** the anterior part of the future alimentary tract. The ventral lip of the fold forms the anterior intestinal portal, whose importance in the formation of the heart has been described. The foregut lengthens and its anterior end fuses with overlying ectoderm to form the **oral plate.** The oral plate eventually breaks through, opening communication between the mouth and pharynx (Figure 11.60B).

After another 24 hours have elapsed, a similar undercutting of the posterior end of the embryo occurs. This undercutting forms the **hind gut,** whose anterior lip is the **posterior intestinal portal.** Partway along its length the endoderm of the hindgut fuses with the ectoderm to form the **cloacal membrane,** which later ruptures. The hindgut pos-

terior to the cloacal membrane is called the **post-anal gut.**

Stalsberg and DeHaan (1968) sought to determine to what extent elongation of the foregut is contributed by the inrolling of yolk-sac endoderm around the fold of the anterior intestinal portal. To do so they removed young embryos at the late head process stage and cultured them ventral side up on a suitable medium. They then carefully positioned fine particles of ferric oxide on the endoderm in the region of the developing anterior intestinal portal. These particles adhered to the endoderm and served as markers for the analysis of the morphogenetic movements involved in elongation of the foregut. These movements, as schematized in Figure 11.61, show that the part of the yolk-sac endoderm nearest the anterior intestinal portal does roll in over the fold, but it contributes only a steadily decreasing fraction of foregut elongation. The major portion of the elongation is contributed by intrinsic elongation of the anterior wall of the first-appearing

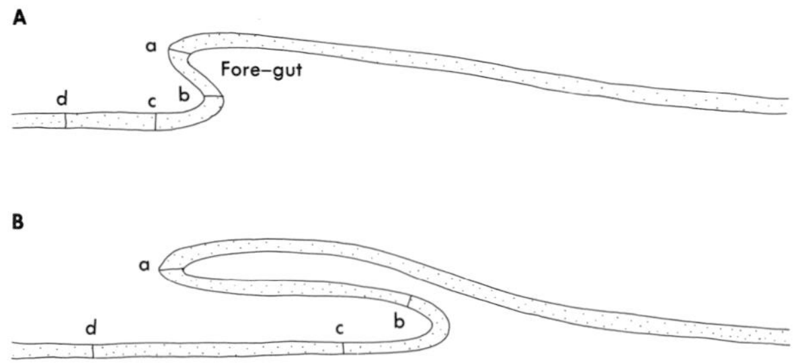

A

B

FIGURE 11.61 Schematic representation of the embryonic endoderm during formation of the foregut. Lettered points indicate marked positions on the endodermal sheet at the head-fold stage (A) and at the stage of about 10 somites (B). The disposition of the marks indicates that the foregut elongates by combined movements of expansion and inrolling. [Adapted with modifications from H. Stalsberg and R. L. DeHaan, *Dev Biol* 18:198–215 (1968).]

fold and the yolk-sac endoderm immediately beneath it.

11.2.3.2 The Derivatives of the Oropharynx.

The combined oral and pharyngeal regions of the embryo are referred to as the **oropharynx.** Often the oral cavity is represented as being lined with ectoderm, but with the breakdown of the oral plate, the boundary between the ectodermal and endodermal lining of the alimentary tract is obliterated. In mammalian embryos epithelial derivatives of the anterior part of the oropharynx, such as the enamel organs of the teeth, are probably all ectodermal in origin. The secretory portions of the salivary glands are also usually regarded as ectodermal derivatives, although their site of origin with respect to the vanished oral membrane is uncertain.

Both teeth and salivary glands originate as solid ingrowths of the oral epithelium onto the subjacent mesoderm. In the case of the tooth, the apex of this ingrowth becomes the goblet-shaped **enamel organ,** which encloses a mass of mesoderm, the **dental papilla.** The inner cells of the enamel organ transform into **ameloblasts,** which secrete the enamel that covers the outer surface of the tooth. Facing the ameloblasts, the **odontoblasts** arise from the dental papilla. The odontoblasts form the dentine which surrounds the pulp. These relationships are diagrammed in Figure 11.62.

There are three pairs of salivary glands in the human embryo (Figure 11.63). Each pair arises as an epithelial bud the end of which branches and

rebranches into a bushwork of solid cords. The ends of these cords round out into the oval secretory parts of the gland called **acini,** and the cords themselves later hollow out to form the ducts. In their growth and branching, the salivary glands invade a dense mesenchyme, which encapsulates the gland and subdivides it into lobules. The sexual dimorphism of the submaxillary salivary gland of the mouse was noted earlier in connection with the discussion of nerve growth factor and epidermal growth factor.

The role of microfilaments and microtubules in the formation of the branching pattern of the submaxillary gland of the mouse was examined by Spooner and Wessels (1972), who cultured the salivary epithelial rudiment in vitro together with its mesenchymal capsule. The new clefts that form at the branch points apparently arise as a result of the contraction of microfilaments (Figure 11.64), for if their organization is disrupted by treatment with cytochalasin B, the initial clefts disappear. They will reorganize into contractile elements at the appropriate locations, however, when the cytochalasin B is washed away. The drug colchicine, which disrupts microtubules, does not prevent re-formation of branch points in such preparations, but it does prevent further morphogenesis. This fact suggests that the microtubules play a role in the stabilization of the pattern of salivary branching, but not its initiation.

Older, well-established points in the branching pattern are not affected by cytochalasin B. Those points are stabilized by a basal lamina rich in glycosaminoglycans. The epithelial rudiments are encompassed by a nearly uniform layer of GAG at their interface with the mesenchyme. Labeling experiments with radioactive glucosamine, a precursor GAG component, shows that the rate of accumulation of newly synthesized GAG is maximal at the morphogenetically active sites (i.e., the buds and lobules that undergo cleft formation). These more active sites appear to host more rapid GAG turnover. Elsewhere, on the surface of established branches and stalks, little GAG turns over (reviewed by Bernfeld et al., 1973). When the glandular epithelium is separated from its mesenchymal envelopment by means of the enzyme collagenase (which hydrolyzes collagen fibers that assist in binding components of the gland), the basal lamina is exposed. If the basal lamina is now removed by

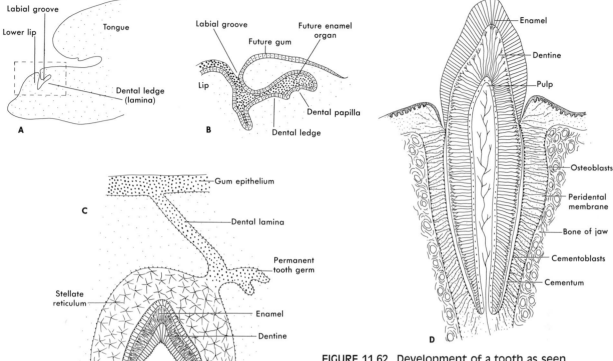

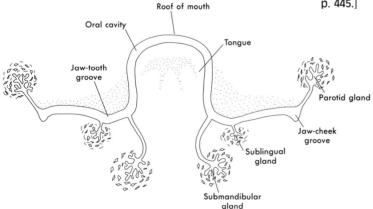

FIGURE 11.62 Development of a tooth as seen diagrammatically in the lower jaw of a human embryo. A, B: Relations of lip, labial groove, and dental lamina in an embryo of about 9 weeks, B representing an enlarged view of the rectangular area bounded by dashed lines in A. C: A human lower incisor at about 7 months. D: The topography of an erupted tooth of the deciduous dentition. (A–C adapted with modifications form L. B. Arey, *Developmental Anatomy*, 7th ed., Saunders, Philadelphia, 1975, pp. 218, 221; D adapted with modifications from B. M. Patten, *Foundations of Embryology*, 2nd ed., McGraw-Hill, New York, 1964, p. 445.]

259

FIGURE 11.63 Diagram of the sites of origin of the salivary glands and their relationships to the tongue and oral cavity as seen in frontal section through the jaws in a human embryo of about 8 weeks. [Adapted with modifications from L. B. Arey, *Developmental Anatomy*, 7th ed., Saunders, Philadelphia, 1965, p. 226.]

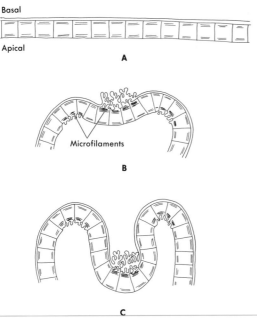

FIGURE 11.64 Scheme of cleft formation during morphogenesis of the salivary gland. A: The site of a future cleft is diagrammed as a flattened sheet of cells bounded basally by mesenchyme and apically by the lumen of the gland. Sets of microfilaments are found at both apical and basal ends of the cells. B, C: Reduction of the surface area as clefts form centrally and at the regions of maximal bending at the lumenal surface. [Adapted with modifications from B. S. Spooner and N.K. Wessels, *Dev Biol* 27:38–54 (1972).]

means of hyaluronidase (an enzyme that hydrolyzes hyaluronic acid, a prominent component of GAGs, as already noted), the epithelium retains its lobular shape and a new lamina is formed. If, however, the epithelium is reassociated with salivary mesenchyme before the lamina is formed it loses its lobular shape within 6 hours, although branching morphogenesis subsequently resumes (Banerjee et al., 1977). This effect of salivary mesenchyme apparently is attributable to the fact that it produces an enzyme that hydrolyzes material of the basal lamina (Banerjee and Bernfeld, 1979). Presumably this enzyme plays a role in the rapid turnover of GAG at the tips of epithelial buds and branches.

The pharynx proper is dorsoventrally flattened (Figure 11.65) and from it five paired pouches, **pharyngeal pouches,** bulge out laterally. These pouches pass through the head mesenchyme to meet corresponding invaginations in the head ectoderm. Here, in lower forms, openings appear such that the pharyngeal cavity now communicates to the exterior. These openings are **gill slits,** through which water exits after entering through the mouth. The mesenchyme between the pouches forms **pharyngeal arches** (Figure 11.66). In fishes and amphibians these arches become bony and fringes of highly vascularized feathery gills develop from them. These gills serve for exchange of respiratory gases.

In higher forms the pouches form, among other things, the eustachian tubes and part of the middle ear, palatine tonsils, thyroid and parathyroid glands, and ultimobranchial bodies. The ultimobranchial bodies are glandular cords that are variously associated with the thyroid and parathyroid glands. The cells of these cords are made up of **calcitonin cells,** which contain large quantities of biogenic amines. In the chick the chief of these amines is **serotonin** (5-hydroxytryptamine); in the quail the principal one is the catecholamine **dopamine** (dihydroxyphenylethylamine), a precursor of epinephrine, which is found in the adrenal medulla, and of the neurotransmitter substance norepinephrine. All of these compounds show a characteristic green fluorescence in properly fixed cells illuminated with ultraviolet light.

Because the calcitonin cells are similar to those of the adrenal medulla, LeDouarin and LeLièvre (1971) suspected that they, too, might be of neural crest origin. To test this possibility they excised a portion of the hindbrain and overlying neural crest cells from a chick embryo of about 36 hours of incubation and replaced it with hindbrain and neural crest of a comparable quail embryo (Figure 11.67). After 2–8 days of further incubation the host chick was killed and prepared for microscopy. The cells of the ultimobranchial bodies contained green fluorescent calcitonin, but examination of their nuclei revealed that they were of quail origin. The epithelium of the fifth pouch is, therefore, simply a target zone for neural crest cells, which infiltrate it and form the true ultimobranchial body (see also LeDouarin et al., 1974).

The epithelium of the floor of the pharynx anteriorly is the source of the thyroid gland (Figure 11.65). In the chick embryo, up to the stage of 13

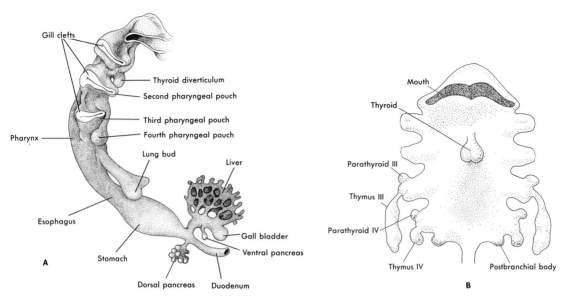

FIGURE 11.65 A: Origin of major endodermal derivatives of the anterior half of the alimentary tract in the embryonic mammal as viewed from the right side. B: Schematic diagram of a ventral view of the pharyngeal endoderm of a mammal showing the pharyngeal pouches and some of their derivatives. The first and second pouches are not labeled. The first pouch contributes to the Eustachian tube and tympanic cavity, the second to the pharyngeal, palatine, and lingual tonsils. Principal derivatives of the other pouches are indicated on the diagram. Note that the post-branchial body, or ultimobranchial body, is derived from the very rudimentary pouch V, which is not visible in A. [A adapted from various sources including G. L. Weller, *Carnegie Contribs Embryol* 24:93–139 (1933), and W. W. Ballard, *Comparative Anatomy and Embryology*, Ronald, New York, 1964, p. 463. B adapted with modifications from B. F. Patten, *Foundations of Embryology*, 2nd ed., McGraw-Hill, New York, 1964, p. 473.]

somites, all the cells of the floor of the pharynx are similar in appearance, being cuboidal and loosely joined to each other. Shortly after that stage, a patch of cells that is different from the rest appears at the level of the second arches. As reported by Shain et al. in 1972, they are packed together with their lateral surfaces tightly apposed, and they show extensive apical blebbing into the pharyngeal lumen (Figure 11.68A, B). Next, these cells become pseudostratified and, in a few hours, form a cup-shaped vesicle extending below the pharyngeal

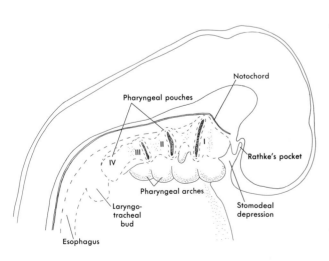

FIGURE 11.66 Diagrammatic view from the right side of a mammalian embryo to show the relationships of pharyngeal arches and pouches. [Adapted with modifications from B. M. Patten, *Foundations of Embryology*, 2nd ed., McGraw-Hill, New York, 1964. p. 472.]

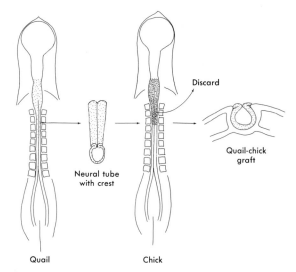

Quail   Chick

FIGURE 11.67   Scheme depicting the replacement of the rhombencephalic region of the chick neural tube and crest with the comparable region of the quail. [Adapted from N. LeDouarin and C. LeLièvre, *Bull Assn Anatomistes*, 56ᵉ Congrès, No. 152:558–68 (1971).]

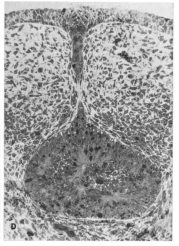

floor. Bundles of microfilaments are present in the cell apices, presumably contributing by their contraction to the formation of the vesicle (see also Smuts et al., 1978).

Constriction of the mouth of the vesicle closes it off (Figure 11.68C), and the vesicle remains connected by a stalk to the floor of the pharynx. The stalk disappers (Figure 11.68D), and the vesicle becomes bilobed and is invaded by mesenchyme, which divides the endoderm into characteristic lobules. Later these hollow out to form the **thyroid follicles.**

Of particular importance to students of embryonic differentiation is the fact that the thyroid hormones, triiodothyronine and thyroxine (Section 17.4.1.3) are detectable in the future thyroid cells from the moment their palisading begins. The genes responsible for enzymes involved in the synthesis of these iodinated amino acids are, therefore, expressed long before the thyroid cell is in its final cytodifferentiated state. A similar phenomenon is seen in the pancreatic rudiment, as described in Section 11.2.3.5.

The lung bud evaginates ventrally from the floor of the pharynx between the fourth pharyngeal pouches (Figure 11.65A). It is surrounded by a mesenchymal capsule. The proximal end of the bud forms the trachea, and its distal end branches to form two **primary bronchi.** From the primary bron-

FIGURE 11.68   The origin of the thyroid rudiment as seen in transverse sections through the pharynx of the chick embryo. A: The floor of the pharnyx prior to the initiation of thyroid morphogenesis as seen at 40–45 hr of incubation. The ventral pharynx consists of a single layer of loosely joined cuboidal cells. B: The same region about 5 hr later. The thyroid primordium now consists of closely packed pseudostratified cells that show blebbing at their apical ends. C: At about 3½ days of incubation the thyroid primordium is a vesicle, now separated from the floor of the pharynx except for a fine channel that remains in the connecting stalk. D: At 4–4½ days of incubation, the thyroid stalk is parted as the vesicle sinks deeper into the encapsulating mesenchyme. [From W. G. Shain et al., *Dev Biol* 28:202–218 (1972), by courtesy of Dr. S. R. Hilfer and permission of Academic Press, Inc., New York.]

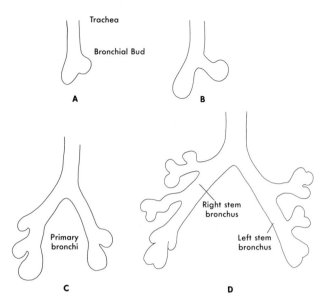

FIGURE 11.69   A–D: Early stages in the development of the right and left bronchi in human lungs as seen in ventral view. The mesenchymal stroma into which budding occurs has been omitted. [Adapted from various sources, including B. F. Patten and B. M. Carlson, *Foundations of Embryology*, 2nd ed., New York, McGraw-Hill, 1974, p. 471.]

chi successive buddings give rise to the **bronchioles** and **air sacs** of the lungs (Figure 11.69). The capsular mesoderm apparently determines the pattern of growth and branching of the endodermal lung primordium. Alescio and Cassini (1962), working on lung buds cultured in vitro, found, for example, that the trachea forms an extra bronchiolar bud if a bit of its covering mesoderm is removed and replaced by mesoderm from later-produced pulmonary buds. Simple excision and replacement of the tracheal covering mesoderm do not have this effect. Neither does supernumerary budding follow replacement of mesoderm of the tracheal region with mesoderm from nonpulmonary sources. These results were confirmed and extended by Wessels (1970), who showed that bronchiolar budding is suppressed if the bronchial covering of mesenchyme is replaced by mesenchyme from the tracheal region.

The thymus (Figure 11.65B) is a pharyngeal derivative which, like the ultimobranchial bodies, has as its effective component a cell population of other than endodermal origin. In it are found thymus-dependent lymphocytes, or "T" cells, that combat viral and fungal infections and that initiate the rejection of tumors and foreign tissues. Using quail marker cells, LeDouarin and Jotereau (1973) sought to determine whether the thymocytes arise from pharyngeal epithelium or have their origin elsewhere. They grafted the third and fourth pharyngeal pouches, free of mesoderm, from quail embryos of 2–3 days of incubation into the splanchnopleure of a 3-day chick embryo and allowed it to develop there until the host was incubated for a total of 14 days. They then examined the grafts histologically, finding that, whereas the reticular and connective tissue framework of the gland was made up of quail cells, the thymocytes were all of chick origin. Apparently they arose from blood-born stem cells originating elsewhere.

The mesoderm of the pharyngeal arches undergoes considerable modification in higher animals. A notable feature is the origin of **maxillary** and **mandibular processes** from the first arch (Figure 11.66), which form the skeleton, bones, teeth, and skin of the upper and lower jaws, respectively. The maxillary process gives rise to medial and lateral **palatine processes.** The fusion of these processes forms the bone of the roof of the mouth, the **hard palate.** The formation of this structure was noted above in connection with the discussion of the epidermal growth factor. The two outermost bones of the middle ear, the **malleus** (hammer) and **incus** (anvil), are formed from material of the mandibular process, and the innermost bone, the **stapes** (stirrup), is formed from mesoderm of the second arch. The malleus, incus, and stapes are the bones that transmit sound-induced vibrations of the tympanic membrane (eardrum) to the inner ear where the vibrations are transduced into nerve impulses. Arches II, III, IV, and V all contribute mesoderm to the **hyoid bone** and associated cartilages and ligaments of the **larynx** ("voice box"). Ectoderm of these arches and associated mesoderm form the integument of the neck.

11.2.3.3   The Digestive Tube.   Posteriorly to the pharynx, the esophagus, stomach, and intestine are formed. The glandular linings of these tubular organs are of endodermal origin, but their muscular walls, their connective tissues, blood vessels, mesenteries, and coelomic epithelial coverings originate from the splanchnic mesoderm. In the early embryo the digestive tract forms a relatively simple tube

263

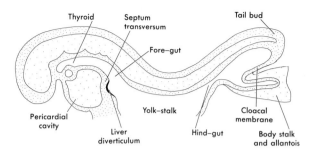

FIGURE 11.70 Highly schematic diagram of a longitudinal section through a human embryo with attention directed particularly to the liver diverticulum and septum transversum (see text).

(Figure 11.70) that follows the curvature of the neural tube and is open to the yolk sac. Its subsequent development is of a remarkable complexity, involving, as it does, elongation to many times the body length, complex foldings to accommodate to the coelomic space and the mass of the other visceral organs, and correlated regional structural and functional differentiations along its length. These events have rarely been the subject of experimental analysis, but are well described in textbooks on development anatomy, such as those of Arey (1965) and of Patten and Carlson (1974). In the remainder of this section we are concerned chiefly with the origin of the major visceral glands of endodermal origin, the liver and pancreas, which have received a good deal of attention from experimentalists and which are considered in other contexts later in this work.

11.2.3.4 Origin of the Liver. The liver arises as a diverticulum of the endoderm at the fold of the anterior intestinal portal. This part of the endoderm is the floor of the future duodenum. From here the endodermal evagination pushes into a mass of mesoderm, the **septum transversum,** which lies between the fold of the anterior intestinal portal and the pericardial cavity (Figure 11.70). The anterior portion of the liver diverticulum forms the glandular tissue of the liver and the common bile ducts, and the posterior portion gives rise to the **gall bladder** (Figure 11.65A) and its duct.

As soon as the liver diverticulum appears, its cranial portion buds off solid epithelial cords that penetrate the **septum transversum** (Figure 11.65A and 11.70), proliferating there into a rapidly ex-

panding spongework heavily infiltrated by blood vessels. The eventual result is a pattern of plate-like epithelial formations enclosing blood sinuses. The liver mass enlarges rapidly, pushing posteriorly into the body cavity as two large lobes, one on either side of the midline. The greater part of the liver is made up of endoderm. Mesodermal cells—largely reticuloendothelial cells, which serve a phagocytic function—line the blood sinuses. Other mesodermal components are the blood vessels and the connective tissue that separates the liver into its characteristic lobular structure.

The location of the prospective endodermal and mesodermal components of the liver in the early chick embryo was determined by LeDouarin in 1964. Her experimental methods involved localized destruction of tissue by means of x-irradiation, marking cell groups with finely divided particles of carbon, and grafts of putative liver endoderm to various regions of the mesoderm. She found that the future liver diverticulum lies exactly in the fold of the anterior intestinal portal (Figure 11.71A) in the embryo of 15 somites. The mesodermal components of the liver, meanwhile, occupy fan-shaped regions of the splanchnopleure on either side of the midline. These regions converge toward the midline as the head fold and lateral body folds deepen, and move into the zone between the prospective liver and heart by the stage when 21 or 22 somites are present. At that stage the liver outpocketing begins and its epithelial buds penetrate the mesenchyme, creating the liver cords.

If, at the 15-somite stage, one places a mechanical block (e.g., a piece of shell membrane from the chick egg) in the blastoderm on one side of the midline (Figure 11.71B), the future liver mesoderm of that side cannot be penetrated by the epithelial cords. The mesoderm will, nevertheless, form reticuloendothelial cells, which will ingest dye particles upon test, and it shows high activities of the enzyme **cholinesterase,** a characteristic liver enzyme. In contrast, if prospective liver endoderm is excised from another embryo and placed posterior to the block, it will populate the splanchnopleure with typical liver cords (Figure 11.71C).

11.2.3.5 Origin of the Pancreas. The pancreas arises from dorsal and ventral outpocketing of the endoderm in close association with the liver diverticulum. In most animals, these rudiments subsequently fuse and their secretions empty by a

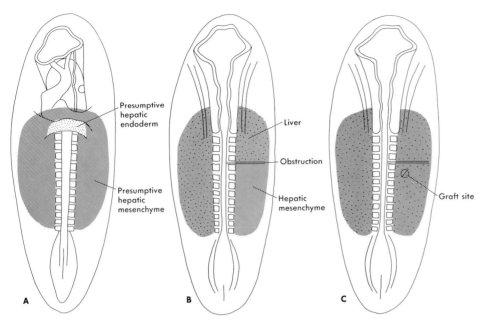

FIGURE 11.71 Dual origin of the functional liver as determined experimentally.
A: Localization of prospective hepatic endoderm and hepatic mesenchyme, respectively, as determined by marking and X-irradiation experiments. B: Insertion of a barrier through the germ layers posterior to the endodermal primordium prevents the epithelial liver cords from penetrating the hepatic mesenchyme. C: Development of epithelial and mesenchymal components posterior to the barrier when a graft of the endodermal component is inserted. [Adapted from N. LeDouarin, *J Embryol Exptl Morphol* 12:141–60 (1964).]

single duct into the common bile duct (cf. Figure 11.65).

The origin of the dorsal pancreas has been studied intensively by Wessels and his collaborators (Wessels and Cohen, 1967; Wessels and Evans, 1968) and by Rutter and his associates (Rutter et al., 1978). The dorsal pancreas in the mouse arises as a bulge of the gut endoderm at about $8\frac{1}{2}$ days (Fig-

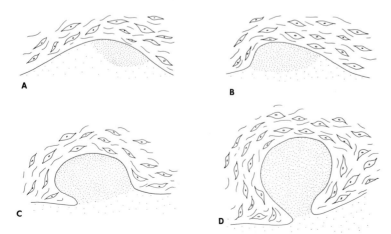

FIGURE 11.72 A–D: Successive stages in the development of the dorsal pancreatic rudiment in the mouse embryo beginning at the ninth day and extending from the 18-somite stage through the 30-somite. The stomach is to the right and the intestine is to the left. [Adapted from N. K. Wessels and J. Evans, *Dev Biol* 17:413–46 (1968).]

ure 11.72A). Already, at this time, distinctive **insulin**-producing cells, beta cells, are recognizable here and there. They will later be found in masses called the **islets of Langerhans** lying between the pancreatic ducts and acini. Other endocrine cells of the islets, alpha cells, which produce **glucagon,** are not seen until after birth.

As the pancreatic rudiment grows out, it develops a stalk and becomes encapsulated in mesenchyme (Figure 11.72B, C, D). Thereafter, the original rudiment begins to form secretory acini, which arise as terminal and side buds of the original rudiment. The acinar cells then begin to show **secretory granules,** the storage form of digestive enzymes later released to the duodenum where they catalyze the breakdown of foodstuffs.

Even before the secretory granules appear, and before the glandular and duct portions of the gland are visibly differentiated, many of the characteristic pancreatic enzymes may be detected in the pancreatic rudiment by means of sensitive biochemical tests (Figure 11.73). It would appear, therefore, that the genes responsible for the synthesis of proteins characteristic of the differentiated state of the pancreas are active in its cells prior to morphogenesis of the gland (Spooner et al., 1977).

266

11.2.3.6 Bursa of Fabricius. An important element in the development of the immune system in birds is the bursa of Fabricius, so called after the anatomist Hieronymus Fabricius of Aquapendente, who described it in the sixteenth century. This element is an endodermal derivative of the roof of the hindgut (Figure 11.74). Stem cells from the embryonic circulation enter it and become lymphocytes known as **B cells.** These are indistinguishable in appearance from thymus-dependent lymphocytes (T cells) described in Section 11.2.3.2. Only B cells, however, can then move to the spleen and become **plasma cells,** which produce **immunoglobulins,** the bloodborne antibodies that combat bacterial infections. If the bursa is prevented from developing in young birds, they have no plasma cells and make no humoral antibodies.

In the chick embryo the first indication of the origin of the bursa occurs at around 66 hours of incubation (Boyden, 1922). At this time the epithelium of the roof of the gut in the region of the future cloaca degenerates, allowing mesenchyme to be exposed to the gut cavity. Endodermal cells in-

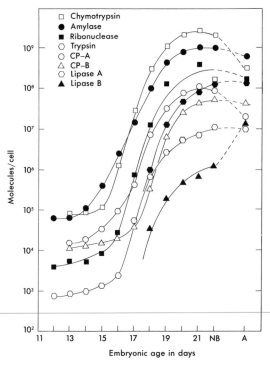

FIGURE 11.73 Developmental profiles of exocrine enzymes specific to the pancreas. Note that an initial period of low and relatively constant specific activity is followed by a rapid rise and then by a leveling-off around the time of birth. In the newborn mouse, NB, the enzymes show different patterns of change in activity (dashed lines). CP–A and CP–B signify procarboxypeptidase A and procarboxypeptidase B activity, respectively. [From data garnered from a number of sources and presented in slightly different form by W. J. Rutter et al., *J Cell Physiol* 72(Suppl 1):1–18 (1968).]

vade the mesenchyme, eventually creating a complex, much-folded gland-like structure that is connected to the cloaca by a narrow lumen. This structure is the bursa of Fabricius. An influx of basophilic stem cells begins during the eighth day of incubation and continues until at least the 15th day. Their numbers increase thereafter by mitosis. In the early stages of bursal development the stem cells remain in the mesenchyme associated with the epithelial folds. Only those that enter the epithelium differentiate as lymphocytes (Houssaint et al., 1976). In birds the bursa is absolutely the only

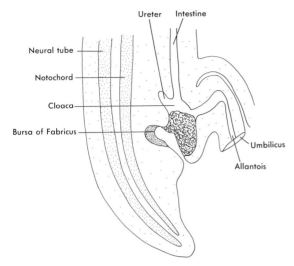

FIGURE 11.74 Origin of the Bursa of Fabricius from the roof of the gut in the chick embryo. [Adapted with modifications from LeDouarin et al., *Proc Nat Acad Sci USA*, 72:2201–05 (1975).]

source of B cells and, hence, of plasma cells. Embryos treated with appropriate doses of testosterone do not form a bursa and have no B cells.

Mammals have B cells, but they do not have a bursa of Fabricius. What induces stem cells to become B cells in mammals is uncertain. Individuals are sometimes born lacking B cells, as reflected in their inability to make circulating immunoglobulins. Such individuals may, however, have T cells which, as noted earlier, differentiate in the thymus (reviewed by Cooper and Lawson, 1974).

11.3 Summary

The molding of the germ layers into the form of organs and organ systems involves a number of morphogenetic processes. Prominent among these are epithelial foldings, including evaginations and invaginations, usually preceded by palisading of cells. These changes involve the contraction of cytoplasmic microfilaments at exposed cell surfaces and the appearance of microtubules preferentially oriented parallel to the long axes of the cells.

The formation and subsequent hollowing of cords of cells, cellular condensations, detachments, and fusions of cellular sheets and aggregates are familiar morphogenetic processes, as are the programmed death of cells and cellular proliferation. Migration of individual cells from source to target areas occurs; this migration involves changes in the arrangement of the microtubules and microfilaments of the migrating cells. Their shapes, moreover, are affected by the character of the substrates over which they move. Cellular locomotion on planar surfaces in tissue culture usually involves a broad ruffled leading lamella behind the edge of which points of very close contact of the cell to its substrate are formed. These zones are plaques, made up of actin filaments and amorphous material. Bundles of microfilaments extend into the cytoplasm away from each plaque toward the rear of the cell. The plaques do not move, but new ones form near the margin of the advancing lamella and those to the rear fade away. Some plaques appear to make tight attachments at the rear of the cell so that, in locomotion, the cell is attenuated. Tensile force, apparently generated by the microfilaments, then causes the cell to detach posteriorly, snapping forward and leaving a part of its membrane behind. This type of locomotion possibly does not occur under normal conditions in vivo. When conditions permit observations in vivo, locomotion usually involves the extension and contraction of long filopodia or, in other cases, the protrusion of blunt lobopodia and cytoplasmic flow.

The major ectodermal derivative is the neural tube, which gives rise to the central nervous system and its derivatives. The neural tube arises from the neural plate. The edges of the neural plate, which consist of palisaded cells, form folds that meet at the dorsal midline. This process involves the contraction of microfilaments at the side of the cells facing the future lumen of the neural tube and the reordering of their microtubules. The neural tube is at first a columnar epithelium whose nuclei migrate to the internal limiting membrane for mitosis. Eventually cells cease mitosis and take up positions laterally to form the mantle layer. Axons and dendrites form the marginal layer toward the external limiting membrane of the neural tube. Neuroblasts are variously displaced, immobilized, or undergo degeneration, creating regional differences in the neu-

267

ral tube along its length. The occurrence of death and proliferation of neuroblasts in both the neural tube and spinal ganglia is correlated with the peripheral load.

Neural crest cells arise at the corners of the folding neural plate. In the head region they form the cranial ganglia (with contributions from ectodermal placodes), all cartilages and bones of the facial and oral region, most connective tissue in the head, odontoblasts, and calcitonin cells. At the trunk level crest cells form the melanocytes, sensory ganglia, all ganglia of the autonomic nervous system and their glia, Schwann cells, and the adrenal medulla. Neural crest cells, for the most part, show considerable lability in the sense that, regardless of their source along the long axis of the body, they differentiate, after experimental exchange, in accordance with their new location.

Sympathetic neuroblasts and large ventrolateral neuroblasts of the spinal ganglia require for differentiation a protein called nerve growth factor, or NGF. Interaction of the factor with receptors on the cell surface induces outgrowth of neurites, but the factor must be internalized and bound to nuclear chromatin in order for the cells to produce their neurotransmitter substance. A variety of tissues and cells in tissue culture have been shown to produce NGF. It is most readily isolated, however, from the submaxillary glands of intact adult male mice where it is secreted by the hypertrophied (under the influence of testosterone) tubular elements of the gland.

Ectodermal placodes give rise to the lens of the eye, middle ear, the sensory epithelium of the nose, and they cooperate with the neural crest cells in forming some of the cranial ganglia.

Neurites growing out from nerve cell bodies are guided to their target organs in unknown fashion, perhaps by subtle chemical gradients. For some kinds of nerves, the first neurite to reach the target organ may be considered a pioneering neurite, whose surface then provides a selective pathway followed by succeeding waves of neurites. The guidance of neurites has been studied extensively in the regenerating optic nerves of fishes and amphibians. It now appears that during embryogenesis, neurites from ganglion cells of the retina grow into the retina in a constant and orderly manner such that they reach appropriate areas of the optic tectum in the midbrain. There they apparently stimulate the for-

mation of regionally specific markers such that, when the optic nerve is severed, regenerating axons reestablish synaptic contact with the same cells of the tectum with which they were originally in contact or, at least, with cells in the same general vicinity.

The remaining major derivatives of the ectodermal layer are the integument and associated structures. Ectoderm forms the epidermal layer of the skin and, in cooperation with mesoderm, forms hairs, feathers, scales, and so on. Maturation and keratinization of the skin are accelerated by an epidermal growth factor (EGF), a protein found in various tissues, especially in the salivary glands of male mice, as in the case of the nerve growth factor.

The chief divisions of the mesoderm are the segmental plate, the intermediate mesoderm, and the lateral mesoderm. The segmental plate is divided into a linear array of somites parallel to the neural tube on each side. These somites contribute to the axial skeleton, trunk and appendicular muscles, and dermis of the back. The intermediate mesoderm is the source of the urogenital system. The functional kidney of the embryo, the mesonephros, forms from this layer and projects into the body cavity. On the medial face of the mesonephros the primorida of the gonads form. The primorida are not the source of the primordial germ cells, but receive them by immigration from elsewhere, usually a source in the endoderm.

The germ cells may be selectively destroyed by x-rays or ultraviolet treatment, resulting in the formation of sterile gonads. They also may be transplanted between individuals of different genotype, and will breed true to their original source. The origin of germ cells is often associated with the presence of ribonucleoprotein-rich zones of cytoplasm, or germinal plasm. Cleavage nuclei that are sequestered into cells containing the germinal plasm become the primordial germ cells.

The lateral mesoderm is first proliferated as a single plate that later splits into somatic and splanchnic layers. The somatic layer forms the muscles of the body wall and the appendages, thoracic and appendicular skeleton, and the peritoneum of the body cavity. The splanchnic mesoderm invests the viscera and forms the mesenteries. Anteriorly two regions of preheart mesoderm converge from bilateral origins in the splanchnopleure, uniting anterior to the fold of the anterior intestinal portal

268

to form the heart. The endothelial blood vessels and the cellular elements of the blood likewise arise from the splanchnic mesoderm. The red blood cells arise in two lines, a primitive, or fetal, line from blood islands in the yolk sac, and a definitive, or adult, line from the liver. The adult line later is proliferated from the bone marrow. Fetal and adult hemoglobins have distinctive chains of polypeptides, which are produced on relatively stable messenger RNAs.

The limbs arise as bulges of the body wall rimmed by a thickening of ectoderm, the apical ectodermal ridge (AER). The mesodermal component of the limb is of dual origin. The skeleton and all connective tissues of the limb arise from the mesoderm of the lateral body wall, but the muscles originate from cells that migrate from the somitic mesoderm into the body wall. The histological differentiation of the limb proceeds in proximodistal sequence. This orderly progression requires the presence of the AER. If it is excised, terminal limb parts are missing. Localized regions of cellular degeneration are important in shaping the contours of the limbs.

The endoderm provides the glandular lining of the digestive tract and the secretory portion of the various glands that discharge their secretions into the digestive tract. It is intimately associated with mesoderm in the construction of all the organs associated with the tract.

From pharyngeal pouches the eustachian tubes, thymus, parathyroids, and ultimobranchial bodies develop. The thymocytes, however, are of mesodermal origin, and the calcitonin cells of the ultimobranchial bodies originate from the neural crest. The thyroid gland and the lungs arise by evagination from the floor of the pharynx. The characteristic hormone of the thyroid gland can be detected in the floor of the pharynx before the palisaded cells of the thyroid rudiment evaginate into the subjacent mesoderm.

The esophagus and stomach form anterior to the fold of the anterior intestinal portal. The liver diverticulum originates from the endoderm of this fold and sends epithelial cords into the mesodermal septum transversum. The epithelial cords then form plates of cells arranged about sinusoids lined with reticuloendothelial cells of mesodermal origin.

11.4 Questions for Thought and Review

1. Recall as many cases as you can in which the initiation of an organ rudiment involves the palisading and subsequent folding, evagination or invagination of a cell layer. What subcellular organelles are likely involved in palisading and folding? How do they affect these processes?

2. Changes in the shape of neural plate cells seem to play a role in the formation of the neural tube. What accompanying changes might be expected to occur in ectodermal cells of regions adjoining the neural plate?

3. Give as many examples as you can of organs or tissues the structure and function of which depend on immigration of cells from other regions of the embryo.

4. Of what evolutionary significance might be the fact that the central nervous system and organs of special sense originate in ectoderm?

5. What is a lamellipodium? What is its relationship to a ruffled membrane?

6. What roles might be played by microfilaments and microtubules in cellular locomotion?

7. What is the relationship between a cell's shape and the macromolecular configuration of its substratum?

8. What are the major kinds of mechanical processes involved in shaping the embryo? Give examples of their occurrence in morphogenesis.

9. Compare and contrast the origin of the neural tube in embryos of urodeles, anurans, and birds.

10. What are neural crest cells? Where do they originate? To what tissues and organs do they contribute?

11. Are neural crest cells from different regions of the neural tube predetermined for their eventual developmental fates? Bring experimental evidence to bear on this question.

12. What is the evidence for a dual origin of neuroblasts in certain cranial ganglia? What alternative explanation might be advanced?

13. What is a pioneering nerve fiber. Nerve fibers usually terminate in specific end organs. Do you suppose that a motor nerve from the spinal cord might be persuaded to make connection with a muscle other than the one in which it usually terminates? What experiment might you devise to answer this question?

14. What is the relationship between the regression of the primitive streak in the chick embryo and the origin of the anterior somites? What is the relationship of the neural plate to these somites? Why is it reasonable to say that the somites are of epithelial origin?

269

15. Describe the involvement of the anterior intestinal portal in the origin of the heart.

16. Distinguish among the pronephros, mesonephros, and metanephros.

17. Compare and contrast the origin of the male and female gonads in higher vertebrates.

18. Compare and contrast the origin and migration of primordial germ cells in birds, anurans, and mammals.

19. What is the relationship between the cytodifferentiation of the primordial germ cells in the bud and their location in the early blastoderm?

20. What is meant by the term germ cell determinant? In what organisms have germ cell determinants been found? How has their function been proved?

21. What is the source of cells for the foregut in the avian embryo? How was this determined experimentally?

22. What is the oropharynx and what adult structures are derived from it in higher vertebrates?

23. How did LeDouarin identify the source of cells for the calcitonin cells found in the ultimobranchial bodies?

24. Describe the origin of the thyroid gland. Of what theoretical significance is the detection of thyroxine in cells of the pharyngeal floor prior to the formation of the thyroid vessels?

25. What is the origin of the thymus? What is its function in the immune system? What is the source of the lymphocyte population of the thymus?

26. Describe the origin of the liver and give the experimental evidence for the separate functions of its endodermal and mesodermal components.

27. Describe the morphogenesis of the pancreas. What is the evidence that the genes responsible for production of the pancreatic enzymes are active prior to morphogenesis of the gland?

28. Describe the origin of the limb in avian embryos. What is the source of muscle cells in the limb? What experiments bear on your answer?

11.5 Suggestions for Further Reading

BEAMS, H. W., and R. G. KESSEL. 1974. The problem of germ cell determinants. *Int Rev Cytol* 39:413–79.

BERNFELD, M. R., and N. K. WESSELS. 1970. Intra- and extracellular control of epithelial morphogenesis. In M. Runner, ed. *Changing Synthesis in Development.* New York: Academic Press, pp. 195–249.

BLACKLER, A. W. 1970. The integrity of the reproductive cell line in the amphibia. *Curr Top Dev Biol* 5:71–88.

CHAN, L.-N., M. WIEDMANN, and V. M. INGRAM. 1974. Regulation of specific gene expression during embryonic development: synthesis of globin messenger RNA during red cell formation in chick embryos. *Dev Biol* 40:174–85.

COULOMBRE, A. J. 1969. Regulation of ocular morphogenesis. *Invest Opthalmol* 8:25–31.

COUNCE, S. J. 1965. Developmental morphology of polar granules in *Drosophila* including observations on pole cell behavior and distribution during embryogenesis. *J Morphol* 112:129–45.

CZOLOWSKA, R. 1969. Observations on the origin of the ''germinal cytoplasm'' in *Xenopus laevis. J Embryol Exp Morphol* 22:229–51.

DEHAAN, R. L. 1968. Emergence of form and function in the embryonic heart. In M. Locke, ed. *The Emergence of Order in Developing Systems.* New York: Academic Press, pp. 208–50.

FRAZIER, W. A., R. R. ANGELETTI, and R. A. BRADSHAW. 1972. Nerve growth factor and insulin. *Science* 176:482–87.

GAZE, R. M., and P. GRANT. 1978. The diencephalic course of regenerating retinotectal fibers in *Xenopus* tadpoles. *J Embryol Exp Morphol* 44:201–16.

HAMBURGER, V. 1962. Specificity in neurogenesis. *J Cell Comp Physiol* 60(Suppl. 1): 81–92.

HEATH, J. P., and G. A. DUNN. 1978. Cell to substratum contacts of chick fibroblasts and their relation to the microfilament system. A correlated interference-reflexion and high-voltage electron-microscope study. *J Cell Sci* 29:197–212.

HUGHES, A. F. 1968. *Aspects of Neural Ontogeny.* London: Logos.

JACOBSON, M., and R. K. HUNT. 1973. The origins of nerve cell specificity. *Sci Am* February:26–35.

MAHOWALD, A. 1971. Origin and continuity of polar granules. In J. Reinert and H. Ursprung, eds. *Origin and Continuity of Cell Organelles.* New York: Springer-Verlag, pp. 159–69.

MONTAGNA, W., and W. C. LOBITZ, eds. 1964. *The Epidermis.* New York: Academic Press.

OHNO, S., U. TETTENBORN, and R. DOFUKU. 1971. Molecular biology of sex differentiation. *Hereditas* 69:107–24.

O'RAHILLY, R. 1966. The early development of the eye in staged human embryos. *Carnegie Contrib Embryol* 625(38):1–42.

O'RAHILLY, R. 1971. The timing and sequence of events in human cardiogenesis. *Acta Anat* 79:70–75.

O'RAHILLY, R. 1978. The timing and sequence of events in the development of the human digestive system and associated structures during the embryonic period proper. *Anat Embryol* (Berl) 153:123–36.

270

O'RAHILLY, R., and E. GARDNER. 1971. The timing and sequence of events of the development of the human nervous system during the embryonic period proper. *Anat Embryol* (Berl) 134:1–12.

O'RAHILLY, R., and E. GARDNER. 1975. The timing and sequence of events in the development of the limbs in the human embryo. *Anat Embryol* (Berl) 148:1–23.

PAPACONSTANTINOU, J. 1967. Molecular aspects of lens cell differentiation. *Science* 156:338–46.

PRATT, R. M., and R. M. GREENE. 1976. Inhibition of palatal epithelial cell death by altered protein synthesis. *Dev Biol* 54:135–45.

PRICE, D. 1972. Mammalian conception, sex differentiation and hermaphroditism viewed in historical perspective. *Am Zool* 12:179–91.

RAWLES, M. E. 1960. The integumentary system. In A. J. Marshall, ed. *Biology and Comparative Physiology of Birds,* Vol. 1. New York: Academic Press, pp. 190–240.

RAWLES, M. E. 1965. Tissue interactions in the morphogenesis of the feather. In A. J. Lynn and B. F. Short, eds. *Biology of the Skin and Hair Growth.* Sydney: Angus and Robertson, pp. 105–28.

SAUNDERS, J. W., JR. 1958. Inductive specificity in the origin of integumentary derivatives in the fowl. In W. D. McElroy and B. Glass, eds. *A Symposium on the Chemical Basis of Development.* Baltimore: Johns Hopkins Press, pp. 239–53.

SAUNDERS, J. W., JR. 1972. Developmental control of three-dimensional polarity in the avian limb. *Ann NY Acad Sci* 193:29–42.

SENGEL, P. 1965. Le développement de la peau et des phanères chez l'embryon de poulet. *Rev Suisse Zool* 72:570–77.

SPOONER, B. S. 1975. Microfilaments, microtubules, and extracellular materials in morphogenesis. *Bioscience* 25:440–51.

SZABO, G. 1967. The regional anatomy of the human integument with special reference to the distribution of hair follicles, sweat glands and melanocytes. *Philos Trans R Soc Lond* [B]252:447–85.

VASAN, N. S., and J. W. LASH. 1979. Monomeric and aggregate proteoglycans in the chondrogenic differentiation of embryonic chick limb buds. *J Embryol Exp Morphol* 49:47–59.

WATTERSON, R. L. 1942. The morphogenesis of down feathers with special reference to the developmental history of melanophores. *Physiol Zool* 55:234–59.

WEISS, P. A. 1972. Neuronal dynamics and axonal flow. V. The semisolid state of the moving axonal column. *Proc Natl Acad Sci* USA 69(3): pp. 620–23.

YAMADA, K. M., B. S. SPOONER, and N. K. WESSELS. 1970. Axon growth: roles of microfilaments and microtubules. *Proc Natl Acad Sci* USA 66:1206–12.

11.6   References

ABERCROMBIE, M., J. E. HAEYSMAN, and S. M. PEGRUM. 1970a. The locomotion of fibroblasts in culture. I. Movements of the leading edge. *Exp Cell Res* 59:393–98.

ABERCROMBIE, M., J. E. HAEYSMAN, and S. M. PEGRUM. 1970b. The locomotion of fibroblasts in culture. II. "Ruffling." *Exp Cell Res* 60:437–44.

ABERCROMBIE, M., J. E. HAEYSMAN, and S. M. PEGRUM. 1970c. The locomotion of fibroblasts in culture. III. Movements of particles on the dorsal surface of the leading lamella. *Exp Cell Res* 62:389–98.

ABERCROMBIE, M., J. E. HAEYSMAN, and S. M. PEGRUM. 1971. The locomotion of fibroblasts in culture. IV. Electron microscopy of the leading lamella. *Exp Cell Res* 67:359–67.

ALESCIO, T., and A. CASSINI. 1962. Induction *in vitro* of tracheal buds by pulmonary mesenchyme grafted on tracheal epithelium. *J Exp Zool* 150:83–94.

ANDRES, R. Y., I. JENG, and R. A. BRADSHAW. 1977. Nerve growth factor receptors: identification of distinct classes in plasma membranes and nuclei of embryonic dorsal root neurons. *Proc Natl Acad Sci* USA 74:2785–89.

AREY, L. B. 1965. *Developmental Anatomy,* 7th ed. Philadelphia: Saunders.

BALLARD, W. W. 1970. Archenteric origin of midgut lumen in amphibia. *Dev Biol* 21:424–39.

BANERJEE, S., and M BERNFELD. 1979. Developmentally regulated neutral hyaluronidase activity during epithelial-mesenchyme interaction. *J Cell Biol* 83:469a (Abstr).

BANERJEE, S. D., R. H. COHN, and M. R. BERNFELD. 1977. Basal lamina of embryonic salivary epithelia. Production by the epithelium and role in maintaining lobular morphology. *J Cell Biol* 73:445–63.

BARD, J. B., and E. D. HAY. 1975. The behavior of fibroblasts from the developing avian cornea. *J Cell Biol* 67:400–18.

BARD, J. B., E. HAY, and S. M. MELLER. 1975. Formation of the avian cornea: a study of cell movement in vivo. *Dev Biol* 42:334–61.

BEDRICK, A. D., and R. L. LADDA. 1978. Epidermal growth factor potentiates cortisone-induced cleft palate in the mouse. *Teratology* 17:13–18.

BELLAIRS, R., A. BOYDE, and J. E. HAEYSMAN. 1969. The relationship between the edge of the chick blastoderm and the vitelline membrane. *Wilhelm Roux' Arch* 163:113–21.

BERNFELD, M. R., R. H. COHN, and S. D. BANGEREE. 1973.

271

Glycosaminoglycans and epithelial organ formation. *Am Zool* 13:1067–83.

BOTERENBROOD, E. D., and P. D. NIEUWKOOP. 1973. The formation of the mesoderm in urodelean amphibians. V. Its regional induction by the endoderm. *Wilhelm Roux' Arch* 173:319–32.

BOYDEN, E. A. 1922. The development of the cloaca in birds, with special reference to the origin of the bursa of fabricius, the formation of the urodeal sinus, and the regular occurrence of a cloacal fenestra. *Am J Anat* 30:163–202.

BRADSHAW, R. A. 1978. Nerve growth factor. *Annu Rev Biochem* 47:191–216.

BRONNER-FRASER, M., and A. M. COHEN. 1980. Analysis of the neural crest ventral pathway using injected tracer cells. *Dev Biol* 77:130–41.

BULL, J. J., and R. G. VOGT. 1979. Temperature-dependent sex determination in turtles. *Science* 206:1186–88.

BURNS, R. K. 1961. Role of hormones in the differentiation of sex. In W. C. Young, ed. *Sex and Internal Secretions,* Vol. 1, 3rd ed. Baltimore: Williams and Wilkins, pp. 76–158.

BURNSIDE, B. 1971. Microtubules and microfilaments in newt neurulation. *Dev Biol* 26:416–41.

BURNSIDE, B., and A. G. JACOBSON. 1968. Analysis of morphogenetic movements in the neural plate of the newt *Taricha torosa. Dev Biol* 18:537–52.

CARPENTER, G., and S. COHEN. 1976. ^{125}I-labeled human epidermal growth factor. Binding, internalization, and degradation in human fibroblasts. *J Cell Biol* 71:159–71.

CARPENTER, G., and S. COHEN. 1978. Biological and molecular studies of the mitogenic effects of human epidermal growth factor. In J. Papaconstantinou and W. J. Rudder, eds. *Molecular Control of Proliferation and Differentiation.* New York: Academic Press, pp. 13–31.

CHEVALLIER, A. 1978. Étude de la migration des cellules somitiques dans le mesoderme somatopleural de l'ébauche de l'aile. *Wilhelm Roux' Arch* 184:57–73.

CHEVALLIER, A. 1979. Role of the somitic mesoderm in the development of the thorax in bird embryos. II. Origin of the thoracic and appendicular musculature. *J Embryol Exp Morphol* 49:73–88.

CHEVALLIER, A., M. KIENY, and A. MAUGER. 1977. Limb-somite relationship: origin of the limb musculature. *J Embryol Exp Morphol* 41:245–58.

CHEVALLIER, A., M. KIENY, and A. MAUGER. 1978. Limb-somite relationship: effect of removal of somite mesoderm on the wing musculature. *J Embryol Exp Morphol* 43:263–78.

CHRIST, B., J. J. JACOB and M. JACOB. 1977. Experimental analysis of the origin of the wing musculature in avian embryos. *Anat Embryol* 150:171–86.

CHUNG, S.-H., and J. E. COOKE. 1975. Polarity of structure and of ordered nerve connections in the developing amphibian brain. *Nature* 258:126–32.

COHEN, S. 1958. A nerve growth-promoting protein. In W. D. McElroy and B. Glass, eds. *Chemical Basis of Development.* Baltimore: Johns Hopkins Press, pp. 665–77.

COHEN, S., R. LEVI-MONTALCINI, and V. HAMBURGER. 1954. A nerve growth-stimulating factor isolated from sarcomas 37 and 180. *Proc Natl Acad Sci USA* 40:1014–18.

COOPER, M. D., and A. R. LAWSON. 1974. III. The development of the immune system. *Sci Am* November:59–70.

COUGHLIN, M. D., M. A. DIBNER, D. M. BAYER, and I. B. BLACK. 1978. Factors regulating development of an embryonic mouse sympathetic ganglion. *Dev Biol* 66:513–28.

COULOMBRE, A. J. 1965. The eye. In R. L. DeHaan and H. Ursprung, eds. *Organogenesis.* New York: Holt, Rinehart and Winston, pp. 219–51.

CRILEY, B. B. 1969. Analysis of the embryonic sources and mechanisms of development of posterior levels of chick neural tubes. *J Morphol* 128:465–502.

DAS, M., T. MIYAKAWA, C. F. FOX, R. M. PRUSS, A. AHARONOV, and H. R. HERSCHMAN. 1977. Specific radiolabeling of a cell surface receptor for epidermal growth factor. *Proc Natl Acad Sci USA* 74:2790–94.

DEHAAN, R. L. 1959. Cardia bifida and the development of pacemaker function in the early chick heart. *Dev Biol* 1:586–602.

DERBY, M. A. 1978. Analysis of glycosaminoglycans within the extracellular environments encountered by migrating neural crest cells. *Dev Biol* 66:321–36.

DIETERLEN-LIÈVRE, F. 1975. On the origin of haemopoietic stem cells in the avian embryo: an experimental approach. *J Embryol Exp Morphol* 33:607–19.

DUBOIS, R. 1968. La colonisation des ébauches gonadiques pars les cellules germinales de l'embryon de poulet, en culture in vitro. *J Embryol Exp Morphol* 20:189–213.

DUBOIS, R. 1969. Données nouvelles sur la localisation des cellules germinales primordiales dans le germe non incubé de poulet. *C R Acad Sci (Paris)* [D] 269:205–208.

EPPERLEIN, H. H. 1978. The ectomesenchymal–endodermal interaction system of *Triton alpestris* in tissue culture. *Differentiation* 11:109–23.

ERICKSON, C. A., K. W. TOSNEY, and J. A. WESTON. 1980. Analysis of migratory behavior of neural crest and fibroblastic cells in embryonic tissues. *Dev Biol* 77:142–56.

ERICKSON, R. P. 1977. Androgen-modified expression compared with Y linkage of male specific antigen. *Nature* 265:59–61.

FALLON, J. F., and J. A. CAMERON. 1977. Interdigital cell death during limb development of the turtle and lizard with an interpretation of evolutionary significance. *J Embryol Exp Morphol* 40:285–89.

272

FARGEIX, N. 1972. Irradiation aux rayons X de blasto-dermes non incubés de cane. Étude de la lignée germi-nale chez les embryons des stades ligne primitive à 14 paires de somites. *C R Acad Sci (Paris)* [D] 275:1011–14.

GAZE, R. N. 1970. *The Formation of Nerve Connections.* New York: Academic Press.

GEIGER, G. 1979. A 130K protein from chicken gizzard: its localization at the termini of microfilament bundles in cultured chicken cells. *Cell* 18:193–205.

GUNDERSON, R. W., and J. N. BARRETT. 1979. Neuronal chemotaxis: chick dorsal-root axons turn toward high concentrations of nerve growth factor. *Science* 206:1079–80.

HAMBURGER, V. 1939. Motor and sensory hyperplasia fol-lowing limb-bud transplantation in chick embryos. *Physiol Zool* 12:268–84.

HAMBURGER, V. 1958. Regression versus peripheral con-trol of differentiation in motor hypoplasia. *Am J Anat* 102:365–410.

HAMBURGER, V. 1962. Specificity in neurogenesis. *J Cell Comp Physiol* 60 (Suppl. 1): 81–92.

HAMBURGER, V. 1975. Cell death in the development of the lateral motor column of the chick embryo. *J Comp Neurol* 160:535–46.

HAMBURGER, V., J. K. BRUNSO-BECHTOLD, and J. W. YIP. 1981. Neuronal death in the spinal ganglia of the chick embryo and its reduction by means of nerve growth factor. *J Neurosci* 1:60–71.

HAMBURGER, V. and R. LEVI-MONTALCINI. 1949. Prolifera-tion, differentiation and degeneration in the spinal ganglia of the chick embryo under normal and experi-mental conditions. *J Exp Zool* 111:457–502.

HARPER, G. P., F. L. PEARCE, and C. A. VERNON. 1980. The production and storage of nerve growth factor in vivo by tissues of the mouse, rat, guinea pig, hamster and gerbil. *Dev Biol* 77:391–402.

HARRIS, A. K., P. WILD, and D. STOPAK. 1980. Silicone rub-ber substrata: a new wrinkle in the study of cell loco-motion. *Science* 298:177–79.

HASSELL, J. R. 1975. An ultrastructural analysis of the inhibition of epithelial cell death and palate fusion by epidermal growth factor. *Dev Biol* 45:90–102.

HASSELL, J. R., J. H. GREENBERG, and M. C. JOHNSON. 1977. Inhibition of cranial neural crest cell development by vitamin A in cultured chick embryo. *J Embryol Exp Morphol* 39:267–71.

HAUSHKA, S., and C. HANEY. 1978. Use of living tissue sec-tions for analysis of positional information during de-velopment. *J Cell Biol* 79:24a (abstr.).

HAY, E. 1968. Organization and fine structure of epithe-lium and mesenchyme in the developing chick em-bryo. In R. Fleischmajer, and R. E. Billingham, eds. *Epithelial-Mesenchymal Interactions.* Baltimore: Williams and Wilkins, pp. 31–55.

HAY, E., and J.-P. REVEL. 1969. *Fine Structure of the Develop-ing Avian Cornea.* Basel, Switzerland: Karger.

HOLLYDAY, M., and V. HAMBURGER. 1976. Reduction of the naturally occurring motor neuron loss by enlargement of the periphery. *J Comp Neurol* 170:311–20.

HOLLYDAY, M., V. HAMBURGER, and J. M. FARRIS. 1977. Localization of motor neuron pools supplying identi-fied muscles in normal and supernumerary legs of chick embryo. *Proc Natl Acad Sci* USA 74:3582–86.

HOUSSAINT, E., M. BELO, and N. M. LEDOUARIN. 1976. In-vestigations on cell lineage and tissue interactions in the developing bursa of fabricius through interspecific chimeras. *Dev Biol* 53:250–64.

HUMPHREY, R. R. 1969. A recently discovered mutant "eyeless" in the Mexican axolotl, *Ambystoma mexi-canum. Dev Biol* 27:365–75.

IKENISHI, K., and P. D. NIEUWKOOP. 1978. Location and ultrastructure of primordial germ cells (PGCs) in *Am-bystoma mexicanum. Dev Growth Differ* 20:1–9.

ISENBERG, G., P. C. RATHKE, N. HULSMAN, W. W. FRANKE, and K. E. WOLFARTH-BOTTERMAN. 1976. Cytoplasmic actomyosin fibrils in tissue culture cells. Direct proof of contractility by visualization of ATP-induced contrac-tion in fibrils isolated by laser microbeam dissection. *Cell Tissue Res* 166:427–43.

IZZARD, C., and L. LOCHNER. 1976. Cell-to-substrate con-tacts in living fibroblasts: an interference reflexion study with an evaluation of the technique. *J Cell Sci* 21:129–59.

IZZARD, C., and L. LOCHNER. 1980. Formation of cell-to-substrate contacts during fibroblast motility: an inter-ference reflexion study. *J Cell Sci* 42:81–116.

JACOB, M., B. CHRIST, and H. J. JACOB. 1978. On the mi-gration of myogenic stem cells into the prospective wing region in chick embryos. A scanning and trans-mission electron microscope study. *Anat Embryol* 153:179–93.

JACOBSON, A. G. 1966. Inductive processes in embryonic development. *Science* 152:25–34.

JACOBSON, A. G. 1978. Some forces that shape the nerv-ous system. *Zoon* 7:13–21.

JACOBSON, A. G., and R. GORDON. 1976. Changes in the shape of the developing vertebrate nervous system analyzed experimentally, mathematically and by com-puter simulation. *J Exp Zool* 197:191–246.

JACOBSON, M. 1966. Starting points for research in the ontogeny of behavior. In M. Locke, ed. *Major Problems of Developmental Biology.* New York: Academic Press, pp. 339–83.

JACOBSON, M. 1970. *Developmental Neurobiology.* New York: Holt, Rinehart, and Winston.

KAHN, C. R., J. T. COYLE, and A. M. COHEN. 1980. Head and trunk neural crest in vitro: autonomic neuron dif-ferentiation. *Dev Biol* 77:340–78.

273

KÄLLEN, B. 1962. Mitotic patterning in the central nervous system of chick embryos studied by a colchicine method. *Anat Embryol* 123:309–19.

KARFUNKEL, P. 1974. The mechanisms of neural tube formation. *Int Rev Cytol* 38:245–72.

KARFUNKEL, P. 1977. SEM analysis of amphibian mesodermal migration. *Wilhelm Roux' Arch* 181:31–40.

KIENY, M., and A. CHEVALLIER. 1979. Autonomy of tendon development in the embryonic chick wing. *J Embryol Exp Morphol* 49:153–65.

LAZARIDES, E. 1975. Tropomyosin antibodies: the specific localization of tropomyosin in nonmuscle cells. *J Cell Biol* 65:549–61.

LAZARIDES, E. 1976. Actin, α-actinin and tropomyosin interaction in the structural organization of actin filaments in nonmuscle cells. *J Cell Biol* 68:202–19.

LAZARIDES, E., and K. BURRIDGE. 1975. α-Actinin: immunofluorescent localization of a muscle structural protein in non-muscle cells. *Cell* 6:289–98.

LAZARIDES, E., and K. WEBER. 1974. Actin antibody: the specific visualization of actin filaments in non-muscle cells. *Proc Natl Acad Sci* USA 71:2268–72.

LEDOUARIN, N. 1964. Isolement expérimental du mésenchyme propre du foie et role morphogène de la composante mésodermique dan l'organogenèse hépatique. *J Embryol Exp Morphol* 12:141–160.

LEDOUARIN, N., J. FONTAINE, and C. LELIÈVRE. 1974. New studies on the avian ultimobranchial glandular cells—interspecific combinations and cytochemical characterization of C cells based on the uptake of biogenic amine precursors. *Histochemistry* 38:297–305.

LEDOUARIN, N., and F. V. JOTEREAU. 1973. Origin and renewal of lymphocytes in avian embryo thymuses studied in interspecific combinations. *Nature New Biology* 246:25–27.

LEDOUARIN, N., and C. LELIÈVRE. 1971. Sur l'origine des cellules a calcitonine du corps ultimobranchial de l'embryon d'oiseau. *Bull Anat* 56e Congrès, Nantes, France, 558–568.

LEDOURAIN, N., and M.-A. TEILLET. 1974. Experimental analysis of the migration and differentiation of neuroblasts of the autonomic nervous system and of neurectodermal mesenchymal derivatives, using a biological cell marking technique. *Dev Biol* 41:162–184.

LEDOUARIN, N., M. A. TEILLET, and C. LE LIÈVRE. 1977. Influence of the tissue environment on the differentiation of neural crest cells. In J. W. Lash and M. M. Burger, eds. *Cell and Tissue Interactions*. New York: Raven Press.

LEE, H.-Y., J. B. SHEFFIELD, R. G., NAGELE, JR. 1978. The role of extracellular material in chick neurulation. II. Surface morphology of neuroloepithelial cells during neural fold fusion. *J Exp Zool* 204:137–54.

LEE, H.-Y., J. B. SHEFFIELD, R. G. NAGELE, JR., and G. W. KALMUS. 1976. The role of extracellular material in chick neurulation. I. Effects of concanavalin A. *J Exp Zool* 198:261–66.

LELIÈVRE, C. S., G. G. SCHWEIZER, C. M. ZILLER, and N. M. LEDOUARIN. 1980. Restrictions of developmental capabilities in neural crest cell derivatives as tested by in vivo transplantation experiments. *Dev Biol* 77:362–78.

LEVI, A., Y. SHECHTER, E. J. NEUFELD, and J. SCHLESSINGER. 1980. Mobility, clustering, and transport of nerve growth factor in embryonal sensory cells and in a sympathtic neuronal cell line. *Proc Natl Acad Sci* USA 77:3469–73.

LEVI-MONTALCINI, R. 1950. The origin and development of the visceral system in the spinal cord of the chick embryo. *J Morphol* 86:253–84.

LEVI-MONTALCINI, R. 1966. The nerve growth factor: its mode of action on sensory and sympathetic nerve cells. *The Harvey Lectures,* Series 60. New York: Academic Press, pp. 217–59.

LEVI-MONTALCINI, R., and V. HAMBURGER. 1951. Selective growth-stimulating effects of mouse sarcoma on the sensory and sympathetic nervous system of the chick embryo. *J Exp Zool* 116:321–62.

LEVI-MONTALCINI, R., H. MEYER, and V. HAMBURGER. 1954. In vitro experiments on the effects of mouse sarcomas 180 and 37 on the spinal ganglia of the chick embryo. *Cancer Res* 14:49–57.

LEVINE, R. L., and M. JACOBSON. 1974. Deployment of optic nerve fibers is determined by positional markers in the frog's tectum. *Exp Neurol* 43:527–38.

LILLIE, F. R. 1917. The free-martin: a study of the action of sex hormones in the foetal life of cattle. *J Exp Zool* 23:371–452.

LILLIE, F. R. 1923. Supplementary notes on twins in cattle. *Biol Bull* 44:47–78.

LIPTON, B. H., and A. G. JACOBSON. 1974a. Analysis of normal somite development. *Dev Biol* 38:73–90.

LIPTON, B. H., and A. G. JACOBSON. 1974b. Experimental analysis of the mechanisms of somite morphogenesis. *Dev Biol* 38:91–103.

MAHONEY, K. A., B. J. HYER, and L.-N. CHAN. 1977. Separation of primitive and definitive erythroid cells of the chick embryo. *Dev Biol* 56:412–16.

MANASEK, F. J. 1975. The extracellular matrix: a dynamic component of the developing embryo. *Curr Top Dev Biol* 10:35–102.

MANASEK, F. J. 1976. Heart development: interactions involved in cardiac morphogenesis. In G. Poste and G. L. Nicolson, eds. *The Cell Surface in Animal Embryogenesis and Development.* Amsterdam: North-Holland, pp. 545–98.

MANASEK, F. J., B. BURNSIDE, and R. E. WATERMAN. 1973. Myocardial cell shape changes as a mechanism of embryonic heart looping. *Dev Biol* 29:349–71.

MANIATIS, G., and V. M. INGRAM. 1971. Erythropoiesis

274

during amphibian metamorphosis. III. Immunochemical detection of tadpole and frog hemoglobins (*Rana catesbeiana*) in single erythrocytes. *J Cell Biol* 49: 390–404.

MARKS, P. A., and R. A., RIFKIND. 1972. Protein synthesis: its control in erythropoiesis. *Science* 175:955–61.

MARTIN, C., D. BEAUPAIN, and F. DIETERLEN-LIÈVRE. 1978. Developmental relationships between vitelline and intraembryonic haemopoiesis studied in avian "yolk sac Chimaeras." *Cell Differ* 7:115–30.

MCLACHLAN, J. C., and A. HORNBRUCH. 1979. Muscle-forming potential of the non-somitic cells of the early avian limb bud. *J Embryol Exp Morphol* 54:209–17.

MCLAREN, A. 1975. Sex chimaerism and germ cell distribution in a series of chimaeric mice. *J Embryol Exp Morphol* 33:205–16.

MCLAREN, A. 1980. Öocytes in the testis. *Nature* 283:688–89.

MCLAREN, A., A. C. CHANDLEY, and S. KOFMAN-ALFARO. 1972. A study of meiotic germ cells in the gonads of foetal mouse chimaeras. *J Embryol Exp Morphol* 27:515–24.

MEIER, S. 1979. Development of the chick embryo mesoblast. Formation of the embryonic axis and establishment of the metameric pattern. *Dev Biol* 73:25–45.

MENKES, B., M. DELEANU, and A. ILIES. 1965. Comparative study of some areas of physiological necrosis in the embryo of man, some laboratory-mammalians and fowl. *Rev Roumaine Cytol* 2:161–71.

MINTZ, B. 1968. Hermaphroditism, sex chromosomal mosaicism and germ cell selection in allophenic mice. *J Anim Sci* 27(Suppl. 1):51–60.

MORAN, D., and R. W. RICE. 1975. An ultrastructural examination of the role of cell membrane surface material during neurulation. *J Cell Biol* 64:172–81.

MUIR, H. 1977. Structure and function of proteoglycans of cartilage and cell-matrix interactions. In J. Lash and M. Burger, eds. *Cell and Tissue Interactions.* New York: Raven Press, pp. 87–99.

MURPHY, R. A., R. H. SINGER, J. D. SAIDE, N. J. PANTAZIS, M. H. BLANCHARD, K. S. BRYON, B. G. ARNASON, and M. YOUNG. 1977. Synthesis and secretion of a high molecular weight form of nerve growth factor by skeletal muscle cells in culture. *Proc Natl Acad Sci* USA 74:4496–4500.

NODEN, D. M. 1978. The control of avian cephalic neural crest cytodifferention. II. Neural tissues. *Dev Biol* 67:313–29.

OHNO, S. 1977. Major regulatory genes for mammalian sex development. *Cell* 7:315–21.

PACKARD, D. S., JR., and A. G. JACOBSON. 1979. Analysis of the physical forces that influence the shape of chick somites. *J Exp Zool* 207:81–106.

PATTEN, B. M., and B. CARLSON. 1974. *Foundations of Embryology.* New York: McGraw-Hill.

PEXIEDER, T. 1975. Cell death in the morphogenesis and teratogenesis of the heart. *Adv Anat Embryol Cell Biol* 51:5–100.

PINTAR, J. E. 1978. Distribution and synthesis of glycosaminoglycans during quail neural crest morphogenesis. *Dev Biol* 67:444–64.

POOLE, T. J., and M. S. STEINBERG. 1977. SEM-aided analysis of morphogenetic movements: development of the amphibian pronephric duct. *Scanning Electron Microsc* 2:43–52.

PRATT, R. M., M. A. LARSON, and M. C. JOHNSTON. 1975. Migration of cranial neural crest cells in a cell-free hyaluronate-rich matrix. *Dev Biol* 44:298–305.

REYER, R. W. 1977. The amphibian eye: development and regeneration. In F. Crescitelli, ed. *Handbook of Sensory Physiology,* Vol. 7. Heidelberg: Springer-Verlag, pp. 309–90.

REYNAUD, G. 1969. Transfert de cellules germinales primordiales de dindon à l'embryon de poulet par injection intravasculaire. *J Embryol Exp Morphol* 21:485–507.

REYNAUD, G. 1973. Contribution a l'étude des relations entre soma et germen chez le poulet au moyen d'une technique de transfert des gonocytes primordiaux. Thesis, University of Provence.

RICE, R. W., and D. J. MORAN. 1977. A scanning electron microscopic and x-ray microanalytic study of cell surface material during amphibian neurulation. *J Exp Zool* 201:471–78.

RIFKIND, R. A. 1974. Erythroid cell differentiation. In J. Lash and J. R. Whittaker, eds. *Concepts of Development.* Stamford, CT: Sinaur Associates, pp. 149–62.

ROGULSKA, T. 1968. Primordial germ cells in normal and transected duck blastoderms. *J Embryol Exp Morphol* 20:247–60.

ROMANOFF, A. 1960. *The Avian Embryo.* New York: Macmillan.

ROSENQUIST, G. C., and R. L. DEHAAN. 1966. Migration of precardiac cells in the chick embryo: a radioautographic study. *Carnegie Contrib Embryol* 38:111–21.

RUGH, R. 1962. *Experimental Embryology,* 3rd ed. Minneapolis: Burgess.

RUTTER, W. J., R. L. PICTET, J. D. HARDING, J. M. CHIRGWIN, R. J. MACDONALD, and A. E. PRZYBYLA. 1978. An analysis of pancreatic development: role of mesenchymal factor and other extracellular factors. In J. Papaconstantinou and W. J. Rutter, eds. *Molecular Control of Proliferation and Differentiation.* New York: Academic Press, pp. 205–27.

SAUER, M. E., and B. E. WALKER. 1959. Radioautographic study of interkinetic nuclear migration in the neural tube. *Proc Soc Exp Biol Med* 101:557–60.

SAUNDERS, J. W., JR. 1948. The proximo-distal sequence of origin of the parts of the chick limb and the role of the ectoderm. *J Exp Zool* 108:363–404.

275

SAUNDERS, J. W., JR. 1966. Death in embryonic systems. *Science* 154:604–12.

SAUNDERS, J. W., JR. 1977. The experimental analysis of chick limb bud development. In D. A. Ede, J. R. Hinchliffe, and M. Balls, eds. *Vertebrate Somite and Limb Morphogenesis.* Cambridge: Cambridge University Press, pp. 1–24.

SAUNDERS, J. W., JR., and J. F. FALLON. 1967. Cell death in morphogenesis. In M. Locke, ed. *Major Problems in Developmental Biology.* New York: Academic Press, pp. 289–314.

SAUNDERS, J. W., JR., M. T. GASSELING, and L. C. SAUNDERS. 1962. Cellular death in morphogenesis of the avian wing. *Dev Biol* 5:147–78.

SCHMIDT, J. 1977. Retinal fibers alter tectal positional markers during the expansion of the half retinal projection in goldfish. *J Comp Neurol* 177:279–300.

SCHOENWOLF, G. C. 1977. Tail (end) bud contributions to the posterior region of the chick embryo. *J Exp Zool* 201:227–46.

SCHROEDER, T. E. 1970. Neurulation in *Xenopus laevis.* An analysis and model based upon light and electron microscopy. *J Embryol Exp Morphol* 23:427–62.

SEARLS, R. L., and M. Y. JANNERS. 1971. The initiation of limb bud outgrowth in the embryonic chick. *Dev Biol* 24:198–213.

SERVER, A. C., and E. M. SHOOTER. 1977. Nerve growth factor. *Adv Protein Chem* 31:339–409.

SHAIN, W. G., S. R. HILFER, and V. G. FONTE. 1972. Early organogenesis of the embryonic chick thyroid. *Dev Biol* 28:202–18.

SHORT, R. V., J. SMITH, T. MANN, E. P. EVANS, J. HALLETT, A. FRYER, and J. L. HAMERTON. 1969. Cytogenetic and endocrine studies of a free-martin heifer and its bull co-twin. *Cytogenetics* 8:369–88.

SIEBER-BLUM, M., and A. M. COHEN. 1980. Clonal analysis of quail neural crest cells; they are pluripotent and differentiate in vitro in the absence of noncrest cells. *Dev Biol* 80:96–106.

SILVERS, W. K., and S. S. WACHTEL. 1977. H-Y antigen: behavior and function. *Science* 195:956–60.

SMITH, L. D. 1964. A test of the capacity of presumptive somatic cells to transform into primordial germ cells in the mexican axolotl. *J Exp Zool* 156:229–42.

SMITH, L. D. 1966. The role of a "Germinal Plasm" in the formation of primordial germ cells in *Rana pipiens. Dev Biol* 14:330–57.

SMITH, L. D., and M. A. WILLIAMS. 1975. Germinal plasm and determination of the primordial germ cells. In C. L. Markert and J. Papaconstantinou, eds. *The Developmental Biology of Reproduction.* New York: Academic Press, pp. 3–24.

SMITH, L. D., and M. WILLIAMS. 1979. Germ plasm and germ cell determinants in anuran amphibians. In D. R.

Newth and M. Balls, eds. *Maternal Effects in Development.* Cambridge: Cambridge University Press, pp. 167–97.

SMUTS, M. S., S. R. HILFER, and R. L. SEARLS. 1978. Patterns of cellular proliferation during thyroid organogenesis. *J. Embryol Exp Morphol* 48:269–86.

SPERRY, R. W. 1965. Embryogenesis of behavioral nerve nets. In R. L. DeHaan and H. Ursprung, eds. *Organogenesis.* New York: Holt, Rinehart and Winston, pp. 161–86.

SPOONER, B. S., H. I. COHEN, and J. FAUBION. 1977. Development of the embryonic mammalian pancreas: the relationship between morphogenesis and cytodifferentiation. *Dev Biol* 61:119–30.

SPOONER, B. S., and N. K. WESSELS. 1972. An analysis of salivary gland morphogenesis: role of cytoplasmic microfilaments and microtubules. *Dev Biol* 27:38–54.

STALSBERG, H., and R. L. DEHAAN. 1968. Endodermal movements during foregut formation in the chick embryo. *Dev Biol* 18:198–215.

SUTASURJA, L. A., and P. D. NIEUWKOOP. 1974. The induction of the primordial germ cells in the urodeles. *Wilhelm Roux' Arch* 175:199–220.

THOENEN, H., M. SCHWAB, and U. OTTEN. 1978. Nerve growth factor as a mediator of information between effector organs and innervating neurons. in J. Papaconstantinou and W. J. Rutter, eds. *Molecular Control of Proliferation and Differentiation.* New York: Academic Press, pp. 101–18.

TOSNEY, K. W. 1978. The early migration of neural crest cells in the trunk region of the avian embryo: an electron microscopic study. *Dev Biol* 62:317–33.

TRINKAUS, J. P. 1973. Surface activity and locomotion of *Fundulus* deep cells during blastula and gastrula stages. *Dev Biol* 30:68–103.

TRINKAUS, J. P. 1976. On the mechanism of metazoan cell movements. In G. Poste and G. L. Nicolson, eds. *The Cell Surface in Animal Embryogenesis and Development.* Amsterdam: North-Holland, pp. 225–329.

VÀN DEUSEN, E. 1973. Experimental studies on a mutant gene (*e*) preventing the differentiation of the eye and normal hypothalamus primordia in the axolotl. *Dev Biol* 34:135–58.

WAKAHARA, M. 1978. Induction of supernumerary primordial germ cells by injecting vegetal pole cytoplasm into *Xenopus* eggs. *J Exp Zool* 203:159–64.

WEBER, K., and U. GROESCHEL-STEWART. 1974. Antibody to myosin: the specific visualization of myosin-containing filaments in nonmuscle cells. *Proc Natl Acad Sci* USA 71:4561–64.

WEISS, P. 1958. Cell contact. *Int Rev Cytol* 7:391–423.

WESSELS, N. K. 1970. Mammalian lung development: interactions in formation and morphogenesis of tracheal buds. *J Exp Zool* 175:455–66.

WESSELS, N. K., and J. H. COHEN. 1967. Early pancreas

organogenesis: morphogenesis, tissue interactions and mass effects. *Dev Biol* 15:237–70.

WESSELS, N. K., and J. EVANS. 1968. Ultrastructural studies of early morphogenesis and cytodifferentiation in the embryonic mammalian pancreas. *Dev Biol* 17:413–46.

WESTON, J. A. 1970. The migration and differentiation of neural crest cells. *Adv Morph* 8:41–114.

WILLIER, B. H. 1921. Structure and homologies of freemartin gonads. *J Exp Zool* 33:63–127.

WILT, F. H. 1965. Regulation of the initiation of chick embryo hemoglobin synthesis. *J Mol Biol* 12:331–41.

WILT, F. H. 1967. The control of embryonic hemoglobin synthesis. *Adv Morph* 6:89–125.

WITSCHI, E. 1967. Biochemistry of sex differentiation in vertebrate embryos. In R. Weber, ed. *The Biochemistry of Animal Development,* Vol. 2. New York: Academic Press, pp. 193–225.

ZUST, B., and K. E. DIXON. 1977. Events in the germ cell lineage after entry of the primordial germ cells into the genital ridges in normal and U.V.-irradiated *Xenopus laevis. J. Embryol Exp Morphol* 41:33–46.

12

Germination of Seeds and the Differentiation of Plant Tissues and Organs

As pointed out in Chapter 6, the seed of the flowering plant arises from a complex that consists of an embryo embedded in endosperm and surrounded by integuments derived from the ovule wall (Figure 6.14). Parts of the embryo are recognizable as prospective root, shoot, and leaves. These parts differentiate to varying degrees as the seed matures. In some cases, notably among the **dicotyledonous** plants (**dicots**—plants with two seed leaves), the material of the endosperm is largely incorporated into the cotyledons as storage material, consisting chiefly of protein and varying amounts of lipid. In others, notably among many **monocotyledonous** plants (**monocots**—plants with only one seed leaf), the bulk of the endosperm, usually containing large amounts of stored starch (as in corn and barley), is not incorporated into the embryo; rather, the embryo remains embedded in the endosperm. In either condition the stored materials are required as sources of energy and structural components for the subsequent germination of the seed and the onset of morphogenesis.

12.1 Germination

In the mature seed the embryo is dormant and is encased in **seed coats** derived from the integuments and the nucellus (Figure 6.15). Germination refers to the resumption of growth by the embryo and the rupture of the seed coats. This is followed by the emergence of the **primary root** from the radicle and the growth of the shoot from the embryonic epicotyl, or plumule. The root turns downward, penetrating the soil, and the shoot pushes upward through the soil.

Seed coats may provide a significant barrier to germination. They are structurally varied in different kinds of plants. They provide mechanical protection to varying degrees, according to their structure and composition. The chief protective elements may arise from either the inner or outer integuments (Figure 12.1). The protective cell walls are often **sclerified**, that is, they become thickened by the deposition of **lignin**, a noncarbohydrate, high-carbon compound characteristic of mature wood. As the seed coats mature on the parent plant, the embryo loses water (Figure 12.2) and becomes dormant before being shed.

Germination involves the imbibition of water followed by the resumption of growth of the embryo and the initiation of differentiation in its parts. This occurs only under appropriate conditions of moisture and (often) light, and may require the intervention of a period of low temperature (**vernalization**). For plants with hard and impermeable seed coats, the imbibition of water and resumption of growth are dependent on mechanical abrasion of the seed coat or its destruction by bacteria of the soil. In some instances inhibitors of germination are present in the seed coat, and these must be leached out by water before the embryo can develop. Examples are the mustard oils in seeds of plants of the family Cruciferae, coumarin (a lactone) in sweet clover, and a variety of organic acids, as found in seeds of some fruits.

During recent years a great deal of research has been devoted to the identification of regulatory mechanisms involved in the maturation of the storage tissue in the seed and to the origin and release of enzymes involved in germination. This research has revealed that plant seeds show patterns of gene

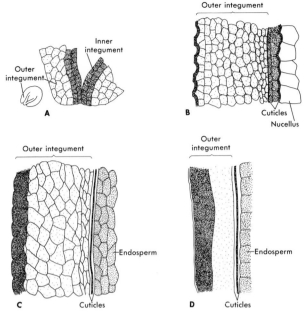

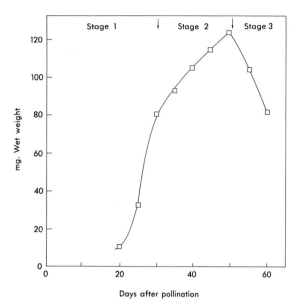

FIGURE 12.1 Origin of the seed coats in *Asparagus*. A: A view of the ovule at low magnification is shown at the left; at the right is a schematic view at higher magnification of the integuments at a slightly later stage. Nucellar tissue and the embryo sac are not shown. B: About 16 days after pollination, showing one side of the developing seed coat. The inner integument is compressed as the expanding endosperm pushes against the nucellus from within the ovule. Cuticles form on either side of the inner integument. C: Expansion of the endosperm compresses the cuticles and essentially eliminates the inner integument. The outer integument begins to desiccate, and its exterior wall becomes greatly thickened. D: The mature seed coat, with endosperm tightly compressed against desiccated remnants of the integuments and their derivatives, these now constituting the seed coat. [Adapted from K. Esau, *Plant Anatomy*, 2nd ed., Wiley, New York, 1965, p. 620.]

FIGURE 12.2 The changes in wet weight of the cotton embryo during embryogenesis. Vertical arrows delimit arbitrary stages. During each stage the embryo can be dissected from the ovule and caused to germinate precociously and will produce protease in its cotyledons. Embryos in stage 3 are undergoing desiccation concomitant with seed formation. [Adapted with modifications from J. N. Ihle and L. Dure, *Biochem Biophys Res Comm* 38:995–1001 (1970).]

279

action and protein synthesis that are quite reminiscent of those found in the fertilized eggs and early embryos of many animal species. Some of the best-studied cases are reviewed briefly in Section 12.1.1 and 12.1.2, after which morphogenesis of root and stem is discussed.

12.1.1 CONTROL OF GENE ACTION AND PROTEIN SYNTHESIS DURING GERMINATION OF COTTON-SEEDS. In the laboratory cottonseeds infiltrated

with water under vacuum for 30 minutes, implanted in the soil, and held at 30°C, germinate and erupt from the soil in about 72 hours. Their wet weight increases rapidly during the first 4 hours and then remains constant for about 20 hours. Thereupon, the root and stems elongate progressively, and their growth is reflected in an increase in both wet and dry weights (Figure 12.3). The cotyledons emerge from the seed coat about 24 hours after planting, but, whereas the wet weight of the cotyledons increases, their dry weight decreases, reflecting the loss of their stored foods and minerals, which support growth of the stem and root (Figure 12.3).

The cotyledons mostly store protein and lipid, but have little starch. Enzymes required for the breakdown and utilization of the stored materials are referred to as **germination enzymes,** as distinct from enzymes involved in morphogenesis of the plant parts. Ihle and Dure (1972) studied the devel-

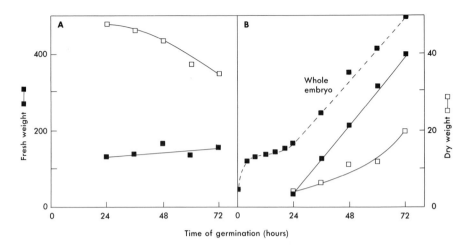

FIGURE 12.3 Changing wet and dry weights of cotton plantlets germinating on moist filter paper after removal of the seed coats: (A) Cotyledons alone; (B) Whole embryo (upper curve) and root plus stem (lower curves). Dry weight is indicated by open squares (□—□), wet weight by filled squares (■—■). [Adapted with modifications from L. C. Waters and L. Dure, *J Mol Biol* 19:1–27 (1966).]

opmental biochemistry of two of the germination enzymes of cottonseed cotyledons. One of these enzymes is a distinctive **carboxypeptidase,** one of a class of enzymes that degrades protein by removing amino acids from exposed —COOH terminal ends. The other one is **isocitratase,** or **isocitrate lyase,** an enzyme not present in cells of higher animals, which is involved in a coordinated set of reactions that results in the incorporation of two-carbon lipid breakdown products into the carbohydrate cycle. This enzyme is thus of great importance in making energy available for growth of the plant from lipid stored in the cotyledons.

Neither of these enzymes can be demonstrated in immature seeds. Nor can they be found in mature seeds until 24 hours after they are placed under conditions for germination. Thereupon, synthesis of the enzymes begins, as demonstrated by two lines of evidence.

1. With respect to the carboxypeptidase, when homogenates of germinating embryos synthesize protein in the presence of radioactive amino acids, the enzyme is much more heavily labeled with radioactivity than are the other proteins of the homogenate.
2. Neither the carboxypeptidase nor the isocitratase is produced if cycloheximide, an inhibitor of protein synthesis, is present in the homogenate.

The question next arises as to whether the onset of synthesis of the germinating enzymes coincides with the onset of transcription of the mRNAs for

these enzymes or whether, as in the case of many proteins of sea urchin eggs, for example (cf. Chapter 10), their synthesis involves the translation of mRNAs produced during an earlier phase of development. The effects of adding actinomycin D, which inhibits transcription of the genome, provide an answer to this question. When the inhibitor is added to homogenates of cotyledons in vitro, the synthesis of the germinating enzymes proceeds normally. Moreover, germination of the intact mature embryo proceeds in a normal fashion in the presence of actinomycin D at concentrations sufficient to suppress mRNA synthesis.

When are the mRNAs for the germinating enzymes produced? The answer to this question was obtained by taking advantage of the fact that an immature cotton embryo, dissected from its ovule and transferred to moist paper, will undergo precocious germination; that is, the root and stem elongate, the cotyledons unfurl, and synthesis of the germination enzymes begins. In embryos of less than 85 mg wet weight (Stage 1, Figure 12.2), however, detectable enzyme synthesis is delayed for 24 hours (Figure 12.4). To determine when the mRNAs for germination of the mature seed are transcribed, embryos of successively younger stages were excised and germinated precociously in the presence of actinomycin D. It was found that the appearance of the germinating enzymes becomes sensitive to the inhibitor at about 32 days after pollination, when the embryo weighs 85 mg, which is about 65% of the maximum weight achieved prior to the onset of desiccation (Figure 12.2). Embryos precociously germinated at this stage, or later, develop normally in the presence of actinomycin D. Embryos excised at younger stages germinate normally in the absence of actinomycin D, but not in its presence.

At the 85-mg stage, therefore, the mRNA for the germinating enzymes normally begins to be synthesized. But the embryo grows for another 20 days before it reaches its full size, after which desiccation and loss of wet weight occur. Why are the mRNAs for the germination enzymes not normally translated during this 20-day interval of growth? A clue to the answer to this question is provided by the fact that embryos at the 85-mg stage, dissected from the ovule and germinated on moist paper, produce the

germination enzymes sooner and at a much more rapid rate if they are carefully washed after dissection than if they are unwashed (Figure 12.4). This fact suggests that unwashed embryos retain some inhibitor, possibly derived from ovular tissue, which is removed by washing.

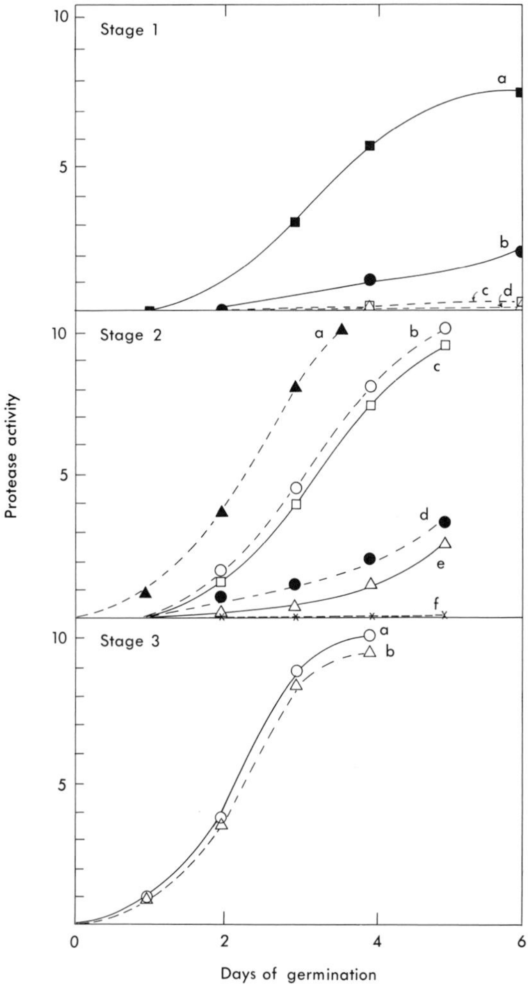

Abscisic acid is a well-known inhibitor of plant growth and an inductor of leaf fall. It is found in considerable quantities in ovular tissue, and its presence is apparently responsible for the nontranslation of mRNAs for the germinating enzymes. If precociously germinating embryos are treated with abscisic acid, their production of the enzymatic proteins is inhibited. The action of abscisic acid is not direct, however, for its effect is abolished by actinomycin D. Its action, thus, possibly involves stimulating the synthesis of an RNA species that ultimately results in production of a translation inhibitor. Alternatively, abscisic acid may act to inhibit translation in conjunction with some substance produced as a result of an RNA synthesis over which it has no direct control.

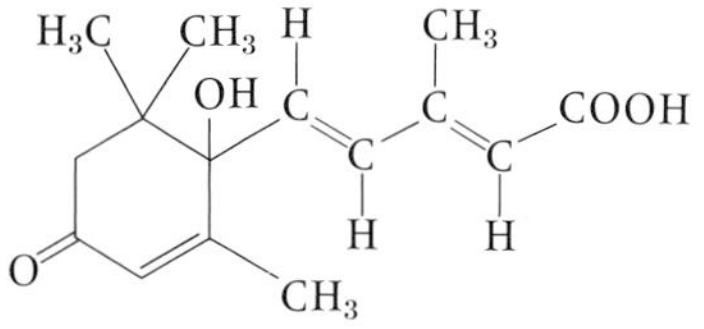

Abscisic acid

281

FIGURE 12.4 Protease activity shown by homogenates of precociously germinated cotton embryos. Protease activity is indicated in arbitrary units per pair of cotyledons. Germination on moist cloth in the presence or absence of test materials. Stage 1 embryos: line a, embryos excised and washed before germination; line b, embryos excised from the ovule and germinated without washing; line c, embryos treated with an extract of ovules or with abscisic acid; line d, embryos germinated under any of the above conditions in the presence of actinomycin D. Stage 2 embryos: line a, enzyme activity of unwashed actinomycin D-treated embryos or of washed embryos treated with ovule extract and actinomycin D; line b, washed embryos treated with actinomycin D or actinomycin D plus abscisic acid; line c, washed embryos; line d, embryos unwashed; line e, washed embryos treated with ovule extract; line f, washed embryos treated with abscisic acid. Stage 3 embryos: line a, unwashed embryos treated with ovule extract or with abscisic acid; line b, embryos similarly treated and germinated in the presence of actinomycin D.

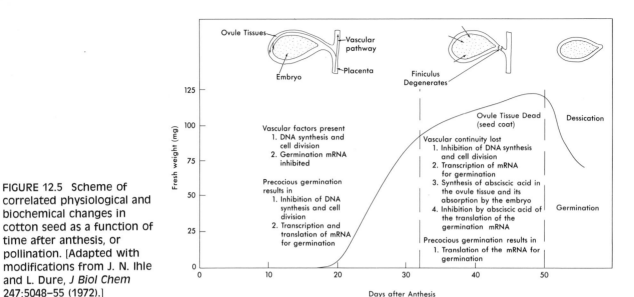

FIGURE 12.5 Scheme of correlated physiological and biochemical changes in cotton seed as a function of time after anthesis, or pollination. [Adapted with modifications from J. N. Ihle and L. Dure, *J Biol Chem* 247:5048–55 (1972).]

After embryos have reached maturity (Figure 12.2), the ovular tissue sclerifies and thus the source of abscisic acid is removed. During the process of desiccation, moreover, embryos are no longer susceptible to abscisin-induced inhibition of enzyme synthesis, nor does actinomycin D affect production of the enzymes (Figure 12.4, Stage 3). Why the translating mechanisms are no longer subject to inhibitory effects of abscisic acid after maturity is reached is not known.

Another question that remains unsolved is why the embryo, clearly capable of synthesizing mRNAs for the germinating enzymes prior to the 85-mg stage, does not do so in the ovule? Interestingly, at the 85-mg stage, the embryo normally loses its connection with the vascular tissue of the ovary and DNA synthesis ceases in the cotyledons. These changes are likewise brought about by the excision of early embryos and their precocious germination in vitro. Conceivably, factors conveyed by the vascular tissue control the pattern of gene transcription in the embryo. The exact nature of these factors and their mode of action remain unknown.

A scheme summarizing the events just described and postulated is presented in Figure 12.5.

12.1.2 SYNTHESIS AND RELEASE OF GERMINATION ENZYMES IN BARLEY SEEDS.

The barley seed is monocotyledonous and is similar in appearance to the wheat seed illustrated in Figure 6.16. The bulk of the seed is an inert starchy mass surrounded by a layer of living cells, the **aleurone.** The aleurone, the only living remnant of the endosperm, is enclosed in seed coats derived from ovular tissue, the wall of the ovule, and the remains of flower parts.

Cells of the aleurone produce the enzymes that digest the food reserves of the starchy endosperm. The appearance of these enzymes is largely under the control of the gibberellins, earlier mentioned in connection with the induction of flowering. Gibberellic acid (GA_3) (page 28) is most frequently used in experimental studies of the hormonal control of germination. Happily, it is easy to isolate the aleurone layer from the endosperm in halved barley seeds and to test directly on these half-aleurones or on their homogenates the effects of GA_3 and inhibitors of RNA synthesis, and to follow the synthesis and release of enzymes from them.

Aleurone layers isolated in vitro slowly synthesize and release to the surrounding medium the enzymes α-amylase, protease, and ribonuclease. The amylase catalyzes the breaking of 1,3 glycoside bonds in starches; the protease catalyzes the breakage of peptide bonds, thus rendering soluble the protein of the endosperm; and the ribonuclease is presumably utilized in solubilizing the nucleotides of the inert endosperm and making them available to the developing embryo.

The synthesis and release of these enzymes from the isolated aleurone are greatly enhanced by incubation in the presence of added GA_3. Aleurones are customarily isolated from the endosperm after a 3-day incubation period in the dark at room temperature (Chrispeels and Varner, 1967). The seeds are then halved and the aleurone layers are separated. At that time there is usually less than 1 μg (10^{-6} g) of α-amylase per half-aleurone layer. When such aleurones are then incubated in the presence of GA_3, there is essentially no synthesis of α-amylase for 6–8 hours, but thereafter, its activity increases linearly for 16–24 hours. Most of the enzyme is secreted into the medium, the cells retaining only a small amount of the enzyme (Figure

FIGURE 12.6 The synthesis of α-amylase by 10 barley aleurone layers incubated in the presence of 1 μM GA_3. Enzyme activity was measured in the medium and in extracts of the aleurone layers. The sum of these two activities is plotted as the total. Activity was measured by the ability of the preparations to hydrolyze a standard starch solution and converted to μg amylase by reference to the action of a standard purified malt amylase. [Adapted with modifications from M. J. Chrispeels and J. E. Varner, *Plant Physiol* 42:308–406 (1967).]

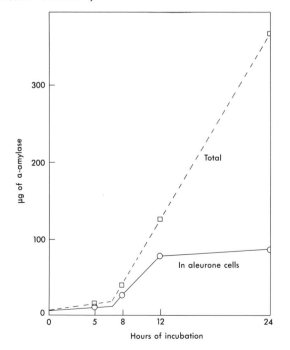

12.6). In the absence of exogenous GA_3, very little α-amylase is synthesized. The requirement for maximal synthesis of α-amylase is satisfied by a concentration of GA_3 in the range of 0.01–0.1 M. The protease shows a similar pattern of synthesis and release in response to GA_3 (Jacobsen and Varner, 1967).

The enhanced activity of α-amylase and of protease in barley aleurone cells after treatment with GA_3 is inhibited by actinomycin D and by puromycin, indicating that transcription and translation of new mRNA are requisite to the synthesis of these enzymes. Moreover, in the case of α-amylase it has been tested and shown that the enzyme is formed from amino acids released by the hydrolysis of preexisting aleurone proteins. Thus, when α-amylase is induced by GA_3 in aleurone cells in the presence of "heavy water" ($H_2^{18}O$), the newly synthesized enzyme has a greater buoyant density (i.e., is "heavier") than enzymes similarly induced in the presence of ordinary water ($H_2^{16}O$). This is because heavy oxygen atoms (^{18}O) are incorporated into amino acids during the hydrolysis of preexisting aleurone proteins. These heavy amino acids then appear in newly synthesized α-amylase molecules (Filner and Varner, 1967).

283

The synthesis and release of small amounts of enzymes by isolated aleurone layers in the absence of exogenous GA_3 probably result from the presence of small amounts of gibberellins in the tissue at the time of its isolation. Indeed, tests show that normally there are small amounts of gibberellin-like substances in barley seeds. Their effects in the intact seed are probably inhibited by abscisic acid during dormancy of the seed. This notion is suggested by the fact that abscisic acid inhibits the production and release of the protease of barley aleurone in the presence of GA_3 (Jacobson and Varner, 1967).

The pattern of synthesis and release of ribonuclease with and without exogenous GA_3 is considerably different from the pattern just described for amylase and protease. Even in the absence of added GA_3, the level of endogenous GA_3 rises steadily in the aleurone layers at a rate considerably greater than that of amylase and protease. Moreover, maximal synthesis of the protease occurs at concentrations of GA_3 as low as $10^{-9} M$ (Figure 12.7), whereas that of amylase requires a concentration 10 to 100 times greater. Another point of difference is found in the pattern of release of ribonuclease

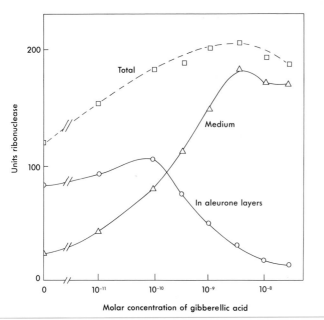

FIGURE 12.7 Production of ribonuclease by barley aleurones incubated for 48 hr in the presence of varying concentrations of GA_3. [Adapted with modifications from M. J. Chrispeels and J. E. Varner, *Plant Physiol* 42:308–406 (1967).]

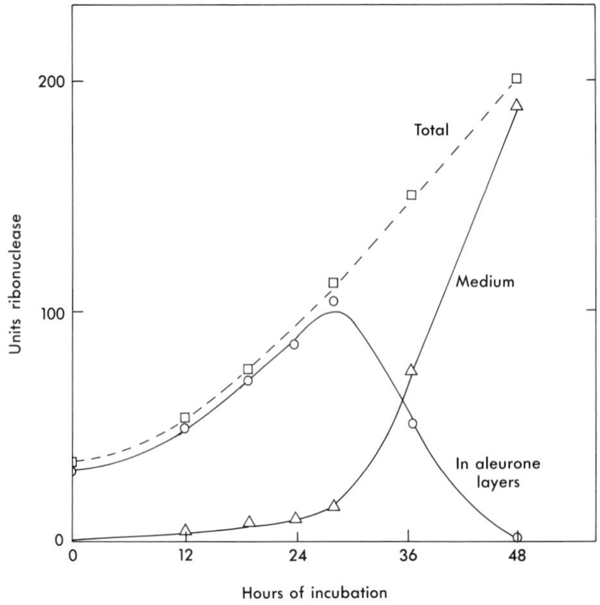

FIGURE 12.8 Delayed release of ribonuclease to the medium from barley aleurones incubated in the presence of $5 \times 10^{-6} M$ GA_3. [Adapted with modifications from M. J. Chrispeels and J. E. Varner, *Plant Physiol* 42:308–406 (1967).]

from the aleurone layer. Most of the ribonuclease synthesized in the presence of low concentrations of GA_3 (10^{-11} to 10^{-10} M) is still present in the aleurone layer after 48 hours. When higher concentrations are used, the amount of enzyme released to the medium is much greater, but this release is delayed for 24 hours nevertheless (Figure 12.8). This evidence suggests that synthesis and release of ribonuclease are independently controlled by GA_3, an interpretation that is supported by the fact that when cycloheximide is added 20 hours after the onset of GA_3 treatment, synthesis of ribonuclease ceases and release of the previously formed enzyme from the aleurone layer is blocked. If, however, the addition of cycloheximide is delayed until 24–28 hours after GA_3 stimulation of the aleurone, then the enzyme is released rapidly. Apparently, therefore, most or all of the release capacity has developed by 28 hours.

It is reasonable to ask whether the coordinated release of amylase and protease are likewise controlled by gibberellin independently of their synthe-

sis. It is difficult to answer this question, for release of amylase into the medium occurs rapidly once its synthesis begins. But small amounts of actinomycin D inhibit the GA_3-stimulated release of amylase more severely than they do its synthesis, thus causing it to accumulate in the aleurone cells.

One way in which gibberellins may facilitate the release of various enzymes from the aleurone is through the production of other enzymes that digest the aleurone cell walls. Treatment of aleurone tissue with GA_3 brings about digestion of the walls of the cells as shown by their loss of reactivity to stains for carbohydrates. Interestingly, this digestion occurs predominantly on the side of each cell facing the endosperm. What brings about the polarized breakdown of aleurone cell walls remains unknown. No clues are found in either the ultrastructure of the aleurone cells or in their pattern of histochemical staining.

One of the enzymes involved in digestion of the aleurone cell walls is β-1,3-glucanase. This enzyme catalyzes the hydrolysis of β-1,3 linkages between

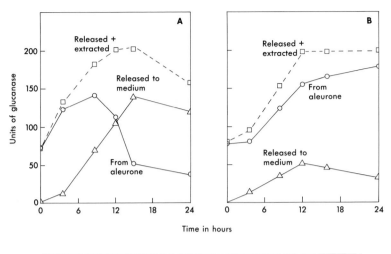

FIGURE 12.9   Release of β-1,3-glucanase from 5 isolated barley aleurone layers in the presence (A) and absence (B) of $5 \times 10^{-7} M$ GA$_3$. Note that GA$_3$ affects the rate of release of the enzyme from the aleurones but not the total amount. [Adapted with modifications from L. Taiz and R. L. Jones, *Planta* 92:73–84 (1970).]

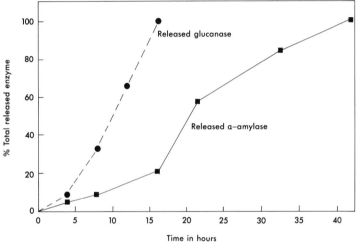

FIGURE 12.10   Comparison of the time course of release of β-1, 3-glucanase and α-amylase from barley aleurones in the presence of $5 \times 10^{-7} M$ GA$_3$. [Adapted with modifications from L. Taiz and R. L. Jones, *Planta* 92:73–84 (1970).]

glucose residues in polysaccharides. Cell walls of the aleurone are rich in carbohydrates containing β-1,3-linked glucose residues (as determined by their fluorescence when stained with the dye aniline blue), but these carbohydrates are lacking in the endosperm. The glucanase can be extracted from dry seeds and in ever-increasing amounts from seeds imbibing water from moistened sterile sand. Treatment of isolated aleurone layers with GA$_3$ brings about the release of β-1,3-glucanase into the medium and a concomitant decrease in the amount of extractable enzyme in the tissue (Figure 12.9). GA$_3$ apparently does not, however, affect the amount of enzyme produced, only the onset and rate of its release into the medium.

When the time course of glucanase release in the presence of GA$_3$ is compared with that of α-amylase (Figure 12.10), important differences are noted. GA$_3$-stimulated release of glucanase begins at 4 hours and is complete in 16–18 hours, whereas that of α-amylase begins at 4 hours, but only slowly, and then, after 20–24 hours, it proceeds more rapidly and continues for 42 hours. This schedule suggests that β-1,3-glucanase exercises a preparatory role in the release of endosperm-digesting enzymes in barley.

12.2 Growth and Differentiation of Seedlings

The growth of plants is initiated in specialized zones of cell division called **meristems.** Among higher plants the cells leaving the division zone usually enlarge considerably and then differentiate into one of the several structural elements. Prominent among these are the complex vascular tissues **xylem** and **phloem, as well as parenchyma, sclerenchyma,** and **cork.** These terms are familiar to students of elementary biology, but some further definition should emerge from the ensuing discussion of the differentiation of structures in the root and stem.

As pointed out in Chapter 6, the parts of the plant embryo are readily distinguishable as to their future fate after relatively few divisions of the zygote have occurred. During maturation of the seed two distinct **apical meristems** become defined, one at the tip of the **radicle,** forming the apex of the **primary root,** and one at the opposite end of the embryo at the base of the cotyledon or cotyledons, which is the apex of the **primary shoot,** called the **epicotyl.** From the latter originate the stem and leaves. In many seeds, such as those of barley, considerable differentiation of structure will have occurred before the onset of dormancy, so that a terminal bud, or **plumule,** consisting of apical meristem and primary leaves, is already present.

12.2.1 DEVELOPMENT OF THE ROOT. Let us begin this discussion by outlining the structural components of the root. These components differentiate from cells that arise by division in the apical meristem. Differentiation of the various tissue types occurs, however, at varying distances behind the tip. A cross section through the root at a level where primary differentiation is complete (Figure 12.11) reveals an outer layer of cells adorned with **root hairs.** This layer is the **epidermis.** It encloses the cortex, a layer of thin-walled, relatively undifferentiated **parenchyma** cells. The parenchyma serves as food storage tissue and also transfers water and minerals, absorbed through the root hairs, to the

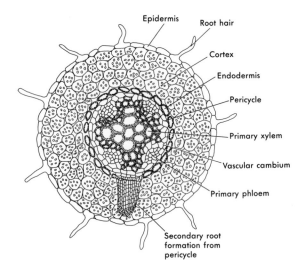

FIGURE 12.11 Cross section through the root of a dicotyledonous plant in the region of maturation.

vascular tissue. The inner boundary of the cortex is the **endodermis,** usually a single layer of thick-walled cells.

Within the endodermis is the **stele** or **vascular cylinder.** Its outermost layer is the **pericycle,** consisting of one or more layers of parenchyma cells. This layer is very important in the production of lateral roots and in the **secondary growth** of the root, which occurs after germination. In the usual dicot pattern the primary xylem occupies the center of the vascular cylinder, often arranged in a star-like pattern as seen in cross section. The first vessels actually differentiate in the peripheral portion of the xylem, and the later differentiating ones appear progressively nearer the center. In the dicotyledonous plants the phloem is peripheral to the xylem and is separated from it at first by layers of thin-walled parenchyma cells. One layer of these cells remains meristematic and is given the name **vascular cambium.** It is important in the production of **secondary tissues,** as occurs when the root enlarges in diameter.

12.2.1.1 Cellular Differentiation in the Primary Root. The organization and pattern of differentiation in the primary root may first be examined in longitudinal sections of the root tip. In such a section (Figures 12.12 and 12.13) one sees parallel files of cells leading from the meristem toward the older part of the root. These files are produced by re-

286

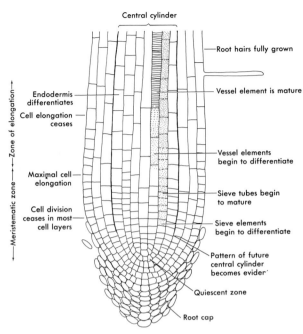

FIGURE 12.12 Diagram of the growing zone of a dicot root as seen in longitudinal section. [Adapted from various sources, including P. M. Ray, *The Living Plant*, Holt, Rinehart and Winston, New York 1963, p. 81.

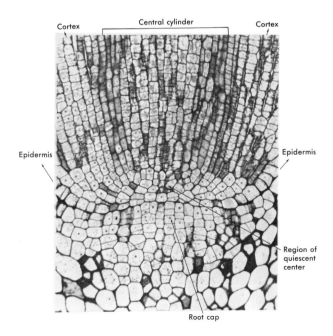

FIGURE 12.13 Median longitudinal section of the root apical meristem of the onion *Allium cepa* showing the cellular relationships in the apical initials, quiescent center, and primary tissues of the root. Compare with Figures 12.12 and 12.14. [Reprinted with permission of Macmillan Publishing Co., Inc, from *Development in Flowering Plants* by J. G. Torrey p. 83. Copyright © 1967, J. G. Torrey. Dr. Torrey's courtesy in providing this photograph, originally given him by Dr. F. A. L. Clowes, is gratefully acknowledged.]

peated division planes parallel to the longitudinal axis. Transverse divisions add new cells to these files proximally and contribute cells to the **root cap** distally. The root cap forms a protective layer the cells of which are sloughed off as the root tip pushes through the soil.

In the columns of cells leading back from the tip both cell division and growth continue to occur until, at a distance of some millimeters from the tip, divisions cease and cells now add to the length of the root by elongation. The meristem ends as the **zone of elongation** begins, but new cells are continually being added to the elongation zone from the meristem. Further back from the tip cells cease elongating and some of those forming the outer layer of the cylinder begin to project **root hairs** into the soil. These are very numerous and provide a greatly increased surface area for the absorption of water and nutrients from the soil.

If one examines the distribution of mitotic figures in the meristem, one finds that cells are not dividing most rapidly at the tip, but rather, at some distance behind it. Moreover, near the tip there is a zone in which mitotic figures do not occur. This is called the **quiescent zone** (Figures 12.13 and 12.14). This zone may involve several hundreds of cells. On its periphery are the **apical initials** whose daughter cells give rise to the root cap on one side and to the files of cells making up the structure of the root on the other.

Patterns of future cellular specialization begin to appear very early. Already in the lower zone of the meristem the arrangement of cells and the distinctive staining characteristics of their nuclei, protoplasts, and cell walls give evidence of the presence of a differentiated outer layer, the epidermis, and a central cylinder, or stele, with a multilayered cortex between.

The visible differentiation of the various kinds of tissue does not occur synchronously. The first-dif-

287

FIGURE 12.14 Median longitudinal section of the root tip of corn, *Zea mays*, showing the quiescent center. The plant was germinated in medium containing ³H-labeled thymidine, which was incorporated into the nuclei of cells synthesizing DNA in preparation for mitosis. This radioautograph reveals dramatically the absence of DNA synthesis and mitosis in the quiescent center. Nuclei of surrounding cells are heavily labeled. [Reprinted with permission of Macmillan Publishing Co., Inc., *Development of Flowering Plants* by J. G. Torrey, p. 86. Copyright © 1967, J. G. Torrey. Gratitude is extended to Dr. Torrey for providing the photograph, originally given him by Dr. F. A. L. Clowes.]

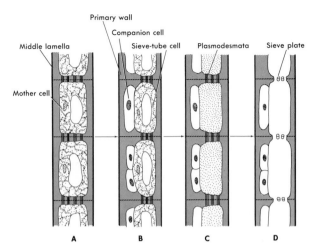

FIGURE 12.15 Schematic illustration of successive stages (A–D) in the development of sieve tubes of the phloem. [Adapted from P. B. Weisz and M. S. Fuller, *The Science of Botany*, McGraw-Hill, New York, 1962, p. 80.]

ferentiated cell types to appear in the future stele are the sieve tubes of the phloem (Figures 12.15 and 12.12). The sieve tubes actually differentiate within the meristematic zone. These join end to end, connecting to progressively older sieve tubes further up the root, meanwhile continuing to elongate. The sieve tubes lose their nuclei at maturity and are associated with special parenchyma cells of the phloem called companion cells. These have cytoplasmic connections with the sieve cells and conceivably aid the latter in regulating conduction. Other parenchyma cells of the phloem are arranged in vertical files among the sieve tubes. Ends of vertically connecting sieve tubes become perforated so that there is continuity between them. The sieve tubes function predominantly in the vertical transport of dissolved organic substances.

Xylem cells, heavier walled than the conducting cells of phloem, begin to appear further behind the tip. The conducting elements of xylem are of two types: **vessels** and **tracheids.** The vessels (Figure 12.16) arise from cells called **vessel elements,** which first appear in the zone of elongation and join end to end with earlier arising elements. Their nuclei die and end plates disappear, creating continuous tubes, the vessels. The wall of the vessel elements that arise in the zone of elongation are thickened by the secondary deposition of lignin in helical array or as separate transverse rings. The turns of the helices and spacing of the rings increase as elongation occurs. In later-differentiating vessels, the

FIGURE 12.16 Diagram of the development of a xylem vessel.

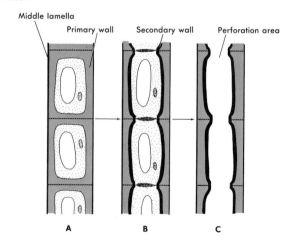

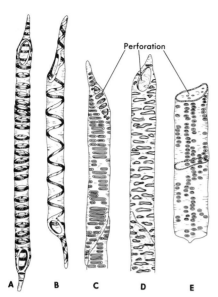

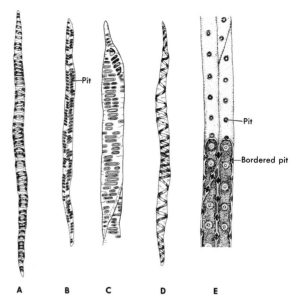

FIGURE 12.17 A–D: Tracheids (E): Portion of a vessel. A and B show the deposition of lignin in annular and spiral patterns, respectively. C and D show the deposition of lignin in pitted patterns and also, at each lower end is shown the manner in which the tracheids are joined to form tubes. D shows two vessel elements joined, forming a portion of a vessel. Perforations permitting communication between vertical xylem elements are labeled in C, D, and E. [From V. E. Greulach and J. E. Adams, *Plants: An Introduction to Modern Botany*, 3rd ed., © 1967, John Wiley & Sons, Inc. Reprinted by permission of John Wiley & Sons, Inc.]

FIGURE 12.18 A–E; Tracheids, showing some of the variety of patterns in which lignin is deposited. In E are shown tracheids with so-called bordered pits, which are paired and specialized for horizontal transport of sap. [From V. E. Greulach and J. E. Adams, *Plants: An Introduction to Modern Botany*, 3rd ed., © 1967, John Wiley & Sons, Inc. Reprinted by permission of John Wiley & Sons, Inc.]

289

secondary walls constitute a continuous structure having pitted walls (Figure 12.17E). Vessels formed by the end-to-end junction of vessel elements may be many feet in length. They function in the upward transport of water and dissolved minerals in the plant body.

The other chief conducting element of xylem is the tracheid. Tracheids, too, are elongate and also form continuous columns. Their secondary walls are thickened in various ways, as shown in Figures 12.17A–D and 12.18. In some cases paired pits form on adjacent tracheids allowing for conduction between them (Figure 12.18E). Tracheids are less prominent in flowering plants than in gymnosperms and other more primitive plants. As xylem matures it comprises a very complex tissue consisting of long woody fibers and other supporting elements and thinner-walled parenchyma cells.

12.2.1.2 Development of Secondary Tissues in Roots. In woody dicots the roots form secondary tissues. These tissues arise from the vascular cambium and from the pericycle, which remains meristematic. When the vascular cambium divides, it produces new cells that are added to the primary xylem and primary phloem, forming secondary xylem and secondary phloem, respectively. The secondary xylem grows much more rapidly and fills out the spaces between the primary xylem cells until the total xylem becomes circular in outline. This movement pushes the primary phloem outward, but the vascular cambium, augmented by new cells from the pericycle, now becomes continuous around the xylem, and produces a continuous circle of phloem external to the xylem and vascular cambium.

Within the pericycle divisions continue. A **phellogen** (Gr. *phellum*, cork) or **cork cambium** forms as a peripheral layer of the pericycle. Outwardly this cambium produces cork cells whose

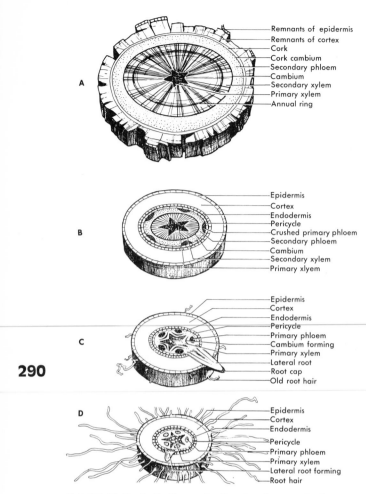

- Remnants of epidermis
- Remnants of cortex
- Cork
- Cork cambium
- Secondary phloem
- Cambium
- Secondary xylem
- Primary xylem
- Annual ring

A

B
- Epidermis
- Cortex
- Endodermis
- Pericycle
- Crushed primary phloem
- Secondary phloem
- Cambium
- Secondary xylem
- Primary xlyem

C
- Epidermis
- Cortex
- Endodermis
- Pericycle
- Primary phloem
- Cambium forming
- Primary xylem
- Lateral root
- Root cap
- Old root hair

D
- Epidermis
- Cortex
- Endodermis
- Pericycle
- Primary phloem
- Primary xylem
- Lateral root forming
- Root hair

290

FIGURE 12.19 A–D: Stages in the development of a root as shown in sections through progressively younger regions A being the oldest and showing two annual growth rings. [Reprinted with permission of Macmillan Publishing Co., Inc., from *Botany: A Functional Approach*, 4th ed., by Walter H. Muller. © 1979, Walter H. Muller.]

secondary walls become thickened and waxy and eventually die. Inwardly the phellogen produces a **phelloderm** whose cells are usually difficult to distinguish from parenchyma derived from the growing pericycle.

The expansion of the vascular cylinder in secondary growth causes the sloughing of the cells of the endodermis, cortex, and epidermis. Sloughing exposes the cork, which becomes wrinkled and furrowed as it ages.

12.2.1.3 Development of Branch Roots.

Most kinds of roots show copious branching through the formation of lateral roots. Lateral roots arise from the parent root during primary growth, as shown in Figures 12.11 and 12.19C, D, usually taking origin from the pericycle. Within the branch root there occurs the same pattern of growth and differentiation as in the root from which it arose. The vascular bundles of the root and branch unite and become confluent.

12.2.2 DEVELOPMENT OF THE SHOOT.

The shoot of a plant is a continuously growing system as a result of the meristematic activity of the shoot apex. The shoot apex is derived from the epicotyl of the seed, which lies at the base of the cotyledons and which may already have the first true leaves developing from it. Stem and leaves continue to develop in orderly relationship to each other, forming what is called a **terminal bud.** Compressed within the bud are successive generations of young leaves at various stages of development, the youngest nearest the apex being enveloped by the older ones below. Leaves are formed in a variety of orderly patterns at the apex, varying in different species. They may arise in pairs opposite each other, in whorls, or singly in various spiral arrangements. The point of leaf attachment is termed a **node** and the stem portion between nodes is an **internode.** In the bud new leaves are formed close upon one another, but later the internodal regions elongate.

12.2.2.1 Origin of the Primary Shoot.

The hemispherical region of the terminal bud that is distal to the base of the youngest leaves (Figure 12.20) is called the **promeristem.** The remainder of the apical meristem extends some distance below and includes the tissue between the forming leaves, **ground meristem,** and the meristems from which the leaves arise. Procambium, from which the vascular vessels arise, appears in the ground meristem.

In angiosperms the two outer layers increase by anticlinal divisions; that is, the cell walls dividing daughter cells are perpendicular to the surface. These layers are the outer and inner tunic layers. They enclose the **corpus,** or body, of the promeristem. In the corpus, mitoses are generally randomly oriented, but tend to occur in an anticlinal direction toward the outside (Figure 12.21A). At the periphery of the promeristem, cell division is most active. Here the new cells that arise in the cor-

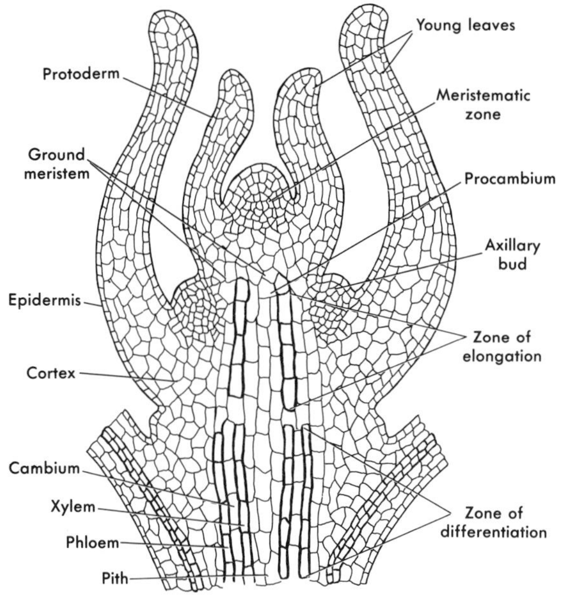

FIGURE 12.20 Growth of the apical meristem and formation of leaf primordia in a flowering plant. [Adapted with modifications from J. G. Torrey, *Development in Flowering Plants*, Macmillan, New York, 1967, p. 116, and from P. M. Ray, *The Living Plant*, Holt, Rinehart and Winston, New York, 1963, p. 89.]

mesophyll layers, and the procambium of the leaf arises from the corpus. A great deal of attention has been paid to the questions of what determines that a leaf shall appear and where it will arise. Despite the fact that reams have been written about these matters, no satisfactory answers have been forthcoming.

The differentiation of vascular tissue in the stem proceeds toward the base of the shoot from the apical meristem (Figure 12.20). Procambium arises in the meristem as groups of elongated cells that lay out the pattern of vascular bundles as one follows longitudinal sections toward the base of the shoot. From these elongated cells sieve tubes and other elements of the phloem (fibers, phloem parenchyma) arise nearer the outside of the stem. At lower levels, in the zone of cell enlargement, vessel elements and tracheids, if any, and other elements of xylem (fibers, xylem parenchyma) appear on the side of the strand toward the inside of the stem.

The pattern of primary vascular tissues in the shoot is related to the pattern of initiation of leaves in the cortex. Just as leaves appear at regular intervals around the base of the promeristem, so do vascular bundles arise at corresponding intervals in a circular pattern in the ground meristem and become continuous with vascular bundles leading into the leaves as the latter appear.

The pattern of differentiation of vascular bundles in the leaf is unlike that found in the root. Primary phloem differentiates in the procambium in strands continuous with the vascular bundles developing below, appearing progressively less differentiated further into the leaf primordium apically. Primary xylem, on the other hand, appears later,

291

pus produce a localized bulge in the periphery at the site of each leaf primordium. The two tunic layers continue over the bulge, and all three components of the leaf primordium (bulge) contribute to the formation of the leaf meristem. The outer tunic forms the upper and lower epidermis of the leaf, the inner tunic contributes the leaf parenchyma and

FIGURE 12.21 Organization of the growth region at the tip of a stem.

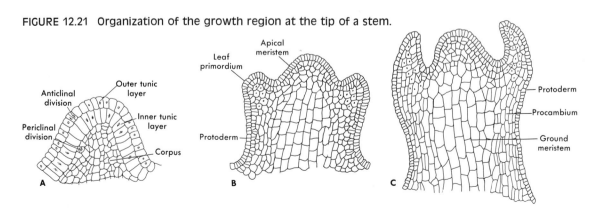

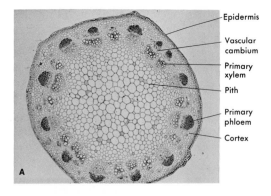

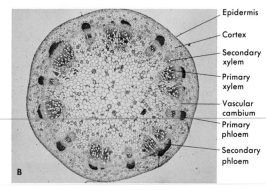

FIGURE 12.22 Cross sections of the stem of an herbaceous dicot, the sunflower. A: Vascular bundles showing primary xylem and phloem. B: The beginning of secondary growth; note the spread of the vascular cambium between the primary vascular bundles. [Courtesy of Carolina Biological Supply Company.]

The upper part expands to form the blade. The vascular tissue differentiates a central vein, or **midrib,** and from this, vascular branches diverge in characteristic fashion through the leaf parenchyma.

The stem of the young seedling consists of primary tissues alone (Figure 12.22A). The outermost layer is the epidermis, one cell in thickness and covered with its own waxy product, **cutin.** Beneath the epidermis is the **cortex,** consisting chiefly of thin-walled parenchyma cells, some with thicker walls, called **collenchyma,** and some fibers. The cortex is chiefly a food-storage tissue that degenerates in older stems. Continuing inward, the first permanent tissue that one encounters is the primary phloem of the vascular bundles. The phloem is separated from the primary xylem by a layer of vascular cambium. The central core of the stem is composed of parenchyma cells that constitute the **pith,** which functions in storage.

12.2.2.2 Development of Secondary Tissues.
The lengthening of the dicot stem occurs entirely through primary growth. Growth in diameter of the stem, however, requires the formation of secondary tissues through the continued division of the vascular cambium. As the cambium divides, cells produced toward the outer portion of the stem develop the secondary phloem, and those that are produced inwardly form the secondary xylem. Meanwhile, the zone of dividing cells spreads into the parenchyma separating the vascular bundles (Figure 12.22B). There the vascular cambium also produces secondary xylem to the inside and secondary phloem to the outside. This process results in the formation of a continuous cylinder of vascular tissues around the circumference of the plant beneath the cortex. The production of secondary xylem is largely responsible for the increase in diameter of a stem or branch.

In woody dicots another group of secondary tissues arises from the cork cambium, a lateral meristem similar to that found in the root. It originates from the outer layers of the cortex and, as its cells divide, it produces outer layers of cork cells and an inner cortical layer called phelloderm. Growth of the cork layer provides a protective tissue over the stem. The cork, together with underlying cork cambium, phelloderm, and the phloem constitutes the bark of the stem (Figure 12.23B).

In woody plants, such as trees and shrubs that

differentiating first at the point of leaf attachment to the axis of the shoot. From this point differentiation proceeds both up the leaf toward the tip and down the stem to connect with the already mature xylem.

At the point where each leaf joins the stem there appears in the angle it makes with the stem (leaf axis) an axial bud (Figure 12.20). These buds remain small and inconspicuous and are, in effect, dormant promeristems that can, under appropriate conditions, form lateral shoots or branches. They usually do so at appropriate distances from the apical bud. The leaf itself first appears as a key-like projection. The lower end of the peg becomes the petiole of the leaf, chiefly occupied by vascular tissue.

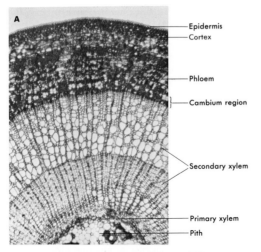

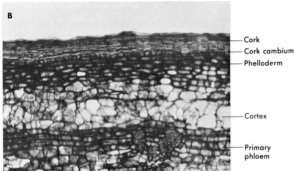

FIGURE 12.23 Characteristics of the stem in a woody dicot, basswood. A: Partial cross section through a stem at the end of its second year of growth. B: Outer layers of 5-year-old stem showing cork, cork cambium, and phelloderm; these constitute the periderm. With advancing age, the cortex will be compressed as secondary phloem pushes the primary phloem outward.

persist from season to season, cambial activity ceases at the end of the growing season. Resumption of activity at the next growing season leaves a clear line of demarcation between the new **spring wood** of larger, thinner-walled xylem cells, and the smaller, thicker-walled xylem elements of the **summer wood** formed the previous season (Figure 12.23A). The amount of tissue formed during one year's growth, as seen in cross section, constitutes the annual growth ring.

12.3 Hormonal Control of Shoot and Root Development

Three major classes of substances are now recognized as serving in the regulation of growth in flowering plants. (Here we do not include "florigens," treated in Chapter 5, for no florigen has yet been clearly identified.) These substances are **auxins, gibberellins,** and **cytokinins.** We have already considered gibberellins in the context of flowering (Chapter 5) and earlier in this chapter with respect to the synthesis and release of enzymes involved in germination. The effects of auxins have been known since 1910, but the discovery of gibberellins and cytokinins and the investigation of their effects on plant growth have been more recent events.

12.3.1 AUXINS. Auxins are growth-promoting substances. The first experiments pointing toward the existence of such a substance had to do with the effects of light on the direction of growth. If a grass seedling is germinated in the dark or under conditions of uniform illumination, its direction of growth is directly upward. If subjected to light coming from only one side, however, it curves toward the light, its chief curvature occurring immediately behind the tip. If the apex is shielded from the source of illumination, growth continues, but it is in the vertical direction. If, however, the base is shielded and the tip is exposed, the tip bends toward the light. Apparently the light stimulus affects the tip of the plant and some substance produced by the tip moves down the stem. The distribution of this substance is affected by light. This interpretation was first suggested by experiments of Charles Darwin (1897), as illustrated in Figure 12.24.

The proof that a growth hormone is actually produced in the tip and transported to lower levels first came from experiments of Boysen-Jansen, of Denmark, in 1910. He studied growth of the **coleoptile** of the oat (*Avena*) seedling. The coleoptile is the modified first leaf of the plant or, possibly, along with the **scutellum,** is derived from the single cotyledon of the embryo (Figures 6.15 and 12.25). The

293

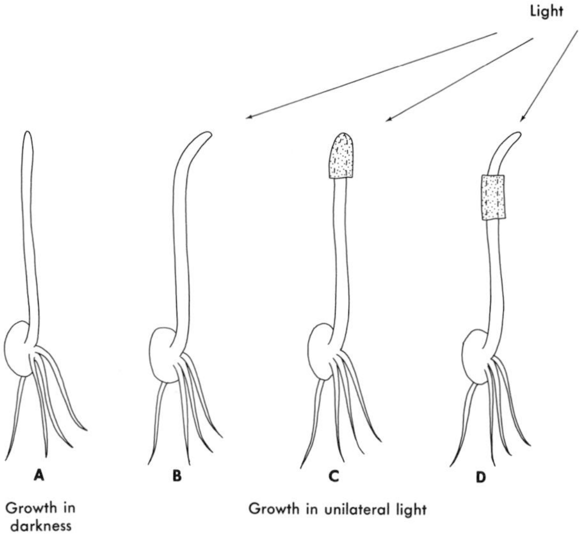

Light

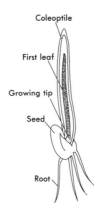

FIGURE 12.24 Growth of a grass seedling under various conditions. (A) Vertical growth in the absence of illumination. (B) The apex curves toward a source of unilateral light. (C) Growth is vertical if the apex is shielded from the light source. (D) The tip curves toward the light source if the coleoptile is shielded below the apex. Thus the effect of light is perceived at the apex. [Adapted with modifications from A. W. Galston and P. J. Davies, *Control Mechanisms in Plant Development,* Prentice-Hall, Englewood Cliffs, NJ, 1970, p. 58.]

A — Growth in darkness
B, C — Growth in unilateral light
D

294

FIGURE 12.25 Diagram of a 3-day oat seedling showing the relationships of the coleoptile to other parts of the plantlet.

Coleoptile
First leaf
Growing tip
Seed
Root

ing, but if the tip is replaced, elongation resumes. This effect is abolished if a flake of impermeable mica (a mineral composed of complex silicates, which crystallizes in thin, translucent, easily separated layers) is inserted between the tip and the rest of the coleoptile. If, however, a thin flake of agar, which is permeable to water and small molecules, is interposed, the coleoptile elongates normally. Moreover, the isolated tip could be placed on a block of agar for some time and then removed, whereupon the bit of agar, when placed on the decapitated coleoptile, would cause it to grow (Figure 12.26A). Clearly a growth substance was captured by the agar and then released.

FIGURE 12.26 Production of growth substance by the apex of the oat coleoptile. A: Normal elongation of the coleoptile in darkness or in uniform illumination. Growth ceases if the coleoptile tip is removed (B), but resumes when the tip is replaced (C). D: Capture and release of growth substance by a block of agar.

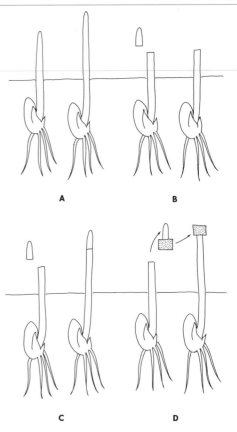

A   B

C   D

coleoptile elongates in advance of the enclosed stem, but it eventually ruptures, releasing the definitive first leaves and the apical meristem.

Growth of the coleoptile results chiefly from cell elongation below the tip. If the tip is cut off, the decapitated coleoptile immediately ceases elongat-

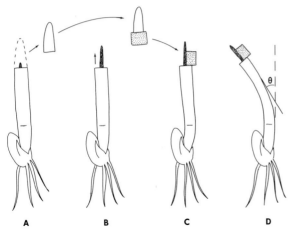

FIGURE 12.27 Went's bioassay for auxin. A: The coleoptile of a 3-day germinated *Avena* (oat) seedling is decapitated. B: The leaf within is pulled upward. C: An agar block containing auxin is then applied, supported by the leaf so that it contacts the cut surface of the coleoptile. D: Auxin transported down the coleoptile from the block causes that side to grow faster, resulting in curvature. The angle of curvature, θ, produced under standard conditions is a measure of the auxin concentration. The auxin source is here shown as supplied by contact with the coleoptile tip. The method is applicable to the measurement of auxin from various sources. [Adapted with modifications from P. M. Ray, *The Living Plant,* Holt, Rinehart and Winston, New York, 1963, p. 99.]

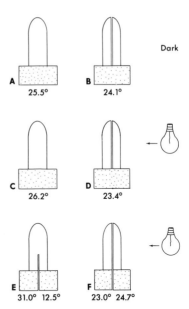

FIGURE 12.28 Redistribution of auxin under the influence of unilateral light. Each figure shows a corn coleoptile tip placed atop an agar block of standard size. In B, D, and F, vertical halves of the coleoptile are shown separated by a thin glass sliver; in E, only the basal portion of the coleoptile is divided, but the glass barrier also divides the block of agar into halves in E and F. Auxin diffusing into the agar during a 3-hr period from coleoptiles held in the dark (A, B) or subjected to unilateral illumination (C–F) was measured by means of Went's *Avena* coleoptile assay as illustrated in Figure 12.27. Results are expressed as degrees of angle θ (Figure 12.27D) and reflect the amount of auxin that diffused into the agar block during the period of test. Because the standard block was divided into equal halves in E and F, two half-blocks from replicate experiments were combined for the *Avena* assay in order to permit the amount of auxin that diffused into them to be compared to the amounts that diffused into standard blocks used in A–D. Figures A–D convey, in sum, that coleoptile tips produce the same amount of auxin whether illuminated or held in the dark and whether split or intact. Figure E shows that when lateral transport of auxin is not barred, as it is in F, more auxin is recovered from the agar below the unilluminated side. [Adapted with modifications from W. S. Briggs et al., *Science* 125:210–13 (1957).]

When an agar block containing the growth substance is placed unilaterally on a decapitated coleoptile in the dark, it causes curvature away from the side on which it was placed. This property was used by Frits Went, around 1930, to develop a method for detecting and measuring the amount of growth substance (Figure 12.27) produced by or extracted from a given source. The growth substance was named auxin.

By Went's method it can be shown that light that impinges unilaterally on a dark-grown seedling produces an asymmetry in the distribution of the auxin in the tip. More auxin is present on the dark side and this causes bending toward the light. Unilateral illumination does not influence the production of auxin, however, for as much auxin can be recovered from the shaded half-tip of a longitudinally split coleoptile as from the illuminated half-tip. What light does is to cause its transport to the unilluminated side (Figure 12.28). The method whereby light induces the lateral transport of auxin toward the unilluminated side of the coleoptile is not known.

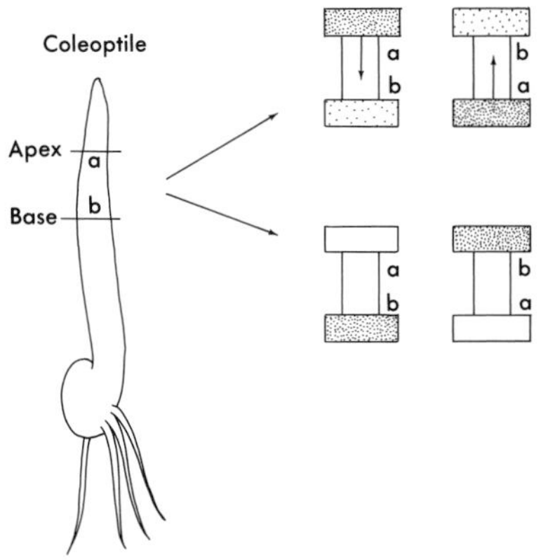

FIGURE 12.29 Polarized transport of auxin through coleoptile tissues. Auxin (indicated by stipples) is transported from the apical to the basal portion of a segment of coleoptile from an agar block regardless of the gravitational orientation of the segment (upper right). It will not diffuse in the opposite direction (lower right). [Adapted with modifications from A. W. Galston and P. J. Davies, *Control Mechanisms in Plant Development*, Prentice-Hall, Englewood Cliffs, NJ, 1970, p. 63.]

296

Of equal interest and likewise puzzling is the fact that auxin is transported strictly in a polarized manner from tip to base. It does not normally move in the opposite direction. Thus, if a subapical section of an *Avena* coleoptile has auxin applied at its apical end, auxin can be collected in an agar block at the basal end. On the other hand, if the auxin is applied at the basal end of the segment, none can be collected at the other end. These same results are found regardless of the orientation of the segment with respect to gravity (Figure 12.29).

Went's method made possible the purification and identification of auxin and auxin-like com-

pounds. The first auxin to be chemically identified was **indole-3-acetic acid** (IAA) (Figure 12.30). It was first extracted from corn seed and has since been obtained from many other plant tissues. Several close relatives of IAA also occur naturally: indoleacetaldehyde, indoleacetonitrile, and indolepyruvic acid, all of which are potential precursors of IAA. A number of related compounds possess some aspects of auxin activity. Among these compounds are α-naphthaleneacetic acid and indolebutyric acid. 2,4-Dichlorophenoxyacetic acid (2,4-D) is another close relative, much used as a herbicide to control broad-leafed plants in lawns. All auxins applied in high concentrations are toxic, roots being the most sensitive. 2,4-D is a particularly potent auxin, exercising toxic effects at relatively low concentrations.

Auxins are responsible for major aspects of plant morphology. IAA (or its relatives) is synthesized in the plant in the terminal bud and in the youngest leaves. Transported down the stem, it controls the rate of growth in the zone of elongation, possibly in cooperation with gibberellin. Thus, the elongation of the internodal regions progresses in proportion to the expansion of the leaves as the shoot grows.

A second major effect of auxin is the suppression of axillary buds. As noted, these buds are promeristems that develop in the angle where leaf meets stem and are the source of side branches. Normally the apical meristem is dominant, in the sense that it contributes to the increasing height of the plant, and side branches develop only at some variable distance behind the apex. This dominance is referred to as **apical dominance.** If, however, one removes the apical meristem from a plant showing apical dominance, then axillary buds near the tip begin to grow and produce their own auxin. Quite often one will achieve dominance over the others,

FIGURE 12.30 Indoleacetic acid and other compounds showing auxin activity.

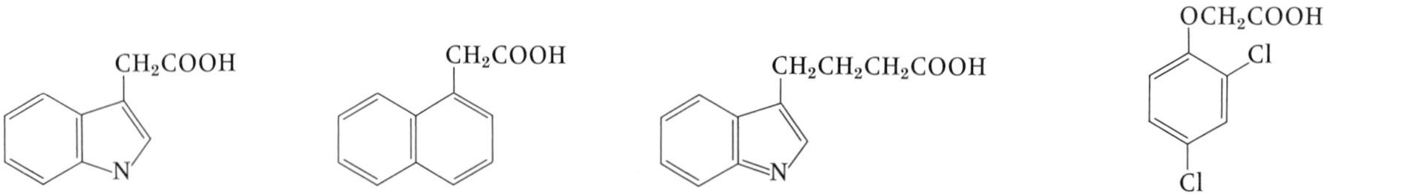

Indoleacetic acid (IAA) α-Naphthaleneacetic acid (NAA)   Indolebutyric acid (IBA)   2,4-Dichlorophenoxyacetic acid (2,4-D)

and become the new apical meristem. The application of auxin to the cut tip, however, prevents growth of the axillary buds. The selective trimming of buds at the tips of shoots and branches can be used to promote bushy outgrowth in plants that would otherwise tend to grow tall and spindly. As branches grow out, their terminal buds, too, produce auxin, thus suppressing the development of the axillary buds more basal to them.

Auxin transported down the stem eventually enters the roots where it is destroyed. The effective agent is probably the enzyme peroxidase, which inactivates IAA.

Many plant stems that are severed and placed in water or moist sand or soil will produce roots near the cut end. Presumably this rooting is the consequence of the accumulation of auxin. Many cuttings that do not root spontaneously can be induced to do so by applying auxin to the cut stem. Indolebutyric acid, an analog of IAA, is especially effective in this regard. The auxin apparently activates the cambium near the cut surface.

Auxin also plays a role in the orientation of a plant with respect to gravity. Emerging from the seed coats, the root grows downward and the shoot upward. If the seedling is laid horizontally, the stem curves upward and the root downward (Figure 12.31). When the seedling is horizontal, auxin is transported to the lower side where it has differential effects on root and stem, inhibiting the elongation of cells in the former and promoting their elongation in the latter. The actual occurrence of auxin transport in response to gravitational force can be demonstrated by observing the movements of radioactively labeled auxin in seedlings. The gravitational field can be substituted or annulled by centrifugal force. Accordingly, if germinating seeds are placed in a centrifuge creating a force of 1× gravity at right angles to the direction of gravitational pull, the roots will grow at an angle of 45° from the verti-

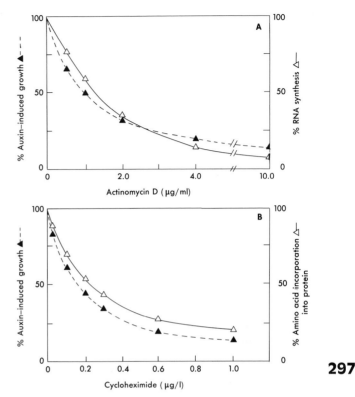

FIGURE 12.32 Effects of inhibitors on growth and macromolecular synthesis of auxin-induced growth of excised soybean hypocotyls. The hypocotyls were incubated in media containing 10 μg/ml 2,4-D. A: Effects of actinomycin D on growth (wet weight) and mRNA synthesis. B: Effects of cycloheximide on growth and incorporation of radioactive leucine. Note that both inhibitors exercise parallel effects. [Adapted with modifications from J. L. Key et al., *Ann NY Acad Sci* 144:49–62 (1967).]

297

cal. They thus respond to the vector of forces created by the earth's gravity and by the centrifuge. Thus evidence suggests that the earth's gravitational field is responsible for the displacement of some dense bodies in cells and that these somehow determine the direction of auxin movement. The exact nature of these bodies and their mode of action are uncertain.

What auxin actually does in order to promote or inhibit cellular elongation remains uncertain, too. It certainly does something to produce changes in extensibility of the rigid cellulose wall of the cell; otherwise, the affected cells could not change in

FIGURE 12.31 Differential effects of auxin under the influence of gravity on root and stem of a seedling.

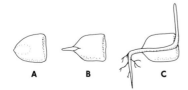

size. Presumably, increased extensibility occurs through the action of enzymes that catalyze reactions breaking cross links between the cellulose microfibrils. Auxin might induce the formation or activation of such enzymes. This possibility is supported by the observations that (1) auxin promotes the incorporation of precursor molecules into RNA and protein and (2) auxin-induced growth is inhibited by drugs that inhibit, respectively, the synthesis of mRNA and the synthesis of protein (Figure 12.32). Unfortunately for this idea, however, is the fact that, at appropriate temperatures, the auxin-induced response occurs almost instantaneously, a fact that precludes attributing its effect on cell elongation to the synthesis of new enzymes.

12.3.2 CYTOKININS. The cytokinins are naturally occurring and synthetic substances, related chemically to adenine, that have profound effects on the growth by cell division of plant cells cultured in vitro and on cells of young fruits and developing seeds. Naturally occurring cytokinins are apparently synthesized in the root and are translocated by the xylem to the shoot, for they appear in the sap bleeding from severed or wounded shoots. Applied to meristems whose growth is suppressed by auxin from the apex, cytokinins counteract the inhibition and stimulate cell division. Thus there is possibly an auxin-cytokinin antagonism operating to determine the form of growing shoots.

In the early 1940s it was found that a factor in **coconut milk** (a liquid form of endosperm) greatly promotes the growth of embryos of *Datura* (Jimson weed) isolated during early stages of seed formation. During the 1950s F. C. Steward investigated the ability of coconut milk to promote the rapid growth of carrot cells in culture, but no single material could be isolated from the coconut milk that would duplicate the effects of the milk, either alone or combined with other plant substances such as auxins. Meanwhile, Skoog and his collaborators (1957) found that adenine or adenosine could promote the growth of tobacco pith cells in culture. A more powerful growth promoter, however, was 6-furfuryladenine, which they isolated after autoclaving DNA from any source under appropriate conditions. It activated cell division when applied with auxin and was named **kinetin.** Later, after other chemical substances with similar activities were identified among synthetic and natural prod-

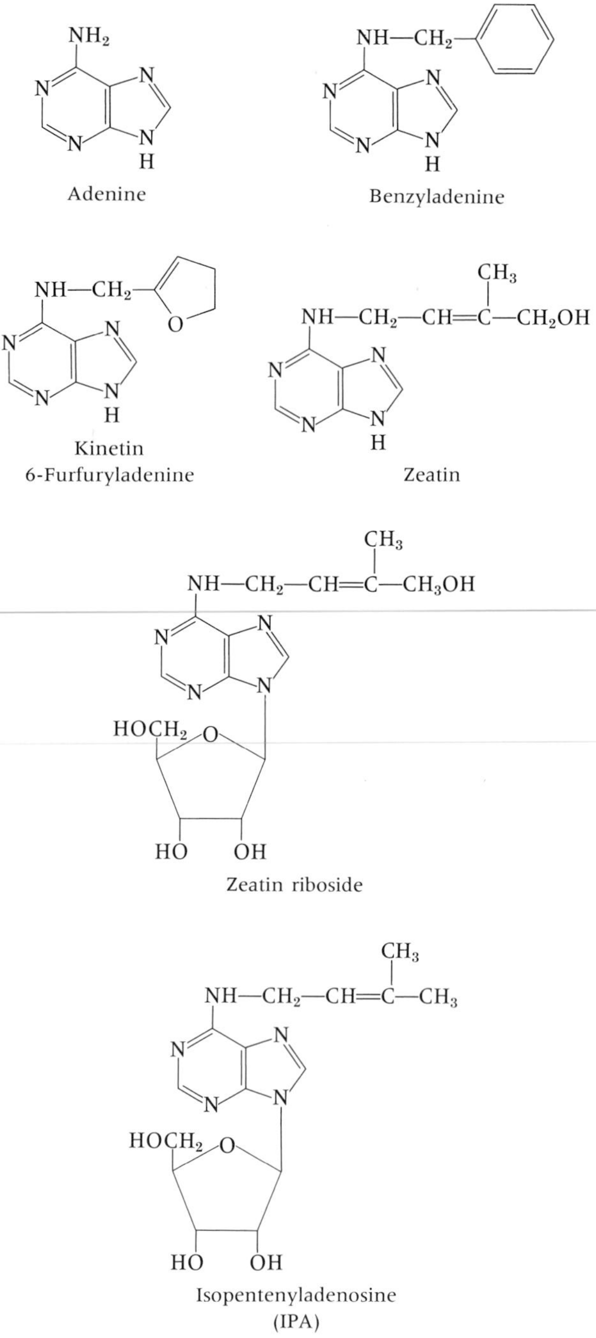

FIGURE 12.33 The structure of adenine and of derivatives showing cytokinin activity. Zeatin and isopentenyl adenosine are naturally occurring cytokinins.

298

ucts, they were collectively called cytokinins (Figure 12.33).

The demonstration that natural substances having growth-promoting activities comparable to the activity of 6-furfuryladenine actually occur awaited the 1960s. In 1964 Letham and his associates identified a powerful cytokinin in young maize seeds. This material, another adenine derivative, bears a five-carbon isoprenoid group instead of the furfuryl group on the nitrogen of the 6-amino group of kinetin; it was called zeatin. About the same time, a similar compound, zeatin riboside, was found in sweet corn. This compound was finally determined in 1967 to be the active compound in coconut milk.

Because the cytokinins constitute a family of compounds that may be classified as adenine derivatives, it is reasonable to inquire whether they naturally occur as parts of the nucleic acids. They have, in fact, been discovered in hydrolyzates of RNA from a number of species, but they occur only in the transfer RNAs of several amino acids and have never been found in hydrolyzates of rRNA. The first cytokinin to be so identified was N^6-(Δ^2-isopentenyl)-adenosine, or IPA, in hydrolyzates of yeast tRNA. IPA was subsequently identified in spinach and sweet peas. Zeatin riboside is its hydroxylated derivative.

Cytokinins bring about both quantitative and qualitative changes in the patterns of protein synthesis. Cultured soybean cells have an absolute requirement for cytokinin. In the absence of the hormone the cells may be held indefinitely in a stationary phase, but when hormone is added to the culture medium, increased polyribosome formation begins within 15 minutes, leading to a three- to fourfold increase in the percentage of ribosomes bound as polyribosomes during the next 6 hours. The hormone-induced formation of polyribosomes does not affect the rate at which uridine is incorporated into messenger or ribosomal RNA and is not blocked by inhibitors of RNA synthesis. Cytokinin does not alter the average size of polyribosomes or the average rate at which protein synthesis is initiated or terminated (cf. Appendix B and Chapter 18). The most likely interpretation of the effect of cytokinin is that it induces the activation of nonpolysomal cytoplasmic mRNAs, some of which are necessary for the synthesis of proteins involved in stimulating cellular proliferation (Tepfer and Fosket, 1978; Fosket and Tepfer, 1978).

12.3.3 GIBBERELLINS. The gibberellins were discussed in Chapter 5 in connection with the induction of flowering in some long-day plants and earlier in this chapter as inducing the de novo synthesis of α-amylase in the aleurone layer of germinating barley seeds and the release of previously synthesized β-1,3-glucanase.

The gibberellins are a family of chemically related compounds that are easily extracted from plants by means of organic solvents. The water-soluble form, gibberellic acid, is usually used experimentally, however. Notably, it produces dramatic elongation of the stems and leaf sheaths of dwarf plants, which are genetically deficient in gibberellins. Here auxin has no effect at all. In contrast, auxins stimulate rapid growth when applied to sections of oat coleoptiles and pea epicotyls, which do not respond at all to gibberellins. In some cases auxins and gibberellins work synergistically. Thus, sections of green pea stems in vitro are stimulated to grow more rapidly in the presence of both auxin and gibberellic acid than with either hormone alone. Interestingly, however, the hormones must be applied together or treatment with gibberellic acid must precede the addition of auxin. Auxin pretreatment followed by gibberellic acid is much less effective (Ockerse and Galston, 1967).

The mechanism whereby gibberellins affect growth of the shoot is, in effect, really unknown. Possibly, they promote the synthesis of auxin from its precursors tryptophan or tryptamine, for increased auxin production follows gibberellin application in some cases. In view of the multiplicity of effects of the hormone on reproductive physiology and development, however, it is likely that it affects a considerable spectrum of biochemical processes, which remain to be revealed by future research.

12.4 Summary

Germination of seeds involves the imbibition of water and resumption of growth by the dormant embryo, rupture of the seed coats, and the emergence of the primary root and shoot.

Energy for germination is provided by reserves stored chiefly in cotyledons and endosperms. Enzymes required for the breakdown and utilization of these reserves are referred to as germination enzymes. In cottonseeds two important enzymes required for breaking down protein and lipid stored in the cotyledons are, respectively, carboxypeptidase and isocitratase. Neither of these enzymes is present in immature seeds, but they appear 24 hours after the embryo is placed under germinating conditions even in the presence of actinomycin D. The mRNAs for these enzymes are synthesized early in embryogenesis, some days before maximum wet weight (followed by desiccation) is attained. Translation of the messages after their transcription is inhibited by abscisic acid.

In barley seeds food reserves stored in the endosperm are digested by enzymes produced in the aleurone. Principally studied enzymes are α-amylase, protease, and ribonuclease, whose synthesis and release by isolated aleurones incubated in the presence of gibberellic acid takes place after a delay of about 8 hours. Most of the enzymes are rapidly released to the incubation medium. Effects of gibberellic acid are inhibited by actinomycin D, which indicates that synthesis and translation of new mRNAs are required for production of these enzymes.

The production of ribonuclease proceeds without exogenous gibberellic acid, and maximal synthesis is elicited by concentrations 10 to 100 times less than that required for synthesis of the amylase. Ribonuclease accumulates in the aleurone for some time before its release. Synthesis and release are differentially affected by gibberellic acid in the presence of cycloheximide.

Gibberellic acid also affects the release, but not the production of β-1,3-glucanase, which catalyzes the digestion of cell wall components in the aleurone layer. The release of this enzyme precedes that of α-amylase and protease, thus facilitating their release to the endosperm.

During germination the root penetrates the soil. During primary growth, the apical meristem produces protective root cap cells in advance of its progress and files of differentiating tissues behind. The meristematic zone is followed by a zone of elongation and a zone of maturation. Cellular differentiation begins in the upper part of the meristematic zone, but is not complete until the files of cells pass into the zone of maturation. There one finds a central cylinder, the stele, surrounding a layer of pericycle cells. Xylem and phloem differen-

tiate from the parenchyma, and the meristematic vascular cambium arises between them. Surrounding the stele are the cortex, made of parenchyma cells, and a covering of epidermis adorned with root hairs.

Secondary growth occurs in some kinds of roots. This growth involves the production of an expanding ring of secondary xylem and secondary phloem by division of the intervening vascular cambium. From the pericycle is formed the cork cambium, a meristematic tissue that produces cork to the exterior and a parenchymous phelloderm to the interior. Expansion of the vascular cylinder causes the sloughing of the cortex and epidermis, leaving the covering periderm composed of cork and underlying cork cambium.

The shoot and leaves arise from an apical meristem which forms a terminal bud consisting of promeristem, meristem, stem segments, and leaves in varying stages of development. Leaves arise from rapidly dividing rudiments appropriately spaced at the edges of the promeristem. Stem segments behind remain meristematic, then cease dividing, and elongate. Within the stem vascular bundles arise at intervals corresponding to the leaf positions and become continuous with the vascular pattern of leaves as they appear.

The outermost layer of the shoot in the seedling consists of epidermis surrounding a cortex consisting of parenchyma and collenchyma. Centrally one encounters vascular bundles consisting of phloem separated from xylem by a layer of vascular cambium. The central core consists of parenchyma cells forming the pith.

Stems lengthen by primary growth. Woody plants increase the diameter of their stems by secondary growth. This growth results from the activity of the vascular cambium and the cork cambium which arises from the cortex. Expansion of the vascular bundles creates a continuous ring of xylem and phloem that expands in annual growing cycles. The cork cambium creates a layer of cork which, together with the cork cambium, phelloderm, and underlying phloem, constitutes the bark of the tree.

Nonwoody plants usually do not show secondary growth. In dicotyledonous plants the vascular bundles are regularly arranged peripherally, but in monocotyledonous plants they are scattered among the parenchyma and lack a cambium.

The pattern of plant growth is largely under the

control of hormones. The principal growth homone is auxin, indole-3-acetic acid, which is produced in apical meristems and young leaves. It controls the elongation of cells below the apex. In growing plants auxin from the terminal bud is transported downward, suppressing the activity of lateral buds. Its movement is polarized basipetally, but light and gravity can affect its distribution transverse to the main shoot axis.

The properties of auxin and its isolation and identification were largely made possible by use of the oat *(Avena)* coleoptile, which is exquisitely sensitive to minute concentrations of auxin and auxin-like compounds.

Cytokinins are compounds that affect patterns of cell division in plant cells cultured in vitro and in cells of young fruits and developing seeds. They also induce division in meristems of buds inhibited by auxins from apical meristems. Natural cytokinins are derivatives of adenine, and one of them, isopentenyl adenosine (IPA) is found adjacent to the anticodon in tRNA. The mechanism whereby exogenous IPA affects plant development is not known.

Gibberellic acid and related compounds (gibberellins), in addition to activating the synthesis and release of germinating enzymes, cause elongation of the stems in genetically dwarfed plants. Their mechanism of action is essentially unknown.

12.5 Questions for Thought and Review

1. What is meant by the expression germination enzymes?

2. At what stage are the germinating enzymes of cottonseed cotyledons synthesized? How was this determined?

3. When are the mRNAs for the cottonseed cotyledons synthesized? What is the evidence on which your answer is based?

4. Are mRNAs for the germinating enzymes in cottonseed cotyledons translated immediately upon being transcribed? If not, what factors are involved in inhibiting their translation?

5. What is the aleurone layer in a grass seed?

6. What are the germinating enzymes chiefly studied

in connection with the germination of barley seeds? What is the role of each?

7. In what respects do the synthesis and release of α-amylase and protease in barley aleurone differ from the synthesis and release of ribonuclease? What is the experimental evidence on which your answer is based?

8. What are the principal zones of morphogenesis in primary root development? Compare and contrast the morphogenetic events in each zone?

9. Compare and contrast secondary growth in the root and shoot of woody plants.

10. Compare and contrast the development of the stem in herbaceous dicotyledonous plants and monocotyledonous plants.

11. What are auxins? Describe critical experiments on the *Avena* coleoptile to show that such a substance exists, where it is produced, and limitation to its transport.

12. What significant morphogenetic functions are served by cytokinins?

12.6 Suggestions for Further Reading

301

BIDDINGTON, N. L., and T. H. THOMAS. 1976. Influence of different cytokinins on the germination of lettuce *(Lactuca sativa)* and celery *(Apium graveolens)* seeds. *Physiol Plant* 37:12–16.

BRIAN, P. W. 1966. The gibberellins as hormones. *Int Rev Cytol* 19:229–66.

CUTTER, E. G., and H.-W. CHIU. 1975. Differential responses of buds along the shoot to factors involved in apical dominance. *J Exp Bot* 26:828–39.

ESAU, K. 1953. *Plant Anatomy,* 2nd Ed. New York: Wiley.

FOUNTAIN, D. W., and J. D. BEWLEY. 1976. Lettuce seed germination. Modulation of pregermination protein synthesis by gibberellic acid, abscisic acid, and cytokinin. *Plant Physiol* 58:530–36.

GALSTON, A. W. 1968. *The Green Plant.* Englewood Cliffs, NJ.: Prentice-Hall.

GALSTON, A. W., and P. J. DAVIES. 1970. *Control Mechanisms in Plant Development.* Englewood Cliffs, NJ: Prentice-Hall.

GOLDSMITH, M. H. 1977. The polar transport of auxin. *Annu Rev Plant Physiol* 28:439–78.

HALPERIN, W. 1978. Organogenesis at the shoot apex. *Annu Rev Plant Physiol* 29:239–62.

HARKES, P. A. 1976. Organization and activity of the root cap meristem of *Avena sativa* L. *New Phytol* 76:367–75.

HOLDER, N. 1979. Positional information and pattern formation in plant morphogenesis and a mechanism for the involvement of plant hormones. *J Theoret Biol* 77:195–212.

JACOBSEN, J. V. 1977. Regulation of ribonucleic acid metabolism by plant hormones. *Annu Rev Plant Physiol* 28:537–64.

KEY, J. L. 1969. Hormones and nucleic acid metabolism. *Annu Rev Plant Physiol* 20:449–74.

KEY, J. L., N. M. BARNETT, and C. Y. LIN. 1967. RNA and protein biosynthesis and the regulation of cell elongation by auxin. *Ann NY Acad Sci* 144:49–62.

MORI, T., Y. WAKABAYASHI, and S. TAKAGI. 1978. Occurrence of mRNA for storage protein in dry soybean seeds. *J Biochem* 84:1103–12.

OBATA, T., and H. SUZUKI. 1976. Gibberellic acid-induced secretion of hydrolases in barley aleurone layers. *Plant Cell Physiol* 17:63–71.

PHILLIPS, J. L., Jr., and J. G. TORREY. 1971. The quiescent center in cultured roots of *Convolvulus arvensis* L. *Am J Bot* 58:665–71.

RAY, P. 1962. *The Living Plant.* New York: Holt, Rinehart and Winston.

SACHS, T., and K. V. THIMANN. 1967. The role of auxins and cytokinins in the release of buds from dominance. *Am J Bot* 54:136–44.

TAYLERSON, R. B., and S. B. HENDRICKS. 1977. Dormancy in seeds. *Annu Rev Plant Physiol* 28:331–54.

TORREY, J. 1965. *Development in Flowering Plants.* New York: Macmillan.

TORREY, J. G., and L. J. FELDMAN. 1977. The organization and function of the root apex. *Am Sci* 65:334–44.

WARNER, J. E., and Y. BEN-TAL. 1975. The role of membrane-bound enzymes in an early response of aleurone tissue to gibberellic acid. *Am Zool* 15:237–40.

WENT, F. W., and K. V. THIMANN. 1937. *Phytohormones.* New York: Macmillan.

12.7 References

CHRISPEELS, M. J., and J. E. VARNER. 1967. Gibberellic acid-enhanced synthesis and release of α-amylase and ribonuclease by isolated barley aleurone layers. *Plant Physiol* 42:398–406.

DARWIN, C. 1897. *The Power of Movement in Plants.* New York: Appleton.

FILNER, P., and J. E. VARNER. 1967. A test for de novo synthesis of enzymes: density labeling with H_2O^{18} of barley α-amylase induced by gibberellic acid. *Proc Natl Acad Sci USA* 58:1520–26.

FOSKET, D. E., and D. A. TEPFER. 1978. Hormonal regulation of growth in cultured plant cells. *In Vitro* 14:63–75.

IHLE, J. N., and L. S. DURE. 1972. The developmental biochemistry of cottonseed embryogenesis and germination. *J Biol Chem* 247:5048–55.

JACOBSEN, J. V., and J. E. VARNER. 1967. Gibberellic acid-induced synthesis of protease by isolated aleurone layers of barley. *Plant Physiol* 42:1596–1600.

OCKERSE, R., and A. W. GALSTON. 1967. Gibberellin-auxin interaction in pea stem elongation. *Plant Physiol* 42:47–54.

SKOOG, F., and C. O. MILLER. 1957. Chemical regulation of growth and organ formation in plant tissues cultured in vitro. *Symp Soc Exp Biol* 11:118–31.

TEPFER, D. A., and D. E. FOSKET. 1978. Hormone-mediated translational control in cultured cells of *Glycine max. Dev Biol* 62:486–97.

FOUR

Major Principles of Morphogenesis

As recognized in Chapter 2, at the end of differentiation the organism consists of discrete types of cells arranged in a species-specific pattern of tissues, organs, and organ systems. The achievement of the final state of differentiation occurs through a sequence of progressive changes, each of which depends on the accomplishment of prior differentiative steps, which are essentially irreversible.

Analysis of the conditions under which these steps occur was long the prerogative of a discipline variously called **experimental embryology, experimental morphogenesis, developmental physiology,** or, as many pioneer German investigators called it, **Entwicklungsmechanik** or **Entwicklungsgeschichte.** Most experimental studies of morphogenesis did and do involve microsurgical procedures on eggs and early embryos, recombinations of cells and tissues, and chemical modifications of the environment of the embryo or its parts. These manipulations are designed to answer questions largely about the developmental potentialities of the parts of the embryo and the interactions between embryonic parts that result in specific morphological patterns. Answers to these kinds of questions are found, not in the test tube, but in the gross and microscopical characteristics of the organisms or their parts that were subjected to manipulation. **Morphogenetic pattern formation** is the currently popular name for this kind of analysis.

Investigations in the field of experimental morphogenesis have produced a great body of information that, over the years, has permitted the formulation of a number of major concepts. These are often called "principles of development." Some major principles were briefly noted in Chapter 2. Most of the information subserving these principles comes from the study of animal development, for the animal embryo is much more accessible to experimental manipulation than is, for example, an apical meristem or an embyro within a seed. Accordingly, most of what is treated in the next four chapters consists of important principles and ideas about morphogenesis that have their bases in observations of animal embryos and their parts as they develop under experimental conditions.

The concept of the morphogenetic field was a particularly controversial one during the early years of this century (see especially Spemann, 1938; Weiss, 1939). It became an almost completely neglected concept of developmental biology with the increasing popularity of studies of the molecular biology of development beginning in the 1950s. This neglect was largely because the morphogenetic field was defined differently by many who used the term, and because the insights to interpret field phenomena at the molecular level were lacking. The lack of insights still prevails, but recent studies, which we consider later, suggest that the field concept should once more occupy a place in our conceptualization of developmental phenomena. In what follows we hope to derive a reasonable definition of a morphogenetic field and possibly assist in the recognition of this concept as being of major importance in modern developmental biology, as well as in classical experimental embryology.

13

Concept of the Morphogenetic Field

13.1 Morphogenetic Fields in Sea Urchin Development

In this discussion a definition of a morphogenetic field will be derived from results of a number of experiments on sea urchin embryos. Their normal development was described in Chapter 9; Figure 9.3 shows the origin and designation of various cell groups that constitute the early blastula.

As shown in Chapter 9, cells derived from the mesomeres at the 16-cell stage form only ectoderm, as do descendents of Veg$_1$, which appear at the 64-cell stage. If we isolate the mesomeres at the 16-cell stage and supply them with only the four micromeres from the vegetal half (Figure 13.1), a small, but normally organized, pluteus develops. Careful observations show that the micromeres form only the skeletal parts of the larva, whereas the gut is differentiated from materials of the animal half that would normally have formed only ectoderm.

Also, if we bisect a blastula through the animal-vegetal axis (Figure 13.2), each half will form a normal, but half-sized, pluteus. In this case, each half-

blastula closes over to form a smaller sphere, and a new center of invagination occurs. Moreover, future ectodermal cells of the body now form apical tuft, and cells that would have contributed to the apical tuft now ring the anus.

On the other hand, if we bisect the blastula by a cut that separates An$_2$ and Veg$_1$, as shown in Figure 10.13, normal plutei do not develop. The animal half forms a sphere with an extensive apical tuft, and the vegetal half usually forms an excessively developed gut and skeletal parts with an insufficient ectodermal covering.

The foregoing exeriments and many others (reviewed by Hörstadius, 1939) reveal that cells and groups of cells in the early embryo have potentialities for morphological differentiation that are far greater than their normal prospective fates. Some

FIGURE 13.1 Mesomeres isolated from a sea urchin embryo at the 16-cell stage and combined with the micromeres in the absence of macromeres (cf. Figure 9.3).

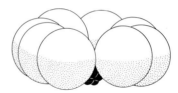

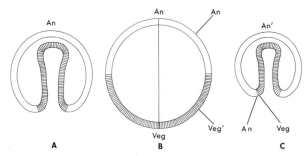

FIGURE 13.2 A whole gastrula from the meridional half of a sea urchin egg: (A) normal gastrula; (B) blastula showing meridional plane of cutting; (C) normal but smaller gastrula from one half. An and Veg, original centers of animal and vegetal fields, respectively; An' and Veg', sites of new field centers that arise after cutting the germ. [Adapted from S. Hörstadius, *Acta Zool* 9:1–191 (1928).]

The field concept is now explored at some length. Perhaps the complexities that are seen tend to distract from the possibility of arriving at some simple conceptual insight that might clarify our understanding of how morphogenetic patterns are formed. It is important, however, for the student of development to realize something of the vast background of factual and conceptual information related to the study of morphogenesis. Only then can he or she appraise more simplistic modern models of field action that are considered in the last chapter of this book.

13.1.1 ANIMAL AND VEGETAL FIELDS. As the sea urchin experiments indicate, a part of the embryo can contain all of the organizing factors required for the morphogenesis of a complete embryo. Meridianal halves have all of these factors. So, too, does the combination of animal halves plus micromeres, as illustrated in Figure 13.2. There are, however, clear differences in the distribution of the organizing factors. Neither those factors present in the animal half nor those in the vegetal half alone are capable of bringing about the differentiation of all parts of the embryo. Factors in the animal half bring about the organization of a ''super blastula'' that dies. Factors in the vegetal half mostly form a ''supergut'' that likewise fails to survive (see Figure 10.13). There are, therefore, two morphogenetic fields in the fertilized sea urchin egg, an **animal** field and a **vegetal** field.

13.1.1.1 Gradients of Organizing Factors in Sea Urchin Fields. The sea urchin egg at the 32- to 64-cell stage can be divided latitudinally into five layers (see Figure 9.3), each of which can be isolated and allowed to demonstrate its developmental potency. In *Paracentrotus lividus,* studied by Hörstadius (1939), An_1 forms a blastula the circumference of which is completely adorned with long stiff cilia like those found in the normal apical tuft. An_2 likewise forms a blastula, but no more than three-fourths of its surface has long cilia. Both of these blastulae die. An isolated Veg_1 layer develops variably, sometimes forming only a blastula with a small tuft of long cilia and other times forming a small gut invagination. Veg_2 produces a larva with an apical tuft of cilia but without the apical organ that the cilia usually adorn; Veg_2 also forms a fairly typical gut and some skeletal parts. The micromeres alone produce a small ball of cells that soon falls

306

cells, indeed, can form any of a variety of parts that are normally seen in a sea urchin larva. But the cells in each of the parts of the blastula do, in fact, differentiate in a particular pattern with a timing and localization appropriate to the morphology of a normal pluteus! From this it follows that there must be present in the embryo organizing factors that see to it that a particular and restricted selection of potencies is put into effect by particular cells. These organizing factors are called **morphogenetic fields.**

What are the organizing factors that constitute a field? Organizing factors and the **organization** of an embryo and its parts are perceived by some as endowing a field with some sort of mysterious properties or with such a degree of complexity as to inhibit imaginative experimentation. As already seen, and further developed in what follows, the morphogenetic performance of a cell within a field depends on its position within the field prior to the occurrence of some differentiative event that fixes its fate. The organization factors may then be looked upon as systems of **positional information** (Wolpert, 1969). Cells in a field have their position specified with respect to a three-dimensional coordinate system of signals within the boundaries of the field and respond with the sequence of cytodifferentiative events appropriate to their positional value in the signaling system. Conceivably, positional signaling could consist of gradients of diffusing molecules, perhaps as simple as cyclic AMP, or could be generated by short-range cell–cell interactions.

apart and dies. But if 1–3% of horse serum is added to the seawater, these cells often secrete skeletal rods (Okazaki, 1975). The Veg$_2$ layer, isolated along with one or more micromeres, produces a larva in which the gut is so large that it cannot invaginate. Instead it turns outward, forming an exogastrula, which soon dies.

The changing patterns of developmental potencies of the four major layers are shown in the first vertical column in Figure 13.3. These changes suggest that the organizing factors in the animal half of the egg progressively lose the ability to organize an apical tuft as one proceeds toward the vegetal pole, and that those of the vegetal half progressively lose their ability to organize gut with increasing distance toward the animal pole. The animal and vegetal fields must, therefore, possess opposing gradients of what we might call "animalizing" and "vegetalizing" factors that have the high points of their gradients at animal and vegetal poles, respectively.

The high point of the vegetalizing factors clearly is in the micromeres. They alone can interact with the animal half to form a normal pluteus. Moreover, if a second set of micromeres is grafted to a normal blastula near the animal pole, a second site of invagination occurs, a supernumerary gut forming from presumptive ectodermal cells at the site of implantation.

The formation of a normal pluteus, therefore, results from an appropriate balance of the organizing factors of the two opposing fields. This idea is reinforced by the further study of Figure 13.3. For example, An$_2$ alone cannot make a pluteus, nor can it do so when supplied with one micromere. Given two or four micromeres, however, it can do so.

The expression of animalizing and vegetalizing forces can be modified by chemical treatment. It has long been known that treatment with lithium ion during cleavage stages leads to a suppression of the animal field and extension of the vegetal field, usually resulting in exogastrulation (that is, evagination of the prospective gut rather than its invagination; see Figure 13.10 for an illustration of exogastrulation in an amphibian embryo). Similar effects are brought about by low concentrations of ammonium ions and a variety of amines (Lallier, 1973b). Animalization of the egg, on the other hand, is brought about by treatment with sodium thiocyanate (NaSCN) and by a variety of surface-active agents such as bile salts and sodium lauryl sulfate (Lallier, 1973a). Diminishing the concentration of potassium ions in the seawater similarly results in suppression of the vegetal field and extension of the animal field. Animalized embryos tend to show excessive spreading of the apical tuft and suppression or elimination of the gut, whereas vegetalized embryos show excessive gut formation and a paucity of ectoderm. In general, it can be said that the chemical modification of the field forces in sea urchin eggs has thrown no light on their nature (but see Lallier's interpretation of the action of lithium and of amines, 1973b).

There have also been a number of attempts to extract animalizing and vegetalizing substances from sea urchin eggs. Hörstadius and his collaborators (see Hörstadius and Josefsson, 1972) have extracted fractions from unfertilized sea urchin eggs and from cleaving embryos, which, applied to vegetal halves, will cause them to develop as normal or almost normal plutei with some frequency. Applied to whole blastulae, they suppress the development of the gut. Other fractions were found to have weak

FIGURE 13.3 The potentialities for differentiation of various tiers of cells of the sea urchin embryo in isolation and combination with micromeres. [Adapted from S. Hörstadius, *Pub Staz Zool Napoli* 14:253–479 (1935), embodying corrections indicated in his footnote on p. 375.]

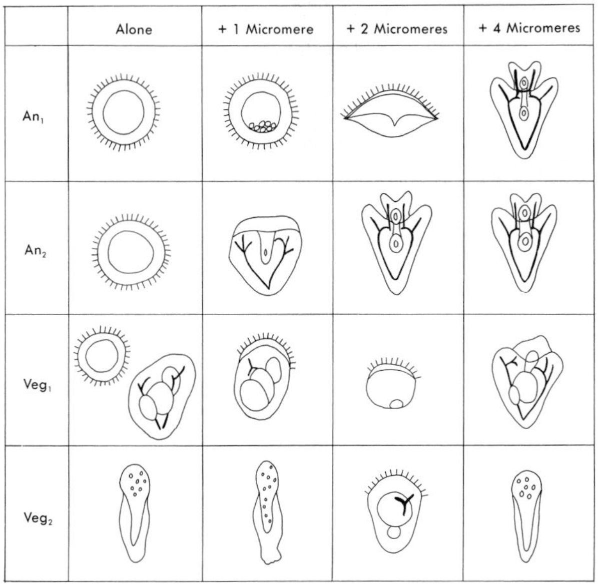

	Alone	+ 1 Micromere	+ 2 Micromeres	+ 4 Micromeres
An$_1$				
An$_2$				
Veg$_1$				
Veg$_2$				

307

vegetalizing activity in that, although they could bring about some gut formation and skeletal parts in animal halves, they could not induce the formation of normal plutei from them and had no effect on intact embryos. The active fractions were prepared from the supernates of homogenized egg after acid precipitation and separation by column chromatography. It is not at all clear whether these fractions act as specific morphogenetic substances in normal development.

13.1.1.2 Inductive Action of Micromeres.

We have seen that an isolated animal half of the sea urchin blastula fails to form a larva and that micromeres alone form, at the most, a few skeleton-producing cells. But, in combination they produce a complete larva to which the micromeres contribute only the skeleton. Hence, the micromeres have somehow caused the cells of the animal half to form the missing vegetal organs, chiefly the gut. Influences of this sort, which cause one group of cells or a tissue to elicit a specific morphogenetic performance in an adjoining tissue, are usually called **inductions.** In the experimental situation we have described we can say that the grafted micromeres have induced the formation of gut in the surrounding animal half. An induction is thus a case of **dependent differentiation,** the topic of Chapter 14. To designate all cases of dependent differentiation as inductions may constitute a problem, but it may be only a semantic one. Thus, the formation of an apical organ in the sea urchin embryo is dependent on the formation of a gut in the presence of an otherwise adequate animal half. This situation is clearly a case of dependent differentiation, but some may not wish to label it as a case of induction.

13.2 The Chordamesoderm Field of the Amphibian Embryo

In Chapter 9 it was pointed out that the bulk of material that involutes around the dorsal lip of the blastopore coincides with the region of the embryo that is marked by the gray crescent. This material is referred to as the chordamesoderm (Figures 9.9 and 9.12). It was also noted in Chapter 9 that the prospective fates of cells in this and other parts of the blastula can be followed into the neurula by marking experiments. Using microsurgery one may, therefore, exchange parts of the embryo of known prospective fate.

In 1923 Otto Mangold reported the effect of grafting a block of prospective belly ectoderm from the early gastrula of the newt *Triton cristatus* in place of an excised block of chordamesoderm at the dorsal lip of the blastopore of *T. alpestris*. In Mangold's experiment, the implant invaginated along with the neighboring chordamesoderm, giving rise to mesodermal structures like those expected of the cells it replaced. Transplanted elsewhere, belly ectoderm would not invaginate or give rise to mesoderm. From this evidence it follows that in the chordamesoderm there are specific influences, organizing factors, that prevail upon the implant so as to impose on it the character of mesoderm.

On the other hand, if one implants a piece of

FIGURE 13.4 Induction of a secondary set of axial organs by a graft of the dorsal lip of the blastopore to the prospective ventral body wall. The host and graft are differently pigmented species so that their cells can be distinguished. [Adapted from H. Spemann and H. Mangold, *Arch Mikr Anat Entw-mech* 100:599–638 (1924).]

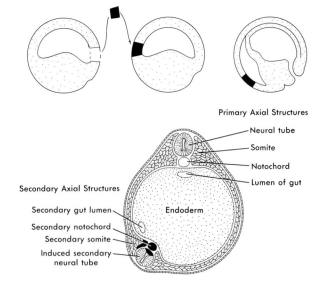

Primary Axial Structures
Neural tube
Somite
Notochord
Lumen of gut

Secondary Axial Structures

Endoderm

Secondary gut lumen
Secondary notochord
Secondary somite
Induced secondary neural tube

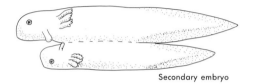

FIGURE 13.5 Secondary embryo formed by grafting a blastoporal lip to the ventral body wall, as described in Figure 13.4

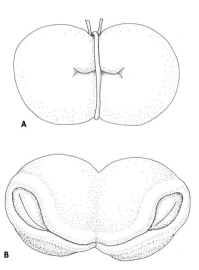

FIGURE 13.6 Effects of weakly constricting a cleaving newt egg in the plane passing through the gray crescent. (A) as seen at the beginning of gastrulation. (B) as seen at the neurula stage. [Adapted with modifications from H. Spemann, *Embryonic Development and Induction*, Yale University Press, New Haven, 1938, pp. 24, 29.]

309

prospective chordamesoderm in place of prospective flank or belly ectoderm (Figure 13.4), the graft always invaginates and, doing so, retains its own character as mesoderm. Furthermore, it induces, or "organizes," its surroundings into a coherent pattern with itself and builds from its own substance and surrounding tissue a small embryo which can reach a high degree of perfection (Figure 13.5). The chordamesoderm is thus a center or organizing factors responsible for determining the fates of parts that come under its influence. Therefore, it is properly called a morphogenetic field.

The foregoing observations, in addition to results of other experiments carried out in his own and other laboratories, led Spemann (summarized by Spemann, 1938) to propose that the chordamesoderm constitutes the **center of organization** on which the organization of all other parts of the embryo depends. Chordamesoderm was, and still is, often referred to simply as the **organizer,** a term that Spemann (1938) explained as simply expressing the fact that it could bring about the organization of materials in and around it into an embryo.

13.2.1 PRECOCIOUS DETERMINATION OF THE CHORDAMESODERM FIELD. The approximate congruence of the gray crescent and the chordamesoderm field and certain experimental evidence long ago suggested that the organizing factors constituting the chordamesoderm field are prelocalized in the gray crescent. Morgan and Boring (1903) found that in newts the first cleavage plane usually isolates the gray crescent in one blastomere and that the second cleavage bisects the gray crescent. In frogs, however, the first cleavage usually passes at or near the center of the crescent. If newt eggs are constricted by a tightened thread along the second cleavage plane, twinned or partially twinned embryos are formed, depending on the tightness of the constriction (Figure 13.6). If, however, the constriction is made in the plane of the first cleavage, gastrulation occurs on only the side of the constriction containing the gray crescent, and a well-formed embryo develops on that side; the other side forms only a disorganized mass of cells (Figure 13.7). With frog eggs twins are obtained when the blastomeres produced by the first cleavage are separated.

In 1960 and 1962 Curtis reported the successful isolation and grafting of fragments of the cortex of *Xenopus* eggs and early embryos. He found that bits of gray crescent excised from one embryo during the first cleavage and grafted to the prospective ventral side of another induced the appearance of a secondary blastopore and the organization of a secondary embryonic axis in the recipient egg. Non-gray crescent cortical material lacked this property. In other experiments, he showed that a bit of gray crescent excised from an eight-cell stage and grafted to the future ventral side of a fertilized egg would likewise induce a secondary axis. Eight-cell embryos would not respond, however, to the putative cortical organizer (Figure 13.8). Moreover, excision of the cortical crescent from an egg during the first cleavage resulted in the formation of a cleaving mass that failed to gastrulate, but excision of the

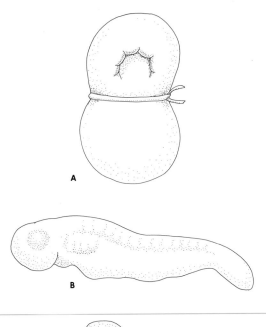

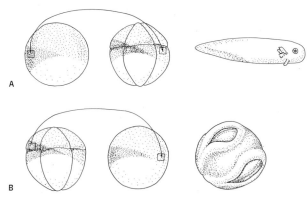

FIGURE 13.8 Effects of grafting gray crescent cortical material in *Xenopus* eggs and embryos. A: Gray crescent cortex from an uncleaved egg grafted to the prospective ventral side of an embryo at the 8-cell stage does not affect subsequent development. B: Gray crescent cortical material from an 8-celled embryo to the prospective ventral side of an uncleaved egg induces formation of a secondary embryonic axis. [Adapted with modifications from A. S. Curtis, *J Embryol Exp Morphol* 10: 410–22 (1968).]

310

FIGURE 13.7 Effects of strongly constricting a newt egg in the cleavage plane that separates the future dorsal side (side of the gray crescent) from the future ventral side. A: As seen at gastrulation. B: Embryo formed from the dorsal side. C: Undifferentiated cell mass formed from the other half-egg. [Adapted with modifications from H. Spemann, *Embryonic Development and Induction*, Yale University Press, New Haven, 1938, pp. 29, 30.]

crescent at the eight-cell stage did not interfere with normal development. Successful inductions by grafts of cortical material were also reported by Tomkins and Rodman in 1971.

It was pointed out in Chapter 9 that the gray crescent arises as a result of a cortical shifting that is marked by a displacement of pigment granules. Because the "organizer" region occupies the equivalent position of the gray crescent, the Curtis experiment makes it seem plausible that the cortical shifting brings to the gray crescent the positional information that establishes the chordamesoderm

field and thus the site of the dorsal blastoporal lip and, accordingly, the plane of bilateral symmetry of the embryo.

Recent studies, however, make this interpretation seem unlikely. Kirschner et al. (1980) have shown that the results of Curtis' gray crescent transplants probably may be attributed to the rotation of the egg that receives the implant of gray crescent. They found that twin embryos occur with high frequency when *Xenopus* eggs are allowed to remain in position to receive a graft of gray crescent without the necessity of making an implant. Similar results were obtained by Malacinski et al. (1980). Moreover, Keller's careful fate maps (Chapter 9) show that the involuting "organizer" region of dorsal mesoderm originates from internal cells of the blastula and not from cells containing the cortical material of the gray crescent.

It is, moreover, likely that there is no precocious determination of the chordamesoderm field but, rather, a progressive organization of the field by virtue of a prior organization in the vegetal portion of the egg. Using eggs of the urodele *Ambystoma mexicanum*, Nieuwkoop (1969a, b; 1970) and Nieuwkoop and Ubbels (1972) arbitrarily divided the blastula

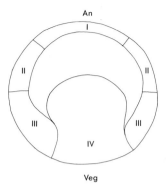

An

I

II   II

III   III

IV

Veg

FIGURE 13.9 Scheme for subdividing the blastula of *Ambystoma mexicanum* into four zones with respect to the animal-vegetal axis. [Redrawn with modification from P. D. Nieuwkoop, *Wilhelm Roux' Arch* 162: 341–73 (1969).]

into four zones (Figure 13.9): Zone I, the animal-pole cap of cells, the prospective ectoneuroderm; Zone II, a ring of cells, consisting of the greater part of the prospective chordamesoderm and the remainder of the ectoneuroderm; Zone III consisting of the remainder of the mesoderm, pharyngeal endoderm, and some intestine; and Zone IV, the yolk mass. He then tested the morphogenetic capacity of each zone in isolation and in a number of combinations. Zone IV, in isolation, formed only a spherical mass of undifferentiated yolky cells; Zone III usually formed an oblong structure with a small amount of ectoderm at one end, and the rest consisted of axial mesoderm such as notochord and somites and pronephric tubules partially enrobed in an endodermal epithelium.

Zones I and II developing alone or in combination formed only a spongy undifferentiated ectodermal mass—with no blastopore and no mesoderm or endoderm—*despite the fact that most of the prospective chordamesoderm, according to the fate maps, is located in Zone II.* It would appear, therefore, that the chordamesoderm field is not prelocalized in the gray crescent region of the egg and early blastula. Nevertheless, field properties are present in this area at the time of gastrulation, as revealed by the experiments of O. Mangold (1923) and of Spemann and H. Mangold (1924) already noted (Figures 13.4 and 13.5) and many others. How do these field properties arise? Nieuwkoop approached this question by combining Zones I and II, which do not gastrulate and make mesoderm, with Zone IV,

which makes only endoderm. In this combination gastrulation occurred, and larvae of almost normal structure developed. By recombining these same zones in known orientation, and using a radioactive label to mark one component, he was able to show that (1) the position of the blastopore is determined by the dorsoventral axis of the endodermal mass regardless of the orientation of the Zone I–II animal cap and (2) both the mesoderm and the anterior portion of the digestive tract, as well as, variably, small dorsal and posterior portions of the tract, arise entirely from the animal cap.

These results suggest that the animal half of the blastula, or earlier embryo, contains no pre-formed morphogenetic potencies, even in the region of the gray crescent, and that it acquires the ability to form nearly all the structures of the embryo from an asymmetric vegetal field. There would thus appear to be no "animal field" in amphibians, as in sea urchins, but only a vegetal field. The vegetal field, present perhaps from the start of development or even in the oöcyte, might then induce Spemann's organizer, the chordamesodermal field, which would set into motion a subsequent sequence of morphogenetic events marked by gastrulation and organ formation.

311

13.2.2 PROGRESSING DETERMINATION: ORGANIZATION OF THE NEURAL FIELD.
Once established, the chordamesoderm field has the power to invaginate and to form notochord, somites, and other mesodermal structures. It will do this even in the situation of grafting to an abnormal location such as the future belly region. Having acquired this capacity, possibly by inductive action of the vegetal field, it is said to be **determined.** The criterion for the determination of the chordamesoderm field is its property of **self-differentiation:** that is, it can realize its normal morphogenetic fate in the absence of its normal surroundings.

During gastrulation the chordamesoderm field moves beneath the ectoderm of the future nervous system, forming, in urodeles, the roof of the archenteron. The ectoderm overlying the archenteron roof now forms the neural plate, and the neural plate next forms the neural tube as described in Chapter 11.

Many lines of evidence point to the fact that the neural plate is determined by inductive action of the roof of the archenteron. Defects in the pattern

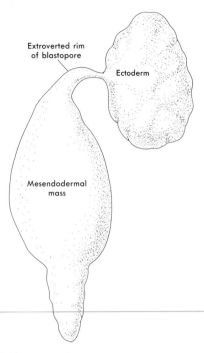

FIGURE 13.10 Exogastrulation of an axolotl embryo. The lower mass, consisting of endoderm and mesoderm turned inside out, remains attached to the upper mass of ectoderm into which it would normally have moved. [Adapted from P. Weiss, *Principles of Development*, Holt, New York, 1939, p. 353.]

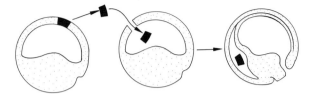

FIGURE 13.11 Blastocoele graft used as a method of testing the differentiation capacities of prospective neural tissue. As illustrated here, the graft would form only epidermal tissue. Had it been previously underlaid by archenteron roof, the graft would have formed neural tissue.

of invagination (occurring, for example, when eggs are compressed or forced to develop in an inverted position) are followed by parallel defects in the neural plate. Moreover, treatment with lithium ions causes exogastrulation in amphibians (Figure 13.10), leaving a disorganized ectodermal mass attached to endodermal and mesodermal structures. If exogastrulation is complete, the neural plate is entirely absent.

If the neural plate is, in fact, determined through induction by the archenteron roof, then prospective neural plate that has been induced should have the capacity to self-differentiate in isolation from its normal surroundings. One way of testing the morphogenetic activity of prospective neural tissue is to insert it in the blastocoele of a host blastula or early gastrula. Gastrulation movements then displace the graft to the future ventral side or flank region (Figure 13.11). O. Mangold (1929) used the blastocoele graft method to test the

determination of the prospective neural plate prior to and during the progress of gastrulation. He found that isolates of neural ectoderm removed in advance of the inrolling archenteron roof formed only epidermal structures when grafted, whereas those excised after being underlaid by the chordamesoderm formed histologically recognizable nervous tissue.

Similar blastocoele cultures were used by Bautzmann (1926) to test the ability of implants to induce neural plate in belly ectoderm. He isolated portions of the beginning gastrula in a systemic plan, using the lip of the beginning blastopore as a reference point. These fragments were separately inserted into the blastocoele of host gastrulae of the same age. The result was that much of the pre-

FIGURE 13.12 Outline of the amphibian blastula as seen from the future left side showing, by means of stipples and degrees of arc, the region capable of inducing secondary axial structures when tested in blastocoele grafts (cf. Figure 13.11). [Adapted with modifications from H. Bautzmann, *Wilhelm Roux' Arch* 108:283–321 (1926).]

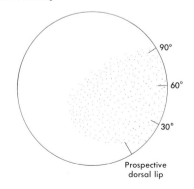

sumptive mesoderm possessed the power of inducing neural plate before invagination (Figure 13.12).

Under normal circumstances, however, it is probably the prerogative chiefly of the notochordal material to carry out the induction. As noted in Chapter 9, the notochord is formed by cells that move forward from the region of the blastopore, converging in the dorsal midline (Figure 9.14), and the anterior end of the neural plate is just beyond the anterior limit of the notochord. In 1926 Lehmann removed the upper lip of the blastopore at various stages during gastrulation. This experiment frequently resulted in partial or, sometimes, total defects in the notochord. These defects brought about corresponding defects in the neural plate. In one particularly clear case in which the notochord was absent partially for some length, the neural plate above it failed completely to form.

13.2.2.1 Hierarchies of Inductions. Since Spemann's time, the chordamesoderm has been referred to by many not only as the organizer or organization center, but also as the **primary inductor,** the latter name by virtue of the fact that the induction of the neural tube by the archenteron roof is the first inductive event that was recognized in amphibian development. As already pointed out, however, Nieuwkoop's investigations and many earlier studies suggest that induction by the chordamesoderm may not be primary, but is actually secondary to a prior induction by the vegetal field. Regardless of which interpretation is correct, the fact that we refer to an induction as primary is to imply that there are **secondary inductor, tertiary inductors,** and so on, in hierarchial array. For example, the brain, having been induced by archenteron roof, forms its various regional differentiations under the influence of head endoderm and mesoderm. The hindbrain induces overlying ectoderm to differentiate the ear placode, which then invaginates and, in turn, induces surrounding mesenchyme to form the cartilaginous auditory capsule. Another sequence that merits special mention is concerned with the origin of the eye (see Figure 11.25). The optic vesicle, an evagination of the forebrain, makes contact with the head ectoderm, which then forms a lens placode. The optic vesicle then invaginates to form the eyecup from which the retina later differentiates. The lens placode then invaginates and forms a lens; and the eyecup-lens complex then induces the overlying tissue to form

the transparent cornea and the surrounding mesenchyme to form the sclerotic coat of the eyeball. These inductions have been examined experimentally; if a step in the sequence is omitted, the next step fails to occur. Thus, if the optic vesicle of the chick embryo is excised prior to its contact with head ectoderm, the lens fails to form and, without lens and optic vesicle, no sclerotic coat arises. The inductions and other events associated with organization of the eye were analyzed by Coulombre (1965).

13.3 An Inventory of Field Properties

As Weiss (1939) remarked, "The field concept has its roots in purely **empirical** ground." Field properties are revealed by experimental evidence, and essentially all phenomena of development show some characteristics of fields. The concept is thus a useful one, and the universality of its application suggests that a section summarizing the properties of fields might be useful at this point. As a matter of fact, we have already noted a number of field properties in the foregoing section. We now summarize these and other field properties in a more formal way, adducing additional experimental evidence as appropriate.

1. **Fields are expressions of physical reality.** That is, fields emerge from the underlying physical and biochemical properties of their component parts. They have no existence independent of cellular activities.

2. **Fields exist in three-dimensional patterns.** Each field pattern is characterized by a zone of highest morphogenetic activity, the **field center.** This activity diminishes along the various axes that extend from the center. On the basis of this information, we can determine that

3. **Fields have polarity,** and that

4. **Fields have strength, or field energy.** Field energy falls off away from the field center, as we saw, for example, in the case of the animal field of the sea urchin egg. Moreover, some fields have more strength than others. Thus, if micromeres are

313

added at the animal pole of an otherwise normal sea urchin blastula, an archenteron will form at the animal pole, as well as at the vegetal pole. The vegetal field, therefore, is the stronger. The fate of cells and cell groups thus depends on the strength of the field in which they come to lie. These examples point up, too, that the field pattern and distribution of field energy coincide.

5. **Fields induce.** This phenomenon has already been illustrated at some length in connection with the induction of neural plate by the archenteron roof. The instance in which a tissue of one kind induces a morphogenetic determination in another is referred to as a **heterotypic** induction. A **homotypic** induction occurs experimentally when, for example, a blastocoele graft of determined neural plate not only forms a neural tube itself but also induces belly ectoderm to form another neural tube overlying it.

Another kind of induction may be called **assimilatory induction.** The experiments of Spemann and H. Mangold illustrated in Figure 13.4 showed that in the formation of a secondary embryo by a grafted "organizer," host cells participated in the formation of notochord and somites. They were thus assimilated into and became a part of the structure of the implant.

6. **Fields have the power of self-differentiation.** This power has already been noted for the chorda-mesoderm and the neural field. There are many other examples among major organ fields. Among the most thoroughly studied self-differentiating systems are the limb fields in amphibians and birds and the fields of the placodal sense organs.

7. **Fields are determined before their parts.** In other words, the field as a whole may have an essentially irreversible morphogenetic commitment, even though the fates of the cells that compose it are not individually determined. As noted earlier, a meridianal half of a sea urchin blastula will form a normal pluteus, a process that involves the production of gut cells from cells that would otherwise have become ectoderm. Moreover, if one fuses two sea urchin embryos during early blastula stages with their animal-vegetal axes oriented in the same direction, they will form a single pluteus. We denote the ability of parts of a field to recreate the whole or the ability of cells within a field to form parts other than those that are their normal fate by saying that

8. **Fields have regulative ability.** This property is obviously a corollary of property 7. The development of regulative eggs described in Chapter 10 provides other examples of regulation in embryos of birds and mammals. Here one may note that regulation of the eye field occurs as an event in normal development. Willier and Rawles (1935) located the eye field in chick embryos by grafting carefully cut and measured fragments of blastoderm to the vascular chorioallantoic membrane of a host chick embryo and observing which fragments formed eyes. They found that at the head process stage the eye field is a single field with a high point of eye-forming potency in its center—actually in the region that normally forms the floor of the forebrain (Figure 13.13). As the chick enters somite-forming stages, the capacity for eye formation (measured as the frequency with which eye structures form from the grafts) increases in the right and left sides of the field while diminishing in the center. By the time eight somites have been formed, the medial region of the original field territory has lost the ability to make eye.

In urodele amphibians the isolated anterior portion of the neural plate cultured in vitro forms only a single eye, but two eyes develop if the same tissue is cultured together with the underlying archenteron roof. Moreover, if the anterior part of the archenteron roof is excised from newt embryos at the early neurula stage, the resulting larva shows a

FIGURE 13.13 The eye-forming area of the chick blastoderm at the stage of the head process. [Adapted from B. H. Willier and M. E. Rawles, *Proc Soc Exp Biol Med* 32:1293–96 (1935).]

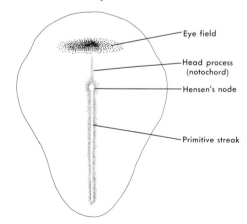

314

single, medially located cyclopean eye. Moreover, anterior neural plate ectoderm substituted for hindbrain ectoderm does not form eyes (reviewed by Lopashov and Stroeva, 1961). These results suggest that the anteromedial part of the mesoderm first induces eye-forming potential in the anterior portion of the neural plate and then inhibits eye formation in the middle of the brain floor. The general presumption seems to be that in birds and mammals the head mesoderm exercises similar controls of eye development (cf. Waddington, 1952, and Arey, 1965).

9. **Fields persist beyond the time required for the exercise of their morphogenetic roles.** As a simple example, a fragment of notochord excised from a urodele neurula and grafted to the blastocoele of a beginning gastrula will induce belly ectoderm to form neural tube. Also, if the limb of a larval or adult newt is amputated, the missing parts are regenerated from cells in the region of the cut (see Chapter 17). Evidently the limb field persists in urodeles into adult stages. In anurans the ability to regenerate limbs is present during tadpole stages but is lost at metamorphosis (Chapter 18). The persistence of fields makes it particularly easy to demonstrate the next property, namely,

10. **Field properties extend beyond the limits of the tissue that contributes to the construction of a specific organ.** Holtfreter (1933) excised blocks of determined epidermis from a series of neurulae of *Triton* at different anteroposterior levels, replacing them in each case with a similar-size piece of undetermined ectoderm from a young gastrula (Figure 13.14). The grafts developed as excrescences from the flank, each giving rise to a variety of ectodermal and mesodermal derivatives, depending on the morphogenetic fields under whose influence they came. As illustrated in Figure 13.14, the various fields have no sharp limit and overlap considerably. Nose and eye can appear at the hindbrain level; hindbrain can develop at anterior trunk levels. The forelimb field is restricted essentially to the level at which forelimb normally arises, but it is overlapped by factors that determine hindbrain and otic vesicles.

In Chapter 11 it was pointed out that the polarization of the anteroposterior and dorsoventral axes of the retina are determined in that order. Prior to their determination the retinal axes can be rotated in situ without affecting either the normal visuo-

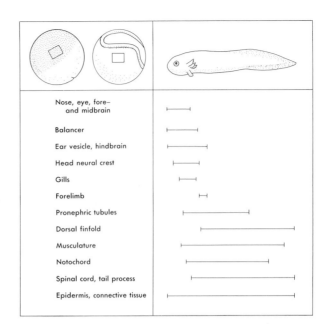

FIGURE 13.14 Field properties extend beyond the limits of tissue normally forming a particular organ. Scheme showing the developmental performance of undetermined gastrula ectoderm grafted to various regions of the flank of a neurula. [Adapted with modifications from J. Holtfreter, *Biol Zentr* 53:404–31 (1933).]

315

motor behavior (see Figure 11.34) or the normal pattern of retinotectal projection as measured electrophysiologically. Determination of polarity in the retina would reasonably be expected to result from field-like influences of the head region. Hunt and Jacobson (1973) showed, however, that polarization of the retinal axes can be affected by factors originating in the trunk. Thus, for example, when an optic primordium, excised after determination of the anteroposterior axis but before the dorsoventral axis is fixed, is grown temporarily on the trunk in inverted dorsoventral orientation, the dorsoventral polarity of the retina develops in accordance with its orientation on the trunk. Grafted in place of a host retina excised from the same side, it will project a normal anteroposterior retinotectal pattern, but its dorsoventral projection will be reversed. These results are quite in line with the notion that organizing factors related to the polarity of embryonic structures are ordered in relation to the major axes of the embryo and are widespread throughout it.

13.4   Summary

Morphogenetic fields are defined as the organizing factors in a region of the germ (egg or embryo) that see to it that a particular and restricted set of developmental capacities, from among a large repertory of potencies, is put into effect by particular cells. The nature of these organizing factors remains essentially unknown, but they are definitely tied to the normal activities of cells and cell groups. They can, moreover, be characterized as systems of positional information; cells within a field presumably have their position in the morphogenetic pattern specified with respect to a three-dimensional set of signals.

In the sea urchin embryo, the pattern of development of the larva is visualized as emerging from the interaction of animal and vegetal fields of opposed polarity. Neither animal nor vegetal halves alone can form complete embryos, but combinations of appropriate cell groups from the two halves can create the appropriate balance of field factors so as to bring about formation of a whole larva from a partial embryo. Selective chemical treatment can modify the action of each field, but little progress has been made in the recognition of normal animalizing and vegetalizing agents.

The chordamesoderm field of the amphibian embryo is perhaps the most widely studied field in developmental biology. It occupies the region of the blastula that normally arises from material of the grey crescent, and it invaginates through the dorsal lip of the blastopore. Its path of invagination determines the anteroposterior and dorsoventral axes of the embryo. The chordamesoderm field is possibly induced by prior action of a strong vegetal field. It, in turn, induces the neural plate in overlying ectoderm.

The properties of morphogenetic fields are arrived at solely on the basis of empirical evidence. This evidence consists in the observed morphogenetic effects of a variety of experimental manipulations of embryonic tissues. The empirical data show that the fields do have physical reality and that they consist of a three-dimensional pattern of organizing factors polarized along gradients of strength.

Fields elicit a particular developmental performance in cells that come under their influence. This process is called embryonic induction. By induction new fields arise, which induce other fields in hierarchical array, so that ultimately there appear fields of very restricted possibilities for differentiation. Once induced, a field has the power of self-differentiation, that is, of proceeding to differentiate without further influence of the inductor. Parts that have self-differentiating power are said to be determined. Development can be looked upon as a complex stepwise process of progressive determination in which each step creates a field of a more restricted potential for differentiation.

Fields have the ability to regulate. A partial field can reorganize itself in the pattern of a whole, but smaller field; moreover, two fields can be fused into a single field if their axial organization coincides. Regulation of fields occurs under both normal and experimental conditions.

Finally, we note that fields persist beyond the time their morphogenetic role has been accomplished and that their influence extends beyond the limits of the material that forms their organ rudiment.

It is one of the most challenging problems of modern developmental biology to discover what the organizing factors—the fields—are which determine what part of a cell's repertory of genetically determined potencies for differentiation it shall exercise.

13.5   Questions for Thought and Review

1. Of what possible use to the developmental biologist is the formulation of the concept of the morphogenetic field?

2. You know from evidence presented in Chapter 10 that all somatic cells have the same complement of genes. How, then, can there be organizing factors that are different in different regions of the blastula?

3. Using material from Chapter 13, formulate a clear and concise argument demonstrating that embryonic development proceeds epigenetically (cf. Chapter 10).

4. Usually a well-formed pluteus larva arises from a sea urchin embryo whose micromeres have been removed. How do you interpret this situation? How can invagination occur? What is the source of cells for forming the skeleton?

5. When a large bit of dorsal blastopore lip is excised from a beginning gastrula and grafted to the ventral side of a host embryo, it induces a secondary embryo showing head parts or even a complete embryonic axis. This chain of events occurs regardless of whether the graft comes to be at the head level or at the trunk level of the host embryo. But, if the isolated dorsal lip originates from an older gastrula, it induces only trunk and tail structures regardless of which anteroposterior level of the host it comes to occupy. From this information, what do you infer about the organization of the chordamesoderm field? You may wish to refer to Chapter 9 for assistance in answering this question. You may also wish to return to this question after reading Chapter 14.

6. What is the reason that the blastocoele culture method has been so widely used in studies of experimental morphogenesis in amphibians? How is it carried out?

7. What is a hierarchy? What do we mean when we refer to a hierarchy of inductions?

8. During formation of the eye in the chick embryo, the tip of the optic vesicle makes contact with the overlying head ectoderm for a period of about 12 hours, then it withdraws. Suppose you now excise the contacted ectoderm and graft it to the flank of the same or another embryo in place of a similar-size bit of ectoderm? What would you expect to happen? Reference to Chapter 9 and 14 may be useful in this connection.

9. Review the properties of fields as outlined in Chapter 13. Find as many examples as you can to show that these properties exist. As you read further in the text, refer to this question to bolster your answer with more data.

13.6   Suggestions for Further Reading

ASASHIMA, M. 1975. Inducing effects of the presumptive endoderm of successive stages in *Triturus alpestris*. *Wilhelm Roux' Arch* 177:301–308.

BARTH, L. J. 1964. The developmental physiology of amphibia. In J. A. Moore, ed. *Physiology of the Amphibia*. New York: Academic Press, p. 469–544.

CHILD, C. M. 1941. Inductors and so-called organizers in embryonic development. In *Patterns and Problems of Development*. Chicago: University of Chicago Press, pp. 435–503.

GROBSTEIN, C. 1956. Inductive tissue interaction in development. *Adv Cancer Res* 3:187–236.

HÖLFTRETER, J., and V. HAMBUGER. 1955. Amphibians. In B. J. Willier, P. Weiss, and V. Hamburger, eds. *Analysis of Development*. Philadelphia: Saunders, pp. 230–96.

JACOBSON, A. G. 1966. Inductive processes in embryonic development. *Science* 152:25–34.

KUHN, A. 1971. *Lectures on Developmental Physiology*. R. Milkman, translator. New York: Springer-Verlag, Lectures 12–17, pp. 166–261.

LOVTRUP, S., U. LANDSTRÖM, and H. LOVTRUP-REIN. 1978. Polarities, cell differentiation and primary induction in the amphibian embryo. *Biol Rev* 53:1–42.

NIEUWKOOP, P. 1967. Problems of embryonic induction and pattern formation in amphibians and birds. *Exp Biol Med* 1:22–36.

NIEUWKOOP, P. 1973. The "organization center" of the amphibian embryo: its origin, spatial organization, and morphogenetic action. *Adv Morph* 10:2–39.

NIEUWKOOP, P. 1977. Origin and establishment of embryonic polarities in amphibian development. *Cur Top Dev Biol* 11:115–32.

SAXON, L., and S. TOIVONEN. 1962. *Primary Embryonic Induction*. London: Logos Press.

WADDINGTON, C. H. 1966. Fields and gradients. In M. Locke, ed. *Major Problems in Developmental Biology*. New York: Academic Press, pp. 105–24.

YAMADA, T. 1962. The inductive phenomenon as a tool for understanding the basic mechanisms of differentiation. *J Cell Comp Physiol* 60(Suppl. 1):49–64.

13.7   References

AREY, L. B. 1965. *Developmental Anatomy*. Philadelphia: Saunders.

BAUTZMANN, H. 1926. Experimentelle Untersuchungen zur Abgrenzung des Organisationszentrums bei *Triton taeniatus*, mit einem Anhang: über Induktion durch Blastula Material. *Wilhelm Roux' Arch* 108:283–321.

COULOMBRE, A. J. 1965. The eye. In R. L. DeHaan and H. Ursprung, eds. *Organogenesis*. New York: Holt, Rinehart and Winston, pp. 219–51.

CURTIS, A. S. 1960. Cortical grafting in *Xenopus laevis*. *J Embryol Exp Morphol* 38:163–73.

CURTIS, A. S. 1962. Morphogenetic interactions before gastrulation in the amphibian, *Xenopus laevis*—the cortical field. *J Embryol Exp Morphol* 10:410–22.

HOLTFRETER, J. 1933. Organisierungsstufen nach regionaler Kombination von Entomesoderm mit Ektoderm. *Biol Zentralbl* 53:404–31.

HÖRSTADIUS, S. 1939. The mechanics of sea urchin development, studied by operative methods. *Biol Rev* 14:134–79.

HÖRSTADIUS, S., and L. JOSEFSSON. 1972. Morphogenetic substances from developing eggs of *Paracentrotus lividus*. *Acta Embryol Exp* 7–23.

HUNT, R. K., and M. JACOBSON. 1973. The origins of nerve cell specificity. *Sci Am* 228:26–35.

KIRSCHNER, M., J. C. GERHART, K. HARA, and G. A. UBBELS. 1980. Initiation of the cell cycle and establishment of bilateral symmetry in *Xenopus* eggs. In S. Subtelny and N. K. Wessels, eds., *The Cell Surface: Mediator of Developmental Processes*. New York: Academic Press, pp. 187–215.

LALLIER, R. 1973a. Animalization de la larve de l'Oursin *Paracentrotus lividus* par traitment de l'oeuf, fecondé invidis, avec un detergent le lauryl sulfate de sodium. *C R Acad Sci* (Paris) [D]276:1033–36.

LALLIER, R. 1973b. Amines et morphogenèse: une nouvelle interprétation du mode d'action des ions lithium, sur la différenciation de l'oeuf de l'Oursin *Paracentrotus lividus*. *C R Acad Sci* (Paris) [D]276:3743–46.

LEHMANN, F. E. 1926. Entwicklungstorungen an der Medullaranlage von *Triton*, erzergt durch Unterlagerungsdefekte. *Wilhelm Roux' Arch* 108:243–83.

LOPASHOV, G. V., and O. G. STROEVA. 1961. Morphogenesis of the vertebrate eye. *Adv Morph* 1:331–77.

MALACINSKI, G. M., H. -M. CHUNG, and M. ASASHIMA. 1980. The association of primary embryonic organizer activity with the future dorsal side of amphibian eggs and early embryos. *Dev Biol* 77:449–62.

MANGOLD, O. 1923. Transplantationsversuche zur Frage der Spezifität und der Bildung der Keimblatter bei *Triton*. *Wilhelm Roux' Arch* 100:198–301.

MANGOLD, O. 1929. Experimente zur Analyse der Determination und Induktion der Medullarplatte. *Wilhelm Roux' Arch* 117:586–696.

MORGAN, T. H., and A. BORING. 1903. The relation of the first plane of cleavage and the gray crescent to the median plane of the embryo of the frog. *Wilhelm Roux' Arch* 16:680–90.

NIEUWKOOP, P. D. 1969a. The formation of the mesoderm in urodelean amphibians. I. Induction by the endoderm. *Wilhelm Roux' Arch* 162:341–73.

NIEUWKOOP, P. D. 1969b. The formation of the mesoderm in urodelean amphibians. II. The origin of the dorsoventral polarity of the mesoderm. *Wilhelm Roux' Arch* 163:298–313.

NIEUWKOOP, P. D. 1970. The formation of the mesoderm in urodelean amphibians. III. The vegetalizing action of the li ion. *Wilhelm Roux' Arch* 166:105–23.

NIEUWKOOP, P. D., and G. A. UBBELS. 1972. The formation of the mesoderm in urodelean amphibians. IV. Quantitative evidence for the purely "ectodermal" origin of the entire mesoderm and of the pharyngeal endoderm. *Wilhelm Roux' Arch* 169:185–99.

OKAZAKI, K. 1975. Spicule formation by isolated micromeres of the sea urchin embryo. *Am Zool* 15:567–81.

SPEMANN, H. 1938. *Embryonic Development and Induction*. New Haven: Yale University Press.

SPEMANN, H., and H. MANGOLD. 1924. Über Induktion von Embryonalanlagen durch Implantation artfremder Organisatoren. *Wilhelm Roux' Arch* 100:599–638.

TOMKINS, R., and W. P. RODMAN. 1971. The cortex of *Xenopus laevis* embryos: regional differences in composition and biological activity. *Proc Natl Acad Sci USA* 68:2921–23.

WADDINGTON, C. H. 1952. *The Epigenetics of Birds*. Cambridge: Cambridge University Press.

WEISS, P. 1939. *Principles of Development*. Chicago: University of Chicago Press.

WILLIER, B. H., and M. E. RAWLES. 1935. Organ-forming areas in the early chick blastoderm. *Proc Soc Exp Biol Med* 32:1293–96.

WOLPERT, L. 1969. Positional information and the cellular pattern of spatial differentiation. *J Theoret Biol* 25:1–47.

The concept of **dependent differentiation** embraces those cases in which the differentiation of one organ or tissue is dependent on the association of its primordium with the primordium of another organ or tissue. In Chapter 13 it was recognized that embryonic inductions are examples of dependent differentiation, but that possibly some researchers would prefer to use the term induction only for cases in which the inductor would seem to specify the nature of the response. A more general expression, which seems to have all the connotations embraced under the concepts of both induction and dependent differentiation, is **tissue interaction.** This term would also include cases in which one tissue simply provides a mechanical support for another or exercises a simple nutritive role in its development. Some cases that may fall into this category are recognized later in this chapter.

First, however, we present some aspects of the relationships between inducing and responding systems that have emerged from classical experiments. Concepts arising from these experiments have generally not yet been integrated with modern molecular models of differentiation. The reason for this situation is that, for most of the "classical" inductions, there is no knowledge about the nature of what passes between inductive and responding systems. Nor is it known for most cases whether the inductive interaction elicits any immediate effect on the synthesis of new macromolecules by either the acting or responding systems. Nevertheless, despite our inability to ascribe particular macromolecular bases to the accomplishment of particular tissue interactions, the concepts that arise from the analysis of these interactions retain their validity and continue to challenge our understanding.

14 Dependent Differentiation

14.1 Competence

The ability of a tissue to react to an inductor by an appropriate morphogenetic response is termed **competence.** We have already observed, for example, that prospective belly ectoderm of an early gastrula, exposed to chordamesoderm, responds by making neural tube. This reaction occurs regardless of whether the exposure is effected by means of the blastocoele culture method described in Chapter 13

or carried out in vitro using gastrula ectoderm wrapped around chordamesoderm. Prospective belly ectoderm of the gastrula is, therefore, competent to form neural tube, just as is prospective neural ectoderm of the gastrula. Any region of the prospective ectoneuroderm, however, will progressively lose its competence to respond to neural induction if isolated in vitro for a prolonged period or in an exogastrula. Scanning electron micrographs of isolates from gastrulae of *Triturus alpestris* held in vitro reveal that they come to resemble epidermal cells of normal neurulae (Grunz et al., 1975).

Competence to respond to neural induction is likewise lost by prospective belly ectoderm in vivo by the time the neurula stage is reached. It is not, however, necessarily restricted to the formation of epidermis. For example, prospective belly ectoderm from an early neurula of the newt *Taricha torosa,* implanted in place of ectoderm that normally forms the sensory placodes (Figure 14.1), forms normal noses, lenses, and ears (Jacobson, 1955, 1963a–c).

A frequently cited case of induction (cf., Chapter 13) is that of the lens of the eye. This induction occurs when the eyecup (presumptive retina) contacts the head ectoderm (Figure 14.2). This interac-

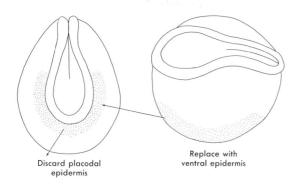

Discard placodal
epidermis

Replace with
ventral epidermis

FIGURE 14.1 Diagram of an operation in which
ectoderm containing prospective sensory placodes is
replaced by prospective belly ectoderm in the newt
embryo. [Adapted with modifications from A. G.
Jacobson, *J Exp Zool* 154:273–84 (1963).]

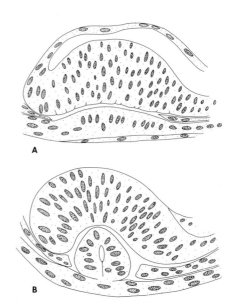

A

B

FIGURE 14.2 Initial stages in the development of the
eyecup and lens of the newt. As the cells of
prospective lens become palisaded in response to the
optic vesicle, the latter responds to the lens by
forming the eyecup. [Adapted with modifications from
H. Spemann, *Embryonic Development and Induction*,
Yale University Press, New Haven, 1938, p. 43.]

320

tion is clearly required for the formation of a properly proportioned and well-differentiated lens.
There is evidence, however, that the competence of
ectoderm to form lens occurs through action of the
endoderm and mesoderm during early neurula
stages. Liedke (1951) found that gastrula belly ectoderm of *Ambystoma punctatum,* transplanted to the
site of the future lens during gastrulation or in the
early neurula stage of a host, formed lens. When
the transplant was made to a host at the late
neurula stage, however, lens did not form from the
graft. In early neurula stages the future lens overlies
the endodermal wall of the archenteron and is bordered by the anterior portion of the lateral plate
mesoderm. Jacobson (1955) isolated prospective
lens epidermis from early neurula stages of another
newt *Taricha torosa* and cultured it: (1) alone;
(2) in combination with the subjacent endoderm;
(3) with the anterior part of the lateral plate mesoderm; and (4) with both endoderm and mesoderm
combined. He observed no lenses in isolates of presumptive lens alone, but found them in a substantial percentage of cases in the other combinations.
Important to these results is the fact that they were
obtained using material from embryos reared at low
temperatures (9–16°C). In other experiments he
reared embryos to the early neurula stage at various
temperatures between 5° and 25°C, removed the
eyecup, and then allowed them to develop at 19°.
From those embryos precultured at 16°C, lenses
formed in the absence of the retina in almost 70%

of cases; from other embryos lenses formed much
less frequently.

Apparently, at colder temperatures the endo-
mesodermal substratum is more effective in pre-
conditioning the future lens epidermis to form lens.
Spemann (1905) routinely reared *Rana esculenta*
embryos at 12°C and reported that they would form
lenses in the absence of the eyecup, a fact not confirmed by Woerderman (1939), who reared embryos of the same species at higher temperatures.

Jacobson (1963b) found that *T. torosa* embryos
deprived of all endoderm will form lenses, but that
the quality of their differentiation is greater the
longer into the neurula stage the endoderm is left
intact. He also observed, however (1963c), that
endoderm of more posterior regions is as effective
as that of anterior regions in promoting lens differentiation. Perhaps the endodermal influence is a
relatively general one. This may explain why, in
some kinds of amphibian embryos, lens competence is present extensively in belly and flank ecto-

derm until relatively late stages. Thus, in the late tail-bud stage of *Rana temporaria* or *Hyla arborea* epidermis from any body region can be grafted in place of prospective lens ectoderm and will respond to the inductive action of eyecup. In *Bombinator pachypus,* however, lens-forming competence is restricted to the head ectoderm at this stage (cf., papers by Spemann, 1912, and von Ubisch, 1927). The general impression is that when lens-forming competence is widespread at early stages, it is progressively restricted to the normal lens-forming area by the late tail-bud stage.

14.2   Genetic Limitation of the Inductive Response

An inductor can elicit from a reacting system only a response that is provided by the latter system's genetic endowment. Particularly pertinent in this regard are experiments carried out in Spemann's laboratory by Schotté and Holtfreter (reviewed by Spemann, 1938). The mouth armament of the larval frog consists of horny jaws, whereas the larval newt's head has calcareous teeth. The newt larva also has on its head ventrolateral outgrowths, the balancers, and the larval frog carries just behind the mouth, in a similar position, suckers. All these structures are ectodermal derivatives that appear to be induced by the head endoderm. If some of the ectoderm of the future head region of an early gastrula of the newt is replaced by ectoderm of an early frog gastrula, the resulting chimera may show horny frog teeth and suckers on the modified newt head. In the reciprocal case, newt ectoderm developing in the head region of a frog embryo may form calcareous teeth and balancers. It is as though each ectoderm understands the inductive language of its own and foreign inductors, but can respond only according to its own genotype. The same principle shows in the fowl, where the structurally distinctive feathers of the silky-chicken mutant are formed from silky ectoderm underlaid by the mesodermal feather inductor from a normal bird (Cairns and

Saunders, 1954). Extensive studies of integumentary derivatives in dermoepidermal recombinations among reptiles, birds, and mammals have been carried out, notably by Dhouailly (1975). These show, in brief, that dermal inductors of integumentary structures are effective in combinations among tissues of these classes of vertebrates. In all cases the spacing and pattern of the derivatives are essentially under dermal control, but the specificity of differentiation, when recognizable, is determined by the ectodermal component. Thus chick epidermis responds to mouse dermis by making feather papillae with characteristic but shorter barb ridges, not by making hairs.

It is possible, however, that the genotype of a responding system may provide for responses to inductions to which they are normally never exposed. For one instructive example, we look at work of Dr. Yujiro Hayashi. He exchanged mesoderm and ectoderm of the prospective upper beak in chick and duck embryos. When the ectodermless beak primordium of the duck is combined with reactive chick ectoderm, the chick ectoderm responds by forming typical tooth ridges of the duck type. These ridges normally do not form in chick ectoderm and are elicited neither in chick–chick combinations nor in ectoderm of duck reacting with upper-beak mesoderm of the chick. These results show that ectoderm of both chick and duck is competent to form tooth ridges, but the ability to elicit this response is possessed only by mesoderm of the duck. In this case the inductive system, not the reaction system, is genetically limited.

Even more dramatically, it has been found by Kollar and Fisher (1980) that epithelium from the floor of the first and second pharyngeal arches of the 5-day chick embryo, combined with mesenchyme from the first submandibular molar of the 16–18 day old mouse embryo, forms an enamel organ characteristic of mammalian teeth. In response to the enamel organ, the mesenchyme organizes a typical tooth papilla within which odontoblasts develop and begin the secretion of dentine. In favorable cases (Figure 14.3) the appearance of dentine is followed by the production of enamel by the enamel organ, which is the normal sequence of events in mammalian odontogenesis. The observations of Kollar and Fisher suggest that chick oral epithelium has retained throughout avian evolu-

321

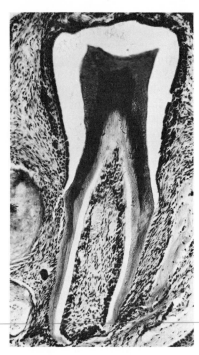

FIGURE 14.3 Well-developed tooth formed from the combination of chick pharyngeal epithelium and submandibular molar mesenchyme of the mouse. The combined tissues were cultured on the chorioallantoic membrane of the chick embryo. The epithelium of the enamel organ is clearly visible as is the clear remnant of enamel left after decalcification of the tooth prior to sectioning. Dentine and pulp of mouse origin are clearly distinguished. [Photograph supplied by Dr. E. J. Kollar, whose generosity is gratefully acknowledged.]

tion the genetic coding for producing an enamel organ. What have been lost are the appropriate tissue interactions required for odontogenesis.

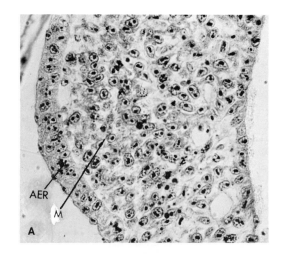

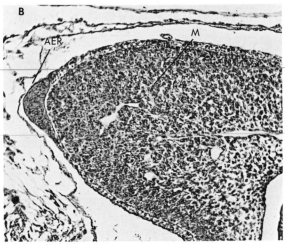

FIGURE 14.4 The apical ectodermal ridge of the chick embryo wing bud. A: Cross section through the somatopleure at the level of the wing primordium in a 60-hr embryo, showing the thickened ectoderm that forms the apical ridge. B: Cross section through the wing bud of the 72-hr embryo showing the growth of the bud and increased thickening of the ectodermal ridge. AER, apical ridge; M, wing-bud mesoderm. [From J. W. Saunders, *J Exp Zool* 108:363–404 (1948) by permission of Alan R. Liss, Inc.]

14.3 The Principle of Reciprocal Action

One usually identifies one component of an inductive interaction as the initiating or action system and the other as the responding or reaction system.

Thus we refer to the induction of lens by optic vesicle, of neural tube by archenteron roof, and so on. Yet when live tissues are in inductive association, it is more accurate to recognize that one is dealing with stimulus-response sequences acting in both directions; that is, there is **reciprocal action** between components of the inductive system. Thus, the lens interacts with the optic cup to determine the geometric pattern of its differentiation. Sometimes, however, the effect of the induced system on the inductor is rather subtle and not readily recognized.

The principle of reciprocal action can be illustrated by the interaction of ectoderm and mesoderm in the morphogenesis of the vertebrate limb. In the chick embryo, when the wing bud appears at 60 hours of incubation, the ectoderm at its apex thickens, forming an anteroposterior ridge (Figure 14.4), which is called the apical ectodermal ridge (cf. Chapter 11). It forms in the ectoderm in response to mesoderm induction, for when prospective wing mesoderm is grafted to a wound in the flank, the ectoderm that heals over it forms a ridge (Figure 14.5). Flank ectoderm loses its competence to give this reponse to mesodermal induction at about 70 hours of incubation. Wing-bud mesoderm loses its capacity to induce at about the same time, and leg-bud mesoderm a little later (Kieny, 1960, 1968; Saunders and Reuss, 1974).

The apical ridge itself is now an inductor. If it is removed from the apex of the mesodermal bulge, limb development ceases (Saunders, 1948). If the mesoderm of the limb primordium is adorned with two ridges by means of suitable microsurgical techniques, two limbs will grow from the originally single primordium (Saunders et al., 1976). The ectodermal ridge, therefore, is said to induce mesodermal outgrowth (Figure 14.6).

The story does not end here, however, for the growing mesoderm now reacts on the ectodermal ridge to keep it in a thick and inductively active configuration (Zwilling and Hansborough, 1956). The mesoderm supplies a maintenance factor, not yet identified, which can pass through filters barring cellular contact (Saunders and Gasseling, 1963). This factor keeps the apical ridge inductively active until the terminal parts of the limb are formed.

The analysis of limb morphogenesis has given considerable insight into the reciprocal nature of tissue interactions during development. It shows that interacting systems maintain not a one-way

323

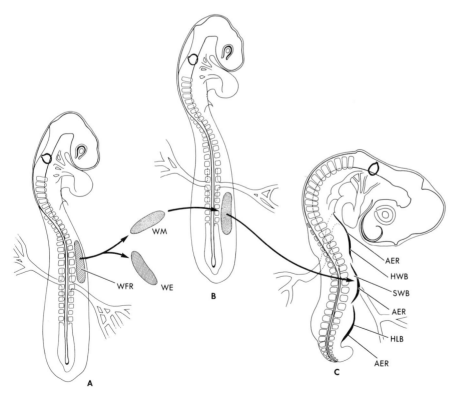

FIGURE 14.5 Induction of apical ectodermal ridge by prospective wing mesoderm. A: Prospective wing-forming region, WFR, of the stage 15 chick embryo is excised, and the ectoderm, EW, is removed and discarded. B: The wing mesoderm, WM, is then grafted to a wound site made by removing a patch of prospective flank ectoderm from a similar host embryo. Host ectoderm then regenerates over the graft. C: The host embryo after 20 hr, showing a supernumerary wing bud, SWB, capped by an apical ectodermal ridge, AER, similar to that on the host wing bud, HWB, and leg bud, HLB.

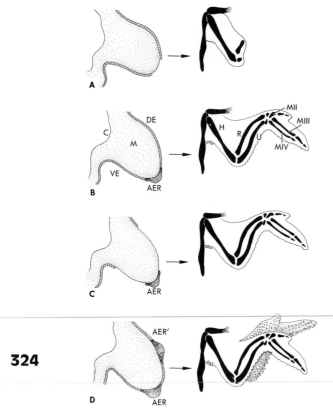

FIGURE 14.6 Induction of limb outgrowth by the apical ectodermal ridge in the chick embryo. *Left:* Schematic cross sections of the wing bud depicting various operations; *right:* the results obtained. A: Excision of the ridge results in deficiencies of terminal limb parts. B: In the presence of a normal ectodermal covering, the wing parts appear in their normal proximodistal sequence and pattern. C: Removal of the dorsal and ventral ectoderm permits outgrowth to continue as long as an apical ridge is present. D: Grafting an extra ridge to the dorsal surface of the wing bud results in formation of two wings. AER, apical ectodermal ridge; AER′, grafted apical ectodermal ridge; C, coelom; DE, ectoderm of the dorsal wing bud; H, humerus; M, mesenchymal core; R, radius; U, ulna; VE, ectoderm of the ventral face of the wing bud. MII, MIII, and MIV, metacarpals II, III, and IV, respectively.

communications system, but a dialogue, transmitting developmental cues in both directions and balancing and adjusting them so as to achieve the construction of a harmonious organ.

14.4 The Question of Inductive Specificity

When inductive interactions have been identified in the embryo there usually exists a notable correspondence in the morphogenetic relationships of the inductive and reacting systems. Thus, nasal placodes are induced by endoderm of the head region and by forebrain, the latter of which receives input from the sensory nasal epithelium; the otic vesicle is induced by head mesoderm and hindbrain, and it subsequently establishes neural connnections with the hindbrain; the lens is induced by eyecup, and so on. The congruence and apparent developmental appropriateness of such relationships suggest strongly that an induction is an instructive event, that is, that the inductor specifies the response that the reacting system will give. It might be so, perhaps, by releasing a specific macromolecule that is taken up by the reacting system and determines what its response will be.

14.4.1 DOES THE INDUCTOR EXERCISE AN INSTRUCTIONAL ROLE? An instructive model of an inductive system would require, moreover, that the responding system should be undifferentiated, in the sense that it is able to respond to any of several different kinds of signals by giving appropriately different kinds of responses. Ectoneuroderm of the urodele gastrula seems to possess this ability, for, as we have already seen (Figure 13.14), it can make essentially any kind of tissue depending on the site to which it is transplanted on a host neurula. The idea that specific instructional molecules mediate inductive events has been further promoted by numerous studies of the differentiation in vitro of ectodermal "sandwiches" enclosing bits of presumed inductive material, isolated from various sources and more or less purified. Similar materials have been tested by implanting them in the blastocoele (Figure 13.11). Such widely divergent materials as killed mouse liver, extracts of bone marrow, kidney, and so on, have inductive effects in such experiments. The inductions include all histological tissue types and, in the case of implants to the blastocoele, sometimes issue in the formation of well-organized body segments. The types of induc-

TABLE 14.1 Regionally Specific Inductions Obtained from Urodele Gastrula Ectoderm

Type of Induction	Structures Formed
Archencephalic	Forebrain Eye Nose
Deuterencephalic	Midbrain Hindbrain Otic vesicle
Trunk-mesodermal	Trunk notochord Trunk somites Pronephros Blood islands
Spinocaudal	Spinal cord Tail notochord Tail somites

tions observed can be classified according to four principal kinds: **archencephalic, deuterencephalic, spinocaudal,** and **trunk-mesodermal.** Each of these types relates to groups of organs and tissues at different body levels, as described in Table 14.1.

In 1958 Yamada reported that all the inductions noted in Table 14.1 could be obtained from an acid-precipitable fraction of the 100,000 $\times G$ supernatant from guinea pig bone marrow extracted in 0.14 M NaCl. The regional effects varied, however, according to the length of time the material was exposed to heat at 90–100°C. Trunk-mesodermal inductions predominated in unheated material, but regional effects changed progressively within a few seconds of exposure to heat in the following sequence:

Trunk-mesodermal
Spinocaudal
Deuterencephalic
Archencephalic

The frequency of inductions was not, however, diminished by the heat treatment.

Beginnning in the 1960s Heinz Tiedemann and his colleagues (reviewed by Tiedemann, 1968) began isolating, from extracts of 9- to 11-day chick embryos, proteins that exercised morphogenetic effects when implanted into the urodele blastocoele or when sandwiched between sheets of undetermined ventral ectoderm of the beginning gastrula. A crude "neuralizing" factor which induces forebrain structures was obtained, and is apparently not

yet purified. A "vegetalizing" factor was also found, which induces the formation of mesodermal and endodermal structures. It was prepared in highly purified form by Born et al. in 1972. Combinations of the neuralizing and vegetalizing factors elicit the formation of deuterencephalic and trunk-mesodermal structures. Heinz Tiedemann et al. (1972) showed that the vegetalizing factor is partially inhibited by binding to chick DNA. Later Hildegard Tiedemann and Born (1978) showed that vegetalizing factor covalently bound to Sepharose beads is inactive. Presumably, it must be internalized by the ectoderm cells in order to induce their differentiation. The crude neuralizing factor, similarly bound to beads, is active in inducing the differentiation of brain structures. Thus, its primary action is apparently on the cell membrane.

What these findings tell us about normal induction is actually very little. They are compatible with the notion that regionalization occurs during "primary induction" (Spemann, 1938; cf. Chapter 13) in response to regional differences in the chorda-mesoderm, but they do not require the interpretation that informational molecules are transferred in the inductive process. Some other experiments are more persuasive that, under some experimental conditions, informational macromolecules can induce specific synthesis, and under other conditions, a specific morphogenetic performance.

For some years a number of investigators, notably Niu and his colleagues, have performed experiments the results of which suggest that exogenous RNA can induce a particular pathway of differentiation in uncommitted cells. As more sophisticated methods for fractionating and purifying RNAs have become available, these results have become quite persuasive. In 1977 Yang and Niu reported that poly(A) RNA from adult livers of bovine, rat, or chick origin, injected into the lumen of the mouse uterus, caused the uterine cells to produce serum albumin immunologically like that produced in the liver of the donor species. In addition, mouse albumin was produced. The synthesis of the albumins was inhibited by cycloheximide, but not by actinomycin D. These results suggest that a polyadenylated mRNA for albumin was taken up and translated by mouse uterine cells. This possibility needs further exploration.

In 1973, Niu and Deshpande reported that both nuclear and cytoplasmic RNA from adult chick

325

heart (very slightly contaminated with protein) could induce the formation of beating heart tissue in posterior sections of the chick blastoderm isolated at the stage of the definitive primitive streak and cultured in vitro. Isolates cultured without heart, or with RNA from other tissues, did not form heart tissue.

Later Deshpande et al. (1977, 1978) isolated poly(A) RNA from total RNA of 16-day embryonic chick hearts. From this they separated a novel RNA which sedimented at 7 S and which induced heart formation in posterior segments of the chick blastoderm that normally do not form heart tissue. RNAs from other sources, isolated under identical conditions, did not induce heart formation. This particular RNA, although of a size comparable to many mRNAs and polyadenylated, as most mRNAs are, is not translated under conditions in vitro in which mRNAs of various other kinds are readily translated. How this untranslatable RNA determines heart differentiation in vitro is not known. It actually inhibits translation of mRNAs isolated from heart. It is not evident that these results have any relationship to the problem of normal heart induction. Jacobson and Duncan (1968) showed that in the newt, heart is induced by anterior endoderm; ectoderm increases the frequency with which hearts differentiate in vitro in the presence of endoderm; and brain tissues delay, or prevent entirely, the onset of heart formation.

14.4.2 IS INDUCTION A PERMISSIVE EVENT? The
early inductive events in the amphibian gastrula are quite suggestive that the regional character of a morphogenetic response is specified by the inductor and is in some sense, therefore, instructive. But, when one turns to the subordinate echelons of inductions (nose, lens, kidney tubules, pancreas, and so on) it is not easy to determine whether a particular inductor specifies the response or merely permits or triggers it. For such inductions the repertory of permitted responses may be very limited. Perhaps the inductor merely throws the balance toward one of two alternative differentiative pathways. As we have seen, in some amphibian embryos lens competence is probably distributed widely in the ectoderm at early stages but is restricted to the lens-placode region as development proceeds. It is questionable whether the prospective

lens has any developmental capabilities other than for lens and head epithelium at the time it is met by the eyecup.

Another difficulty in applying the instructional model to induction is that specific differentiations can often be elicited by subjecting the responding system to abnormal ''inductors.'' The induction of lenses in flank epidermis has been accomplished by grafts of fresh liver and by boiled heart! Particularly in the case of inductive interactions between epithelial and mesenchymal elements, specific reactions can be elicited experimentally by inductors that show no congruence whatsoever with the reacting system. For example, the secretory tubules of the definitive kidney arise in the metanephrogenic mesenchyme (Figure 11.16) in response to induction by the ureteric bud. If the ureteric bud fails to reach the metanephrogenic mesenchyme, as is the case in some lethal mouse mutants, the metanephros fails to form. Once the ureteric bud reaches the metanephrogenic mass, the two elements may be

FIGURE 14.7 Intact metanephric rudiment of the 11-day mouse embryo after 8 days in culture. The branching ureteric bud forming collecting tubules, c, and the secretory tubules, t, formed from the mesenchyme are readily seen. A similar pattern of morphogenesis is seen when the tips of the ureteric bud are separated by means of trypsin and then recombined. [From C. Grobstein, *J Exp Zool* 130:319–40 (1955), by courtesy of Dr. Grobstein and permission of Alan R. Liss, Inc.]

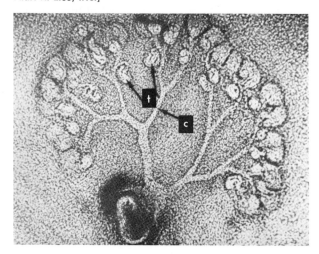

separated by trypsin-digestion and cultured in vitro separately or together. In isolation the mesenchyme of the mouse makes only a sheet of cells, but, if the two components are recombined, the ureteric bud induces the mesenchyme to make collecting tubules (Figure 14.7). Moreover, it can do so even when separated from the mesenchyme by thin filters that bar cellular contact (Figure 14.8). In similar cultures, however, secretory tubules can be induced in the kidney mesenchyme by epithelium of the salivary gland and by fragments of spinal cord. Tubules also form in mouse metanephrogenic mesenchyme grafted to the anterior chamber of the adult mouse eye in the absence of an inductor. This evidence was summarized by Grobstein and Parker (1958).

In Chapter 11 the origin of the dorsal pancreatic rudiment was described. This rudiment can be separated from its mesenchymal capsule and grown in vitro (Grobstein, 1962). Here it fails to form the typical pancreatic acini unless associated with any of several inductors. Its own capsular mesenchyme is inductive, even transfilter (Figure 14.8), but the mesenchymal capsule of the submandibular gland is more effective than the natural inductor. Metanephric mesenchyme and lung capsule are also inductive. Subsequent to these findings, Rutter et al. (1964) prepared cell-free material, difficult to solubilize, from extracts of chick embryos and found that it was an effective inductor of morphogenesis in the pancreatic rudiment of the mouse. Later studies in Rutter's laboratory (reviewed by Rutter et al., 1978) revealed that one may extract from mesenchymal tissues from various sources a carbohydrate-containing protein that elicits pancreatic

FIGURE 14.8 Diagram of a transfilter experiment in a "Grobstein dish." Interacting tissues are placed on opposite sides of a filter whose porosity and thickness may or may not permit transfilter cell contact, depending on the dimensions chosen.

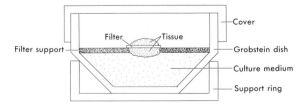

differentiation from mouse or rat pancreatic rudiments. It has been called **mesenchyme factor,** or **MF.** MF has been partially purified and is thought to be a membrane protein.

MF can be bound to Sepharose beads. Fragments of pancreatic rudiment adhere to these beads at their basal surface, which is the surface normally in contact with mesenchyme. The fragments rapidly incorporate ^{3}H-thymidine into DNA and produce zymogen granules. This effect is not observed when albumin-coated beads are used.

Quite evidently, the morphogenesis of the secretory tubules of the metanephros and of the acini of the pancreas are not responses to stimuli that are specific to their normal inductors. Perhaps, in these and similar cases, the "inductors" are simply sources of membrane components provided by contact with mesenchyme, or released from mesenchyme, which interact with recipient cell membrane, setting up a sequence of transmembrane changes required for a cell to express a prior differentiative commitment. Recall that the pancreatic rudiment (Chapter 11) has differentiated specific cellular types and has begun production of digestive enzymes before morphogenesis begins. Grobstein (1962) showed that the pancreatic inductor may act simply to provide conditions under which the pancreatic cells can assemble into proacinar "packages"; thereafter homeotypic interactions between the packaged cells provide appropriate conditions for continued morphogenesis even when the inductor is removed.

From the foregoing examples of epithelial-mesenchymal interactions it seems unlikely that the inductors specify the morphogenetic response of the responding system. Moreover, there are several instances in which interactions once thought to show a measure of specificity have since been shown to be relatively nonspecific. The submandibular salivary gland of the mouse is a case in point. This gland is derived from an epithelial invagination surrounded by a capsule of submandibular mesenchyme. Grobstein (1953) showed that the combined elements isolated from 13-day mouse embryos can undergo characteristic morphogenesis in vitro, the epithelium forming a characteristic compound tubuloalveolar gland. The epithelium alone, however, fails to undergo glandular differentiation. Nor, according to Grobstein's (1953) account, will it

327

do so in association with capsular mesenchyme killed by heat or by freezing or in the presence of mesenchyme from the lung, mandibular arch, or limb bud. Morphogenesis was later reported to occur in the presence of mouse parotid mesenchyme (Grobstein, 1967) or chick submandibular mesenchyme (Sherman, 1960), but Spooner and Wessels (1972) confirmed Grobstein's earlier observation that mouse lung mesenchyme would not support morphogenesis of the mouse submandibular epithelium. Nevertheless, in 1974 Lawson showed that lung mesenchyme of the mouse or rat is effective in bringing about morphogenesis of the mouse submandibular epithelium; the rat mesenchyme is quantitatively more effective. Lawson found that the ability of lung to induce morphogenesis was dependent on culture conditions. Whereas specificity of induction is seen when the recombination of epithelium and mesenchyme is made on a Millipore filter, as carried out by Grobstein (1953, 1967), it is not shown when the recombinants are assembled on a thin agar film supported on a cellulose acetate net in a medium of blood plasma and chick embryonic extract. Lawson ascribed the effectiveness of lung mesenchyme under the latter conditions to increased cellular density of the lung mesenchyme (see also Cunha, 1972).

These and similar cases do lead one to ask whether there are any epithelial-mesenchymal interactions in which the inductor specifies the response of the reacting system. Three groups of examples come immediately to mind. The first of these deals with the induction of teeth. In Section 14.2 it was shown that mouse submandibular mesenchyme can induce the formation of an enamel organ in chick pharyngeal epithelium. The mesenchyme forms a typical dental papilla, and tooth formation follows. The formation of an enamel organ by chick epithelium would seem to be a response to an instructive induction. Other studies on mouse teeth provide further examples. When tooth germs, comprising both epithelial and mesodermal components, are isolated from the mandibles of 14–16-day mouse embryos and cultured in vitro, they develop sufficiently that enamel and dentin are deposited in patterns of molars or of incisors, depending on their regional origin in the mandible. Kollar and Baird (1970) combined mesodermal dental papillae with mouse foot epithelium in vitro and observed the formation of teeth with orderly

deposits of dentin and enamel but unrecognizable as to type. This clearly shows that the dental papilla is inductive and that young, nondental epithelium is able to respond by tooth formation, but in an abnormal manner. Moreover, if the dental epithelium and dental papillae of molars and incisors are separated by trypsin treatment and exchanged, teeth that form from the recombinants develop as incisors or as molars, depending on the source of the papilla. Thus regional differences in the inductive activity of the papillae determine the form of the teeth (Kollar and Baird, 1969).

A second group of examples suggestive of an instructive mode of induction concerns the determination of the ectodermal derivatives of the integument in birds (Figure 14.9). Birds exhibit striking regional differences in their epidermal differentiations. Feathers differ in structure and physiology from one body region to another, and scales and claws are usually restricted to the feet. Determination of these regional differences apparently occurs very early, for if one transplants in the 3-day chick embryo mesoderm from the prospective thigh-forming region of the leg bud to the surface of the wing bud in such a way that it is covered with wing ectoderm, the resulting bird, upon reaching adulthood, may show a patch of thigh feathers on its wing. Upper-wing mesoderm combined with wing ectoderm or leg ectoderm induces wing feathers, and foot-forming mesoderm combined in suitable grafts with wing ectoderm induces the wing ectoderm to form scales and claws. The different inductors—thigh, upper wing, foot—thus have re-

FIGURE 14.9 Specification of response by the inductor. Limb ectoderm of the chick embryo forms derivatives appropriate to the regional origin of different kinds of mesoderm associated with it.

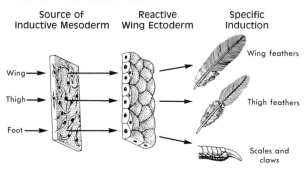

Source of Inductive Mesoderm	Reactive Wing Ectoderm	Specific Induction
Wing		Wing feathers
Thigh		Thigh feathers
Foot		Scales and claws

gionally specific inductive abilities, and early limb ectoderm is competent to respond to each of these inductors with a specific differentiation (Cairns and Saunders, 1954; Saunders et. al., 1957).

A third group of examples suggestive of inductive specificity involves the induction of branching patterns in epithelial rudiments. We have already noted that the tip of the ureteric bud in the mouse embryo branches in a characteristic manner to form the collecting tubules of the kidney when it interacts with the metanephrogenic mass, the latter mass forming secretory tubules. The ureteric bud does not form collecting tubules of the kidney in the absence of the metanephrogenic mass. Using chick embyros Bishop-Calame (1966) examined effects of various mesenchymal tissues on the branching pattern of, and the response of the mesenchymes to, the ureteric bud. She cultured her recombinants on the vascular chorioallantoic membrane of the chick embryo. In brief, she found that the ureter elongated and showed its characteristic branching pattern only in association with metanephrogenic mesenchyme or with undifferentiated mesonephric mesenchyme from very young embryos. In both of these mesenchymes secretory tubules and glomeruli appeared. Surprisingly, however, lung mesenchyme and mesenchyme from the proventriculus likewise formed tubular systems like those of kidney in association with the ureter, but did not promote characteristic ramification of the ureter (Figure 14.10). Thus, the ureter can specify in lung and proventriculus mesenchyme a specific morphogenetic performance appropriate to itself even without undergoing its own typical differentiation.

It does not necessarily follow from these experiments, however, that elicitation of a morphogenetic pattern appropriate to the inductor in a tissue not normally exposed to that inductor, will cause the responding tissue to assume the final cytodifferentiative state appropriate to the inductor. For example, when prospective mammary gland epithelium of the mouse is exposed to salivary gland mesoderm, it forms the branching pattern characteristic of salivary epithelium. It does not produce salivary enzymes, however, and instead can be shown to synthesize α-lactalbumin, a characteristic milk protein, when implanted into a lactating mouse host (Sakakura et al., 1976). These observations are compatible with the idea that membrane components or other substances from the environment can affect the form that cells assume, which may or may not permit them to express the condition of cytodifferentiation called for by their past developmental history, or lineage.

In the environment of migrating cells and cellular layers the most prominent and best characterized materials are collagen, glycosaminoglycans (GAGs), and proteoglycans (cf. Chapter 11). Sclerotome cells, migrating from the somites toward the notochord and neural tube, are synthesizing these substances, as is also the notochord. As a consequence of this medial migration of the sclerotome cells, they become surrounded by a dense matrix which is, in fact, largely collagen and aggregates of proteoglycans. Lash and his colleagues (reviewed by Lash and Vasan, 1977) showed that either exogenous collagen or proteoglycans greatly enhance cartilage production by embryonic chick somites cultured in vitro. They showed, too, that, whereas the presence of notochord in a culture of somites

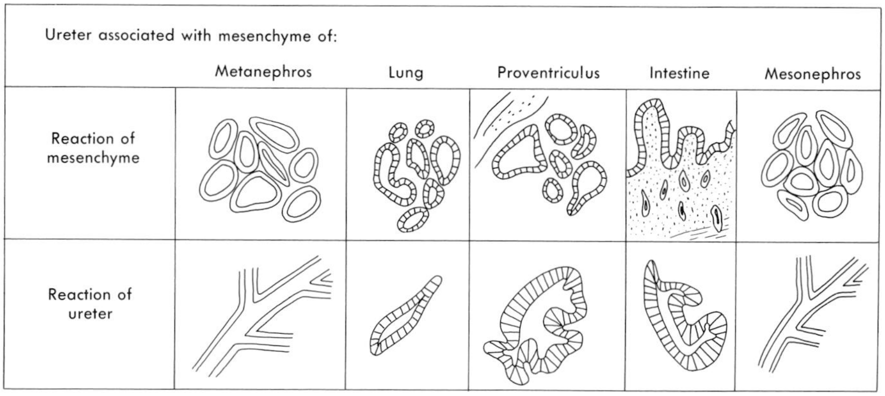

329

Ureter associated with mesenchyme of:					
	Metanephros	Lung	Proventriculus	Intestine	Mesonephros
Reaction of mesenchyme					
Reaction of ureter					

FIGURE 14.10
Morphogenetic performance of the tip of the ureter of the embryonic chick and of various embryonic mesenchymes brought into association with it. This is a diagram of what is typically seen in cross sections of each interacting tissue. See text for details. [Modified from Wolff, *Cur Top Dev Biol* 3:65–94 (1968), after Bishop-Calame.]

enhances their chondrogenesis, it does not do so if the matrix materials associated with the notochord are first removed enzymatically. Proteoglycans form aggregates deposited on and binding together collagen fibers in the matrix. Lash and Vasan (1977) found that production of proteoglycan and proteoglycan aggregates by somite cells is greatly stimulated by their being cultured on a substratum made of collagen.

Another indication that the extracellular materials play a regulatory role in cell and tissue interactions is found in studies of the formation of the corneal stroma. The cornea consists of an epithelium, underlaid by a collagenous stroma containing GAGs (the acronynm as used here does not specify which GAGs or in what form) and an inner limiting endothelium. The stroma is invaded by neural crest cells, which fill in spaces between layers of stroma. Hay and her colleagues (reviewed by Hay, 1977) showed that the stroma is produced by the corneal epithelium and that its production is regulated by components of the stroma. When lens epithelium is cultured in contact with collagen, both new collagen and GAGs accumulate beneath the epithelium twice as rapidly as they do when a collagen substrate is not used. Interestingly, addition of GAGs to the medium in the absence of collagen enhances only the production of GAGs, but not of collagen.

The collagen effect is clearly mediated by contact of the cell membrane with the collagen substratum. When such contact is prevented by porous filters that do not permit penetration of cell processes to the collagen, collagen synthesis by corneal epithelium is not enhanced. If filters with pores of sufficient size are used, however, new collagen is rapidly synthesized and accumulates between the epithelium and the surface of the filter.

14.4.3 CAN WE RESOLVE THE QUESTION OF WHETHER INDUCTIONS ARE INSTRUCTIVE OR PERMISSIVE?

From the foregoing evidence it follows that for some inductive responses the development of the morphogenetic pattern exhibited by a responding system is variable according to the inductor that is used. It is likewise evident that in other inductive responses the morphogenetic specificity of the response arises from properties of the responding system independently of the inductor. Are there, then, two classes of inductions, one in-

structive and the other permissive? The distinction could possibly be more semantic than real! Conceivably all inductions are permissive. Those that appear to be instructive generally issue in fairly complex morphogenetic activities. Possibly they are, in fact, complex and interlocking sequences of permissive inductions.

One can probably discard the notion that the so-called instructional inductions involve the transfer of informational macromolecules that program undifferentiated or multipotential cells for a specific morphogenetic performance. There is simply no clear and unequivocal evidence for this possibility. No inductors are known that will transform a blastula cell into a chondrocyte or myocyte, for example. Cells can differentiate only on the basis of a specific developmental history. A cell of an early blastula will not transform into a cell of a neural tube, but it can do so after a period of proliferation and exposure to appropriate salt solutions (cf., Barth and Barth, 1969, for example) in the absence of chordamesoderm.

Holtzer and his colleagues (see Holtzer et al., 1975, and Dienstman and Holtzer, 1975, for reviews) have developed an ingenious scheme for cell diversification that assigns to inducers only a permissive role in development. Their scheme has drawn on the foregoing considerations, but has largely emerged from their studies of hemopoiesis, myogenesis, and chondrogenesis. These processes terminate in very limited cytodifferentiated states: blood cells, muscle, and cartilage. They have, nevertheless, extended their scheme to embrace the epigenesis of differentiative states throughout all of embryonic development. The essence of this scheme is now presented, but space does not permit a detailed review of the evidence on which it is based or a summary of contrary views (for this kind of summary and appropriate references see Dienstman et al., 1974).

1. All embryonic cells are members of some functional "lineage," or line of descent.
2. Cells within a lineage comprise various "compartments." Each compartment is made up of proliferating cells that have a unique, circumscribed set of metabolic options (program of synthetic activities) distinctive from those found in antecedent and subsequent compartments of the lineage.

3. The sequence of compartments in a lineage (i.e., sequence of metabolic patterns) is obligatory, and a terminal cell type (e.g., myoblast) appears only as the result of an invariable line of precursors.

4. Within a compartment all cells are bipotential in that they have the possibility of yielding, at most, two new cell types as immediate progeny.

5. Within a compartment cells proliferate, but eventually they must undergo a "quantal" mitosis in order to pass into the next compartment.

6. Quantal cell cycles make available for transcription in the daughter cells regions of the genome that were not available for transcription in the mother cell.

7. The decision to move from one compartment to the next or to undergo terminal differentiation is a permissive event that allows one of the two developmental options available to cells in a compartment to be expressed. This event could be exercised by relatively "trivial" (as opposed to "informational") inducer molecules.

The application of this scheme to the differentiation of terminal cell types in the chick limb bud is illustrated in Figure 14.11. This scheme is based on differentiative behavior of cells from chick limb buds, dissociated and cultured in vitro at low cell densities. This scheme describes the early differentiation of limb-bud cells into different lineages. The interpretation has also been made that limb-bud cells are initially homogeneous, but with a bias for producing cartilage in vitro. Presumably, regulatory mechanisms that determine whether cartilage or muscle will differentiate are elicited only in particular microenvironments in vivo and under some conditions in vitro (cf. Medoff, 1967; Searls, 1967; Zwilling, 1968; Caplan and Koutropas, 1973).

The position of Holtzer and his group is strongly supported by results from the laboratories of Kieny and Christ, cited in Chapter 11, which show that the muscle lineage of the limb bud is derived exclusively from cells emigrating into the somatopleure from the somites. Because all other components of the limb arise from somatopleure, however, the scheme shown in Figure 14.11 may not reflect accurately the lineage of the myoblast. There may not be in normal limb buds a cell type that exercises the option of entering either the myoblastic or fibroblastic lineage. Rather, the myoblast is in its termi-

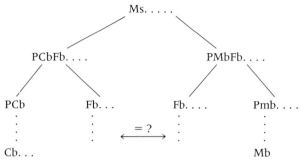

FIGURE 14.11 Scheme depicting postulated lineage relationships among chondrocytes, myocytes, and fibroblasts in the embryonic chick limb bud. Rows of large dots indicate a cell with proliferative potential. Ms is the compartment of hypothetical ancestral mesenchyme cells that can continue to proliferate or to yield cells to the next compartment. From time to time, cells in this compartment enter, respectively, compartments whose metabolic patterns are chondrofibroblastic, (PCbFb), or myofibroblastic, (PMbFb). From the PCbFb compartment, cells may enter either a prospective chondroblastic, PCb, or a fibroblastic, Fb, compartment. From the PMbFb compartment cells enter either the prospective myoblastic compartment, PMb, or the fibroblastic compartment, Fb. Prospective chondroblasts may then become proliferative chondroblasts, Ch; prospective myoblasts, PMb, proliferate, and, from time to time, some enter the ultimate, nonproliferative condition, Mb, and differentiate myofibrils. As the text points out, in vivo there is possibly not a compartment in the limb bud in which the myoblast-fibroblast option occurs. [Adapted with modifications from S. R. Dienstman and H. Holtzer, in H. Ursprung, ed., *Results and Problems in Cell Differentiation*, 7:1–25 (1975).]

331

nal pathway of differentiation as it enters the limb; that is, it is the penultimate compartment. Leaving this compartment it becomes a muscle cell. These considerations do not preclude the possibility that in vitro some of the cells in the penultimate compartment may show a fibroblastic morphology and fail to leave the compartment. Presumably the myogenic cell enters the limb bud, having previously been in a bipotential compartment in the somite, with a set of metabolic options determined, possibly by relatively trivial components of its environment.

It is reasonable to accept the general idea that alterations in the microenvironment of cells and cell groups determine their pathway of differentiation. These changes may involve only the presence of "trivial" as opposed to "informational" molecules. Is it necessary for us, therefore, to discard the concepts of induction and inductive specificity? Not necessarily! They remain as useful concepts about which we can organize a multitude of data, but other considerations we have raised suggest that a redirection of effort is needed in the search for mechanisms of induction and other aspects of tissue interaction.

14.5 Summary

In the field of experimental morphogenesis, competence refers to the ability of a tissue to react to an inductor by an appropriate developmental response. Competence has a transient existence in that prospective tissues acquire, retain, and lose their ability to respond to an inductor during a finite period of development. The prospective lens of the eye apparently acquires its competence to form lens or to react to the optic cup in subsequent induction by prior action of the endomesoderm of the head. Prior to induction competence to react, in many cases, extends beyond the territorial limits of the actual reacting tissue.

Competent tissues may respond to inductors from related members of their own species or foreign species, but they can only differentiate structures permitted by their own genetically determined repertory of responses.

Induction often involves both the action of an inductor in a reaction system and a reciprocal action on the inductive system. The differentiation of a harmonious structure often requires a complex series of inductive interactions between its components.

There is usually an obvious congruence and developmental appropriateness between inductors and reacting systems, which suggests that induction may be an instructive event in which the inducer specifies or determines the nature of the response. Thus ectoneuroderm of the beginning gastrula may form essentially any structure, depending on the inductive field to which it is exposed. Induction in

other cases may be only a permissive event in which the inductor is requisite only to the expression of one of two alternative pathways, such as lens versus head epidermis.

Experimental analysis reveals varying degrees of specificity in interacting systems: kidney tubules and pancreatic acini can be induced by a variety of natural and foreign inductors; collecting tubules can form from the ureter only under the influence of kidney mesenchyme, and submandibular gland can be induced in submandibular epithelium only by submandibular mesenchyme; the tip of the ureter can induce kidney tubules and glomeruli not only in kidney mesenchyme but in mesenchyme of the lung and of the proventriculus; epidermal derivatives in the chick show different regional character depending on the regional origin of their subjacent mesenchyme.

Whereas there is clear evidence in some systems that the inductor specifies the response of the reacting system and in others that the response is determined by the responding system, it does not necessarily follow that induction is ever an instructional event, in the sense that informational macromolecules pass from inductor to responding system and program the genome for a particular differentiative performance. It is possible to construct models of differentiation in which complex differentiations can arise from sequences of permissive events that may result from the action of trivial molecules in cells that, by virtue of their past history, can, at each step in the sequence, show only one of two possible differentiative pathways. Because such schemes have largely been worked out by analyses of cytodifferentiation in vitro, their application to differentiation in vivo requires further verification.

14.6 Questions for Thought and Review

1. Of what value to the study of experimental morphogenesis is the concept of competence?

2. Of what significance in Holtzer's theory of differentiation is the concept of competence? (The answer to this question is implicit in the text but is not presented formally.)

3. The loss of competence by a reacting system is just as important to a harmoniously developing organism as is the acquisition of competence. Evaluate this proposition.

4. Compare and contrast patterns of lens competence and induction in amphibian embryos.

5. Can competent reactive systems respond to living inductors of foreign, but related species? Give examples, including results, of interactions between different orders and classes of vertebrates.

6. What is meant by the expression "genetic limitation of the inductive response"? Illustrate by means of examples.

7. To what aspects of inductive relationships does the term inductive specificity refer?

8. Define and contrast the terms permissive induction and instructive induction. Give examples of each.

9. There are varying degrees of specificity shown by different embryonic inductions. Examine this proposition by bringing appropriate evidence to bear on it.

10. Instructive inductions do not take place in embryonic development. Can you make a case for this proposition? Do so or disprove it.

14.7 Suggestions for Further Readings

BERNFIELD, M., and N. K. WESSELS. 1970. Intra- and extra-cellular control of epithelial morphogenesis. In M. R. Runner, ed. *Changing Syntheses in Development*. New York: Academic Press, pp. 195–249.

DE REUCK, A. V. and J. KNIGHT, eds. 1967. *Cell Differentiation*. Boston: Little Brown.

DEUCHAR, E. 1973. Biochemical aspects of early differentiation in vertebrates. *Adv Morph* 10:175–225.

FLEISCHMAJER, R., and R. E. BILLINGHAM, eds. 1968. *Epithelial-Mesenchymal Interactions*. Baltimore: Williams and Wilkins.

GOETINCK, P. 1966. Genetic aspects of skin and limb development. *Cur Top Dev Biol* 1:253–83.

LEVITT, D., and A. DORFMAN. 1974. Concepts and mechanisms of cartilage differentiation. *Cur Top Dev Biol* 8:103–49.

SAUNDERS, J. W., Jr. 1958. Inductive specificity in the origin of integumentary derivatives in the fowl. In W. D. McElroy, and B. Glass, eds. *The Chemical Basis of Development*. Baltimore: Johns Hopkins Press, pp. 239–53.

SAUNDERS, J. W., Jr., and M. T. GASSELING. 1968. Ectodermal-mesenchymal interactions in the origin of limb symmetry. In R. Fleischmajer, and R. E. Billingham, eds. *Epethelial-Mesenchymal Interactions*. Baltimore: Williams and Wilkins, pp. 78–97.

SENGEL, P. 1971. The organogenesis and arrangement of cutaneous appendages in birds. *Adv Morph* 9:181–230.

TWITTY, V. 1955. Eye. In B. H. WILLIER, P. A. WEISS, and V. HAMBURGER, eds. *Analysis of Development*. Philadelphia: Saunders, pp. 402–14.

WOLFF, E. 1968. Specific interactions between tissues during organogenesis. In A. A. Moscona, and A. Monroy, eds. *Current Topics in Developmental Biology*, Vol 3. New York: Academic Press, pp. 65–94.

YNTEMA, C. L. 1955. Ear and nose. In B. H. Willier, P. A. Weiss, and V. Hamburger, eds. *Analysis of Development*. Philadelphia: Saunders, pp. 415–28.

14.8 References

BARTH, L. G., and L. J. BARTH. 1969. The sodium dependence of embryonic induction. *Dev Biol* 20:236–62.

BISHOP-CALAME, S. 1966. Étude expérimentale de l'organogenèse du système urogenital de l'embryon de poulet. *Arch Anat Microsc Morphol Exp* 55:215–309.

BORN, J., H. P. GEITHE, H. TIEDEMANN, H. TIEDEMANN, and U. KOCHER-BECKER. 1972. Isolation of a vegetalizing factor. *Hoppe Seylers Z Physiol Chem* 353:1075–84.

CAIRNS, J. M., and J. W. SAUNDERS, Jr. 1954. The influence of embryonic mesoderm on the regional specification of epidermal derivatives in the chick. *J Exp Zool* 127:221–48.

CAPLAN, A., and S. KOUTROUPAS. 1973. The control of muscle and cartilage development in the chick limb: the role of differential vascularization. *J Embryol Exp Morphol* 29:571–83.

CUNHA, G. R. 1972. Support of normal salivary gland morphogenesis by mesenchyme derived from accessory glands of embryonic mice. *Anat Rec* 173:205–12.

DESHPANDE, A. K., S. B. JAKOWLEW, H.-H. ARNOLD, P. A. CRAWFORD, and M. A. SIDDIQUI. 1977. A novel RNA affecting embryonic gene functions in early chick blastoderm. *J Biol Chem* 252:6521–27.

DESHPANDE, A. K., and M. A. SIDDIQUI. 1978. Acetylcholinesterase differentiation during myogenesis in early chick embryonic cells caused by an inducer RNA. *Differentiation* 10:133–37.

DHOUAILLY, D. 1975. Formation of cutaneous appendages in dermo-epidermal recombinations between reptiles,

333

birds and mammals. *Roux Arch Entw.-mech* 177:323–40.

DIENSTMAN, S. R., J. BIEHL, S. HOLTZER, and H. HOLTZER. 1974. Myogenic and chondrogenic lineages in developing limb buds grown in vitro. *Dev Biol* 39:83–95.

DIENSTMAN, S. R., and H. HOLTZER. 1975. Myogenesis: a cell lineage interpretation. In J. Reinert and H. Holtzer, eds., *Results and Problems in Cell Differentiation*, Vol. 7, Berlin: Springer-Verlag, pp. 1–25.

GROBSTEIN, C. 1953. Analysis in vitro of the early organization of the mouse sub-mandibular gland. *J Morphol* 93:19–44.

GROBSTEIN, C. 1962. Interactive processes in cell differentiation. *J Cell Comp Physiol* 60(Suppl. 1):35–48.

GROBSTEIN, C. 1967. Mechanisms of organogenetic tissue interaction. In B. B. Westfal, ed. *Cell, Tissue and Organ Culture. Natl Cancer Inst Monogr* 26:279–99.

GROBSTEIN, C., and G. PARKER. 1958. Epithelial tubule formation by mouse metanephrogenic mesenchyme transplanted *in vivo*. *J Nat Cancer Inst* 20:107–19.

GRUNZ, H., A.-M. MULTIER-LAJOUS, R. HERBST, and G. ARKENBERG. 1975. The differentiation of isolated amphibian ectoderm with or without treatment with an inductor. *Roux Arch Entw.-mech* 178:277–84.

HAY, E. 1977. Interaction between the cell surface and extracellular matrix in corneal development. In J. W. Lash and M. M. Burger, eds. *Cell and Tissue Interactions*. New York: Raven Press, pp. 115–37.

HOLTZER, H., N. RUBINSTEIN, S. FELLINI, G. YEOH, J. CHI, J. BIRNBAUM, and M. OKAYAMA. 1975. Lineages, quantal cell cycles, and the generation of diversity. *Q Rev Biophys* 8:523–57.

JACOBSON, A. G. 1955. The roles of the optic vesicle and other head tissues in lens induction. *Proc Natl Acad Sci USA* 41:522–25.

JACOBSON, A. G. 1963a. The determination and positioning of the nose, lens and ear. I. Interactions within the ectoderm, and between the ectoderm and underlying tissues. *J Exp Zool* 154:273–84.

JACOBSON, A. G. 1963b. The determination and positioning of the nose, lens and ear. II. The role of the ectoderm. *J Exp Zool* 154:285–92.

JACOBSON, A. G. 1963c. The determination and positioning of the nose, lens and ear. III. Effects of reversing the antero-posterior axis of epidermis, neural plate and neural fold. *J Exp Zool* 154:293–304.

JACOBSON, A. G., and J. T. DUNCAN. 1968. Heart induction in salamanders. *J Exp Zool* 167:105–15.

KIENY, M. 1960. Role inducteur du mésoderme dans la différenciation précoce du bourgeon de membre chez l'embryon de poulet. *J Embryol Exp Morphol* 8:457–67.

KIENY, M. 1968. Variation de la capacité de mésoderm et de la compétence de l'ectoderme au cours de l'induction primaire du bourgeon de membre, chez l'embryon de poulet. *Arch Anat Microsc Morphol Exp* 57:401–18.

KOLLAR, E. J., and C. FISHER. 1980. Tooth induction in chick epithelium: expression of quiescent genes for enamel synthesis. *Science* 207:993–95.

KOLLAR, E. J., and G. BAIRD. 1969. The influence of the dental papilla on the development of tooth shape in embryonic mouse tooth germs. *J Embryol Exp Morphol* 21:131–48.

KOLLAR, E. J., and G. BAIRD. 1970. Tissue interactions in embryonic mouse tooth germs. II. The inductive role of the dental papilla. *J Embryol Exp Morphol* 24:173–86.

LASH, J. W., and N. S. VASAN. 1977. Tissue interactions and extracellular matrix. In J. W. Lash and M. M. Burger, eds. *Cell and Tissue Interactions*. New York: Raven Press, pp. 101–13.

LAWSON, K. A. 1974. Mesenchyme specificity in rodent salivary gland development: the response of salivary epithelium to lung mesenchyme *in vitro*. *J Embryol Exp Morphol* 32:469–93.

LIEDKE, K. B. 1951. Lens competence in *Amblystoma punctuatum*. *J Exp Zool* 117:573–91.

MEDOFF, J. 1967. Enzymatic events during cartilage differentiation in the chick embryonic limb bud. *Dev Biol* 16:118–43.

NIU, M. C., and A. K. DESHPANDE. 1973. The development of a tubular heart in RNA-treated post-nodal pieces of chick blastoderm. *J Embryol Morphol* 29:485–501.

RUTTER, W. J., R. L. PICTEL, J. D. HARDING, J. M. CHIRGWIN, R. J. MACDONALD, and A. E. PRZYBYLA. 1978. An analysis of pancreatic development: role of mesenchymal factor and other extracellular factors. In J. Papaconstantinou and W. J. Rutter, eds. *Molecular Control of Proliferation and Differentiation*. New York: Academic Press, pp. 205–27.

RUTTER, W. J., N. K. WESSELS and C. GROBSTEIN. 1964. Control of specific synthesis in the developing pancreas. *Natl Cancer Inst Monograph*, No 13, pp. 51–65.

SAKAKURA, T., Y. NISHIZUKA, and C. J. DAWE. 1976. Mesenchyme-dependent morphogenesis and epithelium-specific cytodifferentiation in mouse mammary gland. *Science* 194:1439–41.

SAUNDERS, J. W., Jr. 1948. The proximo-distal origin of the parts of the chick wing and the role of the ectoderm. *J Exp Zool* 108:363–404.

SAUNDERS, J. W., Jr., J. M. CAIRNS, and M. T. GASSELING. 1957. The role of the apical ridge of ectoderm in the differentiation of the morphological structure and inductive specificity of limb parts in the chick. *J Morphol* 101:57–88.

SAUNDERS, J. W., Jr., and M. T. GASSELING. 1963. Transfilter propagation of apical ectoderm maintenance factor in the chick embryo wing bud. *Dev Biol* 7:64–78.

SAUNDERS, J. W., Jr., M. T. GASSELING, and J. E. ERRICK. 1976. Inductive activity and enduring cellular constitution of a supernumerary apical ectodermal ridge grafted to the limb bud of the chick embryo. *Dev Biol* 50:16–25.

SAUNDERS, J. W., Jr., and C. REUSS. 1974. Inductive and axial properties of prospective wing-bud mesoderm in the chick embryo. *Dev Biol* 38:41–50.

SEARLS, R. L. 1967. The role of cell migration in the development of the embryonic chick limb bud. *J Exp Zool* 166:39–50.

SHERMAN, J. E. 1960. Description and experimental analysis of chick submandibular gland morphogenesis. *Wisc Acad Sci Arts Lett* 49:171–89.

SPEMANN, H. 1905. Über Lensenbildung nach experimenteller Entfernung der primären Lensenbildungzellen. *Zool Anz* 28:419–32.

SPEMANN, H. 1912. Zur Entwicklung des Wirbeltierauges. *Zool Jahrb Abt Allg Zool Physiol Tiere* 32:1–98.

SPEMANN, H. 1938. *Embryonic Development and Induction.* New Haven: Yale University Press.

SPOONER, B., and N. K. WESSELS. 1972. An analysis of salivary gland morphogenesis: role of cytoplasmic microfilaments and microtubules. *Dev Biol* 27:38–54.

TIEDEMANN, H. 1968. Factors determining embryonic development. *J Cell Physiol* 72(Suppl. 1):199–244.

TIEDEMANN, HEINZ, J. BORN, and HILDEGARD TIEDEMANN. 1972. Mechanisms of cell differentiation: affinity of a morphogenetic factor to DNA. *Wilhelm Roux' Arch* 171:160–69.

TIEDEMANN, HILDEGARD, and J. BORN. 1978. Biological activity of vegetalizing and neuralizing inducing factors after binding to BAC-cellulose and CNBr-Sepharose. *Wilhelm Roux' Arch* 184:285–99.

VON UBISCH, L. 1927. Beiträge zur Erforschung des Linsenproblems. *Z Wiss Zool* 129:214–52.

WOERDERMAN, M. W. 1939. On lens induction. *Proc. Koninkl. Nederl. Academie van Wetenschappen–Amsterdam* 42:290–92.

YAMADA, T. 1958. Embryonic induction. In W. D. McElroy and B. Glass, eds. *A Symposium on the Chemical Basis of Development.* New York: Academic Press, pp. 184–207.

YANG, S.-F., and M. C. NIU. 1977. Albumin synthesis in mouse uterus in response to liver mRNA. *Proc Natl Acad Sci USA* 74:1894–98.

ZWILLING, E. 1968. Morphogenetic phases in development. In M. Locke, ed. *The Emergence of Order in Developing Systems.* New York: Academic Press, pp. 184–207.

ZWILLING, E., and L. HANSBOROUGH. 1956. Interaction between limb bud ectoderm and mesoderm in the chick embryo. III. Experiments with polydactylous limbs. *J Exp Zool* 132:219–39.

15

Selectivity of Cellular Association

disassembled fragments of embryos or organ primordia, or of suspensions of cells made from embryonic organs.

15.1 The Concept of Affinity

Holtfreter (1939) dissected from gastrulae of urodele embryos a sheet of ectoderm and underlying endoderm cells from the region opposite the blastopore, and was cautious not to include prospective ventral mesoderm. The isolate was then cultured in

FIGURE 15.1 Behavior in culture of pure gastrula ectoderm associated with endoderm. Ectoderm darkly stippled; endoderm lightly stippled. A: The initial isolate, prepared by excising ectoderm and endoderm from the region just opposite the blastopore. B: Formation of a solid sphere of dual composition. In C the components begin self-isolation, and in D they are seen to be almost completely segregated. [Adapted with modifications from J. Holtfreter, *Arch Exp Zellf* 23:169–209 (1939).]

In order for the early embryo to construct tissues and organs, its mass of multiplying cells must be sorted out spatially and assembled into groups that develop different properties. The movements of gastrulation segregate sheets of cells, the germ layers. Within the germ layers cells become sorted into particular organ-forming areas. From these areas cells may separate and affiliate with other cell groups, as neural crest cells and primordial germ cells do. The positional rearrangements that occur in the formation of the germ layers, the segregation of organ-forming areas, and so on, indicate that cells associate in a selective manner in morphogenesis. Early experiments, as cited in this chapter, suggested that cells show changing patterns of affinity for one another as development proceeds. That such affinities occur indicates that mechanisms of cell-cell recognition must be involved in the selectivity with which cells associate, and conversely, in the selectivity with which they sort into particular morphogenetic patterns.

To explore the basis for embryonic cell associations in the intact embryo is clearly impossible. Investigators have, therefore, turned to in vitro experimental systems involving the reassociation of

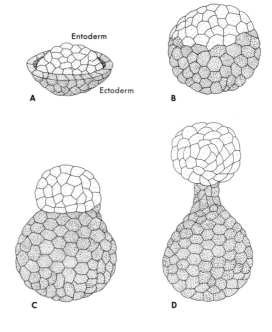

Entoderm

Ectoderm

A

B

C

D

vitro in an appropriate salt solution for several days. As shown in Figure 15.1, the endoderm cells first lie loosely within the curling ectoderm. The ectoderm then proceeds to encase the endoderm, and intimate contact is achieved. But, by 1½ days the endoderm begins to extrude from the ectodermal sac and is thereafter connected to it by a narrow bridge that soon ruptures.

On the other hand, if one includes a bit of prospective mesoderm, separation does or does not occur according to the amount of ectoderm relative to the other components of the preparation. If ectoderm is sufficient in amount to form a vesicle enclosing the endoderm (Figure 15.2A), complete separation of these components occurs with an intervening layer of mesenchyme. When ectoderm is insufficient to form a vesicle by itself (Figure 15.2B), a single vesicle, whose interior is filled with loose mesenchyme, is formed from the ectoderm and endoderm. The two epithelia are delimited from each other but do not separate completely, being held together by a common mesodermal substratum. Holtfreter coined the term **affinity** to indicate the forces, of whatever kind they may be, that determine the processes of attraction and repulsion between different kinds of cells. In the examples already cited one would say that ectoderm and endoderm retain a positive affinity for mesenchymal connective tissue while quickly developing negative affinity for each other. Moreover, whereas young gastrula ectoderm and endoderm initially unite when brought together in vitro in the absence of mesoderm, if they were cultured separately for a day and then brought together, they would show no affinity at all. As another example of changing affinity (Townes and Holtfreter, 1955), one may note that when the floor of the medullary plate of a neurula is associated with neurula endoderm, the neural tissue first sinks within the endoderm, but subsequently reverses direction and returns to the surface.

15.1.1 PERSISTENCE OF AFFINITY CHARACTERISTICS.
Certain affinities become stabilized during development and persist into later life. This fact was shown by experiments on urodele larvae (Chiakulas, 1952). When one makes a wound in the flank of the larva by excising a bit of epidermis, the cells of the cut edge begin to move across the mesodermal

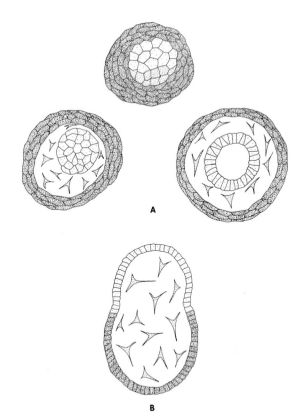

FIGURE 15.2 Segregation of materials of the germ layers of the urodele gastrula, involving different relative amounts of ectoderm in combination with endoderm and mesoderm. A: A relatively large amount of ectoderm (dark stipples) encloses the endoderm (light stipples) with a layer of mesenchyme (intermediate stipple) between. The endoderm forms a gut-like vesicle. B: The result of a similar combination involving a proportionately lesser amount of ectoderm. The ectoderm segregates from the endoderm but is held with it in a vesicle by virtue of the affinity of each layer for mesoderm. [Adapted with modifications from J. Holtfreter, *Arch Exp Zellf* 23:169–209 (1939).]

337

wound bed, towing their neighbors behind them. When the advancing cells from one side of the wound meet those from the other side, their migratory movement ceases and they fuse smoothly. If one places on the wound bed a smaller piece of skin ectoderm from another area of the embryo, the ectoderm pieces of flank and graft move toward each other, meet, and fuse smoothly, so that the grafted

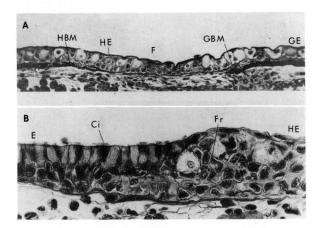

FIGURE 15.3 Tissue recognition in grafts: (A) smooth fusion, F, between flank epidermis of a host salamander, HE, and grafted salamander limb epidermis, GE; (B) failure of fusion between grafted ciliated, Ci, esophageal epithelium, E, and host flank ectoderm, HE. The host ectoderm wedges between the esophageal epithelium and its basement membrane, eventually displacing the graft. A sharp line of demarcation between these tissues is maintained. Host ectoderm piles up at the zone of contact with the graft and shows some cell death as denoted by the fragmentation of nuclei, Fr. [From J. Chiakulas, *J Exp Zool* 121:383–418 (1952), by courtesy of Dr. Chiakulas and permission of Alan R. Liss, Inc.]

338

skin is incorporated into the flank area. It is as though the ectodermal cells from different regions recognize each other and are cued by this recognition to cease their movements and to associate in a stable configuration (Figure 15.3A). Exploratory grafts of this kind show that ectoderm of flank associates smoothly with ectodermal epithelia from any region of the body, even the cornea of the eye and the oral epithelium. This association does not result, however, when the graft is taken from a tissue of endodermal origin, such as esophagus or intestine. In such instances, the spreading ectoderm simply continues its migration, the cells first piling up briefly at the edge of the foreign graft, then tunneling beneath it and displacing it to the outside so that it is shed (Figure 15.3B). The ectodermal edges of the wound continue their migration until they meet, whereupon they fuse smoothly as they come to rest.

15.2 Cell Sorting

Townes and Holtfreter (1955) dissociated different tissues of the urodele neurula into single-cell suspensions and then comingled them in random fashion. They found that the different cell types tended to arrange themselves in a tissue pattern corresponding to their relationships in normal development or in the pattern seen when nondissociated tissue fragments rather than randomized cells were combined. When ectoderm and endoderm cells were mixed, for example (Figure 15.4), the ectodermal cells exteriorized themselves from the endodermal mass and detached. But, when cells of epidermis, mesoderm, and endoderm were dissociated and recombined (Figure 15.4), the epidermis tended to surround the endoderm, with mesoderm sandwiched between.

The tendency of individual cells to segregate with others of their own kind in mixed populations of cells is called **sorting out,** or **cell sorting.** The ability to sort is not restricted to mixtures of cells or tissue fragments from different germ layers of the early embryo. It occurs also when dissociated cells or tissue fragments from different organ primordia are combined, and it has been studied extensively in material from chick and mouse embryos, particularly. Early studies of the sorting of dissociated cells of birds and mammals were carried out using hollow-ground tissue culture slides, in which the mixed cells were aggregated initially by gravity. By this method Moscona (1956) showed with both mouse and chick cells that prospective chonodrogenic cells from the limb bud, mixed with mesonephric cells, would sort out from them.

Other methods for testing sorting behavior are based on a method first used by Moscona (1961) to examine the ability of dissociated embryonic tissue cells to reaggregate and differentiate in a controllable, quantifiable, and reproducible condition. This method involves treating the tissues with trypsin, which hydrolyzes proteins involved in cell binding and permits their separation by means of gentle physical agitation. The cells are washed, suspended in an appropriate nutritive medium, and the suspension is then introduced into a flask placed on a gyratory shaking machine. The machine imparts a gentle swirling motion to the medium, enhancing the rate of collision between cells. Multiple aggregates of rather uniform size (Figure 15.5) appear in the matter of a few hours. The rate of aggregation varies with cell type, history of prior treatment, speed of gyration, temperature, composition of the

FIGURE 15.4 Recombination of dissociated and randomized cells of urodeles at the neurula stage. The left column shows from top to bottom the progressive stages in segregation of ectoderm (darkly stippled) from endoderm. The right column shows, from top to bottom, that the ectoderm tends to surround endoderm where mesoderm intervenes. [Adapted from P. L. Townes and J. Holtfreter, *J Exp Zool* 128:53–120 (1955).]

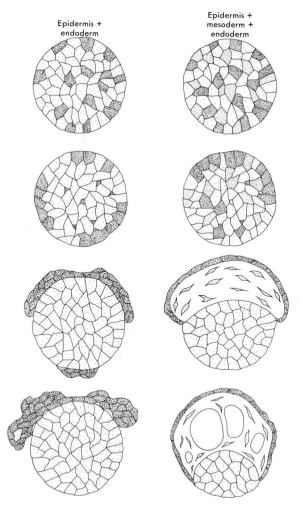

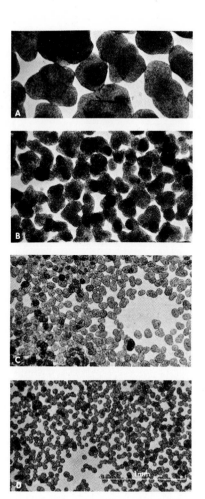

FIGURE 15.5 Size distribution of reaggregates of dissociated 7-day chick embryonic retinal cells rotated on a gyratory shaker at varying speeds for 24 hr at 38°C: (A) 70 rpm, (B) 85 rpm, (C) 100 rpm, (D) 120 rpm. [From A. A. Moscona, *Exp Cell Res* 22:455–75 (1961), by courtesy of Dr. Moscona and permission of Academic Press, Inc., New York.]

medium, and so on. Protein synthesis is required for the aggregation of avian and mammalian cells, for aggregation fails to occur when protein synthesis is inhibited by the drug cycloheximide.

When cells of two different tissue types are mixed in flasks on a gyratory shaker, the aggregates that form are each composed of both cell types. This experiment shows that the initial adhesions of cells need not be type specific. At first the aggregated

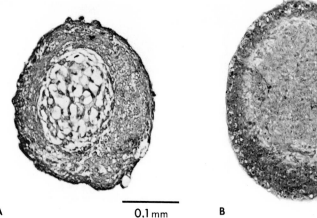
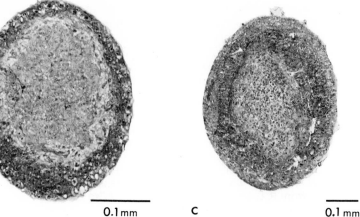

A   0.1 mm   B   0.1 mm   C   0.1 mm

FIGURE 15.6   Sorting of chick embryonic cells from mixtures. Chondrogenic cells of 4-day limb bud, 5-day heart ventricle, and 5-day liver were dissociated by trypsin treatment and allowed to reaggregate in binary combinations on a gyratory shaker. A: Reconstructed heart tissue envelopes limb cartilage. B: Liver cells have reaggregated and envelop heart tissue. C: Reconstructed liver envelops limb cartilage. [From M. S. Steinberg, *Science* 141:401–408 (1963), by courtesy of Dr. Steinberg and with permission. Copyright 1963 by the American Association for the Advancement of Science.]

340   cells move randomly, but very shortly sorting occurs, and the cells stabilize their positions in a characteristic manner. For example, when cells of the 5-day embryonic chick heart are mixed with mesenchymal cells of the limb bud, the latter sort to the interior of each aggregate and form a nodule of cartilage, which is surrounded by a layer of heart tissue (Figure 15.6A). If one similarly mixes embryonic chick liver cells and heart cells, the heart cells sort together and exclude the liver cells, which cohere in a layer around them (Figure 15.6B).

15.2.1   THE BASIS FOR CELL SORTING.   The basis for cell sorting could conceivably be found in cell type-specific adhesivity, in relative rates of adhesion, the relative strengths with which cells adhere to their own kind as compared with other kinds, in forces of mutual attraction (chemotaxis) or repulsion, or in combinations of these factors. There is no substantial evidence to support attraction or repulsion as factors in sorting. There are, however, clear indications that the sorting patterns of cells in mixed aggregates are determined by the relative strengths with which cells cohere with their own kind as

compared with other kinds. This is the basis of the **differential adhesion hypothesis** of Steinberg (Steinberg, 1970, 1978a, b). Steinberg and his students have shown that disaggregated cells of six different kinds of embryonic chick tissues sort from one another in a hierarchical sequence of internal versus external segregation. Listed in the order of their segregation, with the mostly internally segregating tissues given first, these tissues are: epidermal epithelium, prospective limb cartilage, pigmented retinal epithelium, heart, neural tube, and liver. This sequence is transitive in the sense that it predicts the inside-outside position of any combination of tissues in the sequence. First constructed from the behavior of binary combinations, as shown in Figure 15.6A–C, for example, it also serves to predict the sorting of ternary combinations, as shown in Figure 15.7.

That this transitive hierarchy is reflective of differential strengths with which cell types cohere is shown under other circumstances also. Notably, first, when fragments of embryonic tissues of different kinds are brought into association in culture, their cells sort in the same pattern as do mixtures of dissociated cells. Second, when cells of the various

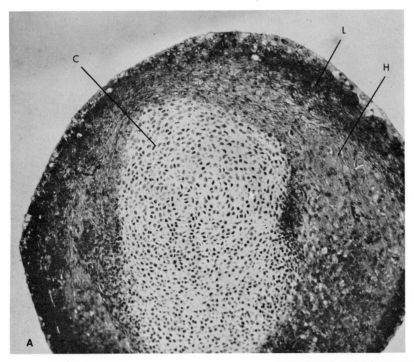

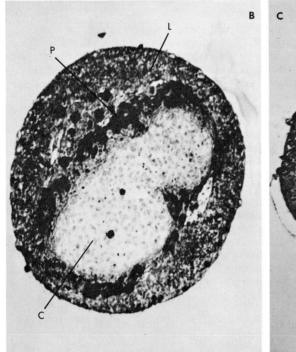

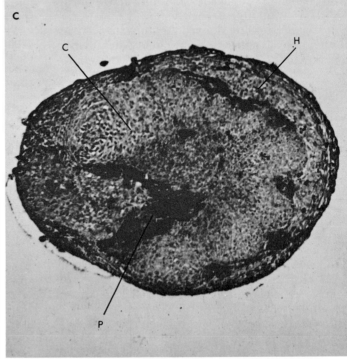

FIGURE 15.7 Sorting of embryonic chick cells in ternary combinations after dissociation and reaggregation in a gyratory shaker. A: Liver, L, envelopes heart, H, which encloses cartilage, C. This sequence is predicted on the basis of results of binary combinations shown in Figure 15.6. In B and C cartilage has segregated internally and is covered by pigmented retina, P. In B the outermost layer is liver, and in C heart is the external tissue. [A from M. S. Steinberg and L. D. Peachy, eds., *Conferences on Cellular Dynamics*, Annals of the New York Academy of Sciences, New York, 1968, p. 351. B, C from photographs supplied by Dr. M. S. Steinberg.]

341

tissue types are separately allowed to form rounded aggregates in gyratory culture and then tested for resistance to deformation when subjected to a centrifugal force of 1000 $\times G$, the aggregates that flatten the most are those that segregate to the exterior. Conversely, flat slices of aggregated cells of various types, when subjected to centrifugal force, tend to round up to a degree dependent on their place in the hierarchy, the more cohesive, internally segregating cells, rounding up to a higher degree than the less cohesive, externally segregating cells. Based on these findings, Steinberg and his associates (Phillips and Steinberg, 1978; and Phillips et al., 1977) have developed an ingenious and highly persuasive analysis of the behavior of embryonic tissues as elastoviscous liquids. Readers are encouraged to pursue this idea by studying the papers cited above and also the 1978 papers of Steinberg shown in the References as the end of this Chapter. A criticism of Steinberg's hypothesis by A. K. Harris (1976) is also of considerable interest.

The differential adhesion hypothesis does not deal with the questions of the speed and possible specificity (Figure 15.8) of cellular adhesions or with changing types of adhesions that develop with time of associations (but see Overton, 1977). Moyer and Steinberg (1976) showed that the rate at which suspended cells adhere to aggregates of their own or of a different kind in gyratory cultures does not reflect the hierarchy of cell sorting. There are, however, good indications that various levels of specificity may be involved in cellular associations. In Moscona's laboratory it was shown that the supernatant medium from monolayer cultures of 10-day chick neural retina cells specifically enhances the aggregation in gyratory cultures of retinal cells freshly dissociated by means of trypsin. This test is made in the absence of serum [blood serum from various sources is often used in cell- and tissue-culture media. Its presence contributes to nonspecific aggregations of cells, but with different kinetics from those shown in specific aggregation (Balsamo and Lilien, 1974a)]. After 24 hours the treated cells form aggregates several times the size of control aggregates, but the addition of the supernatant to dissociated cells of brain or liver is without effect. This datum suggests that the supernatant is involved with cell-specific attachment sites. Moreover, antibodies to the supernatant bind specifically

342

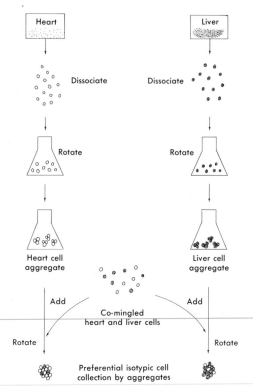

FIGURE 15.8 A method for determining the rate of adhesion between radioactively labeled cells to nonlabeled reaggregates of homotypic or heterotypic cells. Heart and liver tissues were dissociated and allowed to form reaggregates in a gyratory shaker, as shown in sequence from the top. Comingled, dissociated heart and liver cells, labeled with different isotopes that could be differentially counted, were then added to the reaggregates. After a period of further rotation, differential counts of radioactivity were used to determine the number and kind of cells that adhered to the reaggregates. By this means, S. Roth [*Dev Biol* 18:602–31 (1968)] determined that the isotypic cells adhere to aggregates in greater numbers than do heterotypic cells. He used this method to test the relative rates of adhesion for a variety of isotypic and heterotypic-isotypic combinations. His results were confirmed by Steinberg, who showed, however, that rate of adhesion is not necessarily related to position in the cell-sorting hierarchy (see text).

to the cell surface of neural retinal cells but not to cells of other types.

Lilien and Moscona (1967) and Lilien (1968)

reported that the aggregation-promoting factor of neural retina (R–APF) will bind to retinal cells at 4°C or in the presence of cycloheximide, but its aggregation-enhancing effect is not shown in the cold or in the presence of the inhibitory drug. Neural retinal cells, first treated with R–APF and then incubated in the rotary shaker at 38°C, showed enhanced aggregation as compared with non-R-APF-treated controls. These observations suggest that aggregation by specific factors occurs in two steps: (1) uptake by the cells and (2) utilization in a process requiring protein synthesis.

Garber and Moscona (1972a, b) next showed that a cell-free supernatant from 24-hour monolayer cultures of 14-day embryonic mouse cerebrum cells significantly and specifically enhances the histotypic agregation of dissociated cerebrum cells from both mouse and chick embryos. The supernatant is ineffective in promoting aggregation of cells from other brain regions (cerebellum, corpora quadrigemina of the mouse, optic tectum of the chick, or neural retina of chick or mouse). Working with retinal and cerebrum aggregation-promoting factors (R–APF and C–APF), Balsamo and Lilien (1974b) allowed freshly trypsinized cells to pick up their previously prepared APFs and then fixed these cells with glutaraldehyde. Control cells (without APF) were also fixed, and both preparations were tested for rotation-induced aggregation. Neither preparation would adhere. Fixed APF-treated cells would, however, enhance the aggregation of freshly prepared live cells of the same type, but not of heterologous type. Thus, the specific APFs are not alone able to make cells adhere, but require something that is contributed by live cells. This idea is reinforced by the fact that fixed APF-treated cells will aggregate if rotated in flasks in which live isotypic cells are growing as monolayers. Conceivably, what enables fixed APF-treated cells to aggregate is what cells produce in order to utilize APF; APF is not synthesized in the cold or in the presence of inhibitors of protein synthesis. At the time of this writing such a linking protein has not been further identified in cells of higher animals.

In 1975 Hausman and Moscona reported the isolation and partial characterization of a protein from the supernatant of chick retinal cultures. It is a glycoprotein monomer having a molecular weight of 50,000 ± 6000 that shows aggregation-promoting activity 100 times as effective as that of the original supernatant. Later (Hausman and Moscona, 1976; Hausman et al., 1976) the identical factor was isolated from cell membranes of the neural retina. Glycoprotein factors from cerebrum and from spinal cord were also isolated. Each glycoprotein specifically promoted the aggregation of dissociated cells of its tissue source. The cerebrum factor is similar in molecular weight and isoelectric point to the retinal factor and, also like that factor, contains 10–15% carbohydrate.

The fact that cell-aggregating factors of such a high degree of tissue specificity can be isolated from the membrane-rich fraction of cells suggest that they are normally found at the cell surface. This idea is consistent with the postulate that these factors play a significant role in membrane-mediated affinities between cells and the generation of histogenetic cell associations. It is important to note that the aggregation-promoting factors are effective only when tested on freshly trypsinized and dissociated cells. After recovery from trypsin treatment the APFs do not enhance the rate of aggregation.

On the other hand, investigators in Edelman's laboratory at the Rockefeller Institute (Brackenbury et al., 1977; Thiery et al., 1977; Rutishauser et al., 1976; Rutishauser, Gall, and Edelman, 1978; Rutishauser et al., 1978) used immunological methods to isolate from neural retinal cells a protein having a molecular weight of 140,000, which promotes the aggregation of dissociated retinal cells only after they have recovered from the effects of trypsin dissociation, a process that strips away many cell surface proteins. This molecule, designated **cell adhesion molecule** (CAM), was eventually found to be present in all neural tissues examined: retina, brain, optic nerve, spinal cord, and sympathetic and spinal ganglia. When ganglion cells are grown in culture in the presence of monovalent antibodies against CAM (that is, antibodies with only one binding site for CAM), the formation of bundles of neurites is inhibited, so that the ganglion becomes surrounded by a network of fine processes. This evidence suggests that CAM may be involved in the selective adhesion to pioneering neurites that is shown by later-emerging neurites (cf. Chapter 11). Presumably CAM is not involved in earlier, highly specific cellular aggregation, as studies from Moscona's laboratory revealed.

343

15.3 Effects of Intermingling Cells of the Same Tissue Type from a Different Species

What might we expect to result from mingling in culture dissociated chondrogenic (prospective cartilage) cells from mouse and chick embryonic limb buds? This process constitutes a heterospecific association of isotypic cells. On the basis of much of the foregoing material, we might expect that the cells of mouse and chick would segregate in culture, forming separate nodules of cartilage. Before the use of dissociated cells became a common practice Wolff (1954) reported that fragments of mouse and chick embryonic testis unite intimately in culture, cooperating in the formation of testicular tubules. Likewise, fragments of lung epithelium from these species collaborate in the formation of bronchioles. Nuclei of mouse and chick origin are distinguishable histologically, so the determination of their association in heterospecific combination is readily made. Moscona (1956) was apparently the first to test the morphogenetic performance of dissociated and intermingled isotypic cells of mouse and chick. He found that, after several days in culture, prospective chondrogenic cells of both types were completely intermingled in a common cartilaginous matrix. Similarly, mixtures of chick and mouse liver cells formed hepatic cords composed of both cell types in intermingled association.

When heterologous tissues of either isospecific or heterospecific origin are mixed, however, segregation occurs. Mixtures of dissociated chick chondrogenic cells and mouse liver cells form aggregates in which both cartilage and hepatic tissue are present, the cartilage forming one or more central clusters with hepatic cells surrounding them. The same segregation pattern of cartilage is shown in mouse-mouse and chick-chick combinations of prospective liver and cartilage (Moscona 1956, 1957).

These results suggest that properties responsible for cell sorting are quite similar among warm-blooded vertebrates, even those only remotely related. Burdick and Steinberg (1969) and Burdick

(1970) have, however, published results that do not agree completely with those of Moscona. For them, in mouse and chick ventricle cell coaggregates the cells were initially interspersed randomly, but thereafter, the chick cells showed a tendency to segregate internally, although considerable intermingling remained. Chick and mouse limb-bud chondroblasts tended variably to sort during the first 2 days in culture, but by the time chondrogenesis was well advanced the two cell types were found to be rather well intermingled.

It would seem appropriate for other investigators to examine independently the question of sorting in heterospecific combinations of cells. Nevertheless, it is important to recall at this point that the supernatants from mouse cerebrum monolayer cell cultures enhance the aggregation of freshly prepared suspensions of both mouse and chick cerebrum, but are not effective on suspensions of cells from other parts of the nervous system or of cells from nonneural tissue (Garber and Moscona, 1972a, b). Moreover, as already noted, aggregation-promoting factors from membranes of mouse neural retina enhance the aggregation of chick neural retina as well as that of mouse. Certainly, therefore, there would appear to be a commmonality of cellular recognition factors in isotypic cells of mouse and chick.

15.3.1 SEA URCHIN BLASTOMERES SORT WITH SPECIES SPECIFICITY. Sea urchin blastomeres disaggregated at various stages of development through the early gastrula stage can reaggregate and give rise to normal larvae (Giudice, 1962; Giudice and Mutolo, 1970; Spiegel and Spiegel, 1975). Protein synthesis is required for reaggregation to occur, and is possibly required for synthesis of hyalin (cf., Chapter 8) or a hyalin-like matrix (Kondo, 1973) that is involved in cellular adhesion. Using mixtures of cells from embryos of *Paracentrotus lividus* and *Arbacia lixula*, which can be distinguished by pigmentary differences, Giudice (1962) noted that random aggregation occurs first, but that within a few hours cells from the two sources sort out according to species.

The Spiegels (1975) mixed 16-cell stage blastomeres of *Lytechenus pictus* and *Arbacia punctulata*, which are likewise differently pigmented. In this mixture, too, cells at first mutually adhere, but then sort out. Time-lapse photographic studies of the

344

sorting process suggest that sorting does not occur simply by random movements and differential cohesion, as observed in avian and mammalian cells, but that cells and aggregates of each species move directionally toward their own kind.

15.3.2 SORTING OF SPONGE CELLS.

In 1907 H. V. Wilson reported that tissues of marine sponges could be dispersed mechanically into suspensions of single cells by squeezing them through fine-meshed cloth, and that the cells would then readhere and reconstruct many miniature sponges. Later it was shown by Wilson (1910) and others (Galtsoff, 1925) that in mixtures of cells from two species, those of each species preferentially adhere to homospecific cells.

In Moscona's laboratory, T. Humphreys et al. (1960a, b) showed that single-cell suspensions could be obtained by treating sponges with calcium- and magnesium-free seawater (CMF-SW). This treatment is referred to as chemical dissection. They also introduced to the study of sponge-cell adhesion and sorting the method of rotation-mediated aggregation, as described for avian and mammalian cells in Section 15.2. Using this method, T. Humphreys (1970) examined the species specificity of aggregation and sorting in a number of different sponge species. In most combinations, cells would initially adhere in composite aggregates and then sort and separate in a species-specific manner. In a few other combinations the initial adhesion is absolutely species specific, such that aggregates arising from cells mixed in suspension initially form aggregates of two separate types. Among sponges available at the Marine Biological Laboratory in Woods Hole, Massachusetts, only two combinations show absolute specificity. These are *Microciona prolifera* with *Haliclona oculata* and *H. oculata* with *Halichondria panicea*. In contrast, cells of *M. prolifera* and *M. parthena* form mixed aggregates which do not sort out (T. Humphreys, 1963, 1970; Moscona, 1968).

15.3.2.1 Sponge Aggregation Factors.

When sponge cells are dissociated in CMF-SW and tested for aggregability in a variety of artificial seawaters it is found that aggregation requires calcium ions. Magnesium ions can substitute, but not so effectively, and strontium has a very slight enhancing effect on aggregation. Calcium thus plays a significant role in cell binding. It is not sufficient alone, however. Cells dissociated chemically will not reaggregate when rotated at 5°C even in the presence of calcium, whereas they will reaggregate when tested at 22°C. Mechanically dissociated cells, however, will aggregate when rotated at 5°C. This fact suggests that a factor or factors manufactured by the cells are necessary for aggregation and that these factors are produced only at higher temperatures (T. Humphreys, 1963).

Such aggregation factors (AF) have been found in a number of species but have been studied principally in *Microciona prolifera*, *M. parthena*, *Haliclona oculata*, *H. bowerbankii*, and *Terpios zekete*. They can be removed by high-speed centrifugation from supernatants of CMF-SW in which sponge fragments are gently agitated for several hours in the cold, and they require calcium for stability. Added to chemically dissected cells rotated in suspension at 5°C, they promptly promote aggregation. Likewise, they promote the adhesion of formaldehyde-fixed and glutaraldehyde-fixed cells similarly brought together by rotation (Gasic and Galanti, 1966; T. Humphreys, 1965; Margoliash et al., 1965; Moscona, 1968; T. Humphreys et al., 1975; S. Humphreys et al., 1977).

AFs from *Haliclona oculata* and *Microciona oculata*, species whose cells show absolute aggregation specificity, are likewise species specific, each promoting aggregation of only its own kind, even in mixtures of cells from the two species. In pairs in which absolute specificity of aggregation is not shown, AFs share less specificity. Thus, AF from *M. parthena* is equally as effective in promoting aggregation of *M. prolifera* cells as it is with its own cells of origin (T. Humphreys, 1970).

The aggregation factor of *Microciona parthena* was purified and analyzed biochemically in T. Humphreys' laboratory (Henkart et al., 1973; S Humphreys et al., 1977). It is a large glycoprotein, negatively charged, having a molecular weight of about 22,000,000. As seen in the electron microscope, the molecule exists in a sunburst configuration—fibers 45 Å are arranged in a central circle 800 Å in diameter with 15 or 16 radial arms about 1100 Å in length extending from the circle (Figure 15.9). Calcium is required both for the activity and stability of the AF. When treated with ethylenediaminetetraacetate (which chelates divalent cations), the AF is inactivated and the radial side arms separate from the central core. Each side arm disso-

345

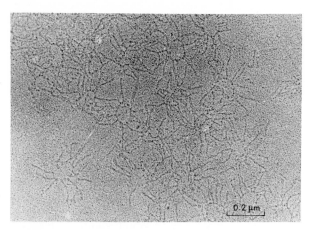

FIGURE 15.9 Electron micrograph of isolated aggregation factor of the sponge *Microciona parthena*. Note the sunburst configuration described in the text. [Reprinted with permission from P. Henkart et al., *Biochem* 12:3045–50 (1973). Copyright 1973 American Chemical Society.]

346

ciates into five to eight subunits having a molecular weight of about 20,000 and the same ratio of protein to carbohydrate as is present in the intact molecule (Cauldwell et al., 1973). When the protein is removed from *M. parthena* sunbursts, two species of polysaccharide remain, one with a molecular weight of about 65,000 and the other with a molecular weight of about 6000. It was calculated by S. Humphreys et al. (1977) that there are four molecules of the larger polysaccharide in each circular core and 100 of the smaller ones. There are approximately 50 of the smaller polysaccharides per radial arm. The AF of *M. prolifera* was isolated by T. Humphreys et al. (1975). It is similar to the AF of *M. parthena*. S. Humphreys et al. (1977) described the AFs of *H. oculata* and of *T. zekete* as having linear backbones with radial arms.

When sponge cells reaggregate after chemical dissection, AF molecules bind to each other, as well as to special attachment sites on the cell surface. That they bridge between cells by attaching to each other is suggested by the fact that when one covalently bonds AF to the surface of minute Sephadex (cross-linked dextran) beads, the beads will show rotation-induced aggregation in the presence of calcium. Their special binding sites on the sponge cell may be removed by subjecting the cells to hypo-

tonic shock. The binding sites, or **base plates,** may then by collected on AF-treated Sepahdex beads. Osmotically shocked cells will not aggregate in the presence of AF unless they are supplied with base plates before AF is added. These studies were summarized by Burger and Jumblatt (1977).

15.4 Summary

The forces that bring about particular associations between cells are related under the term affinity. Cells show different relative affinities for those of their own and other kinds, and these affinities are responsible for the creation of relatively stable cellular associations in tissues and organs.

The tendency of individual cells to associate selectively with others of their own kind in mixed populations of cells is called sorting out.

When fragments of tissue from different organ primordia are mixed, or when dissociated cells from different primordia are comingled, the phenomenon of sorting out occurs. When first mixed, heterotypic cells adhere nonspecifically, but they soon segregate from one another in characteristic patterns. For example, when prospective chondroblasts are dissociated and mixed with dissociated heart cells and liver cells, the resulting pattern is one in which cartilage is surrounded by heart tissue, and heart tissue, in turn, is surrounded by liver tissue. One may presume that chondroblasts cohere more strongly to one another than do heart cells, and heart cells are more coherent than are liver cells. When one examines the sorting of various kinds of tissue cells in different binary and ternary combinations, one can create a transitive hierarchy of cells that describes the internal versus external sorting position of any kind of cell in the hierarchy with respect to any other kind. This hierarchy accurately predicts the resistance to gravitational deformation of aggregates of the different cell types, but does not predict the rate at which cells will adhere with their own kind.

The sorting hierarchy can be interpreted on the basis of differential adhesion (the differential adhesion hypothesis) without respect to the nature of cellular bindings and without regard to elements of specificity in the binding of cell types. There is, however, a large body of evidence suggestive of the fact that like cells may associate by virtue of highly specific configurations of their surface molecules. Cell membranes from the retina and various brain regions yield glycoproteins that enhance, in specific manner, the rate of aggregation of cells of the tissue type from which each specific glycoprotein was obtained.

Dispersed cells of sea urchin embryos from as late as the early gastrula stage can reaggregate and form normal larvae. In mixtures of embryonic cells from different species of sea urchins, species-specific sorting occurs, such that no chimeric larvae are formed from reaggregates. Mature sponges can be dissociated either mechanically or by treatment with calcium- and magnesium-free seawater. Such dispersed cells reaggregate to form new sponges, but they, too, with but few exceptions, sort out according to species from heterospecific mixtures of cells. Dissociated sponge cells require for their reaggregation a large highly acidic glycoprotein having a molecular weight of about 20,000,000. Called aggregation factor, it apparently joins cells by reacting with special membrane components called base plates. The base plates may be separated from sponge cells by means of osmotic shock.

In contrast to the specificity with which heterospecific sponge cells sort is the lack of sorting in heterospecific mixtures of tissues dissociated from embryos of birds and mammals. Isotypic cells in heterospecific association form histologically recognizable tissues in which the cells of both kinds are intermingled. This fact has been shown for a number of tissue combinations, especially between mouse and chick, although there have been reports that some degree of sorting may occur in such mixtures.

15.5 Questions for Thought and Review

1. Review the development of the vertebrate nervous system, and find examples of morphogenetic events that are mediated by or require changing cellular affinities.

2. Cell sorting is usually studied in vitro, using cells that have been dissociated by means of proteolytic en-

zymes and mechanical shearing, and that are then exposed to physical and chemical conditions quite different from those that occur in the embryo. Are we, therefore, justified in concluding that studies of sorting reflect cellular properties and behavior that are operative in morphogenesis? Justify your answer.

3. Compare and contrast the conditions required for reaggregation of dissociated cells of sponges and of embryos of warm-blooded animals.

4. Summarize the evidence that both specific and nonspecific adhesive modalities can operate in the cohesion of cells.

5. Compare and contrast sponges and embryos of sea urchins with respect to their species specificity of reaggregation.

6. Considering the structure of sponges and the various biological and mechanical factors that impinge on them in nature, is there any obvious adaptive value to the species specificity of their cell sorting?

15.6 Suggestions for Further Reading

347

ARMSTRONG, P. B. 1977. Cellular positional stability and intercellular invasion. *Bioscience* 27:803–809.

ARMSTRONG, P. B., and M. T. ARMSTRONG. 1973. Are cells in solid tissues immobile? Mesonephric mesenchyme studied in vitro. *Dev Biol* 35:187–209.

CURTIS, A. S. 1967. *The Cell Surface: Its Molecular Role in Morphogenesis.* London: Logos.

HUMPHREYS, T. 1967. The cell surface and specific cell aggregation. In Davis and Warren, eds. *The Specificity of Cell Surfaces.* Englewood Cliffs, N.J.: Prentice-Hall, pp. 195–210.

KOLATA, G. B. 1975. Embryo development: debate over aggregation factors. *Science* 188:718–19.

LILIEN, J. E. 1969. Toward a molecular explanation for specific cell adhesion. *Cur Top Dev Biol* 4:169–95.

MASLOW, D. E. 1976. In vitro analyses of surface specificity in embryonic cells. In G. Poste and G. L. Nicolson, eds. *The Cell Surface in Animal Embryogenesis and Development.* Amsterdam: North-Holland, pp. 697–745.

MOSCONA, A. A. 1960. Patterns and mechanisms of tissue reconstruction from dissociated cells. In D. Rudnick, ed. *Developing Cell Systems and Their Control.* New York: Ronald, pp. 45–70.

ROTH, S. 1973. A molecular model for cell interactions. *Q Rev Biol* 48:541–63.

TRINKAUS, J. P. 1969. *Cells into Organs.* Englewood Cliffs, N. J.: Prentice-Hall.

TRINKAUS, J. P. 1977. Cell Movements in vivo. Problems and perspectives. Symposium on Biology of Connective Tissue at the 500th Anniversary of the University of Upsala. *Upsala J Med Sci* 82:104.

15.7   References

BALSAMO, J., and J. LILIEN. 1974a. Embryonic cell aggregation: kinetics and specificity of binding of enhancing factors. *Proc Natl Acad Sci USA* 71:727–31.

BALSAMO, J., and J. LILIEN. 1974b. Functional identification of three components which mediate tissue-type specific embryonic cell adhesion. *Nature* 251:522–24.

BRACKENBURY, R., J.-P. THIERY, U. RUTISHAUSER, and G. M. EDELMAN. 1977. Adhesions among neural cells of the chick embryo. I. An immunological assay for molecules involved in cell-cell binding. *J Biol Chem* 252:6835–40.

BURDICK, M. L. 1970. Cell sorting out according to species in aggregates containing mouse and chick embryonic limb mesoblast cells. *J Exp Zool* 175:357–68.

BURDICK, M. L., and M. S. STEINBERG. 1969. Embryonic cell adhesiveness: do species differences exist among warm-blooded vertebrates? *Proc Natl Acad Sci USA* 63:1169–73.

BURGER, M. M., and J. JUMBLATT. 1977. Membrane involvement in cell-cell interactions: a two-component model system for cellular recognition that does not require live cells. In J. Lash and M. M. Burger, eds. *Cells and Tissue Interactions.* New York: Raven Press, pp. 155–72.

CAULDWELL, C. B., P. HENKART, and T. HUMPHREYS. 1973. Physical properties of sponge aggregation factor. A unique proteoglycan complex. *Biochemistry* 12:3051–55.

CHIAKULAS, J. 1952. The role of tissue specificity in the healing of epithelial wounds. *J Exp Zool* 121:383–418.

GALTSOFF, P. S. 1925. Regeneration after dissociation (an experimental study of sponges). I. Behavior or dissociated cells of *Microciona prolifera* under normal and altered conditions. *J Exp Zool* 42:183–221.

GARBER, B. B., and A. A. MOSCONA. 1972a. Reconstruction of brain tissue from cell suspensions. I. Aggregation patterns of cells dissociated from regions of the developing brain. *Dev Biol* 27:183–221.

GARBER, B. B., and A. A. MOSCONA. 1972b. Reconstruction of brain tissue from cell suspensions. II. Specific enhancement of aggregations of embryonic cerebral cells by supernatant from homologous cell cultures. *Dev Biol* 27:235–43.

GASIC, G. J., and N. L. GALANTI. 1966. Proteins and disulfide groups in the aggregation of dissociated cells of sea sponges. *Science* 151:203–205.

GIUDICE, G. 1962. Restitution of whole larvae from disaggregated cells of sea urchin embryos. *Dev Biol* 5:402–11.

GIUDICE, G., and V. MUTOLO. 1970. Reaggregation of dissociated cells of sea urchin embryos. *Adv Morph* 8:115–58.

HARRIS, A. K. 1976. Is cell sorting caused by differences in the work of intercellular adhesion? A critique of the Steinberg hypothesis. *J Theor Biol* 61:267–85.

HAUSMAN, R. E., L. W. KNAPP, and A. A. MOSCONA. 1976. Preparation of tissue-specific cell aggregating factors from embryonic neural tissues. *J Exp Zool* 198:417–22.

HAUSMAN, R. E., and A. A. MOSCONA. 1975. Purification and characterization of the retina-specific cell-aggregating factor. *Proc Natl Acad Sci USA* 72:916–20.

HAUSMAN, R. E., and A. A. MOSCONA. 1976. Isolation of retina-specific cell-aggregating factor from membranes of embryonic neural retinal tissue. *Proc Natl* Acad Sci USA 73:3594–98.

HENKART, P., S. HUMPHREYS, and T. HUMPHREYS. 1973. Characterization of sponge aggregation factor. A unique proteoglycan complex. *Biochemistry* 12:3045–50.

HOLTFRETER, J. 1939. Gewebeaffinität, eim Mittel der embryonalen Formbildung. *Arch Exp Zell* 23:167–209.

HUMPHREYS, S., T. HUMPHREYS, and J. SANA. 1977. Organization and polysaccharides of sponge aggregation factor. *J Supramol Struct* 7:339–51.

HUMPHREYS, T. 1963. Chemical dissolution and in vitro reconstruction of sponge cell adhesions. I. Isolation and functional demonstration of the components involved. *Dev Biol* 8:27–47.

HUMPHREYS, T. 1965. Cell surface components participating in aggregation: evidence for a new particulate. *Exp Cell Res* 40:539–43.

HUMPHREYS, T. 1970. Specificity of aggregation in *Porifera. Transplant Proc* 2:194–99.

HUMPHREYS, T., S. HUMPHREYS, and A. A. MOSCONA. 1960a. A procedure for obtaining completely dissociated sponge cells. *Biol Bull* 119:294.

HUMPHREYS, T., S. HUMPHREYS, and A. A. MOSCONA. 1960b. Rotation-mediated aggregation of dissociated sponge cells. *Biol Bull* 119:295.

HUMPHREYS, T., W. YONEMOTO, S. HUMPHREYS, and D. ANDERSON. 1975. Purification of *Microciona prolifera* aggregation factor. *Biol Bull* 149:430.

KONDO, J. 1973. Cell-binding substances in sea urchin embryos. *Dev Growth Differ* 15:201–16.

LILIEN, J. E. 1968. Specific enhancement of cell aggregations in vitro. *Dev Biol* 17:647–78.

LILIEN, J. E., and A. A. MOSCONA. 1967. Cell aggregation:

its enhancement by a supernatant from cultures of homologous cells. *Science* 157:70–72.

MARGOLIASH, E., J. R. SCHENCK, M. P. BUROKAS, W. R. RITCHTER, G. H. BARLOW, and A. A. MOSCONA. 1965. Characterization of specific cell aggregating materials from sponge cells. *Biochem Biophys Res Commun* 20:383–88.

MOSCONA, A. A. 1956. Development of heterotypic combinations of dissociated embryonic chick cells. *Proc Soc Exp Biol Med* 92:410–16.

MOSCONA, A. A. 1957. The development in vitro of chimaeric aggregates of dissociated embryonic chick and mouse cells. *Proc Natl Acad Sci USA* 43:184–94.

MOSCONA, A. A. 1961. Rotation-mediated histogenetic aggregation of dissociated cells. *Exp Cell Res* 22:455–75.

MOSCONA, A. A. 1968. Aggregations of sponge cells: cell-linking macromolecules and their role in the formation of multicellular systems. *In Vitro* 2:13–21.

MOYER, W. A., and M. S. STEINBERG. 1976. Do rates of intercellular adhesion measure the cell affinities reflected in cell-sorting and tissue-spreading configurations? *Dev Biol* 52:246–62.

OVERTON, J. 1977. Formation of junctions and cell sorting in aggregates of chick and mouse cells. *Dev Biol* 55:103–16.

PHILLIPS, H. M., and M. S. STEINBERG. 1978. Embryonic tissues as elasticoviscous liquids. I. Rapid and slow shape changes in centrifuged cell aggregates. *J Cell Sci* 30:1–20.

PHILLIPS, H. M., M. S. STEINBERG, and B. H. LIPTON. 1977. Embryonic tissues as elasticoviscous liquids. II. Direct evidence for cell slippage in centrifuged aggregates. *Dev Biol* 59:124–34.

RUTISHAUSER, U., W. E. GALL, and G. M. EDELMAN. 1978. Adhesion among neural cells of the chick embryo. IV. The role of the cell surface molecule CAM in the formation of neurite bundles in cultures of spinal ganglia. *J Cell Biol* 79:382–93.

RUTISHAUSER, U., J.-P. THIERY, R. BRACKENBURY, and G. M. EDELMAN. 1978. Adhesion among neural cells of the chick embryo III. Relationship of the surface molecule of CAM to cell adhesion and the development of histiotypic patterns. *J Cell Biol* 79:371–81.

RUTISHAUSER, U., J.-P. THIERY, J. P. BRACKENBURY, R. SELA, B. A. SELA, and G. M. EDELMAN. 1976. Mechanisms of adhesion among cells from neural tissues of chick embryo. *Proc Natl Acad Sci USA* 73:577–81.

SPIEGEL, M., and E. S. SPIEGEL. 1975. The regeneration of dissociated embryonic sea urchin cells. *Am Zool* 15:583–606.

STEINBERG, M. S. 1970. Does differential adhesion govern self-assembly processes in histogenesis? Equilibrium configurations and the emergence of a hierarchy among populations of embryonic cells. *J Exp Zool* 173:395–434.

STEINBERG, M. S. 1978a. Cell–cell recognition in multicellular assembly: levels of specificity. In A. S. Curtis, ed. *Cell-Cell Recognition*. 32nd Symposium of the Society for Experimental Biology. Cambridge: Cambridge University Press, pp. 25–49.

STEINBERG, M. S. 1978b. Specific cell ligands and the differential adhesion hypothesis: how do they fit together? In D. R Garrod, ed. *Specificity of Embryological Interactions*. London: Chapman and Hall, pp. 99–130.

THIERY, J.-P. R. BRACKENBURY, U. RUTISHAUSER, and G. M. EDELMAN. 1977. Adhesion among neural cells of the chick embryo. II. Purification and characterization of a cell adhesion molecule from neural retina. *J Biol Chem* 252:6841–45.

TOWNES, P. L., and J. HOLTFRETER. 1955. Directed movements and selective adhesion of embryonic amphibian cells. *J Exp Zool* 128:53–120.

WILSON, H. V. 1907. On some phenomena of coalescence and regeneration in sponges. *J Exp Zool* 5:245–58.

WILSON, H. V. 1910. Development of sponges from dissociated tissue cells. *Bull Bur Fish* 30:1–30.

WOLFF, E. 1954. Potentialitiés et affinités des tissus révélées par la culture in vitro d'organes en association hétérogènes et xenoplastiques *Bull Soc Zool France* 79:357–68.

16

Regeneration and Reintegration

Once its parts are assembled, the embryo will run its course of development, normally or not, according to favorable or unfavorable circumstances. But either by accident or experimental insult, embryos and adult organisms may be subjected to circumstances that have a deteriorative effect: laceration of the skin, loss of a limb or an eye, bleeding, and so on. Organisms respond to some disturbances of these kinds by activities that we sum up under the heading of **reintegration;** that is, they act to restore their lost unity or carry out some functional adaptation.

Outstanding among the reintegrative activities that some organisms can perform are the regeneration of lost parts, wound healing, and various other forms of adaptive growth. The matter of regeneration is the principal concern of this chapter, but some phases of wound healing and reintegrative growth control are also treated.

16.1 Forms of Regenerative Growth

If a larval salamander loses a limb to a predator, the missing member will be replaced by processes of growth and morphogenesis that originate in the stump. We say that the limb is regenerated from the stump, and the fact that the regenerated appendage is like the one that was lost suggests that the morphogenetic field of the limb persists beyond embryonic life.

The term regeneration embraces activities ranging from restoration, after a traumatic loss, of a limb to the replacement, after a small physiologic loss, of plucked hair. Regeneration may involve local proliferation and differentiation from a germinal layer of cells, as occurs in the replacement of molted feathers and sloughed mucosal cells of the intestine. It may also involve the accumulation at the wound site of a mass of apparently undifferentiated cells, a **blastema**, which proliferates and progressively forms the missing part. The blastema may originate from cells at the cut surface, or it may arise, in part, from special reserve cells (**neoblasts, interstitial cells**) that migrate to the wound site. **Epimorphosis** is the term applied to the reconstruction of missing parts by local proliferation and differentiation from a blastema. Often, however, growth and differentiation of the blastema are accompanied by a more or less extensive reorganization of adjoining tissues, or the restoration of the missing part may be accomplished almost entirely at the expense of the tissues remaining. This kind of regeneration is called **morphallaxis**.

16.2 Regenerative Power Is Widespread

The ability to regenerate lost structure is found in essentially all living things, at least to some degree. Regenerative powers are especially prominent in the sponges, coelenterates, flatworms, annelids, and tunicates, many of which have the ability to reconstitute new organisms from mere fragments of the original body; indeed, for many organisms the usual form of asexual reproduction is fragmentation and regeneration. Among vertebrates the ability of adult animals to regenerate major external body parts is most extensive among urodeles, which can replace some or all of the following parts after loss: tail, limb, jaw, lens, iris, and a portion of the retina. Lizards can regenerate the tail. Fish can usually regenerate distal portions of the severed fin, and tadpoles of anuran amphibians can regenerate the tail and hindlimbs provided that the limbs are severed before metamorphosis is well advanced; however, adult anurans do not naturally regenerate amputated limbs.

Larvae of the clawed toad, *Xenopus,* regenerate the lens from the cornea of the eye, in contrast to the urodeles, in which lens regeneration occurs from the iris. Higher vertebrates have very little power to regenerate appendages, but the physiologic regeneration of blood cells, skin, and integumentary derivatives proceeds constantly throughout life. Bone regenerates after injury, and so do the gall bladder, intestine, and urinary bladder. There often is extensive regeneration of severed axons and dendrites of peripheral nerves, with the subsequent reestablishment of functional connections in many instances. The higher animals show an amazing capacity to regenerate the liver; the rat, for example, can replace the loss of three-fourths of this indispensable organ.

Plants also show extensive powers of reintegration and regeneration. Earlier it was pointed out (Chapter 12) that plants change their patterns of growth in response to light and gravity. In addition, they tend to restore parts that have been removed, and isolated parts can form complete plants once more. Plants heal wounds, and grafted plant parts unite and establish vascular connections; cuttings of shoots and branches regenerate roots; tissues of fallen leaves may form complete new plants.

In this work it is not possible to undertake a detailed analysis of the many well-known regenerating systems of plants and animals. A few of the best-studied cases will be discussed, however, in order to gain the insights they offer into the analysis of regenerative development.

16.3   Regeneration of the Amphibian Limb

In the United States favorite organisms for studying limb regeneration are adult and larval salamanders, notably various species of *Ambystoma* and *Notophthalmus* (formerly known as *Triturus**). Triturus species have been studied to a great extent in Europe, too. Regeneration of the appendages has also been investigated extensively in larval anurans, chiefly of the genus *Rana,* and in the South African clawed toad, *Xenopus laevis.*

16.3.1   LIMB REGENERATION FOLLOWS A BASIC PATTERN. Fundamentally the progress of regeneration after amputation is similar in all these forms (Figures 16.1 and 16.2). The immediate effect of amputation is, of course, the exposure of interior cells of the limb, many of which are killed by the operative trauma; bleeding occurs from the damaged blood vessels and a clot forms over the cut surface (Figure 16.3). Epidermal cells from the stump migrate distally and cover the wound surface. The epithelium is at first only a single layer of cells, but the proximal ectoderm proliferates and shifts toward the wound area so that an **apical ectodermal cap** several cells in thickness accumulates. Meanwhile, subjacent to the cap, the tissues undergo a dramatic change that is referred to as **dedifferentiation.** The intercellular matrix material of cartilage (bone, also, in the case of adult salamanders) and other connective tissues are digested away, releasing the cells from their tissue organization. Sheath cells of Schwann dissociate from their nerve fibers, and multinucleate fibers of skeletal muscle break into

351

---

*The genus name *Notophthalmus* now has priority over *Triturus* and *Diemictylus,* which superseded *Triturus* for a while. The eastern newt of the United States has been named *Triturus* in most of the papers that form the groundwork for present-day studies. In referring to studies conducted on what was called *Triturus,* this work uses the same name. Where *Notophthalmus* is identified as the experimental animal, it will be so termed here. It is worth noting, too, that the western newt *Taricha* was formerly called *Triturus* and that the European genus *Triton* was later called *Triturus.*

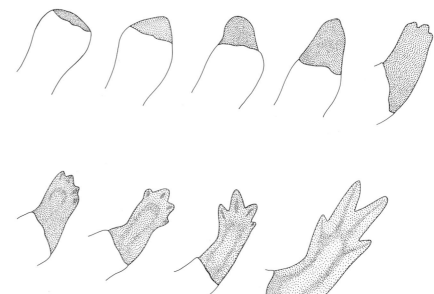

FIGURE 16.1 Regeneration of the forelimb in the adult of *Triturus viridescens*. Amputation at the elbow level is followed by formation of a complete forelimb and hand. [From O. E. Schotte and R. S. Hilfer, *J Morphol* 101:25–56 (1957), by permission of the Wistar Institute.]

352

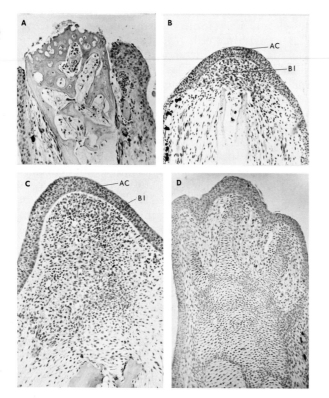

FIGURE 16.2 Histologic changes during regeneration of the limb of the adult *Triturus viridescens*. A: Freshly amputated limb; note retraction of skin and muscles from skeleton. B: A 16-day regenerate, showing thickening of ectoderm and accumulation of blastema cells. C: A 21-day regenerate showing a large blastema that is beginning to redifferentiate. D: A well-developed regenerate 28 days after amputation, showing regenerated skeletal rudiments. AC, apical cap; BL, blastema. [Photographs courtesy of Thomas B. Connelly, T. B. Sprague, and Dr. C. S. Thornton.]

fragments. Nucleated fragments of muscle lose their myofibrils and join the other released cells in forming a mass of proliferating tissue that accumulates beneath the apical epithelial cap as the **regeneration blastema**. Cells of the blastema are now histologically, cytologically, and immunologically indistinguishable. Morphologically they are **dedifferentiated** (see Hay, 1962).

The dissolution of stump tissue that provides cells for the regeneration blastema is referred to as **regression**. Regression ceases as the blastema reaches an appropriate mass. The cells of the blastema then begin to proliferate and, as the blastema increases in mass, the missing limb parts begin to

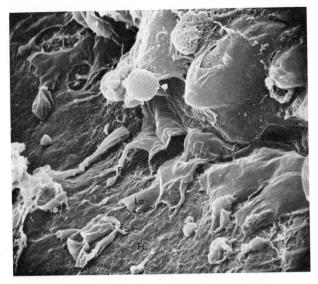

FIGURE 16.3 Scanning electron micrograph of the leading edge of migrating epidermal cells during the wound-healing stage of regeneration in the amputated limb of the adult newt, *Notophthalmus virides-cens*. Leading lamellae, La, and filopodia, F, are recognized in association with the fibrin clot, FC. Refer to the discussion of cellular movement in Chapter 11. [Photograph supplied by Dr. Lillian Repesch, whose generosity is gratefully acknowledged.]

differentiate in proximodistal order beginning from the point at which regression ceased.

16.3.2 THE TIMETABLE VARIES. The rate at which these changes occur and the extent of proximal regression of limb tissues are quite variable, depending on the temperature and on the kind, size, and age of the experimental organism. Regression of stump tissue supplying blastemal cells is much more extensive in larval limbs. Limbs of adult animals regenerate more slowly, and often the quality of the regenerate is inferior to that usually seen when larval limbs are amputated. The adult *Triturus* kept at 20°C requires about 20 days for the typical conical blastema to be formed after amputation of the limb at the elbow level; by 75 days restoration of the limb is essentially complete (Schotté and Hilfer, 1957). In contrast, the amputated limb of a small *Ambystoma* larva regenerates fully in 3 weeks (Steen, personal communication), and an ampu-

tated finger of an *Ambystoma* larva regenerates in 17 days (Hay, 1966).

16.3.3 AXIAL ORGANIZATION OF REGENERATES. When a limb regenerates, the blastema always forms successively more distal parts, starting with those appropriate to the level of amputation or to the level to which the stump regresses before blastema formation. This generalization has been termed by Rose (1962) the **rule of distal transformation.** Because the blastema normally forms only those structures distal to the level of its origin, it is apparent that there exists in the limb level-specific positional information that determines the proximal boundary of the regenerate. Moreover, the anteroposterior and dorsoventral axes of the regenerate correspond to those of the stump. The regenerated limb parts, therefore, have the same pattern of asymmetry (i.e., ''handedness'') as the amputated parts had. There must also exist, therefore, an asymmetric pattern of circumferential positional information, at each limb level, to which the blastemal cells respond.

To discover what information at the amputation surface issues in the three-dimensional polarity of the regenerate is a problem that is challenging most investigators of limb regeneration at this time. A number of formal models of the distribution of positional information have been devised. These models are tested in a variety of ways, involving chiefly the experimental manipulation of blastemata, and are examined in Chapter 19, which is mainly concerned with the formation of morphogenetic patterns in embryonic and regenerative development.

16.3.4 WHERE DO CELLS FOR THE REGENERATE ARISE? To answer this question we make use of the fact that regeneration of the limbs in urodeles is prevented by x-irradiation which, in appropriate doses, inhibits the cellular proliferation required for regeneration. If the entire animal is irradiated, limbs fail to regenerate after amputation. Likewise, irradiation of the limb alone will prevent its regeneration upon subsequent amputation. If the limb is protected from the x-rays by a lead shield, however, and the rest of the body is irradiated, regeneration of the appendage proceeds normally (Butler, 1935). If only a segment of the limb (e.g., the knee-joint region) is irradiated (Figure 16.4), the appendage

353

FIGURE 16.4 Local origin of the cells for the limb regenerate in *Eurycea*, a newt. The left and right hind limbs and pelvic level of the trunk are illustrated to show irradiation of the left knee region and three successive amputation experiments carried out on both limbs of the same animal. The column on the left shows the operation and that on the right the result. A: Amputation through the metatarsal region of right and left limbs is followed by bilateral regeneration. B: Amputation at the knee level is followed by regeneration of only the right (unirradiated) limb. C: Subsequent amputation above the knee results in regeneration of complete limbs on both sides. [Adapted from E. G. Butler and J. P. O'Brien, *Anat Rec* 84:407–13 (1942).]

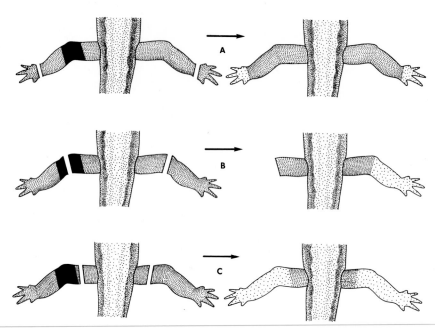

354

will regenerate following amputation if the cut is made above or below the irradiated zone, but not if the cut passes through it (Butler and O'Brien, 1942).

16.3.5 REGENERATED PARTS NEED NOT ARISE FROM STUMP COUNTERPARTS.

The foregoing explanation shows, of course, that cells of the regeneration blastema must arise locally. Does regenerating muscle stem only from cells derived from muscles of the stump, cartilage only from preexisting cartilage cells? What if the stump contained no skeleton; would the skeletal parts form in the regenerate?

If one extirpates the bone from a limb segment of an adult or larval newt, that portion of the limb remains permanently without a skeleton. If one thereafter amputates the appendage through the limb and boneless segment, however, regeneration of distal limb parts, *including their skeletal components,* occurs (Figure 16.5). Clearly, distal skeletal elements can differentiate from a blastema that received no contribution from preexisting skeleton (Weiss, 1925; Thornton, 1938). It must be concluded, therefore, that skeletal tissues of the regenerate need not be contributed by cells from their specific tissue counterparts in the stump.

FIGURE 16.5 Stump counterparts are not required for regeneration of the distal parts of the limb skeleton. A: Forelimb skeleton of an adult newt. B: The humerus is excised and the limb is amputated through the upper arm. C: Regeneration of distal parts, including skeleton, proceeds from the cut surface distally. [Interpretation of results described by P. Weiss, *Wilhelm Roux' Arch* 104:359–94 (1925).]

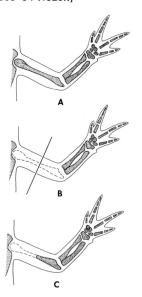

Likewise, it would appear that a normal musculature need not be present at the amputation site in order for distal regeneration of all distal parts. Carlson (1972) amputated the forelimb just above the elbow in *Ambystoma mexicanum* juveniles. He then cleaned out the upper arm of at least 99% of its musculature and observed the subsequent regeneration of grossly normal limbs containing large amounts of anatomically appropriate muscle in forearm and hand.

16.3.6 CAN A "DEDIFFERENTIATED" CELL REDIFFERENTIATE AS ANOTHER CELL TYPE?

Two possibilities are suggested, especially, by the results of, Weiss and Thornton, already cited: (1) that nonskeletal tissues dedifferentiate and then redifferentiate skeletal elements during regeneration or (2) that limbs capable of regeneration contain undifferentiated reserve cells that can differentiate the appropriate tissue types called for in a particular regenerative situation. Although the participation of putative reserve cells cannot be completely ruled out, there is now a large body of evidence that any of the differentiated cell types of the limb can dedifferentiate and then redifferentiate the complete repertory of cell and tissue types of the regenerate under certain conditions.

16.3.6.1 X Rays Help to Provide an Answer.

Appropriate doses of x-rays applied to a urodelean limb prevent its regeneration upon subsequent amputation regardless of the lapse of time between irradiation and operation. In irradiated limbs regeneration is halted at the completion of wound healing, and no blastema accumulates. In larval limbs the stump may then regress completely, or, in the case of partially shielded limbs it may regress as far as the nonirradiated portion and the latter may then regenerate the limb. Irradiated adult limbs show little regression after amputation, but, likewise, no blastema forms.

Regeneration is inhibited by doses of x-rays (800 r locally applied to the limb of a larval *Ambystoma mexicanum* or 2000 r applied locally to the limb of an adult *Triturus*) that do not hinder the normal maintenance functions of the irradiated limb, including the replacement of epidermis at metamorphosis, for example. Irradiation does, however, almost completely inhibit cell division in the regressing tissue beneath the wound epithe-lium, and, therefore, no regeneration blastema forms (Maden and Wallace, 1976).

Suppose that a source of nonirradiated tissue were supplied to an irradiated limb. Would it then regenerate? Using young adults of *Triturus viridescens*, Thornton (1942) x-rayed the right forelimb, amputated it at the elbow, removed the humerus, and then packed into its former space unirradiated muscle taken from the left hind limb. After 2 weeks the composite limb was reamputated through the region containing the implanted muscle. A significant percentage of such limbs regenerated, forming distal parts of varying degrees of perfection complete with cartilaginous elements. Some of the regenerates could be identified as hind limbs. These observations suggest the possibility that

1. The regenerated appendage formed from the nonirradiated implant.
2. The components of muscle (muscle cells and connective tissue) are able to form all tissue elements of a normal limb.
3. The presence of the nonirradiated implant does not "restore" regenerating power to X-rayed limbs; rather, the implant supplies the cells that form the blastema.

355

Full thickness skin (dermis plus epidermis), but not epidermis alone, can contribute blastema cells to an irradiated limb to bring about its regeneration. Umanski (1937) irradiated newt limbs and then excised from each a skin cuff, which was then replaced by a similar cuff of unirradiated skin. After amputation through the implanted cuff regeneration occurred. This observation has been confirmed by several others, including Namenwirth (1974) and Dunis and Namenwirth (1977), who used skin cuffs of triploid axolotl limbs to replace excised irradiated skin of diploid limbs. They found triploid cells in all tissues of the regenerate. When unirradiated epidermis alone was used, no blastema formed and regeneration did not occur.

The latter result makes it clear that unirradiated wound epidermis alone can neither supply cells to the blastema nor evoke blastema formation from irradiated tissues. The skin dermis would seem to contain elements capable of forming a blastema and contributing all the cell types of the regenerate. It should be realized, however, that the unirradiated skin cuffs could be slightly contaminated with scraps of muscle and its associated connective tissue and with connective tissue of nerve sheaths. Dunis and Namenwirth (1977) noted, in fact, that only 10% of the regenerates that they obtained contained muscle. Is it possible that muscle regenerates only from muscle or from **muscle satellite*** cells in regenerates supplied from an irradiated skin cuff?

An earlier observation of Wallace (1972) suggests a negative answer. He implanted a segment of brachial nerve in an irradiated limb of *Ambystoma mexicanum* and, after the wound healed, amputated the limb at or just distal to the implantation site. Regeneration occurred after some delay, but issued eventually in the formation of a well-constructed regenerate. Because the implant consisted only of Schwann cells and fibroblasts (ignoring the axons, which lack nuclei and degenerate) one or both of these cell types must have contributed the diverse types of cells in the regenerate.

Whether cartilage can contribute blastema cells that will redifferentiate to form other tissue components of a limb that is amputated after irradiation has been extensively investigated. Wallace and his collaborators (1974), using axolotl larvae, implanted unirradiated humeral cartilage, carefully freed of all other tissues in place of the humerus excised from an irradiated limb. When larvae of 50–60 mm were used, limbs bearing digits and skeletal muscle, but smaller than normal, were regenerated after amputation through the graft. But when larvae 70–80 mm in length were used as hosts and donors, those regenerates that formed consisted essentially of rods of skin-covered cartilage.

Namenwirth (1974) carried out a similar experiment but used juvenile axolotls 100–180 mm in length. For her experiments, however, the unirradiated implants of cartilage were obtained from triploid donors so that cell types derived from the implant could be identified cytologically. She, too, obtained spike-like regenerates, lacking muscle, but could identify triploid cells in perichondrium, in connective tissue of joints, and in dermal fibroblasts. These results suggest that, whereas differ-

entiated cartilage from younger larvae can form a regeneration blastema from which all other mesodermal cell types arise, the ability to form muscle is lost in cartilage of older larvae. It is unfortunate, however, that triploid or otherwise labeled cartilage was not used as donor tissue in the experiments of Wallace et al. (1974), for limited participation of irradiated cells in the formation of a blastema after amputation through an unirradiated skin cuff has been shown by Desha (1974) and also by Dunis and Namenwirth (1977).

As Steen (1970) pointed out, when cartilage is cleaned of all adherent cells, only one cell type is present. When cells from unirradiated pure cartilage are placed into x-irradiated limb stumps, the chondrocytes are subjected to the extreme test of their differentiative capacities. They can then make all cell types, with the possible exception of muscle— and probably that tissue, too, if sufficiently young donors of cartilage are used. But suppose that the cartilage from the stump of nonirradiated limb were replaced with triploid cartilage or by cartilage labeled with ^{3}H-thymidine. Would regenerates then show distinctively marked nuclei in all tissue types? This experiment was carried out by Steen (1968), who found that when stump tissues were not inhibited from proliferating by means of x-irradiation, the marked chondrocytes contributed essentially only to cartilage. It seems probable, therefore, that during normal regeneration the chondrocytes liberated during regression of the stump retain their chondrocytic character during their incorporation into the blastema.

Muscle cells and fibroblasts are possibly less rigidly specified in the blastema. Steen (1973) minced hindlimb muscle of *Xenopus* tadpoles and prepared clones of cells derived from single myoblasts and single fibroblasts cultured in vitro. Clones were scraped from the culture vessel, and the cells were packed into the stumps of unirradiated freshly amputated *Xenopus* larval hindlimbs. The limbs were then allowed to regenerate and histological sections of them were made. Implant and host cells were distinguished by a nucleolar marker (the *1-nu* mutant of *Xenopus* has only one nucleolus instead of two). Regenerates showed that cells of purely muscle or purely fibroblastic clonal ancestry contributed to all cell types in the regenerate.

These experiments, in sum, seem to provide a rather clear answer to the question as to whether

---

*Muscle satellite cells are small, nonstriated cells located within the connective tissue envelope of striated muscle bundles and known to be capable of differentiating into muscle cells.

cells, once differentiated as a particular type, can, in the regenerative situation, dedifferentiate and then redifferentiate as another type. Yes, they can under experimental circumstances, and with the probable exception of chondrocytes, all cell types possibly do so in normal regeneration. Unfortunately, however, it is still impossible to follow the dedifferentiation and subsequent redifferentiation of any one cell in a normal regenerate.

16.3.6.2 Can Grafts of Nonlimb Tissues Support the Regeneration of an Irradiated Limb?

Grafts of nonirradiated limb skin often promote regeneration of an irradiated limb amputated through the graft to a high degree of perfection. But, suppose that the skin graft were taken from the dorsal fin or tail? In these cases the regenerates are fin-like or tail-like rather than resembling a limb (Glade, 1957). Nevertheless, many researchers have insisted that tissues other than those of limb origin can restore the ability of an irradiated limb to regenerate, either by "healing" the irradiated cells or by providing a source of undifferentiated proliferating cells (reviewed by Lazard, 1967).

Lazard made a comprehensive study of the ability of unirradiated heterotopic implants to promote the regeneration of irradiated limbs. Implants were made in the left hindlimb between knee and foot, and the subsequent amputation plane passed through the implant. When implants of heart or liver were used, there was essentially no regeneration. When various parts of early embryos were implanted, limb-like structures sometimes formed after amputation, especially when prospective mesoderm of the gastrula was used. Grafts of testis, particularly from regions containing spermatogonia, were most effective. Amputations through the testis grafts were followed by regeneration in 40% of the cases. The resulting limb parts were usually hypermorphic (Figure 16.6). In one case, 13 digits formed in the regenerate.

One possible interpretation of these results is that unirradiated implants of various kinds restore to irradiated cells the power to form blastemata that then regenerate limb parts. Another possibility is that a variety of heterotypic tissues contain uncommited cells that can acquire limb characteristics by being exposed to a limb field. Critical evidence bearing on the first possibility is not available and there is no precedent for the second. More reasonably, the grafts of heterotypic tissue that are

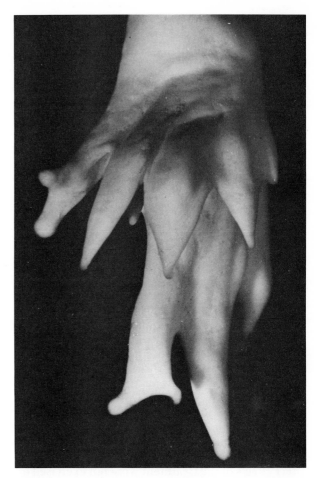

FIGURE 16.6 Hypermorphic regeneration of an irradiated left limb bud of an adult *Ambystoma mexicanum*. A fragment of unirradiated testis was implanted midway between knee and foot, and the limb was later amputated through the site of the implant. [The courtesy of Dr. L. Lazard in providing this hitherto unpublished photograph is gratefully acknowledged.]

most effective in stimulating regeneration are those that cause tissue disruptions that lead to the invasion of the amputation site by unirradiated cells from the base of the limb.

16.3.7 ROLE OF THE APICAL EPITHELIAL CAP.

It was pointed out earlier that the first event in regeneration is the migration of epithelium from the limb stump over the cut surface and the piling up of its cells to form an apical cap. The apical cap remains until regeneration is complete, a fact signaled by the differentiation of a skin dermis at the tips of the fingers and the thinning of the ectoderm. Electron microscopy (Salpeter and Singer, 1960; Norman and Schmidt, 1967; Bryant et al., 1971) has demonstrated that the basal lamina (adepidermal membrane), which normally is closely adherent to the epithelium at its interface with the subjacent mesoderm and matrix materials, is disorganized and partially absent beneath the apical cap. This makes possible direct contact between epithelial and mesenchymal cell processes, which may be of significance in assessing the role of the apical cap.

16.3.7.1 Effect of Absence or Inactivation of the Apical Cap.

Goss (1956a, b) amputated forelimbs of adult *Triturus* at the wrist and promptly inserted the distal end of the stump into the body cavity through a slit in the flank. Epithelium failed to cover the amputation surface and regeneration failed to occur. But if insertion into the body cavity were delayed until an apical cap were well established over a blastema, regeneration of distal parts would take place, albeit somewhat abnormally. Thornton (1957), using young larvae of several species of *Ambystoma*, daily picked away the healing apical epithelium from amputated limbs and observed complete suppression of regeneration. He also showed (1958) that localized daily exposure of the apical cap to ultraviolet light at a wavelength of 253.7 nm suppresses regeneration. The irradiation used penetrated no more than three cell layers, so direct effects of the treatment were not exercised on blastema cells but only on epithelium.

16.3.7.2 Inhibitory Effects of Skin Dermis.

A number of earlier students of limb regeneration in urodeles (e.g., Taube, 1921) showed that the transplantation of full thickness differentiated skin over the wound surface of a freshly amputated limb prevents its regeneration. Later Thornton (1951) showed that treating the cut surface with beryllium ions results in the formation, beneath the wound epithelium, of fibrous dermal scar tissue, which completely inhibits regeneration. Removal of the scar tissue, however, is followed by normal regeneration.

These observations may be correlated with the changing regenerative behavior of anuran hindlimbs. In young larvae a blastema accumulates beneath wound epidermis and regeneration follows. With progress toward metamorphosis (cf. Chapter

17), however, the larva progressively loses the ability to regenerate amputated limbs. This process is correlated with a more adult type of wound-healing behavior, which is characterized by rapid formation of dermis at the wound surface. Rose (1944) found, however, that amputated limbs of adult anurans could be stimulated to show a larval type of healing by treatment of the amputated surface with salt solutions, which led to a greatly enhanced regenerative performance.

More recently, Mescher (1976) prepared amputated legs of adult *Notophthalmus viridescens* in such a way that a flap of skin, continuous with that of the stump, covered the wound surface. He analyzed patterns of regression, mitotic index, and DNA synthesis in these limbs as compared with contralateral controls amputated without further treatment. Both skin-covered and control limbs regressed and showed similar increases in DNA synthesis during the first week. By the end of the second week the skin-covered amputated limbs showed no further regression and no accumulation of blastema cells. The number of cells incorporating ^{3}H-thymidine into DNA and the number of mitotic figures was, however, greatly reduced as compared to controls (Table 16.1). Thereafter controls went on to regenerate completely, whereas there was no further growth in the skin-covered amputated limbs and no distal regeneration.

These observations indicate, in sum, that regeneration requires a special wound epithelium, the apical cap, which must be in contact with regressing mesenchymal tissue of the stump in order for a blastema to form and for regeneration to occur. The

TABLE 16.1 DNA-Labeling Indices and Mitotic Indices of Distal Mesodermal Cells in Amputated Limbs of Adult Limbs of *Notophthalmus viridescens* 1 and 2 Weeks After Operation.[a]

	Stump Covered with Skin Flap	Control Regenerates
Labeling index, 1 week	18.8 ± 2.1	19.2 ± 2.2
Mitotic index, 1 week	0.41 ± 0.05	0.43 ± 0.10
Labeling index, 2 weeks	11.3 ± 2.3	36.9 ± 6.7
Mitotic index, 2 weeks	0.28 ± 0.6	0.69 ± 0.16

[a]Based on data of Mescher (1976). Labeling index based on incorporation of intraperitoneally injected ^{3}H-thymidine during a period of three hours prior to fixation for histological study. Mitotic index based on percentage of cells in mitosis (late prophase to late telophase).

358

effect of the cap is possibly mediated by the absence or disorganization of the basal lamella, thus allowing for contact between epithelial and the blastemal cell processes. Signaling between the cap and blastema could very well involve cell surface components. Pertinent to this possibility is an experiment of Chapron (1974). He injected regenerating newt limbs with radioactive fucose. This carbohydrate is a marker of cell membrane glycoproteins. He then fixed the injected limbs for electron microscope radioautography at 5 minutes and at 4 hours after injection. Label was found in the Golgi region of apical cap cells within 5 minutes and in the cell membranes and intercellular spaces of the cap. After 4 hours radioactivity had diminished in the apical cap and was found in the blastema. Membrane associated glycoproteins are known to be involved in a number of tissue interactions (e.g., induction of cartilage; Kosher and Lash, 1975), so it is possible that the labeled materials Chapron observed are important in the action of the apical cap.

16.3.7.3 Does the Apical Cap Contribute Cells to the Blastema?

It was earlier pointed out that an x-rayed limb supplied with a grafted cuff of non-irradiated limb epithelium does not regenerate after amputation through the region of the graft. A cap of normal wound epithelium apparently cannot, therefore, contribute cells to the formation of a regeneration blastema. It is necessary, however, to determine whether the apical cap of ectoderm contributes cells to the blastema in normally regenerating limbs.

H. J. Anton (1955) exchanged epidermis of the prospective limb region between gastrulae and neurulae of *Triturus vulgaris* and *Triturus alpestris,* species whose cells are sufficiently distinctive to be recognizable in histologic sections. Larvae with chimeric limbs issued from these combinations: the epidermis was derived from the donor, but the mesodermal structures were formed entirely of host-type cells. When the chimeric limbs were amputated, regeneration proceeded in typical fashion, and the blastema of host-like cells was covered by an epithelial cap of cells like those of the donor. The regenerated appendage showed no cells of donor type in its mesodermal tissues but was adorned with a donor kind of epithelium. Apparently, therefore, the regeneration blastema receives no cellular contributions (none that survive or proliferate sufficiently to be detectable, at any rate) from limb ectoderm. A similar conclusion was reached by Hay and Fischman (1961), who studied regeneration in adults *Triturus* limbs whose epidermis was marked with a radioactive label in the nucleus. After limb amputation the blastema formed without contributions from radioactive cells (Figure 16.7). Contrary interpretations of similar experiments have been made, notably by Rose and Rose (1965), but the mass of evidence now indicates that the apical cap does not contribute cells to the regeneration blastema.

16.3.8 NERVES ARE REQUIRED FOR REGENERATION.

When limbs are deprived of their nerve supply and then amputated, regeneration fails to occur until and unless nerves grow back into the denervated tissues. When that occurs, regeneration begins, provided that complete healing of full-thickness of skin over the amputation surface has not yet occurred. The dependence of regeneration on nerve supply has been intensively and cleverly analyzed by a number of investigators, notable among whom is Marcus Singer, who worked principally with adult *Triturus*. He has shown that for regeneration to occur a minimum number of nerve fibers is required (Table 16.2). Either sensory or

Level of Amputation	Median Number of Fibers		
	Total at Cut Surface	per (100 μm)^2 Soft Tissue at Cut Surface	per (100 μm)^2 at Cut Surface Including Bone
Upper arm	1040	10.8	9.6
Forearm	1085	9.9	9.4
Hand	351	4.2	3.2
Digit 2	164	11.4	8.6
Digit 3	225	11.0	8.8
Digit 4	105	11.4	8.1

TABLE 16.2 The Number of Nerve Fibers Required at the Cut Surface for Regeneration to Occur After Amputation of the Forelimb of the Adult *Triturus* Through Different Regions

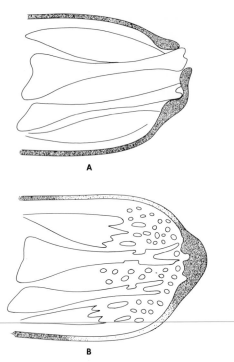

A

B

360 FIGURE 16.7 Expidermal cells apparently do not contribute to the regeneration blastema in the amputated limb of adult *Triturus viridescens*. Prior to operation, the animals were injected with ^{3}H-thymidine, which labels with radioactivity only those cells synthesizing DNA during the brief time the radioactive compound is available. Previously labeled limbs were amputated through the forearm. A: One day after operation the epithelium shows uniform labeling. The epithelium is the only mitotically active tissue in the forelimb of the adult newt. B: Fifteen days after the operation, previously labeled epithelial cells adjacent to the wound have formed a heavily labeled wound epithelium over the regeneration blastema. There is no cell division in the apical cap at this time. In contrast, mitosis has occurred in ectoderm proximal to the wound surface, thus diluting the label. Most significantly, labeled cells are not found in the blastema. Therefore, the blastema apparently receives no cellular contribution from the apical cap of ectoderm. [From E. Hay and D. Fishman, *Dev Biol* 3:26–59 (1961), by courtesy of Dr. Hay and permission of Academic Press, Inc., New York.]

motor nerves are effective, if present in sufficient amount, and sympathetic innervation is not required. Nerves apparently produce a neurotrophic

substance or substances required by limb tissues for their regeneration (see section 16.3.8.3 below). The neurotrophic effect can be supplied by any part of the central or peripheral nervous system and is not species specific: implants of frog ganglion will support regeneration of an amputated newt limb deprived of its nerve supply (A. Kamrin and Singer, 1959).

In amputated denervated limbs, as in x-rayed and amputated limbs, a wound epithelium forms, but an apical cap does not arise and blastema cells do not accumulate at the limb tip. If the operated animal is a young larva, regression will continue unchecked unless nerves are allowed to regrow into the limb. In fact, in the case of the larva, regression will occur without visible external injury to the denervated appendage; pinching the limb with a forceps is sufficient stimulus for complete regression (Thornton, 1953). In adult newts regression is limited to the region of the cut surface and nerve regrowth occurs more slowly.

In nondenervated amputated limbs of *Triturus* adults held at approximately 20%C, cells released from the stump tissues begin to proliferate on the 6th or 7th day, and by the 16th or 17th day a well-established mound-shaped blastema is present. If denervation precedes the formation of the mound stage, the blastema is likely to persist and will undergo morphogenesis. The regenerate will be small and atypical, however, because cell division is inhibited by denervation at all stages of regeneration.

16.3.8.1 Effects of Varying the Nerve Supply. Although regeneration proceeds in newt or axolotl limbs supplied with an amount of nervous tissue above some threshold level, it is found, in general, that the rate of regeneration and the length and volume of the regenerate are all diminished in amputated limbs with a reduced nerve supply. Because there is this quantitative dependence on nerve one may ask what might be the effect of augmenting the nerve supply to a limb normally unable to regenerate. Singer (1954) deviated the sciatic nerve under the skin of the back and into the stump of the freshly amputated forearm of the postmetamorphic (juvenile) anuran *Rana pipiens,* which normally does not regenerate limb parts. The amputated limbs showed some frequency of regeneration, but the limb parts formed were of poor quality. Both the frequency and quality of regeneration were enhanced if Rose's (1944) salt treatment of the stump was used to inhibit skin healing, but complete regenerates were never formed (Singer, et al., 1957).

16.3.8.2 Denervation Affects Patterns of Macromolecular Synthesis.

Mescher and Tassava (1975) examined the number of mitoses and the incorporation of ^{3}H-thymidine in forelimbs of adult *Triturus* during a period of 2 weeks after denervation and amputation. Results were compared with those found in amputated but not denervated contralateral limbs. Radioautographs showed that cells released by regression in both the denervated and control limbs incorporate the label into DNA beginning on day 4. Both stumps were alike in the number and distribution of labeled nuclei until day 10. Thereafter the regenerating controls showed almost twice the number of labeled cells as the denervated limbs. In contrast, mitotic figures were almost entirely absent in the denervated limbs, but began to appear with notable and increasing frequency on day 10 in the controls. Apparently, as cells are released from their parent tissues in the stump, they soon begin DNA synthesis, but do not enter mitosis in the absence of nerve. Similar results have been observed in comparably treated larval limbs of *Ambystoma mexicanum* (Tassava and Mescher, 1976).

In limbs denervated after the onset of regeneration, the blastemata show diminished incorporation of radioactive precursors into DNA, RNA, and protein. This decrease occurs regardless whether the blastema has achieved morphogenetic independence at the time of denervation (Bantle and Tassava, 1974; Dresden, 1973; Dearlove and Stocum, 1974). Major enzymes involved in DNA and RNA metabolism, however, seem to be undiminished in specific activity (Manson et al., 1976). Dearlove and Stocum (1974) raised the question of whether denervation alters the composition of the soluble protein fraction of blastemata in adult *Triturus* limbs denervated at various stages after amputation and sampled 3 days later. Using the technique of polyacrylamide gel electrophoresis, they were able to recognize 24 distinct bands in homogenates of intact limbs. Eight of these are absent from blastemata of limbs denervated 5 days after amputation, some reappearing in blastemata isolated from limbs denervated at later stages. Eight other proteins unique to blastemata also appear, the first of these in blastemata of limbs denervated 10 days after amputation. The stages during which the regenerate is

becoming independent of nerve for its morphogenesis are the ones in which the greatest changes in the protein profiles occur. Morphogenesis can then go on, but not mitosis, in the absence of nerve-dependent proteins.

In 1972 Singer and Caston reported that 10–13 day regeneration blastemata of adult *Triturus* show an enhanced uptake of radioactive precursors of DNA, RNA, and protein in the early hours after denervation (Figure 16.8). Incorporation rates then rapidly drop to about 60% of control values after 2 days. In subsequent studies Singer and his collaborators (Singer and Ilan, 1977; Bast et al., 1979) analyzed nerve-dependent changes in rates of protein synthesis and in content of ribosomes, polysomes, and nascent peptide chains in blastemata from adult *Notophthalmus viridescens* limbs denervated 18–20 days after amputation. When protein synthesis was maximally repressed, 2 days after denervation, ^{35}S-methionine was injected intraperitoneally, and 3 hours later the blastemata from denervated and control limbs were harvested. It was found that the specific activity of ^{35}S-methionine in proteins of blastemata from both denervated and control limbs is the same; therefore, early effects of denervation bring about an absolute reduction in protein synthesis. Further analysis revealed that methionine-labeled nascent polypeptides on polysomes from denervated regenerates was only 48% that of controls. No significant difference in the average molecular weights of proteins of denervated and control regenerates was found. The number of ribosomes and number of translatable mRNAs decreased in a coordinate manner as a result of denervation, but the translation time (cf. Chapter 18) for the average protein was not diminished. In a companion study Geraudie and Singer (1978) found that the depression of protein synthesis occurs in both the epithelium and the blastemal mesenchyme, but is more depressed in the mesenchyme. The specific activity of ^{3}H-thymidine incorporated into DNA was found to be depressed in similar fashion.

16.3.8.3 What Does Nerve Contribute?

In their early studies Singer and his associates made a number of attempts to influence regeneration of limbs denervated after blastema formation by infusing into blastemata characteristic neurotransmitter substances, their inhibitors, and so on. These studies were meticulously carried out and elimi-

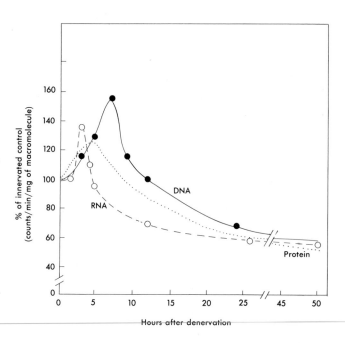

FIGURE 16.8 Changing patterns of macromolecular synthesis as a function of time after denervation of 10–13-day forelimb regenerates of adult *Triturus viridescens*. Specific activities of isotopes incorporated into DNA, RNA, and protein are expressed as percentages of the activities shown by the contralateral nondenervated controls. The operated animals were injected intraperitoneally with the appropriate isotopically labeled precursors three hours before the blastemata were excised and analyzed. Note that RNA, then protein, and then DNA are incorporated in that sequence more rapidly in denervated blastemata than in controls during the first few hours after denervation, then they decline rapidly to plateau at lower levels after 2 days. [Adapted with modifications from M. Singer and J. D. Caston, *J Embryol Exp Morphol* 28:1–11 (1972).]

362

nated from consideration as neurotrophic regenerative agents such products of nerve action as are investigated by the neurophysiologist and pharmacologist.

The more recent studies from Singer's laboratory have been directed to the effects on patterns of macromolecular synthesis of extracts of nervous tissue infused into early regeneration blastemata deprived of innervation. Incorporation of precursors into the major macromolecules of early regeneration blastemata is maximally depressed 48 hours after denervation (Figure 16.8). Singer et al. (1976) isolated from the $140,000 \times G$ supernatant of newt brain homogenates a crude protein preparation that, when infused into blastemata of 9–12-day regenerates denervated 48 hours earlier, caused the blastemata to incorporate radioactive amino acids into protein at a rate equal to or higher than that shown by nondenervated control blastemata. Choo et al. (1978) found similar effects of extracts of both newt and chicken brain on the incorporation of amino acids into protein.

In 1977 Jabaily and Singer measured the specific activity of ^{3}H-thymidine incorporated into DNA in denervated and control blastemata. They found that 2 days after the denervation of 9–12 day regenerates, the specific activity of ^{3}H-thymidine incorpo-

rated into DNA of the blastema after intraperitoneal injection of the isotope was only 48% that of the control. Infusion of the brain factor beginning 24 hours after denervation, however, caused the specific activity of incorporated ^{3}H-thymidine, as measured the next day to be about 90% that of the control.

It is possible that the active factor in nerve extracts used by Singer and his colleagues is **fibroblast growth factor** (FGF). FGF is a basic protein having a molecular weight of about 13,000. The protein is released by proteolytic enzymes from **myelin,** the chief component of nerve sheaths. FGF exercises a strong **mitogenetic** (mitosis-stimulating) effect on many kinds of cells in culture and it also stimulates DNA synthesis when infused into the regeneration blastema of the denervated limb of an adult newt (Mescher and Gospodarowicz, 1979; Gospodarowicz and Mescher, 1980). Amputated limbs, allowed to regenerate until a well-developed blastema is present, show a mitotic rate only 13% that of controls 3 days after denervation. Infusion of 50 µg of FGF into the blastema, however, increases its mitotic rate to as much as 70% of control blastemata, a five-fold increase.

16.3.8.4 A Special Case—The Aneurogenic Limb. It is possible to rear *Ambystoma* embryos in such a way that there is no nerve supply, or only a very limited nerve supply to the limb-forming region. Yntema (1959a, b) joined neurulae of *Ambystoma punctatum* in parabiosis (Figure 16.9), uniting them side by side at the posterior trunk region. After healing and further development the neural tube and placodes of the cranial ganglia were removed from the right parabiont, leaving it parasitic upon the left member. Forelimbs of both host and parasite developed, the latter being only sparsely innervated (Egar et al., 1973) or **aneurogenic** (developing in the complete absence of nerves) limbs. When such limbs are amputated, they regenerate completely and normally except that there may be a delay of 1–7 days in the completion of regeneration.

Aneurogenic limbs can become nerve dependent for regeneration. C. S. and M. T. Thornton (1970) disarticulated aneurogenic forelimbs of *Ambystoma maculatum* at the shoulder and grafted them in place of comparable limbs of neurogenic larvae. If host nerves were prevented from growing into these grafts, they would regenerate after amputation at the forearm level. If, however, host nerves were allowed to grow into the grafts for 13 days before amputation, regeneration failed, sug-

FIGURE 16.9 Parabionts of *Ambystoma punctatum*. A: Embryos at the tail bud stage healed together after being joined posteriorly. In the right member of this pair, the notochord and somites are exposed as a result of removal of the neural tube and adjacent neural crest from the anterior trunk and hindbrain levels. B: Parabionts at a later stage. [From C. L. Yntema, *J Exp Zool* 140:101–24 (1959), by courtesy of the late Dr. Yntema and permission of the Wistar Institute.]

gesting that aneurogenic limbs develop a requirement for nerve in order to regenerate. This requirement, or "addiction," could be banished if grafted aneurogenic limbs that were allowed to be innervated for 2–3 weeks were subsequently denervated and held in the denervated condition for 30 days. Thereupon, they regenerated after amputation in about 50% of the cases.

Other experiments with aneurogenic limbs suggest that the skin plays an important role in mediating the neurotrophic effect on regeneration. If an aneurogenic limb that is provided with a skin sleeve of neurogenic skin in place of its own aneurogenic skin is amputated through the region of the sleeve, a blastema usually does not form and regeneration fails to occur. This experiment suggests that neurogenic skin requires the presence of nerves in order to form a wound epithelium that will promote formation of a blastema in an otherwise regeneration-competent stump. Neurogenic skin is not inhibitory to blastema formation, however, for if amputation is performed through aneurogenic skin distal to a ring of neurogenic skin on an aneurogenic limb, regeneration occurs promptly. The question now arises as to whether the presence of aneurogenic skin might elicit the participation of neurogenic bone and muscle in formation of the regeneration blastema. Steen and Thornton (1963) removed the humerus and musculature from upper arm stumps of aneurogenic limbs and then packed the empty cylinder of skin with cartilage and muscle from a neurogenic upper arm. Regeneration occurred at the same rate as in aneurogenic limbs that were simply amputated. Moreover, if the neurogenic tissues used to pack the aneurogenic skin cylinder were labeled with ^{3}H-thymidine, labeled cells were found in the blastema. Analysis of the dilution of the nuclear label suggested that some of the labeled cells would have divided at least once or even twice by 6–11 days after regeneration began.

It seems, therefore, that the character of the skin determines whether or not a blastema can form under the wound epidermis. Neurogenic skin loses this character in the absence of innervation. Aneurogenic skin has the appropriate property without the presence of nerve. But skin is composed of both epithelial and mesenchymal elements. Is the effect of nerve on one or both of these components? Experiments with neurogenic skin suggest that the epidermal component is not directly affected by

363

nerves, for a normal wound epithelium and blastema are formed in regenerates totally lacking in sensory innervation, which is the only innervation that the epidermis receives (Sidman and Singer, 1960; Thornton, 1960; Thornton and Steen, 1962). Sensory innervation alone suffices for regeneration of neurogenic limbs. Nevertheless, it remains to be determined whether epidermis or dermis is of the greater importance for the neurotrophic effect on regeneration.

There is evidence that x-irradiation interferes with the neurotrophic effect on regeneration. F. C. and S. M. Rose (1974) prepared aneurogenic larvae of *Ambystoma punctatum* and of *A. opacum* by the method of parabiosis. Limbs of both the normal and aneurogenic parabionts were irradiated with doses of x-rays that completely inhibited regeneration in neuralized limbs. The irradiated aneurogenic limbs, however, regenerated in 42 of 71 cases. The Roses next compared the effects of x-irradiation on the regenerative power of limbs in younger versus older larvae. They found that irradiated larvae of between 16 and 24 mm in length regenerated limbs with some frequency, but larger larvae did not. These results suggest that irradiation of the smaller animals came at a time when nerve fiber dependency was being developed, that is, when changes were just beginning to occur in the skin that make regeneration nerve-dependent.

Further study is needed to clarify problems raised by these results. One might infer from the Rose's (1974) work that aneurogenic limb tissues are less susceptible to the antimitotic effects of x-rays. Obviously, a limb cannot regenerate if cells do not divide, and it is quite evident that cell division is clearly compromised, if not inhibited completely, in x-rayed neurogenic tissues. This point requires further examination.

16.3.9 BIOELECTRICITY AND LIMB REGENERATION.

The limbs of adult frogs do not regenerate after amputation, although some regeneration occurs in experimental animals. S. M. Rose (1944, 1945), for example, reported that limbs of adult *Rana clamitans* immersed frequently in strong solutions of NaCl after amputation showed some regeneration. He attributed this regeneration to the fact that the salted limbs maintained a wound epithelium free from an underlying layer of dermis.

In more recent experiments sodium-driven bioelectric currents have been implicated in regeneration. Borgens et al. (1977a) implanted small well-insulated batteries in the dorsal lymph sac of adult *Rana pipiens* and led 0.2 μA of current from the cathode through an insulated silver–silver chloride wick electrode to a forelimb amputation surface. Insulation was removed from the wick where it projected a short distance beyond the wound. A platinum electrode connected to the anode projected into the lymph sac. This preparation constituted a cathode stimulator. Frogs so treated showed some degree of regeneration of bone, muscle, and connective tissue at the amputation surface. When the battery was short circuited, no regeneration occurred; but when anodal stimulators were used, not only was regeneration suppressed, but regression of stump parts occurred. Thus it appears that pulling current out of the stump induces regeneration, whereas drawing current inward induces regression.

Subsequently, Borgens et al. (1977b) used an ultrasensitive vibrating probe to measure the direction and density of currents associated with regenerating and nonregenerating limbs. The probe, vibrating between two external points, measures the minute voltage difference generated by any current that may travel between these two points in the medium in which a biological system is immersed. Current density is then calculated using the measured resistivity of the medium. In the unamputated forelimb of adult newts, *Notophthamus viridescens*, held in pond water, small currents in the range of 0.01–1 μA/cm^2 were found to enter the limb at most locations. Exit currents on the order of 0.2–1.4 μA/cm^2 were located at the base of the first digit and occasionally at the fingertips. These measurements were made at a distance of 340 μm from the skin surface and were calculated to be about 70% higher at the surface.

In amputated limbs relatively intense currents on the order of 10–100 μA/cm^2 left the stump at its cut end, whereas currents of much lower density, on the order of 1–3 μA/cm^2 entered the rest of the limb at the surface. This current pattern persisted, sometimes increasing in strength until the regeneration blastema was fully formed. Thereafter, the current decreased and sometimes reversed.

These currents are apparently sodium driven. It is well known that amphibian skin has the properties of a battery the current of which consists of sodium influx. Borgens and his colleagues found that when sodium in the medium was decreased or increased, stump currents were likewise decreased or increased, respectively. Moreover, when drugs that block sodium entry were introduced into the pond water in which the newts were immersed, stump exit currents decreased dramatically. In subsequent experiments the same investigators (Borgens et al., 1979) found that regeneration is inhibited for some weeks in newts held in sodium-depleted media. Subsequently, however, they recover and regenerate normally.

16.4 Regeneration in Lizards

The tail in most lizards is provided with preformed sites, **autotomy planes,** at which the tail can be snapped off by contraction of the tail musculature. In the vertebrae the autotomy plane passes through the centrum (Figure 16.10) and can be recognized histologically and in dissections. Its position in other tissues is or is not recognizable, according to the species.

The regenerate that forms after autotomy is of special note because of both its structure and its

FIGURE 16.10 Lateral view of caudal vertebrae members 7–10 and 11–18 in the lizard *Anolis carolinensis.* Vertebra 9 has an incomplete autotomy separation. All vertebrae 10–18 show a complete autotomy separation, whereas vertebrae anterior and posterior to this region do not. [Adapted with modifications from P. G. Cox, *J Exp Zool* 171:127–50, (1969).]

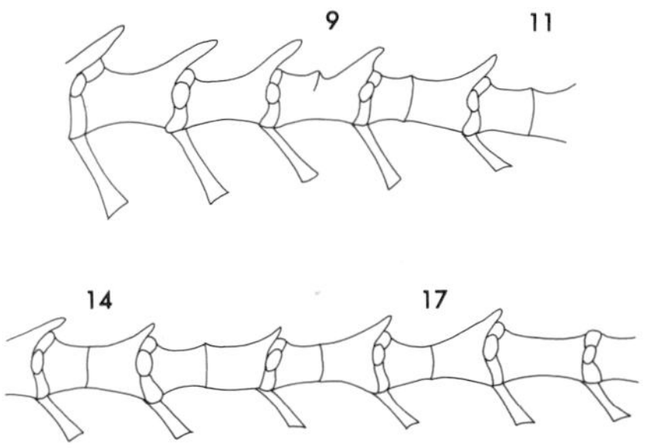

method of origin. Morphologically it is a quite imperfect tail. It lacks vertebrae and, in their place, has an unsegmented cartilaginous tube. This tube contains the regenerated spinal cord, which consists of an extension of the ependymal lining of the central canal of the cord and descending fiber tracts from the old cord. No new neurites are differentiated in the regenerated tail.

The origin of the regenerate, moreover, is not from a proliferating blastema as in amphibians. Very few cells accumulate under the wound epithelium and those that do show very little DNA synthesis. There is essentially no breakdown of muscle and, consequently, no liberation of muscle cells. The principal sources of cells for the regenerate are the ependyma and the various connective tissues comprising muscle septa, dermis, adipose tissue, periosteum, and perhaps osteocytes of the vertebrae (Simpson, 1965; Cox, 1969a, b). The nonnervous elements assemble in precartilaginous and premuscular masses that proliferate behind the apex. These elements differentiate muscle and cartilage in a proximodistal direction. Within the forming tube of cartilage, the ependyma proliferates and gradually extends distally (Figure 16.11).

If the spinal cord is removed for a distance anterior to the autotomy plane, regeneration of the tail fails to occur (R. Kamrin and Singer, 1955). This inability was first thought to be due to failure of nerves to reach the cut surface. Subsequently, however, it was shown that the presence of nerve is not requisite to tail regeneration; rather, the ependyma is the essential ingredient. Destruction of the greater part of the nerve cord at the amputation plane does not inhibit regeneration as long as ependyma is present. Very importantly, small bits of regenerating ependyma within the cartilage tube grafted into the dorsal muscle mass of the tail will induce supernumerary regenerates, whereas implants of cartilage tube alone will not (Simpson, 1964, 1970). The transport of presumed neutrotrophic substances by nerve is thus not required for tail regeneration, for the ependyma produces no neurites in the regeneration situation.

When the limb of a lizard is amputated, it does not regenerate. Singer (1961) and Simpson (1961) both showed, however, that after amputation of one hindlimb, a limited regrowth occurs with low frequency if the sciatic nerve of the opposite limb is deviated surgically to enter the amputated limb.

365

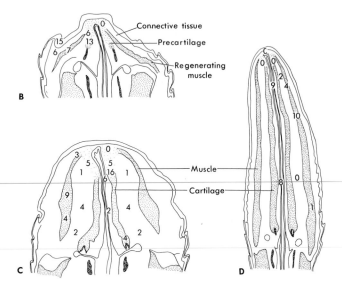

FIGURE 16.11 Stages in the regeneration of the tail in *Anolis carolinensis*, as shown in longitudinal section. The lizards were caused to autotomize by pinching the tail with a forceps or the fingers. At various stages of regeneration the animals were injected with ³H-thymidine and the regenerates were fixed and sectioned for radioautography. In these diagrams the tissues are identified, and relative levels of radioactivity are designated by numbers in different regions of the regenerate. Note especially the absence of labeling (indicated by 0) in the apical zone. [Adapted with modifications from P. G. Cox, *J Exp Zool* 171:127–50 (1969).]

366

The regenerates in these cases are merely short extensions from the stump. Bryant and Wozny (1974) found, however, that an amputated lizard limb will regenerate in a majority of cases and to a greater length if supplied with cartilage and ependyma from a regenerating tail.

16.5 Regeneration of Mammalian Limbs

It is rather clear, as Goss (1968) pointed out, that a warm-blooded animal requires frequent nutrition in order to maintain its heightened metabolic rate and elevated body temperature. Should a bird or mammal lose a limb, it might very well starve, freeze, or be preyed upon before it could regenerate a functional replacement. The ability to regenerate limbs was apparently lost in the course of evolution in exchange for other advantages of the warm-blooded condition.

There have been a number of studies of the sequelae of amputating digits in mice and rats. Amputation surfaces are quickly covered by a thick epidermis, and a definitive dermis then differentiates beneath. The cut bone heals over and no distal outgrowth occurs. Schotté and Smith (1961) found, however, that dermal overgrowth can be considerably delayed by injections of cortisone or adrenocorticotrophic hormone (ACTH). This result is reminiscent of the effects of salt treatment, which promotes limited regenerative outgrowths of anuran limbs (Rose, 1944), as already noted.

When the opossum is born, its hindlimbs are in an immature condition as compared to the forelimbs. The forelimbs are well developed and armed with claws, whereas the hindlimbs show a foot plate with the margins of the digits only just becoming visible. Mizell (1968) found that the forelimbs would not regenerate after amputation. He was able to introduce a column of brain tissue into a channel through the softer hindlimb tissues prior to amputation just above the ankle. Foot parts including digits regenerated, albeit imperfectly, in limbs containing implants, but not in controls in which liver or kidney tissue was introduced. Their augmentation of the nerve supply affects regeneration in newborn opposums as effectively as it does in anurans and lizards, or more so.

The objective of achieving regeneration of limbs or their parts in higher mammals may achieve a limited success, in newborn or young individuals at least, by manipulation of the wound surface. Scharf (1961) amputated the fourth and fifth digits of the forelimbs of newborn rats by cuts through the basal phalanx. Untreated amputations that were simply allowed to heal showed no regeneration, but in cases in which the wound scab was removed by trypsin on the first day after operation, the severed phalanges often regenerated their missing length, but the more distal phalanges and claws did not form. The trypsin treatment delayed wound healing for several days, and this was accompanied by the accumulation of a mesenchymal condensation over the cut end of the bone which was not completely covered by epithelium for some time.

At present there are no indications that regeneration of major limb parts in mammals will be accomplished in the near future. There is now considerable evidence, however, that amputation through the terminal phalanges can be followed by complete regeneration of the fingertip. Its full length may be restored and the nail and fingerprint whorls appear normal (Figures 16.12 and 16.13). There is seldom loss of sensation or mobility. Regeneration occurs, however, only in the absence of surgical intervention. Unfortunately, it is still the practice in many hospitals to trim bone protruding from the amputation site and to cover the exposed surface with a skin graft. This procedure leaves the finger shorter than normal, and the effect is cosmetically unsatisfactory (Douglas, 1972; Illingworth, 1974; Farrell et al., 1977; Pickering, 1978).

Illingworth and Barker (1980) used a vibrating probe similar to that described in Section 16.3.9 to measure electrical currents emanating from regenerating amputated fingertips in children aged 1–15 years. The injured fingers were immersed in saline, and the vibrating probe was held 0.1 mm away

from the cut surface. The investigators found that the exit current rose to a peak of 22 $\mu A/cm^2$ at about the eighth day after injury and usually declined thereafter, disappearing after the wound closed over. These findings suggest that in human beings,

FIGURE 16.12 Regeneration of the tip of the middle finger in a child injured at the age of 22 months. A: Accidental amputation of the third finger proximal to the nail. B: Regeneration of the entire fingertip, including bone and nail as seen 11 weeks after the accident. [From C. M. Illingworth, *J Pediat Surg* 9:853–58 (1974), by courtesy of Dr. Illingworth and permission of Grune & Stratton, Inc.]

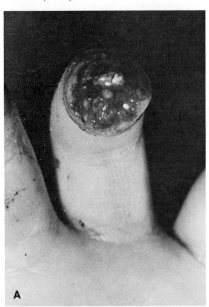

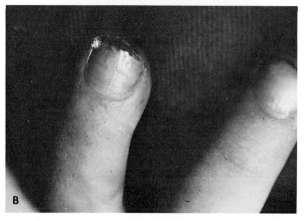

367

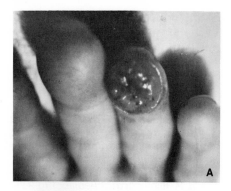

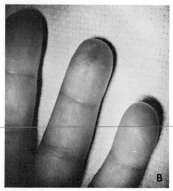

368

FIGURE 16.13 Regeneration of the tip of the ring finger in a 70-year-old man. In this case, no bone was exposed. A: Amputation as a result of lawnmower injury. B: After 9 months, the fingertip is regenerated. There is a small scar, but skin whorls and contours are essentially normal. [From R. G. Farrell et al., *J Am Coll Emergency Physicians* 6:243–46 (1977), by courtesy of Dr. Farrell and permission of W. B. Saunders Co.]

as in salamanders, electrical currents play a role in tissue healing and regeneration.

16.6 Regeneration of the Components of the Eye

In Chapter 11 there was noted the remarkable ability of fish and urodeles to regenerate the optic nerve after the eye has been removed and then replaced

in its own or another emptied orbit. In fact, the regeneration of the nerve is preceded by the degeneration of the entire neural retina (Figure 16.14) and its regeneration from the pigmented layer of the retina. The regenerated ganglion cells of retina then send out their axons to the brain, reconstituting the optic nerve. The iris and lens likewise degenerate after transplantation of the eye (except in larval urodeles). The iris will be regenerated from the pigmented layer of the retina and the lens is then regenerated from the dorsal margin of the iris. The experiments on which these findings are based were reviewed by Goss in 1964 and by Reyer in

FIGURE 16.14 The vertebrate eye: general relationships of its components (A) and enlarged diagram of the layers of the retina (B).

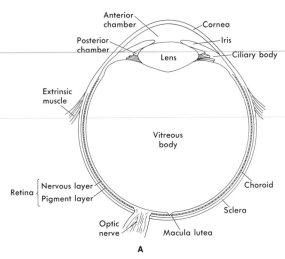

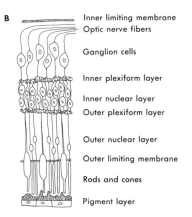

1977, and these references may be consulted for further details.

The rest of this section will be devoted to the regeneration of the lens after its removal through an incision in the cornea (the usual route) or retina. Stone (1967) listed 21 species of urodeles that have the power to regenerate the lens from the margin of the iris. Most of these species belong to the family Salamandridae among which are species of the genera *Notophthalmus* and *Taricha* (most of these were identified under the genus name *Triturus*). A few species in the family Plethodontidae likewise have this property. Members of the Ambystomidae, including those species in which limb regeneration is frequently studied, do not regenerate the lens. In some anuran species (reviewed by Reyer, 1954) lens regeneration from the margin of the iris is possible during tadpole stages, but will not occur on postmetamorphic animals. When the lens is excised from the larval *Xenopus laevis* (Freeman, 1963) it is regenerated, not from the iris, but from the cornea. This property is lost at metamorphosis. Urodeles can regenerate the lens from the iris both as larvae and adults.

16.6.1 THE SEQUENCE OF EVENTS IN LENS REGENERATION.

Lens regeneration provides a clear example of the transformation of one differentiated cellular type, having a distinctive pattern of metabolic activities, to another cellular type, which is morphologically distinctive from the original, and which synthesizes a different array of macromolecules. This process is referred to as **metaplasia.** The iris consists of a bilaminate epithelium whose outer layer (facing the anterior chamber) is covered with a thin connective tissue stroma continuous with the ciliary body. Both inner and outer epithelial lamellae are pigmented (Figure 16.15A) and it is from the

pigmented cells of the dorsal margin of the iris that the lens is regenerated.

The events of regeneration have been divided

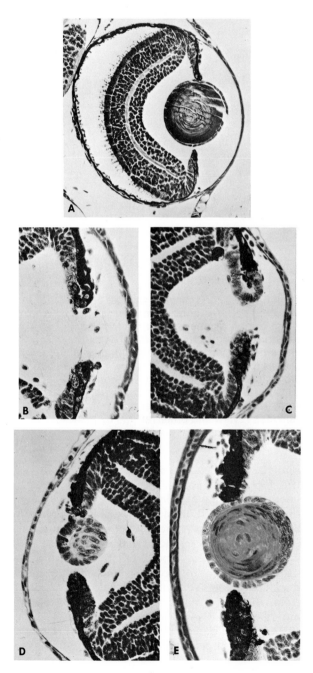

369

FIGURE 16.15 Regeneration of lens from iris in *Triturus viridescens*. A: Section of normal eye showing relationships of lens to iris and to the chambers of the eye. B: Lensectomized eye; 4 days after operation. C, D: Depigmentation of the dorsal margin of the iris and formation of a lens vesicle; 5 and 11 days after operation, respectively. E: Lens regenerate after 20 days. Refer to the text and Figure 16.16 for further details. [From R. Reyer, *Quart Rev Biol* 29:1–46 (1954), by courtesy of Dr. Reyer and permission of the Biological Sciences Department, SUNY at Stony Brook.]

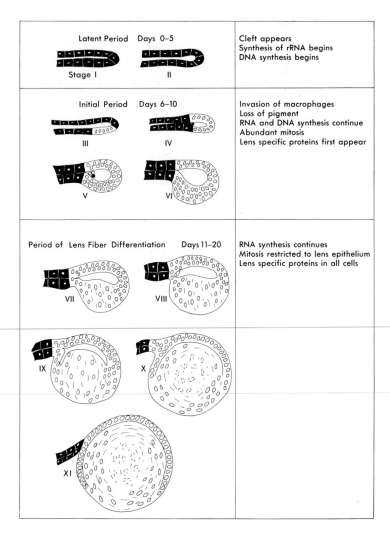

Latent Period Days 0–5	Cleft appears Synthesis of rRNA begins DNA synthesis begins
Stage I II	
Initial Period Days 6–10	Invasion of macrophages Loss of pigment RNA and DNA synthesis continue Abundant mitosis Lens specific proteins first appear
III IV	
V VI	
Period of Lens Fiber Differentiation Days 11–20	RNA synthesis continues Mitosis restricted to lens epithelium Lens specific proteins in all cells
VII VIII	
IX X	
XI	

370

FIGURE 16.16 Morphological changes and changing patterns of macromolecular synthesis during regeneration of the lens from the dorsal margin of the iris in urodeles. Iris and regenerating lens redrawn with modifications from Yamada's (1967) review paper. Stages of regeneration are those described by Sato (1940). All descriptive data were compiled from material presented in Yamada's review and other reviews by Reyer (1959, 1977).

into 12 stages that are grouped in four periods (Figure 16.16). Excision of the lens is followed by a **latent period** of 4 or 5 days. Despite the term latent, this period is a time of considerable change. The border of the dorsal region of the iris thickens and a cleft arises between inner and outer lamellae (Figure 16.15B, C). Ameboid cells move from the stroma on to the surface of the epithelium and into the cleft. This movement is followed by a rapid increase in the rate of incorporation of labeled uridine into ribosomal RNA (Reese et al., 1969). The rate of rRNA production continues to increase until about the tenth postoperative day and then remains ele-

vated until regeneration is complete. About 4 days after operation, the incorporation of labeled thymidine into DNA begins and the first mitotic figures are seen.

The **initial period** occupies postoperative days 6 to 9 or 10. The beginning of this period is signaled by the first appearance of depigmented cells at the dorsal pupillary margin. The pigment is engulfed and carried away by invading amoboid cells. As more cells lose their pigment, they form a hollow epithelial vesicle of high cuboidal cells continuous with the inner and outer lamellae (Figure 16.15C, D). During this period RNA and DNA synthesis continues and more mitotic figures appear. At the end of the initial period cells of the inner wall of the vesicle elongate and protrude into the lumen, forming a few irregularly arranged **primary lens fibers.** Some of these will have left the cell cycle (no longer incorporate thymidine into DNA). At this time also, one may first detect the appearance of lens-specific proteins, the **crystallins,** which are not otherwise present in the iris.

The period of **lens fiber differentiation** now is recognized. Beginning at approximately the 11th day, the primary lens fibers push to the front of the vesicle forming a nucleus behind the **lens epithelium** (Figure 16.15D). The latter then proliferates the **secondary lens fibers** from the equatorial zone where they meet the primary lens fiber nucleus. At about 18 or 20 days after operation the nucleus of primary lens fibers is symmetrically enclosed by secondary lens fibers; DNA synthesis and mitosis now come to an end in the iris but continue in the lens in the epithelium and outer layers of secondary lens fibers (Figure 16.17).

FIGURE 16.17 Diagram showing the location of labeled nuclei in regenerating *Triturus viridescens* at selected stages as identified in Figure 16.16. All regenerates were exposed to ³H-thymidine administered by intraperitoneal injection at stage IX in such concentrations as to be completely bound to DNA in 3 hr. Labeled nuclei are shown as black circles or ovals. A: Stage IX regenerate in animal injected 3 hr earlier: labeled nuclei are seen only in the lens epithelium. B: Stage X regenerate in animal injected 5 days earlier; some of the epithelial cells labeled at stage IX have shifted into lens fibers. C: Stage XI regenerate in animal injected 20 days previously; all epithelial cells labeled at stage IX have now become lens fibers. [Adapted with modifications from R. Reyer, *Dev Biol* 24:533–58, (1971).]

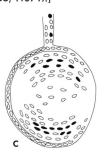

The final period is the **period of growth** (Figure 16.15E), which continues until about the 30th day after operation or later. Mitosis continues in the lens epithelium, but there is a gradual decrease in DNA synthesis elsewhere except in the outermost layers of secondary lens fiber. In the central lens fibers the nuclei eventually degenerate.

16.6.2 THE PRESENCE OF A LENS AFFECTS LENS REGENERATION. The presence of the lens must normally inhibit the iris cells from rechanneling their biosynthetic pathways as just described. When the lens is absent, however, the genetic program leading to lens regeneration is activated in the dorsal iris, but not the ventral. Regeneration is inhibited by actinomycin D, a drug that inhibits transcription of the genome.

If a lens is taken out and immediately replaced, or if the lens of another species is put in its place, the iris fails to form a new lens. When wax or glass spheres similar in size to a normal lens are implanted in the pupil of a lensectomized eye, a lens regenerates from the dorsal iris. Mechanical factors such as the size, shape, and volume of the lens do not, therefore, seem to restrict lens regeneration. Reasonably, therefore, chemical factors that emanate from a differentiated lens inhibit the expression in iris cells of the genetic program responsible for lens differentiation.

Under a variety of experimental conditions (Figure 16.18A, C), two lenses can regenerate simultaneously in one pupil. Moreover, the implantation of an immature regenerate lens (less than 25 days for adult *Triturus*) into a freshly lensectomized eye does not inhibit regeneration (Stone, 1952). It is evident, therefore, that only a fully differentiated lens or an advanced regenerate can inhibit the regenerative activity of the iris. Such a lens fails to inhibit, however, if it is in the process of degenerating owing to injury.

The presence of a mature lens needs not be inhibitory, however, if there is some distance between it and the potentially regenerating dorsal iris or if a barrier is placed between (Figure 16.18D). Eguchi (1961) created tadpoles of *Triturus pyrrhogaster* with a gigantic cyclopean eye by removing the anterior one-fourth of the archenteron roof at the neurula stage. In such eyes the lens is small compared to the size of the pupillary space and can

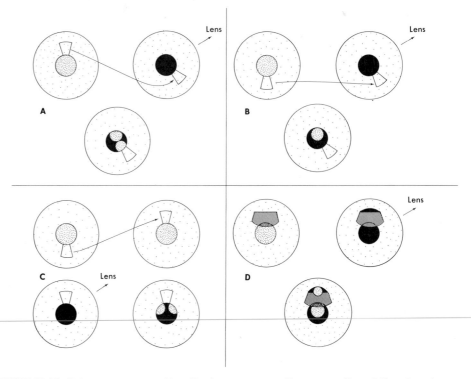

FIGURE 16.18 Schemes representing the lens-regenerative properties of the dorsal margin of the iris and the regeneration-inhibitive properties of an existing lens in the eye of the newt *Triturus viridescens*. In these schemes, the outer large circle represents the outer margin of the iris, which is shown lightly stippled. The lens is shown in heavier stippling and the absence of lens is denoted in black. Excision of lens is denoted by an arrow directed to the word Lens. A: A wedge of dorsal margin of the iris implanted in another eye immediately after lensectomy forms a lens, as does the margin of the dorsal iris of the host. B: A piece of ventral iris margin grafted to the eye of a lensectomized host eye does not regenerate a lens, but the dorsal margin of the iris of the host does. C: Ventral iris grafted to the dorsal iris margin of a lensectomized host does not regenerate lens, but a lens regenerates from dorsal iris tissue on each side of the implant. D: When a sheet of polyethylene film is inserted through a slit in the dorsal part of the iris (upper left), a secondary pupillary space forms above it (upper right). Lensectomy is followed by lens formation of both primary and secondary pupils (lower center). [Adapted with modifications from L. S. Stone, *Am J Ophthalmol* 36:31–39 (1953).]

be displaced to one margin, resulting in the regeneration of a new lens from the opposite margin (Figure 16.19A). Also, Eguchi was able to displace the lens of a normal eye ventrally (Figure 16.19B) without causing its degeneration. A lens then regenerated from the dorsal iris margin so that the pupil contained two lenses. These and related observations by others (Stone, 1963) suggest that if the differentiated lens does indeed exercise its in-

hibitory effect by virtue of a chemical influence, nevertheless, it must be in contact with regeneration-responsive tissue in order to do so.

16.6.3 RETINAL FACTORS ARE NECESSARY FOR LENS REGENERATION. If the iris is separated from the eye of the newt and cultured in vitro, it undergoes the developmental changes characteristic of the early stages of regeneration: enhanced RNA synthesis, DNA synthesis and mitosis, and partial depigmentation. After 4 days, however, the rate of RNA synthesis diminishes, although the synthesis of DNA remains high (reviewed by Yamada et al., 1973). The cultured iris does not form a lens even though it is isolated from the inhibitory effect of a preexisting lens. Its dorsal margin will regenerate a lens, however, if it is implanted in the eye of a lensectomized host, even after prolonged culture in vitro (Figure 16.20). Evidently, removal of inhibition is not sufficient to activate or maintain the complete genetic program for lens differentiation in the dorsal iris. Something that promotes regeneration is required, in addition (Stone, 1958a; Stone and Gallagher, 1958).

The neural retina apparently provides the necessary stimulus, for if disc of impermeable polyethylene film is successfully interposed between the iris and retina, regeneration of the lens fails to occur

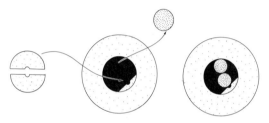

FIGURE 16.20 Lens regeneration by dorsal margin of the iris after prolonged cultivation in isolation from the retina. On the left an iris, cultured in vitro for 34 days and lacking lens, is halved. The dorsal half is then transplanted into the posterior chamber of the eye adjacent to the ventral margin. Lenses regenerate, as seen on the right, from both host and graft. [Adapted with modifications from L. S. Stone, *Anat Rec* 131:151–71, (1958).]

FIGURE 16.19 Lens regeneration in the presence of a preexisting lens that has been displaced. Experiments on *Triturus pyrrhogaster*. A: Lateral displacement of an experimentally created cyclopean eye of a young larva. A second lens forms opposite the original one. B: The lens is displaced ventrally as shown on the left. A lens then forms from the dorsal margin of the iris, as shown in optical section of the eye on the right. The original lens has been drawn disproportionately small in the interests of diagrammatic simplicity. Ac, anterior chamber; Pc, posterior chamber. [Interpretative drawings based on photos in G. Eguchi, *Embryologia* 6:13–35 (1961).]

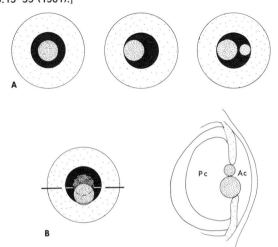

after lensectomy. If, however, the disc is so inserted that some neural retina lies between it and the iris, then a small lens regenerates. The size of such lenses seems to be proportional to the amount of neural retina present (Stone, 1958b). Moreover, the dorsal margin of the iris cultured in vitro in association with neural retina forms well-differentiated lenses with a characteristic lens epithelium and primary and secondary lens fibers (Yamada et al., 1973).

Some tissue environments other than retina can provide stimuli for lens regeneration. Connelly et al. (1973) found that lenses of good quality regenerate from dorsal iris of the newt cultured in vitro in contact with frog or newt pituitary or spinal ganglion. Reyer et al. (1973) obtained lenses from implants of dorsal iris from adult *Notophthalmus veridescens* into regeneration blastemata of the forelimb. Other sources of stimuli to lens regeneration have been found (reviewed by Reyer, 1977), but the regenerates are usually of poor quality.

There is little doubt that the normal stimulus to regeneration comes from the neural retina. The nature of the stimulus is unknown. Also unknown is why ventral retinal cells do not respond regeneratively. Relative to the latter point, it may be noted that ventral retinal cells exposed briefly to the carcinogen (cancer-causing agent) N-methyl-N'-nitro-N-nitrosoguanidine will repeatedly regenerate lenses (Eguchi and Watanabe, 1973) when grafted into a lensectomized host eye at any site around the circumference of the iris and regardless

373

of the simultaneous regeneration of a lens from the dorsal iris. Moreover, the effect of the carcinogen on ventral iris is a stable one, for the treated iris will repeatedly regenerate lenses after the successive removal of the regenerates.

16.6.4 TRANSDIFFERENTIATION OF RETINA.

It seems appropriate to bring the discussion of lens regeneration to an end by introducing the matter of lens formation directly from embryonic cells of the neural or pigmented retina. This phenomenon, which has been called **transdifferentiation,** has been analyzed chiefly in birds and mammals, forms that do not replace an excised lens. Indeed, it has proved to be difficult to find material relating to the direct formation of lens cells from retina in urodeles. Clayton (1970) referred to an observation by Iketa to the effect that lentoids (lens-like groups of cells that lack the characteristic lens epithelium) form in newt retina after neutron irradiation during late embyronic stages. These lentoids contain lens proteins, cyrstallins, as demonstrated by their reaction to antibodies produced against cystallins.

Cell cultures from embryonic neural retina and pigmented retina from birds and mammals are able to redirect their differentiation into lens fiber cells containing the characteristic crystallins, classified as α, β, and δ. This phenomenon, first described by Moscona in 1957, has received a great deal of attention during recent years (see Araki et al., 1979, and Clayton et al., 1979, for references). Of particular interest is the fact that, with respect to the embryonic chick neural retina, cells isolated from $3\frac{1}{2}$-day embryos form more lentoids and at a much more rapid rate than do cells isolated from older embryos. Both rate and frequency of lentoid formation fall off progressively until, by day 19, neural retinal cells show no ability to form lentoids. Moreover, the crystallins formed in lentoids derived from neural retinal cells differ according to the developmental stage at isolation. Thus, in cultures, from $3\frac{1}{2}$-day embryos, δ- and α-crystallins are produced at a ratio of 20:1. This ratio falls as progressively older donors are used. Neural retina cells from 17-day embryos produce only α- and β-crystallins.

Are mRNAs for the crystallins present in retinal cells prior to their transdifferentiation? Clayton et al. (1979) isolated crystallin mRNAs from polysomes of newborn chick lens and then prepared

DNA complementary to the crystallin mRNA (cDNA). The cDNA was then hybridized to neural retinal mRNAs freshly prepared from $3\frac{1}{2}$-day, 8-day, and newborn chicks. It was found that in the $3\frac{1}{2}$-day retina crystallin mRNAs are present at 10 times the level found in the 8-day neural retina. No crystallin mRNA could be detected in the retina of the newborn chick.

Are the crystallin mRNAs translated in the embryonic retina? Very low levels of crystallin proteins have been detected in 8-day embryonic neural retina, but, paradoxically, no crystallins are demonstrated in the $3\frac{1}{2}$-day retinal primordium, even though much more crystallin mRNA is present. Apparently, therefore, as development proceeds there are changes in both the production and translatability of mRNAs for the crystallin proteins in the retina.

16.7 Regeneration in Invertebrates Follows Many Patterns

Great powers of regeneration are found in all major groups of invertebrate animals. Organisms especially favorable for experimental study occur among the sponges, coelenterates, flatworms, annelids, and tunicates.

16.7.1 REGENERATION FROM FRAGMENTS IS COMMON.

Many species among these groups have the ability to reconstitute a new organism from a relatively small fragment of the original body. This ability is expressed not only in the laboratory by organisms that have been insulted experimentally, but also in nature by those that use fragmentation as a regular means of asexual reproduction. Thus there are sponges in which branches of the body normally break off and become new and independent sponges. Several highly efficient regenerates among flatworms and annelids regularly reproduce by reorganizing portions of the body into new worms that are cut off transversely. Among colonial tunicates, all of which can regenerate from relatively

FIGURE 16.21 Regeneration of a missing arm in the starfish *Asterias vulgaris*.

small parts of the original body, there are kinds that regularly bud off segments of the postabdominal region, and each bud, as it separates from the parent, reorganizes itself into a new organism.

16.7.2 LOST APPENDAGES ARE REPLACED.

It is well known that a starfish will regenerate missing arms (Figure 16.21) and that from as little as a single arm, plus part of the central disc, an entire animal can be formed. Starfishes will often autotomize an injured arm, growing a new one in its place, and some kinds will spontaneously separate their bodies into two parts, each thereupon regenerating the lost member.

Among most classes of the arthropods are found organisms that can regenerate limbs, and a few can even replace a considerable portion of the abdomen. By and large, however, the regenerative power of arthropods is not as great as that of the other invertebrate groups. It is interesting to note,

too, that widely distributed among arthropods are species that have special structures for spontaneously snapping off limbs (this is called **autotomy**) that may have become injured or seized by a predator. Such lost limbs are usually regenerated, but not all animals that can autotomize their limbs have the power to regenerate new ones.

16.7.3 REGENERATES USUALLY RETAIN THEIR POLARITY.

In general, when invertebrates regenerate, the axial organization of the new outgrowth corresponds to that of the original organism or part. For example, if you divide an animal such as a flatworm by a transverse cut or isolate a segment of its length, the anterior cut end regenerates anterior parts and the posterior cut end regenerates posterior ones (Figure 16.22). This result can be modified by experimental treatment, however. Thus, Flickinger (1963) showed that regeneration from the anterior cut surface of a middle segment of a flatworm often produces a tail instead of a head if protein synthesis at the anterior end is inhibited by cycloheximide.

375

FIGURE 16.22 Regeneration in a flatworm of the genus *Phagocata*. A: The unoperated animal. B: Anterior and posterior ends separated by a single cut. C: After 10 days the anterior end of the posterior section has regenerated a new head. The regenerating tail of the anteror section is less impressive.

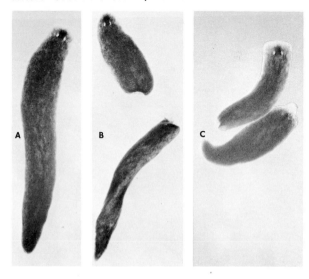

The control of polarity in regenerating coelenterates such as the hydra, has received much attention for many years. The hydra is a two-layered animal of tubular construction (Figure 16.23). At one end, called the distal end, is the **hypostome**, containing the mouth surrounded by a ring of tentacles bearing numerous cnidoblasts (stinging cells). The hydra attaches to the substratum at its proximal end, which consists of a mucus-secreting **foot,** or **basal disc.** Between the hypostome and foot, in distal–proximal order, are the **gastric region,** where digestion takes place, a **budding region** from which new individuals arise asexually, and a **pedicel,** which joins the rest of the body to the foot.

The hydra is rigidly polarized along the distal–proximal axis. When a hydra is divided by a cut through the gastric region, the basal piece forms a new hypostome and tentacles, and the distal piece forms a new base. Moreover, when the animal is sliced transversely into several annular segments, or rings, each segment forms a hypostome and tentacles at the distal end and a basal segment at the proximal end. Occasionally, however, a short seg-

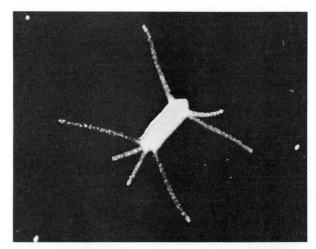

FIGURE 16.24 Bipolar regeneration in *Hydra pirardi.* Regenerate formed from a small central segment, or annulus, showing a hypostome and tentacles at each end. [The courtesy of Dr. Georgia Lesh, who supplied this photograph, is gratefully acknowledged.]

376

FIGURE 16.23 *Hydra pirardi.* The specimen on the right bears a bud. [The courtesy of Dr. Georgia Lesh, who supplied this photograph, is gratefully acknowledged.]

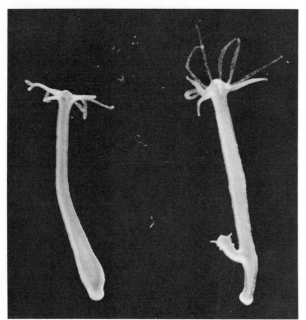

ment isolated just below the hypostome regenerates in a bipolar manner, forming a hypostome and a ring of tentacles at each end (Figure 16.24).

These experiments indicate that all regions of the hydra's body have the capacity to form both hypostome and foot. Why then do not mouths, with rings of tentacles, as well as basal discs, appear at all levels along the longitudinal axis of the animal? Most interpretations of the polarized organization of hydra are based on postulated gradients of activating and inhibiting organizing substances, originating in both the hypostome and foot, the concentration of which determines what will be the morphogenetic significance of cells at each position along the gradient. Both hypostome and foot clearly are sources of axial-organizing factors. Thus, fragments of hypostome grafted to the gastric region induce surrounding cells to join with them in organizing a new hypostome with mouth and tentacles. Fragments of basal disc similarly grafted bring about the formation of a new pedicel and foot (Figure 16.25). Grafts of mixed basal disc and hypostome cells to the gastric region induce no supernumerary parts or, at most, a small foot or a tentacle or two (Newman, 1974). This observation suggests that signals from opposite ends of the hydra tend to counteract each other.

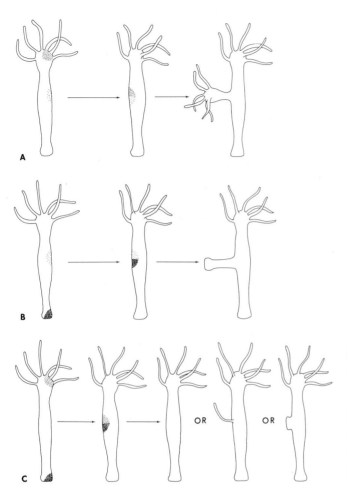

FIGURE 16.25 Axial organizing factors of the hypostome and basal disc in hydra. A: Cells from the hypostome, mixed with cells from the gastric region and grafted to the gastric region, induce the formation of hypostome and tentacles. B: Cells from the basal disc, similarly grafted, induce the formation of pedicel and foot. C: Cells of hypostome and foot, mixed and grafted to the gastric region, induce no outgrowth or only small distal or proximal structures. [Adapted with modifications from S. A. Newman, *J Embryol Exp Morphol* 31:541–55 (1974).]

Morphogenetic signaling in hydra is apparently the function of substances of relatively low molecular weight that are produced at opposite ends of the animal. In 1966 Lesh and Burnett reported that they had extracted from hypostomes a substance, probably a peptide, that would induce the formation of supernumerary hypostomes and tentacles. Annular segments of a hydra, placed in culture medium with this substance, give rise to bipolar regenerates and other bizarre forms, sometimes with 40 or more tentacles. Subsequently, other investigators isolated four distinct substances from hydra that affect the expression of regenerative polarity when added to hydra culture medium in micromolar quantities. The first and best known of these substances, **head activator** (Schaller, 1973), is

a peptide with a molecular weight of about 1000. This polypeptide is possibly the one described by Lesh and Burnett (1966). It accelerates the regeneration of the hypostome and tentacles and promotes budding. Its distribution in the body of the hydra parallels the distribution of nerve cells, which are most concentrated in the hypostome and are progressively more sparse proximally, except that they show a slight increase in the basal region. Nerve cells isolated by means of density gradient centrifugation from a suspension of dissociated hydra cells, contain almost all of the active substance.

Another substance, **head inhibitor,** was described by Berking (1977). It, too, has a molecular weight of 1000 or less, and sediments with particulate components of homogenized hydra cells at a centrifugal force of 10,000 $\times G$. Its distribution in the organism parallels that of head activator, but it exercises the opposite effect. It diminishes the rate of regeneration of the hypostome and tentacles and retards bud formation. Unlike head activator, it is not a polypeptide. It is a positively charged, highly hydrophilic molecule but its chemistry is not otherwise known at this time.

Both head activator and head inhibitor are released into the hydra culture medium when the hypostome and tentacles are removed. They are not released, however, on the same schedule. The inhibitor is released during the first hour after operation. Culture medium collected during this period antagonizes the effects of head activator. In contrast, head activator is released after a short delay and during a time span of several hours.

A **foot activator** (Grimmelikhuijzen and Schaller, 1977) and a **foot inhibitor** (Schmidt and Schaller, 1976) accelerate and retard, respectively, the rate of regeneration of the basal region of the hydra. The concentration of both substances falls off sharply distally to the base. The foot activator, like the head activator, is a peptide. The foot inhibitor is not a peptide, but has not been further characterized, except that it appears to have approximately the same molecular weight as the activator.

Complete biochemical characterization of the activators and inhibitors is hampered by the fact that it has not yet been possible to obtain fractions completely free from contamination. The procedure for isolating these factors from homogenates of hydra was described by Schaller et al. in 1979.

Presumably the regulated release of inhibitors and activators is involved in the stabilization of polarity in hydra and the polarization of its regeneration. The effects of the head and foot morphogenetic substances are specific to head and foot, respectively. Thus, for example, the effect of the head activator is not antagonized by the foot activator or inhibitor.

The fact that the activators and inhibitors are small molecules and that their distribution, particularly as demonstrated for the head factors, parallels the distribution of nerve cells suggests that the activating and inhibitory factors are neurohormones (cf. Chapter 5). If this projection is so, presumably elimination of nerve cells could affect the maintenance of polarity and polarized regeneration.

378

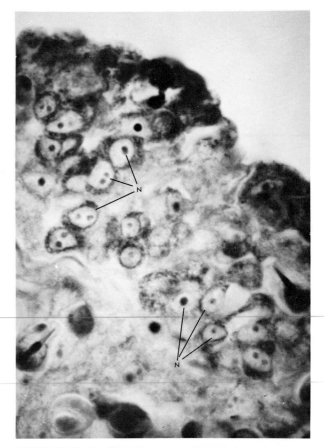

FIGURE 16.26 Section of a hydra showing clusters of interstitial cells, sometimes called neoblasts, N. [The courtesy of Dr. Georgia Lesh, who supplied this photograph, is gratefully acknowledged.]

Nerve-free animals can be obtained by exposing the hydra to low concentration of colchicine, which selectively kills the **interstitial** cells (Figure 16.26) (Campbell, 1976; Marcum and Campbell, 1978). Interstitial cells are the progenitor cells of the nerve cells, germ cells, gland cells of the endoderm, and nematocysts. The latter group constitutes a variety of cells, chiefly found in the tentacles, which are used for the capture and killing of prey. Once interstitial cells are depleted, they are not replaced. Nerve cells, gland cells, germ cells, and nematocysts are constantly lost and depend on interstitial cells for their replacement. Interstitial cell-free animals, therefore, soon consist only of two layers, an outer ectoderm of epitheliomuscular cells and an inner layer of gastric endoderm. They cannot feed, become smaller and smaller, and die. They may be maintained alive, however, if force-fed, but do not attach to the substratum. They lie on their sides and react only sluggishly to very strong mechanical stimuli.

If fed, however, nerve-free animals maintain their polarity, grow, and produce buds at a normal rate. Moreover, they regenerate in a normally polarized manner and at a normal rate. Also, central segments can reverse their polarity, just as central segments of innervated animals do, when they are separated from the hypostome and foot and replaced in reversed orientation. After 36 hours the hypostome and foot may be removed and the central segment will regenerate these structures according to the polarity of its reversed orientation rather than its original polarity (Marcum et al., 1977).

The question now arises as to whether the nerve-free animals contain the head and foot activators. Schaller et al. (1980) made homogenates of nerve-free animals, extracted and fractionated according to the procedure of Schaller et al. (1979) and assayed for activity of the appropriate fractions. Astonishingly, they found that the nerve-free animals showed a specific activity of head activator 8.3 times as great as that of the normal hydra. Specific activities of the head inhibitor and foot activator of the nerve-free animals were, respectively, 3.7 and 1.8 times greater in the noninnervated animals, but the foot inhibitor showed only 80% of the specific activity of that of the normal hydra.

These results indicate that nerve-free hydra contain the same morphogenetic substances as do normal animals. In nerve-free animals these substances obviously must be made by epithelial cells. Yet, in normal animals it is quite certain that essentially all of the head activator, at least, is in nerve cells. Do nerve cells in the normal animal inhibit the production of morphogenetic substances by epithelial cells? Do nerve cells in normal animals store morphogenetic substances produced by epithelial cells? To date these questions have not been resolved.

16.7.4 DO RESERVE CELLS HAVE A REGENERATIVE ROLE?
In many kinds of invertebrates, most notably among coelenterates, flatworms, and oligo-

chaete annelids, there are found distinctive cells, often somewhat teardrop shaped, with large clear nuclei and prominent nucleoli, and with cytoplasm rich in RNA. These cells sometimes occur in nests or clusters; sometimes they are scattered about in mesenchyme; in other cases they are found regularly spaced in segmental array. Such cells have been called by a variety of names, notably **neoblasts** and **interstitial cells.** Many researchers consider these to be reserve cells that are used for regeneration. As such they are presumably **totipotent;** accumulating at a wound site they could differentiate into any of the repertory of cell types needed to restore missing parts. Interstitial cells of hydra were once considered as possibly totipotent. As seen in the foregoing section, however, they are differentiated as stem cells for a limited array of cell types. Hay (1966) reviewed the presumed role of reserve cells in regeneration in a number of invertebrates. She noted that one might reasonably conceive that there is a common basis for the formation of regeneration blastema in various organisms that have the capacity for epimorphic regeneration.

For many years regeneration of missing parts in flatworms has been considered the responsibility of neoblasts. Distinctive RNA-rich cells appear to accumulate at the wound surface after an amputation and appear to form regeneration blastema, which proliferates and then forms the missing parts. After x-ray treatment regeneration fails to occur, but regenerative power may be restored to an irradiated animal by grafting to it a bit of neoblast-containing tissue from an untreated flatworm (Figure 16.27). Amputation at a site remote from the graft is followed by regeneration, usually after a period of delay, which is apparently required for the migration and proliferation of neoblasts from the graft into the depleted host animal (Wolff, 1961).

Hay and Coward (1975), however, concluded on the basis of careful electron microscopic studies that the "neoblasts" seen in flatworms by light microscopists are actually gland cells. They have a prominent juxtanuclear Golgi zone and an elaborate endoplasmic reticulum, as expected of gland cells. This study also revealed that in the parenchyma surrounding the various glandular, muscular, and digestive tissues there are cells, so small as to be poorly resolved by means of the light microscope, that have the characteristics of undifferenti-

379

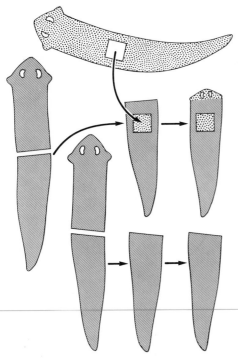

generation are still not solved, as a recent note of Baguña (1981) points out.

Extensive regenerative powers are found among tunicates. Depending on the species, a new and functional animal, or **zooid,** will regenerate from as little as an epidermal vesicle plus a bit of epithelium from the peribranchial or pericardial chamber. In forms that propagate by budding from stolons, epidermis plus a bit of septum that separates the outbound and inbound blood flow in the stolon can regenerate a complete animal. All regenerative fragments probably contain blood cells (of which there are several types in any tunicate). Freeman (1964) raised the possibility that the blood might contain cells that act as regenerative cells comparable to the putative neoblasts of other forms. He studied regeneration in *Perophora viridis,* an organism that reproduces asexually by budding zooids at intervals along a system of growing stolons (Figure 16.28). The zooids arise from epidermal cells of

380

FIGURE 16.27 Restoration of regenerative power to an x-irradiated flatworm. An irradiated fragment that receives a graft from an unirradiated animal can regenerate, whereas, without the graft, it cannot. [An interpretation of results reviewed by E. Wolff, in D. Rudnick, ed., *Regeneration,* Ronald, New York, 1962, pp. 53–84.

FIGURE 16.28 *Perophera viridis.* This is a colonial tunicate that produces individual zooids at intervals along branching stolons.

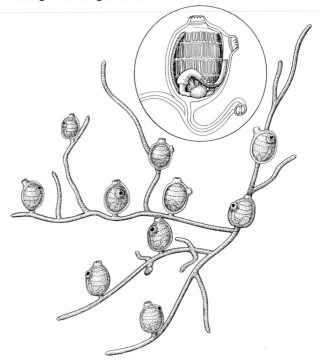

ated cells. Notably, they have a nucleus with small chromatic clumps and no nucleolus and a cytoplasm laden with ribosomes but lacking endoplasmic reticulum. They designated these as **beta cells.** Transitional stages between beta cells and the various differentiated cellular types were found. They suggested that beta cells constitute progenitor cells for various tissue types with which they are associated, much in the same sense that cells of the germinal layer of the skin are progenitor cells for the keratinizing cells or cells of the intestinal crypts are the progenitors of mucosal cells. As such, the beta cells would not necessarily be totipotent. A number of studies of planarian regeneration have been reported since the publication of the paper by Hay and Coward, but the mechanisms of planarian re-

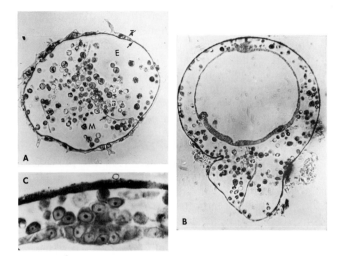

FIGURE 16.29 Origin of a zooid from the stolon in *Perophora viridis*. A: Transverse section of the stolon showing tunic, T, epithelial cell layer, E, and septal mesenchyme, SM. B: Transverse section through stolon (lower) and endoplasmic vesicle (upper); the latter forms the new zooid. Lymphocytes, L, have aggregated on the dorsal side of the endoplasmic vesicle; from these will arise the neural complex and the genital cord. C: Detail from B, showing the aggregation of lymphocytes. [From G. N. Freeman, *J Exp Zool* 156:157–184 (1964) by courtesy of Dr. Freeman and permission of Alan R. Liss, Inc.]

the stolon plus the mesenchymal cells of the septum and its associated blood cells (Figure 16.29).

Freeman isolated on glass slides individual zooids of *Perophora* with attached segments of stolon. When he left such a preparation undisturbed, it showed stolon growth and the formation of new zooids. But when he irradiated the isolate with 5000 roentgens of x-rays, the lymphocytes (one of the kinds of blood cells) disappeared within 2 days and the specimen died without further budding (Figure 16.30). He then injected into x-rayed preparations either the whole blood or various fractions of it from nonirradiated stolons. If either whole blood or the lymphocyte fraction alone was

381

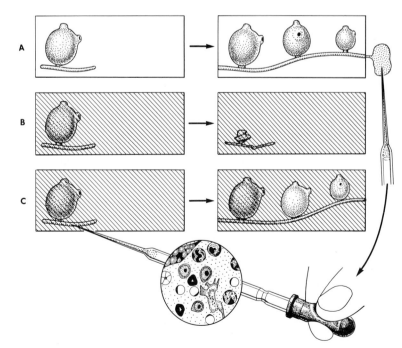

FIGURE 16.30 Effect of blood cells on colony formation by an x-rayed preparation of *Perophora viridis*. A: An untreated isolate of stolon and zooid grows and produces new zooids. B: After x irradiation a similar preparation regresses and dies. C: An irradiated preparation that receives an injection of blood cells from a nonirradiated colony grows and produces new zooids. [An interpretation of data from G. N. Freeman, *J Exp Zool* 156:157–84 (1964).]

used, budding of new zooids occurred with high frequency. The larger the number of lymphocytes injected, the more rapidly did the first bud appear, suggesting that the lymphocytes probably contribute stem cells to the construction of the new zooid.

This survey has shown that regenerative powers are quite widespread among animals of various kinds, degrees of complexity, and habitats, and that mechanisms for regeneration likewise differ from one kind of animal to another. Regenerative growth and morphogenesis clearly are fundamental to the study of the developmental biology of animals and provide numerous challenges to an understanding of the control of developmental processes.

16.8 Regeneration in Plants

Among plants generally there is a great tendency for any part of the organism to restore all parts that are missing or physiologically isolated. As with many of the lower animals, regeneration in plants often serves as a form of asexual propagation, and this property is profitably exploited in agriculture and horticulture. Because their cells are encased in rigid walls, plant tissues undergo relatively little reorganization in the regenerative process. The replacement of parts, therefore, requires that there be present meristematic or potentially meristematic tissue at cut or injured surfaces. Thus the most familiar examples of regeneration are found among the dicots, whose cambium proliferates in response to trauma. Most monocots lack a cambium and this limits their possibilities for regeneration. The following discussion is restricted to regeneration in dicots.

Among various species of plants one finds the regeneration of an entire plant from almost any part of the whole, even as little as a single cell. As any gardener knows, most plants can be propagated by "rooting" stems in moist soil, a process often assisted by the application of an auxin or auxin-like substance (cf. Chapter 12) to the cut surface. Young stems are rooted more easily, as a rule, than older ones, and their new roots are likely to arise from

the pericycle. In older stems the vascular cambium is usually the source of the new roots. Pieces of root can regenerate new stems, too, but this power is more limited. In nature, however, one often finds stems arising from roots that are exposed or that grow close to the surface of the soil. Feldman and Torrey (1976) excised the quiescent center in a variety of *Zea mays* and cultured it on an agar substratum made up in a medium containing zeatin but lacking auxin. Under these conditions the quiescent center regenerated a complete root the polarity of which coincided with that of the center. No shoot appeared, however.

Particularly among succulent plants, but not exclusive to them, are cases in which a single leaf can form roots and stem when the petiole is placed in soil. This property is well known to many *Begonia* fanciers. The petiole undergoes considerable modification in this process. It first becomes anatomically and functionally a stem and then gives rise to roots at the lower end.

16.8.1 REGENERATIVE REPRODUCTION. In many plants regenerative processes have evolved as normal methods of asexual reproduction. Many readers are familiar with the looping canes of blackberry plants, whose drooping stems give rise to new roots and shoots when their apices touch the ground. Some other dicots have a similar method of propagation, "leap-frogging" over considerable areas of ground in a single growing season.

Regenerative reproduction is perhaps most dramatically illustrated in species in which leaves possess or may be induced to develop "foliar embryos." One of the best known of such plants is *Kalanchoe pinnatum*, in which miniature plantlets form in marginal notches of the fleshy leaves. These plantlets, which consist of recognizable leaf and root primordia embedded in the tissue of the leaf, develop as independent plants when the leaf falls to the ground. In other members of the genus *Kalanchoe*, the plantlets may achieve considerable size before the parent leaf matures. In *K. rotundifolia* the leaf bears a meristematic region in the petiole. This meristem forms a shoot bud, but roots do not appear until after the leaf has fallen.

One of the most interesting cases is provided by *Crassula multicava*, the leaf of which forms numerous complete plants, each derived from a single epidermal cell. Nature has anticipated what has only lately been demonstrated in the laboratory, namely, that from a single differentiated plant cell an entire organism can develop.

16.8.2 REGENERATION FROM CALLUS CELLS IN THE LABORATORY.

In nature, regeneration after wounding or cutting often issues first in the formation of a relatively homogeneous mass of parenchymatous cells called **callus.** Later, within the mass of callus, vascular tissues and root or shoot primordia will appear, as appropriate, presumably induced by hormones.

In the laboratory, plant tissues cultured on sterile agar media or in suspension on liquid media produce masses of undifferentiated cells that are also called callus. Tissues of a large number of dicots and representatives of all the major monocot families are now successfully under culture. These results have been made possible by the development of new culture media with high concentrations of mineral salts, especially nitrates, and by the use of 2,4-dinitrophenoxyacetic acid, a synthetic auxin that stimulates proliferation of the cultured tissues and makes callus available in quantity.

A major research goal that has been achieved in the past approximately 20 years has been the regeneration of entire plants from cells in culture. The first notable success in this regard issued from the successful growth of tobacco pith (normally nondividing) as callus culture in the presence of auxin, and the demonstration by Skoog and Miller (1957) that proper ratios of auxin and cytokinin in the medium would result in the differentiation of roots or shoots or both from the callus tissue, as desired. Regenerated plants were cultured to reproductive maturity. The differentiation of plant organs from callus may be referred to as **regeneration by organogenesis** (Figure 16.31).

Regeneration can also occur from callus cells by a process of **embryogenesis.** Shortly after the successful regeneration of tobacco plants was accomplished, F. C. Stewart and his associates at Cornell grew callus cells derived from secondary phloem of wild carrot in rotating cultures in liquid media containing coconut milk. Free cells frequently detached from the rotating callus and divided to form structures remarkably like carrot embryos. Such "embryoids," which could be obtained by the hundreds of thousands, could be continued in culture, transplanted to semisolid media, and eventually to soil, where they formed adult flowering plants (Stewart et al., 1958).

At this time several dozen plant species have been regenerated from callus cells by organogenesis or by embryogenesis. Many ornamental plants are now being propagated commercially by these tech-

383

FIGURE 16.31 Regeneration by organogenesis. A: Long-term callus culture of *Lilium* induced to form roots and shoots by hormone treatment. B: Mature lily plant derived from organ primordia induced in cultured callus. [From W. F. Sheridan in L. Ledoux, ed., *Genetic Manipulations with Plant Material,* 1975, by courtesy of Dr. Sheridan and permission of Plenum Publishing Corporation New York.]

niques, and it is to be expected that many food plants will be propagated initially in this manner also. It was earlier pointed out (cf. Chapter 6) that the vegetative cell of the pollen grain can form a complete haploid plant. Very importantly, a number of important food plants such as barley, rye, rice, and corn have yielded haploid plants from anther cultures. The creation of haploid plants enables the rapid stabilization of new genotypic strains. Haploid plants can be bred if they spontaneously become diploid or if the floral parts of mature haploids are caused to be diploid by treatment with colchicine, which inhibits cytokinesis by disrupting the mitotic spindle.

16.9 Control of Reintegrative Growth

384

When an organ or part that is capable of undergoing complete regeneration does so, it grows until it achieves its proper proportionate size relative to the rest of the organism. This process also occurs in the normal growth of body parts, each of which grows to a size that stands in some proportion to other parts. These proportions change during development, for growing parts usually have different growth rates relative to the overall growth of the body. But eventually, growth stops, and in adult life organs will have achieved a stable size and cellular population.

The complex of factors that causes an organ system, organ, or part to cease growth at a certain size relative to other body parts is little understood. Hormones certainly play a role in normal development of plants and animals, but factors intrinsic to the growing part have a determining role, also. For example, an isolated tail vertebra of an 8-day-old mouse, transplanted beneath the renal capsule of an adult mouse, grows as rapidly and to the same size as does a control vertebra that remains undisturbed on the tail of a normal mouse (Figure 16.32).

The control of organ size in animals has been extensively investigated in liver, which has great regenerative power, some paired organs such as the kidney, which show compensatory hypertrophy

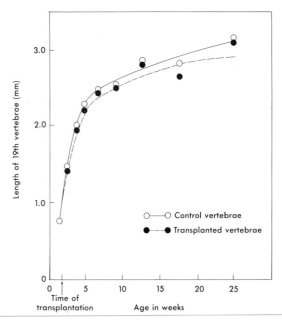

FIGURE 16.32 A comparison of the mean length at various ages of 19th caudal vertebrae of young mice transplanted beneath the capsule of the kidney of older hosts, and control 19th vertebrae growing normally in the tail. [Adapted with modifications from J. F. Noel, *J Embryol Exp Morphol* 29:53–64 (1973).]

when one member of the pair is lost, and epidermis. From such investigations has emerged the concept that tissues contain and secrete substances that specifically inhibit DNA synthesis in their own cells but not in those of other tissues. These substances are referred to as **chalones** (pronounced *kay-lones*).

16.9.1 IS THERE A LIVER CHALONE? In the liver of the young adult rat very few cells are in mitosis and they are found around the central vein in each of the polygonal lobules of which the liver is composed. If one removes 70% of the liver, cells at the periphery of each lobule begin DNA synthesis and mitosis, and these activities then spread as a wave toward the center of the lobule. The peak of DNA synthesis is achieved after 24 hours. The liver will have recovered its normal mass 6 or 7 days later, and at least 95% of its cells will have undergone mitosis.

Conceivably, the release of mitotic block upon operation could be caused by the removal of an inhibitor to mitosis, or to the production or release of a mitotic stimulant, or to both. When rats are joined side by side with crossed circulations and one member of the pair undergoes partial hepatectomy, the partner with the intact liver shows, after 48 hours, a great increase in the number of mitoses in its own liver, as judged by comparison with the number of mitoses in samples of the liver removed from its parabiont at the time of operation. Also, in tissue cultures of rabbit liver, made in media containing blood plasma from partially hepatectomized rabbits, there is more rapid outgrowth of connective tissue and epithelial elements than is found in similar cultures made with plasma from a nonhepatectomized animal. Thus there would seem to be present in the blood factors that are involved in liver regeneration. Does normal serum have an inhibitor of liver growth or does it lack a growth stimulating factor?

Glinos (1958) withdrew upwards of 30% of the blood from intact rats every 12 hours for 36–96 hours, replacing the blood cells after suspending them in saline after each withdrawal. This process, termed **plasmapheresis,** effectively diminishes the normal components of plasma and presumably would remove any inhibitors or stimulators of liver mitosis. If the former were removed, the mitotic rate of the liver should increase as compared to controls. This result proved to be the case. Subsequently, Verly (1973) isolated from intact rabbit liver a small polypeptide that inhibits DNA synthesis in slices of liver removed from an animal 24 hours after hepatectomy. The chalone is not species specific, but it affects DNA synthesis only in liver.

A hypothetical mechanism for control of liver regeneration cannot safely be based only on deficiency of a chalone, however. Hepatectomized animals show an increased concentration of fatty acids in the blood. Infusion of an appropriate fatty acid mixture into the blood of an intact animal leads to as intense an increase in the rate of DNA synthesis in intact liver as occurs following partial hepatectomy (Short et al., 1972).

16.9.2 EPIDERMAL MITOTIC INHIBITOR. Skin cells from the outer layers of the epidermis are constantly being lost and are replaced by products of mitosis in the lowermost layer of the skin, the stratum germinativum (Figure 11.35). Mitosis goes on in this layer at a slow rate and in a diurnal rhythm. The mitotic rate steps up greatly in response to a wound, providing a population of epidermal cells

that covers the wound surface. In past years the enhanced mitotic rate adjacent to a wound was explained as reaction to a stimulating "wound hormone," but no such hormone has been demonstrated. In 1960 Bullough and Lawrence pointed out that epidermis only reacts to wounding when it, itself, is injured; trauma applied to dermis and other tissues subjacent to the epidermis does not enhance epidermal mitosis. Moreover, they noted that the region of mouse skin epidermis stimulated to enhanced mitosis by wounding was confined to a narrow band about 1 mm wide in a gradient with highest activity near the cut edge (Figure 16.33). Enlarging the wound size did not cause a wider band of mitosis, which might be expected if wounding produced mitotic stimulating substances. They proposed instead, therefore, that wounding the epidermis removes mitotic inhibitors produced by the epidermis, and that the gradient in mitotic activity adjacent to a wound may be inversely proportional

FIGURE 16.33 Relative mitotic activity in mouse epidermis as a function of distance from the edge of an incision 1 cm in length through the middorsal body skin. Biopsy material was taken 12, 18, and 24 hr after wounding. [Adapted with modifications from W. S. Bullough and E. B. Laurence, *Proc Roy Soc Lond* [B] 151:517–36 (1960).]

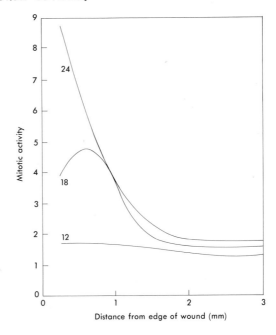

385

to a gradient in mitotic inhibitor, which ceases to be produced within damaged cells and which may also diffuse away from the wound.

In subsequent years it has been possible to extract from epidermis of several mammalian species substances that act as mitotic inhibitors. They are not species specific but are specific for epidermis. Rather astonishingly, too, they do not inhibit the mitotic activity of sebaceous glands, which are derived from epidermis. In 1976 Thornley and Laurence reported the inhibitory effect of two fractions from pig skin on mitosis in epithelium of the mouse ear. One fraction, the **G_1-inhibitor,** prevents the onset of DNA synthesis; the other, the **G_2-inhibitor,** prevents cytokinesis in cells that have already synthesized DNA in preparation for division. The G_1 and G_2 inhibitors are possibly glycoproteins having molecular weights of 2×10^4 and $3-4 \times 10^4$, respectively.

16.9.3 GROWTH IN RESPONSE TO FUNCTIONAL DEMAND.

When the availability of atmospheric oxygen is reduced, as it is for mammals residing at high altitudes, for example, the bone marrow responds by accelerating the production and release of red blood cells into the circulatory system. This increase enhances the oxygen transport capacity of the blood, compensating for the diminished oxygen tension of the inspired air. Conversely, when oxygen tension is increased, the rate of red cell production is decreased. In a growing rat fed on a higher than normal amount of protein, the liver increases in size as compared to that of normally fed controls. Restriction of dietary protein is followed by liver atrophy. When one kidney is removed, the remaining kidney increases the number of geomeruli in use at any one time and increases the diameter and length of the tubules in each nephron.

What the various modes of compensatory and regenerative growth have in common is that they tend to adjust to appropriate levels the functions that various stressed or depleted organs carry out. This and the above considerations lead to what is referred to as the "functional demand theory" of growth regulation. In essence this thesis states that changes in the mass of an organ occur relative to the physiological need for its products.

The identification of the component of function that is responsible for regenerative or other compensatory growth is often difficult. Which of the various functions of liver is the one through which its growth is controlled? Liver is the site of synthesis of many plasma proteins including α- and β-globulins and fibrinogen. It produces heparin, prothrombin, and hypertensinogen. It is the source of bile. The liver is essentially, however, a digestive organ in the sense that all blood from the digestive tract enters it through the hepatic portal vein. It also has an arterial blood supply, and both sources of blood empty ultimately into the vena cava. If the portal supply to some liver lobes, but not the arterial supply, is cut off, those lobes gradually atrophy, whereas the undisturbed lobes undergo hypertrophy. Is it loss of digestive function that provides the stimulus to liver regeneration? If so, what, if any, is the relationship of liver chalones to this physiological loss?

What is the physiological component of kidney function that affects kidney size? It filters blood, selectively adjusts concentrations of glucose and electrolytes, and eliminates urea. It has an overall role in maintaining water balance in the various fluid compartments of the body and therefore is of great importance in regulating blood pressure. Possibly it is the latter that is most important. Reduction of blood pressure decreases compensatory growth of the kidney remaining after unilateral nephrectomy—but changing blood pressure also affects other aspects of kidney function, so that it is difficult to assign responsibility for size regulation to a single physiological process.

The regulation of production of red blood cells is somewhat better understood. The blood serum of a previously bled, and thus anemic, rabbit injected into a normal rabbit causes rapid **erythropoiesis** (Gr., *erythros,* red; Gr., *poiesis,* creation; i.e., red-cell production) in the bone marrow of the recipient. The bloodborne substance appears under conditions of oxygen deprivation as well as anemia. It is erythropoietin, a sialic acid-containing protein, that has a molecular weight of 5000–60,000.

The principal source of erythropoietin is the kidney. It is released from kidneys held at low oxygen tensions or perfused with cobalt ions. The latter have long been known to mimic effects of hypoxia in the induction of polycythemia (the condition of elevated production and release of cells in the erythrocytic series). The kidney is not the only source of erythropoietin, however, for human patients who have suffered bilateral nephrectomy (removal of both kidneys) have some ability to regulate erythropoiesis in response to anemia.

The effect of erythropoietin is diminished in animals with a functionless liver. It has been suggested that an α-globulin originating in the liver is needed to stabilize erythropoietin.

Erythropoietin is finding use in medical practice. It is obtained commercially from the blood plasma of anemic sheep and is secreted to some extent in the urine of anemic patients. Erythropoietin stimulates globin synthesis in bone marrow. This effect is blocked by actinomycin D, indicating that transcription of the genome is involved. Whether altered transcription is a direct or indirect effect of erythropoietin is not clear, however.

16.10   Summary

Growth-related activities that act to restore functional integrity after postnatal disturbance are considered under the heading of reintegration. Prominent among reintegrative activities are various forms of regeneration, the replacement of parts that are lost. The ability to restore a whole organism from a part, sometimes only a very small part, is widespread among plants and animals.

Amphibians, notably the urodeles, have the ability to regenerate complete limbs. Epidermis heals over the cut stump; beneath this epidermis blastema cells arise by the dedifferentiation and proliferation of cells at the cut surface. All types of mesodermal cells of the stump contribute to the regeneration blastema, but it is not clear to what extent morphologically dedifferentiated cells form cells other than those of their type of origin during regeneration. X-irradiated limbs do not form a blastema and fail to regenerate. If, however, at the amputation site the limb is provided with unirradiated limb skin, limb muscle, or a segment of brachial nerve, a blastema will form and all tissue types of a normal limb will regenerate. Unless it is from a very young larva, unirradiated cartilage, will not give rise to muscle cells in this experimental situation, however.

The wound epithelium, or apical cap, is essential for the formation of the regeneration blastema. If skin dermis intervenes between epithelium and the amputation plane, a blastema fails to form and regeneration does not take place. In nonregenerating appendages (e.g., adult anuran, rat), dermis normally covers the wound surface quickly and regeneration fails.

The urodele limb does not regenerate if it has been previously denervated and maintained free of regenerating nerves. The source of the nerve supply is not critical, but the amount of nerve tissue at the cut surface is. After amputation denervated limbs initiate DNA synthesis, but few of the cells enter mitosis, in contrast to the controls wherein the blastemata show numerous mitotic figures. The neural contribution to normal regenerates is not known, but a factor isolated from brain enhances the incorporation of amino acids into protein in freshly amputated denervated limbs.

Urodele limbs prevented from receiving a nerve supply during embryogenesis are called aneurogenic. Such limbs will regenerate when amputated. When an aneurogenic limb is supplied with a cuff of neurogenic limb skin, and then amputated at the distal end of the cuff, the wound epithelium fails to promote blastema formation and regeneration fails to occur. The requirement for nerve in amphibian limb regeneration is, therefore, apparently exercised in the skin.

When a limb regenerates, the blastema gives rise to those parts that normally lie distal to the cut surface, regardless of the level of amputation. This is called the rule of distal transformation. Also, the anteroposterior and proximodistal axes of the regenerated parts correspond to those of the stump. Therefore, there must be present at the amputation site both level-specific and circumferential patterns of positional information that determine the three-dimensional polarity of the regenerated parts.

Higher vertebrates show only limited powers to regenerate appendages. Lizards regenerate the tail, provided that the ependymal layer of the cut nerve cord is present at the amputation site. Lizard limbs with an augmented nerve supply regenerate spike-like outgrowths. The hindlimbs of the newborn opossum, too, show partial regeneration in the presence of additional nervous tissue. Higher mammals can regenerate at least terminal parts of a phalanx if a wound epithelium without dermis can be maintained.

The lens regenerates from the dorsal margin of the iris in urodeles. The iris cells lose their pigmen-

387

tation and begin DNA and RNA synthesis, eventually detaching as a lens. Lens will not regenerate in close proximity to another lens or in the absence of factors supplied by the retina. In *Xenopus* tadpoles lens regenerates from the cornea rather than from the iris.

Among invertebrates regeneration of the whole animal from fragments is commonly seen as a means of asexual reproduction, and it occurs readily under experimental conditions. Regenerates usually retain their original polarity. For many invertebrates regeneration is considered to take place at the expense of reserve cells, or neoblasts, although this conclusion has been seriously questioned.

In plants, as in many invertebrates, regenerative reproduction is quite common, notably among plants whose succulent leaves bear foliar embryos in various stages of development and which develop when the leaf falls. As little as a single epidermal cell from a leaf can form a complete plant in some cases.

In the laboratory entire reproductive plants can be reared from callus cultures induced to undergo organogenesis by appropriate hormone treatment. Also, haploid plants derived from anther cultures have been created in a number of species of cereal grains. These plants promise to be of great importance in the rapid development of crop plants with desirable agricultural and nutritive properties.

For regenerating organs in mammals inhibitory substances called chalones are thought to limit growth. Liver and epidermal chalones, tissue specific but not species specific, inhibit DNA synthesis and mitosis in their respective regenerating and healing tissues. These chalones have been partially purified and characterized as to molecular weight.

When the functional mass of a tissue is diminished with respect to the physiological output required of it, it tends to undergo a compensatory increase in size and output. Regeneration can be considered as a response to functional demand as can the compensatory growth and physiological response of one kidney after removal of the other. The factors of physiological demand that elicit compensatory, or adaptive, growth are not well understood in most cases. The output of red blood cells by bone marrow is apparently regulated by erythropoietin, produced chiefly in the liver in response to anemia and to lowered oxygen tensions.

16.11 Questions for Thought and Review

1. What are the major roles of the apical epithelial cap in regeneration of the urodele limb? What experiments demonstrate that the cap is required for these roles?

2. Cells in the regeneration blastema in the urodele limb are not distinguishable as to their tissue of origin, resembling embryonic cells rather than those of differentiated tissues. What is their origin in normally regenerating limbs? Do all cells have the same potentiality for regenerating limb parts? On what experimental evidence do you base your answer? Can you completely eliminate the possibility that regenerated limb parts arise only from hitherto undifferentiated reserve cells?

3. By what means does x-irradiation inhibit regeneration of an amphibian limb?

4. The question has often been raised as to whether nonlimb cells can support the regeneration of irradiated limbs. How do you answer this question and on the basis of what evidence?

5. Does the apical epithelial cap require an input from nerve to exercise its normal effect on limb regeneration in amphibians? Describe the appropriate experimental evidence.

6. Conceivably, the apical epithelial cap could contribute cells to the regeneration blastema. Does it do so? Describe the appropriate experimental evidence.

7. Compare and contrast the patterns of DNA, RNA, and protein synthesis in amputated control limbs and in amputated denervated limbs. Consider cases denervated before and after amputation.

8. A tremendous mass of evidence attests to the fact that normally reared urodele larvae will not regenerate amputated limbs that lack a nerve supply. How do you reconcile this with the fact that larval limbs that have never experienced a nerve supply will regenerate?

9. There are several aspects of the loss and regeneration of the lizard tail that are different from these processes in urodeles. Describe these differences and cite experimental data as appropriate.

10. Describe the sequence of morphological changes in the regeneration of the lens in the eye of the newt. Correlate these changes with changing patterns of macromolecular synthesis.

11. What are the principal factors that determine whether a lens can be regenerated in a newt eye? Describe the experimental evidence that supports your answer.

12. What kind of evidence suggests that neoblasts serve to form the regeneration blastema in flatworms?

What evidence is there to the contrary? Can you devise a critical experiment to test the hypothesis that there are reserve cells that serve for regeneration in flatworms?

13. Review Freeman's experiments as reported in the text. All of his controls were not indicated in this text. What controls are required to support the proposition that unirradiated lymphocytes alone restore regenerative power to an irradiated stolon of *Perophora?*

14. The sea anemone *Metridium* often sheds fragments of its foot as it glides across a rock. These fragments may reorganize new, miniature anemones which then feed and grow. What name is given to this kind of regeneration?

15. What is the evidence that nerve cells are involved in the polarized regenerative ability of hydras? How can one account for the regeneration of nerve-depleted hydras?

16. Compare and contrast regenerative reproduction, regeneration by organogenesis, and regeneration by embryogenesis in plants.

17. Define the term *chalone.* What is compensatory growth? How is it different from regenerative growth? What is meant by functional adaptation in reference to growth? How are the concepts of functional demand and of chalones interrelated?

16.12 Suggestions for Further Readings

ARGYRIS, T. S. 1972. Chalones and the control of normal, regenerative, and neoplastic growth of skin. *Am Zool* 12:137–49.

BAGUÑÀ, J. 1976. Mitosis in the intact and regenerating planarian *Dugesia mediterranea* n. sp. I. Mitotic studies during growth, feeding and starvation. *J Exp Zool* 195:53–64.

BAGUÑÀ, J. 1976. Mitosis in the intact and regenerating planarian *Dugesia mediterranea* n. sp. II. Mitotic studies during regeneration and a possible mechanism of blastema formation. *J Exp Zool* 195:65–80.

BERRILL, N. J. 1976. Regeneration and budding in worms. *Biol Rev* 27:401–38.

BULLOUGH, W. S. 1973. The chalones: a review. *Natl Cancer Inst Monogr* 38:5–15.

BULLOUGH, W. S. 1975. Chalone control mechanisms. *Life Sciences* 16:323–30.

CARLSON, B. M. 1978. Types of morphogenetic phenomena in vertebrate regenerating systems. *Am Zool* 18:869–82.

CARLSON, P. S., H. H. SMITH, and R. D. DEARING. 1972. Parasexual interspecific plant hybridization. *Proc Natl Acad Sci USA* 69:2292–94.

DEUCHAR, E. M. 1976. Regeneration of amputated limb-buds in early rat embryos. *J Embryol Exp Morphol* 35:345–54.

JOHANSSON, L., and T. ERIKSSON. 1977. Induced embryo formation in anther cultures of several *Anemone* species. *Physiol Plant* 40:172–74.

KIORTSIS, V., and H. A. L. TRAMPUSCH. 1965. *Regeneration in Animals and Related Problems.* Amsterdam: North-Holland.

LENDER, T. 1962. Factors in morphogenesis of regenerating fresh-water planaria. *Advan Morphogen* 2:305–31.

LESH-LAURIE, G. E. 1973. Expression and maintenance of organismic polarity. In A. L. Burnett, ed. *Biology of Hydra.* New York: Academic Press, pp. 143–66.

NOME, O. 1975. Tissue specificity of the epidermal chalones. *Virchows Archiv* [*Cell Pathol*] 19:1–25.

ROSE, S. M. 1964. Regeneration. In J. A. Moore, ed. *Physiology of the Amphibia.* New York: Academic Press, pp. 545–622.

RUDNICK, D., ed. 1961. *Regeneration.* New York: Ronald.

SANGWAN, R. S., and B. NORREEL. 1975. Induction of plants from pollen grains of *Petunia* cultured in vitro. *Nature* 257:222–24.

SCADDING, S. R. 1977. Phylogenic distribution of limb regeneration capacity in adult *Amphibia. J Exp Zool* 202:57–68.

STOCUM, David L. 1979. Stages of forelimb regeneration in *Ambystoma maculatum. J Exp Zool* 209:395–416.

TASSAVA, R. A., and W. D. MCCULLOUGH. 1978. Neural control of cell cycle events in regenerating salamander limbs. *Am Zool* 18:843–54.

THORNTON, C. D., ed. 1959. *Regeneration in Vertebrates.* Chicago: University of Chicago Press.

THORNTON, C. S. 1970. Amphibian limb regeneration and its relation to nerves. *Am Zool* 10:113–18.

THORNTON, C. S., and S. C. BROMLEY, eds. 1973. *Vertebrate Regeneration.* Stroudsburg, Penn.: Dowden Hutchinson and Ross.

WEBSTER, P. L., and H. D. LANGENAUER. 1973. Experimental control of the activity of the quiescent centre in excised root tips of *Zea mays. Planta* 112:91–100.

389

16.3 References

ANTON, H. J. 1955. Die Regeneration heteroplasticher keimblattchimärischer Extermitäten bei *Triturus vulgaris* und *Triturus alpestris. Z Naturforsch* 10B:723–25.

ARAKI, M., M. YANAGIDA, and T. S. OKADA. 1979. Crystallin

synthesis in lens differentiation in cultures of neural retinal cells of chick embryos. *Dev Biol* 69:170–81.

BAGUÑA, J. 1981. Planarian neoblasts. *Nature* 290:14–15.

BANTLE, J. A., and R. A. TASSAVA. 1974. The neurotrophic influence on RNA precursor incorporation into polyribosomes of regenerating adult newt forelimbs. *J Exp Zool* 189:101–13.

BAST, R. E., M. SINGER, and J. ILAN. 1979. Nerve-dependent changes in content of ribosomes, polysomes, and nascent peptides in newt limb regenerates. *Dev Biol* 70:13–26.

BERKING, S. 1977. Bud formation in *Hydra:* inhibition by an endogenous morphogen. *Wilhelm Roux' Arch* 181:215–25.

BORGENS, R. B., J. W. VANABLE, Jr., and L. F. JAFFE. 1977a. Bioelectricity and regeneration. I. Initiation of frog limb regeneration by minute currents. *J Exp Zool* 200:403–16.

BORGENS, R. B., J. W. VANABLE, Jr., and L. F. JAFFE. 1977b. Bioelectricity and regeneration: large currents leave the stumps of regenerating newt limbs. *Proc Natl Acad Sci USA* 74:4528–32.

BORGENS, R. B., J. W. VANABLE, Jr., and L. F. JAFFE. 1979. Reduction of sodium dependent stump currents disturbs urodele limb regeneration. *J Exp Zool* 209:377–86.

BRYANT, S. V., D. FYFE, and M. SINGER. 1971. The effects of denervation on the ultrastructure of young limb regenerates in the newt, *Triturus*. *Dev Biol* 24:577–95.

BRYANT, S. V., and K. J. WOZNY. 1974. Stimulation of limb regeneration in the lizard *Xantusia vigilis* by means of ependymal implants. *J Exp Zool* 189:339–52.

BULLOUGH, W. S., and E. LAURENCE. 1960. The control of epidermal mitotic activity in the mouse. *Proc R Soc Lond* [B]151:517–36.

BUTLER, E. G. 1935. Studies on limb regeneration in x-rayed *Amblystoma* larvae. *Anat Rec* 295–307.

BUTLER, E. G., and J. P. O'BRIEN. 1942. Effects of localized x-radiation on regeneration of the urodele limb. *Anat Rec* 84:407–13.

CAMPBELL, R. F. 1976. Elimination of *Hydra* interstitial and nerve cells by means of colchicine. *J Cell Sci* 21:1–13.

CARLSON, B. M. 1972. Muscle morphogenesis in axolotl limb regenerates after removal of stump musculature. *Dev Biol* 28:487–97.

CHAPRON, C. 1974. Mise en évidence du rôle, dans la régénération des amphibiens, d'une glycoprotéine sécrétée par la cape apicale: étude cytochimique et autoradiographique en microscopie électronique. *J Embryol Exp Morphol* 32:133–45.

CHOO, A. F., D. M. LOGAN, and M. P. RATHBONE. 1978. Nerve trophic effects: an in vitro assay for factors involved in regulation of protein synthesis in regenerating amphibian limbs. *J Exp Zool* 206:347–354.

CLAYTON, R. M. 1970. Problems of differentiation in the vertebrate lens. *Cur Top Dev Biol* 5:115–80.

CLAYTON, R. M., I. THOMSON, and D. I. DE POMERAI. 1979. Relationship between crystallin mRNA expression in retina cells and their capacity to re-differentiate into lens cells. *Nature* 282:628–29.

CONNELLY, T. G., J. R. OTIZ, and T. YAMADA. 1973. Influence of the pituitary on Wolffian lens regeneration. *Dev Biol* 31:301–15.

COX, P. G. 1969a. Some aspects of tail regeneration in the lizard, *Anolis carolinensis*. I. A description based on histology and autoradiography. *J Exp Zool* 171:127–50.

COX, P. G. 1969b. Some aspects of tail regeneration in the lizard, *Anolis carolinensis*. II. The role of peripheral nerves. *J Exp Zool* 171:151–60.

DEARLOVE, G. E., and D. L. STOCUM. 1974. Denervation-induced changes in soluble protein content during forelimb regeneration in the adult newt, *Notophthalmus viridescens*. *J Exp Zool* 190:317–28.

DESHA, D. L. 1974. Irradiated cells and blastema formation in the adult newt, *Notophthalmus viridescens*. *J Embryol Exp Morphol* 32:405–16.

DOUGLAS, B. S. 1972. Conservative management of guillotine amputation of the finger in children. *Aust Paediatr J* 8:86–89.

DRESDEN, M. H. 1973. Denervation effects on newt limb regeneration: autoradiography with ³H-thymidine. *Cell Differ* 2:255–59.

DUNIS, D. A., and M. NAMENWIRTH. 1977. The role of grafted skin in the regeneration of x-irradiated axolotl limbs. *Dev Biol* 56:97–109.

EGAR, M., C. L. YNTEMA, and M. SINGER. 1973. The nerve fiber content of *Ambystoma* aneurogenic limbs. *J Exp Zool* 186:91–96.

EGUCHI, G. 1961. The inhibitory effect of the injured and the displaced lens on the lens-formation in *Triturus* larvae. *Embryologia* 6:13–35.

EGUCHI, G., and K. WATANABE. 1973. Elicitation of lens formation from the 'ventral iris' epithelium of the newt by a carcinogen, *N*-methyl-*N*'-nitro-*N*-nitrosoguanidine. *J Embryol Exp Morphol* 30:63–71.

FARRELL, R. G., W. A. DISHER, R. S. NESLAND, T. H. PALMATIER, and T. D. TRICHLER. 1977. Conservative management of fingertip amputation. *J Am Coll Emergency Physicians* 6:243–46.

FELDMAN, L. J., and J. G. TORREY. 1976. The isolation and culture in vitro of the quiescent center of *Zea mays*. *Am J Bot* 63:345–55.

FLICKINGER, R. A. 1963. Control of cellular differentiation by regulation of protein synthesis. *Am Zool* 3:209–21.

FREEMAN, G. 1963. Lens regeneration from the cornea in *Xenopus laevis*. *J Exp Zool* 154:39–66.

FREEMAN, G. N. 1964. The role of blood cells in the process of asexual reproduction in the tunicate *Perophora viridis*. *J Exp Zool* 156:157–84.

GERAUDIE, J., and M. SINGER. 1978. Nerve dependent macromolecular synthesis in the epidermis and blastema of the adult newt regenerate. *J Exp Zool* 203:455–60.

GLADE, R. W. 1957. The effects of tail tissue on limb regeneration in *Triturus viridescens*. *J Morphol* 101:477–522.

GLINOS, A. D. 1958. The mechanism of liver growth and regeneration., In W. D. McElroy and B. Glass, eds. *The Chemical Basis of Development*. Baltimore: Johns Hopkins Press.

GOSPODAROWICZ, D., and A. L. MESCHER. 1980. Fibroblast growth factor and the control of vertebrate regeneration and repair. *Ann NY Acad Sci* 339:151–74.

GOSS, R. J. 1956a. Regenerative inhibition following limb amputation and immediate insertion into the body cavity. *Anat Rec* 126:15–27.

GOSS, R. J. 1956b. The regenerative responses of amputated limbs to delayed insertion into the body cavity. *Anat Rec* 126:283–97.

GOSS, R. J. 1964. *Adaptive Growth*. New York: Academic Press.

GOSS, R. J. 1968. *Principles of Regeneration*. New York: Academic Press.

GRIMMELIKHUIJZEN, C. J. P., and H. C. SCHALLER. 1977. Isolation of a substance activating foot formation in hydra. *Cell Differ* 6:297–305.

HAY, E. 1962. Cytological studies of dedifferentiation and differentiation in regeneration amphibian limbs. In D. Rudnick, ed. *Regeneration*. New York: Ronald, p. 177–210.

HAY, E. 1966. *Regeneration*. New York: Holt.

HAY, E. D., and S. J. COWARD. 1975. Fine structure studies on the planarian, *Dugesia*. I. Nature of the "neoblast" and other cell types in noninjured worms. *J Ultrastruct Res* 50:1–21.

HAY, E. D., and D. A. FISCHMAN. 1961. Origin of the blastema in regenerating limbs of the newt *Triturus viridescens,* an autoradiographic study using tritiated thymidine to follow cell proliferation and migration. *Dev Biol* 3:26–59.

ILLINGWORTH, C. M. 1974. Trapped fingers and amputated fingertips in children. *J Pediatr Surg* 9:853–58.

ILLINGWORTH, C. M., and A. T. BARKER. 1980. Measurement of electrical currents during the regeneration of amputated fingertips in children. *Clin Phys Physiol Meas* 1:87–89.

JABAILY, J. A., and M. SINGER. 1977. Neurotrophic stimulation of DNA synthesis in the regenerating forelimb of the newt, *Triturus*. *J Exp Zool* 199:251–56.

KAMRIN, A. A., and M. SINGER. 1959. The growth influence of spinal ganglia implanted into the denervated forelimb regenerate of the newt, *Triturus*. *J Morphol* 104:415–40.

KAMRIN, R. P., and M. SINGER. 1955. The influence of the spinal cord in regeneration of the tail of the lizard, *Anolis carolinensis*. *J Exp Zool* 128:611–27.

KOSHER, R. A., and J. W. LASH. 1975. Notochordal stimulation of in vitro somite chondrogenesis before and after enzymatic removal of perinotochordal materials. *Dev Biol* 42:362–78.

LAZARD, L. 1967. Restauration de la régénération de membres irradiés d'axolotl par des greffes hétérotopiques d'origines diverses. *J Embryol Exp Morphol* 18:321–42.

LESH, G. E., and A. L. BURNETT. 1966. An analysis of the chemical control of polarized form in hydra. *J Exp Zool* 163:55–78.

MADEN, M., and H. WALLACE. 1976. How x-rays inhibit amphibian limb regeneration. *J Exp Zool* 197:105–14.

MANSON, J., R. TASSAVA, and M. NISHIKAWARA. 1976. Denervation effects on aspartate carbamyl transferase, thymidine kinase, and uridine kinase activities in newt regenerates. *Dev Biol* 50:109–21.

MARCUM, B. A., and R. D. CAMPBELL. 1978. Development of hydra lacking nerve and interstitial cells. *J Cell Sci* 29:17–33.

MARCUM, B. A., R. D. CAMPBELL, and J. ROMERO. 1977. Polarity reversal in nerve-free hydra. *Science* 197:771–73.

MESCHER, A. L. 1976. Effects on adult newt limb regeneration of partial and complete skin flaps over the amputation surface. *J Exp Zool* 195:117–27.

MESCHER, A. L., and D. GOSPODAROWICZ. 1979. Mitogenetic effect of a growth factor derived from myelin on denervated regenerates of newt forelimbs. *J Exp Zool* 207:497–503.

MESCHER, A. L., and R. A. TASSAVA. 1975. Denervation effects on DNA replication and mitosis during the initiation of limb regeneration in adult newts. *Dev Biol* 44:187–97.

MIZELL, M. 1968. Limb regeneration: induction in the newborn opossum. *Science* 161:283–86.

MOSCONA, A. A. 1957. Formation of lentoids by dissociated retinal cells of the chick embryo. *Science* 125:598–99.

NAMENWIRTH, M. 1974. The inheritance of cell differentiation during limb regeneration in the axolotl. *Dev Biol* 41:42–56.

NEWMAN, S. A. 1974. The interaction of the organizing regions in hydra and its possible relation to the role of the cut end in regeneration. *J Embryol Exp Morphol* 31:541–55.

NORMAN, W. P., and A. J. SCHMIDT. 1967. The fine structure of tissues in the amputated-regenerating limb of the adult newt, *Diemictylus viridescens J Morphol* 123:271–312.

PICKERING, R. D. l1978. Conservative management of fingertip amputations. *J Am Coll Emergency Physicians* 7:34–36.

REESE, D., E. PUCCIA, and T. YAMADA. 1969. Activation of ribosomal RNA synthesis in initiation of Wolffian lens regeneration. *J Exp Zool* 170:259–68.

REYER, R. W. 1954. Regeneration of the lens in the amphibian eye. *Quart Rev Biol* 29:1–46.

REYER, R. W. 1977. The amphibian eye: development and regeneration. In F. Cressetalli, ed. *Handbook of Sensory Physiology,* Vol. VII/5. *The Visual System in Vertebrates.* Berlin: Springer-Verlag, pp. 309–90.

REYER, R. W., R. A. WOOLFITT, and L. T. WITHERSTY. 1973. Stimulation of lens regeneration from the newt dorsal iris when implanted into the blastema of the regenerating limb. *Dev Biol* 32:258–81.

ROSE, F. C., and S. M. ROSE. 1965. The role of normal epidermis in recovery of regenerative ability in x-rayed limbs of *Triturus. Growth* 29:361–93.

ROSE, F. C., and S. M. ROSE. 1974. Regeneration of aneurogenic limbs of salamander larvae after x-irradiation. *Growth* 38:97–108.

ROSE, S. M. 1944. Methods of initiating limb regeneration in adult *Anura. J Exp Zool* 95:149–447.

ROSE, S. M. 1945. The effect of NaCl in a stimulating regeneration of limbs of frogs. *J Morphol* 77:119–39.

ROSE, S. M. 1962. Tissue-arc control of regeneration in the amphibian limb. In D. Rudnick, ed. *Regeneration.* New York: Ronald, pp. 153–76.

SALPETER, M. M., and M. SINGER. 1960. The fine structure of mesenchymatous cells in the regenerating forelimb of the adult newt *Triturus. Dev Biol* 2:516–34.

SATO, T. 1940. Vergleichende Studien über die Geschwindigkeit der Woffschen Linsenregeneration bei *Triton taeniatus* und bei *Diemyctylus pyrrhogaster. Rous Arch Entw-mech* 140:570–613.

SCHALLER, H. C. 1973. Isolation and characterization of a low-molecular weight substance activating head and bud formation in hydra. *J Embryol Exp Morphol* 29:27–38.

SCHALLER, H. C., T. RAU, and H. BODE. 1980. Epithelial cells in nerve-free hydra produce morphogenetic substances. *Nature* 283:589–91.

SCHALLER, H., C. SCHMIDT, and C. J. P. GRIMMELIKHUIJZEN. 1979. Separation and specificity of action of four morphogens from hydra. *Roux Arch Entw-mech* 186:139–49.

SCHARF, A. 1961. Experiments on regenerating rat digits. *Growth* 25:7–23.

SCHMIDT, T., and H. C. SCHALLER. 1976. Evidence for a foot-inhibiting substance in hydra. *Cell Differ* 5:151–59.

SCHOTTÉ, O. E., and S. R. HILFER. 1957. Initiation of regeneration in regenerates after hypophysectomy in adult *Triturus viridescens. J Morphol* 101:25–55.

SCHOTTÉ, O. E., and C. B SMITH. 1961. Effects of ACTH and of cortisone upon amputational wound healing processes in mice digits. *J Exp Zool* 146:209–30.

SHORT, J., R. F. BROWN, and A HUSKOVA. 1972. Induction of deoxyribonucleic acid synthesis in the liver of the intact animal. *J Biol Chem* 247:1757–66.

SIDMAN, R. E., and M. SINGER. 1960. Limb regeneration without innervation of the apical epidermis in the adult newt, *Triturus. J Exp Zool* 144:105–10.

SIMPSON, S. B. 1961. Induction of limb regeneration in the lizard, *Lygosoma laterale,* by augmentation of nerve supply. *Proc Soc Exp Biol Med* 107:108–11.

SIMPSON, S. B. 1964. Analysis of tail regeneration in the lizard *Lygosoma laterale.* I. Initiation of regeneration and cartilage differentiation: the role of the ependyma. *J Morphol* 114:425–35.

SIMPSON, S. B. 1965. Regeneration of the lizard tail. In V. Kiortsis and H. A. L. Trampusch, eds. *Regeneration in Animals and Related Problems.* Amsterdam: North-Holland, pp. 431–43.

SIMPSON, S. B., Jr. 1970. Studies of regeneration of the lizard's tail. *Am Zool* 10:157–65.

SINGER, M. 1954. Induction of regeneration of the forelimb of the postmetamorphic frog by augmentation of the nerve supply. *J Exp Zool* 126:419–72.

SINGER, M. 1961. Induction of regeneration of body parts in the lizard, *Anolis. Proc Soc Exp Biol Med* 107:106–108.

SINGER, M., and J. D. CASTON. 1972. Neurotrophic dependence of macromolecular synthesis in the early limb regenerate of the newt, *Triturus. J Embryol Exp Morphol* 28:1–11.

SINGER, M., and J. ILAN. 1977. Nerve-dependent regulation of absolute rates of protein synthesis in newt limb regenerates. Measurement of methionine specific activity in peptidyl-tRNA of the growing polypeptide chain. *Dev Biol* 57:174–87.

SINGER, M., R. P. KAMRIN, and A. ASHBAUGH. 1957. The influence of denervation upon trauma-induced regeneration of the forelimb of the postmetamorphic frog. *J Exp Zool* 136:35–52.

SINGER, M., C. E. MAIER, and W. S. MCNUTT. 1976. Neurotrophic activity of brain extracts in forelimb regeneration of the urodele *Triturus. J Exp Zool* 196:131–50.

SKOOG, F., and C. O. MILLER. 1957. Chemical regulation of growth and bud formation in plant tissues cultured in vitro. *Symp Soc Exp Biol* 11:118–31.

STEEN, T. 1968. Stability of chondrocyte differentiation and contribution of muscle to cartilage during limb regeneration in the axolotol *(Siredon mexicanum). J Exp Zool* 167:49–78.

STEEN, T. 1970. Origin and differentiative capacities of cells in the blastema of the regenerating salamander limb. *Am Zool* 10:119–36.

STEEN, T. 1973. The role of muscle cells in *Xenopus* limb regeneration. *Am Zool* 13:1349–50 (abstract).

STEEN, T., and C. S THORNTON. 1963. Tissue interaction in

amputated aneurogenic limbs of *Ambystoma* larvae. *J Exp Zool* 154:207–21.

STEWART, F. C., M. MAPES, and K. MEARS. 1958. Growth and organized development of cultured cells. *Am J Bot* 45:705–08.

STONE, L. S. 1952. An experimental study of the inhibition and release of lens regeneration in adult eyes of *Triturus viridescens viridescens*. *J Exp Zool* 121:181–224.

STONE, L. S. 1958a. Inhibition of regeneration in newt eyes by isolating the dorsal iris from the neural retina. *Anat Rec* 131:151–72.

STONE, L. S. 1958b. Lens regeneration in adult newt eyes related to retina pigment cells and the neural retina factor. *J Exp Zool* 139:69–84.

STONE, L. S. 1963. Experiments dealing with the role played by the aqueous humor and retina in lens regeneration in adult newts. *J Exp Zool* 153:197–210.

STONE, L. S. 1967. An investigation recording all salamanders which can and cannot regenerate a lens from the dorsal iris. *J Exp Zool* 164:87–104.

STONE, L. S., and S. B. GALLAGHER. 1958. Lens regeneration restored to iris membrane when grafted to neural retina environment after cultivation in vitro. *J Exp Zool* 139:247–62.

TASSAVA, R. A., and A. L. MESCHER. 1976. Mitotic activity and nuclei acid precursor incorporation in denervated and innervated limb stumps of axolotl larvae. *J Exp Zool* 195:253–62.

TAUBE, E. 1921. Regeneration mit Beteiligung ortsfremder Haut bei Tritonen. *Wilhelm Roux' Arch* 49:269–315.

THORNLEY, A. L., and E. B. LAURENCE. 1976. The specificity of epidermal chalone action: the results of in vivo experimentation with two purified skin extracts. *Dev Biol* 51:10–22.

THORNTON, C. S. 1938. The histogenesis of the regenerating forelimb of larval *Amblystoma* after exarticulation of the humerus. *J Morphol* 62:219–42.

THORNTON, C. S. 1942. Studies on the origin of the regeneration blastema in *Triturus viridescens*. *J Exp Zool* 89:375–89.

THORNTON, C. S. 1951. Beryllium inhibition of regeneration. III. Histological effects of beryllium on the amputated forelimbs of *Amblystoma* larvae. *J Exp Zool* 118:467–93.

THORNTON, C. S. 1953. Histological modifications in denervated injured forelimbs of *Amblystoma* larvae. *J Exp Zool* 122:119–50.

THORNTON, C. S. 1957. The effect of apical cap removal on limb regeneration in *Amblystoma* larvae. *J Exp Zool* 134:357–82.

THORNTON, C. S. 1958. The inhibition of limb regeneration in urodele larvae by localized irradiation with ultraviolet light. *J Exp Zool* 137:153–80.

THORNTON, C. S. 1960. Regeneration of asensory limbs of *Ambystoma* larvae. *Copeia* 4:371–73.

THORNTON, C. S., and T. P. STEEN. 1962. Eccentric blastema formation in aneurogenic limbs of *Ambystoma* larvae following epidermal cap deviation. *Dev Biol* 5:328–43.

THORNTON, C. S., and M. T. THORNTON. 1970. Recuperation of regeneration in denervated limbs of *Ambystoma* larvae. *J Exp Zool* 173:293–302.

UMANSKI, E. 1937. Untersuchung des Regenerationsvorganges bei Amphibien mittels Ausschaltung der einselnen Gewebe durch Röntgenbestrahlung. *Biol Zhurn* 6:739–56.

VERLY, W. G. 1973. The hepatic chalone. In *Chalones: Concepts and Current Researches. Natl Cancer Inst Monogr* 38:175–84.

WALLACE, H. 1972. The components of regrowing nerves which support the regeneration of irradiated salamander limbs. *J Embryol Exp Morphol* 18:419–35.

WALLACE, H., M. MADEN, and B. M. WALLACE. 1974. Participation of cartilage grafts in amphibian limb regeneration. *J Embryol Exp Morphol* 32:391–404.

WEISS, P. 1925. Unabhängigkeit der Extremitätenregeneration vom Skelett (bei *Triton cristatus*). *Wilhelm Roux' Arch* 104:359–94.

WOLFF, E. 1961. Recent researches on the regeneration of planaria. In D. Rudnick, ed. *Regeneration*. New York: Ronald, pp. 53–84.

YAMADA, T. 1967. Cellular and subcellular events in Wolffian lens regeneration. *Curr Top Dev Biol* 2:247–83.

YAMADA, T., D. H. REESE, and D. S. McDEVITT. 1973. Transformation of iris into lens in vitro and its dependency on neural retina. *Differentiation* 1:65–82.

YNTEMA, C. L. 1959a. Regeneration in sparsely innervated and aneurogenic forelimbs of *Amblystoma* larvae. *J Exp Zool* 140:101–24.

YNTEMA, C. L. 1959b. Blastema formation in sparsely innervated and aneurogenic forelimbs of *Amblystoma* larvae. *J Exp Zool* 142:423–40.

393

FIVE

Metamorphosis

The topic of Chapter 16, regeneration, provides examples of postembryonic reactivation of morphogenesis, induced by loss that leads to restoration of the original form of the organism. In most major animal groups there is another kind of postembryonic change that occurs as a normal part of the life cycle. This kind of change is called **metamorphosis.** It is seen in organisms in which embryonic development produces a larva, a stage intermediate between embryo and adult. Metamorphosis almost always involves a dramatic change in body form and life style. Witness, for example, the transformation of a tadpole swimming in a pond, into a land-dwelling, hopping toad. Or observe the transformation of a pluteus larva, propelled by cilia and carried by currents, into a sea urchin that creeps along over the ocean bottom. Observe, too, the tadpole of the tunicate, free swimming until it attaches to a wharf piling or dock and becomes a sessile organism.

For animals that are sessile—most tunicates, mussels, giant clams, sponges, and so on—larvae serve as a means of dispersal of the species, enabling the exploitation of suitable ecological niches elsewhere. Dispersal also minimizes crowding and competition for nutrient sources. Adapted as they are for moving, feeding, attaching, or exploiting environments other than those of their parents, larvae often bear little resemblance to

395

adults of the species. Indeed, it is difficult for anyone other than a person especially trained to classify, even as to phylum, most of the larval forms that can be found at the surface of the sea. Because there are so many kinds of larvae, Chapter 17 can present only a brief overview of some larval forms. Thereafter, only some details of metamorphosis among insects and some vertebrates, principally amphibians, are considered. Emphasis is placed on hormonal controls of the morphological and physiological changes and the altered patterns of gene action that occur during metamorphosis in these forms.

Postnatal development in reptiles, birds, and mammals includes a more or less prolonged period of growth, which may proceed smoothly or in spurts. The growth period is accompanied by changes in body proportions. As a general rule, the adult body form is achieved shortly after the gonads mature, but growth usually continues for a time thereafter. Animals that show this pattern of postnatal change are said to have **direct development.**

Among the fish, amphibians, and all major invertebrate groups are species whose postnatal development includes one or more larval stages. Such animals have **indirect development.** Table 17.1 identifies larval forms found in major animal groups.

17

Larval Forms and Metamorphosis

TABLE 17.1 Examples of Larval Forms in Major Animal Groups

Animal Group	Larval Form
Lower Mesozoa	
Sponges	Amphiblastula
Coelenterates	Planula, actinula, syphistoma, ephyra
Prostomia	
Flatworms	
Turbellarians	Müllers larva
Trematodes	Miricidium, cercaria, redia, sporocyst
Cestodes	Oncosphere
Nemertines	Pilidium
Annelids	Trochophore
Molluscs	Trochophore, veliger
Crustaceans	Nauplius, metanauplius, cypris, zoea, mysis, megalops
Insects	
with incomplete metamorphosis	Nymphs, naiads
with complete metamorphosis	Grubs, caterpillars, maggots, pupae, etc.
Ectoprocts	Cyphonautes
Phoronids	Actinotroch
Deuterostomia	
Echinoderms	Pluteus, auricularia, bipinnaria
Hemichordates	Tornaria
Tunicates	Tadpole
Cyclostomes	Ammocoetes
True eels	Leptocephalus
Salmon	Alevin, parr, smolt
Amphibians	Tadpole, eft

17.1   Larvae of Some Mesozoa

Among sponges is found a larva called an **amphiblastula.** Initially a free-swimming form, it soon attaches to a solid surface and inverts, forming a sedentary sponge (Figure 17.1). Coelenterates that have a polyp stage in their life cycle, such as *Hydra* (Figure 16.23) and *Campanularia* (*C. flexuosa* is shown in Figure 8.1) usually have a free-swimming **planula** larva (Figure 17.2), which soon attaches to the substrate and forms a polyp. The planula of *Aurelia* forms a specialized polyp called a **syphistoma.** The syphistoma may propagate by means of stolons, from which new syphistomas arise (Figure 17.3A). Constrictions along the length of the syphistoma create segments, each of which then forms a disc-like structure with eight notched peripheral lobes. Each disc is an **ephyra** larva (Figure 17.3B), which becomes detached and swims about, gradually transforming into a mature jellyfish, or **medusa** (Figure 17.3C), which reproduces sexually, thus giving rise to planulae.

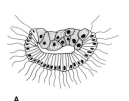

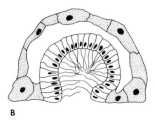

A

B

FIGURE 17.1 Schematized stages in the development of a simple sponge. A: Section through the amphilblastula; nonflagellated cells and flagellated cells are distinctively distributed. B: Section through the gastrula; the flagellated cells invaginate, and the animal attaches to the substratum at the blastoporal end.

A

17.2 Larvae of Some Invertebrate Bilateria

In contrast to sponges and coelenterates, which have a basic radial symmetry, the later evolving animal phyla show bilateral symmetry, and are referred to as the **Bilateria.** The members of the Bilateria may be divided into two groups, the **Prostomia** and the **Deuterostomia.** The name of the first group is derived from the fact that the blastopore, the first body opening to appear, gives rise to the mouth (Gr., *pro,* before; *stoma,* mouth). The lar-

FIGURE 17.2 Schematized section through a planula larva such as is found in the life cycle of many coelenterates. After a period of swimming about, the ciliated larva attaches to the substratum by its anterior end and forms a polyp. The solid endodermal mass shown here is characteristic of most members of the Hydrozoa. Among Scyphozoa and Anthozoa, the endoderm of the planula is hollow.

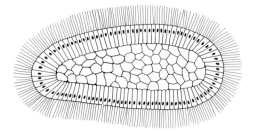

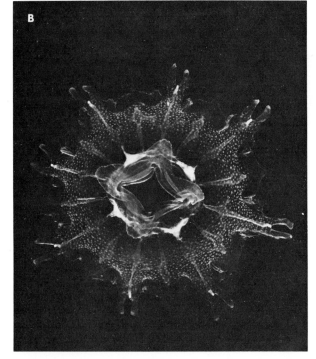

B

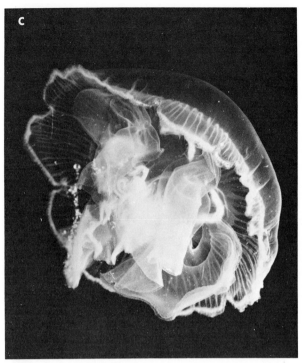

FIGURE 17.3 Strobilization and development of the medusa in the hydroid *Aurelia aurita*. A: Polyps, called scyphistomae, interconnected by stolons, are undergoing strobilization, a process of terminal constriction whereby ephyra larvae are detached as they mature. B: Early stage in the transformation of the ephyra into a medusa. C: Early medusoid stage, showing the complex folding of oral arms; planula larvae develop from fertilized eggs in pockets formed by frills of the oral arms. [From D. B. Spangenberg, *J Exp Zool:* A, 178:183–94 (1971), B–D, 159:303–18 (1965), courtesy of Dr. Spangenberg and permission of Alan R. Liss, Inc.]

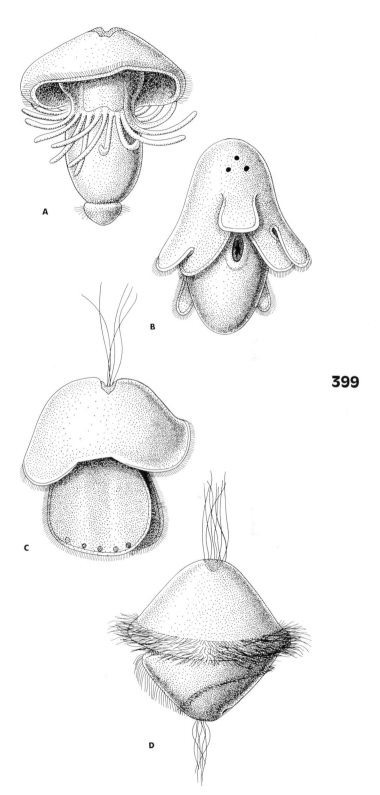

399

vae of prostomes arise from spirally cleaving eggs that show mosaic development and have a number of morphological features in common (Figure 17.4). Prostome larvae are thought to resemble ancestral

FIGURE 17.4 Representative larvae of members of the Prostomia. A: Actinotroch larva of a phoronid worm (phylum Phoronida). B: Müller's larva of a polyclad flatworm (phylum Platyhelminthes). C: Pilidium larva of a nemertean (phylum Nemertinea). D: Trochophore larva of a polychaete annelid (phylum Annelida).

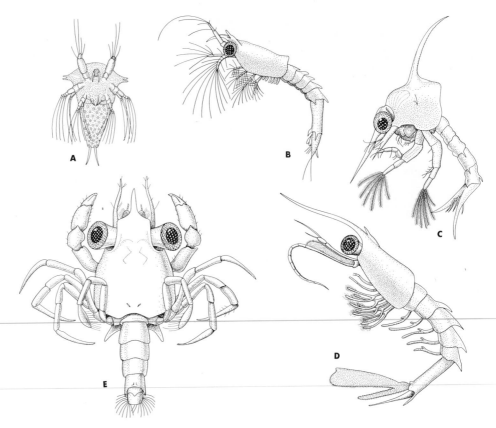

FIGURE 17.5 Crustacean larvae. A: Nauplius, the basic larval form, which appears at hatching in a large number of crustacean groups. Among higher crustaceans larval molts lead to successively more complex larval stages: protozoea (B), zoea (C), mysis (D), and megalops (E). This succession is suggestive of evolutionary relationships. Adults of the genus *Mysis* pass through the protozoea and zoea stages before achieving the adult form. In the case of *Peneaus,* a third larval stage, the mysis, precedes the final molt. The mysis larva of *Peneaus* resembles adult shrimp of the genus *Mysis,* suggesting that peneid shrimps have a *Mysis*-like ancestor. The true crabs (Brachyura, or "short tails") have a forward-directed and much reduced abdomen. Crabs pass through the nauplius and protozoea stages in the egg, hatching as zoea larvae. The zoea of the crab then molts to form a mysis larva that, upon molting, becomes a megalops. The megalops has a large tail such as is found in lobsters and other Macrura ("large tails"). When the megalops molts, the adult form of the crab appears. The occurrence of the megalops stage suggests that the crabs arose from ancestors that had long and strong abdomens such as those of lobsters.

forms that were in the main line of evolution leading to the annelids, mollusks, and arthropods. The arthropods departed radically from the developmental patterns shown by their presumed ancestral forms, however. Insect larvae and their metamor-

phic changes are described in Section 17.3. Some crustacean larvae and aspects of possible evolutionary relationships among crustaceans are illustrated in Figure 17.5.

During development of the Deuterostomia (Gr.,

deuteros, secondary; *stoma,* mouth) the mouth arises as a secondary opening into the archenteron. The origin of the mouth is illustrated for the sea urchin larva in Figure 9.4 and for a chick embryo in Figure 11.60B. The blastopore or its associated cells transforms into the anus. Deuterostomes generally show indeterminate cleavage and tend to form mesoderm from two sources, first from the blastoporal region and later as outpocketings of the archenteron. These outpocketings form one or more coelomic cavities. Some deuterostome larvae are illustrated in Figure 17.6. Of particular note in

this figure is the resemblance between the echinoderm larval types, the **auricularia** and the **bipinnaria,** and the **tornaria** larva of *Balanoglossus. Balanoglossus* is one of the hemichordates, which are burrowing, worm-like organisms that have gill slits and a hollow dorsal nervous system. But for the lack of a notochord they would be classified as chordates. The tadpole larva of ascidians (Figure 10.1), which are urochordates whose ancestors were in the line of vertebrate evolution, has a notochord in addition to a dorsal hollow nervous system and gill slits. The hemichordates thus show relationships to both the echinoderms and primitive chordates.

FIGURE 17.6 Representative larvae of members of the Deuterostomia. A: Auricularia larva of a holothurian (sea cucumber). B: Tornaria larva of *Balanoglossus* (acorn worm). C: Pluteus larva of an echinoid, such as a sea urchin or sand dollar. D: Young stage in the bipinnaria larva of an asteroid (starfish and similar echinoderms).

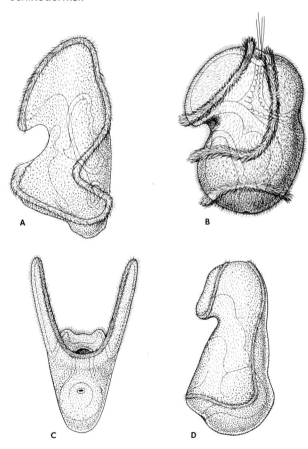

17.3   Insect Metamorphosis and Its Determinants

401

In a textbook of developmental biology considerable attention might be legitimately addressed to the analysis of metamorphosis in all of the animal groups. With only limited time and space for treating these matters, however, it is important to concentrate on those organisms whose larvae and patterns of metamorphosis have received the most rewarding analytical attention, namely, the insects and the amphibians.

17.3.1 PATTERNS OF INSECT METAMORPHOSIS.
When the young insect hatches it is covered by a firm cuticle; growth and change in form require the molting of this cuticle and its replacement with a larger one. The loss of the cuticle is called **ecdysis.** Development is marked off by a series of molts, the number of which is usually predetermined, and the form of the insect changes with each molt in a precise pattern characteristic of the species.

Certain insects, such as the springtails and bristletails, have direct development. They are referred to as **ametabolous.** In the winged insects, however, very marked changes in the body form occur during the posthatching period. These changes occur gradually in some kinds of insects and abruptly in others. Insects that undergo gradual metamorphosis are described as **hemimetabolous.** In this group the

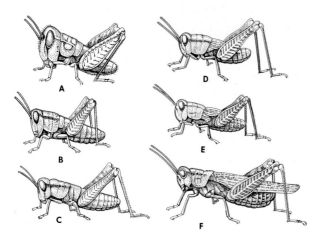

FIGURE 17.7 Gradual metamorphosis in a hemimetabolous insect. In this grasshopper there are five nymphal stages before the adult condition is reached. (Reprinted with permission of Macmillan Publishing Co., Inc., from *College Zoology*, rev. ed., by R. W. Hegner. Copyright © 1912, 1926, 1931, 1936, 1951, 1959 by Macmillan Publishing Co., Inc.)

402

wings are present only as buds at the time of hatching and the body form is disproportionate to that of the adult. As the molts ensue, the configuration of the insect progressively approaches that of the adult; the wings become fully developed and sexual maturity is achieved at the last molt (Figure 17.7). A juvenile insect undergoing this gradual metamorphosis is called a **nymph** or, if it adapted to an underwater existence during its immature phase, a **naiad.**

Holometabolous insects hatch from the egg in forms popularly known as grub, caterpillar, maggot, cutworm, and so on. After a period of feeding and a variable number of molts, depending on the species, the larva reaches a size often hundreds of times greater than its original size (Figure 17.8). The stages between larval molts are called **instars.** At the end of the last instar the larva undergoes a metamorphic molt to form a **pupa.** This molt may occur within a special **cocoon** (Figure 17.9A) woven by the larva in its final instar. Pupation may involve the formation of a special pupal cuticle (Figure 17.9B), as occurs in butterflies and silkmoths, or the pupa may remain encased in the old larval skin, which becomes tanned and hardened to form a **puparium.**

The pupa is a nonfeeding larval stage during which a complex series of internal changes occurs. These changes involve the wholesale destruction of many, if not all, of the larval organs and the formation of the adult body from nests of organ-specific cells called **imaginal discs** (Figure 17.10).

Within the pupal cuticle the adult, or **imago,** is formed. When its development is complete, a final molt, the **imaginal molt,** occurs and the adult emerges. The adult is essentially a new organism, different in form and function from both larva and

FIGURE 17.8 Reproduction in crane flies. A: Mating, female on the left. B: Eggs, larvae and pupa. On the left are two newly laid eggs, on the right is a pupa, and in the middle are two of the larval stages. Note the tremendous size increase from egg to pupa. [Photographs courtesy of Dr. Florian Muckenthaler.]

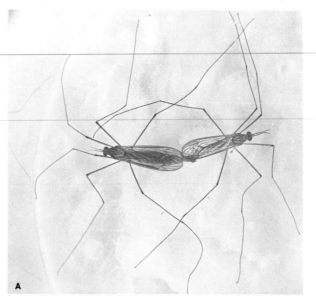

A

B

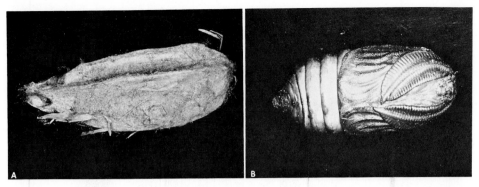

FIGURE 17.9 Silkmoth pupa. A: Pupal case spun by the larva at the end of the feeding period. B: Diapausing pupa excised from pupal case; note the sculptured pupal cuticle, showing outlines of antennae and wings. The large size of the antennae marks this pupa as a male.

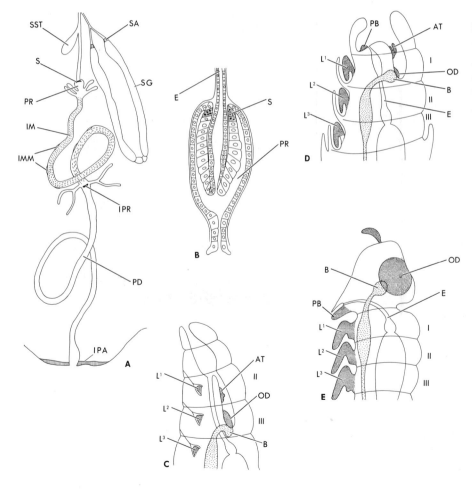

FIGURE 17.10 Imaginal discs in larvae and pupae of *Musca*, a fly. A: The imaginal discs of the larval digestive tract. B: Close-up of the larval proventriculus showing imaginal cells of the stomodaeum. C–E: imaginal discs of the legs and head parts seen from the left side for the larva (C) and for the pupal stages (D,E). AT, antennal imaginal disc; B, brain; E, esophagus; IM, imaginal cells of mid-gut epithelium; IMM, imaginal cells of the mid-gut muscles; IPA, posterior abdominal imaginal disc; IPR, imaginal ring of proctodaeum; L^1, L^2, L^3, discs for the first, second, and third legs; OD, optic disc; PB, proboscis; PD, larval proctodaeum; PR, proventriculus; S, stomodael imaginal ring; SA, imaginal ring of salivary gland; SG, larval salivary gland; SST, sucking stomach; I, II, III, first, second, and third thoracic segments. [From E. Korschelt and K. Heider, *Embryology of the Invertebrates*, Vol. III Engl. trans., Macmillan, New York, 1899.]

pupa. Imagos of some species do not feed, functioning exclusively as flying reproductive machines. Adults of other species may feed and produce several generations of offspring. They do not, however, molt again in nature.

17.3.1.1 The Origin and Development of Imaginal Discs.

Imaginal discs are found in all holometabolous insects. In some insects, however, only certain parts of the imago originate from the imaginal discs, for example, wings, antennae, and genitalia. In Lepidoptera the larval alimentary tract, mouth parts, and tracheal system are remodeled to fit the adult pattern. In the higher Diptera, such as *Drosophila*, the entire adult integument, except for that of the abdomen, arises from imaginal discs, of which there are 10 major pairs, plus a genital disc. Three major pairs of discs form the head: the labial discs, the imaginal cells of the clypeo-labrum, and the eye-antenna discs. The adult thorax arises from three pairs of discs ventrally, which form the legs, and from three pairs of dorsal discs, which form the wings, halteres, and the dorsal part of the integument. An unpaired genital disc on the last abdominal segment forms the genital apparatus. The integument of the abdomen forms from small groups of imaginal cells called **histioblasts.** Nests of imaginal cells in the larval gut and elsewhere form the primordia of the internal organs (Figure 17.10).

Imaginal discs of the principal body parts first appear in the embryo or early larva as thickenings of the epidermis, each in the form of a circumscribed zone of simple columnar epithelium. During larval life these thickenings invaginate, remaining attached to the larval epithelium by a stalk (Figure 17.10). Meanwhile, their cells proliferate and the epithelium becomes greatly folded. The folds become telescoped together in a form resembling the packing of the rings of a collapsible drinking cup (Figure 17.11A). The cells meanwhile are undifferentiated and remain so until metamorphosis, at which time the number of cells in a disc may range from a few hundred in the smaller discs to many thousands in larger ones. At metamorphosis the epithelium of the discs undergoes a dramatic evolution called **eversion** (Figure 17.11) and projects from the body. The cells at the bases of each disc proliferate the epidermis of the adjoining body regions as the larval epidermis is displaced and degen-

404

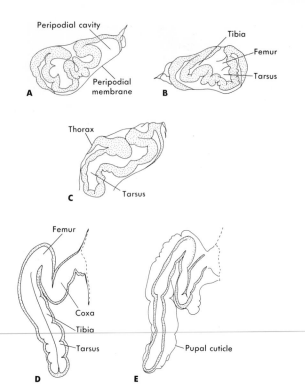

FIGURE 17.11 Successive stages in the eversion of the imaginal leg disc in *Drosophila*. [Adapted from P. Mandaron, *Bull Soc Zool France* 97:382–400 (1972).]

erates. The cells of the appendages and body epidermis then differentiate and secrete a cuticle in the pattern specific to each body region.

The developmental potentialities of imaginal discs might reasonably be pursued in considerably more detail. Interested readers are invited to consult reviews by Gehring (1968), Hadorn (1968), and Postlethwait and Schneiderman (1974).

17.3.2 MOLTING, PUPATION, AND ADULT DIFFERENTIATION CONTROLLED BY HORMONES.

Larval and pupal molts in holometabolous insects and nymphal molts of the hemimetabolous forms are controlled by hormones (Figure 17.12). Periodically, neurosecretory cells in the brain release a brain hormone called **ecdysiotropin.** This hormone, probably a polypeptide, stimulates glands in the **prothorax** (the segment of the thorax that bears the anteriormost of its three pairs of legs). The **pro-**

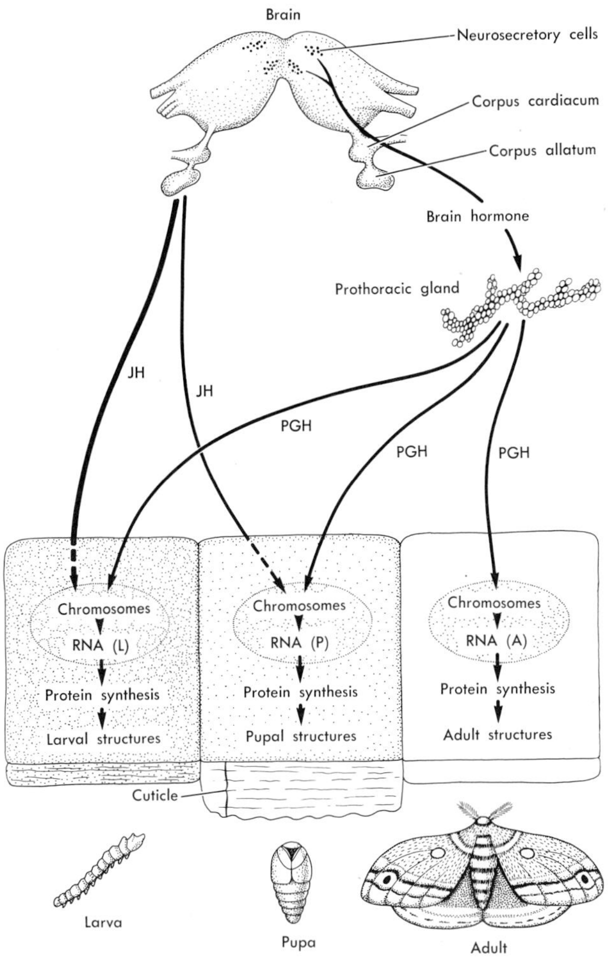

FIGURE 17.12 Scheme of hormone action in metamorphosis of the cecropia silkmoth. The larval molts (left) are apparently initiated by brain hormone, which stimulates the prothoracic glands to secrete ecdysone, the prothoracic gland hormone, PGH. At the same time, the corpora allata secrete juvenile hormone, JH, which affects the portions of the genome that code for synthesis of larval protein patterns (see Chapter 18, and Appendix B). After four (sometimes five) larval molts, the corpora allata cease secreting juvenile hormone, so that the next molt (center) occurs in response to PGH, with only a low concentration of JH. This combination leads to the production of pupal cuticle, formation of pupal structures, and breakdown of larval ones. At the final molt (right), no juvenile hormone remains, and an adult cuticle forms in the presence of PGH alone; the other pupal structures either break down or develop into adult ones [From H. A. Schneiderman and L. I. Gilbert, *Science* 143:325–333 (1964), by courtesy of the authors and with permission. Copyright 1964 by the American Association for the Advancement of Science.]

405

thoracic glands respond to the brain hormone by secreting a steroid hormone **ecdysone,** which stimulates growth and causes the epidermis to secrete a new cuticle, initiating the molting process. If the brain of a larva is removed microsurgically, ecdysone is not produced and growth and molting cease.

Paired glands, the **corpora allata,** which lie near the brain, secrete another hormone, **juvenile hormone,** an aliphatic hydrocarbon (Meyer et al., 1968, 1970) that is completely unlike other hormones known in the animal kingdom. The corpora

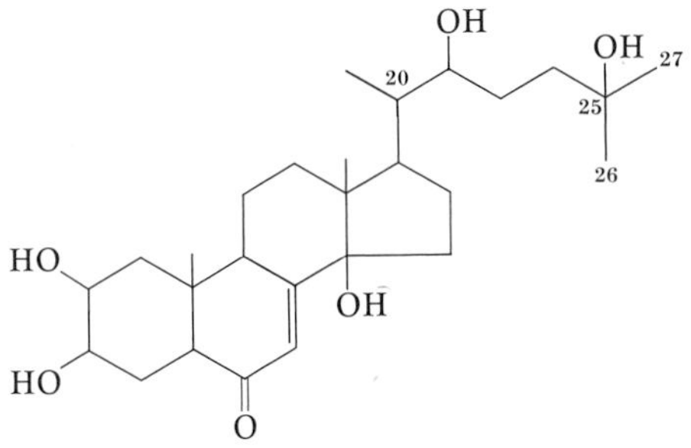

α-Ecdysone

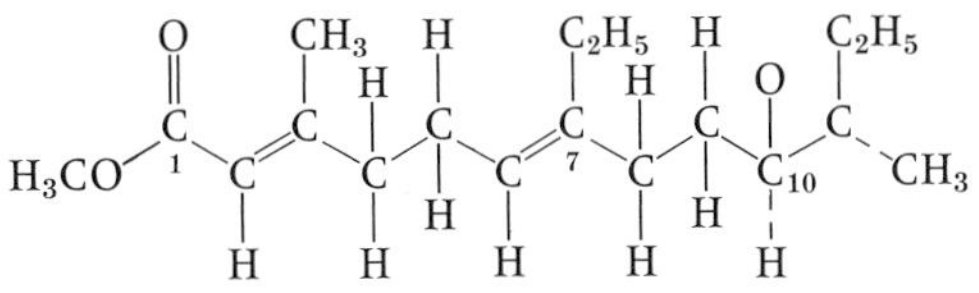

Juvenile hormone

allata produce their secretion throughout larval life and, as long as juvenile hormone is present, molts that take place under the influence of ecdysone lead to the production of an additional larval stage. In the last larval instar of holometabolous insects the secretion of juvenile hormone is reduced, and the next molt is the pupal molt. The production of juvenile hormone then ceases, and at the next molt the insect differentiates the adult form. If a pupa is treated with juvenile hormone, however, it will undergo another molt, producing a second pupal cuticle.

Ecdysone continues to be produced until the adult molt is completed. It is required for the eversion of the imaginal discs. In fact, isolated discs of wing (Fristrom et al., 1973) and leg (Mandaron, 1971) from third instar *Drosophila* larvae show spontaneous eversion in vitro in the presence of ecdysone (Figure 17.11). Ecdysone stimulates the synthesis of both RNA and protein, which are required for eversion of the imaginal discs.

When the imago emerges, the prothoracic glands degenerate. Therefore, there is no further source of ecdysone and additional molts do not occur. The corpora allata, however, resume secretion of juvenile hormone after the adult emerges. Juvenile hormone affects protein and fat metabolism and is required for the production of vitellogenic proteins (cf. Chapter 7).

17.3.3 CHANGING PATTERNS OF GENE ACTION.
In the holometabolous insects changing patterns of protein synthesis accompany the succession of larval molts and, especially, the larval–pupal and pupal–adult transformations. Prominent among the changing protein populations are those of the salivary glands, various hydrolases—especially those involved in eliminating larval tissues—enzymes involved in forming the puparium or in producing components of the cocoons, and the amino acyl synthetases. These changes involve changing gene activity under hormonal control. Some aspects of the altered transcriptional patterns that occur during metamorphosis are presented in Chapter 18, in which the general problem of control of gene action in development is considered in more detail. In this section attention is particularly directed to those aspects of gene action revealed by the study of giant chromosomes (Figure 17.13) that are found in cells of certain organs of larval and

406

adult flies. These chromosomes are best known from studies of the larval salivary glands, but similar giant chromosomes also occur in large cells of the Malpighian tubules, seminal vesicles, heart, pericardium, and foot pads.

The large size of these chromosomes results from three factors. First, the homologous chromosomes are permanently paired. Second, they are uncoiled over most of their length, so that each stretches out as a strand many times the metaphase length. Third, each strand is present in multiple copies (thus they are polytene, as are the nurse cells in some insect ovaries). At irregular but frequent intervals along each chromosomal strand there are regions that are coiled, or condensed, and these appear as dark-staining areas. The interband regions appear to be unstained in contrast. The condensed areas of the replicate strands are in register, so that a pattern of transverse bands is created across the polytene structures. The bands are of such constant and characteristic pattern that each can be given a coded designation that indicates which chromosomes it is on and its precise location on the chromosome. These bands have been studied particularly well in the larval salivary glands of such flies as *Sciara coprophilia, Chironomus thumni, Drosophila melanogaster* and *D. hydei*. Correlated genetic and cytological studies permit the identification of the various banded segments with specific gene loci. Approximately 1900 bands have been recognized in *D. melanogaster* and more than 2000 in *C. thumni*.

In *D. melanogaster* the polytene, banded configuration is achieved by the middle of the third instar. At this time the chromatin of some of the bands begins to show structural modifications. The band becomes swollen and the arrangement of the chromatin is less ordered and more lightly staining. The strands of chromatin extend beyond the margins of the chromosome, forming a **puff.** Particularly large puffs, found in chironomid flies, give the appearance of surrounding the chromosome like a ring. These large puffs are called Balbiani rings after the scientist who discovered them in 1881. The appearance of a puff is believed to mark the exposure to transcription of DNA that was hitherto condensed and unavailable. RNA polymerase II, which transcribes mRNA, can be demonstrated on the puffs and, to a lesser extent, in the interband regions, but not on the bands (Jamrich et al., 1977). Moreover,

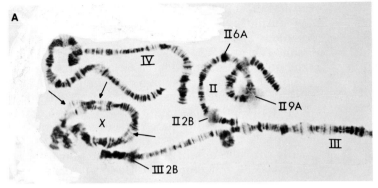

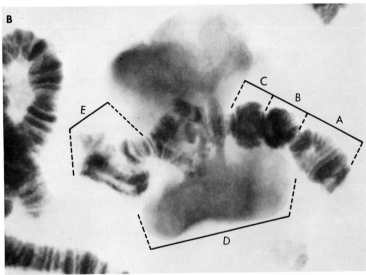

FIGURE 17.13 A: The distribution of DNA in bands and puffs of the four giant chromosomes of a salivary gland cell of a larva of the fly *Sciara coprophilia,* as revealed by staining with the Feulgen reagent. The autosomes are identified by roman numerals, the sex chromosome by X. This preparation was made from a female larva, and the permanent pairing of homologous chromosomes is shown in the X chromosome at points indicated by arrows. Certain bands that are puffed in chromosomes II and III B: The small chromosome IV of a salivary gland cell of a larva of *Chironomus thumni,* showing puffing of segments B and C and of the nucleolar region D. The puffs are sites of intensive RNA synthesis. The banding pattern and the dual nature of the chromosome are clearly seen. (Photographs for A and B were supplied, respectively, by Dr. Ellen M. Rasch and Dr. Hans Laufer. The courtesy of these colleagues is gratefully acknowledged.]

407

the puffs are sites of rapid RNA synthesis, as revealed by their rapid incorporation of radioactive uridine (Pelling, 1964). The interbands, too, may be sites of RNA synthesis, but at a relatively low level (Zhimulev and Belyaeva, 1975). Lambert et al. (1972) isolated from the nuclear sap of *Chironomus tentans* a mRNA of relatively large size that binds preferentially to a large Balbiani ring on chromosome IV. Moreover, heat shocks (37°C for a short period) induce new puffs that are shown unequivocally to be sites for production of mRNAs that code for specific heat-shock proteins (Berendes et al., 1973; Ashburner and Bonner, 1979).

Numerous bands undergo puffing and regression in a pattern that is controlled by ecdysone and precisely correlated with the progress of the larval–pupa molt. In the strain of *Drosophila melanogaster*

used by Ashburner (1967) the eggs, hatched as first instar larvae and reared at 25°C, complete the third instar after about 120 hours. At this point the larva becomes a soft, white immobile mass, having discharged the content of its salivary glands and everted its **spiracles** (openings to its respiratory system). The accomplishment of this condition is called **pupariation,** or **puparium formation,** and the larva is referred to as a **prepupa.** After 4 more hours it undergoes a prepupal molt (the occurrence of this molt is disputed by some, e.g., Whitten, 1957), but it is still referred to as a **prepupa.** It molts to form a true **pupa** 8 hours later. These changes are paralleled by a remarkable increase in puffing activity in the salivary gland chromosomes. There are only about 10 prominent puffs in the entire chromosomal complement 10 hours prior to

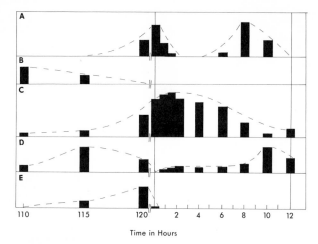

FIGURE 17.14 Histograms showing relative sizes of puffs in selected band on the left arm of chromosome 3 of *Drosophila melanogaster*. The bands are, in order from top to bottom: 63E, 68C, 71CE, 75B and 78D. The abscissa shows time in hours during the late third instar larval and prepupal stages. The vertical line at 0 hr indicates the time of puparium formation, and that at 12 hr the time of pupation. [Adapted from M. Ashburner, *Chromosoma (Berl.)* 21:398–428 (1967).]

408

pupariation. But as pupariation begins, and for the next 12–14 hours, at least 125 puffs appear. In *D. melanogaster* as many as 200 bands form large puffs at some time during the development of the third instar and the prepupa. At any particular time only certain bands are puffed, but the pattern of puffs changes continuously and in a predictable manner throughout this period. The correlation of puffing activity late in the third instar and during the prepupal period is illustrated for some loci in the left arm of chromosome III in Figure 17.14. Ashburner (1967) mapped the occurrence of puffs in chromosomes 2 and 3, finding a total of 108 puffs, of which 83 were strictly stage dependent for their appearance. Very importantly, the normal pattern of puffing in salivary gland chromosomes can be induced by ecdysone in glands isolated in vitro (Ashburner, 1972; Ashburner and Richards, 1976; Richards, 1976a, b).

The relevance of puffs to the physiology of development of Diptera is not at all clear. If cycles of high puffing activity are correlated with molting, why should there be so much activity in the salivary gland, which degenerates at the pupal molt? The main function of the larval gland appears to be the production and extrusion of a protein secretion that hardens and attaches the pupal case to a substratum. The production of secretion begins in about the middle of the third larval instar, and extrusion of the secretion into the lumen of the gland begins about 3 hours before puparium formation. After puparium formation production of the secretion ceases, yet it is at this time that puffing is most active under the influence of ecdysone! Correlation of puffs with the production of specific salivary proteins has been demonstrated for *Drosophila melanogaster*, *Chironomus tentans*, and *C. paladivitattus* (Korge, 1975, 1977a, b; Grossbach, 1973). But, as Wobus et al. (1970) pointed out, there is no correlation between the large number of puffed bands in salivary gland chromosomes and the small number of salivary gland specific proteins that can be isolated from extracts of the glands or their secretions. Perhaps some of the puffing activity, however, reflects transcription that is related to the synthesis of lysosomal enzymes involved in the ultimate histolysis of the cells (cf. Henrikson and Clever, 1973) or with enzymes involved in the sequestration of hemolymph proteins in the glands as described for *Chironomus thumni* by Laufer and his colleagues (Laufer and Nakase, 1965; Laufer et al., 1964; Doyle and Laufer, 1969a, b).

As a final note on the topic of changing gene activity during insect metamorphosis, one may leave the Diptera and turn to the well-studied beetle *Tenebrio molitor,* in which an interesting change in gene action begins during the first days of pupal life. At this time the pupa begins to produce a new mRNA that is translated as a tyrosine-rich protein that is specific to the adult cuticle (Ilan et al., 1966). This mRNA is not translated, however, until 5–7 days later, whereupon there is a sharp rise in cuticular tyrosine. Later (1970) Ilan and colleagues showed that the translation of the message requires populations of isoaccepting tRNA species and of aminoacylsynthetases that are different from those that are available at the beginning of pupation. The application of juvenile hormone prevents the appearance of these components of the translating system, and a second pupal rather than adult cuticle is formed. These findings, which are examined further in Chapter 18, indicate that changes in the translational machinery for protein may also be brought about by the hormones of metamorphosis.

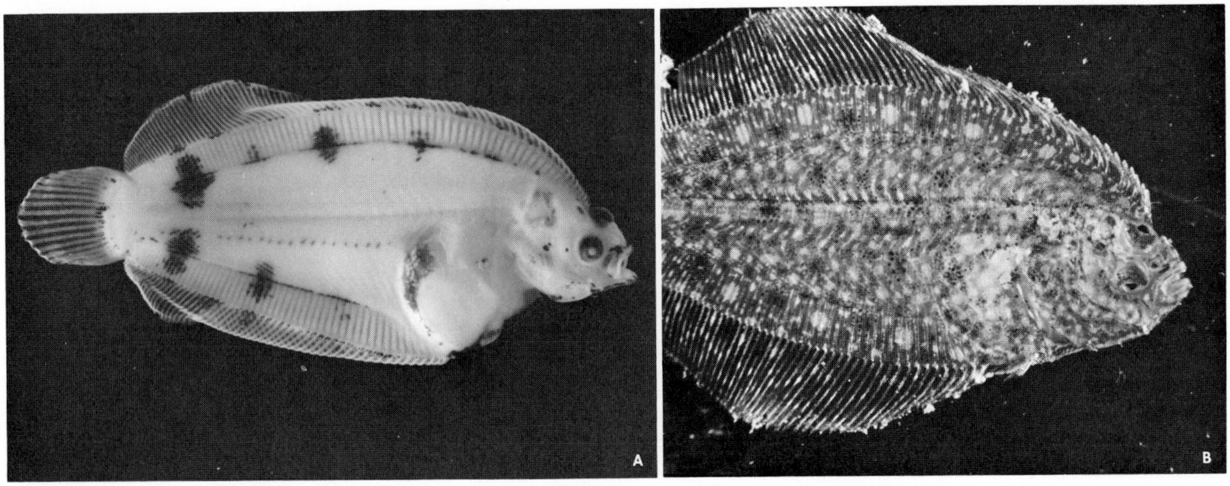

FIGURE 17.15 A: larval form of the sand sole *Psettichthys* at the onset of metamorphosis (preserved specimen). B: Living sand sole approaching the end of metamorphosis; note that the eyes have shifted so that both are on the right side of the head. [Photographs kindly supplied by the late Mac V. Edds, Jr.]

17.4   Metamorphosis in Vertebrates

When one thinks of vertebrates that undergo metamorphosis, the transition of tadpole to frog or toad is what usually comes first to mind. This metamorphic event is what is most frequently observed by nature lovers, and it has been more intensively studied in the laboratory than the metamorphosis of other vertebrates. But many other vertebrates undergo forms of metamorphosis, some less drastic than those of the anuran amphibian, others involving more complex changes both in morphology and physiology. Many of our marine food fish undergo a metamorphosis early in life that adapts them to a bottom-dwelling existence. Starting postembryonic life as bilaterally symmetrical animals, many fish, including sole, flounder, plaice, and so on, undergo structural changes in such a way that they come to lie on the ocean bottom on their right or left sides, depending on the species. The skin and scales of the lower side become smooth and depigmented and those of the upper side become rough. Pigment cells of the upper side, under complex control, expand

and contract in patterns that tend to mimic the pattern of the ocean floor. A most spectacular change occurs in the head (Figure 17.15), where the eyes shift their position so that they both occupy the upper surface. The jaws meanwhile, however, retain essentially their original position with respect to the dorsoventral axis. In subsequent sections, metamorphic changes in several other kinds of vertebrates are noted briefly, but first, a more complete description of metamorphosis in anurans is presented.

17.4.1 FROM TADPOLE TO FROG. The most familiar kinds of anurans are frogs whose eggs develop into tadpoles, which are limbless aquatic animals with an oval torso and long, finned tail (Figure 17.16). Metamorphosis transforms tadpoles into creatures like their parents, four-legged, hopping, carnivorous animals. After metamorphosis most kinds of anurans take up a land-dwelling existence, returning to the water only to breed. Toads show this life habit. Others, for example, bullfrogs, spend a good deal of time in the water or on the banks of streams. Anurans of the genus *Xenopus* remain in water as adults.

Tadpoles have many adaptations, both anatomical and physiological, that differentiate them from

409

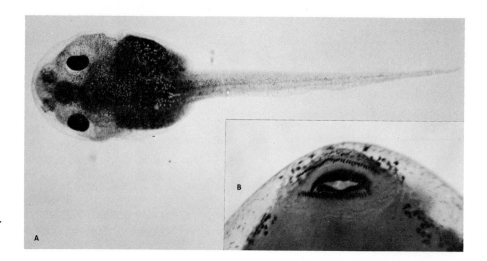

FIGURE 17.16 Tadpole of
Rana pipiens. A: Dorsal view.
B: Closeup of the mouth-
parts on the ventral side.

410

adults. The mouth is adorned with two horny beaks
and rows of horny teeth (Figure 17.16B). On each
side of the head a fold of skin, the operculum, covers gills that grow from the lower ends of the visceral arches. Lungs are rudimentary and nonfunctional. The epidermis is populated with large
pigment cells of different kinds and is underlaid by a
jelly-like dermis rich in hyaluronic acid. The larva
feeds on plants, digesting the food in a long, coiled
alimentary tract. The eyes, recessed in the head, utilize the visual pigment **porphyropsin*** in the retina.

---

*Porphyropsin is the visual pigment utilized by animals that
live in fresh water or by animals during the freshwater phase
of a life cycle that may otherwise include a marine or land-
dwelling phase. Porphyropsin is a purple pigment formed by
the combination of a protein, **opsin**, with the aldehyde of
vitamin A_2 or **$retinal_2$**. Another visual pigment, **rhodopsin**, is
found in the retina in marine or land-dwelling animals or during the marine or land-inhabiting phase of animals that spend
a part of the life cycle in fresh or brackish water. Rhodopsin is
formed by the combination of **vitamin A_1** or **$retinal_1$** with an
opsin slightly different from that found in porphyropsin.

Retinal₁

Nitrogenous waste products are excreted chiefly as
ammonia. The red blood cells in the circulation are
chiefly produced in the kidney. Their hemoglobin
(HbF) is different from that of the adult (HbA).

**17.4.1.1 Anatomical Changes Associated
with Growth and Metamorphosis.** In the common grass frog, *Rana pipiens,* the first postembryonic
period is one of much growth with little change in
body form, it is designated as the **growth period** or
premetamorphic period. The growth period lasts
about 7 weeks at laboratory temperatures of
22–25°C. During the last 2 weeks of the growth
period the hindlimbs appear and total body length
increases from about 35 mm to 55 mm. By the end
of the growth period the hindlimbs are about 4 mm
in length and the torso is about 20 mm long.

The growth period is followed by the **prometamorphic period,** which lasts about 3 weeks. During
this period the length of the hindlimbs increases
very rapidly relative to the growth of the torso.
About 5 days before the end of the prometamorphic
period, the tadpole will have reached a total length

Retinal₂

FIGURE 17.17 Metamorphosing larva of *Rana pipiens.* This is a hypophysectomized larva that has been treated with thyroxine, seen from the dorsal side. It shows harmonious metamorphic changes comparable to those of normal froglet during metamorphic climax, except that the snout is slightly less pointed. [Photograph courtesy of Dr. J. Kollros.]

of about 70 mm (Figure 17.17). Thereupon, a number of changes begin, notably a shift in the position of the anus and a thinning of the operculum on each side, a process that forms translucent "skin windows" through which the forelimbs will subsequently erupt.

With the accomplishment of these changes metamorphic climax ensues. Within as little as 24 hours the horny beaks of the oral region are lost, the forelegs emerge, and the mouth widens. Within a week the mouth becomes much wider and muscular jaws

develop. The eyes are repositioned to a higher level. These changes involve a complete remodeling of the head skeleton and are adaptive to a predatory life style requiring an aerial sensory input. The gills and the tail are completely resorbed. The skin hardens as the epidermis thickens, and the jelly-like dermis is replaced by a tougher and more fibrous tissue. Within the skin the pigment cells become arranged so as to give the adult pattern of coloration.

Internally, a muscular tongue used in grasping prey is formed, and, as the lungs enlarge, the hyoid cartilages differentiate and muscles develop for the pumping of air into the lungs. The cells of the larval alimentary tract are almost completely sloughed off, and essentially a new and shorter digestive tract develops.

17.4.1.2 Biochemical Changes. Patterns of macromolecular synthesis based on changing patterns of gene activity accompany metamorphosis. During metamorphic climax porphyropsin is replaced almost entirely by rhodopsin as the visual pigment. Also, in the eye the larval form of α-crystallin in the lens is replaced by an adult α-crystallin that differs considerably in electrophoretic mobility (Polansky and Bennett, 1970). An adult form of skin keratin replaces the larval form at metamorphosis, too (Reeves 1975), and the mRNA for adult epidermal keratin has been isolated and translated in a cell-free preparation (Reeves, 1977). During metamorphosis large quantities of hyaluronidase are produced. This enzyme eliminates the hyaluronic acid of larval skin. In the post metamorphic skin hyaluronic acid is almost absent, having been replaced by a mixture of other glycosaminoglycans. Hyaluronidase is not present in appreciable quantities in adult skin (Lipson et al., 1971). Other aspects of changes in the skin at metamorphosis involve altered patterns of collagen synthesis and deposition resulting in a tougher skin more appropriate to terrestrial life.

Upon metamorphosis the frog begins to excrete the bulk of its nitrogenous waste as urea rather than ammonia. Urea, like ammonia, is very soluble, but it is less toxic. Thus urea can be formed and retained in the blood after metamorphosis and then eliminated by the kidneys with less water loss than would be required for the elimination of an equivalent amount of nitrogen in the form of ammonia. Production of urea requires the activation of the

411

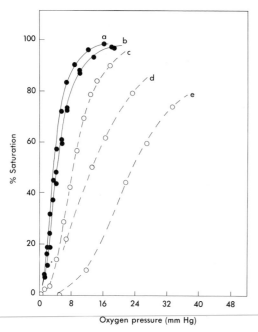

FIGURE 17.18 Oxygen saturation curves for hemoglobin of tadpoles (filled circles) and adults (open circles) of *Rana catesbeiana*. Curve a, pH 7.62; on curve b are shown points plotted at both pH 9.0 and pH 7.32; acidification produces essentially no change in the oxygen affinity. Curves c, d, and e show saturation curves plotted at pH 8.4, 7.22, and 6.94, respectively. Note that the oxygen-binding power of adult hemoglobin is less than that of tadpole hemoglobin and that the affinity of adult hemoglobin for oxygen is much reduced upon acidification. [Adapted from G. Wald, *Science* 128:1481–90 (1958).]

ornithine–urea cycle in the liver. In this cycle carbon dioxide and nitrogen are eliminated in the form of urea. Metamorphic changes thus require a reorganization of metabolic patterns in the liver for production of the appropriate enzymes of the ornithine–urea cycle. This reorganization begins very early in metamorphosis, even before there is any recognizable change in body form (Munro, 1939; Cohen, 1970).

At metamorphosis the site of erythropoiesis shifts to the spleen and the bone marrow. This shift is marked by the production of hemoglobin of different physiological (Figure 17.18) and electrophoretic properties. In *Rana catesbeiana* the shift from production of tadpole hemoglobin (HbF) to adult

hemoglobin (HbA) is essentially complete just before the tail is resorbed. There has been some uncertainty as to whether the onset of HbA production involves the differentiation of a new erythropoeitic cell line or whether cells of the larval erythropoietic lineage suppress gene loci for HbF production and promote gene activity at loci involved in the production of HbA. Jurd and MacLean (1970) showed that during the metamorphosis of *Xenopus laevis* many of the individual circulating erythrocytes contain hemoglobins that react to antibodies against both HbF and HbA. Maniatis and Ingram (1971), however, found that during metamorphosis of *Rana catesbeiana* the hemoglobin of individual circulating erythrocytes reacts with antibodies against HbF or HbA, but not against both.

Also, during metamorphosis there is considerable synthesis of hydrolytic enzymes involved in the resorption of the larval gut and the tail. This synthesis almost certainly involves new gene activity, which we now discuss in the context of hormonal controls.

17.4.1.3 Hormonal Control of Metamorphosis in Anurans.
Metamorphosis of anuran amphibians proceeds in an orderly manner, as just shown. Regulation of the orderly pattern of metamorphic events that transform the tadpole to a froglet is primarily attributed to changing patterns of secretion of thyroid hormone. This hormone is released as a mixture of **triiodothyronine** (T_3) and **thyroxine** (T_4), T_3 being the most effective agent. The thyroid gland depends for its development and hormone production on **thyroid stimulating hormone** (TSH) released by the anterior pituitary gland.

3,5,3′-Triiodothyronine (T_3)

3,5,3′,5′-Tetraiodothyronine (T_4)
(Thyroxine)

During the premetamorphic period the thyroid gland grows under the influence of TSH. The growing tadpole does not initiate metamorphic changes, however, for the pituitary gland, while releasing TSH, likewise produces prolactin. Prolactin stimulates body growth but inhibits thyroid-induced metamorphosis. Meanwhile, however, under the influence of low thyroid hormone levels, the larval hypothalamus becomes active and releases TSH-releasing factor (TSH–RF), which stimulates increased production and release of TSH. (As isolated from higher animals, this TSH–RF is a tripeptide: pyroglutamic acid–histidine–proline—NH_2). The hypothalamus also produces a substance that inhibits production and release of prolactin by the pituitary. Increased levels of TSH now trigger more rapid growth of the thyroid, which now releases more of its hormone. At this point prometamorphic changes are initiated, as already described. Meanwhile, thyroid hormone exercises a positive feedback on the hypothalamus, thus stimulating the release of more TSH via TSH–RF. Production of thyroid hormone is, therefore, further enhanced. When the level of thyroid hormone is sufficiently high, metamorphic climax ensues. This occurrence results not only in the drastic morphological and biochemical changes just described, but also in the partial regression of the thyroid gland. The hypothalamus now diminishes its output of factors affecting the pituitary hormones, and a hormonal balance appropriate to the life of the growing froglet ensues.

Although essentially all aspects of amphibian metamorphosis are attributable to changes in the level of thyroid hormone, the story of metamorphic control is considerably more complex. Larval tissues differ in the time at which they become competent to respond to thyroid hormone (Moser, 1950), in the rate of their response, and in the level of hormone (threshold) required to elicit a response. Moreover, threshold requirements for thyroid hormone in a responding tissue may change during the course of the response. Kollros (1961), for example, showed that complete development of the hindlimb in *Rana pipiens* requires a hormone level more than 20 times as great as that required to initiate growth of the limb. On the other hand, the regressing tail becomes increasingly sensitive to the hormone as metamorphosis proceeds.

An understanding of the molecular mechanisms of thyroid hormone action has been slow in coming. Studies to this end have been directed principally to the liver and tail. In the liver, as metamorphosis begins, the originally rounded euchromatic nuclei of the hepatocytes become indented and heavily heterochromatic, the mitochondria enlarge and increase in number, and the rough endoplasmic reticulum, previously found only in the periphery of the cells, increases throughout their cytoplasm (Bennett and Glenn, 1970). These changes require accelerated production of rRNA and of ribosomal and membrane proteins. The metamorphic transformation of the liver occurs without loss of larval hepatocytes or their proliferation. The response of the liver to thyroid hormone is thus unique among tissues of the animal undergoing metamorphosis, for other target tissues respond by growth involving mitosis or by remodeling that is achieved by death and resorption of larval cells and the subsequent proliferation of adult tissue (Morris and Cole, 1978).

There is no unequivocal evidence that the thyroid hormone stimulates the production of novel mRNAs in the liver. It affects the rate of transcription of both rRNA and mRNA and the rate of protein synthesis, but to what extent new kinds of proteins are made is not clear. HbA is almost certainly a protein that first appears under the influence of thyroid hormone. Very likely, too, cytoplasmic and nuclear receptor proteins for estrogens are not present prior to the action of the thyroid gland, for estrogen-dependent synthesis of the vitellogenic proteins (Chapter 7) can only be demonstrated after metamorphosis begins (Huber et al., 1980). The most dramatic hormone-induced change in the protein synthetic pattern of the liver is the coordinated and very rapid increase in production of the enzymes involved in urea synthesis. Treatment with actinomycin D is not immediately effective in preventing enhanced synthesis of these enzymes, however. This observation suggests that mRNAs for the ornithine–urea cycle enzymes are already present in liver cells prior to hormone treatment and that thyroid hormone enhances their translation (Cohen, 1970).

The analysis of effects of T_3 and T_4 on the regression of the tadpole tail has been greatly facilitated by the fact that the amputated tail can be held in organ culture in a healthy condition for prolonged periods. Cultured tails exposed to thyroid hormone undergo regression with a timing and magnitude

413

similar to regression in vivo. Tata (1966) showed that in the hormone-induced isolated tail tip there is a burst of RNA and protein synthesis prior to or coincident with the onset of regression. Regression is completely abolished by actinomycin D, puromycin, or cycloheximide, which suggests that new transcription and protein synthesis are required in order for regression to occur. Many hydrolytic enzymes, including cathepsins and acid phosphatase, increase in total activity during tail regression. These enzymes are identified as lysosomal enzymes in other tissues and species and, in regressing tadpole tails, are largely located within the phagocytic vacuoles of macrophages. The destruction of the tail cannot be attributed to macrophage invasion, however, or to secretion of degradative enzymes by macrophages. Disintegration of muscle, for example, begins well in advance of any evidence of tissue digestion by macrophages, and the spatiotemporal distribution of degenerative change is not readily correlated with the secretion of enzymes from phagocytic cells.

414

The primary action of thyroid hormone on tail regression still remains a puzzle. Reasonably, however, it induces the synthesis or activation of histolytic enzymes in a manner involving genetic transcription. Collagenase, shown by Gross and Lapiere (1962) to be released during tail regression in vitro, might be one such enzyme. After histolysis is under way, phagocytes developing in the mesenchyme ingest tissue debris. This ingestion creates phagocytic vacuoles whose presence in the cells stimulates synthesis of lysosomal enzymes (Beckingham Smith and Tata, 1976).

17.4.1.4 Receptors for Thyroid Hormone.
The step that initiates the hormone-induced biochemical changes in target tissues is probably the binding of the hormone to cellular receptor molecules. Frieden and his associates have determined that the cytoplasm of a number of tadpole tissues contains acidic proteins that bind T_3 with different affinities. Binding is cation dependent (Yoshizato et al., 1975a). In the tail fin the receptors show an affinity for T_3 that is 250 times greater than that for T_4 (Kistler et al., 1977), which suggests that T_4 is probably not bound under physiological conditions.

Liver cells in vitro show rapid binding of T_3 and T_4 to both cytoplasmic and nuclear receptors. Kistler et al. (1975) determined that the maximum number of binding sites per liver nucleus is about 12,300 for T_3 and about 2300 for T_4. There are, however, tissue-specific differences in the number and proportion of nuclear binding sites for T_3 and T_4. Yoshizato et al. (1975b) found, for example, that in tail nuclei the maximum number of binding sites per nucleus is 1500 for T_3 and 800 for T_4.

Whether the cytoplasmic receptors play a role in the translocation of the hormone to the nucleus is not known. Rat liver cells, which respond to thyroid hormone by enhanced production of certain globulins, have cytoplasmic receptors similar to those found in frog cytoplasm. These receptors, too, show a greater affinity for T_3 than for T_4. Nevertheless, isolated rat liver nuclei readily take up T_3 and T_4, and just as do frog nuclei, they preferentially bind T_3 (Spindler et al., 1975). Nuclear binding, therefore, apparently does not require prior binding to a cytoplasmic receptor. The nuclear receptor in rat liver is a nonhistone acidic chromosomal protein that is bound to DNA in regions that are rich in RNA polymerase and that are transcriptionally active (Spindler et al., 1975). Nuclear receptors show a very high affinity for T_3, but they rapidly lose this affinity when they are solubilized and separated from other chromosomal proteins. Their affinity for T_3 is restored, however, by addition of the histone proteins H2A, H2B, H3, and H4.* This information suggests that T_3 promotes a histone–nonhistone protein interaction that is involved in the regulation of gene expression (Eberhardt et al., 1979).

17.4.2 METAMORPHOSIS IS LESS DRASTIC FOR URODELES.
Most larval salamanders (Figure 17.19) have external gills and a long tail with dorsal and ventral fins. The forelimbs appear before the hindlimbs do, and both pairs of appendages grow gradu-

---

*These histones are components of **nucleosomes**, which consist of two molecules of each of the named histones with about 140 nucleotide pairs of DNA wrapped around them. Nucleosomes are self-assembling units of chromosome structures; that is, DNA and the histones automatically assume the nucleosome structure when combined. When nuclei are caused to swell and burst by water, the chromosomes unravel and nucleosomes are seen as "beads on a string." Each nucleosome is linked to its neighbor by 20–75 base pairs of DNA with which histone H1 is associated.

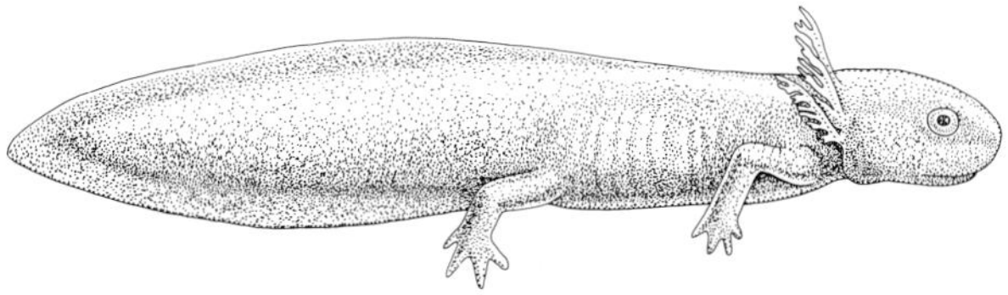

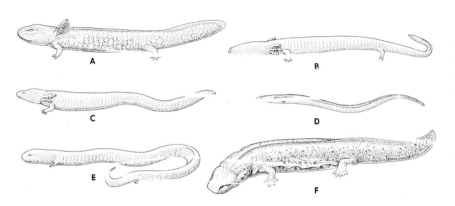

FIGURE 17.19 Larva of *Ambystoma*, a salamander.

ally, independent of any metamorphic stimulus. The mouth of the larva is broad, essentially as it is in the adult. Metamorphosis involves loss of the external gills and the tail fins, the formation of lungs, and disappearance of the gills; eyelids appear, the maxillary bones form, the skeleton begins to ossify, the skin cornifies, and skin glands are differentiated. After metamorphosis urodeles may reduce the time spent in water or they may become land dwellers, depending on the species, but most return to the water to breed.

17.4.3 NOT ALL AMPHIBIANS HAVE FREE-LIVING LARVAL STAGES.
Among both anurans and salamanders there are species in which the egg develops into a gilled, tadpole-like stage within the jelly membranes and then goes through metamorphosis before hatching. Other species, moreover, show direct development, bypassing the larval configuration completely. Such developmental patterns occur especially among amphibian groups that live permanently on land (see Lynn, 1961, and Dent, 1968, for reviews).

17.4.4 SEXUAL MATURITY OF SOME LARVAL FORMS.
Many urodeles show little metamorphic change in nature, retaining larval characteristics,

even as reproductive individuals, throughout their life history. There is considerable variation, however, among different families in the degree to which metamorphic change occurs (Figure 17.20). The Proteidae, of which the mudpuppy, *Necturus*, is an example, form simple lungs, but retain their gills and other larval characteristics. Among the Sirenidae are species whose external characteristics are entirely larval except for the skin, which acquires the glandular character of the adult. *Cryptobranchus* and *Amphiuma*, members of other families, lose their gills but otherwise remain larval in form. All of these organisms achieve sexual maturity while existing as permanent larvae. This condition is referred to as **neoteny.**

Neotenous races are also found among families whose members characteristically show metamorphosis. Races of *Ambystoma tigrinum* living in certain high cold lakes of the Rocky Mountains of the United States achieve sexual maturity as larvae. Generations regularly pass without the appearance of an adult. Yet many of these organisms do metamorphose in typical fashion when brought to warmer temperatures of the laboratory. The axolotl *Amblystoma mexicanum* is another neotenous salamander. It does not show external metamorphic changes in nature but does so in the laboratory in response to the administration of thyroid hormone (Prahlad and DeLanney, 1965; Taurog, 1974). Nevertheless, in the absence of thyroid stimulation, the neotenous axolotl produces serum proteins of the adult type and the adult form of hemoglobin (Ducibella, 1974). The neotenous animal also gradually increases the amount of metabolic nitrogen excreted as urea. If it is forced to go through metamorphosis, the animal excretes 90% of its nitrogenous waste as urea (Schultheiss, 1977).

415

FIGURE 17.20 Some neotenous urodeles. These retain larval features throughout life, but they are sexually mature. A: *Necturus*. B: *Protein*. C: *Siren*. D: *Pseudobranchus*. E: *Amphiuma*. F: *Cryptobranchus*. [From W. G. Lynn, *Am Zool* 1:151–61 (1961), by courtesy of Dr. Lynn and permission of *American Zoologist*.]

17.5 Second Metamorphosis in Vertebrates

It is seldom noted in developmental biology texts that many higher animals show more than one major change in form, physiology, and life habit between birth (or hatching), and adulthood.

The familiar eastern spotted newt, *Notophthalmus viridescens*, exhibits two prominent metamorphic changes during the life cycle. This newt begins its posthatching life as an olive-green larva equipped with gills, a keeled tail, and a well developed **lateral-line organ** (a system of receptors sensitive to local displacement of water). After a few months of growth metamorphosis occurs: the gills and tail fins are resorbed, the lateral-line system becomes nonfunctional, and the skin becomes rough and dry and orange-red in color. The visual pigment, predominantly porphyropsin, in the larva changes to a mixture of porphyropsin and rhodopsin. The newt is now called an **eft** (Figure 17.21), and it becomes a woodland dweller for 2 or 3 years, remaining on land and growing to full size. The eft phase is terminated through the action of prolactin and pituitary gonadotropins. These hormones inaugurate a drive

FIGURE 17.21 Red eft stage of *Notophthalmus viridescens*. [Photograph kindly supplied by Dr. Margaret Stewart, whose courtesy is gratefully acknowledged.]

toward water and initiate physiological changes resulting in the functional reinstatement of the lateral-line system, the reversion of the skin to a mucus-secreting, wet, and shiny organ, the restoration of a finned tail, the maturation of the gonads, and the adoption of porphyropsin as the sole visual pigment. In this condition the eastern newt returns to the water to spawn and to remain throughout its life, which lasts for several breeding seasons (Dent, 1975).

Multiple metamorphic changes are especially prominent in aquatic organisms that undertake spawning migrations between fresh and salt water, or the reverse. The spawning runs of Pacific salmon are probably familiar to most citizens, for the news media frequently feature illustrations of the adults fighting their way against raging currents and up "fish ladders" to bypass flood control dams of the Pacific Northwest. These salmon return to spawn in the headwaters of streams where they, themselves, were hatched years before. The young hatch and go through a series of larval forms before proceeding down the river and out to sea where they spend several years feeding and developing to maturity. With development of the gonads, the adults return to the rivers and undergo a final metamorphosis. They cease feeding, their bodily pigmentation changes, the digestive tract degenerates, and the males develop great hooked jaws. By the time they have reached the spawning grounds both males and females are moribund, their final metamorphosis marking the onset of irreversible processes leading to death, which occurs a few hours after spawning.

Organisms, such as the salmon that move upstream from the sea or large lakes to spawn are referred to as **anadromous** (*ana-*, a prefix from a Greek root meaning upward; *dromeus*, to run). Another anadromous organism showing a second metamorphosis is the sea lamprey, *Petromyzon marinus*, a cyclostome that attaches to a host fish by way of its spherical, rasping mouth and literally sucks the blood out of the host's body (Figure 17.22A). It is found in the waters off the Atlantic coast of North America, the coasts of Europe, the west coast of Africa, and in the Great Lakes of the United States. *P. marinus* became landlocked in the eastern Great Lakes after the last glaciation, about 7000 years ago. With the opening of the Welland Ship Canal in 1829, the species spread further westward, reaching Lake Superior by 1946. In subse-

FIGURE 17.22 *Petromyzon marinus*, the sea lamprey. A: Adult prior to second metamorphosis. B: Similar specimen parasitizing moribund host. [Photographs kindly supplied by Dr. P. J. Manion, whose courtesy is gratefully acknowledged.]

417

quent years the lampreys severely reduced the population of rainbow trout and almost eliminated the harvest of whitefish for some years. Control measures have now reduced the lamprey population, and valuable food fish are now more plentiful in the Great Lakes.

Lampreys spawn in freshwater tributaries of the oceans and Great Lakes. From the egg there emerges a blind **ammocetes** larva about 5 mm in length. It burrows in the mud by day, feeding on diatoms and desmids, and emerges at night. The ammocetes remains in the river for as much as 8 years reaching a length of 130–150 mm. Upon reaching a size for metamorphosis the animal ceases feeding and undergoes a profound metamorphosis (Figure 17.23), that adapts it for a parasitic

existence. It develops a rasping, suctorial mouth and eyes that utilize rhodopsin, as though in anticipation of its forthcoming marine existence. The journey to the sea may take several weeks or months without a feeding period. Leaving the river the lamprey begins to parasitize fish (Figure 17.22B), and during a period of upwards of 1½ years it grows to a length of about 720 cm in the sea (430 cm in the Great Lakes). It then ceases feeding and undergoes a second metamorphosis: the gonads mature, the sexes become distinct, and both acquire a golden mating tint. The visual pigment changes to porphyropsin and the digestive tract degenerates. The animals then migrate to their natal stream or another tributary to spawn and die (Manion and Stauffer, 1970; Potter et al., 1978).

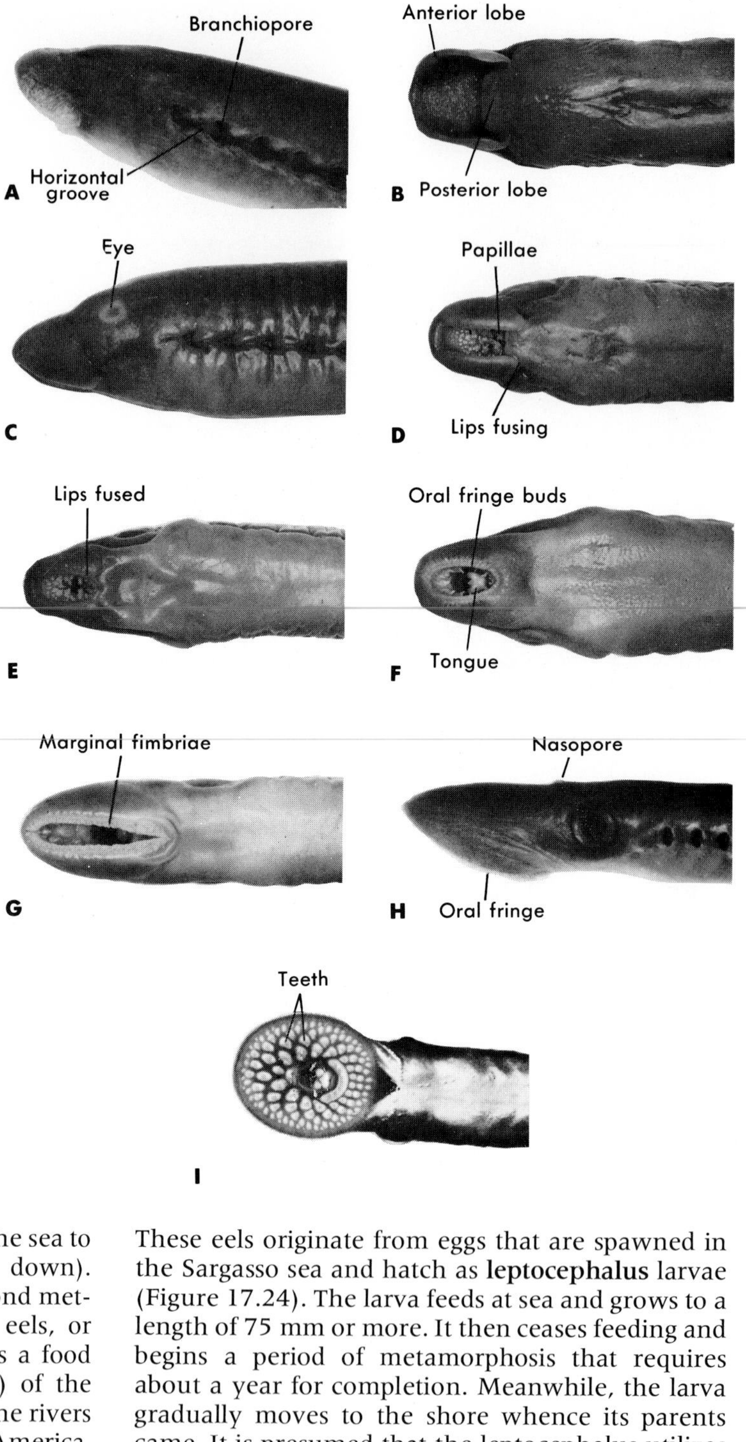

FIGURE 17.23 Changes in head region during metamorphosis of the ammocetes of *Petromyzon marinus*. A, B: Lateral and ventral views of the premetamorphic larva. C, D: Onset of metamorphosis (stage 1) as seen in lateral and ventral views, respectively; note onset of eye develoment and fusion of the lips of the oral lobes. E, F: 2nd and 3rd stages of metamorphosis as seen in ventral views. G: Ventral view of mouthparts as seen during the 4th stage of metamorphosis. H, I: Lateral and ventral views of the head of the newly metamorphosed lamprey; openings of the spiracles are seen caudal to the large eye in H, and the impressive rasping apparatus of mouth and tongue is seen in I. [From P. J. Manion and T. M. Stauffer, *J Fisheries Res Bd Canada* 27:1735–46, (1970), by courtesy of Dr. Manion and with permission.]

418

Organisms that return from streams to the sea to spawn are called **catadromous** (Gr., *kata,* down). Many of these organisms also exhibit a second metamorphosis, for example, the freshwater eels, or river eels, which are regarded by many as a food delicacy. They are teleost fish (bony fish) of the family Anguillidae and are found in estuarine rivers along the shores of Europe and North America.

These eels originate from eggs that are spawned in the Sargasso sea and hatch as **leptocephalus** larvae (Figure 17.24). The larva feeds at sea and grows to a length of 75 mm or more. It then ceases feeding and begins a period of metamorphosis that requires about a year for completion. Meanwhile, the larva gradually moves to the shore whence its parents came. It is presumed that the leptocephalus utilizes

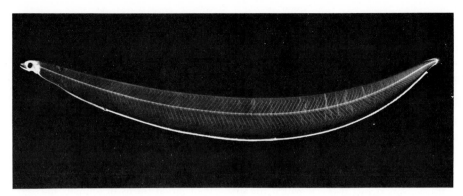

FIGURE 17.24 Leptocephalus of the eel *Ariosoma balearicum*. This is a photograph of a specimen fixed in formaldehyde. If unfixed, it would be completely transparent except for the eyes. [From W. H. Hulet, *Phil Trans Roy Soc Lond* [B]282:107–38 (1978), by courtesy of Dr. Hulet and permission of The Royal Society.]

a deep-sea variety of rhodopsin as visual pigment. As it reaches the shore, metamorphosis to the familiar adult eel form is completed, and the animal enters brackish or fresh water for a prolonged period of growth during which it may grow to a meter in length. Specimens from the rivers have a mixture of porphyropsin and rhodopsin in the retina. This rhodopsin, however, shows an absorption peak for light at a different wavelength from the rhodopsin of deep-sea fish.

When the larva is ready for its spawning run downstream, it ceases feeding, the digestive tract deteriorates, its color changes, and the eyes double in size. Porphyropsin is lost from the retina and the rhodopsin changes to the deep-sea form. The gonads begin to develop and the adults now progress thousands of miles to the Sargasso sea. There they spawn and die.

17.6 Summary

The development of many animals is indirect in the sense that one or more larval forms intervene between the embryonic and adult stages. Larvae usually differ greatly from the adult in habitat and mode of life. The process of change that occurs between larval forms or between larva and adult is termed metamorphosis. Animals that have direct development do not have larval forms but may, as in the case of many arthropods, undergo a series of stages separated by molts.

The study of larval forms of invertebrates has assisted in the identification of the principal branches of the evolutionary tree leading to modern animals, particularly the two principal divisions, the Prostomia and the Deuterostomia; the latter provide the line of descent to the vertebrates.

Insects are said to be ametabolous if they show direct development, hemimetabolous if they show a gradual metamorphosis, and holometabolous if the (usually) winged, sexually mature adults are separated from a grub-like or maggot-like larva by an intervening pupal stage. The pupa, in many kinds of insects, suffers the disintegration of all or most of its larval organs, forming the major adult structures from nests of previously undifferentiated cells called imaginal discs.

Molting in the hemimetabolous and holometabolous insects is controlled by hormones. Larval molts occur at appropriate titers of ecdysone in the presence of juvenile hormone. Ecdysone, a steroid hormone, is produced and released by the prothoracic glands under the stimulus of a brain hormone. Juvenile hormone, a complex aliphatic hydrocarbon, is secreted by the corpora allata. The pupal molt, which requires ecdysone, occurs when juvenile hormone is absent, at the end of the last larval instar.

The larval–pupal transformation is marked by changing patterns of gene activity. In dipteran insects giant, transversely banded polytene chromosomes in large cells of the salivary gland give visual evidence of gene activity when certain banded regions form puffs. The puffs are sites of mRNA synthesis. As many as 200 puffs appear in the chromosomes of each salivary gland cell during the larval–pupal transformation. Various bands undergo puffing and regression in a pattern that is controlled by the level of ecdysone and is precisely

419

correlated with the progress of the larval–pupal molt. Certain puffs are evidently sites of mRNA synthesis for the production of proteins of the salivary secretion, but the relevance of the great number of puffs to the physiology of transformation of larva to pupa is not clear.

Hormones also control metamorphosis in amphibians, most of which have free-living larval stages. Among urodeles, however, many kinds realize reproductive maturity while in a larval form and retain larval characteristics throughout life. Some urodeles may be caused to undergo metamorphosis by treatment with the thyroid hormones thyronine and triiodothyronine or certain of their analogues. The thyroid normally controls the onset of metamorphosis, which involves degeneration and death and growth in others (e.g. jaws, limbs). The different target organs also show differential thresholds for response.

Whereas many invertebrates may show a succession of larval forms, most vertebrates have only one obvious metamorphosis. Exceptions are found, however, among urodeles and a number of marine organisms. These show a prominent second metamorphosis when they return to their natal environment to spawn (and then to die, in many cases) after a prolonged existence in another habitat. Pacific salmon, freshwater eels, and sea lampreys are familiar examples. Some newts, too, undergo a second metamorphosis before reaching sexual maturity after a prolonged life on land.

The anatomical and biochemical changes that occur during vertebrate metamorphosis have been studied most intensively in the transformation of tadpole to frog. Remodeling of the body form and preparation for the adult life habit involve degeneration of larval organs and changing patterns of protein synthesis, notably in the eyes, skin, and liver. Adult specific hemoglobin is produced, and enzymes of the ornithine–urea cycle are greatly increased. It is not certain that amphibian metamorphosis involves translation of novel mRNAs. Stimulation by thyroid hormone, however, certainly alters the availability and rate of translation of mRNAs involved in production of proteins characteristic of the adult.

Triiodothyronine is the most effective form of thyroid hormone. It binds to cytoplasmic and to nuclear receptors. In the nucleus it binds to a non-histone chromosomal protein, but it requires for stable binding the presence of the histone proteins of nucleosomes.

17.7 Questions for Thought and Review

1. Illustrate by means of several examples ways in which the analysis of larval forms of animals may give insights into evolutionary relationships.

2. Distinguish between direct and indirect development as illustrated by different orders of insects. Distinguish among ametabolous, hemimetabolous, and holometabolous forms of insects.

3. Imaginal discs from a larval fly, such as *Drosophila melanogaster,* can be transplanted into the abdominal cavity of an adult and can be maintained indefinitely by serial transplants from one adult to another. Why would you not expect such discs to undergo metamorphosis? How might you induce them to undergo metamorphosis?

4. Among insects there are found hormones that resemble those of vertebrates in some ways. What are these hormones? In what ways do they resemble vertebrate hormones? One hormone important to larval life is unlike any known vertebrate hormone. What is it, and what is its chemical nature? In what organs are the hormones involved in metamorphosis produced?

5. Compare and contrast the morphological changes of metamorphosis in anurans and in urodeles.

6. Describe the sequence of hormonal action during metamorphosis of an anuran tadpole. Characterize each hormone chemically and indicate its target organ or organs.

7. What is meant by the term second metamorphosis as applied to vertebrates? Describe the morphological and physiological changes that take place during primary and secondary metamorphosis in the eastern spotted newt.

8. What is the correlation between visual pigment and spawning behavior in anadromous and catadromous animals?

9. Discuss the effect of thyroxine on the kinds of proteins synthesized as the tadpole begins metamorphosis. Correlate these changes with preparation for a land-dwelling existence.

10. Does thyroxine initiate transcription of hitherto unused genes in anuran metamorphosis? Adduce as much evidence as you can in support of your answer.

11. What adaptive value may be attributed to the facts that tadpole hemoglobin becomes saturated with oxygen at low tensions and that the oxygen dissociation curve of tadpole hemoglobin hardly changes over a wide pH range?

12. Puffs in giant polytene chromosomes found in secretory organs of many flies appear in various patterns that change in preparation for pupation and in response to heat shock and ecdysone. Newly arising puffs are considered to be sites of new gene activity. Review the evidence that indicates such puffs are sites of mRNA synthesis that is translated into protein.

13. Are changes in the translating machinery of the cell possibly brought about by the hormones of metamorphosis? What is the evidence? Return to this question once more after reading Chapter 18.

17.8 Suggestions for Further Reading

ASHBURNER, M., C. CHIHARA, P. MELTZER, and G. RICHARDS. 1974. Temporal control of puffing activity in polytene chromosomes. *Cold Spring Harbor Symp Quant Biol* 38:655–62.

BARRINGTON, E. J. W. 1961. Metamorphic processes in fishes and lampreys. *Am Zool* 1:97–106.

BENBASSAT, J. 1974. The transition from tadpole to frog haemoglobin during natural amphibian metamorphosis. I. Protein synthesis by peripheral blood cells in vitro. *J Cell Sci* 15:347–57.

BERENDES, H. D. 1973. Synthetic activity of polytene chromosomes. *Int Rev Cytol* 35:61–116.

BIESSMANN, H., B. W. LEVY, and B. J. MCCARTHY. 1978. In vitro transcription of heat-shock-specific RNA from chromatin of *Drosophila melanogaster* cells. *Proc Natl Acad Sci USA* 75:759–63.

BULTMANN, H., and U. CLEVER. 1969. Chromosomal control of foot pad development in *Sarcophaga bullata*. I. The puffing pattern. *Chromosoma* 28:120–35.

CLONEY, R. A. 1969. Cytoplasmic filaments and morphogenesis: the role of the notochord in ascidian metamorphosis. *Z Zellforsch Mikrosk Anat* 100:31–53.

ETKIN, W. 1966. How a tadpole becomes a frog. *Sci Am* (May)76–88.

ETKIN, W., and L. I. GILBERT, eds. 1968. *Metamorphosis: A Problem in Developmental Biology*. New York: Appleton-Century-Crofts.

FELL, H. B. 1948. Echinoderm embryology and the origin of chordates. *Biol Rev* 23:81–107.

FRAENKEL, G., A. BLECHL, J. BECHL, P. HERMAN, and M. J. SELIGMAN. 1977. 3':5'-Cyclic AMP and hormonal control of puparium formation in the fleshfly *Sarcophaga bullata*. *Proc Natl Acad Sci USA* 74:2182–86.

FRIEDEN, E. 1968. Biochemistry of amphibian metamorphosis. In W. Etkin and L. I. Gilbert, eds. *Metamorphosis*. New York: Appleton-Century-Crofts, pp. 349–98.

GORELL, T. A., L. I. GILBERT, and J. B. SIDDAL. 1972. Studies on hormone recognition by arthropod target-tissues. *Am Zool* 12:347–57.

GRANT, W. C., Jr. 1961. Special aspects of the metamorphic process: second metamorphosis. *Am Zool* 1:163–71.

HARDISTY, M. W., and I. C. POTTER, eds. 1971. *The Biology of Lampreys*, Vols. 1 and 2. New York: Academic Press.

JUST, J. J., J. SCHWAGER, R. WEBER, H. FEY, and H. PFISTER. 1980. Immunological analysis of hemoglobin transition during metamorphosis of normal and isogenic *Xenopus*. *Wilhelm Roux Arch* 188:75–80.

KAFATOS, F. and C. M. WILLIAMS. 1964. Enzymatic mechanism for the escape of certain moths from their cocoons. *Science* 146:538–40.

LAUFER, H. 1963. Hormones and the development of insects. *Proc Int Cong Zool XVI, Wash* 4:215–20.

LAUFER, H. 1965. Chromosomal puffing and its relation to cell function and development. *Arch Anat Microsc Morphol Exp* 54:648–51.

MCAVOY, J. W., and K. E. DIXON. 1977. Cell proliferation and renewal in the small intestinal epithelium of metamorphosing and adult *Xenopus laevis*. *J. Exp Zool* 202:129–38.

OHTSU, K., K. NAITO, and F. H. WILT. 1964. Metabolic basis of visual pigment conversion in metamorphosing *Rana catesbeiana*. *Dev Biol* 10:216–32.

PASSANO, L. M. 1961. The regulation of crustacean metamorphosis. *Am Zool* 1:89–95.

SPANGENBERG, D. B. 1974. Thyroxine in early strobilation in *Aurelia aurita*. *Am Zool* 14:825–31.

SPANGENBERG, D. B. 1971. Thyroxine induced metamorphosis in *Aurelia*. *J Exp Zool* 178:183–94.

SURKS, M. 1., W. KOENER, W. DILLMANN, and J. H. OPPENHEIMER. 1973. Limited capacity binding sites for L-triiodothyroxine in rat liver nuclei. Localization to the chromatin and partial characterization of the L-thiiodothyroxine-chromatin complex. *J Biol Chem* 248:7066–72.

TATA, J. R. 1968. Early metamorphic competence of *Xenopus* larvae. *Dev Biol* 18:415–40.

TAUROG, A., C. OLIVER, R. L. PORTER, J. C. McKENZIE, and J. M. McKENZIE. 1974. The role of TRH in the neoteny of the Mexican axolotl (*Ambystoma mexicanum*). *Gen Comp Endocrinol* 24:267–79.

VAN DER KLOOT, W. G. 1961. Insect metamorphosis and its endocrine control. *Am Zool* 1:3–9.

WHITTEN, J. M. 1969. Coordinated development in the foot pad of the fly *Sarcophaga bullata* during metamor-

421

phosis: changing puffing patterns of the giant cell chromosomes. *Chromosoma* 26:215–44.

WRIGHT, SISTER MARY L. 1977. Regulation of cell proliferation in tadpole limb epidermis by thyroxine. *J Exp Zool* 202:223–34.

ZDAREK, J., and G. FRAENKEL. 1972. The mechanism of puparium formation in flies. *J Exp Zool* 179:315–24.

17.9   References

ASHBURNER, M. 1967. Patterns of puffing activity in the salivary gland chromosomes of *Drosophila*. I. Autosomal puffing patterns in a laboratory stock of *Drosophila melanogaster*. *Chromosoma* 21:398–428.

ASHBURNER, M. 1972. Patterns of puffing activity in the salivary gland chromosomes of *Drosophila*. VI. Induction by ecdysone in salivary glands cultured in vitro. *Chromosoma* 38:255–81.

ASHBURNER, M., and J. J. BONNER. 1979. The induction of gene activity in *Drosophila* by heat shock. *Cell* 17:241–54.

ASHBURNER, M., and G. RICHARDS. 1976. Sequential gene activation by ecdysone in polytene chromosomes in *Drosophila melanogaster*. III. Consequences of ecdysone withdrawal. *Dev Biol* 54:241–55.

BECKINGHAM SMITH, K., and J. R. TATA. 1976. Cell death. Are new proteins synthesized during hormone-induced tadpole tail regression? *Exp Cell Res* 100:129–46.

BENNETT, T. P., and J. S. GLENN. 1970. Fine structural changes in liver cells of *Rana catesbeiana* during natural metamorphosis. *Dev Biol* 22:535–59.

BERENDES, H. D., C. ALONSO, P. J. HELMSING, H. J. LEENDERS, and J. DERKSEN. 1973. Structure and function of the genome of *Drosophila hydei*. *Cold Spring Harbor Symp Quant Biol* 38:645–54.

COHEN, P. P. 1970. Biochemical differentiation during amphibian metamorphosis. *Science* 168:533–43.

DENT, J. N. 1968. Survey of amphibian metamorphosis. In W. Etkin and L. I. Gilbert, eds. *Metamorphosis: A Problem in Developmental Biology*. New York: Appleton-Century-Crofts, pp. 271–311.

DENT, J. N. 1975. Integumentary effects of prolactin in the lower vertebrates. *Am Zool* 15:923–935.

DOYLE, L., and H. LAUFER. 1969a. Sources of larval salivary gland secretion in the dipteran *Chironomus tentans*. *J Cell Biol* 40:61–78.

DOYLE, D., and H. LAUFER. 1969b. Requirements of ribonucleic acid synthesis for the formation of salivary gland specific proteins in larval *Chironomus tentans*. *Exp Cell Res* 57:201–10.

DUCIBELLA, T. 1974. The occurrence of biochemical metamorphic events without anatomical metamorphosis in the axolotl. *Dev Biol* 38:175–86.

EBERHARDT, N. L., J. C. RING, L. K. JOHNSON, K. R. LATHAM, J. W. APRILETTI, R. N. KITSIS, and J. D. BAXTER. 1979. Regulation of activity of chromatin receptors for thyroid hormone: possible involvement of histone-like proteins. *Proc Natl Acad Sci USA* 76:5005–5009.

FRISTROM, J. W., W. R. LOGAN, and C. MURPHY. 1973. The synthetic and minimal culture requirements for evagination of imaginal discs of *Drosophila melanogaster* in vitro. *Dev Biol* 33:441–56.

GEHRING, W. 1968. The stability of the determined state in cultures of imaginal discs in *Drosophila*. In H. Ursprung, ed. *The Stability of the Differentiated State*. New York: Springer-Verlag, pp. 137–54.

GROSS, J., and C. M. LAPIERE. 1962. Collagenolytic activity in amphibian tissues: a tissue culture assay. *Proc Natl Acad Sci USA* 48:1014–22.

GROSSBACH, U. 1973. Chromosome puffs and gene expression in polytene cells. *Cold Spring Harbor Symp Quant Biol* 38:619–27.

GUILLEMIN, R., and R. BURGUS. 1972. The hormones of the hypothalamus. *Sci Am* (November)24–33.

HADORN, E. 1968. Dynamics of determination. In M. Locke, ed. *Major Problems in Developmental Biology*. New York: Academic Press, pp. 85–104.

HENRIKSON, P. A., and U. CLEVER. 1973. Protease activity and cell death during metamorphosis in the salivary gland of *Chironomus tentans*. *J Insect Physiol* 18:1981–2004.

HUBER, S., G. U. RYFFEL, and R. WEBER. 1980. Thyroid hormone induces competence for oestrogen-dependent vitellogenin synthesis in developing *Xenopus laevis* liver. *Nature* 278:65–67.

ILAN, J., J. ILAN, and J. H. QUASTEL. 1966. Effects of actinomycin D on nucleic acid metabolism and protein biosynthesis during metamorphosis of *Tenebrio molitor* L. *Biochem J* 100:441–47.

ILAN, J., J. ILAN, and N. PATEL. 1970. Mechanism of gene expression in *Tenebrio molitor*. Juvenile hormone determination of translational control through transfer ribonucleic acid and enzyme. *J Biol Chem* 245:1275–81.

JAMRICH, M., A. L. GREENLEAF, and E. K. F. BAUTZ. 1977. Localization of RNA polymerase in polytene chromosomes of *Drosophila melanogaster*. *Proc Natl Acad Sci USA* 74:2079–83.

JURD, R. D., and N. MACLEAN. 1970. An immunofluorescent study of the haemoglobins in metamorphosing *Xenopus laevis*. *J Embryol Exp Morphol* 23:299–309.

KISTLER, A., K. YOSHIZATO, and E. FRIEDEN. 1975. Binding of thyroxine and triiodothyronine by nuclei of isolated tadpole liver cells. *Endocrinology* 97:1036–42.

KISTLER, A., K. YOSHIZATO, and E. FRIEDEN. 1977. Preferential binding of tri-substituted thyronine analogs by bullfrog tadpole tail fin cytosol. *Endocrinology* 100:134–37.

KOLLROS, J. J. 1961. Mechanisms of amphibian metamorphosis: hormones. *Am Zool* 1:107–14.

KORGE, G. 1975. Chromosome puff activity and protein synthesis in larval salivary glands of *Drosophila melanogaster. Proc Natl Acad Sci USA* 72:4500–54.

KORGE, G. 1977a. Larval saliva in *Drosophila melanogaster:* production, composition and relationship to chromosome puffs. *Dev Biol* 58:339–55.

KORGE, G. 1977b. Direct correlation between a chromosome puff and the synthesis of a larval saliva protein in *Drosophila melanogaster. Chromosoma* 62:155–74.

LAMBERT, B., J. WIESLANDER, B. DANEHOLT, E. EGHAZI, and U. RINGBORG. 1972. In situ demonstration of DNA hybridizing with chromosomal and nuclear sap RNA in *Chironomus tentans. J Cell Biol* 53:407–18.

LAUFER, H., Y. NAKASE, and J. VANDERBERG. 1964. Developmental studies of the dipteran salivary gland. I. The effects of actinomycin D on larval development, enzyme activity, and chromosomal differentiation in *Chironomus thumni. Dev Biol* 9:367–81.

LAUFER, H., and Y. NAKASE. 1965. Salivary gland secretion and its relation to chromosomal puffing in the dipteran *Chironomus thumni. Proc Natl Acad Sci USA* 53:511–16.

LIPSON, M. J., R. A. CERSKUS, and J. E. SILBERT. 1971. Glycosaminoglycans and blycosaminoglycan-degrading enzyme of *Rana catesbeiana* back skin during late stages of metamorphosis. *Dev Biol* 25:198–208.

LYNN, W. G. 1961. Types of amphibian metamorphosis. *Am Zool* 1:151–61.

MANDARON, P. 1971. Sur le mécanisme de l'évagination des disques imaginaux chez la Drosophile. *Dev Biol* 25:581–605.

MANIATIS, G. M., and V. M. INGRAM. 1971. Erythropoiesis during amphibian metamorphosis. III. Immunological detection of tadpole and frog hemoglobins in single erythrocytes. *J Cell Biol* 49:390–401.

MANION, P. J., and T. M. STAUFFER. 1970. Metamorphosis of the landlocked sea lamprey *Petromyzon marinus. J Fish Res Bd Canada* 27:1735–46.

MEYER, A. S., H. A. SCHNEIDERMAN, E. HANZMANN, and J. H. KO. 1968. Hormones from the cecropia silk moth. *Proc Natl Acad Sci USA* 60:853–60.

MEYER, A. S., E. HANZMANN, H. A. SCHNEIDERMAN, L. I. GILBERT, and M. BOYETTE. 1970. The isolation and identification of the two juvenile hormones from the cecropia silk moth. *Arch Biochem Biophys* 137:190–213.

MOSER, H. 1950. Ein Beitrag zur Analyse der Thyroxinwirkung im Kaulquappenversuch und zur Frage nach dem Zustandekommen der Fruhbereitschaft des Metamorphosereaktionssystems. *Rev Suisse Zool* 57 (Suppl. 2) 1:1–144.

MORRIS, S. M., Jr., and R. D. COLE. 1978. Histone metabolism during amphibian metamorphosis. Isolation, characterization and biosynthesis. *Dev Biol* 62:52–64.

MUNRO, A. F. 1939. Nitrogen excretion and arginase activity during amphibian development. *Biochem J* 33:1957–65.

PELLING, C. 1964. Ribonucleinsäure-synthese der Reisenchromosomen. Autoradiographische Untersuchungen an *Chironomus tentans. Chromosoma* 15:71–122.

POLANSKY, J. R., and T. P. BENNET. 1970. Differences in the soluble lens proteins from tadpole and adult bullfrogs. *Biochem Biophys Res Commun* 38:450–57.

POSTLETHWAIT, J. H., and H. A. SCHNEIDERMAN. 1974. Developmental genetics of *Drosophila* imaginal discs. *Annu Rev Genet* 7:381–433.

POTTER, I. C., G. M. WRIGHT, and J. H. YOUSON. 1978. Metamorphosis in the anadronous sea lamprey, *Petromyzon marinus* L. *Can J Zool* 56:561–70.

PRAHLAD, K., and L. DELANNEY. 1965. A study of induced metamorphosis in the axolotl. *J Exp Zool* 160:137–45.

REEVES, O. R. 1975. Adult amphibian epidermal proteins: biochemical characterization and developmental appearance. *J Embryol Exp Morphol* 34:55–73.

REEVES, R. 1977. Hormonal regulation of epidermis-specific protein and messenger RNA synthesis in amphibian metamorphosis. *Dev Biol* 60:163–79.

RICHARDS, G. 1976a. Sequential gene activation of ecdysone in polytene chromosomes of *Drosophila melanogaster*. IV. The mid prepupal period. *Dev Biol* 54:256–63.

RICHARDS, G. 1976b. Sequential gene activation by ecdysone in polytene chromosomes of *Drosophila melanogaster*. V. The later prepupal puffs. *Dev Biol* 54:264–75.

SCHULTHEISS, H. 1977. The hormonal regulation of urea excretion in the Mexican axolotl (*Ambystoma mexicanum* Cope). *Gen Comp Endocrinol.* 31:45–52.

SPINDLER, B. F., K. M. MACLEOD, J. C. KING, and J. D. BAXTER. 1975. Thyroid hormone receptors. Binding characteristics and lack of hormonal dependency for nuclear localization. *J Biol Chem* 250:4113–19.

TATA, J. R. 1966. Requirement for RNA and protein synthesis for induced regression of the tadpole tail in organ culture. *Dev Biol* 13:77–94.

TAUROG, A. 1974. Effect of TSH and long-acting thyroid stimulator on thyroid[131] I-metabolism and metamorphosis of the Mexican axolotl (*Ambystoma mexicanum*). *Gen Comp Endocrinol* 24:257–66.

WHITTEN, J. M. 1957. The supposed pre-pupa in Cyclorhaphous diptera. *Q J Microsc Sci* 98:241–50.

WOBUS, U., R. PANITZ, and E. SERFLING. 1970. Tissue specific gene activities and proteins in the *Chironomus* salivary gland. *Mol Gen Genet* 107:215–23.

YOSHIZATO, K., A. KISTLER, and E. FRIEDEN. 1975a. Metal ion dependence of the binding of triiodothyronine by

423

cytosol proteins of bullfrog tadpole tissues. *J Biol Chem* 250:8337–43.

YOSHIZATO, K., A. KISTLER, and E. FRIEDEN. 1975b. Binding of thyroid hormones by nuclei of cells from bullfrog tadpole tail fins. *Endocrinology* 97(4):1930–35.

ZHIMULEV, I. F., and E. S. BELYAEVA. 1975. ^{3}H-Uridine labeling patterns in the *Drosophila melanogaster* salivary gland chromosomes X, 1R and 3L. *Chromosoma* 49:219–31.

S I X

Genetic and Epigenetic Control of Development

Much emphasis has been placed earlier on the fact that **425** differentiation is marked by variable patterns of protein synthesis. Differentiated cells and tissues owe their distinctive characteristics to the functional attributes of their proteins. Because the genetic DNA determines the structure of the various proteins found in a cell, it is evident that the differentiated state of a cell depends on the transcription and translation of information in its genome.

We now know a great deal about the mechanism of genomic transcription and translation and can identify many points at which these processes can be regulated during development. What remain as serious problems, however, are the mechanisms whereby morphological patterns of high orders of complexity emerge from variable patterns of protein synthesis. What factors determine, for example, that the pattern of muscular and skeletal elements shall have the form of a human leg at the lower extremity and that of an arm at the upper extremity? These factors must largely be epigenetic, having their basis in macromolecular charge distributions and configurations, interactions between cells and between cells and cell products, and so on.

We have few insights into the origin of morphogenetic patterns, but we treat this problem briefly after
first considering the evidence for patterns of variable
gene activity and the regulation of protein synthesis
and processing as the basis for cellular differentiation.

It is now quite certain that, with few exceptions, every nucleus in a multicellular plant or animal contains the same genetic information that was present in the zygote nucleus. (Some evidence for this derives from studies of regulation and regeneration as outlined in Chapters 10 and 15; other evidence is adduced in later sections.) It follows from this fact, therefore, that the achievement and persistence of various differentiated conditions must issue from variable patterns of transcriptional activity on identical genomes, the controlled translation of specific transcripts into protein, and the processing of these proteins into their functional form. In this chapter we first present the basic evidence for the identity of genetic information in the variously differentiated cells of an organism and then present further evidence that development involves stage and tissue-specific variations in the pattern of cytoplasmic gene transcripts. The remainder of the chapter is devoted to a discussion of mechanisms that can be invoked to bring about these variations and that can control the kind and amount of the ultimate gene products, the proteins.

18

Control of Gene Expression in Realization of the Phenotype

18.1 Each Genome Contains a Complete Construction Plan

It is important to realize that the sequences of nucleotides that constitute the genes are responsible both for synthesis of the building materials of the organism and for the order and timing of their production and the pattern in which they are assembled. This discussion begins therefore with emphasis on the fact that various kinds of nuclei in an organism, probably even those of tumor cells, can direct normal patterns of morphogenesis.

The first indications of this fact were provided by experiments of Hämmerling (1943), who exchanged nuclei between species of the algal genus *Acetabularia*. Each organism consists of a single cell that is regionally differentiated into a **stalk, rhizoids,** and a **reproductive cap.** The rhizoids anchor the plant to its substratum, and the nucleus resides in one of them at the base of the stalk. At the other

end of the stalk is the umbrella-shaped reproductive cap. When this cap is mature, the single nucleus makes numerous daughter nuclei, which migrate to the cap, forming cysts filled with haploid ($1n$) spores. Germination of the cysts releases flagellated swimming cells that fuse in pairs to form the zygote ($2n$), which starts the next generation.

When a developing cap is severed from the stalk, a new cap regenerates, and regeneration can be forced repeatedly. If the nucleus is removed, however, a new cap may be regenerated, but repeated regenerations cannot occur. It is as though the nucleus-bearing rhizoid provides a substance necessary for cap formation, which is used up in its absence.

The kind of cap formed varies from species to species. Hämmerling raised the question of whether the nucleus controls the kind of cap produced and sought an answer by means of transplantation experiments involving two species, *A. mediterranea* (*med*) and *A. crenulata* (*cren*). Severing the rhizoid and cap of *cren* he produced an anucleate stalk to which he then grafted the nucleated rhizoid of *med*. The stalk regenerated a cap intermediate in form between *cren* and *med*, but forced by decapitation to regenerate again and again, it thereafter produced

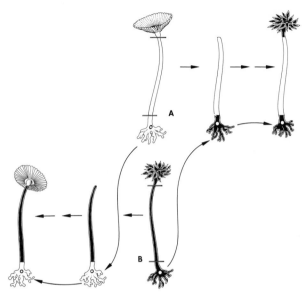

FIGURE 18.1 Nuclear transplantation between *Acetabularia mediterranea* (A) and *A. crenulata* (B). [After J. Hämmerling, *induk Abstum Vererbungsl* 81:114–80 (1943).]

only caps of *med* type, that is, of the kind genetically determined by the grafted nucleus. The overall result of this experiment and its reciprocal are diagrammed in Figure 18.1.

Even more dramatic evidence for the genetic equivalence of somatic cell nuclei in the control of morphogenesis in plants was provided by studies of F. C. Stewart and his colleagues, as noted in Chapter 16. They showed that secondary phloem cells of the carrot can be held in culture, forming embryo-like bodies that give rise to reproductively mature plants.

Evidence that nuclei of animal cells likewise contain a complete construction plan awaited the development of techniques for transplanting nuclei from differentiated cells into the cytoplasm of uncleaved eggs. In 1952 Briggs and King reported that they had successfully transplanted into activated but enucleated eggs of *Rana pipiens* the functional nucleus from embryos of the blastula stage. About one-third of such eggs gave rise to normal embryos. With increasing frequency as older donors were used, the transplanted nuclei were less able to bring about normal embryogenesis. Moreover, with advancing donor age, the recipient eggs showed dif-

ferent patterns of abnormal development correlated with the regional origin of the donor nucleus (King and Briggs, 1955; Briggs and King, 1957; Hofner and DiBerardino, 1980). Nevertheless, in some cases normal larvae, capable of metamorphosis, resulted from the transplantation of neurula nuclei into eggs. In 1963, Gurdon reported results of injecting nuclei from intestinal epithelial cells of *Xenopus laevis* tadpoles into activated oöcytes of *X. laevis*, the pronuclei of which had been inactivated by ultraviolet irradiation. Such eggs often cleaved and, with very low frequency, developed into normal larvae and, eventually, into sexually mature adults. Similar results were obtained by Gurdon and Laskey (1970) using nuclei from cultured cells of an epithelial type derived from minced tadpole tissues, dissociated by means of trypsin and grown in glass vessels. More recently, Gurdon et al. (1975) placed fragments of hindfoot interdigital cells from adult *Xenopus* into tissue culture under such conditions that 99% of cells growing out from the explant along the glass surface of the culture vessel became keratinized, as is typical of skin epidermis, within 10 days. This observation suggested to these investigators that such cells are fully determined as keratinizing skin cells at the time of their emigration from the explant. Injecting nuclei of 6-day cultured cells into activated *Xenopus* oöcytes resulted in formation of cleaving eggs in a high percentage of cases, and a number of those formed tadpoles that had a beating heart and were capable of muscular swimming movements. Regardless of the fact that such tadpoles apparently did not develop further, they showed an essentially normal tissue organization. Thus it is highly probable that nuclei of relatively specialized cells of adult *Xenopus* contain all the information requisite to development of a normal embryo. This is probably the case in nuclei of adult *Rana* cells, also, but they show a greater tendency to develop chromosomal anomalies as a result of the transplantation procedure. DiBerardino and Hoffner (1971) transplanted spermatogonial cell nuclei from juvenile and adult *Rana pipiens* into enucleated activated eggs and obtained some cases of development to the tadpole stage. They found, however, that in most cases severe chromosomal anomalies occurred and development was abnormal. Hennen (1970) reported that lower temperature and the addition of the polyanion spermine to the medium in which the transplant nucleus is prepared significantly enhance the developmental performance of the recipient eggs.

The very precise control of morphogenesis that can be exercised by transplanted nuclei is well illustrated by cases in which the transplanted nucleus is of a different genotype than the one it replaces. Signoret et al. (1962) studied nuclear control of pigmentation pattern in the Mexican axolotl, *Ambystoma mexicanum,* by means of the nuclear transplantation technique. In this urodele amphibian a white variant form results from a simple recessive muta-

tion (*dd*), which causes the dark melanin pigmentation to be restricted to a narrow band on each side close to the dorsal fin. In the wild type, or dark form (*DD* or *Dd*), the pigment cells are distributed over most of the body. Signoret and co-workers transplanted the diploid nucleus from a cell of the early embryo of a *dark* animal into an activated *white* egg whose own nucleus had been inactivated by means of ultraviolet irradiation. The egg thus developed with a functional *dark* nucleus, and the larva that formed showed the *dark* pigmentary pattern characteristic of the nuclear-donor strain (Figure 18.2).

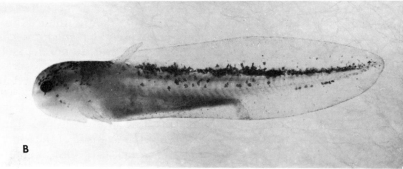

FIGURE 18.2 Nuclear control of pigmentary pattern in the larva of *Ambystoma mexicanum.* A: Larva developed from egg or an ultraviolet-irradiated white (*dd*) female injected with diploid nucleus from the early embryo of a dark, or wild-type, genotype (*DD* or *Dd*). B: Control larva of genotype *DD.* C: Control larva of genotype *dd.* Note that the *D* gene in *dd* cytoplasm (A) duplicates exactly the pigmentary pattern that it shows in its natural environment. [From J. Signoret, R. Briggs, and R. R. Humphrey; *Dev Biol* 4:134–64 (1962), by courtesy of the authors and permission of Academic Press, Inc., New York.]

429

18.1.1 REPROGRAMMING GENOMES OF CANCER CELLS. Most cancerous growths consist of tumors that show uncontrolled growth of a differentiated tissue type, histologically recognizable and exercising the principal biosynthetic activities of that tissue type. They often have an abnormal chromosome constitution and thus may lack some of the genetic information of normal cells. One such tumor is a **renal adenocarcinoma** (a tumor originating from epithelial cells of the kidney tubules), which occurs spontaneously with high frequencey in some populations of the grass frog, *Rana pipiens,* in the northeastern United States. King and DiBerardino (King and DiBerardino, 1965; DiBerardino and King, 1965) showed that a single tumor cell nucleus, transplanted into an enucleated host egg, could promote normal cleavage and, rarely, the development of larvae, all somewhat abnormal. These larvae exhibited all cell types found in differentiated organ systems of control larvae, however, and showed principally a diploid, but cytologically abnormal, chromosome constitution. Most nuclear transplant embryos revealed a wide range of deficiencies in organ development and chromosome constitution. Embryos that arrested at blastula and gastrula stages were variously aneuploid and had gross chromosomal abnormalities. Embryos that developed to later stages showed less severe defects. To the degree, therefore, that the tumor nuclei approached a normal chromosome constitution, they were better able to promote the cytodifferentiative and histodifferentiative states appropriate to normal larvae. Some tumor cells in these experiments clearly could exercise the range of cytodifferentiative controls possessed by a normal zygote nucleus, but not, apparently, the totality of morphogenetic controls.

In the situation somewhat comparable to a nuclear transplant, certain mouse **teratoma** cells have shown a remarkable ability to show normal genetic expression. A teratoma is a complex tumor that may or may not be malignant and is composed of several, usually variable, tissue types such as teeth, hair, nervous tissue, cardiac and striated muscle, lung, and so on. These tissues are derived from **teratocarcinoma**, or **embryonal carcinoma**, stem cells. Single cells from such teratocarcinomas, grafted to suitable hosts, can give rise to solid tumors showing a number of tissue types in variously organized array (Kleinsmith and Pierce, 1964).

Mice are particularly susceptible to the experimental induction of teratocarcinomata by tissue grafts consisting of embryos or their genital ridges grafted beneath the capsule of the adult testis. One such teratoma, designated OTT6050, was derived by Stevens (1970) from a 6-day chromosomally male embryo from a strain carrying known genetic markers for a number of enzymes and coat color characteristics. It became malignant and has been carried since 1967 by serial transplants from the body cavity of one mouse to the next. The tumor is a modified **ascites** form, that is, it exists in the form of free groups of cells floating in the coelomic fluid. The cell groups constitute **embryoid bodies,** small cores of undifferentiated cells surrounded by an epithelial rind. Single cells from embryoid bodies, when injected subcutaneously into young mice, form solid tumors containing various tissues. Whether individual core cells could make all adult tissue types, however, was long unknown, for the solid tumors consistently failed to show some tissue types.

Illmensee and Mintz (1976) dissociated the core cells of embryoid bodies of the OTT6050 line after they had been carried for 8 years by mouse-to-mouse transfer. Isolated cells were introduced singly into blastocysts recovered from inseminated females, and maneuvered next to the inner cell mass. These blastocysts were then introduced into the uteri of females made pseudopregnant by mating with vasectomized males. Normal development ensued in a high percentage of cases. Tissues from embryos, newborns, and adults were examined for distinctive electrophoretic allelic variants of the enzyme glucose phosphate isomerase and for coat color. In a number of cases all tissues showed the slow-migrating enzyme characteristic of the inbred embryoid donor strain, as well as the fast-migrating species characteristic of the host blastocyst strains used. Moreover, patches of skin showed structural and pigmentary patterns specific to the embryoid donor strain. Genes for these characteristics were probably not expressed for over 8 years, during which time embryoid bodies were propagated by transfer from one mouse host to the next almost 200 times. Yet the genes were capable of normal expression when introduced into the proper developmental situation. Reproductive effectiveness of mice derived from the injected blastocysts was shown by the finding that a male mouse host blastocyst of a genetically marked strain, injected with five embryoid core cells, developed into a mature male that bred for the genetic characteristic of the embryoid donor line (Mintz and Illmensee, 1975).

These results suggest that the conversion to cancerous growth, at least for teratocarcinoma cells, does not involve structural changes in the genome, but rather changes in gene expression. Genetic mutation is thus not necessarily involved in neoplastic growth. Whether other types of tumors have a nonmutational basis for their origin is not so certain, but, as experiments with transplanted renal adenosarcoma nuclei suggest, essentially all structural gene functions must be unaltered in cancer cells, as they are in differentiated normal cells.

18.2 Genomic DNA Sequences in Differentiated Cells Are Probably Identical

From the foregoing information it would seem to follow that differentiated cells would have identical genomic DNA sequences. But the successful testing of this concept awaited the development of competitive DNA–DNA hybridization techniques. If DNA double helices are sheared in a high-speed blender or in a special pressure cell to lengths of 400–500 nucleotides, then caused to separate by heat treatment (melting) and allowed to cool under appropriate conditions, the strands will reunite (anneal) preferentially with their complementary strands. If they are immobilized as separate strands by being trapped in cooling agar or on a special filter, for example, they cannot reunite with each other, but are available to unite with added mobile nucleotide sequences in solution that may be brought to them and that show some degree of complementarity. Such union is referred to as hybridization. When a known amount of immobilized unlabeled DNA is exposed to radioactive labeled single-stranded DNA in solution, the amount of hybridization can be calculated by measuring the amount of bound radioactivity after the removal of unbound label. When unlabeled (''cold'') DNA is added to the labeled

DNA in the hybridizing solution, it will compete with the labeled DNA for binding to the immobilized DNA to a degree dependent on the identity of its nucleotide sequences. This process is called competitive hybridization. The best competitors are those having nucleotide sequences identical to those of the labeled DNA.

McCarthy and Hoyer (1964) tested the ability of unlabeled DNA from whole mouse embryo, mouse brain, kidney, thymus, spleen, and liver to compete with labeled DNA from mouse L cells (a line of cells derived from a single normal fibroblast that has been carried in vitro for many years and which is now malignant when tested by injection into normal mice) for binding to whole mouse embryo DNA. The results of their study (Figure 18.3) were interpreted as indicating that each kind of cell contains the same DNA nucleotide sequences and in the same proportion. Actually given the amounts of DNA used and the short annealing times, the results

FIGURE 18.3 Competition by unlabeled DNA for binding of ^{14}C-labeled mouse DNA to 60 μg of mouse embryo DNA trapped in 0.5 g of agar. 1 μg of ^{14}C-labeled DNA from mouse L-cells (a cultured cell line) was incubated for 15 hr at 60°C with varying amounts (abscissa) of DNA from various mouse sources, including L-cells, whole embryos, brain, kidney, thymus, spleen, liver, and from a bacterial source, *Bacillus subtilis*. DNA from the latter source did not compete with ^{14}C for binding. All mouse tissue did compete and to precisely the same degree. The plot (filled circles) shows, for simplicity, only the percent of label bound by mouse embryo DNA. Points for all other competing DNAs showed the same plot. [Adapted from B. J. McCarthy and B. H. Hoyer, *Proc Natl Acad Sci USA* 52:915–22 (1964).]

431

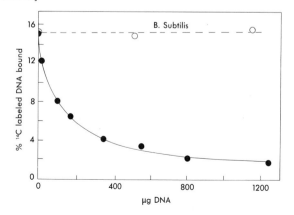

of McCarthy and Hoyer show the identity, or great similarity, of only those nucleotide sequences present in very many copies in the genome. Because genes that encode for most proteins are represented by only one or a very few copies in the genome, the data shown in Figure 18.3 do not relate directly to the genes for most structural proteins. The initial concept of McCarthy and Hoyer has been, however, very influential in the development of modern ideas relating to the role of genes in development.

Since 1964 new and exceedingly more sensitive methods have been developed for measuring the complementarity of single-stranded DNA sequences. These methods usually involve a large excess of unlabeled DNA (**driver** DNA) and a trace amount of radioactive DNA. Kinetics of annealing in liquid solution can then be measured by sampling at intervals and analyzing the radioactivity of double-stranded DNA that can be absorbed to special columns of the mineral hydroxyapatite, or by assaying DNA that is resistant to nucleases that digest only single-stranded DNA. Results of kinetic studies show that the major portion of the genome is constituted by nucleotide sequences represented only once in each haploid genome, and this evidence is consonant with the fact, to be considered further, that most of the structural genes (as many as 70% in the mouse, for example) exist as single copies in the haploid genome. These are referred to as **single-copy** sequences. For *Xenopus* DNA Davidson et al. (1973) estimated that single-copy sequences account for 54% of the nucleotides, 6% of the nucleotides are involved in sequences present about 20 times per haploid genome, 31% are present in sequences reiterated about 1600 times; and 6% are in sequences repeated about 32,000 times. The remaining nucleotides are found in sequences reiterated with such a high frequency that their number cannot be estimated; that is, they anneal almost instantaneously.

The question is thus left unsettled by the experiments of McCarthy and Hoyer (1964) as to whether the single-copy portion of the genome is identical from cell to cell. More recent studies provide some confidence that they are. When single-stranded DNA from a cell type is allowed to anneal until the highly and moderately repetitive sequences are renatured, the remaining sequences, representing chiefly single-copy sequences, can then be tested in competition with single-copy preparations from

other cell types. For the relatively small number of organisms and tissues in which such studies have been made (reviewed by Davidson, 1976) it would appear that single-copy DNA from each tissue type in an organism is identical, although very small differences would be undetectable.

From the foregoing considerations it is reasonable to conclude that all cells of an organism, regardless of the tissue or organ in which they occur, contain the same genetic information. Some interesting exceptions to this general conclusion were discussed in Section 10.1.5 and are treated further in Section 18.4.2.

18.3 Stage- and Tissue-Specific Differences in Cytoplasmic mRNA

Given that the characteristics of different cells and tissues are dependent on the genome, and recognizing the genomic equivalence of cells in an organism, one must look to differential gene expression as the basis for functional differences that appear from one developmental stage to the next and from one cell type to another. The general premise is, and has been, that differential gene expression, however controlled, derives from the translation on polyribosomes of diverse and specific sets of messenger RNAs.

It has been only in recent years that parallel increases in technology and analytical insights have permitted the unequivocal determination that there are stage- and tissue-specific populations of mRNAs present in embryos and in differentiated tissues. DNA complementary to polyribosomal mRNA can now be prepared with relative ease. This DNA is referred to as **cDNA** and is usually prepared from radioactive deoxynucleotide precursors. The mRNA released from polyribosomes is incubated with the nucleotides in the presence of RNA-dependent DNA polymerase. This polymerase is often called **reverse transcriptase,** for it catalyzes the transcription of DNA from single-stranded RNA. The use of reverse transcriptase thus permits one to prepare precise copies of the DNA nucleotide sequences represented in mRNAs from a particular source. mRNAs from embryos of different stages or from different tissues or organs can then be compared with respect to their ability to bind to cDNA prepared from different sets of mRNA.

Because polysomal mRNAs are principally derived from nucleotide sequences present in only one copy per genome, it is also reasonable to compare mRNAs from different tissues as to their ability to bind single-copy genomic DNA. DNA is melted and allowed to anneal, as described in Section 18.2. Annealing is permitted to continue until all sequences present in multiple copies have reassociated. Single-copy sequences are the last to anneal and so can be isolated and tested for binding to mRNA.

The use of these and related methods show that, whereas cells from embryos at different stages and tissues of different types utilize many of the same mRNAs, they also show considerable diversity in their mRNA populations, as well as considerable differences in the frequencies with which individual mRNAs appear on polyribosomes. In sea urchins 56% of the mRNAs on the polysomes of embryos at the blastula stage are distinct from those of the gastrula and 44% are homologous. In the mouse brain and kidney 22% of the mRNAs are distinctive and 78% are homologous. In chicken liver and oviduct about 85% of the mRNA species detected are homologous and 15% are distinctive (Figure 18.4). References to the various works on which this material is based and discussion of other aspects of specific differences in polysomal mRNAs are available in an important paper of Davidson and Britten (1979).

FIGURE 18.4 Evidence that chicken liver and oviduct utilize both distinctive and homologous mRNAs. A trace amount of ^{125}I-labeled, highly labeled (10^6 cpm), was hybridized with 550 μg/ml of oviduct mRNA (□– –□), with 550 mg/ml liver mRNA (■——■) or with 550 mg/ml each of both (●——●). The great excess of RNA assures that each DNA sequence corresponding to the tissue mRNA will be hybridized. Each point shown is corrected for self-annealing of DNA. Note that the reaction with oviduct mRNA is saturated at 1.8% and that the reaction with liver DNA is saturated at 2.05% of single copy genomic DNA. The reaction is saturated at 2.4%, however, when mRNAs from both sources are used. This is an increase of 17% of the hybridization shown by liver mRNA alone. If liver and oviduct hold all their mRNA sequences in common, the extent of hybridization obtained using both mRNAs would have been the same. [Adapted from R. Axel, P. Fiegelson, and G. Schutz, *Cell* 7:247–54 (1976).]

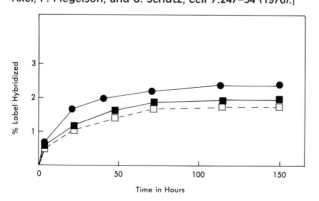

18.4 Transcriptional Control of Gene Expression

433

The first line of control of the ultimate product of gene expression, namely, protein, is at the level of transcription—that is, at the level of production of gene copies for various messenger RNA species and for the ribosomal and transfer RNAs required for their translation. In order for transcription to occur, the appropriate DNA sequences must be accessible to the appropriate enzymes, namely, the RNA polymerases. The supply of polymerases must be sufficient to produce transcripts in amounts adequate to the functional needs of each cell type. The polymerases, as components of the transcriptional control system, are discussed first. Some aspects of the availability of the genome for transcription can be treated in a relatively simple manner, and are taken up next. These aspects involve the elimination or inactivation of large blocks of the genome. The most difficult aspects of the question of transcriptional control of gene expression have to do with the relationship between the RNA species that are transcribed in the nucleus and those that appear in the cytoplasm. Data that reveal cell- or tissue-specific differences in cytoplasmic RNA populations, especially certain of the mRNAs, do not reliably reflect specific controls of transcription. They may, rather, reflect differences in the posttranscriptional processing of primary gene transcripts. The control of initial transcripts is considered separately in Section 18.4.3.

18.4.1 ROLE OF RNA POLYMERASES IN TRANSCRIPTIONAL CONTROL. Three distinctive classes of RNA polymerases are involved in transcription of genomic DNA. RNA polymerase I is found in nucleoli and is responsible for transcription of 28 and 18 S rRNA. RNA polymerase II is responsible for the transcription of the various mRNAs, and RNA polymerase III catalyzes the transcription of 5 S ribosomal RNA and the various transfer RNAs (tRNAs). Selectivity of the enzymes for these classes of RNA is now fairly well established (Chambon, 1975; Parker and Roeder, 1977). In *Rana pipiens* RNA polymerases I and II accumulate in the germinal vesicle during oögenesis and are conserved there. They are later used during early embryonic development (Hollinger and Smith, 1976). Roeder et al. (1970) examined the activities of the three polymerases during development of *Xenopus laevis* up to the tadpole stage and found no correlation between the enzyme activity and the amount or kind of RNA present at any stage. It is particularly striking that during gastrulation, when the pattern of RNA synthesis changes rapidly (cf. Chapter 10), little or no change is seen in the amount and proportion of polymerases I and II. It seems unlikely, therefore, that the polymerases are rate limiting in transcription during embryogenesis.

Nor do the polymerases seem likely to be rate-limiting factors in tissues stimulated to undergo specialized protein synthesis. For example, induction of synthesis of egg white proteins in the immature chick oviduct by means of estrogen is preceded by an enhanced production of the polymerases. In the nucleoli, for example, polymerase I rapidly rises to twice the amount found in untreated controls well in advance of any increases in rRNA synthesis or of total RNA synthesis (Bieri-Bonniot et al., 1977).

In some instances the activity of polymerases is controlled by inhibitors. A case in point is the brine shrimp, *Artemia salina,* which is commonly collected as a dormant, encysted gastrula in which polymerases I and II are present but inactive. They begin transcriptional activity when the cyst is hydrated, and development of the nauplius larva (cf. Chapter 17) then ensues (Hentschel and Tata, 1977). Extracts of the dormant embryos contain a factor that inhibits the activity of RNA polymerases I and II (D'Alessio and Bagshaw, 1977). Hydration apparently removes or inactivates the inhibitor(s) and makes available the polymerases necessary to genetic activity and new protein synthesis. Another instance is provided by the plasmodial slime mold, *Physarum* (sp.), in which an inhibitor of RNA polymerase I, but not of RNA polymerase II, appears when the organism begins to differentiate resistant spores, or spherules, under adverse conditions. The inhibitor, a high molecular weight organic phosphate, appears when food is withdrawn, binding to polymerase I in the nucleoli. When the spherules are allowed to germinate to form new plasmodia, the inhibitor disappears and RNA polymerase I begins to transcribe the ribosomal genes once more. The phenotypic control mechanism in *Physarum* thus involves changes in the activity of polymerase I mediated by changes in the level of an inhibitor (Hildebrandt and Sauer, 1977).

Conceivably, specific stimulators of the activity of preexisting RNA polymerases could have developmental significance. Such a stimulator has been described in the calf thymus. From this tissue a protein, designated *S,* specifically stimulates transcription of native DNA by ribonuclease II (Brand et al., 1977). From lentil root nuclei Teissere et al. (1975) isolated an auxin-induced protein that specifically stimulates the activity of RNA polymerase I. Another factor, not under auxin control, enhances the activity of RNA polymerase II. These factors do not affect the rate at which transcription proceeds, but rather the frequency with which transcription is initiated.

18.4.2 TRANSCRIPTION CONTROLLED BY ELIMINATING OR SILENCING LARGE PARTS OF THE GENOME. We now consider briefly some examples of relatively coarse mechanisms of transcriptional control. One such mechanism is the elimination of chromosomes from nuclei, which was discussed in Chapter 10. To reiterate, during cleavage many kinds of organisms eliminate a good deal of genetic material from blastomeres that are to contribute to the **soma,** or body, as distinct from progenitors of the

434

germ line, or prospective gametes. Apparently in such organisms the future somatic cells eliminate those genes not needed for formation of body parts. The germ line, however, requires these genes in order to carry out the specialized activities of gametogenesis.

Another mechanism for coarse control is that of **heterochromatization** of extensive segments of the genome. Interphase nuclei of most cell types contain regions of densely packed chromatin, usually referred to as **heterochromatin.** Heterochromatin is relatively inactive in RNA synthesis, which is shown by its diminished ability to stimulate incorporation of labeled uridine. Distribution patterns of heterochromatin are different in nuclei of different cell types, and this presumably reflects the variable gene activity in different cell types. In some cases, however, heterochromatization involves an entire chromosome or an entire haploid genome.

In female mammals one of the X chromosomes is inactivated shortly after implantation of the blastocyst (Aitken, 1974). It appears as a heterochromatic patch called a Barr body (Barr and Bertram, 1949). The Barr body is attached to the nuclear membrane during interphase and it does not replicate its DNA until late in the synthetic phase (S phase) of the cell cycle. As males do not have a Barr body, the cytological examination of cells sloughed from the embryo into the amniotic fluid allows the opportunity for prenatal sex determination in human beings and other mammals.

Lyon (1961) analyzed the expressions of several phenotypes in female mice heterozygous for X-chromosome-linked mutations. These mutations have visible effects on the surface of the body, such as skin color and eye pigmentation. The mutant females often showed a variegated effect in random patterns, indicating most reasonably that random inactivation of one of the X chromosomes occurred early in development. Thus heterozygous females would have two kinds of cells: one with the maternally derived X (X^m) active and the other with an active paternally derived X (X^p). Among human females a number of X-chromosome-linked mutated genes affecting the activities and electrophoretic properties of several enzymes have been studied (R. G. Davidson et al., 1963; Migeon et al., 1968; Romeo and Migeon, 1970; Deys et al., 1972; Riciutti et al., 1976). In general, the distribution of cells showing altered enzyme activities usually indicates that the X chromosomes are inactivated at random in human beings, but for some cells, such as lymphocytes or reticulocytes, there is the possibility that either nonrandom inactivation occurs, or that random inactivation occurs, followed by clonal selection against the population carrying one variant form (Yen et al., 1978).

In kangaroos X^p is preferentially inactivated in most tissues (Cooper et al., 1975), and in mice and rats it is preferentially inactivated in the extraembryonic membranes, but not in the embryonic tissue (Wake et al., 1976; West et al., 1977; Takagi et al., 1978). It has been demonstrated in females of mice and human beings that both X chromosomes are active throughout the growth period of the oöcyte. In both forms the gene for glucose 6-phosphate dehydrogenase is located on the X chromosome. Mangia et al. (1975) analyzed the activity of this enzyme in oöcytes of XX and X0 mice at several stages of follicular development. They found the activity of the enzymes of XX females to be twice that of X0 females. The activity of lactic dehydrogenase, which is determined by an autosomal gene, showed no difference in the two kinds of oöcytes, however. In female human fetuses heterozygous for a variant allele of glucose 6-phosphate having altered electrophoretic properties, Migeon and Jelilian (1977) found both forms of the allele to be expressed in the germinal cells, but not in somatic tissues of the ovary. They suggested that possibly the cells of the germ line do not undergo X-inactivation. On the other hand, Ohno (1963; cited by Mangia et al., 1975) reported that female primordial germ cells resemble somatic cells in having a Barr body. If so, then the inactive X of prospective oöcytes must be reactivated early in oögenesis.

Apparently, some of the genetic material remains active in each X chromosome homologue. This conclusion is suggested by several lines of evidence, two of which are cited here.

1. Red blood cells from females heterozygous for a particular blood group antigen designated Xg^a cannot be separated into two types by use of antibodies to the variant antigens; all cells show both forms of the antigen.
2. Human females with only one X chromosome (X0) exhibit bodily abnormalities, especially of the accessory sexual characteristics (Turner's syndrome); if the second X chromosome were

435

completely inactivated in normal females, the absence of one of two X chromosomes should have no effect in X0 individuals.

Moreover, whereas cells of females with three X chromosomes show two Barr bodies, such individuals likewise tend to show abnormal sexual development and often a degree of mental retardation.

In some insects an entire haploid chromosome set becomes heterochromatic and hence genetically inactive or extremely limited in ability to be transcribed (reviewed by Brown and Nur, 1964, and Nur, 1967). In one group of coccids (mealybugs) both maternal and paternal sets of chromosomes are euchromatic as cleavage begins and they remain so in female embryos. During the fifth to eighth cleavages, however, all genetic males show heterochromatization of the entire paternally derived haploid set of chromosomes. At meiosis in the male chromosomes do not pair, so no genetic recombination occurs, and both sets go to opposite poles of the spindle and the paternal set disintegrates. The male thus contributes to his sperm only the chromosome set derived from his female parent, which was euchromatic throughout his ontogeny. In the next generation, however, this set of chromosomes becomes heterochromatic.

When the wild type allele of a recessive mutant is inherited from a mother in these coccids, it is expressed in both sons and daughters. But when the mother contributes the recessive allele and the wild type is present in the father's sperm, only the daughters show the wild phenotype, the sons showing the recessive—evidence that the chromosome set contributed by the male parent to his sons is the one that is heterochromatically inactivated. In some of the tissues of males, varying according to species, heterochromatization of the paternal set of chromosomes reverses in some tissues later in development. This reversal occurs notably in the intestine, Malpighian tubules, some skeletal muscle, testis sheath cells, and auxiliary glands of the testis. Sons of irradiated fathers are often sterile, which is possibly the consequence of damage to genetic material destined to show reversal of heterochromatization in tissues involved in reproduction.

18.4.3 SPECIFIC GENE ACTIVATION. Recognizing that there are circumstances under which large segments of the genome are not available for transcrip-

tion, we now turn to a more difficult problem. Do the particular populations of mRNAs that characterize cells of embryos in different developmental stages or in different tissue types reflect differences in the activation of specific segments of the genome for transcription? Much of the research that has influenced thinking about mechanisms of transcriptional control has been carried out on differentiated cell types that respond to specific stimuli by the production of a large amount of mRNA that is translated as a very limited spectrum of distinctive proteins. Reticulocytes, for example, produce abundant transcripts of the genes encoding for α- and β-globulins, the hemoglobin proteins. Another example is provided by the portion of the chick oviduct known as the **magnum.** The tubular gland cells of the magnum produce large quantities of mRNAs specific for the egg-white proteins in response to stimulation by estrogen and progesterone. Studies of such specialized cells, especially those in which the protein-synthetic pattern is hormonally induced, have led to the general concept that the immediate control of transcription involves interactions between RNA polymerase, DNA, and some control molecule or molecules which come to occupy sites on the chromosomes adjacent to the specific coding region.

This concept may possibly express a relationship that occurs during the terminal differentiation of particular cell types in which certain DNA sequences, present as single copies or only a few copies in each genome, are represented by great numbers of mRNA copies in the cytoplasm. Such mRNAs are called the **superprevalent class** (Davidson and Britten, 1979). The mRNA for the egg-white protein ovalbumin, which may be present in as many as 78,000 copies per tubular gland cell in the magnum of the oviduct of the laying hen, is an example of RNA of the superprevalent class. One may question whether mechanisms that control the production of superprevalent mRNAs are applicable to the control of mRNA species that are represented by fewer than the superprevalent class. Polysomal mRNAs of eukaryotes can be divided into naturally occurring, although somewhat arbitrary, abundance classes that do not show a marked continuum between groupings (Hastie and Bishop, 1976; Axel et al., 1976). There is, first, the **complex** class of mRNA; members of the complex class appear in only 1–15 copies per cell. Members of the complex

436

class are so called because they contribute to most of the complexity of the genome, that is, to the total length of diverse DNA sequences, counted as base pairs. Next, are the moderately prevalent mRNAs that occur in 15–300 copies per cell. The super-prevalent messages are present in 10^4–10^5 copies per cell but they represent transcripts of the structural DNA that contributes least to the complexity of the genome. The superprevalent messages, how-ever, represent transcripts of but a minute fraction of all the diverse structural genes required by the organism at any point in its developmental history. In sea urchins cytoplasmic mRNA of the complex class in the oöcyte, in various embryonic stages and in adult tissues, differs by an amount of single-copy RNA sequences equivalent to thousands of different structural genes (Galau et al., 1976). These differ-ences in polysomal mRNAs must be regulated, and these mRNAs are presumed to be translated into protein. The quantities of their translation products, however, are too small to be detected by present methods. Changes in the pattern of protein synthe-sis that may be detected when the sea urchin em-bryo begins gastrulation and pluteus formation (Westin, 1969; Brandhorst, 1976) are possibly re-lated to changes in transcript prevalence. Lasky et al. (1980) showed that whereas most of the preva-lent class of messages present in the pluteus are also present in the unfertilized egg and zygote, a minor-ity of sequences nevertheless show sharp stage-specific changes in representation in the cytoplasm. Conceivably, the stage-specific changes in the pat-tern of protein synthesis become detectable as some of the complex class of mRNAs shift into the moder-ately prevalent class.

Evidence is accumulating that the changes in the population of cytoplasmic mRNAs that are seen during early developmental stages do not necessar-ily result from differential transcription of the ge-nome. Thus Kleene and Humphreys (1977) showed that nuclear RNA complementary to single-copy genomic DNA sequences shows no detectable change from the blastula to the pluteus stage in sea urchins, whereas 45% of the complex class of cyto-plasmic mRNAs changes between these stages (Galau et al., 1976). These observations make it clear that factors other than those that control tran-scription determine what gene products ultimately appear as cytoplasmic mRNA. There must be, in other words, posttranscriptional controls that deter-

mine what portions of the genome shall ultimately be available for translation. The post transcriptional control of cytoplasmic RNA is the topic of Section 18.5. Before proceeding to that topic, however, it is appropriate to consider controls of certain mRNAs that appear to operate at the level of transcription.

18.4.3.1 Chromosomal Proteins and the Question of Specific Gene Transcription. For some years research has been directed to the ques-tion of whether the chromosomal proteins exercise a measure of control over the specificity of genetic transcription. Histones have long been known to be prominently associated with chromosomes. Some time ago Huang and Bonner (1962) found that the ability of DNA to act as a template for RNA synthe-sis in a cell-free system is progressively inhibited as it is increasingly complexed with histones. Alfrey et al. (1963) then showed that the ability of isolated nuclei to synthesize RNA is progressively increased as histones are removed from their binding to DNA. It is unlikely, however, that histones are involved in regulating the specificity of gene transcription. They remain relatively constant from tissue to tissue within the organism, and amino acid sequences in some of them show considerable resemblance in plants and animals.

The nonhistone chromosomal (NHC) proteins, on the other hand, may play a significant role in determining transcriptional specificity. The NHC proteins are highly acidic, quite variable from tissue to tissue, and have a wide range of stabilities. Their half-lives vary from several minutes to several cell generations. Importantly, the proportion of NHC proteins is much higher in transcriptionally active regions of the genome than in regions not being transcribed. Moreover, the NHC proteins are differ-ent in different kinds of tissue and are possibly in-volved in the realization of tissue-specific patterns of transcription. An experiment from Paul's labora-tory (Paul et al., 1973) illustrates this. He and his collaborators isolated chromatin from both fetal mouse liver, which produces the globins of hemo-globin, and mouse brain, which does not. Tested in a cell-free transcribing system, liver chromatin pro-duced globin mRNAs, but brain chromatin did not. Chromatin from fetal liver and brain was then sepa-rately dissociated and fractionated. When brain DNA was reconstituted into chromatin in the pres-ence of brain nonhistone chromosomal proteins, it did not make globin messages, but when it was re-

437

constituted in the presence of fetal liver nonhistone chromosomal proteins, it did. The production of globin mRNA was measured by hybridization of the transcripts to highly labeled cDNA prepared by means of reverse transcriptase. This method permits the detection of very minute quantities of globin message. Unfortunately, in this experiment and in similar experiments carried out in other laboratories, RNA polymerases of bacterial origin, rather than endogenous RNA polymerases, were used in the transcribing system (endogenous RNA polymerases are almost impossible to obtain in sufficient quantities to carry out the experiment). If one assumes that bacterial RNA polymerases faithfully copy the available genome of mammalian chromatin, the results of Paul and his collaborators suggest that the role of specific NHC proteins is to "unmask" particular genes. If, on the other hand, bacterial RNA polymerases cannot adequately replace endogenous mammalian RNA polymerases, the putative unmasking role of NHC proteins must be questioned. It now seems, in fact, that mouse brain nuclei normally transcribe the globin gene and export globin mRNA to the cytoplasm. Humphries et al. (1976) made cRNA complementary to mouse α- and β-globin mRNA and used it as a probe to determine whether polyadenylated RNA isolated from brain, liver, and several mouse cell lines contains complementary sequences. Their hybridization data enabled the calculation to be made that in adult brain there are approximately 80 globin gene transcripts in the nucleus and 300 copies in the cytoplasm of each brain cell. The adult liver, no longer an erythropoietic organ, has about 75 copies in the nuclear compartment and 70 in the cytoplasm. These numbers are small as compared to those obtained from measurements of RNA in fetal liver. (In the 14-day fetus there are about 4600 nuclear sequences complementary to the probe and about 66,000 in the cytoplasm of each liver cell.) The data show, nevertheless, that the globin genes are transcribed and exported to the cytoplasm in nonerythroid cells. The globin genes thus are not masked in brain cells and possibly not in any cells of the organism.

These considerations cast considerable doubt on the interpretation that the NHC proteins play a role in transcriptional specificity. That there are differences in NHC proteins correlated with functionally differentiated cells is, however, a reality that must

438

be reckoned with in any interpretation of the control of gene expression.

18.4.3.2   The Action of Some Steroid Hormones in Control of Gene Expression. The effects of steroid hormones on the pattern of gene expression seem, superficially at least, to provide examples of direct control of specific gene action. In general, the pattern of action of steroid hormones has been viewed somewhat as follows:

1. The hormone is taken up by the target cell.
2. In the target cell the hormone binds to a specific receptor.
3. The hormone-receptor complex is translocated to the nucleus where it binds to an acceptor site on chromatin.
4. Binding activates transcription of DNA adjacent to the binding site.

This interpretation is simplistic and thus somewhat appealing. In general, one does find that the administration of hormone leads to increased activity of RNA polymerases, increased nuclear RNA, changing patterns of repetitive sequences of nuclear RNA, increased transcription of single-copy DNA, and an increase in the capacity of the chromatin to serve as a template.

The role of steroids in controlling gene expression has been studied principally in the reproductive tracts of birds and mammals, among which the most studied organs have been the uterus of the rat or calf and the magnum of the oviduct in birds. Early stages in the development of these organs are marked by the differentiation of specific estrogen receptor proteins, which are necessary in order for estrogen-induced maturation of these organs. In the immature female rat, for example, when estrogen is administered, it becomes concentrated in the uterus to an extent 500 times its concentration in the blood (Jensen and DeSombre, 1973). In the cytoplasm it unites with a receptor protein, a monomer, that sediments at 4 S. In an estrogen-dependent reaction the receptor associates with a second monomer to form a 5 S complex, which is rapidly translocated to the nucleus where it binds to chromatin. Binding is followed by the accelerated synthesis of RNA polymerases, rRNA, and mRNA, and by an increased rate of cell division. Uterine glands then develop and new proteins appear, notably a progesterone receptor protein (Leavett et al., 1977).

In the newly hatched chick the various regions of the oviduct are recognizable, but the lining of the duct is a uniform single-layered epithelium. Administration of estrogen as early as 4 days after hatching and continuing for about 10 days causes the epithelium of the magnum to differentiate tubular glands, which invaginate into the underlying connective tissue, and ciliated cells and goblet cells, which are interspersed over the surface of the lumen. The tubular gland cells are the source of the principal proteins of the egg white. These proteins are **ovalbumin, conalbumin, ovomucoid,** and **lysozyme,** which together account for 86% of the protein produced in the gland. Each tubular gland cell produces all four proteins. Their synthesis begins within 24 hours of estrogen administration, prior to the onset of the morphogenetic movements of the gland cells out of the surface epithelium (Palmiter and Gutman, 1972). Another well-known protein, **avidin** (which binds vitamin B_{12}), is produced by the goblet cells, but only after stimulation of the estrogen-primed oviduct by progesterone. If estrogen treatment is continued for 18 days, about 50,000 sequences complementary to cDNA made from ovalbumin mRNA are present in each tubular gland cell (McKnight et al., 1975).

A great many investigations dealing with effects of steroid hormones on the synthesis of egg-white proteins are carried out on "withdrawn" chicks. Typically, a withdrawn chick is prepared by administering a standard dose of estrogen for 10 days and then ceasing (withdrawing) treatment for 10–15 days. During withdrawal the production of the specific proteins rapidly diminishes; ovalbumin, which is the major protein, cannot be detected in the magnum of the withdrawn chick. Ovalbumin mRNA apparently occurs to the extent of about 60 copies per cell. It can be detected by means of a cDNA probe. 80% of the complementary sequences are in the cytoplasm (McKnight et al., 1975).

During primary stimulation by estrogen immature chicks differentiate progesterone receptors. Thus, whereas progesterone cannot induce synthesis of the egg-white proteins in the immature chick, it can do so alone in the withdrawn chick, as does estrogen. With either hormone the chromatin sites for binding to the steroid-receptor complex are saturated in less than 30 minutes. Alone or in combination the hormones cause an increase in the concentration of translatable mRNA and a concomitant increase in protein synthesis. The syntheses of the various proteins are not coordinated, however. For example, conalbumin mRNA begins to accumulate about 30 minutes after secondary stimulation, whereas there is a lag of 3 hours before ovalbumin mRNA accumulates. The onset and rate of synthesis of the corresponding proteins are determined by their mRNAs. It is also appropriate to note that the relative proportions in which the different egg-white proteins are synthesized change during both primary and secondary stimulation.

Investigators in O'Malley's laboratory (Schwartz et al., 1977; Schrader et al., 1977) reported the isolation of a steroid receptor protein from the withdrawn oviduct. It is a **dimer** consisting of two dissimilar subunits, designated A and B. Both subunits bind a molecule of hormone. The B subunit binds to the nonhistone chromosomal protein of oviduct chromatin, but not to DNA. The A subunit, on the other hand, binds to DNA, but only poorly to chromatin. These properties suggest that A might be a regulatory protein determining that RNA polymerase could transcribe, and that the B protein could determine where the A protein would localize. The presumption is that both estrogen and progesterone affect protein synthesis in the withdrawn chick oviduct by creating new initiation sites for the binding of RNA polymerase by interacting with the receptor dimer. The number of binding sites was determined in chromatin of withdrawn oviducts and oviducts that received a second stimulation by either estrogen or progesterone alone or in combination. It was reported that either of the two hormones increases the number of binding sites for RNA polymerase in oviduct chromatin threefold in an hour's time. Used in combination their effects are not additive, so it has been assumed that the receptor is the same for both steroids. Because the method used in the assay measures the binding of RNA polymerase from *Escherichia coli*, conclusions about the role of the steroids and the presumed identity of their receptors must be questioned.

There is, however, other evidence that casts doubt both on the identity of the estrogen and progesterone receptors formed upon secondary stimulation and on the identity of their mode of action. Estrogen and progesterone show different patterns of dose-dependent effects on the rates of transcription of mRNAs for the different egg-white proteins (Palmiter, 1972a, b; Palmiter and Haines, 1973).

439

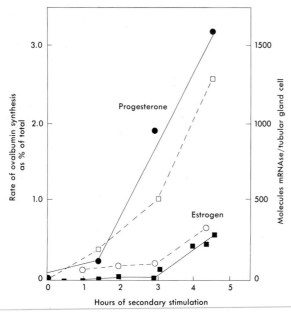

FIGURE 18.5 A comparison of the differential effects of estrogen and progesterone on the rate of ovalbumin production (●—● and ■—■) and the number of molecules of ovalbumin mRNA per tubular gland cell (○– –○ and □– –□) in the magnum of the withdrawn chick. The chicks received 3 mg of the respective hormones intramuscularly in the thigh. [Adapted from R. T. Schimke, P. Penneguin, D. Robins, and G. S. McKnight, *Ann NY Acad Sci* 286:116–24 (1977).]

440

Also the two hormones affect differently the time that the proteins appear in the withdrawn oviduct following secondary stimulation. For example, the production of ovalbumin mRNA increases rapidly beginning $1\frac{1}{2}$ hours after secondary stimulation with progesterone, but ovalbumin begins to appear only after 4 hours following secondary stimulation with estrogen, and accumulates less rapidly (Schimke et al., 1977) (Figure 18.5). Conalbumin synthesis, on the other hand, begins about 1 hour after administration of estrogen, but shows a 2–3-hour lag after the injection of progesterone. McKnight (1978) showed that minced oviductal tissues of the withdrawn chick respond to estrogen and progesterone in vitro. The production of ovalbumin and conalbumin in response to the presence of estrogen or progesterone in the medium followed the same pattern as was earlier shown in vivo. The observations recorded here make it clear that the regulation of mRNA production in the chick oviduct is not easily explained by a transcriptional control model in which hormone-receptor complexes bind at promoter sites and thereby increase transcription of specific mRNAs.

18.4.3.3 Are There Multiple Copies of Genes Coding for Rapidly Produced Products?

When it was discovered some years ago that the ribosomal RNA genes in *Xenopus* are highly reiterated and that the reiterated sequences are amplified during oögenesis (Chapter 7), it seemed reasonable to expect that cells engaged in intensive specialized activities might use one or both of these stratagems to provide for rapid production of appropriate mRNAs. Thus far, however, except for oöcytes, gene amplification has been found only rarely as a possible mechanism for facilitating genetic transcription. Moreover, in essentially all cases in which they are examined, cells producing large amounts of specialized proteins do so by transcription of only one or possibly two or three copies per haploid genome. Evidence for this idea is obtained in some cases by examining the kinetics of reannealing of a labeled DNA probe complementary to a particular mRNA with an excess of total nuclear DNA. The ovalbumin gene in the chick oviduct is apparently represented only once per haploid genome (Woo et al., 1975). Duck and mouse reticulocytes have only one gene for each globin chain (reviewed by Davidson and Britten, 1973). Human beings have two α-globin genes, which can mutate independently, but the β-globin gene has only one locus. These examples suggest that rates of transcription and translation are sufficient to allow single genes to produce transcripts for large amounts of protein.

In silkmoths the gene for silk fibroin is probably present as a single copy in each haploid genome (Suzuki et al., 1972), but the silk gland cells are highly polytenic, so there are large members of the fibroin gene available for translation when silk is being made rapidly. Polyteny also provides nurse cells in insect ovaries with multiple copies of genes whose ultimate expression is in the form of products added rapidly to the egg cytoplasm.

The follicle cells of the egg chamber of *Drosophila* are polytenic, but they also show selective amplification of parts of the genome. In *Drosophila* the proteins that form the chorionic coverings of the eggs are synthesized by the follicle cells during a period of only 5 hours. During this time, the structural genes for several of the chorionic proteins are expressed at very high levels. The production programs for the various proteins are temporally regulated so that a follicle cell produces the mRNA for a particular protein for only 1–2 hours. Prior to the onset of synthesis of chorionic proteins the follicle cells, first diploid, replicate their DNA, in the absence of cell division, reaching a DNA content 16 times that of the haploid genome. Further increases in DNA then take place by the selective amplification of regions of the genome that code for chorionic proteins. Eventually each follicle cell contains about 45 times the DNA found in a haploid genome. Spradling and Mahowald (1980) found that genes for three of the chorionic proteins are specifically amplified tenfold in the follicle cells of the egg chambers. These genes code for proteins having molecular weights of 38,000, 36,000 and 18,000 respectively. The work of Spradling and Mahowald provides the first clear evidence of the specific amplification of structural genes in a normal developmental process.

Specific gene amplification also occurs under certain conditions in which cultured mammalian cells are selected for resistance to the deleterious effects of the drug *methotrexate*. This analogue of folic acid inhibits the activity of dihydrofolate reductase, an enzyme that has an essential role in the biosynthesis of purines. The ability of cell lines to become resistant to the drug is correlated with an increase in the cellular content of dihydrofolate reductase, and this increase, in turn, is correlated with an increase in cellular levels of translatable mRNA for the enzyme. Alt et al. (1978) characterized the methotrexate resistance of a line of cells derived from mouse sarcoma 180. In this line dihydrofolate reductase is present in concentrations 200 times as great as in cells that are sensitive to the drug. When methotrexate is removed, high levels of resistance are lost, and there is a corresponding decrease in dihydrofolate reductase synthesis and in the enzyme-specific mRNA. They also analyzed a line of mouse lymphoma cells, L1210, that retains its resistance to methotrexate even when grown in the absence of the drug. L1210 shows high stable levels of the dihydrofolate reductase and a correspondingly large population of mRNA for the enzyme.

Purified DNA sequences complementary to mouse dihydrofolate reductase mRNA were used to quantitate both mRNA and numbers of dehydrofolate reductase genes in both cell lines. Alt and collaborators (1978) found through the analysis of hybridization kinetics that increased levels of dehydrofolate reductase and its mRNA are associated with a proportionate increase in the number of dihydrofolate reductase gene copies. Thus a 200-fold increase in enzyme is correlated with a 200-fold amplification of the dihydrofolate reductase gene. When unstable resistant lines are grown in the absence of methotrexate, loss of resistance is associated with a decrease in the number of copies of the gene. Subsequently it was found that in the lines that are unstably resistant to methotrexate the amplified genes are located in small paired chromosomal elements that lack spindle-fiber attachments. These elements are lost during successive mitoses and are not replaced in the absence of methotrexate. Kaufman et al. (1979) thereupon drew the inference that in stably resistant lines the amplified genes are located on chromosomes and segregate equally during mitosis. Cytological demonstration of the chromosomal location of the amplified dehydrofolate reductase gene has been made in a line of Chinese hamster ovary cells that is stably resistant to methotrexate (Nunberg et al., 1978). The amplified gene is found on one arm of one of the pair of chromosomes designated as number 2. The arm of the chromosome bearing the amplified gene is larger than the corresponding arm of the homologous number 2 chromosome. After denaturation of the amplified DNA in situ on a microscope slide, it binds specifically to isotopically labeled DNA complementary to dehydrofolate reductase mRNA.

441

18.5   Processing of Gene Transcripts

The term **processing** refers to modifications of the initial gene transcripts that enable them to function in the synthesis of protein. Processing of transcripts begins in the nucleus and is usually continued in the cytoplasm. All RNA, as it is transcribed, is promptly associated with protein; most RNA used in translation is excised by nucleases from larger primary transcripts, and some is apparently formed

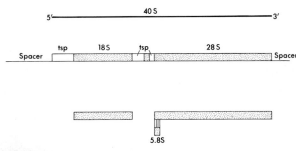

FIGURE 18.6 Processing of the ribosomal precursor gene. Spacers separate gene copies. The primary transcript is a molecule sedimenting at 40 S. Processing removes a transcribed leader sequence and intervening sequences (tsp), as shown, leaving 18, 5.8, and 28 S remnants. The 5.8 S fragment is bonded to the 28 S subunit by complementary base pairs. [Adapted from R. P. Perry, *Annu Rev Biochem* 45:605–92 (1976).]

442

of sequences spliced together after excision of intervening sequences. Methylation of nucleotide bases and sugars occurs in nucleus and cytoplasm, and methylation can affect transport of RNA from nucleus to cytoplasm. Special modifications of both the 5' end (first transcribed) and the 3' end of mRNA occur chiefly in the nucleus and to some degree in the cytoplasm. Factors controlling all aspects of processing constitute the first line of **post-transcriptional controls** of gene expression. In what follows we describe briefly some aspects of the processing of the ribosomal and transfer RNAs and then turn to the more complex question of how cytoplasmic mRNA originates.

Some aspects of the processing of rRNA were noted in connection with the discussion of oögenesis in amphibians in Chapter 7. The 28 S and 18 S components of rRNA are components of the same primary transcript. In mammals the primary transcript sediments at 45 S and in *Xenopus*, where it has been most extensively studied, it sediments at 40 S. The organization of the *Xenopus* precursor is shown in Figure 18.6. Processing of this precursor occurs almost entirely within the nucleus and includes the methylation of some of the bases and the excision of the 18 S and 28 S RNA segments. When methylation is inhibited, transcription of the rRNA gene is not inhibited, nor is the processing of the transcript into 28 S and 18 S segments prevented. Exit of the subunits to the cytoplasm is greatly inhibited, however (Caboche and Bachellerie, 1977).

In mammalian cells the final trimming to yield the 28 S fragment from a 32 S intermediate form is often rate limiting in the processing pathway (reviewed by Perry, 1976). The processing of the principal rRNA precursor in yeast is somewhat exceptional in the sense that the final maturation of the RNA of the small ribosomal subunit occurs in the cytoplasm in which a 20 S RNA segment is cleaved to a segment sedimenting at 18 S (Udem and Warner, 1973). In *Drosophila* the processing of rRNA involves the excision of an internal sequence of bases from the primary gene transcript and the splicing of the cut ends to form the 40 S precursor (Wellauer and Dawid, 1977). The internal sequence codes for 5.8 S RNA somewhat in the manner illustrated for the *Xenopus* rRNA in Figure 18.6 (Pellegrini et al., 1977).

One will recall from Chapter 7 that the 5 S component of active ribosomes in *Xenopus* is transcribed from moderately repetitive sequences found on several chromosomes. It is apparently not excised from a higher molecular weight transcript (Brown et al., 1976).

The initial tRNA transcripts are also larger than the mature tRNAs. The average difference in length between pre-tRNA and mature tRNA is about 20 nucleotides, with a range of about 15–35 nucleotides. Studying two isoaccepting alanine tRNAs from the fibroin-secreting posterior silk gland of *Bombyx mori*, Garber et al. (1978) found that there are 14–17 extra nucleotides in the pre-tRNAs and that some of the extra nucleotides are trimmed from each end of the 76 nucleotides of the mature tRNA^{Ala} species. In yeast DeRobertis and Olson (1979) described the sequence of nucleotides and their processing in the isoaccepting species of tyrosine tRNAs of baker's yeast. They described the processing of one yeast tRNA^{Tyr} as shown in Figure 18.7. Processing includes base modification, addition of the nucleotide sequence —C—C—A at the 3' end, deletion of a leader sequence at the 5' end, and excision of a 15-nucleotide sequence adjacent to the coding region. Splicing of the cut ends exposed by removal of the intervening sequence occurs in a two-stage reaction in which half-molecules accumulate and are then spliced in a reaction involving ATP (Peebles et al., 1979). Examination of a variety of yeast tRNAs revealed intervening sequences of 14–60 nucleotides (Peebles et al., 1979; Knapp et al., 1979).

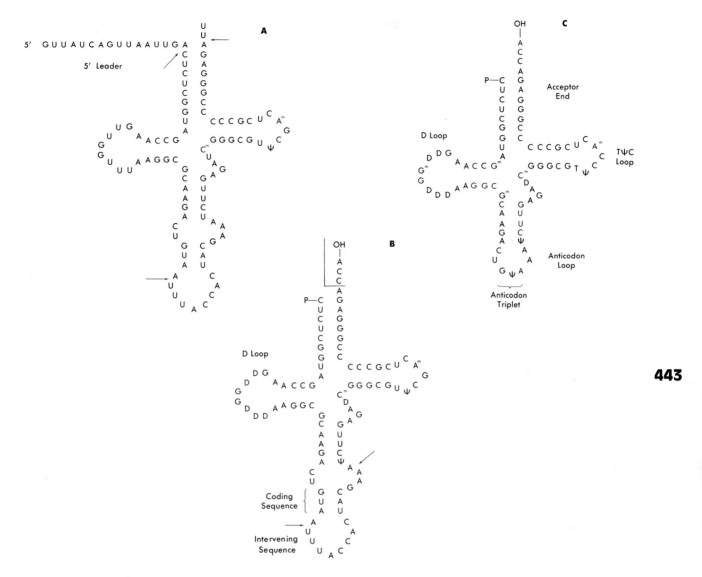

FIGURE 18.7 Processing of one of the species of tyrosine tRNA from yeast. The primary transcript (A) has a leader sequence at the 5′ end and a two-nucleotide trailing sequence at the 3′ end of the molecule. B: Both leading and trailing sequences have been cleaved, and the —CCA sequence has been added at the 3′ end. C: Mature tyrosine tRNA is formed when the intervening sequence is excised and the cut ends spliced. Note that the primary transcript has three modified nucleotides: pseudouridine (ψ), which is a compound similar to uridine, but with atoms N and C at positions 3 and 5 interchanged; methylated adenine (Aᵐ); and methylated cytosine (Cᵐ). Other modifications are seen in B, including the formation of dihydrouridylic acid in the D loop. Further methylation occurs as maturation of the molecule proceeds, including the formation of thymine by methylation of uridine. [Adapted from E. M. DeRobertis and J. B. Gurdon, *Sci Am*, 74–82 (Dec, 1979), and E. M. DeRobertis and M. V. Olson, *Nature* 278:137–43 (1979).]

Processing of tRNA occurs in both the nucleus and cytoplasm. Addition of the —C—C—A nucleotide sequence to the 3′ end occurs chiefly in the cytoplasm, but modification of the nucleotide bases by methylation takes place principally in the nucleus. When methylation is inhibited, tRNA is translocated to the cytoplasm and is found on polysomes. The rate at which undermethylated tRNA appears in the cytoplasm is diminished, and it has a reduced probability of being incorporated in polyribosomes and of participating in protein synthesis (Amalric et al., 1977).

Transfer RNAs exit rapidly from the nucleus. Studies of the relative rates of export to the cytoplasm of tRNA and of ribosomal subunits of *Xenopus* neurula cells show that newly processed tRNAs appear first, the 40 S subunit bearing 18 S RNA appear next, and the 60 S subunit, with 28 S RNA, appears considerably later. The ribosomal subunits contain large numbers of different proteins, and their assembly apparently accounts for their slow processing and transport. By 1977 38 different proteins had been isolated from the 60 S subunit and 32 from the 40 S subunit (Collatz et al., 1977).

18.5.1 ORIGIN OF CYTOPLASMIC mRNA. The principal problem in understanding the origin of cytoplasmic mRNA stems from the fact that only a small fraction of the single-copy sequence represented in nuclear RNA of a cell or tissue is also represented in the mRNA. In sea urchin embryos, depending on the stage, this fraction may be only 10–20%, in rat liver it is about 11%, and in mouse brain about 18%. What determines which species of nuclear RNA are to become cytoplasmic mRNAs? What is the significance of the nuclear RNAs that are transcribed and broken down without being exported? These are complex questions, and answers are highly speculative. Before approaching them, therefore, let us start with the nuclear RNAs that are somehow already destined to be represented in mRNA, examining first the structure of mRNA itself, and then the processing reactions involved in its formation.

Most mature mRNAs, of mammals at least, fall within a size range of 400–4000 nucleotides, with a median of about 1200. Taking the β-globin mRNA of rabbit as an example, we may describe its parts in sequence from the 5′ end as follows.

1. "Cap": a modification of the 5′ end of mRNA as found in most eukaryotes; it consists of 7-methylguanosine (m^7G) added posttranscriptionally to the first transcribed nucleoside by a 5′ to 5′ triphosphate linkage. In rabbit β-globin mRNA this nucleoside is adenosine. It and the

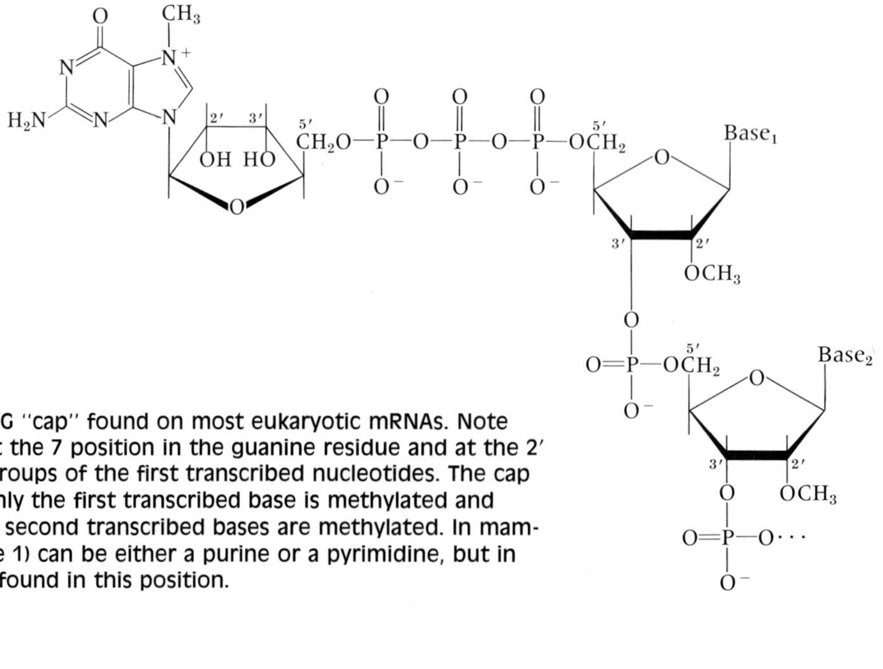

444

FIGURE 18.8 Structure of the m^7G "cap" found on most eukaryotic mRNAs. Note the characteristic methylation at the 7 position in the guanine residue and at the 2′ position on each of the ribose groups of the first transcribed nucleotides. The cap is referred to as "Cap I" when only the first transcribed base is methylated and "Cap II" when both the first and second transcribed bases are methylated. In mammals, the penultimate base (Base 1) can be either a purine or a pyrimidine, but in lower forms purines are usually found in this position.

FIGURE 18.9 Organization of mature β-globin mRNA from rabbit reticulocytes. The 56-base nontranslated leader sequence includes the AUG initiator codon. The coding region is translated as 146 amino acid residues. [Adapted from A. Efstratiadis, F. C. Kafatos, and T. Maniatis, *Cell* 10:571–85 (1977).]

next nucleoside, cytidine, are methylated in the 2′ position of the ribose residue. (This arrangement is known as "Cap II"; Figure 18.8.)

2. Leader segment: 54 untranslated nucleotides, including the initiator codon AUG.

3. Initiator codon: AUG signals the start of translation; it codes for methionine, a residue that is later removed from the terminal end.

4. Translated sequence: 146 triplet codons that are translated as the β-hemoglobin molecule.

5. Terminator codon: UGA in the rabbit.

6. Untranslated sequence: an untranslated trailer sequence of 95 bases. A similar sequence is characteristic of most eukaryotic mRNAs.

7. Polyadenylated terminal sequence: as noted in Chapter 7, most eukaryotic mRNAs have poly(A) sequences varying from 15–200 or more nucleotides.

The arrangement of the parts of the β-globin message is schematized in Figure 18.9. The coding of the entire nucleotide sequence of the rabbit β-globin message was determined by Efstratiadis et al. (1977). Their paper may be consulted for further details.

18.5.1.1 Processing of Prospective mRNA.
We now discuss modifications of the initial transcripts that are involved in the production of mature mRNA, that is, mRNA that is exported to the cytoplasm and essentially ready to become associated with polyribosomes and to act as a template for protein synthesis.

Most eukaryotic mRNAs have a m⁷G cap similar to that shown in Figure 18.8. The Gppp component of the cap is added at the initiation of transcription when the first phosphodiester bond is formed between the 5′ terminal bases in the presence of RNA polymerase (Furuichi and Shatkin, 1976; Wei and Moss, 1977). A methyl group is then added at the 7 position on the guanine and, depending on the spe-

cies, to the 2′ position of the terminal base sugar of the transcript. If additional methyl groups are to be added to adjoining nucleotides, as in Figure 18.8, the additional methylation occurs in the cytoplasm (Perry and Kelley, 1976). Methylation of the cap could conceivably be a posttranscriptional control point, for mRNAs that are normally capped are inhibited in translation if they are produced in the presence of inhibitors of S-adenosylmethionine transferase. This point, however, remains uncertain.

Most mRNAs are adorned at the 3′ end with a nontranscribed poly(A) sequence. Not all mRNAs have this poly(A) tail, however, Histone mRNAs from mammals, sea urchins and birds lack the terminal poly(A) sequence, whearease 75% of histone mRNA in amphibian oöcytes is polyadenylated. For messages that are polyadenylated, addition of the poly(A) sequences begins in the nucleus. Sawicki et al. (1977) estimated for the human cell line (HeLa) that nuclear RNA in each cell acquires about 200 poly(A) residues per minute. As messages enter the cytoplasm, further polyadenylation takes place, but end-group addition proceeds more slowly than loss of (A) residues occurs; so that the net effect is the progressive diminution of the poly(A) tail of the mRNA. Gradual loss of poly(A) with age seems to be characteristic of messages that bear this 3′ appendage, but translation of such messages is not inhibited.

Polyadenylation changes under a number of conditions, but the significance of these changes for protein synthesis is obscure. In rat liver cells cultured in vitro, the proportion of poly(A) containing mRNA moving from nucleus to cytoplasm is three times as great in cells taken from 50-day-old rats as in those from 180-day-old rats (Yannarell et al., 1977). When starfish oöcytes are stimulated by 1-methyladenine (cf. Chapter 7), polyadenylation of newly synthesized mRNA produces 3′ termini of about 170 nucleotides, in contrast to the 120 nucleotide tails present on previously synthesized messages, which acquire no additional (A) residues (Jeffrey, 1977). Similarly, in sea urchin eggs fertilization is accompanied by an increase in the posttranscriptional polyadenylation of cytoplasmic mRNA sequences.

As mRNA is transcribed in the nucleus, protein is bound to it. In the cytoplasm mRNA that is not being transcribed exists as a **messenger ribonucleo-**

445

protein (mRNP) complex. mRNA that is being transcribed is also complexed with proteins. When polyribosomes are isolated and treated so as to release the mRNA from them, an array of proteins remains bound to the mRNA. For example, Jeffrey and Peters (1977) found four principal proteins bound to rabbit globin mRNA obtained from polyribosomes. The largest of these, having a molecular weight of 79,000, bound specifically to the poly(A) sequence. Other investigators have reported finding poly(A)-binding proteins of approximately the same size in HeLa cells (Shiokawa and Pogo, 1974), and rat liver cells (Mazur and Schweiger, 1978). This protein is presumably involved along with the polyadenylated sequence in transport of the mRNA from the nucleus to cytoplasm. Schwartz and Darnell (1976) showed that for HeLa cells the poly(A)-binding protein is not produced when polyadenylation is specifically inhibited by 3′-deoxyadenosine. Protein synthesis and binding are thus important aspects of the processing of gene transcripts that appear as mRNAs.

The genomic transcript that is "capped," complexed with protein, and polyadenylated in the nucleus requires further processing before being exported to the cytoplasm, for mRNAs that are sufficiently abundant to be studied readily, it is now clear that the nucleotide sequences coding for one peptide chain are arranged in noncontiguous blocks within an initial transcript that may be three or four times the size of the mature mRNA. Ross (1978) isolated from nuclei of mouse erythroid cells a capped and polyadenylated sequence of about 1860 nucleotides (including a poly(A) sequence of 145 bases) that hybridizes to cDNA prepared from cytoplasmic β-globin mRNA. The cDNA, however, consists of only about 610 nucleotides, corresponding to the coding portion of mature mRNA. The processing of the initial transcript requires about 20 minutes and involves the excision of two internal segments, one of approximately 125 nucleotides and another of about 780 nucleotides (Kinniburgh et al., 1978). Mature β-globin mRNA fails to hybridize to genomic DNA corresponding to the excised regions of the initial transcript (Figure 18.10).

Among other transcripts whose processing involves the excision of intervening sequences and splicing of the cut ends are those for rabbit β-globin, which has two intervening sequences (van den Berg et al., 1978), mouse immunoglobins (Schibler

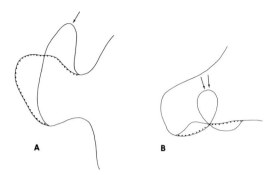

FIGURE 18.10 Annealing of 15 S precursor β-globin mRNA (A) and of mature β-globin mRNA (B) with partially denatured genomic DNA containing the entire β-globin gene. The gene was "cloned" along with flanking DNA base pairs as a plasmid in *E. coli*. In A, 15 S precursor β-globin mRNA, represented as a string of dots, has displaced one strand (arrow) of DNA. In B, the 10 S mature mRNA has displaced one strand on either side of a loop of double-stranded DNA (double arrow). [Interpretative drawings of electron micrographs by S. M. Tilghman et al., *Proc Natl Acad Sci USA* 75:1309–13 (1978).]

et al., 1978), rat liver albumin (Strair et al., 1978), and chick vitellogenin (Jost et al., 1978). As might be expected, the mRNAs for the egg-white proteins have been intensively examined. The initial ovalbumin transcript is coded by about 6000 base pairs of DNA, which corresponds to a length three times that of the translated ovalbumin mRNA. Excision of six intervening sequences and splicing of seven segments are required in processing mature ovalbumin mRNA (Dugaiczyk et al., 1978; Mandel et al., 1978). The chick conalbumin gene transcript contains about 10,000 bases, but the mature message is only about 2400 nucleotides long. Processing of the primary transcript requires the excision of 16 intervening segments. The formation of the mature message involves the splicing of sequences comprising 60–200 nucleotides (Cochet et al., 1979).

It is appropriate to close this subsection by remarking on the unusual case provided by the transcription and processing of genes for histones. The histone genes have been best studied in sea urchins and in *Drosophila melanogaster*. In the sea urchins the genes for H1, H4, H2B, H3, and H2A are located close together in the sequence given. A short transcribed sequence separates each of them. The cluster is repeated in linear series with nontranscribed

446

spacers between. It now appears that each cluster is transcribed as a single molecule and then processed to give the individual histone messages. There do not seem to be intervening sequences within the individual histone coding regions, however.

18.5.1.2 Nuclear RNA and Control of Cytoplasmic mRNA.

We turn, finally, to questions raised at the beginning of Section 18.5.1., where it was noted that only a small fraction of nuclear RNA (nRNA) appears as cytoplasmic mRNA. What determines which of the genomic transcripts found in nRNA will serve as templates for structural proteins, and what is the significance of genomic transcripts that do not form mRNA? The material presented in the following discussion is extensively documented in the 1979 paper of Davidson and Britten; therefore, attribution to individual researchers is omitted in what follows.

In the late 1960s and early 1970s it was found that radioactive uridine incorporated into nuclear RNA appears in transcripts of a heterogeneous size range extending from several times that of cytoplasmic mRNA down to and overlapping the size of mRNA. The large transcripts, referred to as **heterogeneous nuclear RNA (HnRNA)**, were found to show m^7G "cap" and a degree of polyadenylation and, also, to be complementary to genomic DNA. In addition, cDNA complementary to specific mRNAs was found to bind to HnRNA. Nevertheless, as noted, very little HnRNA is represented in cytoplasmic mRNA, most of it being degraded in the matter of 20–30 minutes. The loss of HnRNA is far greater than can be accounted for by the excision of internal segments during the processing of mRNA.

These considerations led to intensive investigations, during the past several years, of the relationship between nucleus and cytoplasm with reference to the generation of cytoplasmic mRNA. These investigations have led to the recognition that differences in thousands of complex and of moderately prevalent classes of mRNAs characterize differentiated cell types. Thus the regulation of the classes of structural genes corresponding to these classes of mRNA is fundamental to the accomplishment of cellular differentiation; more so, probably, than is the regulation of genes transcribed as the superprevalent classes. The appearance of superprevalent classes of mRNA is usually a feature of the terminal differentiation of cells already specialized for a particular performance, the formation of egg-white

proteins, for example. Also, in many cases superprevalent mRNAs account for only a small portion of the total cytoplasmic mRNA of a cell.

Some of the salient facts that have emerged from the analysis of nuclear RNA and its relationship to cytoplasmic RNA are outlined here. These facts form the basis of a theory of the control of gene expression that has yet to be thoroughly tested, but which is expected to stimulate numerous investigations during years to come.

1. HnRNA consists of many sequences of nucleotides that are highly repeated and that have no counterpart in cytoplasmic mRNA.

2. The repetitive sequences in HnRNA may be divided into families, members of which are sufficiently alike that they can form stable base–pair duplexes.

3. The repetitive-sequence families are highly represented in nuclei of mature oöcytes and young embryos, but, as cells become differentiated, there is a gradual shutdown of the transcription of certain families such that each differentiated cell type has its own characteristic pattern of repetitive-sequence families. Importantly, in each cell type the complementary sequences of each family are both represented. This knowledge suggests that the formation of complementary duplexes could play a role in the regulation of gene expression.

4. In genomic DNA repeat sequences are interspersed with single-copy DNA. Moreover, the transcripts of the single-copy DNA are represented in nuclear RNA in a tissue-specific manner. This observation would seem to be at odds with the evidence (still a bit tenuous, to be sure) cited in Section 15.5.1.1, that in all cells all potential mRNAs are transcribed continually, although they may possibly not appear in the cytoplasm.

5. If potential mRNAs are transcribed in all cells, however, then the fact that there are cell-type-specific differences in single-copy gene transcripts indicates that not all genomic single-copy sequences are structural genes and that some single-copy transcripts may have a role other than the production of mRNA. This role could be a regulatory one.

The foregoing considerations led Davidson and Britten (1979) to formulate a model for the regulation

447

of differential gene activity involving both types of single copy sequences, that is, those that are and those that are not the source of prospective mRNAs (Figure 18.11). In the genome the structural genes, including sequences coding for leading, intervening, and trailing portions of the gene, are flanked by repetitive DNA sequences. Each such complex comprises a **constitutive transcriptional unit** (CTU). In all cells the CTUs are transcribed at a basic rate. The transcript is a single-stranded RNA copy of the CTU and is called **constitutive transcript** (CT). Each CT is potentially capable of being processed and exported to the cytoplasm, but as we have seen, different cell types have cytoplasmic mRNAs that are distinctive from those found in other cell types, as well as mRNAs that are common to all cell types. Some potential mRNAs, therefore, must be degraded before being processed and exported to the cytoplasm.

It is proposed that the processing of structural gene transcripts in the nucleus is controlled by portions of the genome that are transcribed in cell-specific fashion and that do not contain coding regions for structural genes. These genomic regions are called **integrating regulatory transcriptional units** (IRTU). The transcription of an IRTU gives RNA that controls the expression of functional genes. The IRTUs consist of interspersed repetitive and single-copy DNA sequences or of clusters of repetitive sequences. Each IRTU is under the control of a nucleoprotein "sensor" element, which is responsive to internal or external signal molecules. The sensor determines whether its associated IRTU will be transcribed. The transcript of the IRTU is referred to as an **integrating regulatory transcript**, or IRT. The IRTs should constitute the nRNA sequences that are cell specific (as indicated in the foregoing item 4), whereas the CTs are the ubiquitous nRNAs. Together IRTs and CTs constitute the HnRNA of the nucleus, except as part of the HnRNA may be constituted by superprevalent mRNA precursors.

In the model gene expression is regulated by the formation of intranuclear RNA–RNA duplexes between the repeated sequence regions of IRTs and complementary sequences on CTs. The RNA–RNA duplexes are required for the survival and processing of cell-specific sets of mRNAs. In the absence of duplex formation, both CTs and IRTs are rapidly degraded, as suggested by the rapid turnover of HnRNA. For CTs bearing repeat sequences comple-

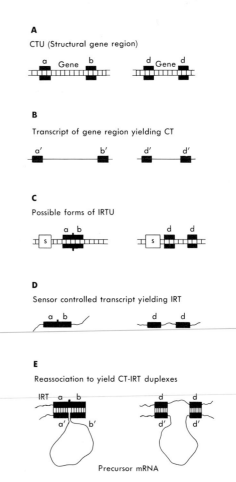

FIGURE 18.11 Model for the regulation of cytoplasmic mRNA. A: Two regions of the genome, each including a constitutive transcriptional unit (CTU) for a structural gene; such units are transcribed in all cell types along with flanking repetitive sequences, here designated by means of lowercase letters. B: mRNA transcripts (CTs) of the gene regions and their flanking sequences. C: Some possible forms of integrating regulatory transcription units (IRTUs) bearing sensor nucleoproteins that control the transcription of the IRTUs in response to external signals. D: Integrating regulatory transcripts (IRTs) transcribed from the regions shown in C. E: Two possible forms of intranuclear mRNA duplexes resulting from the complementary base pairing between flanking repetitive sequences of IRTs and complementary sequences on CTs. Formation of such duplexes is presumably required to prevent the degradation of the gene transcripts and to permit further processing of the mRNA [Adapted from E. Davidson and R. J. Britten, *Science* 204:1052–59 (1979).]

mentary to large families of repeat sequences on IRTs, it would be expected that opportunities for duplex formation would be enhanced and that such CTs, being produced at a basic rate, should be processed frequently and, hence, give rise to the moderately prevalent class of mRNAs. In contrast, CTs that are destined to form the complex class of mRNAs would have repeat sequences complementary to relatively small families of repetitive sequences on IRTs. The frequency of degeneration of these CTs could be such that only a small fraction of them would be processed into mRNAs.

This model requires that appropriate sets of sensor structures would characterize each cell type. These sets would necessarily result from developmentally controlled activities in other parts of the genome at an earlier time. These activities might be greatly different in highly regulative eggs, as contrasted with the mosaic eggs. In mosaic eggs one might imagine that regional differentiations in the cytoplasm arising during oögenesis or in postfertilization cytoplasmic movements would precociously bring about differentiation of sensor structures in nuclei that are relegated to these differentiated cytoplasmic regions by the cleavage pattern. In regulative eggs differentiation of nucleoprotein sensors would presumably occur more progressively by virtue of conditions arising epigenetically during cleavage.

18.6 Translational Controls of Gene Expression

Translational controls are those controls of the pattern of protein synthesis that are exercised in the cytoplasm. Points at which the control of protein synthesis may occur are many and varied. They include the stability of mRNAs, the translational capacity of the cellular machinery, the competitive fitness of different mRNAs for forming peptide-initiating complexes with the ribosomes, the elongation of the peptide chain and its termination, and so on. These and other points of control are taken up in what follows.

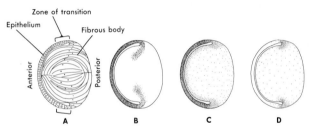

FIGURE 18.12 Long-lived mRNA in lens of the 12-day chick embryo. A: The normal lens. Cells multiplying in the epithelial layer migrate into and contribute to the fibrous body. B–D: The pattern of incorporation of radioactivity by isolated lenses incubated in labeled media. B: The pattern of incorporation of labeled uridine into RNA during 3 hr of exposure; label is restricted to cells of the epithelium and to those that have just entered the fibrous body. C: The incorporation of radioactive leucine in a lens exposed for 1 hr to the amino acid; uptake is greatest in the epithelium and that portion of the fibrous body subjacent to it. D: The distribution of the radioactivity in a lens similarly exposed to a labeled leucine, but after 8 hr of pretreatment with actinomycin D, which presumably prevents RNA transcription. Note that incorporation of leucine continues in the central zone of the fibrous body, just as in the controls, i.e., in C, but fails to occur elsewhere; it is most intense at the point where cells are just entering the fibrous body. [Adapted from R. Reeder and E. Bell, *Science* 150:71–72 (1965).]

449

18.6.1 STABILITY OF mRNA. The stability of mRNA in the cytoplasm is measured by its half-life, that is, the time required for one half of the message existing at any one time to decay. Eukaryotic mRNAs are, for the most part, relatively stable. Whereas mRNAs of prokaryotes have half-lives of only a few seconds or minutes, those of eukaryotes range from minutes to days. For example, there is a class of mRNAs in human lymphocytes that has a half-life of 17 minutes (Berger and Cooper, 1975), but mRNAs for lens fiber proteins (Figure 18.12) have half-lives of 8 hours or more (Reeder and Bell, 1965). In estrogen-treated immature chicks ovalbumin mRNA has a half-life of 4–5 days, and each message is probably translated about 50,000 times during its lifetime.

When the half-life of a species of mRNA is changed, the number of times it can be translated is also altered. Such changes affect the pattern of protein output in a cell. In general, when the develop-

FIGURE 18.13 Life cycle of *Dictyostelium discoideum*. The mature fruiting body releases spores that germinate to form amoeba-like organisms that feed and divide mitotically. When the dividing population is deprived of nutrients, its members aggregate to form a pseudoplasmodium, which forms a slug-like grex. After some hours the grex settles on its base and forms a stalked fruiting body.

450

mental situation no longer calls for the synthesis of a particular protein, the existing mRNAs for that protein rapidly decay. In the cellular slime mold *Dictyostelium discoideum* there is a phase in its life cycle (Figure 18.13) during which vegetative ameboid cells, if deprived of food, begin to aggregate to form a cohesive migratory body called a **grex.** Actin is the major protein synthesized by aggregating cells, and its rate of production increases rapidly during the first 2 hours after aggregation begins. The rate peaks and then begins to decline an hour later. The changing rate of actin synthesis is precisely paralleled by changes in the amount of translatable actin mRNA (Figure 18.14), shown when whole-cell RNA is added to an in vitro translating system (Alton and Lodish, 1977a, b).

Histone mRNAs show remarkable changes in half-life, and these changes are correlated with the cell cycle. In cultured mammalian cell lines histone mRNAs can be detected in the cytoplasm only during the DNA synthetic period (S phase) of the cell cycle (Detke et al., 1978). This period lasts for several hours, during which histone mRNA shows a long half-life. When the DNA synthetic period is over, or when DNA synthesis is blocked by drugs,

FIGURE 18.14 Parallel changes in the rate of actin synthesis and the content of actin mRNA in *Dictyostelium discoideum* after removal from growth medium. The data on which this plot was based were normalized with respect to the content of actin mRNA and the relative rate of actin synthesis in vegetative cells on growth medium. Aggregation begins about 6 hr after removal from growth medium; the grex forms after 10 hr. [Adapted from T. H. Alton and H. F. Lodish, *Dev Biol* 60:180–206 (1977).]

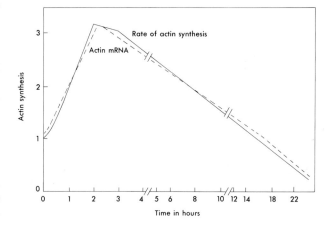

the half-life of histone mRNA is shortened to a matter of minutes and soon the message can no longer be detected. Interestingly, however, if both DNA and protein synthesis are simultaneously blocked with appropriate concentrations of NaCl during S phase, the polysomes release mRNA for histones into the cytoplasm in a relatively stable form (Stahl and Gallwitz, 1977).

When the estrogen-treated immature chick is withdrawn, production of the egg-white proteins drops dramatically, and the appropriate messages rapidly disappear from the polyribosomes. Palmiter and Carey (1974), assuming that the loss of polysomal message is the result of accelerated degeneration of the message, calculated for ovalbumin mRNA that its rate of decay is increased 10-fold during the first 20 hours after the last estrogen treatment.

Because some messages that have been processed but are no longer being translated tend to decay at accelerated rates, the question has been raised as to whether metabolic turnover of mRNA is linked to its translation. If it is, treatment of cells with drugs that prevent the attachment of mRNAs to ribosomes or that cause them to detach after initial binding would decrease their half-lives. This theory was tested for the mRNAs of two very similar enzymes, alanine aminotransferase (AAT) and tyrosineaminotransferase (TAT). These enzymes are of about the same size, have the same intracellular localization, and their mRNAs are translated at the same rate. The enzymes are induced in hepatoma cell cultures by hydrocortisone or related steroids. When hormone is withdrawn, the specific mRNAs for these enzymes decay, but at strikingly different rates. The mRNA for AAT decays with a half-life of 12– 14 hours, whereas mRNA for TAT does so with a half-life of 2 hours. It was found that neither fluoride ion, which inhibits attachment of mRNA to ribosomes, nor puromycin, which detaches nascent peptide chains from ribosomes, changes the decay rates of messages. It is quite evident, therefore, that for these hormone-induced transcripts, at least, the rate of message decay is the same whether they are being translated or not. Presumably, the half-lives of these messages are determined by the different primary structures of the two mRNAs. Cycloheximide, which slows chain elongation but not the initial binding of mRNA to the ribosome, prolongs the half-life of TAT mRNA, but not that of the more stable ATT mRNA. This evidence suggests that the half-life is determined by the character of the mRNA near or at the site of ribosomal binding, and that possibly the shorter lived TAT mRNA, which

451

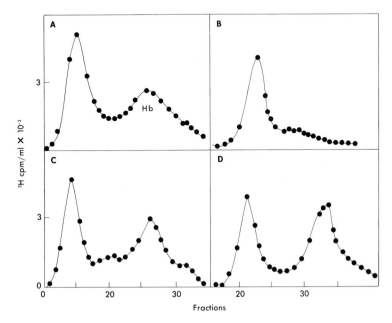

FIGURE 18.15 Effects of the 3′ poly(A) terminus on persistence of rabbit globin mRNA being translated in *Xenopus* oöcytes. Batches of eggs were injected with rabbit globin mRNA from which the poly(A) sequence was removed (A, B) or with equal amounts of native globin mRNA (C, D). Immediately after injection (A, C) or 51 hr later (B, D) ^{3}H-histidine was added to the culture medium. After 5 hr, the eggs were homogenized and the proteins fractionated by a molecular sieving technique that resolved proteins synthesized using endogenous mRNAs from those made on rabbit globin templates. In each plot, the endogenous protein peak is on the left and the hemoglobin peak on the right. Eggs receiving poly(A)-free mRNA (C) show no incorporation of label into hemoglobin after 51 hr, whereas eggs injected with native globin mRNA (D) show a large peak of activity in hemoglobin. [Adapted from G. Marbaix et al., *Proc Natl Acad Sci USA* 72:3065–67 (1975).]

would be stearically protected during translation, is more susceptible to attack by nucleases at this point (Stiles et al., 1976).

Polyadenylation has been suggested as contributing to the stability of messages that terminate in the 3' poly(A) sequence, but this matter is uncertain. Polyadenylated α- and β-globin mRNAs from rabbit reticulocytes injected into fertilized eggs of *Xenopus laevis* are translated for several days as development of the egg proceeds (Gurdon et al., 1974; Froehlich et al., 1977), showing remarkably little decay. Likewise, these messages are translated in large unovulated oöcytes. If, however, one removes the 3' poly(A) sequence, the message decays rapidly (Marbaix et al., 1975). Presumably, the stability of the injected polyadenylated mRNA is promoted by poly(A)-mediated association of the message with one or more *Xenopus* proteins. The contrasting template activities of polyadenylated and poly(A)-free rabbit globin mRNAs are shown in Figure 18.15.

18.6.2 REGULATION BY COMPONENTS OF THE TRANSLATIONAL MACHINERY.
Stimuli leading to the expression of a new pattern of protein synthesis at advanced stages of differentiation seem, in general, to induce the elaboration of a translational system quantitatively and qualitatively adapted to the developmental situation. As noted in Chapter 7, when the liver of vertebrates with large-yolked eggs is stimulated by estrogen, it prepares for the increased transcription and translation of the vitellogenin gene by an enhanced production of the translational apparatus: ribosomes, tRNAs, aminoacyl synthetases, factors promoting initiation, elongation and termination of amino acid chains, and so on. One gains the impression that the translational apparatus is geared rather precisely to the production job at hand.

Whereas the foregoing conclusion seems to be true, it is also evident that for a given stage of development the translational machinery may, in a sense, be overloaded or at least saturated with messages capable of being translated. During early cleavage stages of sea urchin embryos after a 1-hour labeling period with ^{3}H-uridine, only 50% of labeled (i.e., newly synthesized) mRNA is found in polyribosomes, and the rest is located in cytoplasmic RNP particles. These particles contain mRNA, as determined by translation in a cell-free system. As

development proceeds to the mesenchyme blastula stage, 80% of the newly labeled mRNA is found on polysomes and 20% is present in cytoplasmic mRNP. Some of the cytoplasmic mRNP, however, never appears on polysomes and decays with a half-life of about 40 minutes. These observations suggest that potential mRNAs are transcribed and exported to the cytoplasm in excess of what the translational machinery can handle. Presumably, newly synthesized mRNA, much of which is identical to maternal mRNA, is produced without regard to the maternal messages, and the two compete for binding to polysomes. With advancing development and progressive decay of maternal messages, more translational machinery becomes available for handling newly synthesized messages, so that their percentage on polyribosomes increases (Dworkin and Infante, 1976; Dworkin et al., 1977).

In the mature oöcyte of *Xenopus* the translational machinery is apparently saturated. Laskey et al. (1977) injected mouse or rabbit globin mRNA

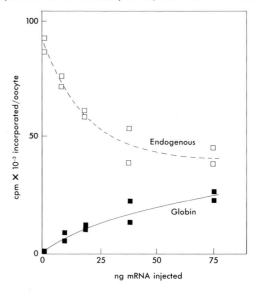

FIGURE 18.16 Depression of endogenous protein synthesis in *Xenopus laevis* oöcytes by injection of extraneous mRNA. Oöcytes were injected with varying concentrations of rabbit globin mRNA and labeled with a mixture of amino acids. Samples were homogenized after 2 hr and the amount of label incorporated into endogenous protein and into globin was determined. [Adapted from R. A. Laskey et al., *Cell* 11:345–51 (1977).]

into *Xenopus* oöcytes and found that any concentration of injected mRNA that elicited appreciable globin synthesis caused a concomitant depression in the synthesis of endogenous proteins (Figure 18.16). Injection of reticulocyte polyribosomes bearing globin mRNAs, however, resulted in globin synthesis without appreciable effect on the synthesis of *Xenopus* proteins. Thus it is evident that, whereas the translational machinery of the oöcyte is effectively saturated, an overall increase in protein synthesis can occur within the oöcyte when additional translational factors are supplied. Factors controlling initiation of the polypeptide chains on the ribosomes could then be rate limiting, a possibility that we now consider.

18.6.2.1 Chain Initiation as a Site for Translational Control. The initiation of translation requires that the mRNA achieve a stable initiation complex with the ribosome. The formation of this complex involves a special initiator methionyl tRNA (Met-tRNA$_i$), the ribosomal subunits, energy sources in the forms of adenosine triphosphate (ATP) and guanosine triphosphate (GTP), and a number of proteins and protein complexes known as **initiation factors.** For eukaryotes the initiation factors are designated by a prefix, eIF, which means

eukaryotic initiation factor. The prefix is followed by a numerical designation. The various eIFs currently recognized and their apparent roles in the initiation of translation are listed in Table 18.1. The first step in the formation of the **initiation complex** is the assembly of a **ternary complex** consisting of Met-tRNA$_i$, eIF-2, and GTP. The binding of Met-tRNA$_i$ to the 40 S ribosomal subunit is then catalyzed by eIF-2. mRNA is next bound to the 40 S subunit. Then eIF-3 and members of the eIF-4 group are involved in the stabilization of the binding of mRNA to the 40 S subunit, a process in which eIF-1 seems to play a minor role. Hydrolysis of ATP, catalyzed by eIF-4b, is essential to the binding process. Finally, eIF-5 catalyzes the union of the 40 S and the 60 S ribosomal subunits; GTP, which was involved in the formation of the ternary complex, is hydrolyzed in this reaction, and the eIFs are released. This step completes the formation of the stable 80 S initiation complex, and translocation of the message through the ribosome begins.

Because translation begins at the 5′ end of mRNA, it seems reasonable that the m^{7}G cap and the leader sequence of the message are important in initiation of translation. The three-dimensional configuration of the cap and leader sequence or of

453

TABLE 18.1   Eukaryotic Initiation Factors[a]

Designation	Molecular Weight		Probable Functions
eIF-1	15,000		Binding of Met-tRNA to 40 S ribosomal subunit Binding of mRNA to 40 S subunit
eIF-2	3 proteins	34,000 48,000 52,000	Formation of ternary complex [Met-tRNA$_1$–GTP–eIF-2] ↓ [Met-tRNA$_1$–40 S]
eIF-3	9–11 proteins	total 500,000 or more	*stabilize*
eIF-4a		50,000	Binding of mRNA to 40 S subunit
eIF-4b		80,000	catalyzes hydrolysis of ATP to ADP + P$_i$
eIF-4c		19,000	
eIF-5		160,000	Joining of 60 S ribosomal subunits to form 80 S initiation complex Hydrolysis of GTP to GDP + P$_i$
eIF-2a		65,000	Formation of initiation complex with artificial mRNAs
eIF-4d		165,000	Diminishes Mg^{2+} requirement for globin synthesis in absence of spermine in vitro

[a] Chiefly from M. H. Schreier, B. Erni, and T. Staehelin. 1977. *J Mol Biol* 116: 727–53. Whereas eIF-2a and eIF-4d can be purified from rabbit reticulocyte lyzates, they are not necessary to the synthesis of proteins using natural mRNAs.

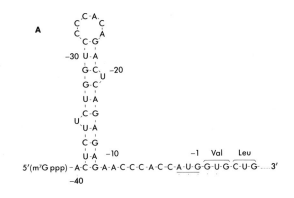

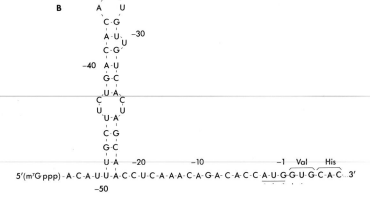

FIGURE 18.17 Possible secondary structures relating the 5′ m⁷G cap to the untranslated leader sequence of Human α-globin mRNA (A) and β-globin mRNA (B). [Adapted from J. C. Chang et al., *Proc Natl Acad Sci USA* 74:5145–49 (1977).]

the leader sequence alone, in the case of a non-capped message, might determine the efficiency of binding of mRNA and hence the speed with which translation begins. A possible relationship of the cap to the leader sequence is shown in two-dimensional representation for the mRNAs of human α-hemoglobin and β-hemoglobin in Figure 18.17. Under some conditions in vitro messages that normally bear the m⁷G cap are inhibited in translation by analogues of the normal cap (Hickey et al., 1976, 1977). Also, messages that are decapped enzymatically fail to be translated in some systems and are rapidly degraded (Shimotohno et al., 1977). Messages that are not normally capped, such as some eukaryotic viruses, are not readily degraded, nor are they affected in translation by analogs of the normal cap.

The normally uncapped encephalomyocarditis virus (ECM) binds to ribosomes at a rate similar to that of normally capped messages in the cell, but when tested in a cell-free system, ECM is a more effective competitor for initiation than capped messages, by a factor of about 100 times. In the cell-free system ECM shows a very high affinity for eIF-4b, but its competitive advantage for initiation can be diminished by the addition of large amounts of eIF-4b (Baglioni et al., 1978).

A cap-binding protein (CBP) having a molecular weight of 24,000 has been found by Sonenberg et al. (1980) to co-purify with eIF-3 and eIF-4b. CBP specifically recognizes capped messages, whether of viral origin or normal to the viral host cell, and facilitates the initiation of protein synthesis. It does not promote the initiation of normally uncapped viral messages, but if such messages are provided experimentally with a 5′ m⁷G terminus, their rate of initiation is increased. The addition of CBP also relieves translational competition between capped and naturally uncapped mRNAs in favor of the capped species, apparently acting in collaboration with eIF-3 and eIF-4b.

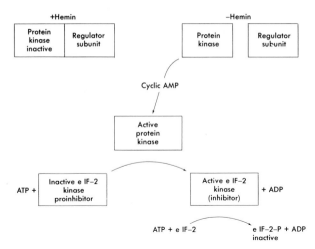

FIGURE 18.18 Control by hemin of an inhibitor of chain initiation. In the presence of hemin (upper left) a cyclic-AMP-dependent protein kinase is rendered enzymatically inactive by binding to its regulator subunit. In the absence of hemin, however, the regulator subunit is detached from the protein kinase, which then becomes an active enzyme catalyzing the phosphorylation of another inactive enzyme, namely eIF-2 kinase proinhibitor. Phosphorylation transforms the proinhibitor into an active kinase that inactivates eIF-2 by catalyzing the phosphorylation of the smallest subunit protein of eIF-2, rendering the initiation factor inactive. [Adapted from a scheme of A. Datta et al., *Proc Natl Acad Sci USA* 75:1148–52 (1978).]

Perhaps the best-known pathway for the regulation of gene expression through control of chain initiation involves eIF-2. The activity of eIF-2 is regulated in many cells and cell-free translating systems by **hemin.** Hemin is the oxidized form of the tetrapyrolle heme, which is found in hemoglobin, myoglobin, cytochrome c, and other enzymes. Through an indirect process that is illustrated in Figure 18.18, hemin controls the enzymatic phosphorylation of the smallest (molecular weight of 38,000) of the three protein subunits of eIF-2. Phosphorylation of this subunit renders eIF-2 inactive in catalyzing the binding of Met-tRNA$_i$ to the 40 S subunit of the ribosome and, accordingly, chain initiation is inhibited (Clemens et al., 1974; Datta et al., 1977a, b; Datta et al., 1978).

Translational control by hemin may be of general significance. A control system identical to that illustrated in Figure 18.18 has been described for embryos of *Artemia salina,* the brine shrimp. Hemin prevents the inactivation of eIF-2 in rat liver, and in reticulocytes it regulates the initiation of translation of the mRNA for carbonic anhydrase (Delaunay et al., 1977), in addition to regulating chain initiation for globin synthesis. Added to cultures of intact ascites cells, it stimulates endogenous protein synthesis, and in lysates of ascites cells, it promotes the translation of exogenously supplied mRNA.

Chain initiation is inhibited in the presence of double-stranded RNA (dsRNA) that is more than 50 base pairs in length. Lenz and Baglioni (1978) showed that dsRNA causes phosphorylation of the smallest subunit protein of eIF-2. Interestingly, however, very high concentrations of dsRNA are not inhibitory. Finally, one may note that phosphorylation of the other subunits of eIF-2 and phosphorylation of the other eIF proteins has not yet been shown to affect chain initiation (Benne et al., 1978). There has been some suggestion, however, that phosphorylation of various eIFs may possibly affect mRNA selection (see review by Ochoa and de Haro, 1979).

18.6.2.2 Competition Among Messages for Chain Initiation. When initiation factors are not rate limiting, mRNAs, even closely related ones, may differ considerably in their ability to initiate translation. Thus in rabbit reticulocytes the mRNA for α-globin initiates translation only 60% as frequently as does β-globin mRNA (Lodish and Jacobsen, 1972). These messages have the same m⁷G cap at the 5′ end and several subsequent nucleotides in common, but differ somewhat in nucleotides of the leader sequence. Conceivably, these differences account for the different rate constants for initiation of the globin messages. Once initiation is achieved, however, the rate of chain elongation is the same for both globin messages. Both globins are produced in equal amounts, coordination of their production being achieved by the fact that the ratio of mRNA for α-globin to that of β-globin is about 1.4 : 1 (Lodish, 1971).

Changing nutritional or hormonal conditions may affect differentially the translocation to ribosomes of messages with different rate constants for initiation. In slices of bovine pancreatic tissue or in isolated pancreatic islets of Langerhans incubated in vitro, addition of glucose to the medium stimulates synthesis of all proteins, but under conditions in

455

which total protein shows a twofold increase in production, that of proinsulin increases threefold. These increases occur independently of new RNA synthesis and in face of no change in the overall rate of chain elongation (Lomedico and Saunders, 1977). In the case of a partial hemin deficiency in reticulocytes, there is an overall decrease in the rate of chain initiation because of effects on eIF-2, as we have just described. This decrease results in the disproportionate reduction in synthesis of α-globin as compared to β-globin. In the estrogen-primed and withdrawn chick oviduct the initiation rates of all proteins are slow. When the oviduct is restimulated with estrogen, however, whereas there is an overall increase in the rate of chain initiation and protein synthesis, the increase in initiation of ovalbumin is disproportionately large in comparison to that of mRNAs for the other egg-white proteins and for proteins generally. But when progesterone is used to stimulate the withdrawn chick, the proportional enhancement of initiation and translation of ovalbumin mRNA is the same as that of the other oviduct proteins (Schimke et al., 1977). Conceivably, estrogen—but not progesterone—selectively increases the rate constant for initiation of ovalbumin mRNA.

The fact that different mRNAs have different initiation rate constants makes it possible that factors promoting overall changes in initiation rates could be responsible for radically different patterns of protein synthesis in different cell types of an organism. Cells that are undergoing mitosis, for example, initiate protein synthesis at a rate considerably lower than postmitotic cells, even when preexisting mRNAs remain intact (Fan and Penman, 1970). Thus during early embryogenesis regional differences in mitotic rate could result in regional differences in the translation of similar species of mRNAs because of message-specific differences in rate constants for chain initiation.

18.6.2.3 Chain Elongation and Termination as Translational Controls. Once the mRNA has formed a stable initiation complex with the ribosomal subunits, the rate of protein synthesis depends on the rate at which amino acids are added to the growing peptide chains and the efficiency with which they are released. In eukaryotes these processes are mediated by the **elongation factors** EF-1 and EF-2 and a single **releasing factor** RF. As described more fully in Appendix B, EF-1 catalyzes

the binding of aminoacyl tRNA (aatRNA) to aminoacyl site A on the ribosome, and EF-2 is responsible for translocating the incoming aatRNA to the **peptidyl** site B, after its amino acid has joined in peptide linkage to its antecedent amino acid in the growing chain. **Termination** occurs when the tRNA bearing the terminal amino acid of the chain is translocated to the P site. Termination requires the presence of a **termination codon**, UAA, UAG, or UGA, and the releasing factor.

The action of each of the elongation factors and the releasing factor requires its binding to GTP and to the ribosome, and each step in elongation and the process of termination involves the hydrolysis of the terminal high-energy phosphate of GTP to form GDP and inorganic phosphate. This reaction is presumably the signal for release of the bound factor from the ribosome. As noted, GTP is also hydrolyzed in connection with the formation of the initiation complex. Clearly the initiation, growth, and termination of the nascent protein chain require that GTP be constantly regenerated from GDP. This regeneration occurs at the expense of ATP generated by the anaerobic and oxidative phosphorylation reactions of the cell. Reduction in cellular metabolism would thus reduce the availability of ATP for the recharging of GDP and could thereby depress protein synthesis.

The availability and activity of the EFs and of RF would also greatly affect the rate of protein synthesis. Twardowski et al. (1977) described EF-1 of *Artemia salina* as existing in two forms: (1) EF-1$_H$ with a molecular weight of about 200,000 and (2) EF-1$_L$, which has a molecular weight of about 50,000. Presumably GTP disperses EF-1$_H$ into its subunits, which are EF-1$_L$. EF-1$_L$ and GTP then unite and react with aminoacyl tRNA, binding it to the ribosome. In dormant gastrular cysts of *A. salina*, EF-1 exists chiefly as EF-1$_H$, but in the hatched larva EF-1$_L$ predominates. From nauplii a protease has been isolated that disaggregates EF-1$_H$ at low concentrations without altering the size or activity of the resulting EF-1$_L$. Possibly, in this organism and in others a proteolytic modification of the subunits is involved in their disaggregation.

The relationship of components of EF-1 outlined here for *A. salina* may be over simplistic. Nagata and his associates (1978) isolated EF-1 from pig liver and then purified from it three subunits, EF-1α, EF-1β and EF-1γ. They proposed that EF-1α medi-

456

ates the mRNA stimulated binding of aatRNA to ribosomes in the presence of GTP. GTP is hydrolyzed to GDP + P; and then EF-1β or EF-1γ promotes the replacement of GDP bound to EF-1α with exogenous GTP.

When ribosomes from cysts are extracted at high-salt concentrations, a ribonucleotide is released that blocks the binding of EF-1 to aminoacyl tRNA. The extracted nucleotide has a molecular weight of about 6000, and uracil constitutes 40% of its bases. In high-salt ribosomal extracts from developing nauplii there is found another ribonucleotide, essentially absent in cysts, which counteracts the effect of the inhibitor. The activator has a molecular weight of approximately 9000, and its predominant base is guanine. The inhibitor is sensitive to a nuclease, similar to pancreatic RNAase, that can also be extracted from ribosomal washes. The activity of the RNAase increases significantly in developing embryos and may thus be another factor in regulating chain elongation (Lee-Huang et al., 1977).

It has been shown that the elongation of α- and β-globin chains proceeds at the same average rate, namely, about 1.5 amino acids per second at 25°C and that it requires about 15 additional seconds for chain termination in each case (Lodish and Jacobsen, 1972). About 5% of the total transit time of the message through a ribosome is involved in termination in these cases. Transit times would be expected to differ, depending on the length of message to be translated, but conceivably, the amino acids of each mRNA might be added at the same average rate. This seems not always to be so. In rat liver or cultured hepatoma cells the average transit time for all soluble proteins is about 2 minutes, but that of tyrosine aminotransferase (TAT), is 5–8 minutes. Production of the enzyme is greatly stimulated by glucocorticoid hormones, as already noted, but this enhanced synthesis results from increased transcription of TAT mRNA, not from diminution of the transit time. When an analogue of cyclic AMP, namely, dibutyryl cyclic AMP, is added to the induced system, the transit time for TAT may be diminished to as little as 45 seconds, whereas the overall transit time for soluble proteins remains unchanged at around 2 minutes. (Dibutyryl cyclic AMP mimics the action of cyclic AMP mediating a number of cellular processes. Unlike cyclic AMP, it is not hydrolyzed rapidly by phosphodiesterases

and its concentration can thus be controlled better.) This dramatic reduction in transit time for TAT is almost certainly the result of enhanced activity of the elongation factors. Unfortunately, however, the contribution of a possible change in termination time to total transit time for the enzyme cannot be determined with present methodologies until the terminal amino acids of the enzyme are identified (Roper and Wicks, 1978).

In germinating pea epicotyls the application of auxin (Chapter 12) enhances dramatically and without appreciable lag the accumulation of mRNA for the enzyme cellulase. There is a 24-hour lag, however, before an increase in the activity of cellulase is observed. Appropriate tests show that the epicotyl is not deficient in initiation factors during the lag period, but that there is a deficiency in elongation factors. Cellulase production appears to be regulated at two levels. One level may be that of the transcription, processing, or stabilization of cellulase mRNA. The other level of control involves the factors necessary for chain elongation (Verma et al., 1975).

18.6.2.4 Transfer RNA and Aminoacyl Synthetases in Translational Control. Transfer RNA plays a central role in protein synthesis, for it is the adaptor that directs the positioning of amino acids in the elongating peptide chain. The position in which a particular amino acid is inserted is determined by the sequence of trinucleotide codons in the mRNA that is on the ribosomes. Particular tRNAs are charged with the appropriate amino acids in a reaction catalyzed by a corresponding aminoacyl synthetase before binding to ribosomes. The role of tRNA in chain elongation is described in Appendix B for the benefit of readers who may wish to review this phase of protein synthesis.

When released from the ribosome, a particular tRNA is recycled with a frequency that depends on its rate of aminoacetylation, the frequency with which its specific amino acid is called for by messages being translated, the affinity of the aminoacetylated form of tRNA for the appropriate codon, and the duration of its attachment to the ribosome. Limiting amounts of tRNA charged with a particular amino acid could delay translation of a protein at the point where the amino acid is required. Conversely, the synthesis of special proteins characteristic of the differentiated state of a cell might be facilitated by the functional adaptation of tRNA

457

population to the frequencies with which amino acids appear in the special proteins.

Possibilities for regulation of protein synthesis during development are also provided by the fact that there are several synonymous mRNA codons for each of the 20 amino acids, some of which are more readily recognized by charged tRNAs than others. Moreover, for each of the 20 amino acids in protein, there are two or more tRNAs that can be specifically charged. Different tRNAs that recognize the same amino acids are referred to as **isoaccepting species** of tRNA.

Isoaccepting species usually differ in at least some base sequences and in the degree to which they have methyl groups added after being transcribed. Different isoaccepting species for an amino acid are, for the most part, transcribed from separate clusters of genes in linear array separated by spacer sequences (recall the condition of the ribosomal genes in the nucleolar organizing region discussed in Chapter 7). For example, in the *Xenopus* genome four isoaccepting species of the tRNA for valine (tRNA^{Val}) can be distinguished by hybridization to single-stranded DNA. Each of these species is transcribed from separate clusters of DNA sequences, which are, on the average, reiterated 90-fold (Clarkson et al., 1973a, b).

Isoaccepting species of tRNA may differ in their rates of aminoacetylation and in their affinities for their specific aminoacyl synthetases and may show differential affinities for their synonymous codons. The aminoacyl synthetases also exist in multiple forms for the different amino acids and tend to show preferential charging of particular isoaccepting tRNA species. These variables together with those noted in the foregoing paragraphs offer almost infinite possibilities for the functional adaptation of tRNA to particular developmental situations. This statement is not intended to imply that specific tRNA levels or their charging enzymes are programmed as a part of the process of cell differentiation. In fact, they may represent physiological adaptations to the pattern of protein synthesis that is called for by a developmental event. A few examples of the physiological adaptation of tRNA populations are presented in the material that follows.

First, let us note that intracellular concentrations of specific tRNAs are highly correlated with the needs of the cell for their corresponding amino acids in protein synthesis. As an example, in hemo-

globin A (HbA) of sheep there are no isoleucine residues and six methionine residues; in sheep hemoglobin of type B there are likewise no isoleucine residues and eight methionine residues; in hemoglobin C there are two isoleucine residues and two methionine residues. Correspondingly, in sheep reticulocytes synthesizing HbC the tRNA accepting capacity for isoleucine is two or three times higher than for reticulocytes producing HbA or HbB, and the HbA- and HbB-producing reticulocytes have a greater tRNA-accepting capacity for methionine (Litt and Kabat, 1972).

How the synthesis and degradation of a particular tRNA in a cell is regulated is not clear. There is some evidence, however, that deprivation of a particular amino acid required for protein synthesis tends to increase the concentration of the tRNA for that amino acid relative to other tRNAs. Litt and Weiser (1978) cultured a mouse leukemia cell line in histidine-depleted media and in media containing L-histidinol, a molecule that competitively inhibits histidyl tRNA synthetase. Under both of these conditions total tRNA synthesis and protein synthesis are inhibited. But likewise under both conditions the proportion of tRNA-binding histidine was increased proportionately to that of tRNA binding to other amino acids. It is not evident whether changes in specific tRNA levels are controlled by varying their transcription, processing, or degradation.

Changing patterns of protein synthesis are usually accompanied by changes in the patterns of isoaccepting species of tRNAs, especially for the most frequently utilized amino acids. During the last larval instar of the silk moth the posterior part of the silk gland begins to produce **fibroin,** the major protein of silk. At this time, also, the cells of the middle portion of the gland begin production of another protein **siricin.** Fibroin is made up principally of glycine (44%), alanine (29%), and serine (12%). The amino acids of siricin are chiefly serine (37%), glycine (17%), and alanine (6%). The relative abundances of the tRNAs for each amino acid in active glands exist in corresponding proportions. Delaney and Siddiqui (1975) found that prior to the onset of fibroin synthesis in the posterior part of the gland two isoaccepting species of glycine tRNA (tRNA^{Gly}) are present. When fibroin production begins, a third tRNA^{Gly} appears and all three isoaccepting species increase dramatically in activity,

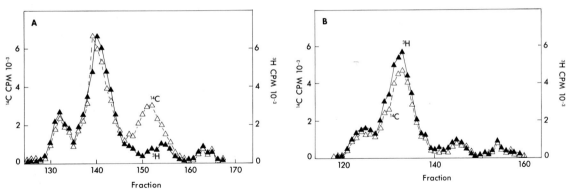

FIGURE 18.19 Hormone-elicited changes in isoaccepting tRNAs and their charging enzymes (aminoacyl synthetases). A: (△– –△) Charging of tRNA^{Leu} with ^{14}C-leucine by rat liver tRNA synthetases 3 hr after injection of hydrocortisone; (▲—▲) charging of tRNA^{Leu} with ^{3}H-leucine by rat liver synthetases 3 hr after saline injection. The two preparations were mixed and co-chromatographed, and the activities of ^{14}C and ^{3}H were differentially counted. Note that there are four peaks of activity, two large and two small, representing 4 isoaccepting tRNA^{Leu}s in the controls. The injection of hormone elicits a new peak after 3 hr. B: (△– –△) Charging of tRNA^{Leu} obtained from rat liver 12 hr after hydrocortisone injection by means of synthetases isolated from livers 3 hr after hydrocortisone treatment; (▲—▲) charging of tRNA^{Leu} obtained from rat liver 3 hr after hydrocortisone injection by means of tRNA synthetases isolated 12 hr after hormone treatment. Note that the synthetases obtained 3 hr after hormone treatment do not elicit the new charging peak shown in A when used with tRNA^{Leu} extracted 12 hr after treatment. Also, the synthetases present after 12 hr do not charge the new tRNA^{Leu} that is present 3 hr after treatment. [Adapted from K. Altman et al., *Proc Natl Acad Sci USA* 69:3567–69 (1972).]

459

but in differing proportions. Meza et al. (1977) analyzed the isoaccepting species of tRNA^{Ala} in both the middle and posterior portions of the gland. They found changes in isoaccepting tRNA^{Ala} species in both glands at the onset of special protein synthesis, but the pattern of changes differed in cells producing siricin and those producing fibroin.

Differences in isoaccepting species of tRNAs have been reported for early stages in sea urchin development (Yang and Comb, 1968), in liver in response to thyroxine treatment (Yang and Sanadi, 1969), in erythroid cells of embryonic and adult chicks (Lee and Ingram, 1967), and in insects during embryogenesis (Smith and Forrest, 1973). Differences are found among different organs of the same animal (Rogg et al., 1977), between cancerous and the normal tissues of their origin, and as a result of viral infection. The list could be extended to some length, but these and the foregoing examples should suffice to show the almost universal correla-

tion between diversity of isoaccepting tRNA species and diversified states of cellular differentiation.

The aminoacyl synthetases for individual amino acids also vary from one developmental stage to the next and among differentiated cell types. When one isolates total aminoacyl synthetases from newly hatched larvae of *Rana pipiens* they aminoacetylate very efficiently tRNAs isolated from similar larvae, but are quite ineffective in charging tRNAs isolated from unfertilized, fertilized, and cleaving eggs. Conversely, enzymes from unfertilized eggs have high charging potential for tRNAs of eggs and cleaving embryos, but show little capacity to aminoacylate those from larvae (Caston, 1971). Bick and Strehler (1971) showed that during the first 21 days of germination of soybeans there are time-related changes in six isoaccepting leucyl tRNAs and in three leucyl tRNA synthetases that occur in the cotyledons. Leucyl-tRNA synthetases also change in rat liver in response to hydrocortisone and related compounds. When the isoaccepting species of tRNA^Leu from livers of untreated rats are charged with labeled amino acids, using aminoacyl synthetases from normal liver and displayed chromatographically, they show four peaks of activity corresponding to four isoaccepting species of tRNA^Leu (Figure 18.19A). Within 3 hours after the intra-

FIGURE 18.20 Changed charging properties for tRNA^Leu between the 1st and 7th day after the larval–pupal molt in *Tenebrio molitor*. The abscissa indicates micrograms of synthetases from day 1 (■—■) or day 7 (□—□). A: Addition of day 7 enzymes to day 7 tRNA increases charging tRNA^Leu as compared to day 1 enzymes; day 7 enzymes recognize either more or different day 7 tRNA^Leu than day 1 enzymes do. B: Charging of day 7 tRNA^Leu by day 7 enzymes is not enhanced by addition of day 1 enzymes. C: Day 1 enzymes charge day 1 tRNA^Leu as effectively as they charge day 7 tRNA^Leu, but no additional charging occurs when day 7 enzymes are added. D: Day 7 enzymes apparently recognize and charge all species of day 1 tRNA^Leu, for the addition of day 1 enzymes does not enhance charging above that carried out by day 7 enzymes. [Adapted from J. and Judith Ilan, *Dev Biol* 42:64–74 (1975).]

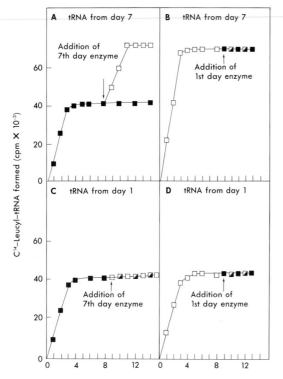

peritonial injection of hydrocortisone, a new leucine-accepting tRNA and a new leucyl-tRNA synthetase appear in liver cytoplasm. The new isoaccepting species can be acylated only with the synthetase derived from livers of hormone-treated animals (Figure 18.19B). Both components of the new charging system are transient; 12 hours after administration of hormone they disappear from livers of treated animals (Altman et al., 1972). During days 1–7 after the larval-pupal molt in the beetle *Tenebrio molitor* a change in the leucine-charging system occurs also. During this period a new isoaccepting species of tRNA^Leu appears and possibly a new leucyl tRNA synthetase. The leucyl tRNA synthetase activity at day 1 and day 7 is identified in each case with seemingly identical proteins with a molecular weight of about 9900. The two enzymes could be differentiated only by virtue of their charging potential for day 7 tRNA, as shown in Figure 18.20. Possibly, therefore, they are modified forms of the same enzyme, but could conceivably be translated from different transcripts (Ilan and Ilan, 1975).

18.6.2.5 Control of Chain Elongation by Functionally Coordinated Proteins.

The products of translation are often metabolically active only when associated as subunits of a larger protein complex. Recall that eIF-2, for example, is a complex of three proteins of different molecular weights. Many common enzymes are active as complexes of identical or closely related subunits. A familiar example is lactic dehydrogenase, which is effective only in the form of any of five possible tetramers formed by the association of A and B subunits, which are separate gene products. It would seem reasonable that the production of separately translated but functionally interacting subunits should be coordinated. Two examples of translational control of one subunit by the other are given here.

Hemoglobin is made up of two α-globin and two β-globin subunits. Wolf et al, (1973) showed that in adult rabbit bone marrow, the principal blood-forming organ, the synthesis of β-globin is depressed when the synthesis of α-globin is specifically inhibited. The investigators used rabbits with a variant strain in which the single isoleucine residue of the β-globin chain had been replaced by valine. The α-globin unit, however, contains three isoleucine residues. An antagonist of isoleucine, Lmethylthreonine, administered to bone marrow preparations would thus directly inhibit globin synthesis—but not α-globin synthesis—by diminishing the availability of isoleucine. Nevertheless, the synthesis of β-globin was depressed also. Stevens and Williamson (1973) examined the synthesis of immunoglobulin in a mouse plasmacytoma (lymphocyte tumor). The final product is a tetrameric structure consisting of two heavy chains and two light chains. They found that the light chain exists as a pool 5 times larger than the heavy chain. It binds specifically to the heavy chain in mRNA and suppresses its synthesis. This type of repression is not likely to occur in the case of coordinated repressors of β-globin synthesis in rabbit erythroid cells, for there is no pool of α-globin in them.

18.6.2.6 Is the Translational Machinery Cell Type Specific?

From time to time evidence has been presented that cellular specialization is marked by changes in the translational machinery such that messages appropriate to the differentiated state of a cell are preferentially translated. The possibility that muscle-specific initiation factors may be required for translation of myoglobin mRNA was raised in Heywood's laboratory (Thomson et al., 1973; Heywood and Kennedy, 1974), but this idea requires further testing. By and large, the preponderance of evidence suggests that the translational apparatus in all cells, even highly specialized ones, has the ability to translate mRNAs from a wide variety of sources. Reticulocytes, for example, are highly specific cells, 90% of whose output of protein consists of globins (Lodish and Desalu, 1973). Despite the specialization of reticulocytes, their lysates will translate with great efficiency mRNAs for proteins of kinds that reticulocytes never produce. Moreover, they will do so in the absence of initiation factors or other components from the cells that normally translate the foreign mRNAs. A partial list of mRNAs that are effectively translated by rabbit reticulocyte preparations are those for ovalbumin, conalbumin, lysozyme, and ovamucoid, which are components of chick egg white, α-crystallins of lens, myosin, and many viral mRNAs. Lodish (1976) has reviewed these and many other instances that show the nonspecificity of the reticulocyte translating system. Brain- and liver-translating factors, moreover, can translate globin mRNAs in vitro (Schreier and Staehlin, 1973; Crystal et al., 1972).

Do tests that are carried out using cellular extracts or lysates reflect a true lack of translational specificity for different mRNAs in living cells? The fact that *Xenopus* oöcytes and fertilized eggs will translate injected messages provides an opportunity to answer this question. Gurdon and his colleagues (Gurdon et al., 1973, 1974; Woodland et al., 1974) injected globin mRNA from mouse and rabbit reticulocytes into oöcytes and fertilized eggs of *Xenopus*. The fertilized eggs were allowed to develop—some as late as the swimming tadpole stage, which requires 8 days at 19°C. At appropriate intervals embryos or isolated fragments of early larval stages were put in medium supplemented with ^{3}H-histidine for labeling of newly synthesized protein, and after 14 hours they were sacrificed, fractionated, and examined for labeled globins. It was found that mouse globins were translated at the same rate in oöcytes, fertilized eggs, and all tissues of the developed embryo and that the α-globin/β-globin ratio was about equal at all stages if hemin was included in the material injected (Lingrel and Woodland, 1974). Rabbit globins were synthesized at a low rate in oöcytes, but much more rapidly in advanced embryos. Again, however, all tissues were equally effective in translating the globin messages, but for the rabbit globins α- and β-chains were translated in approximately equal amounts regardless of addition of hemin. Clearly cells do not have to be in the erythropoietic line in order to translate globin mRNAs. Muscle and nerve cells from the dorsal axial part of the embryo, a region far removed from presumptive or actual erythropoietic tissue, produced globins as effectively as did cells from the ventral part of the embryo or larva. In similar experiments carried out by Froelich and his collaborators (Froelich et al., 1977), rabbit globin mRNA was injected into fertilized eggs of *Xenopus* and globin synthesis was measured at the gastrula stage. The investigators also extracted globin mRNA from different regions of the gastrula and determined its content by hybridization with globin-complementary DNA. They found that globin synthesis, as a percentage of endogenous protein synthesis, decreases in a gradient from the small animal pole cells of the gastrula to the large cells of the vegetal pole. Measurements of the distribution of rabbit globin mRNA in the gastrula showed that it, too, declined along the same gradient. Presumably, therefore, the greater proportion of globin synthesis at the animal pole is a reflection of a greater globin mRNA content per cell. In contrast, Gurdon's group concluded that injected mRNA was evenly distributed and equally well translated throughout the embryo. They did not, however, directly measure the regional distribution of globin mRNA in their embryos. These differences notwithstanding, the results from both laboratories are compatible with the conclusion that translation of messages occurs with little specificity or selectivity in all cell types.

18.7 Posttranslational Regulation of Gene Expression

The control of gene expression continues beyond the fabrication of linear sequences of amino acids on the ribosomal mRNA complex. Posttranslational modifications of the peptide chains determine their functionally active configuration; additional controls adjust their life spans as called for by the metabolic situation.

One level of control is constituted by changes that occur spontaneously in the chain by virtue of its specific sequences of amino acids. Anfinsen (1968) has brilliantly illustrated for bovine pancreatic ribonuclease that the three-dimensional structure of the molecule emerges rapidly and automatically from the interaction of the amino acid side-chains with one another, with the completed peptide backbone, and with the environment, without needing further genetic information. The formation of disulfide bonds between cysteine residues is particularly important in stabilizing molecular configuration, and, together with weaker noncovalent bonds, it is important in determining the association of proteins with one another and with various ligands.

Another very important but often ignored aspect of processing is the modification of individual amino acids in a chain by covalent bonding to various reactive groups. Acetylation, methylation, and hydroxylation of residues occur in all proteins. Notably, too, tRNAs mediate the covalent bonding of their specific amino acids to reactive sites on inter-

nal members of the chain. Uy and Wold tabulated for all 20 amino acids the then-known (1977) posttranslational modifications that occur to each of them in peptide chains. Lysine and tyrosine are the amino acids most frequently modified by covalent bonding to reactive groups, in effect, creating 33 different derivatives of lysine and 19 of tyrosine. Many of these modifications, as well as the initiation of folding into the three-dimensional configurations, occur while the nascent protein is still being translated on the ribosome. Translation and post-translational processing thus go on simultaneously.

In Chapter 7 it was noted that amphibian vitellogenin* is secreted into the blood by the liver under the influence of estrogen. Before being secreted, however, the vitellogenin is highly modified by the addition of lipids, carbohydrates, and phosphates. In the blood of *Xenopus laevis* it appears as a protein dimer with a molecular weight of 450,000 and is transported to the ovary where it is further processed in the oöcytes. Within the oöcyte each dimer is cleaved into three polypeptide chains. Two of the three, with molecular weights of 120,000 and 31,000 respectively, contain essentially all of the lipid and are noncovalently bonded to form lipovitellin. The other subunit has a molecular weight of about 35,000 and is heavily phosphorylated. This subunit is phosviten (Bergink and Wallace, 1974).

Concomitant translation and post translational assembly of unlike polypeptide subunits is required in some cases for the assembly of functional protein aggregates. The α-crystallin of calf lens, for example, is a large aggregate having a molecular weight of about 800,000. This aggregate is formed by the

---

*Vitellogenin was once thought to result from the transcription of a single gene, translation of the transcript, and intracellular processing of the resulting protein. It now appears that *Xenopus laevis* has four vitellogenin genes in each haploid genome (Jaggi et al., 1980; Ryffel et al., 1980; Wahli et al., 1981). After extensive nuclear processing of primary transcripts, four different mRNAs, each about 6300 nucleotides in length, can be isolated. The different messages show extensive similarities in their nucleotide sequences. The four mRNAs can be translated in vitro by a cell-free reticulocyte lyzate preparation to produce four different proteins, each with a molecular weight of about 200,000. At this time it is not known to what extent each of these products of gene expression is represented in circulating vitellogenin.

462

assembly of two different gene products, α-A crystallin and α-B crystallin, in the ratio of 2:1. Asselbergs et al. (1978) showed that the translation product of α-A mRNA forms aggregates in the absence of α-B mRNA, but the incorporation of α-B into aggregates requires that its mRNA be translated along with α-A mRNA.

Proteins that are to be secreted are synthesized on ribosomes attached to the rough endoplasmic reticulum (RER). As the elongating peptide emerges from the ribosome, it is directed through the membranes of the RER into the cisternal space. Here it begins its three-dimensional folding, and, in the case of secreted glycoproteins, modifications such as the addition of N-acetylglucosamine and mannoal are incurred. Hydroxylation of lysine and protein residues in collagen subunits likewise occur in the cisternae of the RER. Further modifications take place in the Golgi complex. Here the glycoproteins receive their complete complement of carbohydrate residues. Here also, proteins that are to be stored and later secreted, such as insulin and digestive enzymes, are packaged in an inactive form.

Most secreted proteins are usually synthesized as transient precursors containing 15–30 additional amino acid residues at the NH_2 terminal end, that is, the first-translated portion of the peptide chain. These transient forms are usually detectable when the appropriate mRNAs are translated in vitro, but are now known for light and heavy immunoglobulin chains, serum albumin, a number of pancreatic enzymes, several polypeptide hormones (e.g., prolactin, growth hormone, parathormone, corticotropin, glucagon, insulin), honey bee venom (mellitin), and the egg-white proteins conalbumin, lysozyme, and ovomucoid. The prefix pre is usually used to designate the appropriate precursor: Thus one refers to preglucagon, preprolactin, and so on.

The NH_2 terminal sequences of the precursor molecules differ for each protein, but have in common the fact that they are rich in hydrophobic residues. This property renders them highly suitable for binding to and passing through membranes of the RER and has led to the formulation of the **signal hypothesis** for the synthesis and segregation of secretory proteins (Blobel and Dobberstein, 1975a, b). The essence of the hypothesis is that the initiation and elongation of secretory proteins occur on free ribosomes in the cytoplasm. As the hydrophobic NH_2 terminal end of the amino acid and chain

emerges from the ribosome, it serves as a signal that causes the ribosome to attach to the membrane of the RER. The signal sequence then passes through the membrane into the cisternal space. The signal sequence is removed by a membrane-bound endopeptidase. Meanwhile, the rest of the chain enters the cisternal space as it is synthesized. In 1977 Jackson and Blobel reported the successful extraction of the endopeptidase from the RER of dog pancreas. They tested its activity in the wheat-germ translating system, which is dependent on exogenous mRNA. When this system is supplied with mRNAs for prolactin and for growth hormone, the hormones isolated from the preparation are in their precursor forms. In the presence of the endopeptidase, however, cleavage of the leader sequences occurred and the mature forms of these hormones were isolated. (See also, related experiments of Shields and Blobel, 1978; in addition, see Wickner's 1979 review for some arguments against the signal hypothesis.)

Some proteins require further cleavage after removal of the leader sequence in order to achieve their functional form. One example is the case of honey bee venom, mellitin. It is a polypeptide composed of only 23 residues. In the venom gland it is formed slowly from promellitin, which is about twice as large. When mellitin mRNA is translated in a wheat-germ system, however, a prepromellitin composed of 70 residues can be isolated; the first 21 of these residues constitute the hydrophobic signal sequence and the last 23 constitute the active venom (Suchanek et al., 1978).

Insulin also exists in pro- and prepro- forms. Insulin consists of two amino acid chains, an A chain and a B chain, derived by the translation of a single mRNA. In the rat the template codes for a chain of 110 amino acids that constitute preproinsulin. The first step in processing of the chain is the elimination of a signal sequence of 24 amino acids. Proinsulin, the remainder of the chain, now folds as it enters the cisternal lumen of the RER, and disulfide bridges form between cysteine residues, stabilizing the three-dimensional configuration of the molecule. As proinsulin enters the Golgi complex, it undergoes cleavage at two points, eliminating a central segment of 33 amino acids. This cleavage leaves A and B chains of 22 and 31 residues, respectively, connected by disulfide bridges (Lomedico et al., 1979). Many other examples of

463

the posttranslational processing of protein precursors may be found in a publication of the New York Academy of Sciences edited by Zimmerman et al. (1980).

It is appropriate to say a final word about the persistence of the expressed gene product. Like the RNAs, the protein composition of a cell is regulated in accordance with the particular developmental stage and metabolic situation. Very little is known about the way in which the persistence of a protein species is regulated. For enzymes, however, it appears that when the presence of a particular enzyme is no longer appropriate, it is degraded rapidly. Thus, for liver enzymes such as tryptophan pyrrolase, tyrosinaminotransaminase, arginase, and others, the activity of the enzyme is enhanced by the administration of the appropriate substrate, for example, tryptophan, tyrosine, arginine. The effect of the substrate in these cases is to inhibit the rate of decay of the enzymes, which are normally held at a steady state by a balanced rate of production and degradation (Schimke, 1967).

464

18.8   Summary

Results of many experiments indicate that, with few exceptions, nuclei of cells in different regions of embryos and adult organisms each contain the same genetic information that was present in the zygote. A nucleus from a cell at an advanced stage of development transplanted into an activated compatible egg whose nucleus has been removed or inactivated, can foster development of a complete organism, expressing genetically controlled phenotypic differences found in the donor organism. Certain mouse teratocarcinoma cells, carried for many generations as undifferentiated embryoid cells, are also able to express the complete range of normal differentiated states in a fully integrated manner when introduced into mouse blastocysts that implant and develop to adulthood.

Techniques of competitive DNA–DNA hybridization and kinetic analyses of DNA–DNA reannealing after thermal melting reveal with a high degree of confidence that all cells are alike in their composition of DNA base pairs. Similar techniques reveal, moreover, that there are stage- and tissue-specific differences in the mRNA species transcribed from the genomic DNA and that these differences involve transcripts of both single-copy and reiterated genes. These observations suggest the general interpretation that variable gene activity is at the basis of cellular differentiation. This interpretation leads to the more difficult problem of how this variability is controlled. What determines which messages will be transcribed and processed, and at what rates? What controls the lifetime of the transcripts and the rate of their translation? And, what is involved in controlling the ultimate aspect of gene expression, the functionally active, three-dimensional protein product?

The genome is transcribed in the presence of adequate amounts of ribonucleotide triphosphates by the RNA polymerases I, II, and III, which catalyze transcription of rRNA, mRNA, and tRNA, respectively. The activity of polymerases is generally not rate limiting except as inhibitors of polymerase activity are sometimes found in dormant embryos. Very crude controls in genomic transcription are exercised by mechanisms that inactivate whole chromosomes, such as the X chromosome in female mammals or the paternal genome in certain insects, and by mechanisms that eliminate from somatic cells large numbers of chromosomes of which some or all are requisite to the differentiation of the germ line, but not of the body cells generally.

A serious question is whether or not the particular population of mRNAs found in the cytoplasm in embryos at different stages of development, or in the cytoplasm of cells from different kinds of tissues, reflects differences in the activation of specific segments of the genome for transcription. Conceivably, all of the genome could be transcribed in the nucleus of each cell, but the population of cytoplasmic mRNAs would reflect the operation of post-transcriptional controls. This question has not been settled as yet, but there is increasing evidence that the cytoplasmic mRNAs involved in the differentiation of particular cell types are controlled by intra-nuclear factors that determine which mRNAs will be exported to the cytoplasm for transcription.

With the exception of histone mRNAs, most cytoplasmic mRNAs are transcribed from DNA sequences that are represented only once or, at most, a few times in genomic DNA. Such DNA sequences are called single-copy genes. In cells that synthesize large quantities of specialized proteins a few single-copy genes may be represented in the cytoplasm by 10^5–10^6 mRNA copies per cell. Such mRNAs are referred to as the superprevalent class. In all cells, however, there are mRNA species that are present in only one or a few copies per cell. These mRNAs are referred to as the complex class. They represent single-copy DNA the complexity of which (that is, the sum of diverse sequences of base pairs of which) is greater than that of the other single-copy DNAs. There is another natural group of cytoplasmic mRNAs referred to as the moderately prevalent class. The mRNAs in this class are represented by 15–300 copies per cell.

Most studies of the control of protein synthesis have been concerned with the generation, processing, and translation of superprevalent mRNAs. Results of these studies have led to the general concept that whether a gene is transcribed or not involves interaction between RNA polymerase and some control molecule or molecules which come to occupy sites adjacent to a particular coding region. Nonhistone chromosomal proteins, which are different in different specialized cells, have been thought of, for example, as unmasking particular gene loci for transcription. Cytoplasmic receptors, translocated to the nucleus under the influence of steroid hormones, have been held responsible for the initiation of specific patterns of transcription. It is not certain, however, that specific control molecules actually initiate transcription of hitherto inactive genes. They may, rather, affect their rate of transcription or rate of intranuclear decay.

Gene transcripts include those for rRNA and tRNA, in addition to potential control transcripts and mRNAs. All gene transcripts undergo some processing in order to become functional. All RNA becomes associated with protein as it is transcribed, and apparently all RNAs exist in precursor forms that require modifications that begin in the nucleus in order for the RNAs to become functional. Both rRNA and tRNA are transcribed as larger precursors that must be trimmed and methylated before being exported. Cytoplasmic mRNAs as found in the cytoplasm are considerably modified from the original transcript. Except for certain histone mRNAs, cytoplasmic mRNA consists of a special "cap" of 7-methylguanosine (m^7G) attached to the first transcribed nucleotide by a 5' to 5' triphosphate linkage. After the cap comes a leader sequence, which is not translated; then the initiator codon, AUG; the

translated sequence; a terminator codon, UGA; an untranslated trailer sequence; and a "tail" consisting of a variable number of residues of polyadenylic acid, poly(A).

The m^7G cap and poly(A) tail are added post-transcriptionally. The remaining part of the mature mRNA represents, in most cases, only a small part of the initial nuclear transcript, for intranuclear processing removes large intervening segments of the initial transcript, followed by splicing of cut ends to form the message that is exported to the cytoplasm. Processing of the primary transcript of the conalbumin gene requires the excision of 16 intervening segments and the splicing of 17 fragments 60–200 nucleotides long.

The properties of mRNAs and their processing as described here have been determined by studying the production of superprevalent mRNAs. Yet during development different cell types show differences in thousands of complex and moderately complex classes of mRNA. Presumably, the regulation of the structural genes corresponding to these classes of mRNA is fundamental to cellular differentiation, but because the mRNAs of these classes exist in so few copies, it is difficult to identify these mRNAs with particular protein products. Thus one cannot carry out studies parallel to those involving genes that are transcribed in the production of superprevalent mRNAs. Moreover, the regulation of gene activity involved in the production of complex and moderately prevalent mRNAs may be different from regulation of genes for superprevalent mRNAs.

On the basis of a number of facts cited in the text, but not repeated here, Davidson and Britten have proposed that in all cells the structural genes consist in single-copy DNA sequences that are flanked by repetitive DNA sequences. Each such complex is a constitutive transcriptional unit (CTU). In all cells all CTUs are transcribed at a basic rate to form a constitutive transcript (CT). Each CT is a potential candidate for processing into mRNA. Also in the genome are other single-copy DNA sequences flanked by repetitive DNA. Each of these sequences is an integrating regulatory transcriptional unit (IRTU), and its transcript is an integrating regulatory transcript (IRT). IRTs differ in different kinds of cells, for only those IRTUs are transcribed that possess an activated sensor element that is responsive to external signal molecules. The sets of sensor

465

structures would have arisen as a result of developmentally controlled activities at an earlier time. The IRTs determine which of the CTs will survive to be processed to mRNA by virtue of the degree of complementarity between their flanking repetitive sequences and those of the CTs. For CTs bearing repeat sequences complementary to large families of repeat sequences on IRTs, complex formation and survival for processing to mRNA are enhanced.

Once the determination is made as to which sets of mRNA will be processed and exported to the cytoplasm, other factors determine the extent of their participation in protein synthesis. One factor is the stability, or half-life, of the message. The m^7G cap and the poly(A) tail sequence of the message may be involved in message stability, but a major factor is the requirement for its translation product.

In general, the translational capacity of the cell is geared to the demand for synthesis of a particular set of proteins. There are a number of sites at which the rate of translation may be controlled, however. The rate at which mRNA achieves a stable initiation complex with the 80 S ribosome is critical, but thereafter, the rate at which the message is translated and terminated are also subject to regulation. The first step in chain initiation involves formation of a ternary complex between Met-tRNA$_1$, which is specified by the initiator codon of the mRNA, with initiation factor eIF-2 and GTP. This complex binds to the 40 S ribosomal subunit, which then binds to mRNA. Other eIFs stabilize this binding and catalyze the union of the 40 and 60 S subunits of the ribosome; thereupon, elongation of the chain begins as the components of the ternary complex are discharged. Different mRNAs vary in their rate constants for initiation, probably by virtue of differences in their three-dimensional configuration at the 5' end, which is most closely associated with the various initiation factors.

Regulation of protein synthesis through control of initiation factors is best known for eIF-2. Its activity depends on the presence of hemin, which inhibits the phosphorylation of one of the proteins of eIF-2, a reaction which would inactivate the factor.

Elongation of the protein chain requires the action of elongation factors EF-1 and EF-2. A releasing factor RF is required for termination of translation. Control of these factors is but little understood, except for EF-1, which has known inhibi-

tors and activators, especially in the brine shrimp embryo.

Transit time of a message through the ribosome is different for different proteins. It varies not only with the length of the mRNA molecule, but also with the properties of the message. Moreover, metabolic conditions that alter the overall rate of protein synthesis do so differentially with respect to the transit times for some mRNAs.

The transfer RNAs and their aminoacyl synthetases offer some possibilities for regulation of gene expression. There are two or more isoaccepting species of tRNA for each amino acid and a variable number of amino acid specific synthetases, which differ in their ability to charge the different isoaccepting species of tRNA. The pattern of isoaccepting species of tRNA changes from one developmental stage to the next and from one differentiated tissue to another. Likewise, synthetases from one stage or tissue type show a diminished charging effectiveness with the tRNAs of another. Similarly, the pattern of isoaccepting tRNAs and their synthetases changes when immature or unstimulated organs alter the pattern of their protein synthesis, as in seed germination, metamorphosis of insects, and, in vertebrates, in response to hormones.

Regulation of chain elongation is also affected by the translation of functionally coordinated proteins in an as yet unknown manner. Thus the synthesis of β-globin in bone marrow cells is depressed when that of α-globin is specifically inhibited, and immunoglobulin light chains bind to immunoglobulin heavy-chain mRNA and suppress its synthesis.

Cell-free synthesizing systems can translate mRNAs from whatever source, faithfully producing products like those of the mRNA donor. This fact is also true of the fertilized amphibian egg injected with foreign mRNAs. The introduced mRNAs may persist and be translated in all tissues of the body. Thus it would appear that different tissues do not have a cell- or tissue-specific translating system, although there is some puzzling evidence to the contrary.

The full measure of gene expression is not seen until a functionally active protein is produced. This production requires posttranslational modifications of the translated product. Some of these modifications proceed spontaneously by virtue of the properties of the side-chains of the polypeptides and involve folding of the chain and association between chains by virtue of noncovalent bonding and disulfide bridges. Essentially all proteins undergo further modifications involving covalent bonding to internal residues of methyl, acetyl, and hydroxyl groups, particularly, and the binding of additional amino acids to side-chains of internal residues promoted by tRNA. Other modifications involve enzymatic cleaving of inactive precursor proteins into an active form. Proteins that are to be secreted usually have a hydrophobic leader sequence, which precedes the remainder of the peptide chain into the cisterna of the rough endoplasmic reticulum. The leader is there cleaved by an endopeptidase as the rest of the molecule enters. Within the RER the molecule undergoes its initial folding, formation of disulfide bonds, initial binding of carbohydrate residues, and so on. Further modifications take place in the Golgi complex as the proteins are prepared for export or storage prior to subsequent export.

Finally, the expression of a gene in development is terminated when its product is no longer appropriate to a particular stage or developmental condition. The termination of gene expression is but little understood except in the case of many liver enzymes that are constantly being produced and degraded. The rate of degradation of these enzymes is diminished, however, by the presence of the substrates whose reactions they catalyze.

18.9 Questions for Thought and Review

1. Present a rational argument for the proposition that tissue differentiation results from differential gene activation. What evidence suggests the contrary position, namely, that the production and availability of mRNA may be regulated primarily at the posttranscriptional level.

2. Many tumor cells and teratocarcinoma cells apparently have sufficient genetic information in their nuclei for producton of all adult tissue types. What is the evidence that supports this statement? What might be some of the reasons that mouse embryoid bodies derived from a teratocarcinoma do not form mouse embryos in the coelomic fluid in which they are carried?

3. What is meant by a "cell-free translation system"?

What advantages does such a system offer for the analysis of transcription and translation? What disadvantages?

4. What types of RNA polymerases are present in nuclei, and what is the function of each? Do they exercise a role in transcriptional control? Bring appropriate evidence to bear on your reply.

5. Discuss the significance of heterochromatin in the control of gene expression during development.

6. Nonhistone chromosomal proteins show major differences among different tissues. What is the evidence that they are involved in the control of tissue-specific gene action? Discuss the validity of this evidence.

7. Hormones bring about significant changes in transcriptional patterns in their target organs. What are the mechanisms whereby these changes are thought to be brought about?

8. All gene transcripts require a certain amount of processing before they can be utilized in translation. Outline the major elements in the processing of the various classes of RNAs as they occur in the nucleus and the cytoplasm.

9. What is the difference between gene reiteration and gene amplification?

10. What role or roles might be served by the 5' cap of 7-methylguanosine in the control of message translation? Present the appropriate evidence.

11. What are some possible functions of the 3' poly(A) terminus found on most mRNAs?

12. Eukaryotic mRNAs are relatively stable. Of what selective advantage might this be? Are all eukaryotic messages equally stable? What are some factors that affect the stability of messages? Do these factors affect stability differentially? Give examples.

13. Is there a correlation between the functional requirement for a particular mRNA and its stability? Give evidence in support of your answer.

14. The translational capacity of cells seems generally to be adjusted to the supply of mRNA and the functional demands of the cell. Give an example that suggests that there are cases in which there is more mRNA in a cell than can be handled by its translational machinery.

15. What is the mechanism whereby hemin presumably affects initiation of translation in eukaryotes?

16. Different mRNAs show different rate constants for the initiation of protein synthesis under standard conditions. What is the effect of these differences under conditions that limit the overall rate of protein synthesis? Are there circumstances under which extrinsic factors differentially alter initiation rates for different mRNAs? Explain.

17. With reference to message translation, what is meant by the term *transit time*? Are transit times the same for all proteins? What are the components that determine transit time?

18. Are amino acids added at the same rate during

elongation of different kinds of peptide chains? Explain, giving examples. Do conditions that alter transit times affect all peptide chains equally? Illustrate your answer by means of examples.

19. Discuss and give examples of the apparent functional adaptation that occurs during development between the isoaccepting tRNAs and cellular specialization. Likewise, review the relationships among specific aminoacyl synthetases, populations of isoaccepting species of tRNA, and developmental stage.

20. The concentration of a protein may affect the rate of translation of another protein with which it is functionally coordinated. Discuss the evidence for this proposition.

21. The ultimate expression of a gene is the presence of a functionally active protein. This expression usually requires that the peptide chain undergo a series of modifications. Describe the types of posttranslational modifications that occur spontaneously, by cleavage, by covalent bonding, and so on.

22. Construct a flow sheet, based on the material in this chapter and your additional reading, depicting in sequence the events that are involved in the expression of a gene, the points at which control of gene expression may be exercised, and the operative factors at these points.

468

18.10 Suggestions for Further Readings

ARCECI, R. J., and P. R. GROSS. 1977. Noncoincidence of histone and DNA synthesis in cleavage cycles of early development. *Proc Natl Acad Sci USA* 74:5016–20.

ASSELBERGS, A. M., E. MEULENBERG, W. J. van VENROOIJ, and H. BLOEMENDAL. 1980. Preferential translation of mRNAs in an mRNA-dependent reticulocyte lysate. *Eur J Biochem* 109:159–65.

BAGLIONI, C. 1974. Nonpolysomal messenger RNA. *Ann NY Acad Sci* 241:183–90.

BAGSHAW, J. C., R. S. BERNSTEIN, and B. H. BOND. 1978. DNA-dependent RNA polymerases from *Artemia salina*. Decreasing polymerase activities and number of polymerase II molecules in developing larvae. *Differentiation*. 10:13–21.

BRIMACOMBE, R., G. STÖFFLER, and H. G. WITTMAN. 1978. Ribosome structure. *Annu Rev Biochem* 47:217–49.

BROWN, D. D. 1976. Genome organization in higher organisms. *Fed Proc* 35:11–12.

BROWN, S. W. 1969. Developmental control of heterochromatization in coccids. *Genetics* 61:191–98.

CASE, R. M. 1978. Synthesis, intracellular transport and discharge of exportable proteins in the pancreatic acinar cell and other cells. *Biol Rev* 53:211–354.

CHAN, L., S. E. HARRIS, J. M. ROSEN, A. R. MEANS, and B. W. O'MALLEY. 1977. Processing of nuclear heterogenous RNA: Recent developments. *Life Sci* 20:1–16.

DIBERARDINO, M. A. 1980. Genetic stability and modulation of metazoan nuclei transplanted into eggs and oocytes. *Differentiation* 17:17–30.

FESSLER, J. H., and L. FESSLER. 1978. Biosynthesis of procollagen. *Annu Rev Biochem* 47:129–62.

FILIPOWICZ, W., J. M. SIERRA, C. NOMBELA, S. OCHOA, W. C. MERRICK, and W. F. ANDERSON. 1976. Polypeptide initiation in eukaryotes: initiation factor requirements for translation of natural messengers. *Proc Natl Acad Sci USA* 73:44–48.

GAREL, J. P. 1974. Functional adaptation of tRNA population. *J Theor Biol* 43:211–25.

ILAN, J., and J. ILAN. 1975. Regulation of messenger RNA translation during insect development. *Curr Top Dev Biol* 9:89–136.

KEDES, J. H. 1976. Histone messengers and histone genes. *Cell* 8:321–31.

KLEINSMITH, L. J., J. STEIN, and G. STEIN. 1976. Dephosphorylation of nonhistone proteins specifically alters pattern of gene transcription in reconstituted chromatin. *Proc Natl Acad Sci USA* 73:1174–78.

KORN, L. J., and J. B. GURDON. 1981. The reactivation of developmentally inert 5 S genes in somatic nuclei injected into *Xenopus* oöcytes. *Nature* 289:461–65.

LEVIN, D., and I. M. LONDON. 1978. Regulation of protein synthesis: activation by double-stranded RNA of a protein kinase that phosphorylates eukaryotic initiation factor 2. *Proc Natl Acad Sci USA* 75:1121–25.

LI, H. J. 1976. Chromatin structure. *Int J Biochem* 7:181–85.

LUCCHESI, J. C. 1977. Dosage compensation: transcription-level regulation of X-linked genes in *Drosophila*. *Am Zool* 17:685–93.

LYON, M. F. 1968. Chromosomal and subchromosomal inactivation. *Annu Rev Genet* 2:31–52.

O'MALLEY, B. W., H. C. TOWLE, and R. J. SCHWARTZ. 1977. Regulation of gene expression in eucaryotes. *Annu Rev Genet* 11:239–75.

MOHANDAS, T., R. S. SPARKER, and L. J. SHAPIRO. 1981. Reactivation of an inactive human X chromosome: evidence for X inactivation by DNA methylation. *Science* 211:383–86.

PARK, W., R. JANSING, J. STEIN, and G. STEIN. 1977. Activation of histone gene transcription in quiescent WI-38 cells of mouse liver by a nonhistone chromosomal protein fraction from HeLa S_3 cells. *Biochem USA* 16:3713–21.

PAUL, J. 1970. DNA masking in mammalian chromatin: a molecular mechanism for determination of cell type. *Curr Top Dev Biol* 5:317–52.

PEDERSON, T. 1981. Messenger RNA biosynthesis and nuclear structure. *Am Sci* 69:76–84.

PELHAM, H. R. B., and D. D. BROWN. 1980. A specific transcription factor that can bind either the 5 S RNA gene or 5 S RNA. *Proc Natl Acad Sci USA* 77:4170–74.

PTASHNE, M., and W. GILBERT. 1970. Genetic repressors. *Sci Am* 222:36–44.

RAFF, R. A. 1977. The molecular determination of morphogenesis. *Bioscience* 27:394–401.

RAZIN, A., and A. D. RIGGS. 1980. DNA methylation and gene function. *Science* 210:604–10.

ROA, S. R. V., and S. C. JHANWAR. 1975. Is late replication of the inactive X chromosome irreversible in all cells of mammals? *Cytogenet Cell Genet* 14:140–49.

REVEL, M., and Y. GRONER. 1978. Post-transcriptional and translational control of gene expression in eukaryotes. *Annu Rev Biochem* 47:1079–1126.

SCHIMKE, R. T. 1980. Gene amplification and drug resistance. *Sci Am* (November):60–69.

SCHMIDT, O., and D. SÖLL. 1981. Biosynthesis of eukaryotic transfer RNA. Bioscience 31:34–9.

SCHWARTZ, R. W., M.-J. TSAI, S. Y. TSAI, and B. W. O'MALLEY. 1975. Effect of estrogen on gene expression in the chick oviduct. *J Biol Chem* 250:5175–82.

SHATKIN, A. J. 1976. Capping of eucaryotic mRNAs. *Cell* 9:645–53.

STEIN, G., and J. STEIN. 1976. Chromosomal proteins: their role in the regulation of gene expression. *Bioscience* 26:488–98.

STEIN, H., and P. HAUSEN. 1970. Factors influencing the activity of mammalian RNA polymerase. *Cold Spring Harbor Symp Quant Biol* 35:709–25.

TATA, J. R. 1976. The expression of the vitellogenin gene. *Cell* 9:1–14.

UMBARGER, H. E. 1978. Amino acid biosynthesis and its regulation. *Annu Rev Biochem* 47:533–606.

WAHLI, W., I. B. DAVID, G. U. RYFFEL, and R. WEBER. 1981. Vitellogenesis and the vitellogenin gene family. *Science,* 212:298–304.

WEATHERALL, D. J., and J. B. CLEGG. 1976. Molecular genetics of human hemoglobin. *Annu Rev Genet* 10:157–78.

WEISSBACH, H., and S. OCHOA. 1976. Soluble factors required for eukaryotic protein synthesis. *Annu Rev Biochem* 45:191–216.

18.11   References

AITKEN, R. J. 1974. Sex chromatin formation in the blastocyst of the roe deer. (*Capreolus careolus*) during delayed implantation. *J Reprod Fertil* 40:235–39.

ALFREY, V. G., V. C. LITTAU, and A. E. MIRSKY. 1963. On the role of histones in regulating ribonucleic acid synthesis in the cell nucleus. *Proc Natl Acad Sci USA* 49:414–21.

ALT, F. W., R. E. KELLEMS, J. R. BERTINO, and R. T. SCHIMKE. 1978. Selective multiplication of dihydrofolate-resistant variants of cultured murine cells. *J Biol Chem* 253:1357–70.

ALTMAN, K., A. L. SOUTHERN, S. C. URETZKY, P. ZABOS, and G. ACS. 1972. Hydrocortisone induction of rat-liver leucyl-transfer RNA and its synthetases. *Proc Natl Acad Sci USA* 69:3567–69.

ALTON, T. H., and H. F. LODISH. 1977a. Developmental changes in messenger RNAs and protein synthesis in *Dictyostelium discoideum*. *Dev Biol* 60:180–206.

ALTON, T. H., and H. F. LODISH. 1977b. Synthesis of developmentally regulated proteins in *Dictyostelium discoideum* which are dependent on continued cell-cell interaction. *Dev Biol* 60:207–16.

AMALRIC, F., J. P. BACHELLERIE, and M. CABOCHE. 1977. RNA methylation and control of eukaryotic RNA biosynthesis. Processing and utilization of undermethylated tRNAs in CHO cells. *Nucleic Acids Res* 4:4257–70.

ANFINSEN, C. B. 1968. Spontaneous formation of the three-dimensional structure of proteins. In M. Locke, ed. *The Emergence of Order in Developing Systems.* New York: Academic Press, pp. 1–20.

ASSELBERGS, F. A. M., M. KOOPMANS, W. J. VAN VENROOIJ, and HANS BLOEMENDAL. 1978. Post-transcriptional assembly of lens α-crystallin in the reticulocyte lysate and in *Xenopus laevis* oöcytes. *Eur J Biochem* 91:65–72.

AXEL, R., P. FEIGELSON, and G. SCHUTZ. 1976. Analysis of the complexity and diversity of mRNA from chicken liver and oviduct. *Cell* 7:247–54.

BAGLIONI, C., M. SIMILI, and D. A. SHAFRITZ. 1978. Initiation activity of ECM virus RNA, binding to initiation factor EIF-4B and shut-off of host cell protein synthesis. *Nature* 275:240–43.

BARR, M. L., and E. G. BERTRAM. 1949. A morphological distinction between neurons of the male and female, and the behavior of the nucleolar satellite during accelerated nucleoprotein synthesis. *Nature* 163:676–77.

BENNE, R., J. ERDMAN, R. R. TROUT, and J. W. B. HERSHEY. 1978. Phosphorylation of eukaryotic protein synthesis initiation factors. *Proc Natl Acad Sci USA* 75:108–112.

BERGER, S. L., and H. L. COOPER. 1975. Very short-lived stable mRNAs from resting human lymphocytes. *Proc Natl Acad Sci USA* 72:3873–77.

BERGINK, E., and R. A. WALLACE. 1974. Precursor-product relationship amphibian vitellogenin and phosvilen. *J Biol Chem* 249:2899–2903.

BICK, M. D., and B. L. STREHLER. 1971. Leucyl transfer RNA synthetase changes during soybean cotyledon senescence. *Proc Natl Acad Sci USA* 68:224–28.

BIERI-BONNIOT, F., U. JOSS, and C. DIERKS-VENTLING. 1977.

Stimulation of RNA polymerase I activity by 17 β-estradiol-receptor complex on chick liver nucleolar chromatin. *FEBS Lett* 81:91–96.

BLOBEL, G., and B. DOBBERSTEIN. 1975a. Transfer of proteins across membranes. I. Presence of proteolytically processed and unprocessed nascent immunoglobulin light chain on membrane-bound ribosomes of murine myeloma. *J Cell Biol* 67:835–51.

BLOBEL, G., and B. DOBBERSTEIN. 1975b. Transfer of proteins across membranes. II. Reconstitution of functional rough microsomes from heterologous components. *J Cell Biol* 67:852–62.

BRAND, J., E. SPINDLER, and H. STEIN. 1977. The role of basic proteins in the DNA-dependent RNA polymerase reaction. Evidence that RNA polymerase subunits are distinct from fraction S protein. *FEBS Lett* 80:173–76.

BRANDHORST, B. P. 1976. Two-dimensional gel patterns of protein synthesis before and after fertilization of sea urchin eggs. *Dev Biol* 52:310–17.

BRIGGS, R., and T. J. KING. 1952. Transplantation of living nuclei from blastula cells into enucleated frogs' eggs. *Proc Natl Acad Sci USA* 38:455–63.

BRIGGS, R., and T. J. KING. 1957. Changes in the nuclei of differentiating endoderm cells as revealed by nuclear transplantation. *J Morphol* 100:269–312.

BROWN, D. D., J. DOERING, S. EMMONS, N. FEDEROFF, and E. JORDAN. 1976. Developmental genetics by gene isolation: the dual 55 DNA system in *Xenopus. Carnegie Institution Year Book* 75:12–15.

BROWN, S. W., and U. NUR. 1964. Heterochromatic chromosomes in the coccids. *Science* 145:130–36.

CABOCHE, M., and J.-P. BACHELLERIE. 1977. RNA methylation and control of eukaryotic RNA biosynthesis. Effects of cyclolencine, a specific inhibitor of methylation, on ribosomal RNA maturation. *Eur J Biochem* 74:61–129.

CASTON, J. D. 1971. Studies on tRNA and aminoacyl-tRNA synthetases during the development of *Rana pipiens. Dev Biol* 24:19–36.

CHAMBON, P. 1975. Eukaryotic nuclear RNA polymerases. *Annu Rev Biochem* 44:613–38.

CLARKSON, S. G., M. L. BIRNSTIEL, and V. SERRA. 1973a. Reiterated transfer RNA genes of *Xenopus laevis. J Mol Biol* 79:391–410.

CLARKSON, S. G., M. L. BIRNSTIEL, and I. F. PURDOM. 1973b. Clustering of transfer RNA genes of *Xenopus laevis. J Mol Biol* 79:411–29.

CLEMENS, M. J., E. C. HENSHAW, H. RAHAMIMOFF, and I. M. LONDON. 1974. Met-tRNA$_f^{Met}$ binding to 40 S ribosomal subunits: a site for the regulation of initiation of protein synthesis by hemin. *Proc Natl Acad Sci USA* 71:2946–50.

COCHET, M., F. GANNON, R. HEN, L. MAROTEAUX, F. PERRIN, and P. CHAMBON. 1979. Organization and sequence studies of the 17-piece chick conalbumin gene. *Nature* 282:567–74.

COLLATZ, E., N. ULBRICH, K. TSURUGI, H. N. LIGHTFOOT, W. MACKINLAY, A. LIN, and I. G. WOOL. 1977. Isolation of eukaryotic ribosomal proteins. Purification and characterization of the 40 S ribosomal subunit proteins Sa, Sc, S3a, S5', S9, S10, S11, S12, S14, S15, S15', S16, S17, S18, S19, S20, S21, S26, S27' and S29. *J Biol Chem* 252:9071–80.

COOPER, D. W., P. G. JOHNSTON, C. E. MURTAGH, G. B. SHARMAN, J. L. VANDEBERG, and W. E. POOLE. 1975. Sex-linked isozymes and sex chromosome evolution in kangaroos. In C. L. Markert, ed. *Isozymes,* Vol III. New York: Academic Press, pp. 559–73.

CRYSTAL, R. W., A. W. NEINHEUS, P. M. PRICHARD, D. PICICANO, N. A. ELSON, W. C. MERRICK, H. GRAF, D. A. SHAFRITZ, D. G. LAYCOCK, J. A. LAST, and W. F. ANDERSON. 1972. Translation by rabbit reticulocytes of globin mRNA using initiation factors from liver. *FEBS Lett* 24:310–14.

D'ALESSIO, J. M., and J. C. BAGSHAW. 1977. DNA-dependent RNA polymerases from *Artemia salina.* IV. Appearance of nuclear RNA polymerase activity during preemergence development of encysted embryos. *Differentiation* 8:53–56.

DATTA, A., C. deHARO, J. M. SIERRA, and S. OCHOA. 1977a. Role of 3':5'-cyclic-AMP-dependent protein kinase in regulation of protein synthesis in reticulocyte lysates. *Proc Natl Acad Sci USA* 74:1463–67.

DATTA, A., C. deHARO, J. M. SIERRA, and S. OCHOA. 1977b. Mechanism of translational control by hemin in reticulocyte lysates. *Proc Natl Acad Sci USA* 74:3326–29.

DATTA, A., C. deHARO, and S. OCHOA. 1978. Translational control by hemin is due to binding to cyclic AMP-dependent protein kinase. *Proc Natl Acad Sci USA* 75:1148–52.

DAVIDSON, E. H. 1976. *Gene Activity in Early Development,* 2nd ed. New York: Academic Press.

DAVIDSON, E. H., and R. J. BRITTEN. 1973. Organization transcription and regulation in the animal genome. *Q Rev Biol* 48:565–613.

DAVIDSON, E. H., and R. J. BRITTEN. 1979. Regulation of gene expression: possible role of repetitive sequences. *Science* 204:1052–59.

DAVIDSON, E. H., B. R. HOUGH, E. S. AMENSON, and R. J. BRITTEN. 1973. General interspersion of repetitive and nonrepetitive sequence elements in the CNA of *Xenopus. J Mol Biol* 77:1–23.

DAVIDSON, R. G., H. M. NITOWSKY, and B. CHILDS. 1963. Demonstration of two population of cells in the human female heterozygous for glucose-6-phosphate variants. *Proc Natl Acad Sci USA* 50:481–85.

DELANEY, P., and M. A. Q. SIDDIQUI. 1975. Changes in in vivo levels of charged transfer RNA species during development of the posterior silkgland of *Bombyx mori.* *Dev Biol* 44:54–62.

DELAUNAY, J., R. S. RANU, D. H. LEVIN, V. ERNST, and J. M. LONDON. 1977. Characterization of a rat liver factor that inhibits initiation of protein synthesis in rabbit reticulocyte lysates. *Proc Natl Acad Sci USA* 74:2264–68.

DEROBERTIS, E. M., and M. V. OLSON. 1979. Transcription and processing of cloned yeast tyrosine tRNA genes microinjected into frog oöcytes. *Nature* 278:137–377.

DETKE, S., J. L. STEIN, and G. S. STEIN. 1978. Synthesis of histone messenger RNAs by RNA polymerase II in nuclei from S phase HeLa S3 cells. *Nucleic Acids Res* 5:1511–28.

DEYS, B. F., K. H. CRZESCHIK, A. CRZECHIK, E. R. JAFFE, and M. SINISCALO. 1972. Human phosphoglycerate kinase and inactivation of the chromosome. *Science* 175:1002–03.

DIBERARDINO, M., and N. HOFFNER. 1971. Development and chromosomal constitution of nuclear transplants derived from male germ cells. *J Exp Zool* 176:61–72.

DIBERARDINO, M. A., and T. J. KING. 1965. Transplantation of nuclei from the frog renal adenocarcinoma. II. Chromosomal and histologic analysis of tumor nuclear-transplant embryos. *Dev Biol* 11:217–242.

DUGAICZYK, A., S. L. C. WOO, E. C. LAI, M. L. MACE, JR., L. MCREYNOLDS, and B. W. O'MALLEY. 1978. The natural ovalbumin gene contains seven intervening sequences. *Nature* 274:328–33.

DWORKIN, M. B., L. M. RUDENSEY, and A. A. INFANTE. 1977. Cytoplasmic nonpolysomal ribonucleoprotein particles in sea urchin embryos and their relationship to protein synthesis. *Proc Natl Acad Sci USA* 74:2231–35.

DWORKIN, M. B., and A. A. INFANTE. 1976. Relationship between the mRNA of polysomes and free ribonucleoprotein particles in the early sea urchin embryo. *Dev Biol* 53:73–90.

EFSTRATIADIS, A., F. C. KAFATOS, and T. MANIATIS. 1977. The primary structure of β-globin mRNA as determined from cloned DNA. *Cell* 10:571–85.

FAN, H., and S. PENMAN. 1970. Regulations of protein synthesis in mammalian cells. II. Inhibition of protein synthesis at the level of initiation during mitosis. *J Mol Biol* 50:655–79.

FROEHLICH, J. P., L. W. BROWDER, and G. A. SCHULTZ. 1977. Translation and distribution of rabbit globin mRNA in separated cell types of *Xenopus laevis* gastrulae. *Dev Biol* 56:356–71.

FURUICHI, Y., and A. J. SHATKIN. 1976. Differential synthesis of blocked and unblocked 5' termini in reovirus mRNA: effect of pyrophosphate and pyrophosphatase. *Proc Natl Acad Sci USA* 73:3448–52.

GALAU, G. A., W. K. KLEIN, M. M. DAVIS, B. WOLD, R. J. BRITTEN, and E. H. DAVIDSON. 1976. Structural gene sets active in embryos and adult tissues of sea urchin. *Cell* 7:487–505.

GARBER, R. L., M. A. Q. SIDDIQUI, and S. ALTMAN. 1978. Identification of precursor molecules to individual tRNA species from *Bombyx mori.* *Proc Natl Acad Sci USA* 75:635–39.

GURDON, J. B. 1963. Nuclear transplantation in amphibia and the importance of stable nuclear changes in promoting cellular differentiation. *Q Rev Biol* 38:54–78.

GURDON, J. B., and R. A. LASKAY. 1970. The transplantation of nuclei from single cultured cells into enucleate frogs' eggs. *J Embryol Exp Morphol* 24:227–48.

GURDON, J. B., R. LASKEY, and O. R. REEVES. 1975. The developmental capacity of nuclei transplanted from keratinized skin cells of adult frogs. *J Embryol Exp Morphol* 34:93–112.

GURDON, J. B., J. B. LINGREL, and G. MARBAIX. 1973. Message stability in injected frog oocytes: long life of mammalian α and β globin messages. *J Mol Biol* 80:539–51.

GURDON, J. B., H. R. WOODLAND, and J. B. LINGREL. 1974. The translation of mammalian globin mRNA injected into fertilized eggs of *Xenopus laevis.* I. Message stability in development. *Dev Biol* 39:125–33.

HAMMERLING, J. 1943. Ein- und zweikernige Transplantante zwischen *Acetabularia mediterranea* und *A. crenulata.* *Z Ind Abst Verer* 81:114–80.

HASTIE, N. D., and J. O. BISHOP. 1976. The expression of three abundance classes of messenger RNA in mouse tissues. *Cell* 9:761–74.

HENNEN, S. 1970. Influence of spermine and reduced temperature on the ability of transplanted nuclei to promote normal development in eggs of *Rana pipiens.* *Proc Natl Acad Sci USA* 66:630–37.

HENTSCHEL, C. C., and J. R. TATA. 1977. Differential activation of free and template-engaged RNA polymerase I and II during the resumption of development of dormant *Artemia* gastrulae. *Dev Biol* 57:293–304.

HEYWOOD, S. M., and D. S. KENNEDY. 1974. The control of myoglobin synthesis during muscle development. *Dev Biol* 38:390–93.

HICKEY, E. D., L. E. WEBER, and C. BAGLIONI. 1976. Inhibition of initiation of protein synthesis by 7-methylguanosine-5'-monophosphate. *Proc Natl Acad Sci USA* 73:19–23.

HICKEY, E. D., L. E. WEBER, C. BAGLIONI, C. H. KIM, and R. H. SARMA. 1977. A relation between inhibition of protein synthesis and conformation of 5-phosphorylated 7-methylguanosine derivatives. *J Mol Biol* 109:173–83.

HILDEBRANDT, A., and H. W. SAUER. 1977. Transcription of ribosomal RNA in the life cycle of *Physarum* may be regulated by a specific nucleolar inhibitor. *Biochem Biophys Res Commun* 74:466–72.

HOFFNER, N. J., and M. A. DIBERARDINO. 1980. Develop-

471

mental potential of somatic nuclei transplanted into meiotic oöcytes of *Rana pipiens*. *Science* 209:517–19.

HOLLINGER, T. G., and L. D. SMITH. 1976. Conservation of RNA polymerase during maturation of the *Rana pipiens* oöcyte. *Dev Biol* 51:86–97.

HUANG, R. C., and J. BONNER. 1962. Histone, a suppressor of chromosomal RNA synthesis. *Proc Natl Acad Sci USA* 48:1216–22.

HUMPHRIES, S., J. WINDASS, and R. WILLIAMSON. 1976. Mouse globin gene expression in erythroid and non-erythroid tissues. *Cell* 7:267–77.

ILAN, J., and J. ILAN. 1975. Similarities in properties and a functional difference in purified leucyl-tRNA synthetase isolated from two developmental stages of *Tenebrio molitor*. *Dev Biol* 42:64–74.

ILLMENSEE, K., and B. MINTZ. 1976. Totipotency and normal differentiation of single teratocarcinoma cells cloned by injection into blastocysts. *Proc Natl Acad Sci USA* 73:549–53.

JACKSON, R. C., and G. BLOBEL. 1977. Post-transcriptional cleavage of presecretory proteins with an extract of rough microsomes from dog pancreas containing signal peptidase activity. *Proc Natl Acad Sci USA* 74:5598–5602.

JAGGI, R. B., B. K. FELBER, S. MAURHOFER, and G. R. RYFFEL. 1980. Four different vitellogenin proteins of *Xenopus* identified by translation in vitro. *Eur J Biochem* 109:343–47.

JEFFREY, W. R. 1977. Polyadenylation of maternal and newly-synthesized RNA during starfish oocyte maturation. *Dev Biol* 57:98–108.

JEFFREY, W. R., and C. PETERS. 1977. Polypeptide composition of the globin poly(A)-protein complex from rabbit reticulocytes. *Mol Biol Rep* 3:379–86.

JENSEN, E. V., and E. R. DESOMBRE. 1973. Estrogen-receptor interaction. *Science* 182:126–34.

JOST, J.-P., G. PEHLING, T. OHNO, and P. COZENS. 1978. Identification of a large precursor of vitellogenin mRNA in the liver of estradiol-treated chicks. *Nucleic Acids Res* 5:4781–94.

KAUFMAN, R. J., P. C. BROWN, and R. T. SCHIMKE. 1979. Amplified dehydrofolate reductase genes in unstably methotrexate-resistant cells are associated with double minute chromosomes. *Proc Natl Acad Sci USA* 76:5669–73.

KING, T. J., and R. BRIGGS. 1955. Changes in the nuclei of differentiating gastrula cells, as demonstrated by nuclear transplantation. *Proc Natl Acad Sci USA* 41:321–25.

KING, T. J., and M. A. DIBERARDINO. 1965. Transplantation of nuclei from the frog renal adenocarcinoma. I. Development of tumor nuclear-transplant embryos. *Ann NY Acad Sci* 126:115–26.

KINNIBURGH, A. J., J. E. MERTZ, and J. ROSS. 1978. The

precursor of mouse *β*-globin messenger RNA contains two intervening RNA sequences. *Cell* 14·681–93.

KLEENE, K. C., and T. HUMPHREYS. 1977. Similarity of HnRNA sequences in blastula and pluteus stage sea urchin embryos. *Cell* 12:143–55.

KLEINSMITH, L. J., and G. B. PIERCE. 1964. Multipotentiality of single embryonic carcinoma cells. *Cancer Res* 24:1544–51.

KNAPP, G., R. C. OGDEN, C. L. PEEBLES, and J. ABELSON. 1979. Splicing of yeast tRNA precursors: structure of the reaction intermediates. *Cell* 18:37–45.

LASKEY, R. A., A. D. MILLS, J. B. GURDON, and G. A. PARTINGTON. 1977. Protein synthesis in oöcytes of *Xenopus laevis* is not regulated by the supply of messenger RNA. *Cell* 11:343–51.

LASKY, L., Z. LEV, J.-H. XIN, R. J. BRITTEN, and E. H. DAVIDSON. 1980. Messenger RNA prevalence in sea urchin embryos measured with cloned cDNAs. *Proc Natl Acad Sci USA* 77:5317–21.

LEAVETT, W. W., T. J. CHEN, and T. C. ALLEN. 1977. Regulation of progesterone receptor formation by estrogen action. *Ann NY Acad Sci* 286:210–25.

LEE-HUANG, S., J. M. SIERRA, R. NARANJO, W. FILLIPOWICZ, and S. OCHOA. 1977. Eucaryotic oligonucleotides affecting mRNA translation. *Arch Biochem Biophys* 180:276–87.

LEE, J. C., and V. M. INGRAM. 1967. Erythrocyte transfer RNA: change during chick development. *Science* 158:1330–32.

LENZ, J. R., and C. BAGLIONI. 1978. Inhibition protein synthesis by double-stranded RNA and phosphorylation of initiation factor. *J Biol Chem* 253:4219–23.

LINGREL, J. B., and H. R. WOODLAND. 1974. Initiation does not limit the rate of globin synthesis in message-injected *Xenopus* oocytes. *Eur J Biochem* 47:47–56.

LITT, M., and D. KABAT. 1972. Studies of transfer nucleic acids and of hemoglobin synthesis in sheep reticulocytes. *J Biol Chem* 237:6659–64.

LITT, M., and K. WEISER. 1978. Histidine transfer RNA levels in Friend leukemia cells: stimulation by histidine deprivation. *Science* 201:527–29.

LODISH, H. F. 1971. Alpha and beta globin messenger ribonucleic acid. Different amounts and rates of initiation of translation. *J Biol Chem* 246:7131–38.

LODISH, H. F. 1976. Translational control of protein synthesis. *Annu Rev Biochem* 45:39–72.

LODISH, H. F., and O. DESALU. 1973. Regulation of synthesis of non-globin proteins in cell-free extracts of rabbit reticulocytes. *J Biol Chem* 248:3520–27.

LODISH, H. F., and M. JACOBSEN. 1972. Regulation of hemoglobin synthesis. Equal rates of translation of *α*- and *β*-chains. *J Biol Chem* 247:3622–29.

LOMEDICO, P., N. ROSENTHAL, A. EFSTRATIADIS, W. GILBERT, K. KOLODNER, and R. TIZARD. 1979. The structure and evolution of two nonallelic rat preproinsulin genes. *Cell* 18:545–58.

LOMEDICO, P. T., and G. F. SAUNDERS. 1977. Cell-free modulation of protein synthesis. *Science* 198:620–22.

LYON, M. F. 1961. Gene action in the X-chromosome of the mouse *(Mus musculus* L.) *Nature* 190:372–73.

MANDEL, J. L., R. BREATHNACK, P. GERLINGER, M. LEMEUR, F. GANNON, and P. CHAMBON. 1978. Organization of coding and intervening sequences in the chick ovalbumin split gene. *Cell* 14:641–53.

MANGIA, F., G. ABBO-HALBASCH, and C. J. EPSTEIN. 1975. X chromosome expression during oögenesis in the mouse. *Dev Biol* 45:366–68.

MARBAIX, G., G. HUEZ, A. BURNY, Y. CLEUTER, E. HUBERT, M. LECLERQ, H. CHANTRENNE, H. SOREQ, U. NUDEL, and V. Z. LITTAUER. 1975. Absence of polyadenylate segment in globin messenger RNA accelerates its degradation in *Xenopus* oöcytes. *Proc Natl Acad Sci USA* 72:3065–67.

MAZUR, G., and A. SCHWEIGER. 1978. Identical properties of an m-RNA-bound protein and a cytosol protein with high affinity for polyadenylate. *Biochem Biophys Res Commun* 80:39–45.

MCCARTHY, B. J., and B. H. HOYER. 1964. Identity of DNA and diversity of messenger RNA molecules in normal mouse tissues. *Proc Natl Acad Sci USA* 52:915–22.

MCKNIGHT, G. S. 1978. The induction of ovalbumin and conalbumin mRNA by estrogen and progesterone in chick oviduct explant cultures. *Cell* 14:403–413.

MCKNIGHT, G. S., P. PENNEQUIN, and R. T. SCHIMKE. 1975. Induction of ovalbumin mRNA sequences by estrogen and progesterone in chick oviduct as measured by hybridization to complementary DNA. *J Biol Chem* 250:8105–10.

MEZA, L., A. ARAYA, G. LEON, M. KRAUSKOPF, M. A. Q. SIDDIQUI, and J. P. GAREL. 1977. Specific alanine-tRNA species associated with fibroin biosynthesis in the posterior silk-gland of *Bombyx mori* L. *FEBS Lett* 77:255–60.

MIGEON, B. R., and K. JELILIAN. 1977. Evidence for two active X chromosomes in germ cells of female before meiotic entry. *Nature* 269:242–43.

MIGEON, B. R., V. M. DER KALOUSTIAN, W. L. NYHAN, W. J. YOUNG, and B. CHILDS. 1968. X-linked hypoxanthineguanine phosphoribosyl transferase deficiency: heterozygote has two cloned populations. *Science* 160:425–27.

MINTZ, B., and K. ILLMENSEE. 1975. Normal genetically mosaic mice produced from malignant teratocarcinoma cells. *Proc Natl Acad Sci USA* 72:3585–89.

NAGATA, S., K. MOTOYOSHI, and K. IWASAKE. 1978. Interaction of subunits of polypeptide chain elongation factor 1 from pig liver. *J Biochem* 83:423–29.

NUNBERG, J. H., R. J. KAUFMAN, R. T. SCHIMKE, G. URLAIRB,

and L. A. CHASIN. 1978. Amplified dehydrofolate reductase genes are localized to a homogeneously staining region of a single chromosome in a metholrexate-resistant Chinese hamster ovary cell. *Proc Natl Acad Sci USA* 75:5553–56.

NUR, V. 1967. Reversal of heterochromatization and the activity of the paternal chromosome set in the male mealy bug. *Genetics* 56:375–89.

OCHOA, S., and C. DE HARO. 1979. Regulation of protein synthesis in eukaryotes. *Annu Rev Biochem* 48:549–80.

PALMITER, R. D. 1972a. Regulation of protein synthesis in chick oviduct. I. Independent regulation of ovalbumin, conalbumin, ovomucoid, and lysozyme induction. *J Biol Chem* 247:6450–61.

PALMITER, R. D. 1972b. Regulation of protein synthesis in chick oviduct. II. Modulation of polypeptide elongation and initiation rates by estrogen and progesterone. *J Biol Chem* 247:6770–80.

PALMITER, R. D., and N. H. CAREY. 1974. Rapid inactivation of ovalbumin messenger ribonucleic acid after acute withdrawal of estrogen. *Proc Natl Acad Sci USA* 71:2357–61.

PALMITER, R. D., and G. A. GUTMAN. 1972. Fluorescent antibody localization of ovalbumin, conalbumin, ovamicord and lysozyme in chick oviduct magnum. *J Biol Chem* 247:6459–61.

PALMITER, R. D., and M. E. HAINES. 1973. Regulation of protein synthesis in chick oviduct. IV. Role of testosterone. *J Biol Chem* 248:2017–2116.

PARKER, C. S., and R. G. ROEDER. 1977. Selective and accurate and transcription of the *Xenopus laevis* 5 S RNA genes in isolated chromatin by purified RNA polymerase III. *Proc Natl Acad Sci USA* 74:44–48.

PAUL, J., R. S. GIHMOUR, N. AFFARA, G. BIRNIE, P. HARRISON, A. HELL, S. HUMPHRIES, J. WINDLASS, and B. YOUNG. 1973. The globin gene: structure and expression. *Cold Spring Harbor Symp Quant Biol* 38:885–90.

PEEBLES, C. L., R. C. OGDEN, G. KNAPP, and J. ABELSON. 1979. Splicing of yeast tRNA precursors: a two-stage reaction. *Cell* 18:27–35.

PELLEGRINI, M., J. MAMMING, and N. DAVIDSON. 1977. Sequence arrangement of the rDNA of *Drosophila melanogaster*. *Cell* 10:213–24.

PERRY, R. P. 1976. Processing of RNA. *Annu Rev Biochem* 45:605–29.

PERRY, R. P., and D. E. KELLEY. 1976. Kinetics of formation of 5' terminal caps in mRNA. *Cell* 8:433–42.

REEDER, R., and E. BELL. 1965. Short- and long-lived messenger RNA in embryonic chick lens. *Science* 150:71–72.

RICIUTTI, R. C., T. D. GELEHRTER, and L. G. ROSENBERG. 1976. X-chromosomes inactivation in human liver: confirmation of X-linkage of ornethine transcarbamylase. *Am J Hum Genet* 28:332–38.

473

ROEDER, R. G., R. H. REEDER, and D. D. BROWN. 1970. Multiple forms of RNA polymerase in *Xenopus laevis:* their relationship to RNA synthesis in vivo and their fidelity of transcription in vitro. *Cold Spring Harbor Symp Quant Biol* 35:727–35.

ROGG, H., P. MULLER, G. KEITH, and M. STAEHELIN. 1977. Chemical basis for brain-specific serine transfer RNAs. *Proc Natl Acad Sci USA* 74:4243–47.

ROMEO, G., and B. R. MIGEON. 1970. Genetic inactivation of the α-galactosidase locus in carrier's of Fabry's disease. *Science* 170:180–81.

ROPER, M. D., and W. D. WICKS. 1978. Evidence for acceleration of the rate of elongation of tyrosene aminotransferase nascent chains by dibutyryl cyclic AMP. *Proc Natl Acad Sci USA* 75:140–44.

ROSS, J. 1978. Purification and structural properties of the precursors of the globin messenger RNAs. *J Mol Biol* 119:21–35.

RYFFEL, G. U., T. WYLER, D. B. MUELLENER, and R. WEBER. 1980. Identification, organization and processing intermediates of the putative precursors of *Xenopus* vitellogenin messenger RNA. *Cell* 19:53–61.

SAWICKI, S. G., W. JELINEK, and J. E. DARNELL. 1977. 3'-Terminal addition to HeLa cell nuclear and cytoplasmic poly(A). *J Mol Biol* 113:219–35.

SCHIBLER, U., K. B. MARCY, and R. P. PERRY. 1978. The synthesis and processing of the messenger RNAs specifying heavy and light chain immunoglobulins in MCP-11 cells. *Cell* 15:1495–1510.

SCHIMKE, R. T. 1967. Protein turnovers and regulation of enzyme levels in rat liver. In M. P. Stulberg. ed. *Enzymatic Aspects of Metabolic Regulation. Natl Cancer Inst Monogr* 27:301–14.

SCHIMKE, R. T., P. PENNEQUIN, D. ROBINS, and G. S. MCKNIGHT. 1977. Effects of estrogen and progesterone on ovalbumin synthesis in the withdrawn oviduct. *Ann NY Acad Sci* 286:116–24.

SCHRADER, W. T., W. A. COTY, R. G. SMITH, and B. W. O'MALLEY. 1977. Purification and properties of progesterone receptors from chick oviduct. *Ann NY Acad Sci* 286:64–80.

SCHREIER, M. H., and T. STAEHELIN. 1973. Translation of rabbit hemoglobin messenger RNA *in vitro* with purified and partially purified components from brain or liver of different species. *Proc Natl Acad Sci USA* 70:462–65.

SCHWARTZ, H., and J. E. DARNELL. 1976. The association of protein with the polyadenylic acid of HeLa cell messenger RNA: evidence for a "transport" role of a 75,000 molecular weight polypeptide. *J Mol Biol* 104:833–51.

SCHWARTZ, R. J., C. CHANG, W. T. SCHRADER, and B. W. O'MALLEY. 1977. Effect of progesterone receptors in transcription. *Ann NY Acad Sci* 286:147–60.

SENGEL, P., and M. RUSAOUËN. 1968. Aspects histologiques de la différenciation précoce des ébauches plumaires chez le poulet. *C R Acad Sci Paris* [D]266:795–97.

SHIELDS, D., and G. BLOBEL. 1978. Efficient cleavage and segregation of nascent presecretory proteins in a reticulocyte lysate supplemented with microsomal membranes. *J Biol Chem* 253:3753–56.

SHIMOTOHNO, K., Y. KODAMA, and J HASHIMOTO, and K.-I. MIURA. 1977. Importance of 5'-terminal blocking structure to stabilize mRNA in eukaryotic protein synthesis. *Proc Natl Acad Sci USA* 74:2734–38.

SHIOKAWA, K., and A. O. POGO. 1974. The role of cytoplasmic membranes in controlling the transport of nuclear messenger RNA and initiation of protein synthesis. *Proc Natl Acad Sci USA* 71:2658–62.

SIGNORET, J., R. BRIGGS, and R. R. HUMPHREY. 1962. Nuclear transplantation in the axolotl. *Dev Biol* 4:134–64.

SMITH, R. L., and H. S. FORREST. 1973. Variations in isoaccepting species of transfer RNA during embryogenesis of *Oncopeltus fasciatus* (Dallas). *Dev Biol* 33:123–29.

SONENBERG, N., H. TRACHSEL, S. HECHT,. and A. J. SHATKIN. 1980. Differential stimulation of capped mRNA translation in vitro by cap binding protein. *Nature* 285:331–33.

SPRADLING, A. C., and A. P. MAHOWALD. 1980. Amplification of genes for chorion proteins during oogenesis in *Drosophila melanogaster. Proc Natl Acad Sci USA* 77:1096–1100.

STAHL, H., and D. GALLWITZ. 1977. Fate of histone messenger RNA in synchronized HeLa cells in the absence of initiation of protein synthesis. *Eur J Biochem* 72:385–92.

STRAIR, R. K., S. H. YAP, B. NADAL-GINARD, and D. A. SHAFRITZ. 1978. Identification of a high molecular weight presumptive precursor to albumin mRNA in the nucleus of rat liver and hepatoma cell line $H_4A_zC_2$. *J Biol Chem* 253:1328–31.

STEVENS, L. 1970. The development of transplantable tertocarcinomas from intratesticular grafts of pre- and postimplantation mouse embryos. *Dev Biol* 21:364–82.

STEVENS, R. H., and A. R. WILLIAMSON. 1973. Translational control of immunoglobulin synthesis. I. Repression of heavy chain synthesis. *J Mol Biol* 78:505–16.

STILES, C. D., K. L. LEE, and F. T. KENNEY. 1976. Differential degradation of messenger-RNAs in mammalian cells. *Proc Natl Acad Sci USA* 73:2634–38.

SUCHANEK, G., G. KREIL, and M. A. HERMODSON. 1978. Amino acid sequence of honey-bee prepromelittin synthesized in vitro. *Proc Natl Acad Sci USA* 75:701–704.

SUZUKI, Y., L. P. GAGE, and D. D. BROWN. 1972. The genes for silk fibroin in *Bombyx mori. J Mol Biol* 70:637–40.

TAKAGI, N., N. WAKE, and M. SASAKI. 1978. Cytologic evidence for preferential inactivation of the paternally

474

derived X chromosome in XX mouse blastocysts. *Cytogen Cell Genet* 20:240–48.

TEISSERE, M. P. PENON, R. B. VAN HUYSTEE, Y. AZOU, and J. RICHARD. 1975. Hormonal control of transcription in higher plants. *Biochim Biophys Acta* 402:391–402.

THOMSON, W. C., E. A. BUZASH, and S. M. HEYWOOD. 1973. Translation of myoglobin messenger ribonucleic acid. *Biochemistry* 12:4559–65.

TWARDOWSKI, T., J. M. HILL, and H. WEISSBACH. 1977. Studies on the disaggregation of EF-1 with carboxypeptidase A. *Arch Biochem Biophys* 180:444–51.

UDEM, S. A., and J. R. WARNER. 1973. The cytoplasmic maturation of a ribosomal precursor ribonucleic acid in yeast. *J Biol Chem* 248:1412–16.

UY, R., and F. WOLD. 1977. Posttranslational covalent modification of proteins. *Science* 198:890–96.

VAN DEN BERG, J., A. VAN OOYEN, N. MANTEI, A. SCHAMBÖCK, G. GROSVELD, R. A. FLAVELL, and C. WEISSMANN. 1978. Comparison of cloned rabbit and mouse β-globin genes showing strong evolutionary divergence of two homologous pairs of introns. *Nature* 278:37–44.

VERMA, D. P. S., G. A. MACLACHLAN, H. BYRNE, and D. EWING. 1975. Regulation and in vitro translation of messenger ribonucleic acid for cellulase from auxin-treated pea epicotyls. *J Biol Chem* 250:1019–26.

WAHLI, W., I. B. DAVID, G. U. RYFFEL, and R. WEBER. 1981. Vitellogenesis and the vitellogenin gene family. *Science*, 212:298–304.

WAKE, N., N. TAKAGI, and M. SASAKI. 1976. Non-random inactivation of X chromosome in the rat yolk sac. *Nature* 262:580–81.

WEI, C.-M., and B. MOSS. 1977. 5'-Terminal capping of RNA by guanylyltransferase from HeLa cell nuclei. *Proc Natl Acad Sci USA* 74:3758–61.

WELLAUER, P., and I. DAWID. 1977. The structural organization of ribosomal DNA in *Drosophila melanogaster*. *Cell* 10:193–212.

WEST, J. D., W. I. FRELS, V. M. CHAPMAN, and V. E. PAPAIOANNOU. 1977. Preferential expression of the maternally-derived X chromosome in the mouse yolk sac. *Cell* 12:873–82.

WESTIN, M. 1969. Effect of actinomycin D on antigen synthesis during sea urchin development. *J Exp Zool* 171:297–304.

WICKNER, W. 1979. The assembly of proteins into biological membranes: the membrane trigger hypothesis. *Annu Rev Biochem* 48:23–45.

WOLF, J. L., R. G. MASON, and G. R. HONIG. 1973. Regulation of hemoglobin β-chain synthesis in bone marrow erythroid cells by α-chains. *Proc Natl Acad Sci USA* 70:3405–3409.

WOO, S. L. C., J. M. ROSEN, C. D. LIAROKOS, Y. C. CHOC, N. BUSCH, A. R. MEANS, B. W. O'MALLEY, and D. L. ROBBERSON. 1975. Physical and chemical characterization of purified ovalbumin messenger RNA. *J Biol Chem* 250:7027–39.

WOODLAND, H. R., J. B. GURDON, and J. B. LINGREL. 1974. The translation of mammalian globin mRNA injected into fertilized eggs of *Xenopus laevis* II. The distribution of globin synthesis in different tissues. *Dev Biol* 39:134–40.

YANG, S. S., and D. G. COMB. 1968. Distribution of multiple forms of lysyl transfer RNA during early embryogenesis of sea urchin, *Lytechinus variegatus*. *J Mol Biol* 31:139–42.

YANG, S. S., and D. R. SANADI. 1969. Changes in the distribution of transfer ribonuclei acid species specifically induced by thyroxine. *J Biol Chem* 224:5081–86.

YANNARELL, A., D. E. SCHUMM, and T. E. WEBB. 1977. Age-dependence of nuclear RNA processing. *Mech Ageing Dev* 6:259–64.

YEN, R. C. K., W. B. ADAMS, C. LAZAR, and M. A. BECKER. 1978. Evidence for X-linkage of human phosphoribosylpyrophosphate synthetase. *Proc Natl Acad Sci USA* 75:482–85.

ZIMMERMAN, M., R. A. MUMFORD, and D. F. STEINER, eds. 1980. *Precursor Processing in the Biosynthesis of Proteins*. *Ann NY Acad Sci* 343, 449 pp.

475

19

Epigenesis of Morphogenetic Patterns

and measured at various stages of development?

In Chapter 18 we pointed out that the polypeptide chain, as it is being translated, immediately and automatically assumes a three-dimensional structure that is determined by the sequence of its component amino acids. Necessarily, the emerging protein assumes and fixes its structure in the thermodynamically most favorable free-energy conformation. Moreover, such molecules will associate, under conditions that can be defined, with molecules of the same or of different kinds, into three-dimensional arrays determined by the distribution of their electrical charges, configurational opportunities for the formation of hydrogen bonds, and so on. Thus does organization of a higher level of order emerge from lower levels of order—thus does epigenesis occur—as the biochemist or biophysicist can see it. The rules for predicting the three-dimensional patterns that will emerge at the macromolecular level or at the level of macromolecular assemblies are rapidly becoming clear as increasingly sophisticated physical techniques are applied to these problems.

Now consider, for example, the groups of cells that constitute the early wing bud or leg bud of the chick embryo. Each, isolated and grown as a graft in an abnormal site, makes a wing or leg, as the case may be, of right or left asymmetry, depending on the way it was taken from the right or left side of the donor embryo. Two important points must now be considered: first, the cells of wing bud and leg bud are cytologically indistinguishable, have identical histological arrangements, and, as far as one can tell (Saunders, 1972), are making the same array of proteins; second, the kinds of tissues found in the final products, wing and leg, are likewise identical and are composed of the same proteins. Thus it follows that the array of structural genes operating in the anterior and posterior appendages are identical. Nevertheless, the skeletal, muscular, vascular, and connective tissues and the integumentary derivatives of wings and legs are arranged quite differently. What are the rules that govern that cells will produce cartilage and bone in the upper arm in the shape of a humerus, and that of a femur in the upper leg? Or, for that matter, that fingers will arise at the tip of a human hand and toes at the tip of the leg?

A number of formal models for morphogenetic patterning in developing limbs and other systems

476 The development of an embryo may be seen as the emergence of systems of order at successively higher levels of organizational complexity. The order, represented as form and function, which accrues to each successive level, has its foundations on, and emerges automatically and progressively from, the properties of the system at the preceding level, until the final form and function of the phenotype are achieved. This development is epigenesis.

Underlying the end-product of epigenesis is the genome. The genome determines the repertory of proteins that cells can synthesize, and thus the kinds of structural and catalytic compounds that they can produce. But what determines which ones will be synthesized by cells at a particular organizational level? And from these syntheses how does a morphogenetic pattern emerge? What, in effect, are the conditions existing in a group of cells, an embryo, or an embryonic field at any cross-section of time that determine the characteristics of the system at a succeeding moment? Or, is this question naïve? Should it even be asked in the context of the present state of our knowledge or should we be looking for new conceptual approaches that will illuminate the parameters that must be identified have been proposed and have stimulated serious experimental tests. Some are considered here, but first some simpler systems of patterned macromolecules are examined.

19.1   Self-assembly of Macromolecular Systems

There are a number of biological macromolecules that spontaneously come together in a predictable three-dimensional array under appropriate conditions. This joining together is referred to as **self-assembly.** Keratin, actin, and myosin are notable examples of self-assembly systems, as well as **tubulin,** the component of microtubules, and **collagen,** a major component of connective tissues. The self-assembly of microtubules and collagen fibrils is discussed in what follows.

Microtubules play a role in a number of morphogenetic processes, as noted in earlier discussions of fertilization, cleavage, gastrulation, and various cell movements. They are hollow, cylindrical structures, about 24 nm in diameter and often of great length. In columnar cells they may extend the entire length of the cell, and in nerve cell processes they are probably very long. The diameter of the inner hollow core is 14 nm and the wall is 5 nm thick. The latter is made of tubulin, a **heterodimer** (having two nonidentical subunits) that sediments at 6 S and has a molecular weight of 110,000. The α and β subunits are of the same molecular weight, 55,000, but differ slightly in amino acid composition. Each subunit is coded by different mRNAs, as shown by Bryan et al. (1978) for tubulin of the embryonic chick brain. Miocrotubules are probably formed by the polymerization of dimers in helical array, as depicted in Figure 19.1, in a reaction that involves the hydrolysis of one molecule of GTP to GDP per heterodimer added. The guanidine nucleotides remain associated with the polymer (Maccioni and Seeds, 1977). When mechanically prepared microtubule fragments are used as "seeds" in the presence of purified tubulin, elongation of the fragment proceeds by the consecutive association of the

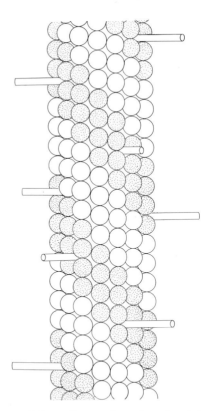

FIGURE 19.1 Schematic drawing of a microtubule showing the relationships of its subunits and high molecular weight projections. [Adapted from R. P. Vallee and G. G. Borisey, *J Biol Chem* 253:2834–45 (1978).]

477

heterodimers onto the existing free ends. Depolymerization occurs by the dissociation of the dimers from the ends of microtubules.

In the absence of microtubule fragments polymerization requires the presence of a high molecular weight protein (HMW) doublet (Figure 19.2). The subunits, having molecular weights of 286,000 and 271,000, respectively, subsequently appear as filamentous projections on the microtubule surface (Murphy et al., 1977). In the presence of HMW, 30 S oligomer rings of tubulin polymerize and then can serve as nucleation sites for microtubule elongation even in the absence of further HMW. The net rate of elongation of the microtubules is greater in the presence of HMW, for HMW reduces the depolymerization rate constant, thus shifting the equilibrium of association–dissociation in favor of assembly (Murphy et al., 1977; Johnson and Borisy,

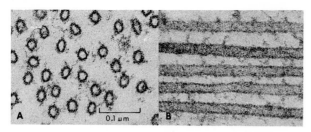

FIGURE 19.2 Electron micrographs of transverse (A) and longitudinal (B) sections of microtubules formed by reassembly of microtubule subunits in the presence of high molecular weight protein. [Electron micrographs kindly supplied by Dr. D. D. Murphy; similar to micrographs published by D. D. Murphy and G. G. Borisey, *Proc Natl Acad Sci USA* 72:2696–2700 (1975).]

1977). Lower molecular weight proteins designated *tau* are also involved in polymerization of tubulin, but their function is not yet clear (Vallee and Borisy, 1978).

How the assembly and disassembly of microtubules are controlled in the cell still offers many puzzles. They have been studied principally in cleaving eggs or other cells where they make up the spindle apparatus, including astral rays when present (cf. Chapter 9), in organelles such as the **axopodia** of certain protozoans (Figure 19.3), in the cytoskeleton (Figure 19.4A), of tissue culture cells, and in cells undergoing morphogenetic movements (cf. Chapters 9 and 11). When disassembled by means of drugs or low temperatures, they reassemble from preexisting subunits, apparently using the original nucleation centers.

In tissue-culture cells, as exemplified by those of the 3TC mouse line, cytoplasmic microtubules make up an extensive array that is demonstrable by means of fluorescent antibodies produced against

478

FIGURE 19.3 A: Idealized segment of the multinucleate heliozoan protozoan *Actinosphaerium*. The projections, called axopodia, contain a central core of microtubules, the axoneme, and have a rim of cytoplasm in which mitochondria are seen. B: Cross section through an axoneme near the axopodial base; note that it is made up of two spiral rows of microtubules and that these rows are coiled about each other. [From L. G. Tilney and K. R. Porter, *Protoplasma* 60:317–44 (1965), by courtesy of the authors and permission of Springer-Verlag.]

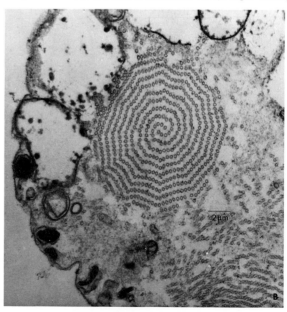

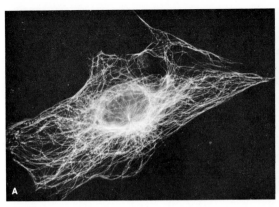
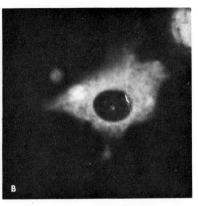

FIGURE 19.4 A: Microtubular network in a mouse tissue cell treated with a fluorescent antibody to tubulin. B: Similar cell in which the microtubules have been disassembled by means of the drug colchicine; note that the microtubule organizing center has persisted (arrow). [From M. Osborn and K. Weber, *Proc Natl Acad Sci USA* 73:867–71 (1976), through the courtesy of Dr. Osborn, who generously supplied original photographs.]

purified tubulin (Figure 19.4A). When the microtubular system is disassembled by means of drugs, there remains adjacent to the nucleus a small structure that reacts with the antibody (Figure 19.4B) and that serves as the nucleation center for the microtubular array of the entire cell when the cell is restored to conditions favoring reassembly of the tubulin subunits (Osborn and Weber, 1976).

Centrioles have been considered as nucleation centers for microtubules, but it is not known whether a centriole is present in the adnuclear body just described. Centrioles are prominent in the basal bodies of flagella, which contain ordered arrays of microtubules, and the fertilizing sperm in many invertebrate eggs provide a centriole from which microtubules appear to emanate (cf. Chapter 8). Centrioles are sometimes considered to be organizing centers for mitotic spindles. But many kinds of spindles do not have centrioles, notably spindles in most plant cells and the meiotic spindles in most animals. Moreover, parthenogenetic stimulation of many marine eggs, including enucleate fragments, can induce the formation of many **cytasters,** foci of microtubular arrays, in numbers far in excess of possible preexisting centrioles. In addition, there is now substantial evidence that the formation of centrioles occurs subsequent to the formation of asters. In 1975 Weisenberg and Rosenfeld prepared ho-

mogenates of eggs of the surf clam *Spisula solidissima* in the cold and then observed these homogenates under a cover-slip as the temperature was raised to 28°C. If the eggs were previously activated by treatment with excess KCl in seawater, microtubules, assembled from preexisting tubulin, polarized about dense structures here and there throughout the homogenate, forming numerous aster-like structures. Electron micrographs revealed no centriolar structures in such dense structures initially, but centrioles did form within them after 4.5 minutes. Homogenates of unactivated eggs formed few microtubules and no asters. Apparently, KCl treatment activated or brought about formation of microtubule organizing centers (MTOC). The MTOCs from homogenates of activated eggs could be partially purified by centrifugation and could induce aster and centriolar formation in homogenates of unactivated egg cytoplasm.

In sum, self-assembly of relatively complex three-dimensional patterns occurs in living cells or in vitro in the presence of microtubule subunits and suitable nucleation factors. How the distribution of nucleation factors is controlled in the cell has yet to be learned.

The idea of the automatic emergence of higher levels of order from the properties of lower echelons may be further illustrated by the formation of

479

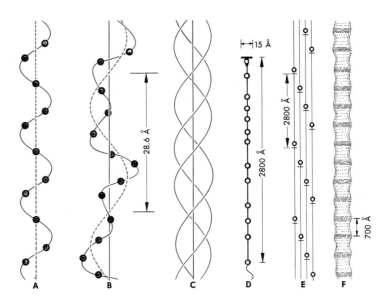

FIGURE 19.5 Structure of collagen. The basic single-chain molecular helix of collagen is schematized in A. It is depicted as thrown into a coiled helix in B. Tropocollagen, formed by the association of three helices, is shown in C and is represented at lower magnification in D. In E tropocollagen molecules are shown in parallel array with quarter-length overlap. This leads to the 700Å spacing seen in electron micrographs, as diagrammed in F and shown in Figure 19.6A. [Adapted with modifications from J. Gross, *Sci Am*, May 1961, pp. 120–30.]

480

macromolecular fabrics made from collagen. Collagen is the major fibrous component of skin, tendon, ligament, cartilage, and bone (Gross, 1961). The basic molecular unit of collagen is a chain of amino acids that, looked at from one end, makes a left-handed helix. Three such chains held together by hydrogen bonds and twisted about each other in a right-handed coiled helix make up a stiff rod about 280 nm in length and 1.5 nm in diameter. This structure has a molecular weight of about 340,000 and is called **tropocollagen;** it is the unit from which collagen fibrils are made (Figure 19.5). Molecules of tropocollagen are synthesized in cells called fibroblasts and then emptied into the intercellular spaces where they polymerize in the form of bundles showing a periodic banding with a repeat pattern of about 70 nm. In living systems, these bundles become distributed in various complex arrays, depending on the kind of tissue: in skin they make a dense feltwork; in tendon they form in tight parallel array; in bone they are arranged like the struts and girders of a bridge; and in cartilage they are randomly and sparsely distributed in a matrix of chondroitin sulfate.

It has long been known that if collagen is dissolved in acid, reconstituted fibers appear automatically when the solution is neutralized. The reconstituted fibers show precisely the same periodic banding as do those in the living systems. Coming

out of solution, the tropocollagen molecules polymerize in parallel array, facing the same direction and overlapping about one fourth of their length. Heads and tails of the molecules thus meet across the fiber array at about 70-nm intervals, giving the characteristic repeat pattern. The electric charges along the length of the tropocollagen molecules are such that this pattern emerges automatically (Figures 19.5D and 19.6A). But association of tropocollagen molecules can be changed to a nonoverlapping one if large, negatively charged molecules are added to the solution in which polymerization of the tropocollagen takes place. Thereupon, heads and tails of the tropocollagen line up in register and consequently a band-repeat pattern of 280 nm, the length of the molecule, is shown by the fibers (Figure 19.6B).

These considerations show that molecules of tropocollagen, being endowed with a particular set of properties, form structures of a higher level of organization whose new properties are determined by physiochemical conditions of the milieu. Looking back to a level of organization lower than that of tropocollagen, we note that the component proteins of tropocollagen are produced on ribosomes and then become entwined in a triparite structure. However, it is not the fact that these proteins are synthesized, but rather that there are built-in properties that determine that they will twist together

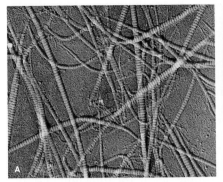

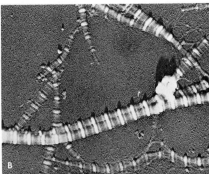

FIGURE 19.6 Electron micrographs of reconstituted collagen of the swim bladder of a fish; preparations shadowed with chromium. A: Fibers showing the normal period of 700Å; prepared by dialyzing acid-solubilized collagen against a solution of NaCl. B: Fibers showing long spacing of about 2800 Å; prepared by adding serum glycoprotein to the collagen solution and dialyzing against water. [Photographs generously supplied by Dr. Jerome Gross.]

FIGURE 19.7 Electron micrograph showing the arrangement of collagen fibrils in multiple plies in the skin of the tadpole. The stacked fibers in each ply are ordered at right angles to those in the plies above and below. Approximately 33 layers of stacked fibrils separate the epidermis (upper left) from the dermis (lower right). [Photograph supplied by Dr. Norman Kemp.]

to form tropocollagen under appropriate conditions within the cell (or in a test-tube system). Automatically, a structure is formed that is of a higher order of organization than its components. Furthermore, the fact that tropocollagen is emptied by the cell into the tissue spaces does not determine that it will form into associations showing a certain repeat pattern. Rather, the attributes of the tropocollagen are such that its molecules will associate or not and make certain repeat patterns or not, depending on the milieu.

The collagen that is produced in tissues approaching their final differentiated state is exposed to a milieu created by the physical and chemical circumstances that issue from the arrangements and activities of the cells that compose the tissue. The milieu is varied and complex, depending on the tissue, and it determines the order of emerging collagenous fabrics. Now, indeed, the analysis becomes difficult. What conditions, for example, determine that collagen fibers will arrange themselves in tight parallel array to form a tendon? Or how can one account for the fact that in tadpole skin they are laid down in stacked parallel plies, each ply having its fibers at right angles to those in the layers above and below (Figure 19.7)?

482

19.2 Epigenesis of Integumentary Patterns in the Chick

When one proceeds from the tissue level, complexities increase. We can possibly see our way clear to deriving a geometric pattern of collagen molecules from the known basic physiochemical properties of the molecule and its environment. But accounting for the geometry that is superimposed on tissues by their integration into organs poses problems of much greater magnitude and complexity.

There is, indeed, no rational way in which complex morphogenetic patterns can be thought of as arising from the self-assembly of molecules. Rather, complex patterns arise under strict temporal and spatial constraints. Obviously some parts of the sequence of events in organ development may have a self-assembly component. Certain kinds of cells, as we illustrated in Chapter 15, do automatically associate with other kinds of cells in predictable patterns, but such association patterns change when cells of different ages are brought together. Thus the origin of morphogenetic patterning can be understood only as one learns how temporal and spatial orderliness is controlled.

As an example of the emergence of order at higher levels of complexity, let us look at the case of a (deceptively) simple organ, the integument of the fowl, and focus attention on the development of the patterns in which its derivatives, the feathers, are distributed. Relatively early in development ($4\frac{1}{2}$–6 days of incubation, depending on location) a primitive kind of dermis differentiates in the superficial mesenchyme. It is closely associated with the overlying epidermis and is slightly demarcated from the premuscle masses beneath. Feathers arise from **feather germs,** or **feather papillae,** to which both ectoderm and mesoderm contribute. The origin of a feather germ is first signalled by a localized ectodermal thickening beneath which a dermal condensation, the dermal papilla arises (Sengel and Rusaouën, 1968). Over the dermal condensation the epidermis proliferates, forming a conical projection, the feather papilla (Figure 19.8). The dermal condensation proliferates a mesenchymal **pulp,** but remains chiefly at the base of the feather germ, receiving blood vessels that supply the pulp and ectodermal covering of the germ. The base of the feather germ then sinks downward into an ectodermal cyclinder, the **feather follicle,** the walls of which are continuous with the covering ectoderm of the germ. The epidermal component of the feather germ breaks into longitudinal ridges, each of which becomes a fluffy barb of the down feather, which is entirely made of keratin, the same kind of protein that is found on most epidermal derivatives, such as hoofs, claws, and nails. When the down feather is molted after hatching, the dermal papilla remains at the base of the follicle and is quickly covered by epidermal cells that migrate over its exposed surface. From these epidermal cells the first juvenile feathers arise. In contrast to the down feather, the juvenile feather has a shaft, called a rachis, and bilaterally arranged barbs. After a variable number of molts, the adult feather is formed, differing in size

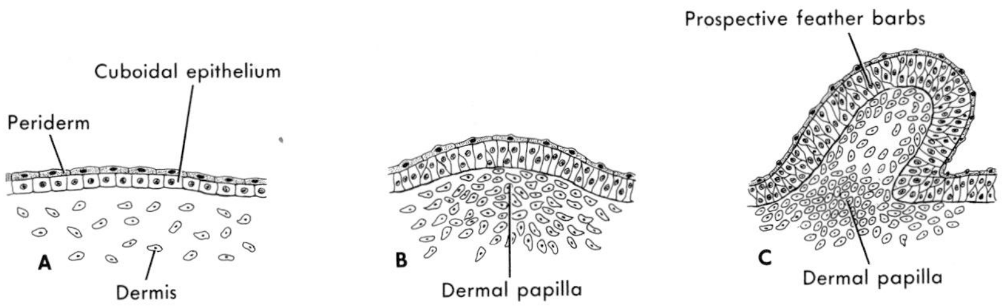

FIGURE 19.8 Origin of a feather papilla. A: components of the prospective integument prior to the onset of feather development. B: Thickening of the ectoderm and formation of a dermal condensation, the dermal papilla. C: The feather papilla prior to formation of the follicle and down-feather barbs.

and form from the previous juvenile feather. After each molt the new feather, entirely epidermal, originates from cells of the follicular wall that migrate over the dermal papilla.

This is as far as we shall go in pursuing the developmental history of the feather in this book. Feather development has been described in great detail in reports published by Lillie and his co-workers (Lillie, 1942; Lillie and Juhn, 1932; Holmes, 1935) and by Rawles and Watterson (Rawles, 1960; Watterson, 1942), especially. Their works may be consulted for further information. We now direct attention to the patterns in which feather germs arise and to the unique morphogenetic performance that occurs in each feather follicle. First, note that feather germs arise only in certain regions of the skin, called **feather tracts** or **pterylae.** These are

separated by featherless regions called **apteria.** Within each pteryla the feather germs arise in precisely prescribed order and form rigidly specific geometric patterns (Holmes, 1935). There first appears a row of primordia, running parallel to the head–tail axis of the embryo. This is the primary row, or **row of origin.** Secondary rows arise on either side of the first, the new papillae occupying positions corresponding to the spaces between those of the primary row. Tertiary and quaternary rows subsequently appear and the alternating positions taken by the primordia determine that feather positions will occur at the intersections of diagonal coordinates with longitudinal rows (Figure 19.9).

This precise geometric relationship of feather positions is maintained throughout the life of the bird (Figure 19.10A, B). New follicles do not form,

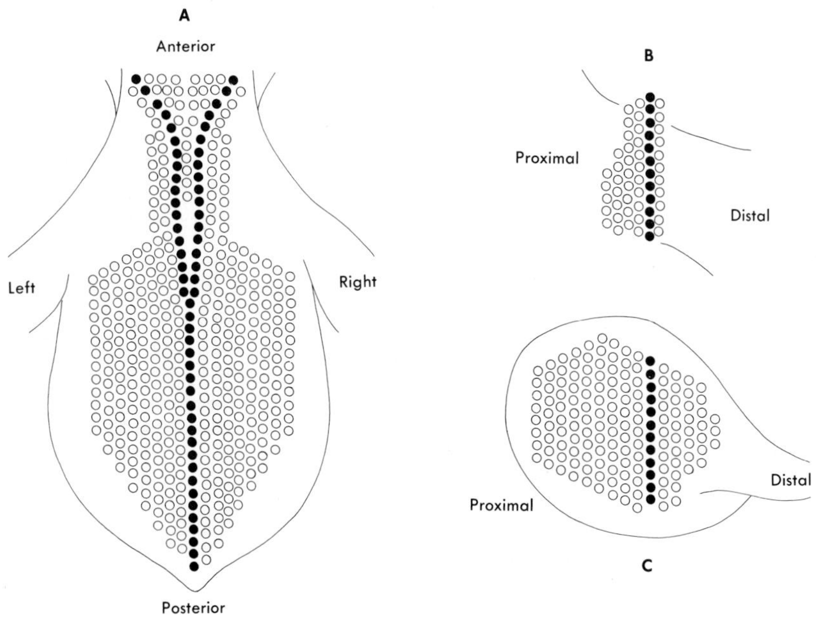

483

FIGURE 19.9 Origin of feather tracts in the chick embryo. In each tract certain feather germs arise almost simultaneously, constituting a row of origin (filled circles). Other rows then arise in succession on either side of the first row (open circles). A: The back and saddle tracts. B: The shoulder tract. C: The thigh tract. [Adapted from A. Holmes, *Am J Anat* 56:513–37 (1935).]

and after a feather is plucked from an adult bird a new feather forms from the same follicle. Very importantly, when the definitive feathers of the adult plumage are carefully examined, it is found that feathers of each tract are distinctively different from those of other tracts (Figure 19.10B). Moreover, within a tract feathers show gradients in structure, growth rates, response to hormones, and so on, along each of the coordinate axes that define the tract. Thus each feather in each tract is a unique structure having, however, its mirror-twin on the opposite side of the bird (Landauer, 1930; Juhn and Fraps, 1934).

The foregoing discussion implies that there must be systems of positional information that determine where a feather tract will form, where individual feathers will arise in a tract and the degree to which each feather must differ from its neighbor in morphology and physiology. Nothing is known about the character of this positional information, but there is a body of data relating to the time at which various determinative steps in the execution of tract pattern and specificity occur.

First, it is very clear that the feather tract fields and the regional characteristics of feather structure within a tract are established very early. Experiments involving the shoulder tract and its adjacent apterous region, and other experiments involving

484

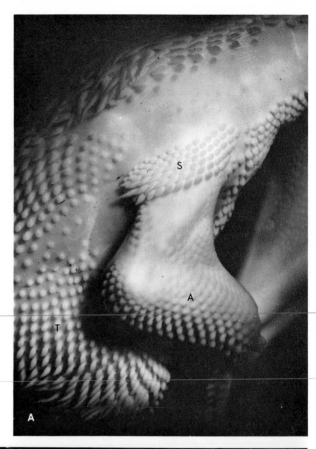

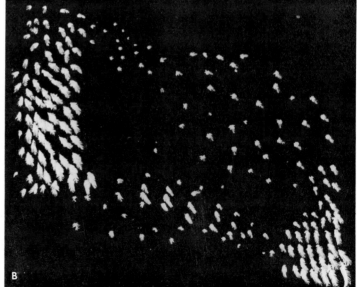

FIGURE 19.10 A: A portion of the right side of an 11-day chick embryo showing the precise geometric patterns in which the feather germs are arranged in shoulder, S, alar, A, and thigh, T, tracts. B: Feathers of the shoulder tract and part of the alar tract of the right wing of an adult White Leghorn fowl plucked and mounted in an expanded facsimile of their distribution on the wing. The shoulder tract is at the left and a prominent apterium occupies the center of the photograph. [B from J. M. Cairns and J. W. Saunders, *J Exp Zool* 127:221–48 (1954), by permission of Alan R. Liss, Inc.]

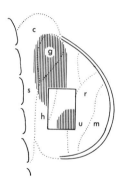

FIGURE 19.11   Fate map for the wing bud of a chick embryo at about 3½ days of incubation (adapted from Saunders, 1948) illustrating the effects of reorienting material of the future shoulder tract and adjacent apterium. The position of the future shoulder tract is indicated by vertical hatching. The approximate limits of various skeletal areas are enclosed by dotted lines: c, corocoid; g, glenoid region; h, humerus; m, manus; r, radius; s, scapula; u, ulna. [From J. W. Saunders and M. T. Gasseling, *J Exp Zool* 135:503–28 (1957), by permission of Alan R. Liss, inc.]

exchanges of materials between thigh and feather tracts in very young embryos illustrate this fact. In the 3-day chick embryo there are no feathers and a dermis is not yet organized. At this time, as established by means of marking experiments, the prospective shoulder tract occupies the zone of ectoderm and superficial mesoderm at the base of the wing bud anteriorly, as shown in Figure 19.11. There is a prospective apterium posterolateral to this zone. If one excises a block of superficial wing bud tissue involving the posterolateral quadrant of the shoulder tract and adjacent apterium and rotates it 180 degrees, retaining the ectodermal side upward, one effectively replaces a portion of the shoulder tract with prospective apterium, and, reciprocally, prospective shoulder feather material is shifted into the apterium with its anteroposterior polarity reversed.

Such experiments were carried out on embryos incubated for 72–96 hours. Effects of the operations were analyzed when the pattern of feather germs became visible at 11 days of incubation. Four classes of effects were found, as shown in Figure 19.12. These classes were analyzed according to arbitrary developmental stages covering the period of the experiment. Some embryos operated on at all

stages showed a normal shoulder tract and adjoining apterium. Others showed a shoulder tract deficient in feather germs in the posterior quadrant, but, again, with the normal adjacent apterium. In others, especially those resulting from operations of embryos of older stages, not only was the humeral tract deficient, but there was a group of feathers in the normally apterous area. These were directed posteriorly in some cases, and the longer feather germs were found posteriorly in the group, as is normal for the shoulder tract. Thus their direction and pattern or outgrowth had regulated. In other cases, but only from the oldest group of operated embryos, the feathers in the normally apterous area were directed anteriorly, having failed to regulate their anteroposterior polarity. These showed the larger feather germs on the anterior side. Some birds were allowed to hatch after operation and to achieve the adult plumage. Of those showing supernumerary feathers in the apterous area, examination revealed that they were of shoulder-tract type and of a size and structure expected of the lower left quadrant of the shoulder tract. These results indicate that positional values and polarity of feather germs within the shoulder tract field are progressively fixed in very early wing-bud stages, long before the first morphological signs of feather-germ formation.

Other experiments indicate that the properties of the future feather tract are initially determined in the mesoderm. Thus when superficial mesoderm of

485

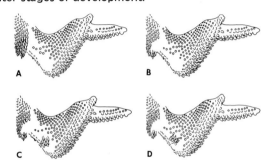

FIGURE 19.12   Four classes of results obtained after reorienting the tissues of the prospective shoulder tract in the 3–4-day chick embryo. A–D: Typical effects on the position and orientation of feather germs of the shoulder tract after operations at successively later stages of development.

the leg bud is excised from the future thigh feather tract in the 3-day chick embryo and grafted to the wing bud (Figure 19.13), it induces feathers of thigh type in the wing ectoderm (cf. Chapter 14). These thigh feathers are seen, in favorable cases, to be arranged in patterns identifiable with their regional origin in the thigh tract (Cairns and Saunders, 1954). Importantly, similar results are obtained when a block of deeper leg-bud mesoderm is grafted to the wing bud. Thus the system of positional information that specifies feather pattern in the limb is not necessarily confined to the prospective dermis, but is present in deeper layers of mesenchyme, which can take over the function of the original prospective dermis.

The steps that intervene between the early stages in prepatterning of the integumentary derivatives and the realization of the morphological pattern have been analyzed chiefly in the feathers of the **saddle tract** (Figure 19.9). This tract is bilaterally symmetrical, and its row of origin is in the dor-

FIGURE 19.14 The origin of dermal condensations, precursors of the dermal papillae of feather germs, in a hexagonal lattice.

FIGURE 19.13 Scheme of operation for replacing prospective dermis of the wing bud with prospective dermis of the region of thigh feathers of the leg bud. Wing bud epidemis heals over the graft of leg bud tissue and is induced to form leg feathers. [Redrawn with modifications from J. M. Cairns and J. W. Saunders, *J Exp Zool* 127:221–148 (1954).]

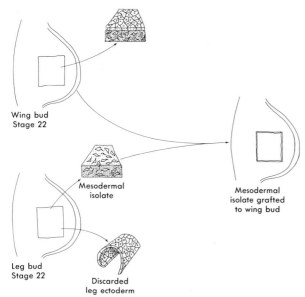

Wing bud
Stage 22

Mesodermal
isolate

Mesodermal
isolate grafted
to wing bud

Leg bud
Stage 22

Discarded
leg ectoderm

sal midline; additional rows of feather primordia are added symmetrically to the right and left. The dermis progressively thickens in advance of the prospective dermal condensations that become the dermal papillae. These condensations consist of a core of rounded, closely packed cells. Peripherally, the cells become elongate, are arranged circumferentially around the central core, and are continuous with a longitudinal band of cells connecting to adjacent condensations. Each condensation thus appears at the intersection of a hexagonal lattice of oriented elongate cells (Figure 19.14).

How are the dermal condensations realized? They could conceivably originate as foci of localized cell divisions, or by cell migration or by both mechanisms. Wessels (1965) labeled DNA in developing back skin by means of tritiated thymidine, finding that label was diluted more rapidly at the center of newly forming dermal condensations than it was in further developed dermal papillae. This finding would suggest that focal centers of higher mitotic rate initiate dermal condensation. Stuart et al. (1972) reported that in back skin explanted after the initial row of condensations appeared, subsequent rows formed even in the presence of the drug colchicine, which blocks mitosis at metaphase. Blocked metaphases were found in no greater a proportion of cells in the dermal condensations than elsewhere. The latter authors suggested, therefore, that the dermal condensations arise by the migration of cells along a preformed collagenous lattice whose intersections are the sites of the dermal condensations. In support of this thesis, they pointed out that isolated back skin containing several rows of condensations, when viewed under the microscope with polarizing optics, shows a pattern of birefringence, indicative of an ordered array of macromolecules, connecting each dermal condensation in the typical hexagonal array already described. When the skin was treated with collagenase the birefringence disappeared. Skin cultivated in vitro in the presence of the enzyme collagenase failed to form additional condensations, although control skins in vitro did. They reasoned, therefore, that a collagenous fabric serves as a pathway of migration for condensing cells. They were not, however, able to provide unequivocal evidence that an oriented fabric of collagen forms in advance of successive rows of feather germs, although they did find that a line of birefringence is found in the midline immediately in advance of the formation of the initial row of papillae. Wessels and Evans (1968) found no evidence for the preferential orientation of collagenous fabrics either in or between developing dermal condensations when they examined back skin by means of the electron microscope. Goetinck and Sekellick (1972) confirmed the existence of a pattern of birefringence in normal back skin, as described by Stuart et al. (1972), but found no evidence of patterned birefringence in the skin of the **scaleless** mutant, which lacks feather germs. As matters now stand, we have no certain information as to how the dermal condensations arise. Perhaps both division and migration are involved in the origin of the dermal condensations, but it is still unclear which process, if either, is more important.

The experiments of Saunders and Gasseling (1957) and of Cairns and Saunders (1954), referred to in the text and in Figures 19.11 and 19.13, show that, in the absence of subsequent experimental interference, the feather pattern that is expressed in the embryo is determined by the mesoderm long before the pattern is visible. Nevertheless, the emer-

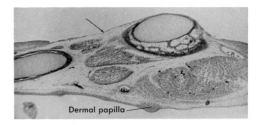

FIGURE 19.15 Cross section through the forearm of a 9½-day chick embryo with an epidermal wound that failed to heal after an operation at 3 days of incubation that removed most of the ectoderm of the dorsal side of the wing bud. Note the absence of ectoderm and of dermis and dermal papillae in the wound area (arrow).

gence of the pattern requires further ectoderm–mesoderm interactions in which both components share responsibility. Thus, as Saunders and Weiss (1950) showed, when the healing of an ectodermal wound in a feather-forming area is delayed beyond the normal time of feather-germ initiation (Figure 19.15), the exposed mesoderm neither organizes a normal dermis nor forms dermal papillae. The critical period for later phases of the ectoderm–mesoderm interaction was analyzed for the saddle tract of the chick embryo by Sengel (1958). The initial row of feather germs in this tract appears in the dorsal midline of the lumbosacral region at about $6\frac{1}{2}$ days of incubation, and adjacent rows are then added laterally. Using an agar-based medium supplemented with an extract of young chick embryos, Sengel analyzed the ability of isolates of the prospective saddle tract to self-differentiate in vitro when explanted with dermis and epidermis intact or separated and reassociated in heterochronous combinations. He distinguished as Stage 0 saddle skins from embryos of $5\frac{3}{4}$–$6\frac{1}{2}$ days of incubation, which showed no signs of feather initiation and which would not initiate feather formation on his particular culture medium in vitro. Embryos of $6\frac{1}{2}$–7 days of incubation showed saddle tracts with one, two, or three rows of feather germs on either side of the midline. These tracts were designated as Stage 1. When isolated in vitro, samples of such saddle tracts formed feather germs, but only after the original dermal condensations and epidermal placodes disappeared and reorganized. Explants of back skins at Stage 2 (7–$8\frac{1}{2}$ days) contained four or more rows of feather germs lateral to the initial

487

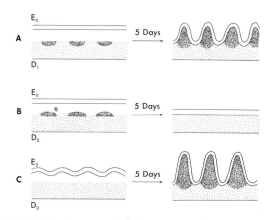

FIGURE 19.16 Reciprocal action of ectoderm and mesoderm in the formation of feather papillae in the chick embryo. A: Dermis that has formed prospective dermal papillae can interact with undifferentiated ectoderm and give rise to feather germs. B: Dermal papillae of a slightly later stage, however, are unable to participate in feather germ formation when associated with undifferentiated ectoderm. C: Epidermis, having been induced to initiate feather formation, can cause the formation of dermal papillae and will then cooperate in the formation of typical feather papillae. [Adapted from P. Sengel, *Ann Biol* 34:29–52 (1958).]

488

row, and these proceeded to differentiate a typical feather pattern without a period of reorganization. The same results were obtained when epidermal (E) and dermal (D) components were separated, recombined in isochronous combinations (E_0/D_0, E_1/D_1, and E_2/D_2, respectively). In heterochronous combinations, however, as illustrated in Figure 19.16, feather germs form in vitro from the combination E_0/D_1, but do not do so in the combination E_0/D_2. Evidently, therefore, the dermal papilla loses its capacity to induce the epidermal component to form a feather papilla after Stage 1. The epidermis, having been induced, however, is capable of eliciting the feather forming reaction in dermis, for feather germs arise from the combination E_2/D_0.

The necessity of two-way communication between ectodermal and mesodermal components at the stage of feather-germ formation is further shown by means of experiments involving combination between ectoderm and mesoderm of normal and of scaleless mutant fowl. Scaleless fowl are deficient both in scales and in feather germs. Sengel and Abbott (1963) combined components of the

saddle tract of scaleless and normal skin of the saddle region in vitro at Stage 1, as just defined. They found that, whereas scaleless epidermis underlaid with normal dermis failed to form feather germs, the combination of normal epidermis and scaleless dermis did so. In the scaleless mutant, therefore, the epidermis is not competent to interact with dermis at the normal time that feather-germ formation is initiated. Could this deficiency have arisen, however, from a failure of a prior action of scaleless dermis? Apparently not, for Goetinck and Sekellick (1972) found that the mesodermal core of the 3–4 day scaleless limb bud, covered with normal limb ectoderm and grafted to a scaleless host, formed characteristic wing feather patterns, whereas the reciprocal combination did not. The prospective dermis of the scaleless mutant is thus not directly affected by the genetic lesion that results in a deficiency of integumentary derivatives. Rather, it is the ectoderm that is affected, and deficiencies in the ectoderm result in abnormal morphogenesis in the dermis, as reflected in the lack of organization of dermal papillae and oriented hexagonal patterns of cells, as just described.

19.2.1 SIGNIFICANCE OF THE ROW OR ORIGIN OF THE FEATHER TRACT. The fact that successive rows of feather germs arise sequentially with reference to the origin of an initial row raises the question of whether the row of initiation is requisite for further feather formation and for the temporal sequence in which feathers arise in a tract. Linsenmayer (1972) pursued this question in studies of the thigh feather tract. In this tract the primary row is approximately parallel to the long axis of the embryo and arises near the lateral border of the tract. The principal body of the tract arises as more proximal rows appear in distoproximal order, so that at any one time successively smaller feather germs are seen in that same order (Figures 19.9 and 19.10A). The initial row arises about day 7 of incubation, but will appear essentially on schedule when the prospective thigh tract is excised as early as 4.5– 5 days of incubation and grafted to the vascular chorioallantoic membrane. Grafts to this membrane made at the stage of feather germ initiation do not undergo reorganization as does back skin isolated and cultured in vitro (Sengel, 1958; Novel, 1973). Moreover, on the chorioallantoic membrane the initial and subsequent rows of feather germs arise in normal se-

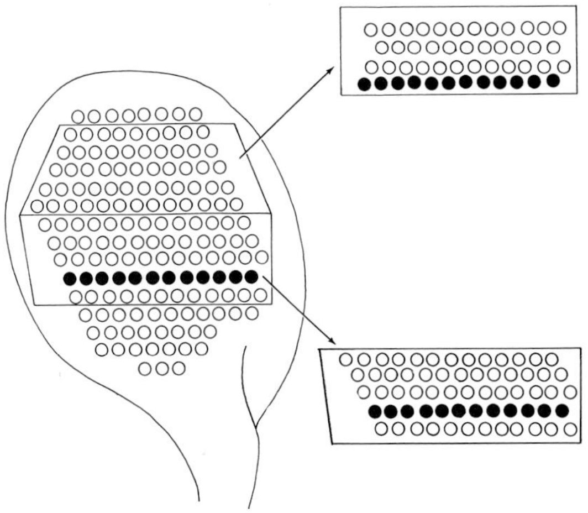

FIGURE 19.17 An interpretative drawing of the results of Linsenmayer (1972), who showed that isolates of the prospective thigh tract, growing on the chorioallantoic membrane, do not require the presence of the "row of origin" in order to initiate feather formation.

quence. Do the later-appearing feather germs depend for their origin on the presence of the initial row? Linsenmayer, 1972, excised lateral and medial halves of the prospective thigh tract and grafted them to the chorioallantoic membrane prior to the appearance of feather primordia. He found that both lateral and medial halves of the separated explants initiated feather formation in longitudinal rows at their lateral margins, with successive rows appearing medially (Figure 19.17). Feather germs were initiated later in the medial isolate and, for a given length of incubation, were smaller than those arising from the lateral isolate. These results suggest that the initiator row is not requisite to the appearance of subsequent rows of feather germs in a tract.

Some experiments of Sengel (1958) and of Novel (1973) are suggestive of a contrary interpretation, however. These investigators prepared explants of lateral halves of the prospective saddle tract from embryos of $6\frac{1}{2}$–$7\frac{1}{2}$ days of incubation. In some cases the cut was made parallel to the row of origin on the contralateral side, so as to include the row of origin undamaged in the explant. In other cases the cut was made through the row of origin, effectively destroying the original primordia (Figure 19.18). In grafts to the chorioallantoic membrane or in explants in vitro to a special culture medium (commercially available Medium 199), both kinds of isolates formed rows of feather germs sequentially and in the usual hexagonal pattern. On agar-based media supplemented with extracts of whole

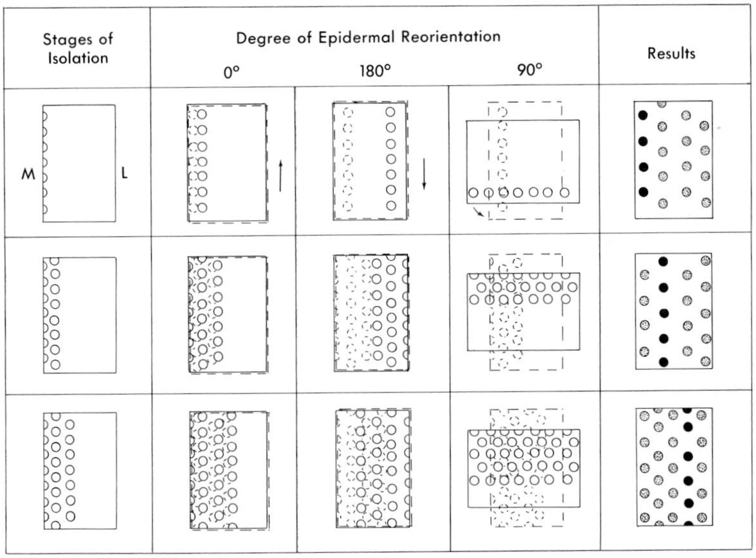

489

FIGURE 19.18 Effects of isolating and recombining right halves of the dermal and upidermal components of back skin cut through the row of origin of the saddle tract. The left column shows the rows of feather germs visible at the time of operation (6½–7½ days). The three central columns show the degrees of epidermal rotation, and the right column indicates the positions in which feather germs subsequently formed. Dashed circles indicate dermal papillae that became disorganized as a result of the operation. Open circles denote epidermal thickenings of feather germs, filled circles (right column) indicate the row of origin of the feather-germ pattern in recombinants, and stipple-filled circles signify successive rows arising secondarily. [Adapted from P. Sengel in R. Porter and J. Rivers (eds.), *Cell Patterning*, Ciba Foundation Symposium 29 (new series), North-Holland, Amsterdam, pp. 51–70, 1975.]

chick embryo or of embryonic chick brain, however, the pattern of feather germs developed normally only if the row of initiation was undamaged. If it was destroyed by cutting along its length, the entire pattern of feather-germ organization existing at the time of isolation was disrupted, as described for Sengel's earlier experiments, and a new pattern appeared in which an initial longitudinal row of feather germs arose lateral to the original medial edge. Subsequent rows then appeared medial and lateral to the new initial row. Interestingly, the new row of initiation formed farther laterally the further the feather pattern had developed at the time of isolation (Figure 19.18). These observations led Sengel (1975) to propose that, whereas the position of the first row is defined by unknown mechanisms in normal development, the feather germs in the next row have their positions defined by the first one, and rudiments of each newly defined row in turn control the positions occupied by feather germs in the next row. This suggestion is obviously at variance with the interpretation of Linsenmayer's experiments. It still remains unclear, therefore, what factors determine the foci of feather-germ formation in a tract.

490

19.2.2 ECTODERMAL CONTROL OF FEATHER PO-LARITY.

The experiments of Sengel and his collaborators, as well as many others not described here (e.g., Rawles, 1963), make it quite evident that some properties of the integumentary derivatives are capable of modification after morphogenesis begins. One of the early determined properties that is relatively resistant to modification, however, is that of the anteroposterior polarity of feather germs. As Saunders and Gasseling (1957) showed, their polarity can be reversed by reversing the orientation of prospective shoulder-tract feathers at stages prior to the origin of the dermal papillae.

Normally, the feather germ is directed caudally over most regions of the body, a fact that is signaled very early by the configuration of cells of the epidermal placode (Sengel and Mauger, 1976) and by the asymmetric distribution of pigment cells in the feather germs of pigmented breeds of fowl (Watterson, 1942). The definitive feathers show a characteristic dorsoventral organization. The ventral side, which faces the skin, is identified by a longitudinal groove in the rachis and by the direction of curvature of the vane and barbs. In the experi-

ments of Saunders and Gasseling (1957) the dorso-ventral organization of feathers resulting from reversal of the anteroposterior axis of the inductive mesoderm always corresponded to the feather slope; that is, the ventral side of the definitive feather always faced the skin, regardless of feather orientation.

This property of the ectoderm seems to be fixed permanently during the earliest stages of feather tract organization. When feather-forming epidermis and dermal papillae are separated by digestion with trypsin and recombined in vitro (Sengel, 1958; Novel, 1973), or on the chorioallantoic membrane after rotation of the epidermal component either 90 or 180 degrees with respect to the mesodermal component, feather germs arise in longitudinal rows as previously described, but the direction of orientation of the feather germs conforms to the original polarity of the epidermal component (Figure 19.18).

19.3 Formal Models for Predicting Patterns Formed During Regeneration and Embryonic Development

The immediately foregoing analysis has tended to emphasize the complexity of organizational patterns achieved in developing systems, but has provided little insight into the problem of how specific patterns of differentiated states can arise within an ensemble of apparently identical cells. We now turn attention to formal models that seek to establish conceptual frameworks within which the development and regeneration of spatial patterns may be discussed. These models are designed particularly to predict the effect on morphogenetic patterns of the experimental perturbation of developing systems.

Prominent among efforts to develop a satisfactory working model are those of Wolpert and his students. They have sought to formulate a more insightful definition of the embryonic field (cf.

Chapter 13) that will eventually be productive of more meaningful interpretations of cellular behavior in the differentiation of morphogenetic patterns. The chief points of this formulation are as follows.

1. There are mechanisms whereby cells in a developing system are specified as to their position with respect to a set of points or boundaries in the system. Cells that have their positional information determined by the same set of points or boundaries constitute an embryonic field.

2. The positional information that each cell receives specifies the nature of the molecular differentiation that it will undergo, with due allowance to its genome and previous developmental history.

3. Polarity (i.e., axial order) with respect to spatial coordinates is defined with respect to the direction in which positional information is specified and perceived.

4. Regulation of pattern in an embryonic or regenerative field (cf. Chapters 13, 14, and 16) is dependent on the ability of cells to adjust their positional information content with respect to their altered relationships with sets of points or boundaries that are caused by perturbation (i.e., if a cell cannot perceive or react to altered patterns of signals, it cannot engage in regulative development).

5. Positional information in different kinds of organisms may be specified by the same mechanisms (Wolpert, 1978; French et al., 1976; Bryant et al., 1977).

19.3.1 SPECIFICATION OF POSITIONAL INFORMATION IN MORPHALLACTIC REGULATION. In Chapter 16 the term morphallaxis was introduced in connection with the analysis of the regenerative regulation of morphological patterns. The term morphallactic regulation can be applied to any case in which the field pattern is restored after perturbation by virtue of the reorganization of its remaining components. An excised central segment of a hydra, for example, forms a miniature hydra before it can once more feed and grow. Also recall from Chapter 13 that a meridianal half of a sea urchin

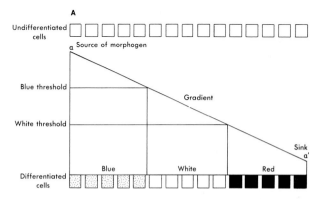

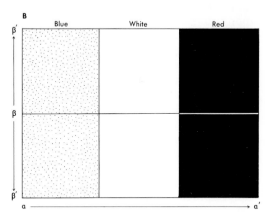

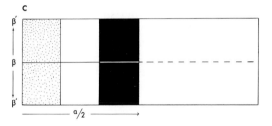

FIGURE 19.19 The "French-flag" model for morphallactic regeneration. A: For generation of the French flag sequence of blue, white, and red vertical stripes, positional information is presumed to be delivered by a gradient of diffusible morphogen whose concentration is fixed at a level α at the left and kept at a lower level α' by a "sink" on the right. Files of cells from left to right have, by virtue of prior developmental experience, thresholds that enable the gradient of information to be interpreted. Thus, at a high concentration of the morphogen the interpretation "blue" is made. B: If one imposes a second gradient on the pattern of A, a pattern of vertical stripes is generated. C: If the $\beta'-\beta'$ axis shown in B is shortened, then the "cells" reassess their positional input and adjust their performance to create a smaller version of the flag. [Adapted from various papers of L. Wolpert, especially *J Theoret Biol* 25:1–27,(1969).]

491

blastula regulates to form a normally proportioned pluteus of half size. This, too, is a case of morphallaxis.

Wolpert applied his concept of the role of positional information to the problem of pattern regulation by creating his now well-known "French flag" model. He generated the pattern of the flag of France from a hypothetical file of "cells," each of which can differentiate as blue, white, or red (Figure 19.19). In the French flag these colors appear as three vertical stripes originating in the same sequence from the left side (i.e., from the flagpole or masthead). Now one assumes that positional signals (possibly in the form of a diffusible substance, or **morphogen**), originate at the masthead border and are delivered as a gradient toward the right, and that cells have thresholds of response to the morphogen. A source keeps the concentration high on the left, and a sink keeps the concentration fixed on the right. Cells in files from left to right thus are exposed to gradients of positional information and have thresholds that enable this information to be interpreted. Thus the threshold for the blue response is highest, and only cells nearest the left margin differentiate as blue. Thresholds for white and red are successively lower. Onto this one-dimensional system one could now impose a second coordinate system at right angles to the first so that an appropriate number of parallel files of cells would differentiate so as to produce a flat sheet with vertical stripes (Figure 19.19B). Regulation in this hypothetical field after surgical intervention would lead to a reassessment by each of the cells of its position in the now-smaller whole, leading to the formation of a miniature but complete pattern (Figure 19.19C).

Conceivably, the same system of positional information could give rise to different patterns, depending on how the positional information is interpreted. Consider, for example, that the American flag contains the same colors as the French flag. If "development" of the two flags is based on positional information specified by the same coordinate system, then the graft of an undifferentiated piece of one flag to the other should give rise to a composite pattern. For example, if a small piece of "undifferentiated' American flag were grafted to the upper left corner of a French flag, it should develop white stars on a blue field whereas the excised piece of French flag grafted to the donor site of the American flag should form vertical stripes (Figure 19.20).

Regulation in a number of biological systems is amenable to interpretation by the French flag analogy. An example that is sometimes cited is the case in which proximal mesoderm of the prospective thigh region of the leg bud is grafted subjacent to the ectodermal tip of the wing bud in the $3-3\frac{1}{2}$ day chick embryo (Figure 19.21). In its new location the graft makes toes and induces scales and claws, which suggests that its cells acquired positional information that elicited distal differentiation in their new location, but that they responded according to a prior developmental history that determined them to be leg as opposed to wing.

The nature of signals emanating from sets of points or boundaries in a field that gives a cell positional information is not known for these cases or, for that matter, any others. The notion that posi-

FIGURE 19.20 Differential interpretation of identical positional information. The assumption is that the patterns of American and French flags are both generated by identical systems of positional information, but the components of the flag are programmed to respond differently. [Adapted from L. Wolpert, *Sci Am*, May 1978, pp. 154–164.]

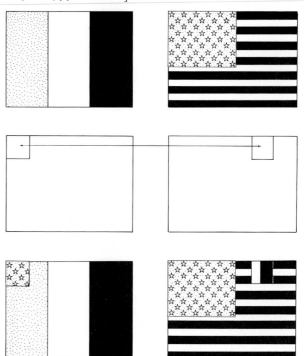

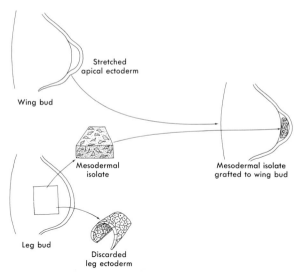

FIGURE 19.21 Scheme depicting the method of grafting prospective thigh mesoderm beneath the apical ectoderm of the wing bud in the 3½-day chick embryo. [Redrawn with modifications from J. M. Cairns and J. W. Saunders, *J Exp Zool* 127:221–48 (1954).]

tional information has a reality has given rise, however, to a number of other formal models for predicting the pattern of morphogenesis in regenerating and developing systems, especially the appendages, in both vertebrates and invertebrates. In what follows we discuss a new model that has been applied to regenerative and regulative patterns in regenerating limbs of cockroaches and amphibians and to imaginal discs of *Drosophila*. Thereafter, the question of positional signaling in the embryonic development of amphibian and chick limbs is discussed. The concept of positional signaling has been applied to the regeneration of *Hydra* by Wolpert and his colleagues (Wolpert et al., 1972), but is not treated here. Some aspects of regeneration in *Hydra* were treated in Chapter 16.

19.3.2 SPECIFICATION OF POSITIONAL INFORMATION IN EPIMORPHIC FIELDS.
Epimorphic fields are regenerative fields in which growth is required. The zone of growth is provided by a blastema, which, in regenerating amphibian limbs, at least, is derived almost exclusively from cells at the cut surface. The blastema reproduces the lost structures in proper polarity and symmetry if left undisturbed, but may,

under appropriate conditions, provide supernumerary regenerates, sometimes of mirror-twin asymmetry and sometimes not.

Rules for predicting regenerative patterns in epimorphic fields have been formulated by French and his collaborators (1976) and proposed as having universal application. These rules are based on a simple model in which the regenerating system is presented in two dimensions by a system of polar coordinates. This system can be represented either as a clock-faced disc or, if desired, as a cone (Figure 19.22). The center of the disc or the apex of the

FIGURE 19.22 The polar coordinate model for interpreting regulation in epimorphic fields. It is shown in A as a clock-face model and in B as a cone. The boundary of the field (e.g., base of the limb) is represented by the outermost circle of the clock face and by the left end of the cone, as illustrated. Recognize that the cone (B) would appear in the form of concentric circles as in the clock face if viewed end on after truncation. As visualized in A, each cell receives positional information with respect to its distance along a radius (A–E) and also with respect to its position on a circle (0–12). Positions 0 and 12 are identical so that the sequence is continuous. [From S. V. Bryant in D. E. Ede et al. (eds.), *Vertebrate Limb and Somite Morphogenesis*, Cambridge University Press, Cambridge, 1977, pp. 311–327.]

493

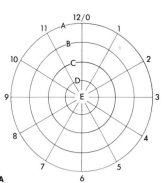

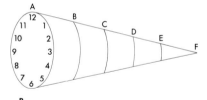

cone represents the distal end of the regenerate. The circumference of the disc or the circumference at the base of the cone represents the proximal boundary of the epimorphic field. Each cell in the field has a unique positional value, which is defined by two components. One of these relates to the circumferential position of the cell with respect to coordinates arbitrarily numbered as a clock face with numerals from 1 to 12. The other component relates to the proximodistal level of the cell with respect to circular coordinates at levels A, the proximal boundary, through E, the tip.

Two rules describe the behavior of cells in the epimorphic field after surgical intervention. The first one is called the **rule of shortest intercalation,** and the second one is called the **rule for distalization.** We use simple examples to illustrate these rules and to define the terms used in them. Then the rules are applied to the generation of a variety of epimorphic patterns.

19.3.2.1 Pattern Formation During Limb Regeneration.

We begin this section by developing first the notion of intercalation. The most important part of the polar coordinate model is the proposition that when cells from any two normally nonadjacent radial (i.e., proximodistal) or circumferential positions are brought together as a result of wound healing or grafting, a discontinuity of positional values at the site of confrontation results. This discontinuity stimulates local growth, and during growth those positional values that normally separate the confronted positions are generated. The generation of the missing positional values is called intercalation. Intercalation of both proximodistal and of circumferential positional values occurs, as we illustrate next.

First recall that when the limb of a newt, for example, is amputated, it regenerates only those parts distal to the level of amputation. Because the regenerate arises from cells at the amputation site, it follows therefore, that cells of different proximodistal levels of the limb have a different content of positional information. Now consider the case in which one grafts the regeneration blastema formed after amputation of the newt limb through the forearm to the stump of a similar limb freshly amputated through the upper arm (Figure 19.23). The operation brings into apposition cells of different positional value with respect to the proximodistal axis, and intervening values are absent. As a conse-

494

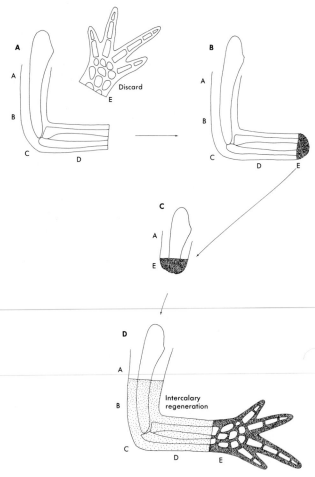

FIGURE 19.23 Intercalary regeneration in the limb of a newt or salamander. A: A limb is amputated at the level of the wrist, whose positional value is designated E. B: A regeneration blastema forms at the amputation site. C: The regeneration blastema shown in E is now grafted to the freshly severed stump of a limb amputated through the humerus; this brings tissue of proximodistal value, A, in apposition to tissue of positional value, E. D: Growth is stimulated and the missing positional values B, C, and D are intercalated between A and E.

quence, growth is stimulated and brings about the intercalation of positional values intermediate between those of the cut surfaces. This intercalation results in the regeneration of a limb with a complete proximodistal sequence of positional values.

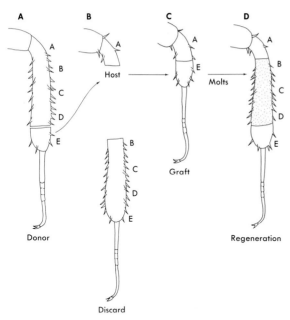

FIGURE 19.24 Intercalary regeneration of the tibial segment in a young cockroach. The lower end of the tibia of a donor foreleg (A) is amputated at positional level E and then grafted (C) to a host tibia (B) amputated at positional level A. After several molts, the missing positional values are filled by intercalary regeneration (D).

Similarly, grafting the distal part of a segment of a cockroach limb to a proximal amputation site on the same segment (Figure 19.24) results, after the next molt, in the intercalation of the intermediate positional values and regeneration of a complete segment.

The reciprocal experiment of grafting terminal limb parts of proximal positional value to the stump of a limb segment severed distally gives different results when amphibian and insect materials are compared. When the regeneration blastema of *Ambystoma*, amputated through the humerus, is grafted to the stump of a host forelimb amputated through the forearm or wrist, tissues of different positional value are brought together, but the intermediate values are seldom intercalated, and then only incompletely. According to Maden (1980a), the intercalated structures originate from the blastema in these cases, whereas when distal blastemata are grafted to proximal stumps, the interca-

lated materials are chiefly of stump origin. In cockroaches, on the other hand, the intercalation of missing positional values occurs readily when terminal parts of proximal positional value are approximated to parts of more distal positional value in the same limb segment.

In the cockroach the proximodistal organization is repeated in each limb segment. Therefore, grafting the middle part of the tibia, a distal segment, to the middle part of the femur, a proximal segment, does not lead to intercalary regeneration, for tissues of the same positional values are apposed. But intercalation can occur between tissues of different segments if different levels of segments are confronted. Bohn (1976), for example, severed the prothoracic leg of a cockroach at the distal level of the femur and grafted it to the stump of the metathoracic leg (third leg) severed through the proximal region of the tibia. The resulting appendage after two molts characteristically showed an intercalary regenerate whose bristles and pigmentary characteristics were appropriate to the femur of the metathoracic leg, but were ordered in normal proximodistal sequence (Figure 19.25).

Regeneration can also result in the intercalation of missing circumferential positional values, as represented by the numbered sequences shown in Figure 19.22. This process is most easily illustrated by regeneration of cockroach limb segments from which longitudinal strips of integument (cuticle plus overlying epidermis) have been removed. In cockroach limbs patterning in the circumferential sense is clearly indicated in the integument by the distribution of longitudinal rows of cuticular bristles secreted by the epidermis. Because these rows are nonuniform in circumferential disposition, it is clear that the content of positional information is different around the circumference of the limb segment as seen in cross-section (Figure 19.26A). When a longitudinal strip of integument is excised from a limb segment (Figure 19.26B), wound healing results in the confrontation of groups of epidermal cells that were originally some distance apart around the circumference and that differ in positional information. Interaction between the apposed surfaces results in growth and intercalary regeneration; that is, the missing circumferential structures are regenerated (Figure 19.26C).

Importantly, the positional values that are intercalated are the shorter of two possible sets of inter-

495

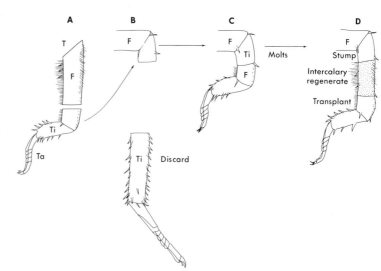

FIGURE 19.25 Intercalary regeneration between the distal end of the femur of the foreleg and the proximal end of the tibia of the third leg in the cockroach *Blaberus cranifer*. A: Donor foreleg. B: Host third leg severed through the proximal part of the tibia. C: Host-graft combination. D: Intercalary regenerate showing spines characteristic of the proximal end of the foreleg femur. F, femur; T, trochanter; Ta, tarsus; Ti, tibia. [Adapted from H. Bohn, *Dev Biol* 53:285–93, (1976).]

496

mediate values that separate the apposed surfaces. Thus, when values 4 and 8 are confronted, as shown in Figure 19.26B, three intermediate values (5, 6, and 7) are intercalated, instead of the seven intermediate values (3, 2, 1, 12/0, 11, 10, 9) that separate 4 and 8 in the other direction. This chain of events leads to what is meant by the **rule of shortest intercalation.**

We now turn to the rule for distalization, which was first presented as the **complete circle rule for**

distal transformation. The rule originally proposed that cells at the proximodistal level can give rise to new cells with all the positional values distal to their own values, but only when a complete sequence of circumferential positional values is exposed by amputation or by intercalation. When a normal limb of a newt or cockroach is amputated, all circumferential positional values are presumed to be present at the cut surface, and distal regeneration occurs, regardless of whether one deals with regeneration from the proximal or distal cut surface (Figure 19.27). The complete circle rule would lead to the prediction that if composite limbs are formed surgically (Figure 19.28) by combining two half-limbs (e.g., two half-dorsal, half-ventral, half-anterior, or half-posterior upper or lower limbs) in such a way that each contains a mirror image of the circumferential information of the other, formation of distal parts should not occur after amputation (Bryant, 1976). In fact, artificially constructed half-limbs in newts and axolotls do not show distal outgrowth if healing of the symmetrical halves is allowed to proceed for 30 days or more before amputation (Bryant, 1976; Bryant and Baca, 1978; Tank, 1978a). If, however, amputation is carried out at earlier stages, distal outgrowth occurs to an extent that is inversely proportional to the time between grafting (Figure 19.27) and amputation. Such outgrowths tend to be hypermorphic, usually showing only a single symmetrical digit. But, especially in the case of double-posterior combinations

FIGURE 19.26 Intercalation of circumferential positional values in the leg of the cockroach. The outlines show positional values as seen in cross section. A: Positional values before excision of a longitudinal strip of ectoderm and cuticle. B: Apposition of cells of different positional value by healing. C: Intercalary regeneration of positional information by the shortest route. [Adapted from P. J. Bryant et al., Sci, Am, July 1977, pp. 66–81.]

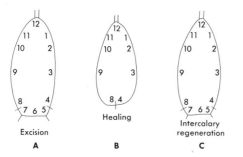

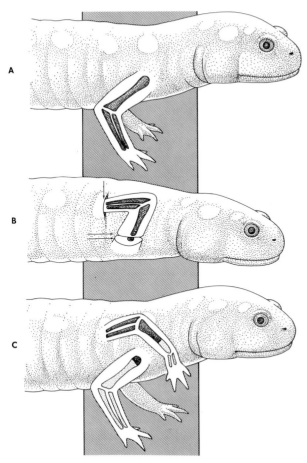

FIGURE 19.27 A classic experiment showing that both cut ends of an amphibian limb can regenerate terminal limb parts. A: *Ambystoma,* showing partial right forelimb skeleton. B: The hand is amputated (arrow), and the forelimb stump is inserted into the back musculature; after the forelimb heals to the back, the limb is again amputated below the shoulder (double arrow). C: The right forelimb regenerates typical distal parts of right asymmetry; the dorsal implant regenerates a portion of the humerus, the forearm, and hand; the regenerate is of left asymmetry. [An interpretative drawing based on results obtained by J. N. Dent, *Anat Rec* 118:841–56 (1954), and E. G. Butler, *J Morphol* 96:265–82 (1955).]

amputated as early as 5 days after grafting, terminal parts comprising two to seven digits may form, and these usually show a striking mirror-image symmetry about the anteroposterior axis. They tend to

form, in effect, double posterior limb tips (Stocum, 1978; Bryant and Baca, 1978; Tank and Holder, 1978; Holder et al., 1979; Holder et al., 1980; Krasner and Bryant, 1980).

The latter observations have led to a refinement of the rule of distalization. At the amputation surface new cells are generated during circumferential intercalation. These new cells, as a result of strictly local interactions, must adopt positional values more distal than those of the cells preexisting at the wound edge. In normal regeneration repeated rounds of circumferential intercalation and distalization will give an outgrowth that is both circumferentially and distally complete. Applying the refined

FIGURE 19.28 Construction of double anterior and double posterior upper arm segments in the newt. A: Left and right upper arms showing position of cuts isolating, respectively, anterior and posterior portions of the limb; arbitrary positional values are indicated by numbers. B: Scheme of positional values at the amputation surfaces of double anterior and double posterior upper arms. [Adapted from S. V. Bryant, *Nature* 263:676–79 (1976).]

497

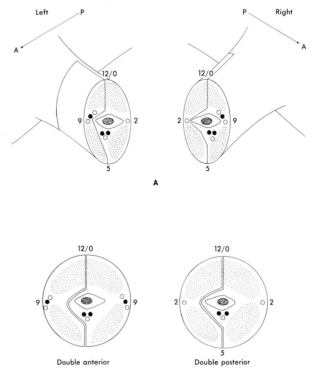

rule of distalization to regeneration fields made symmetrical by surgical means, as in the case of double half-limbs in amphibians, the model would predict that distalization can occur from double half-circumferences, but that its extent will depend on the pattern of healing at the wound site. Thus, amputation of a fully healed double half-limb allows for little or no distalization. Cells at the cut surface intercalate only with those neighbors that have different positional value. Wound healing would tend to bring into apposition cells of identical positional value across the midline, and thus intercalation and distalization would soon cease.

In contrast, in limbs amputated shortly after grafting, presumably the mirror-twin surfaces behave independently of one another. They would not intercalate across the midline. Thus cells of identical positional value are not brought into confrontation. In each separate half, therefore, intercalation and distalization can proceed to a degree depending on the number of different positional values present at the amputation surface and the number of distal values to be replaced. Regenerates formed after amputation of double half-posterior limb halves usually lose distal values in the midline so that double posterior sets of digits of opposite handedness occur (Bryant et al., 1981).

Slack (1977a) found that *Ambystoma mexicanum* embryos form double posterior forelimbs if a bit of flank tissue is grafted immediately anterior to the site of origin of the future forelimb (Figure 19.29). Such limbs show normal dorsoventral differentials, but are completely symmetrical otherwise in the sense that posterior limb structures are present in mirror-twin symmetry on both anterior and posterior edges of the appendages. When such limbs are amputated, they usually give rise to double posterior limbs, but occasionally they form relatively normal regenerates. Interestingly, after repeated amputation and regeneration such double posterior limbs tended to show a diminished number of digits. In contrast, the number of digital elements in double posterior regenerates studied by Holder et al. (1980) either increased the number of digits upon regeneration after further amputation or showed no decrease.

Slack (1977b) and Slack and Savage (1978a, b) explained the origin and regeneration of double posterior limbs on the basis of the distribution of a diffusible morphogen which normally originates

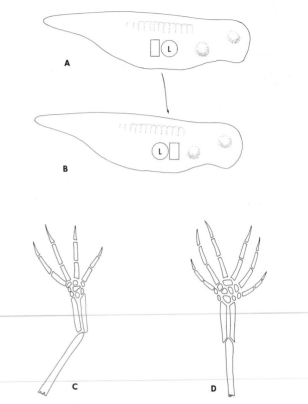

FIGURE 19.29 Scheme of operation leading to the embryonic development of double posterior forelimbs in the axolotl. A, B: Flank somatopleure is excised from the region posterior to the prospective limb bud and placed anterior to it. C: The skeleton of a normal forelimb. D: The skeleton of a double posterior limb resulting from the operation described. [Adapted from J. M. W. Slack, *Nature* 261:44–46 (1976).]

posterior to the limb bud and persists after limb development. Such a morphogen, having higher concentrations both anteriorly and posteriorly after the transplantation operation, is presumed to direct the formation of posterior structures. (See Section 19.3.2 for a similar interpretation of the development of the anteroposterior axis in chick embryos.) Holder et al. (1980) pointed out, however, that the occasional distal development of regenerates from double anterior and double dorsal limb halves suggests that an explanation of results other than the one offered by Slack and Savage should be invoked.

The original clock-face model was invoked by Bryant and Iten (1976) to account for the appearance of supernumerary limbs that appear after a regeneration blastema is axially rotated with respect to the stump. Using adults of *Notophthalmus viridescens*, they showed that when the regeneration blastema, formed after amputation of the left forearm, is removed and grafted to the freshly amputated stump of the right forearm of the same or another newt, one or two supernumerary limbs, both of right asymmetry, arise at the host–graft junction. If the anteroposterior axis of the blastema is reversed with respect to that of the stump, the supernumerary regenerates may form at either or both the anterior or posterior margins of the limb (Iten and Bryant, 1975). If the dorsoventral axis of the

FIGURE 19.30 Results of grafting the regeneration blastema from a left forelimb to the stump of an amputated right forelimb. Adults of *Notophthalmus viridescens* were used, and amputations were carried out at the midlevel of the upper arm. The grafts were placed with dorsoventral axes reversed with respect to the stump. A, B: A regenerate and a dorsal supernumerary limb (arrow) before and after staining the skeleton. C, D: A regenerate with both dorsal and ventral supernumerary limbs before and after staining. Arrows identify the supernumerary limbs. Note that C and D are oppositely oriented. V, ventral; D, dorsal. [From S. V. Bryant and L. E. Iten, *Dev Biol* 50:212–34 (1976), through the courtesy of the authors and permission of Academic Press, Inc., New York.]

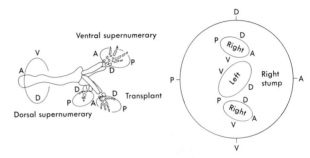

FIGURE 19.31 At the left is a diagram of the skeletal parts of the limb shown in Figure 19.30D. On the right is a scheme in which the host stump is represented as a circle and the graft and supernumerary limbs are shown as ellipses to indicate orientation. A, anterior; D, dorsal; P, posterior; V, ventral. [Redrawn from S. V. Bryant and L. E. Iten, *Dev Biol* 50:212–34 (1976), through the courtesy of the authors and permission of Academic Press, Inc., New York.]

FIGURE 19.32 Effects of rotating a regeneration blastema 180° on its own stump. A: Skeletal parts of a supernumerary left limb arising dorsocaudally. B: A supernumerary right limb arising anteroventrally in relation to the host axis. Other aspects of the diagrams are as described in Figure 19.31. [Redrawn from S. V. Bryant and L. E. Iten, *Dev Biol* 50:212–34 (1976), by courtesy of the authors and permission of Academic Press, Inc., New York.]

499

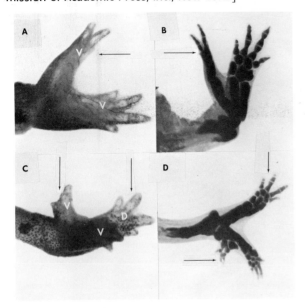

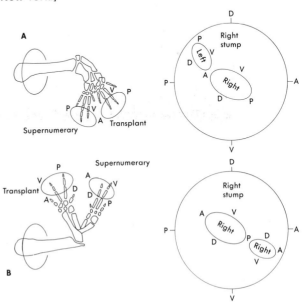

blastema is reversed, the supernumerary regenerates arise either dorsally or ventrally or in both positions (Figures 19.30 and 19.31). When the regeneration blastema, formed after amputation of the right forearm, is severed and replaced on the stump with both dorsoventral and anteroposterior axis

reversed, a supernumerary regenerate right asymmetry may form at the host–graft junction anteroventrally. Another supernumerary may arise dorsocaudally, and, if it does, it will be of left symmetry (Figure 19.32). Bryant and her collaborators did not observe supernumerary regenerates when other

FIGURE 19.33 Schemes for predicting the occurrence and handedness of supernumerary limbs regenerated in *Notophthalmus viridescens* after axial rotation of a regeneration blastema with respect to the amputation surface. The outer circle represents the distribution of positional information in the stump, and the central circle the positional information in the blastema. The small circles represent the position and handedness of supernumerary regenerates that may occur. They are located at the points of greatest incongruity between positional values of stump and blastema and are thus points where a complete circle of circumferential positional information can be generated by intercalation between host and graft. They are, therefore, possible sites for the origin of supernumerary regenerates. The two upper diagrams depict the rotation of a left blastema in a right stump, and the lower diagram the rotation of a right blastema on its own stump. [Redrawn with slight modifications from S. V. Bryant and L. E. Iten, *Dev Biol* 50:212–34 (1976), by courtesy of the authors and permission of Academic Press, Inc., New York.]

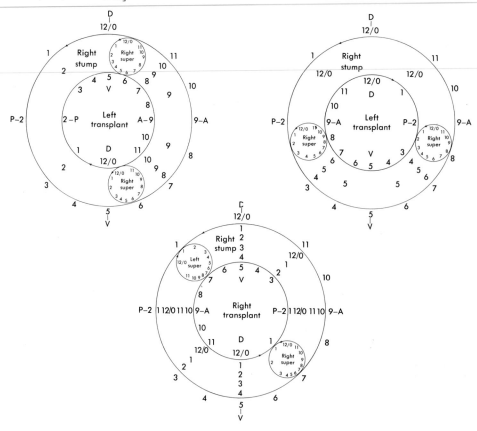

rotational positions of the grafted blastema were used. They showed, however, that the position and handedness of supernumerary regenerates can be predicted by the clock-face model, provided that positional values are assigned asymmetrically, as schematized in Figure 19.33. Their scheme predicts points at which there should exist maximal incongruity of positional information content at the host–graft junction. At these points a complete circle of positional information can be intercalated between host and graft. Regeneration at these points should thus permit distalization of a complete set of terminal limb parts. The scheme and arguments used in applying the polar coordinate model to the occurrence of supernumerary regenerates in the forelimb of *Notophthalmus viridescens* are detailed in the caption of Figure 19.33 and in an important paper of Bryant (1977).

Some others who have investigated the effects of axial rotation of the blastema have disagreed with the results and interpretation of Bryant and Iten (1976) and have argued that the polar coordinate model has but little predictive value. Maden and his collaborators (Maden and Turner, 1978; Maden, 1980b; Maden and Goodwin, 1980) reported that rotation of the axolotl blastema at a variety of angles between 45 and 315 degrees resulted in supernumerary outgrowths. Moreover, in contrast to results of Bryant and Iten (1976), they reported that supernumerary regenerates appeared in no favored quadrant after 180-degree rotation of a blastema on the ipsilateral (same side of origin) stump. In cases in which two regenerates formed, they were seldom positioned oppositely. Wallace and Watson (1979) made similar studies using juvenile axolotls and reported that supernumerary regenerates formed as a result of rotating the blastema 90 or 290 degrees and noted that this would not be predicted by the polar coordinate model. Wallace and Watson suggested that these discordant results indicate that the polar coordinate model for epimorphic regeneration should be abandoned. It should be recognized, however, that the circumferential distribution of positional information in axolotl limbs may be different from that of *Notophthalmus*, and also that it may be differently distributed at different levels along the proximodistal axis. Moreover, the refined rule of distalization permits the possibility that some regeneration of supernumerary limb parts might occur from regions in

which less than a complete set of circumferential positional values is present. These arguments notwithstanding, it should be pointed out that Tank (1978b) found that grafts of blastemata, from regenerating left forelimbs of the axolotl to the stump of the right forelimb, gave supernumerary regenerates that correspond to the Bryant and Iten (1976) model for *Notophthalmus*. The regenerates were not diagnosable in all cases, however.

Finally, we may remark on the fact that the polar coordinate model is a two-dimensional one, lacking the inside–outside component. The amphibian limb, however, is relatively complex, having a central core of bone surrounded by muscle and covered with skin, which consists of dermis overlayed by epidermis. Is it possible that positional information is restricted to only two dimensions in view of this complexity? An affirmative answer to this question is suggested by experiments of Carlson

FIGURE 19.34 Scheme showing the rotation of a cuff of skin of the upper arm (A) in a salamander and subsequent amputation (B) through the region of the graft. The amputation surface (C), thus shows skin and inner tissues oppositely polarized with respect to both anterioposterior and dorsoventral axes.

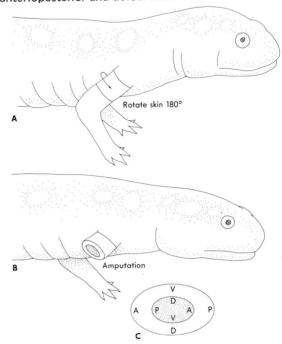

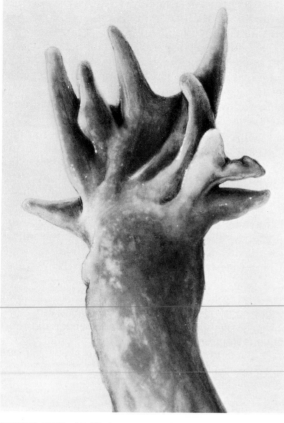

FIGURE 19.35 Multiple regenerate formed by the left limb of an axolotl treated as described in Figure 19.34. [From B. M. Carlson, *Dev Biol* 39:263–85 (1974), by courtesy of Dr. Carlson and permission of Academic Press, Inc., New York.]

(1974), who showed that whereas the reorientation of bone or of epidermis in salamanders does not lead to the regeneration of supernumerary limb parts, the manipulation of muscle and dermis does. It is not unreasonable, therefore, to assume that positional values in amphibian limbs are arranged as a cone (Figure 19.22B) and that interior tissues lack positional information. In fact, a great deal of positional information resides in skin alone, as can be shown by reorienting a portion of skin and amputating through the regions of reoriented skin (Figures 19.34 and 19.35). This procedure usually results in multiple regenerates. Moreover, as shown in Chapter 16, amputation of an x-irradiated limb

through that portion of the limb bearing an unirradiated cuff of skin can lead to normal regeneration. If a hindlimb skin cuff is used in normal orientation to replace an excised skin cuff of an irradiated forelimb, typical hindlimb structures frequently form (Clark, 1978). This gives further evidence that a great deal of positional information is found in the two-dimensional pattern of skin.

19.3.2.2 Epimorphic Regulation of Pattern in the Imaginal Wing Disc of *Drosophila*.

The imaginal wing disc of *Drosophila* is clearly a two-dimensional structure from which a three-dimensional wing arises. Its ability to regenerate has been extensively exploited in experiments the results of which are largely responsible for the original formulation of the polar coordinate model of epimorphic regeneration (French et al., 1976). Recall

FIGURE 19.36 The imaginal wing disc of *Drosophila melanogaster.* [From a photograph kindly supplied by Dr. P. J. Bryant, whose courtesy is gratefully acknowledged.]

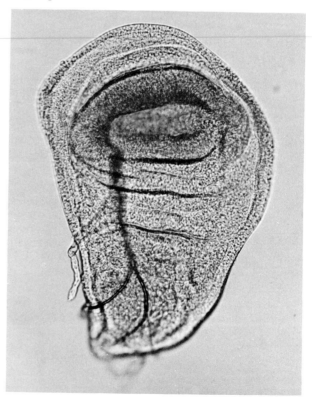

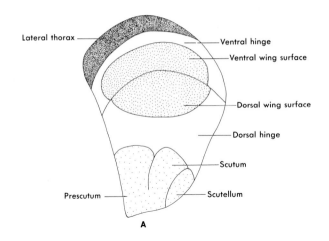

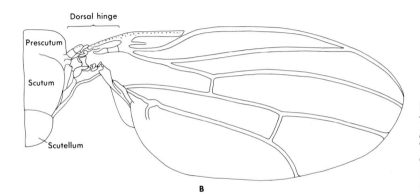

FIGURE 19.37 A: Simplified fate map for the right imaginal wing disc of *Drosophila*. B: Principal parts of the right wing and adjacent thoracic structures of *Drosophila* as seen in dorsal view. [Drawings simplified from various papers of P. J. Bryant.]

from Chapter 17 that the imaginal disc of the wing and other appendages is part of a hollow epithelial sac formed by invagination of the larval epidermis and attached to it by a stalk. One part of the sac, the peripodial membrane, is made up of squamous cells that largely disintegrate during metamorphosis. The base of the sac is a columnar epithelium that enlarges by cell division and is thrown into folds such that the future integumentary epithelium of the appendage consists of parts arranged in telescoped fashion (Figure 19.36). At metamorphosis, under the influence of ecdysone, the epithelium undergoes eversion, a process that converts the folded plate of cells into the extended epithelium of the adult appendages. The epithelium then secretes the cuticle with its typical sculpturing and its adornment of bristles and sensory hairs.

When an imaginal disc is extirpated from a larva during its last instar and transplanted into the body cavity of a host that is undergoing metamorphosis, it will cease growing and will undergo metamorphosis simultaneously with the host. When a disc is dissected into fragments by cuts along known coordinates and the fragments are implanted as described into separate hosts, each forms a discrete part of the pattern of the whole adult appendage, with little or no overlapping or duplication of parts. The structure recovered after metamorphosis can then be used to construct a fate map of the imaginal disc. Such a fate map for the imaginal wing disc of *Drosophila* is shown in Figure 19.37. By referring to the fate map, one can precisely locate the future margins of the wing, the integument of its upper and lower surfaces, the attachments of wing to body wall, the pattern of wing veins, the position of future bristles, and so on.

Fragments of the wing disc thus behave as parts of a mosaic pattern when subjected to immediate metamorphosis. If, however, such fragments are given an opportunity to grow in the body cavity of an adult host (which does not provide the metamorphic hormones), they then show both regenerative and regulative behavior that is demonstrable by retransplanting them subsequently into a metamorphosing host. Quite characteristically, when various sectors of the wing disc are excised, the larger sector tends to regenerate the missing parts of the expected pattern and the smaller makes the expected mapped part plus its mirror-twin duplicate, at least to some degree. The origin of other patterns is amenable to interpretation by the short-

est intercalation rule and can be illustrated most readily by the behavior of isolated 90- and 270-degree sectors of the wing disc (Figure 19.38). Isolates growing in the adult body cavity bring their cut edges into approximation, thus allowing for the confrontation of tissues of different positional values. The shortest intercalation between confronted values calls for regeneration of a complete wing pattern from a 270-degree sector and duplication of the pattern normally formed from the 90-degree sector. The patterns formed by these isolates when retransplanted to a metamorphosing larva verifies these expectations. Of special interest in the application of the polar coordinate model is the case in which a central sector of the disc is removed and

FIGURE 19.38 Application of the polar coordinate model to predict the morphogenetic performance of fragments of the imaginal wing disc of *Drosophila*. A: The position of cuts producing 90° and 270° isolates of the wing disc shown in outline. The model predicts that the 90° segment, D, will duplicate its pattern and that the 270° segment, R, will regenerate missing parts of the pattern. Note that any part of the adult wing pattern can be recognized by virtue of the distribution of integumentary bristles and of supporting structures. B: Excision of a 90° segment from the clock face. C: Approximation of cut edges and intercalation of missing positional values by the 90° isolate. D: Approximation of cut edges and regeneration of missing positional values by the 270° isolate. [Adapted from various sources, including P. J. Bryant et al., *Sci Am*, July 1977, pp. 66–81.]

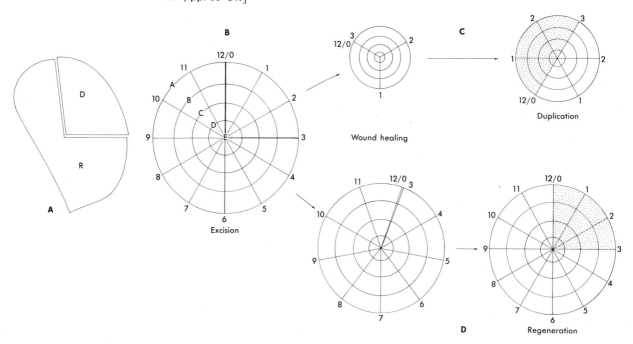

cultured separately from the periphery. In this experiment both isolates have a complete set of circumferential positional information. According to the rules, therefore, each should undergo distal transformation. Retransplantation to a metamorphosing host shows that the peripheral sector tends to regenerate the pattern of a wing tip, which would normally form from the central sector, and thus forms a complete wing. The central sector makes a wing tip, as expected on the basis of the fate map, but it also regenerates the distal wing parts in mirror-image symmetry to itself, which is expected, based on the complete circle rule for distal transformation. These and other examples of the application of the polar coordinate model to regeneration of the wing disc in *Drosophila* were described and schematized most lucidly by P. J. Bryant et al. in 1977.

19.3.3 PATTERN FORMATION AND REGULATION IN THE DEVELOPMENT OF THE VERTEBRATE LIMB.

Another model based on the concept of positional information has enjoyed a considerable vogue as applied to the development and regulation of morphogenetic patterns in the vertebrate limb, especially in the wing of the chick embryo (Figure 19.39). Wolpert and his co-workers (Wolpert, 1969; Tickle et al., 1975) have proposed that positional information is specified with reference to two orthogonal axes, the proximodistal and anteroposterior ones (the model does not consider the dorsoventral axis). For the specification of limb structures along the proximodistal axis, it has been proposed

that the apical ridge of ectoderm (AER), which rims the apex of the limb bud (Figure 14.4), exercises an action on the subjacent mesoderm that marks out an apical "progress zone." As the limb bud elongates, cells in the progress zone autonomously change their positional values to those of more distal levels, while they remain under the influence of the AER. Overflowing the progress zone proximally by virtue of proliferation, cells then become stabilized in the positional values they had achieved. Thus, cells first emerging from the progress zone would have proximal positional values and later emerging ones would have more distal values (Figure 19.40).

The mechanism proposed for the specification of limb structures along the anteroposterior axis is that there is a localized zone at the posterior junction of the limb bud and body wall (Figure 19.41), the zone of polarizing activity (ZPA) (Saunders and Gasseling, 1968), which is the source of a diffusible morphogen. The hypothetical morphogen is broken down as it moves away from its source, thus resulting in an exponential decrease in its concentration from posterior to anterior. A cell's positional value along the anteroposterior coordinate would thus depend on its distance from the ZPA.

The model is attractive, especially with respect to the origin of positional values along the proximodistal axis, for it requires only that cells autonomously determine their positional value in proximodistal sequence as long as an AER is present. When the AER is experimentally removed or replaced by other ectoderm, the limb lacks its terminal parts, as shown by Saunders (1948). The AER does not provide positional information with respect to the proximodistal axis, for the exchange of AERs between limb buds of different developmental stages does not affect the generation of normal proximodistal positional values in the mesoderm (Rubin and Saunders, 1972). This model permits differentiation of terminal limb parts in the absence of continued input from the proximal part of the limb bud. Experimentally it is shown that the apex of a limb bud, isolated from its stump, forms a normal limb tip when grafted elsewhere on the embryo.

The postulated source of positional information with respect to the anteroposterior axis, the ZPA, was suggested to Wolpert and his colleagues by experiments of Saunders and Gasseling (1968). These

505

FIGURE 19.39 The wing skeletal pattern in the chick. Digits are numbered according to the usual convention, which omits numbers I and V. H, humerus; M_2, M_3, and M_4, metacarpals; R, radius; RC, radial carpal; U, ulna; UC, ulnar carpal.

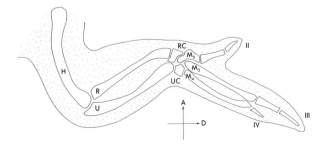

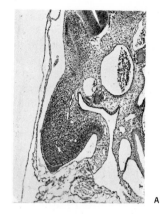
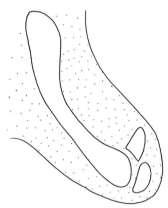

A

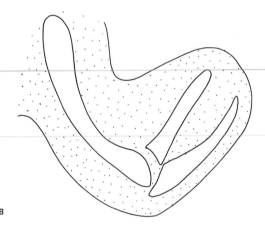

B

FIGURE 19.40 Effects of excising
the apical ectodermal ridge of the
chick wing bud on the differentiation of terminal wing parts. A–C
(left): Sections parallel to the long
axis of the wing bud at successively later stages of development.
On the right are depicted the skeletal deficiencies resulting from
removal of the apical ectodermal
thickening at each stage (cf. Figure 19.39). According to the thesis
of Wolpert and his colleagues, as
described in the text, the longer
cells are present in a "progress
zone," which is determined by the
apical ectodermal ridge, the more
distal their positional value
becomes.

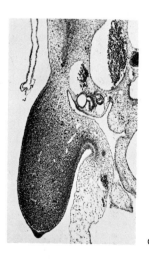
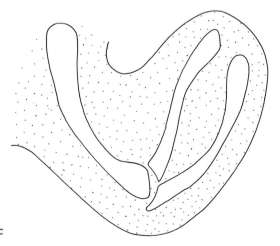

C

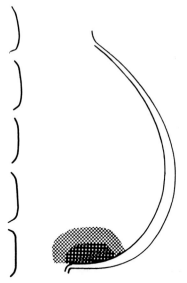

FIGURE 19.41 Approximate distribution of the putative zone of polarizing activity in an embryonic chick limb bud at about 3½ days of incubation. The darker stippling indicates the region of highest activity as originally tested. [From J. W. Saunders and M. T. Gasseling in R. Fleischmajer and R. P. Billingham (eds), *Epithelial-Mesenchymal Interactions,* © 1968, The Williams and Wilkins Co. Baltimore.]

investigators reported that when a notch of tissue is removed from the rim of the early right wing bud at its anterior junction with the body wall and replaced by a similar bit of tissue from the posterior edge of a donor limb at its junction with the body wall, the anterior side of the host limb bud gave rise to a supernumerary limb whose posterior edge faced the graft (Figure 19.42B), and was thus of left asymmetry. A similar implant placed at the apex of the host wing bud likewise caused a supernumerary limb to form from anterior limb-bud tissue. It also developed with its posterior side facing the graft, and was thus of right asymmetry (Figure 19.42A). The power of inducing and polarizing a supernumerary limb was first reported as being confined to the ZPA of the wing bud and to the homologous region of the leg bud. It is a mesodermal property. Similar results were reported for the developing forelimb of *Xenopus* by Cameron and Fallon (1977) and, as noted in Section 19.3.1.1, Slack (1977a) found that flank tissue grafted to the anterior edge

of the forelimb of *Ambystoma mexicanum* induces the formation of twin posterior limbs.

In Wolpert's laboratory Tickle et al. (1975) tested the morphogenetic affects of varying the position of the grafted ZPA along the margin of the early limb bud from anterior to posterior. The patterns of supernumerary digits that resulted changed correspondingly in such a manner as to be compatible with the assumption that digits IV, III, and II of the wing differentiate in response to thresholds of morphogen diminishing in that order. Figure 19.43 illustrates the distribution of the putative morphogen in a normal limb bud and in one to which an additional ZPA was added. In the normal limb the origin of digits is principally from the posterior half of the limb bud, although digit II receives preaxial contributions. As the figure shows, the concentration of presumed morphogen is high near the ZPA, above the threshold for signaling the position for digit IV. Further anterior, the concentration falls to a level below the thresholds successively for digits IV, III, and II, and correspondingly, digits arise in the anterioposterior order II, III, and IV. Addition of another source of morphogen anterior to the normal source

507

FIGURE 19.42 Effects of grafting tissue from the posterior side of the wing bud to the apex (A) and to the anterior side (B) of the bud. In A, one sees the formation of a supernumerary limb tip of right asymmetry from the preaxial tissue of the bud. In B, one sees the formation of a limb tip of left asymmetry preaxially. Numbering of the digits is as described in Figure 19.39 [Redrawn from various original papers of the author.]

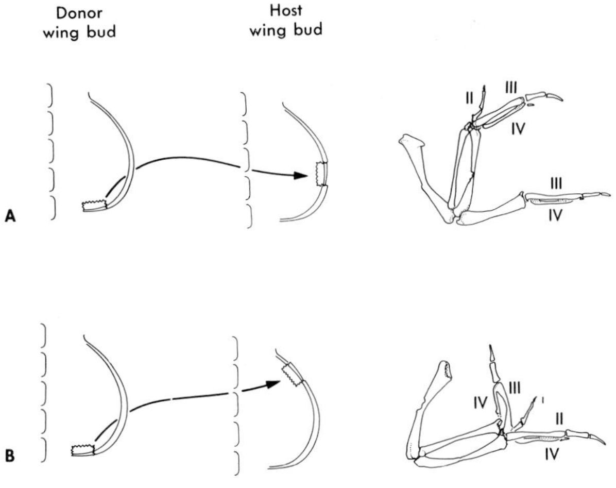

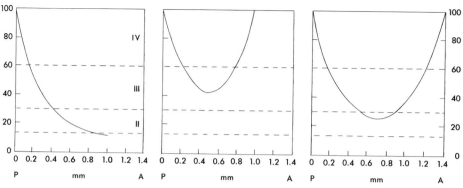

FIGURE 19.43 The concentration profile for a presumptive morphogen produced by the ZPA and affecting the differentiation of positional values along the anterioposterior axis of the chick limb bud is shown for a normal bud on the left. The concentration of morphogen at its source, posterior, P, is fixed on the left at 100, and, for this hypothetical case, the rate of diffusion and breakdown is fixed such that the value of the morphogen at 1 mm from its source is 10. (1 mm is approximately the length of the limb bud along its anteroposterior axis at its junction with the body wall.) Horizontal lines in the model suggest threshold values of the morphogen required for the three wing digits, IV, III, and II, in descending order. In the central scheme, the hypothetical concentration of morphogen is plotted for the case in which an additional ZPA is grafted to the wing bud at its anterior junction with the body wall. If the assumption is made that the length of the anterioposterior axis of the limb is not altered by the operation, the resulting distribution of morphogen would suggest that digits IV, III, III and IV might be formed in anteroposterior order without an intervening digit II. In fact, however, the presence of an anterior ZPA implant brings about an increase in the anterioposterior dimension of the distal region of the wing bud. As suggested by the scheme at the right, this change would effectively increase the distance between the two sources of morphogen and thus permit the concentration of morphogen to fall below the threshold for forming digit III. Therefore one or more digits II might arise, resulting in anteroposterior digital sequences of IV, III, II, III, IV or IV, III, II, II, III, IV, both of which have actually been obtained by investigators in several laboratories. [Adapted from C. Tickle et al., *Nature* 254:199–202 (1975).]

508

would give varying levels of morphogen, corresponding to different threshold patterns. Thus, when an extra ZPA is added preaxially, two complete limb tips could be formed showing, from anterior to posterior, the digital sequence IV, III, II, II, III, IV, corresponding to a mirror-twinned wing tip.

A final aspect of the model requires that only cells in the progress zone can respond to the morphogen from the ZPA. The positional values of limb parts are thus specified in both anteroposterior and proximodistal axes as they emerge from the progress zone. This conclusion is experimentally supported by the fact that the effects of grafting an additional ZPA are restricted progressively to more terminal parts the later the developmental stage of the limb bud at the time of operation.

This model is also compatible with the notion that signaling mechanisms involved in the generation of morphogenetic patterns may be similar in the development of diverse groups of organisms. The effect of adding an additional ZPA of chick origin to a chick wing bud is effectively imitated by isolates homologous to the ZPA originating from anterior or posterior forelimbs of mouse, quail, duck, hamster, ferret, human, painted turtle, and snapping turtle (Tickle et al., 1976; Fallon and Crosby, 1977; Saunders, 1977). With respect to the differentiation of the proximodistal axis, the AER of the chick may be substituted by that of the duck (Hampé, 1957; Saunders and Fallon, 1966; Pautou, 1968), quail (Saunders et al., 1976), mouse (Cairns, 1959, 1965), and rat (Jorquera and Pugin, 1971).

The foregoing model has stimulated a good deal of experimental work. Many of the results are highly compatible with it, as evidence presented herein suggests. Other evidence, however, is difficult to reconcile with it. An important aspect of the model is its implication that a cell, having emerged proximally from the progress zone, has its fate fixed with respect to its positional value along the proximodistal axis. Thus it would be expected that no regulation of morphological pattern would follow upon the removal or addition of presumptive limb segments along the proximodistal axis. It has been shown, however (Kieny and Pautou, 1976; Kieny, 1977), that limb buds can regulate for intercalary defects proximal to the progress zone. Regulation for intercalary excesses can also occur, but it is less complete (Kieny and Pautou, 1976, 1977; Kieny, 1977).

With respect to the ZPA as a source of a morphogen that determines positional value, the situation is now quite confusing. Saunders and his colleagues (Saunders, 1977) showed that tissues from a variety of nonlimb sources implanted at the anterior border of the wing bud and body wall can elicit the formation of supernumerary limbs oriented with posterior side toward the graft. Fragments of chick somites, flank tissue, tail bud, mesonephros, and even the turtle notochord, exercise the effect when grafted in intimate contact with both mesoderm and AER at the anterior margin of the chick wing bud (Figure 19.44). In the limb bud only the anterior tissue lacks the capacity. These considerations, together with the facts that normal limbs can result from the removal of the originally designated ZPA (Fallon and Crosby, 1975) and that no morphological or biochemical marker of ZPA cells has been found, suggest that the ZPA does not exist as a specialized entity, nor do its counterparts in the embryonic limb of *Xenopus* or the flank of *Ambystoma*.

Unfortunately, the idea that a morphogen diffusing from the ZPA determines positional values with respect to the anteroposterior axis of the limb bud has proved so attractive (See Summerbells, 1979, paper and reference list) that few investigators have tried to design experiments that give a critical test of the thesis. One exception is found in the case of Iten and her students, whose experimental results cast considerable doubt that the ZPA plays the role popularly attributed to it. Iten and

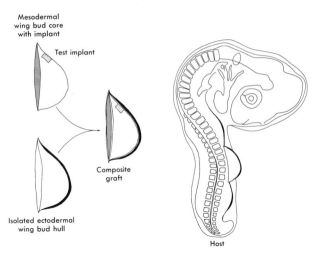

FIGURE 19.44 Scheme for testing the ability of various nonlimb tissues to induce supernumerary limb outgrowth from a single limb bud. An isolated limb bud is stripped of its ectoderm by means of a cation-chelating agent (the sodium salt of ethylenediaminetetraacetic acid). The test implant is then inserted into a notch cut in the anterior edge of the denuded mesoderm (upper left). The combined tissues are then encased in an ectodermal hull prepared by trypsin treatment of a donor wing bud. The whole is then grafted to the flank or to the dorsal side of a host chick embryo and allowed to develop.

509

Murphy (1980) made a slit in the ZPA region of the wing bud in chick embryos of about $3\frac{1}{2}$ days of incubation. The slit was oriented radially with respect to the curvature of the AER and passed deep toward the body wall (Figure 19.45). Next a pie-shaped wedge of anterior tissue was excised from one or another location on the anterior rim of a donor wing bud. This wedge consisted of the AER, dorsal and ventral ectoderm, and the full thickness of the mesoderm. The donor wedge was then inserted into the posterior slit, forcing the cut edges apart. The AERs of host and implant were aligned and development was allowed to proceed until all wing cartilages were formed. Presuming that the ZPA region is the source of a morphogen, the cells of the grafted wedge would now be exposed to a high concentration of it and would become respecified to form extra skeletal elements. This scenario, indeed, is what occurred. But importantly, the frequency with which supernumerary parts formed depended

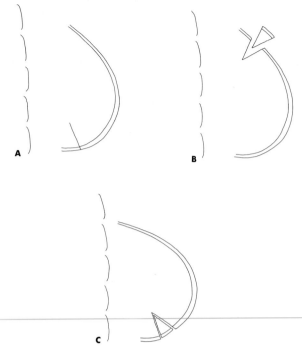

510   FIGURE 19.45 Scheme of the operation used by Iten and her collaborators to insert a wedge of wing bud tissue into the region of the presumed zone of polarizing activity. A: A slit is made in the posterior margin of a prospective host wing bud. B: A pie-shaped wedge of tissue is excised from the anterior margin of a donor wing bud. C: The wedge is inserted into the slit made in the host wing bud.

on the precise source of the wedge. Wedges from the anteriormost level of the donor wing bud formed more supernumerary limb parts and with greater frequency than did wedges of anterior tissue taken from the region immediately anterior to the apex of the wing. Yet, presumably, each type of implant was exposed to the same concentration of putative morphogen.

In another experiment, Javois and Iten (1981) made similar slits in the ZPA region of host embryos and inserted wedges of tissue from the ZPA region of donor embryos. Because no tissue was removed from the host, the addition of ZPA material should have created a more potent source of morphogens, resulting in the formation of additional digits, but this was not the case. Hosts to such implants formed completely normal limbs.

The results of Iten and her collaborators are more compatible with a model in which local cell-to-cell interactions, rather than diffusible morphogens, control pattern formation. A particularly persuasive argument that this is the case is provided by experiments in which a wedge of tissue from the anterior rim of the wing bud is inserted into a slit made in a host wing bud at the same level. If the implant is normally oriented with respect to the dorsoventral axis, the limb develops normally, but if the graft is implanted with the dorsoventral axis reversed, a supernumerary humerus often forms (Iten and Murphy, 1980).

It is possible that certain aspects of the polar coordinate model for epimorphic regeneration are applicable to the embryonic development of the chick limb. Recall that when the imaginal wing disc of *Drosophila* is divided, one portion tends to regenerate the complete pattern of the wing and the other tends to duplicate its pattern in mirror-twin fashion. Iten (1977) analyzed the regulative capacities of complementary anterior and posterior portions of the early limb bud. She found that, depending on the position of the isolating cut, one member of the complementary pair appears to regulate an essentially complete and normally polarized limb, whereas the other member tends to form a bilaterally symmetrical structure.

It remains to be seen to what extent the polar coordinate model can be used to predict morphogenetic patterns in the chick wing bud. One complication, not found in the newt, is that the ectoderm apparently plays a more significant role in limb outgrowth and pattern formation in chick limbs than it does in urodeles. Terminal limb parts can form from randomized limb mesodermal cells covered with an intact limb bud ectoderm (MacCabe et al., 1973). Moreover, such outgrowths show a pattern of dorsoventral polarity corresponding to that of the ectoderm. Similarly, terminal limb parts consisting of intact limb-bud mesodermal cores encased in limb-bud ectoderm having reversed dorsoventral polarity, form their terminal parts in reversed handedness with respect to the origin of the mesoderm (MacCabe et al., 1974; Pautou, 1977).

Some years ago Turing (1952) pointed out that the coupling of chemical reactions to diffusion can lead to stable heterogeneous patterns of chemical concentration. Inspired by the Turing hypothesis, Newman and Frisch (1979) constructed a model for the generation of the pattern of the skeletal elements of the chick wing. They pointed out that a molecule that encourages cell-to-cell contact could be responsible for the initiation of foci of chrondrogenesis; they then created a mathematical model for generating the pattern of wing cartilages. Their molecule M is considered to be loosely bound to the surfaces of all wing bud cells, all of which produce it. It is presumed to be necessary in order for cells of the cartilage lineage to form cartilage. Wherever a critical concentration of M is reached, a precartilaginous condensation appears. M is considered to be kept at a high concentration at the interface of the AER and subjacent mesenchyme by virtue of being produced rapidly in that region. The concentration of M throughout the limb bud is subject to change by virtue of its synthesis and breakdown by the cells and by its diffusion in the intercellular space. A spatial distribution of M is thereby created wherein the concentration of M exceeds the threshold for triggering cartilage differentiation only at certain points.

The process of patterning is presumed to occur in a diffusion chamber in the limb bud. This chamber consists of an extracellular matrix of dilute hyaluronate gel. The chamber enlarges through the replication of mesenchyme and contracts through the recruitment of cells into premuscle and precartilage. The latter processes occur at the proximal end of the chamber. The diffusion chamber is presumed to involve the entire volume of the limb bud when the bud first develops, but it is ultimately confined to the apex, where the cells are under the influence of the AER.

Newman and Frisch devised equations which, under the assumption of certain reasonable boundary conditions, admit of solutions that are standing waves of concentrations of M along the anterioposterior and dorsoventral axes of the model limb bud. The equations are rather complex in their derivation and are not presented here. They do, however, enable one to predict the number of chondrogenic foci and to show that their pattern changes discontinuously in a way that depends on the proximodistal length of the diffusion chamber. The shorter the length of the diffusion chamber, the greater will be the number of parallel cartilaginous elements. One of the presumptions is that contraction of the region subjacent to the AER occurs in abrupt jumps, which is suggested by the morphogenetic effects of

excising the ridge in different limbs at closely spaced intervals (Summerbell, 1974). For example, the time required for the cartilages of the wrist region to become capable of differentiating upon removal of the AER is greater than that required for the entire limb proximal to the wrist. The observation would suggest an abrupt contraction of the subridge diffusion chamber after the forearm cartilages are determined. If such abrupt changes do occur, then the equations enable one to predict the origin of humerus, forearm, and hand, in that order.

Newman and Frisch suggested that a suitable candidate for M is **fibronectin,** which is a glycoprotein with a molecular weight of about 250,000. It is ubiquitously distributed on cell surfaces, in extracellular spaces, and in blood plasma. Tomasek (1980) working in Newman's laboratory showed that fibronectin is heavily concentrated in the subridge space at the apex of the limb bud. Also it appears in the extracellular space of the precartilaginous condensations, but it disappears from that location when chondrogenesis begins. Fibronectin inhibits the formation of cartilage by prospective chrondrocytes in vitro, but, from Tomasek's morphological evidence, it seems to be involved in the formation of cell-to-cell contacts that characterize the precartilaginous condensations.

511

19.4 Concluding Remarks

This chapter has dealt with but a few aspects of the major problem of how the genetic information in individual cells is expressed in the supracellular domain of morphogenetic pattern formation. We have examined a few of the developmental strategies that might underlie pattern formation and have neglected many that deserve our attention. Among the latter are the studies of developmental compartmentalization in clonal lines within insect imaginal discs (Garcia-Bellido et al., 1973) and possible epigenetic switching mechanisms that control future pathways of development in the insect blastoderm (Kaufmann et al., 1978).

In this chapter we have ignored the molecular domain that deals with the way in which positional information is interpreted in development. This issue has been ignored because we know essentially nothing about it. The examples that we have examined in this chapter, however, should be sufficient to set the challenge for tomorrow to the student of today. The challenge is to learn how to ask questions of the developing system in such a way as to reveal the mechanisms whereby progressively higher orders of integration emerge from systems at lower orders. What indeed is the nature of positional information in a developing system and how is it read by the genome?

19.5   Summary

The epigenetics of morphogenetic patterns may be analyzed at all organizational levels from the assembly of orderly arrays of macromolecules to the emergence of form and function in organs and organ systems. The assembly of microtubules from preexisting subunits and of collagen fibrils from tropocollagen illustrate the necessary and automatic creation of morphological order from disorder under defined conditions.

It is much more difficult to account for the geometry that is imposed on cells and their products as they are integrated into tissues and organs. This difficulty is well illustrated by the problems attending the organization of feather tracts in fowl. Each feather is structurally and physiologically unique, and its characteristics and topological relationships are prepatterned in the early embryo several days before the feather germs actually appear. Some regulation of the tract pattern after surgical interference is possible, however, as the feather germs begin to emerge.

The nature of positional information specifying the arrangement of feathers in tracts in unknown. Transplantation experiments indicate that this information resides initially in the future dermis, but is subsequently shared with the ectoderm. Feather germs first appear as localized thickenings of ectoderm covering dermal condensations. These thickenings enlarge to form feather papillae. These papillae emerge in characteristic rows, an initial row being followed by formation of secondary and tertiary rows on either side, usually parallel to the long

axis of the embryo. The feather germs appear in a pattern of hexagonal packing. There is disagreement concerning the factors determining this origin of sequential rows and the mechanisms whereby the initial mesenchymal condensations arise.

During recent years a number of formal models have been devised in the attempt to establish conceptual frameworks within which the development and regeneration of spatial patterns may be considered. Basic to several of these models is the concept of positional information, which refers to the mechanisms operating within an embryonic field whereby each cell has its position specified with respect to the boundaries of the field and to the cells around it. The positional information that a cell receives determines its position in the three coordinates of space and its future molecular differentiation. In response to the experimental alteration of field boundary conditions in both developmental and regenerative fields, cells readjust their positional value and hence their subsequent differentiation. Conceivably, also, positional information might have an element of universality in the sense that the same mechanisms of positional signaling are employed by different systems in the same or different organisms. Cells would respond to this information in accordance with their genetic endowment and prior developmental history.

The concept that cells perceive and adjust to changes in positional information has been applied to the analysis of morphogenetic patterns that emerge in epimorphic fields of regeneration. From this analysis has emerged a model or set of rules, based on a system of polar coordinates, that is moderately successful in predicting patterns of normal and abnormal regeneration, particularly in surgically insulted amphibian limbs and in imaginal discs of insects. This model is often referred to as the clock-face model. The circumference of the clock represents the base of the regenerating limb or circumference of an embryonic field, such as an imaginal disc, and the center of the circular face represents the distal end of the appendage. Each cell in the field has a unique positional value, which is defined by two components: its circumferential position with respect to coordinates defined by numbers arbitrarily spaced around the clock face, and its proximodistal level as defined by concentric circles between the rim of the "clock" and its center. The chief rules appreciable to this model, espe-

512

cially as applied to the regeneration of limbs and the imaginal discs of appendages, are that (1) when cells having a different content of positional information are brought together either by deletion or grafting experiments, growth is stimulated and the absent positional information between the apposed cells is intercalated by the shortest route (rule of shortest intercalation); and (2) after amputation intercalation of positional information and distal growth proceed to a degree dependent on the number of positional values present at the cut surface and the number of distal values to be replaced (rule of distalization).

These rules have been applied with some success to the occurrence of regeneration of limb parts and of supernumerary limb parts under a variety of conditions in urodele amphibians. Conflicting results from different laboratories, however, suggest that the complete circle rule for distal transformation may require considerable modification. Likewise, the rules predict with some accuracy the regenerative and regulative behavior of fragments of imaginal wing discs in *Drosophila.* Exceptions to the rules are seen with sufficient frequency as to suggest that they probably require refinement or replacement in these cases, also.

A different kind of model has been proposed to interpret the normal and regulative behavior of the embryonic chick limb with respect to its proximodistal and anteroposterior axes. The ectodermal ridge (AER), which rims the apex of the limb bud, presumably marks out an apical progress zone in which cells autonomously change their positional value to successively more distal levels. Overflowing the progress zone proximally by virtue of proliferation, they are stabilized in the positional value previously achieved. For the specification of positional value along the anteroposterior axis, a putative zone of polarizing activity (ZPA), located at the posterior margin of limb and body wall, provides a source of diffusible morphogen to which limb elements in anteroposterior sequence respond to different thresholds. Grafting an additional ZPA anteriorly causes the anteroposterior pattern of the limb bud to be duplicated in a pattern of mirror symmetry. This model suffers from the fact that cells that have overflowed the progress zone should be incapable of regulation, and yet they seem to do so under some experimental conditions. As far as the ZPA is concerned, it appears, on the basis of certain kinds

of tests, that it is the source of universal morphogen in organisms from the three highest classes of vertebrates. Its activity, however, is a property shown by a variety of nonlimb tissues that are grafted to the limb bud under slightly modified conditions.

Some aspects of regulative behavior after surgical interference in development of the chick limb bud are interpretable on the basis of the polar coordinate model. Complementry anterior and posterior portions of the limb bud tend to either form a relatively complete limb, normally polarized with respect to the anteroposterior axis, or to form an incomplete but bilaterally symmetrical limb structure, as is the case for the imaginal wing disc in *Drosophila.* In the case of the chick wing bud, the problem of applying models derived from other systems is complicated by the fact that control of the dorsoventral axis is, at least partially, controlled by the ectoderm, whereas the other axes are determined by mesodermal mechanisms.

Another model treats the apex of the chick limb bud as a diffusion chamber. Mathematical equations have been devised which, under assumed boundary conditions, admit of solutions that are standing waves of concentration of a molecule such as fibronectin. The spatial distribution of this molecule is presumed to determine patterns of prechondrogenic foci.

513

19.6 Questions for Thought and Review

1. The term self-assembly is applied to the spontaneous generation of three-dimensional macromolecular structures from their subunits. Might not the same term be applied to the construction of tissues and organs? Document your answer. Reference to Chapter 15 will be helpful in this regard.

2. Outline the roles of nucleation centers and of high molecular weight proteins in the assembly of microtubules.

3. Systems at higher levels of organizational complexity emerge automatically and necessarily from systems at lower levels of complexity under appropriate conditions. Discuss the applicability of this proposition with respect to problems of embryonic development.

4. Feathers are integumentary derivatives that occur in

precise geometric arrays and are distinctive from one body region to the next and from feather to feather within each array or tract. In what respects does the development of a feather tract illustrate the principle of progressiveness in development (cf. Chapter 13)?

5. What are the respective roles of ectoderm and mesoderm in the emergence of feather tract patterns and at what stages are each of the components capable of dominant roles?

6. Recent years have witnessed the attempt to formulate the embryonic field with respect to systems of positional information. What are the salient features of the new formulation?

7. Apply the concept of positional information to the generation of the "French flag" pattern. Use the flag analogy to suggest how different morphological patterns may arise based on the same sets of positional information.

8. The polar coordinate model, or clock-face model, has been used to predict the morphogenetic patterns produced by regenerative growth in a variety of epimorphic fields. What is the form of this model, that is, what sets of positional information are involved in determining the positional value of cells in the field?

9. With reference to the clock-face model, define and illustrate the rule of shortest intercalation using examples of the intercalation of positional information in both the circumferential and proximodistal senses. Use examples involving regeneration of newt and cockroach limbs, and the imaginal wing disc of *Drosophila*.

10. Likewise, with reference to the above model, define and illustrate the rule for distalization. Apply this rule to the case of regeneration of the limb of the newt and to the formation of supernumerary regenerates.

11. Describe models that have been used to predict the regulative behavior of embryonic vertebrate limbs after experimental disturbance. Address evidence in favor of and contradictory to these models.

12. Of what significance to the understanding of epigenesis is the design and testing of formal models of normal, regulative, and regenerative development?

19.7   Suggestions for Further Reading

BERNS, M. W., and S. M. RICHARDSON. 1977. Continuation of mitosis after selective microbeam distruction of the centriolar region. *J Cell Biol* 75:977–82.

BOHN, H. 1975. Tissue interactions in the regenerating cockroach leg. In P. A. Lawrence, ed. *Insect Development*. Oxford: Blackwell, pp. 170–85.

BRYANT, P. J. 1975. Regeneration and duplication in imaginal discs. *Cell Patterning*, Ciba Foundation Symposium 29. Amsterdam: ASP, pp. 71–93.

GEHRING, W. J. 1976. Developmental genetics of *Drosophila*. *Annu Rev Genet* 10:209–52.

GIERER, A., and H. MEINHARDT. 1972. A theory of biological pattern formation. *Kybernetik* 12:30–39.

GOODWIN, B. C., and M. H. COHEN. 1969. A phase-shift model for the spatial and temporal organization of developing systems. *J Theor Biol* 25:49–107.

MADEN, M. 1977. The regeneration of positional information in the amphibian limb. *J Theor Biol* 69:735–53.

MOSCONA, A. A., and B. B. GARBER. 1968. Reconstruction of skin from single cells and integumented differentiation in cell aggregates. In R. Fleischmajer and R. P. Billingham, eds. *Epithelial–Mesinchymal Interactions*. Baltimore: Williams and Wilkins, pp. 230–43.

MACWILLIAMS, H. K., and S. PAPGEORGIOU. 1978. A model of gradient interpretation based on morphogen binding. *J Theor Biol* 72:385–411.

PRIGOGINE, I., and G. NICOLIS. 1971. Biological order, structure and instabilities. *Q Rev Biophys* 4:107–48.

SENGEL, P. 1971. Organogenesis and arrangement of cutaneous appendages in birds. *Adv Morphogenesis* 9:181–230.

SLACK, J. M. W. 1978. Determination of polarity in the amphibian limb. *Nature* 261:44–46.

SPOONER, B. S. 1975. Microfilaments, microtubules and extracellular materials in morphogenesis. *Bioscience* 25:440–51.

STOCUM, D. L. 1978. Organization of the morphogenetic field in regenerating amphibian limbs. *Am Zool* 18:883–96.

19.8   References

BOHN, H. 1976. Regeneration of proximal tissues from a more distal regeneration level in the insect leg (*Blaberus cranifer, Blatteridae*). *Dev Biol* 53:285–93.

BRYAN, R. N., G. A. CUTTER, and M. HAYASHI. 1978. Separate mRNAs code for tubulin subunits. *Nature* 272:81–83.

BRYANT, P. J., S. V. BRYANT, and V. FRENCH. 1977. Biological regeneration and pattern formation. *Sci Am* (July) 66–81.

BRYANT, S. V. 1976. Regenerative failure of double half limbs in *Notophthalmus viridescens*. *Nature* 263:676–79.

BRYANT, S. V. 1977. Pattern regulation in amphibian

limbs. In D. A. Ede, J. R. Hinchliffe and M. Balls, eds. *Vertebrate Limb and Somite Morphogenesis.* Cambridge: Cambridge University Press, pp. 312–27.

BRYANT, S. V., and B. A. BACA. 1978. Regenerative ability of double-half and half upper arms in the newt, *Notophthalmus viridescens. J Exp Zool* 204:307–24.

BRYANT, S. V., V. FRENCH, and P. J. BRYANT. 1981. Distal regeneration and symmetry. *Science* 212:993–1002.

BRYANT, S. V., and L. E. ITEN. 1976. Supernumerary limbs in amphibians: experimental production in *Notophthalmus viridescens* and a new interpretation of their formation. *Dev Biol* 50:212–34.

CAIRNS, J. M. 1959. Evidence for the operation of a growth control system in the limb bud of the mouse embryo similar to that of the chick embryo. *Anat Rec* 134:543 (abstr.).

CAIRNS, J. M. 1965. Development of grafts from mouse embryos to the wing bud of the chick embryo. *Dev Biol* 12:36–52.

CAIRNS, J. M., and J. W. SAUNDERS. 1954. The influence of embryonic mesoderm on the regional specification of epidermal derivatives in the chick. *J Exp Zool* 127:221–48.

CAMERON, JoAnn, and J. F. FALLON. 1977. Evidence for polarizing zone in the limb buds of *Xenopus laevis. Dev Biol* 55:320–30.

CARLSON, B. M. 1974. Morphogenetic interactions between rotated skin cuffs and underlying stump tissues in regenerating axolotl forelimbs. *Dev Biol* 39:263–85.

CLARKE, B. J. 1978. Restoration of regenerative ability of x-rayed newt limbs after grafting proximal or distal skin. *Growth* 42:275–95.

FALLON, J. F., and G. M. CROSBY. 1975. Normal development of the chick wing following removal of the polarizing zone. *J Exp Zool* 193:449–55.

FALLON, J. F., and G. M. CROSBY. 1977. Polarizing zone activity in limb buds of amniotes. In D. A. Ede, J. R. Hinchliffe, and M. Balls, eds. *Vertebrate Limb and Somite Morphogenesis.* Cambridge: Cambridge University Press, pp. 55–69.

FRENCH, V., P. BRYANT, and S. V. BRYANT. 1976. Pattern regulation in epimorphic fields. *Science* 193:969–81.

GARCIA-BELLIDO, A. P. RIPOLI, and G. MORATA. 1973. Developmental compartmentalization of the wing disc of *Drosophila. Nature New Biol* 245:251–53.

GOETINCK, P. F., and M. J. SEKELLICK. 1972. Observations on collagen synthesis, lattice formation, and morphology of scaleless and normal embryonis skin. *Dev Biol* 28:636–48.

GROSS, J. 1961. Collagen. *Sci Am* (May) 121–30.

HAMPÉ, A. 1957. Sur le rôle du mésoderme et de l'ectoderme du bougeon de patte dans les éxhanges entre le poulet et le canard. *C R Acad Sci* [D] (*Paris*) 244:3179–81.

HOLDER, N., S. V. BRYANT, and P. W. TANK. 1979. Interactions between irradiated and unirradiated tissues during supernumerary limb formation in the newt. *J Exp Zool* 208:303–10.

HOLDER, N., P. W. TANK, and S. V. BRYANT. 1980. Regeneration of symmetrical forelimbs in the axolotl *Ambystoma mexicanum. Dev Biol* 74:301–14.

HOLMES, A. 1935. The pattern and symmetry of adult plumage units in relation to the order and locus of origin of the embryonic feather papillae. *Am J Anat* 56:513–37.

ITEN, L. 1977. Pattern regulation in complementary anterior and posterior portions of the chick wing bud. *Am Zool* 17:963 (abstr.).

ITEN, L. E., and S. V. BRYANT. 1975. The interaction between blastema and stump in the establishment of the anterior–posterior and proximal–distal organization of the limb regenerate. *Dev Biol* 44:119–47.

ITEN, L. E., and D. J. MURPHY. 1980a. Pattern regulation in the embryonic chick limb: supernumerary limb formation with anterior (non-ZPA) limb bud tissue. *Dev Biol* 75:373–85.

ITEN, L. A., and D. J. MURPHY. 1980b. Supernumerary limb structures with regenerated posterior chick wing tissues. *J Exp Zool* 213:327–35.

JAVOIS, L. C., and L. E. ITEN. 1981. Position of origin of donor posterior chick wing bud tissue transplanted to an anterior host site determines the extra structures found. *Develop Biol* 82:329–42.

JOHNSON, K. A., and G. G. BORISY. 1977. Kinetic analysis of microtubule self-assembly in vitro. *J Mol Biol* 117:1–31.

JORQUERA, B., and E. PUGIN. 1971. Sur le comportement du mésoderme et de l'ectoderme du bourgeon de membre dans les éxhanges entre le poulet et le rat. *C R Acad Sci Paris* [D]272:1522–25.

JUHN, M., and R. M. FRAPS. 1934. Pattern analysis in plumage. IV. Order of asymmetry in the breast tracts. *Proc Soc Exp Biol Med* 31:1187–90.

KAUFMANN, S. A., R. M. SHYMKO, and K. TRABERT. 1978. Control of sequential compartment formation in *Drosophila. Science* 199:259–70.

KIENY, M. 1977. Proximodistal pattern formation in avian limb development. In D. A. Ede, J. R. Hinchliffe, and M. Balls, eds. *Vertebrate Limb and Somite Morphogenesis.* Cambridge: Cambridge University Press, pp. 87–103.

KIENY, M., and M.-P. PAUTOU. 1976. Régulation des excédents dans le développement du bourgeon de membre de l'embryon d'oiseau. *Wilhelm Roux' Arch* 179:327–38.

KIENY, M., and M.-P. PAUTOU. 1977. Proximo-distal pattern regulation in deficient avian limb buds. *Wilhelm Roux' Arch* 183:177–91.

KRASNER, G. N., and S. V. BRYANT. 1980. Distal transfor-

515

mation from double-half forearms in the axolotl, *Ambystoma mexicanum*. *Dev Biol* 74:315–25.

LANDAUER, W. 1930. Studies on the plumage of the silver spangled fowl. III. The symmetry conditions of the spangled pattern. *Bull Storrs Agric Exp Sta* 163:71–82.

LILLIE, F. R. 1942. On the development of feathers. *Biol Rev* 17:247–66.

LILLIE, F. R., and M. JUHN. 1932. The physiology of development of feathers. I. Growth rate and pattern in the individual feather. *Physiol Zool* 5:124–84.

LINSENMAYER, T. F. 1972. Control of integumentary patterns in the chick. *Dev Biol* 27:244–71.

MADEN, M. 1980a. Intercalary regeneration in the amphibian limb and the rule of distal transformation. *J Embryol Exp Morphol* 56:201–209.

MADEN, M. 1980b. Structures of supernumerary limbs. *Nature* 286:803–805.

MADEN, M., and B. C. GOODWIN. 1980. Experiments on developing limb buds of axolotl, *Ambystoma mexicanum*. *J Embryol Exp Morphol* 57:177–87.

MADEN, M., and R. N. TURNER. 1978. Supernumerary limbs in the axolotl. *Nature* 273:232–35.

MACCABE, J. A., J. ERRICK, and J. W. SAUNDERS. 1974. Ectodermal control of the dorsoventral axis of the leg bud of the chick embryo. *Dev Biol* 39:69–82.

MACCABE, J. A., J. W. SAUNDERS, and M. PICKETT. 1973. The control of the anteroposterior and dorsoventral axes in embryonic chick limbs constructed of dissociated and reaggregated limb-bud mesoderm. *Dev Biol* 31:323–35.

MACCIONI, R., and N. W., SEEDS. 1977. Stoichiometry of GTP hydrolysis and tubulin polymerization. *Proc Natl Acad Sci USA* 74:462–66.

MURPHY, D. B., K. A. JOHNSON, and G. G. BORISY. 1977. Role of tubulin-associated proteins in microtubule nucleation and elongation. *J Mol Biol* 117:33–52.

NEWMAN, S. A., and H. L. FRISCH. 1979. Dynamics of skeletal pattern formation in developing chick limb. *Science* 205:662–68.

NOVEL, G. 1973. Feather pattern stability and reorganization in cultured skin. *J Embryol Exp Morphol* 30:605–33.

OSBORN, M., and K. WEBER. 1976. Cytoplasmic microtubules in tissue culture cells appear to grow from an organizing structure towards the plasma membrane. *Proc Natl Acad Sci USA* 73:867–71.

PAUTOU, M.-P. 1968. Rôle déterminant du mésoderme dans la différenciation spécifique de la patte de l'oiseau. *Arch Anta Microsc Morphol Exp* 57:311–28.

PAUTOU, M.-P. 1977. Establissment de l'axe dorso-ventral dans le pied de l'embryon de poulet. *J Embryol Exp Morphol* 42:177–94.

RAWLES, M. E. 1960. The integumentary system. In A. J. Marshall, ed. *Biology and Comparative Physiology of Birds*, Vol. 1. New York: Academic Press, pp. 189–240.

RAWLES, M. E. 1963. Tissue interactions in scale and feather development as studied in dermal-epidermal recombinations. *J Embryol Exp Morphol* 11:765–89.

RUBIN, L., and J. W. SAUNDERS, Jr. 1972. Ectodermal–mesodermal interactions in the growth of limb buds in the chick embryo: constancy and temporal limits of the ectodermal induction. *Dev Biol* 28:94–112.

SAUNDERS, J. W., Jr. 1948. The proximo–distal sequence of origin of the parts of the chick wing and the role of the ectoderm. *J Exp Zool* 108:363–404.

SAUNDERS, J. W. 1972. Developmental control of three-dimensional polarity in the avian limb. *Ann NY Acad Sci* 193:29–42.

SAUNDERS, J. W. 1977. The experimental analysis of chick limb bud development. In D. A. Ede, J. R. Hinchliffe, and M. Balls, eds. *Vertebrate Limb and Somite Morphogenesis*. Cambridge: Cambridge University Press, pp. 1–24.

SAUNDERS, J. W., Jr., and J. F. FALLON. 1966. Cell death in morphogenesis. In M. Locke, ed. *Major Problems in Developmental Biology*. New York: Academic Press, pp. 289–314.

SAUNDERS, J. W., Jr., and M. T. GASSELING. 1957. The origin of pattern and feather germ tract specificity. *J Exp Zool* 135:503–28.

SAUNDERS, J. W., Jr., and M. T. GASSELING. 1968. Ectodermal–mesodermal interactions in the origin of limb symmetry. In R. Fleischmajer and R. E. Billingham, eds. *Epithelial-Mesenchymal Interactions*. Baltimore: Williams and Wilkins, pp. 78–97.

SAUNDERS, J. W., Jr., M. T. GASSELING, and J. M. ERRICK. 1976. Inductive activity and enduring cellular constitution of a supernumerary apical ectodermal ridge grafted to the limb bud of the chick embryo. *Dev Biol* 50:16–25.

SAUNDERS, J. W., Jr., and P. WEISS. 1950. Effects of removal of ectoderm on the origin and distribution of feather germs in the wing of the chick embryo. *Anat Rec* 108:581 (abstr.).

SENGEL, P. 1958. La différenciation de la peau et des germes plumaires de l'embryon de poulet en culture in vitro. *Ann Biol* 34:29–52.

SENGEL, P. 1975. Feather pattern development. In R. Porter and J. Rivers, eds. *Cell Patterning Ciba Foundation Symposium 29*. Amsterdam: North Holland, pp. 51–70.

SENGEL, P. and U. K. ABBOTT. 1963. In vitro studies with the scaleless mutant: interactions during feather and scale differentiation. *J Hered* 54:254–62.

SENGEL, P., and A. MAUGER. 1976. Peridermal cell patterning in the feather-forming skin of the chick embryo. *Dev Biol* 51:166–71.

SLACK, J. M. W. 1977a. Determination of anteroposterior polarity in the axolotl forelimb by an interaction between limb and flank rudiments. *J Embryol Exp Morphol* 39:151–68.

SLACK, J. M. W. 1977b. Control of anteroposterior pattern

in the axolotl forelimb by a smoothly-graded signal. *J Embryol Exp Morphol* 39:169–82.

SLACK, J. M. W., and S. SAVAGE. 1978a. Regeneration of reduplicated limbs in contravention of the complete circle rule. *Nature* 271:760–61.

SLACK, J. M. W., and S. SAVAGE. 1978b. Regeneration of mirror symmetrical limbs in the axolotl. *Cell* 14:1–8.

STOCUM, D. L. 1978. Regeneration of symmetrical hindlimbs in larval salamanders. *Science* 200:790–93.

STUART, E. S., G. GARBER, and A. A. MOSCONA. 1972. An analysis of feather germ formation in the embryo and in vitro, in normal development and in skin treated with hydrocortisone. *J Exp Zool* 179:97–118.

SUMMERBELL, D. 1974. A quantitative analysis of the effect of excision of the AER from the chick limb bud. *J Embryol Exp Morphol* 32:651–60.

SUMMERBELL, D. 1979. The zone of polarizing activity: evidence for a role in normal chick limb morphogenesis. *J Embryol Exp Morphol* 50:217–33.

TANK, P. W. 1978a. The failure of double-half forelimbs to undergo distal transformation following amputation in the axolotl, *Ambystoma mexicanum. J Exp Zool* 204:325–36.

TANK, P. W. 1978b. The occurrence of supernumerary limbs following blastemal transplantation in the regenerating forelimb of the axolotl, *Ambystoma mexicanum. Dev Biol* 62:143–61.

TANK, P. W., and N. HOLDER. 1978. The effect of healing time on the proximodistal organization of double-half forelimb regenerates in the axolotl, *Ambystoma mexicanum. Dev Biol* 66:72–85.

TICKLE, C., G. SHELLSWELL, A. CRAWLEY, and L. WOLPERT. 1976. Positional signalling by mouse limb polarizing region in the chick wing bud. *Nature* 259:396–97.

TICKLE, C., D. SUMMERBELL, and L. WOLPERT. 1975. Positional signalling and specification of digits in chick limb morphogenesis. *Nature* 254:199–202.

TOMASEK, J. 1980. Organization of selected extracellular matrix materials in relation to limb development. Doctoral Dissertation, State University of New York at Albany.

TURING, A. M. 1952. The chemical basis of morphogenesis. *Philos Trans R Soc* London. [B]237:32–72.

VALLEE, R. B., and G. C. BORISY. 1978. The non-tubulin component of microtubule protein oligomers. Effect of self-association and hydrodynamic properties. *J Biol Chem* 253:2834–45.

WALLACE, H., and A. WATSON. 1979. Duplicated axolotl regenerates. *J Embryol Exp Morphol* 49:243–58.

WATTERSON, R. L. 1942. The morphogenesis of down feathers with special reference to the developmental history of melanophores. *Physiol Zool* 15:234–59.

WEISENBERG, R. C., and A. C. ROSENFELD. 1975. In vitro polymerization of microtubules into asters and spindles in homogenates of surf clam eggs. *J Cell Biol* 64:146–58.

WESSELS, N. K. 1965. Morphology and proliferation during early feather development. *Dev Biol* 12:131–53.

WESSELS, N. K., and J. EVANS. 1968. The ultrastructure of oriented cells and extracellular materials between developing feathers. *Dev Biol* 18:42–61.

WOLPERT, L. 1969. Positional information and the spatial pattern of cellular differentiation. *J Theor Biol* 25:1–47.

WOLPERT, L. 1978. Pattern formation in biological development. *Sci Am* (October) 154–64.

WOLPERT, L., M. R. B. CLARKE, and A. HORNBRUCH. 1972. Positional signalling along hydra. *Nature New Biol* 239:101–05.

517

The term **cell cycle** refers to the complex sequence of nuclear and cytoplasmic events that occurs between successive cellular divisions. The purpose of this appendix is to provide a brief outline principally of the nuclear events that precede and accompany the division of a cell into two daughter cells. This appendix is only an outline, designed chiefly to refresh the memories of students who will already have been exposed to similar materials in an introductory course in high school or college. Readers who wish to make a detailed study of mitosis and of the nuclear and cytoplasmic changes during the cell cycle should consult textbooks of cytology, cytogenetics and cell physiology, and special review articles.

A

The Cell Cycle and Mitosis

A.1 Phases of the Mitotic Cycle

Because a population of dividing cells shows repetitive nuclear changes related to the event of mitosis, one speaks of the **mitotic cycle.** The phases of the mitotic cycle are illustrated in Figures A.1 and A.2. The first phase of the cycle to be described is strictly not a part of mitosis, per se, but it is a part of the cell cycle, often rather prolonged, called **interphase.** The appearance of the nucleus during interphase is described first.

A.1.1 INTERPHASE. The nucleus of a cell that has not initiated the process of division contains long threads of material called **chromatin.** Chromatin consists of large molecules composed of **deoxyribonucleic acid** (DNA), **ribonucleic acid** (RNA), and protein. Ribonucleic acid makes up the bulk of special organelles called **nucleoli.** The nucleoli are the site of origin of ribosomes, which are later exported to the cytoplasm where they are involved in protein synthesis. The structures of the nucleic acids and an outline of the process of protein synthesis are treated in Appendix B.

When fixed and stained cells are examined by means of the light microscope, the chromatin is visible only where it is aggregated in strands or clumps, for the ultimate strands of chromatin are too small in diameter to be seen. Nucleoli are usually recog-

nized, but are not present in very early stages of most embryonic cells.

FIGURE A.1 Mitotic cell division. Six chromosomes, or three pairs, compose the diploid complement of this hypothetical cell. Each chromosome duplicates its DNA during interphase and is seen as double in prophase. The double chromosomes separate at anaphase, and at telophase each daughter cell receives six chromosomes.

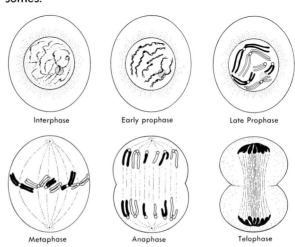

Interphase   Early prophase   Late Prophase

Metaphase   Anaphase   Telophase

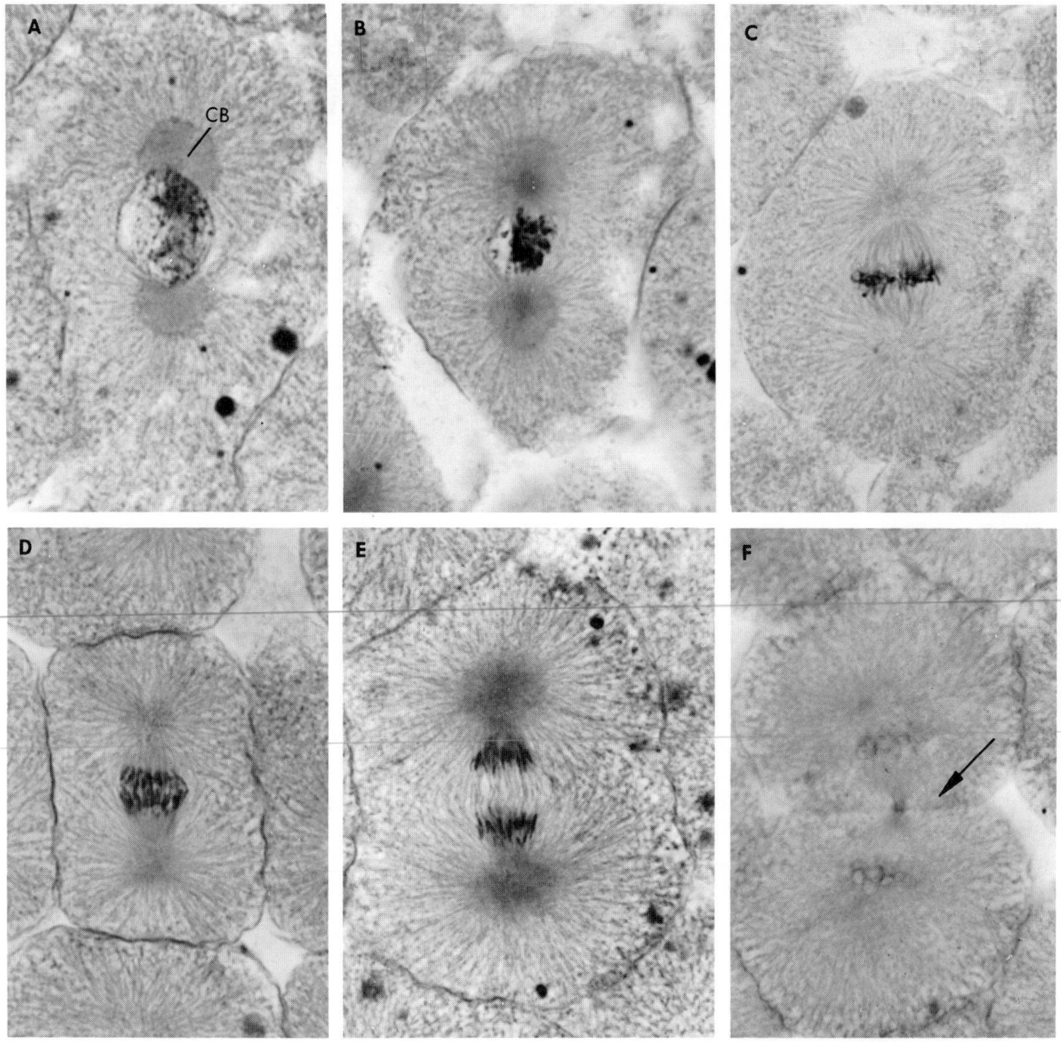

FIGURE A.2 Mitotic stages in the whitefish blastula. A: early prophase showing early appearance of chromosomes and large central bodies, CB, from which astral rays and spindle fibers arise. B: Late prophase. Chromosomes are more condensed, and nuclear membrane is breaking down. C: The end of metaphase, with daughter chromosomes just beginning to separate. Note well-defined spindle. D: Early anaphase. E: Late anaphase. F: Telophase. The daughter nuclei appear to arise by the coalescence of chromosomal vesicles, a phenomenon found in several kinds of embryos, notably those of fish. The midbody, probably a condensed spindle fiber remnant, is found at the center of the cytoplasmic constriction between daughter cells (arrow).

A.1.2 PROPHASE. As the time for mitosis nears, the chromatin gradually becomes organized into shorter and thicker strands that can be resolved visually as **chromosomes,** each of which is double in structure. The chromosomes become visible in the prophase stage of mitosis as a consequence of the progressive coiling of the chromatin strands in an ever-tightening pattern. Also, during prophase in most animal cells, but not in plant cells, structures called **centrioles** take up positions at opposite sides of the nucleus. The centrioles are often a part of a dense **central body** (Figure A.2). As the centrioles assume their positions, fibers connecting the central bodies appear, thus creating a **mitotic spindle.** The nuclear membrane breaks down leaving the chromosomes held within the spindle. The chromosomes reach their ultimate degree of coiling at the end of the prophase and are thus quite prominent as double structures. Each subunit is called a **chromatid.** Once the events of prophase have occurred, the chromosomes quickly assume positions midway between the ends of the spindle. They are connected to the centrioles by spindle fibers attached to specialized regions of the chromosomes called **centromeres.** The plane passing through the centromeres at right angles to the long axis of the spindle is called the **equatorial plane.**

A.1.3 METAPHASE AND ANAPHASE. With the chromosomes at the equatorial plane, mitosis is at the stage of **metaphase.** This phase may be more or less prolonged, but it is followed eventually by **anaphase.** Anaphase is the stage during which the chromatids separate to form daughter chromosomes, each with its own centromere. The daughter chromosomes then move to opposite poles of the spindle with their centromeres leading the way. As sets of daughter chromosomes move apart, the cell and spindle elongate in preparation for the **telophase** stage, which follows.

A.1.4 TELOPHASE. During telophase the cytoplasm constricts between the two sets of daughter chromosomes, thus dividing the original cell into two daughter cells. Within each daughter cell, the chromosomes aggregate at the poles of the spindle, and a new nuclear membrane forms around them as they uncoil and assume the elongated, entangled configuration that is typical of the interphase cell.

A.2 Phases of the Cell Cycle

The phase of the cell cycle that is occupied by mitosis is relatively brief and is referred to as the M phase. At some time during interphase sufficient

DNA and histones are synthesized to replicate the chromosomes. This period of synthesis is called the **DNA synthetic phase** or, simply, the **S phase** of the cell cycle. The portion of interphase after mitosis but before S is the **first gap phase,** which is called G_1. The interphase period after S but prior to mitosis is the **second gap** G_2.

During interphase the cells grow actively and carry out metabolic activities appropriate to their function in the embryo or differentiated organism. These functions continue during M, also, but transcription of the genes is inhibited in tightly coiled chromosomes.

In many plants and animals certain highly specialized cells leave the cell cycle as they approach the final stages of their differentiation. Human brain cells undergo their final division around the time of birth and never replicate their DNA again. Cells permanently arrested in the G_1 period are often said to be in a phase of G_0. G_1 varies in duration from a few minutes to many years in cells under different conditions. S varies between minutes and several hours. G_2 is somewhat less variable but can be very prolonged. Many liver cells, for example, replicate their DNA and remain in G_2, possibly for years.

521

A.3 Suggestions for Further Readings

BURNS, G. W. 1980. The cytological bases of inheritance. *The Science of Genetics,* 4th ed. New York: Macmillan, pp. 37–90.

CANDE, W. Z., E. LAZARIDES, and J. R. McINTOSH. 1977. A comparison of the distribution of actin and tubulin in the mammalian mitotic cycle as seen by direct immunofluorescence. *J Cell Biol* 72:552–67.

DeROBERTIS, E. D. P., W. W. NOWINSKI, and F. A. SAEZ. 1970. *Cell Biology,* 5th Ed. Philadelphia: Saunders.

DUSTIN, P. 1980. Microtubules. *Sci Am* (August) 67–76.

MAZIA, D. 1961. Mitosis and the physiology of cell division. In J. Brachet and A. E. Mirsky, eds. *The Cell,* Vol 3. New York: Academic Press, pp. 77–412.

MITCHISON, J. M. 1971. *The Biology of the Cell Cycle.* Cambridge: Cambridge University Press.

PRESCOTT, D. M. 1976. *Reproduction of Eukaryotic Cells.* New York: Academic Press.

PRESCOTT, D. M. 1976. The cell cycle and the control of cellular reproduction. *Adv Genet* 18:100–77.

B
Mechanism of Protein Synthesis

The genetic material comprises a set of instructions for protein synthesis. These instructions are encoded in sequences of molecules called **purines** and **pyrimidines,** each of which is a part of a **deoxyribose nucleotide.** The nucletides are polymerized in long chains called deoxyribose nucleic acid, or DNA. The DNA is a part of each chromosome of a cell, and each time a cell divides the genetic instructions are duplicated so that each daughter cell acquires a complete set. Between divisions appropriate portions of the genetic instructions are copied by sequences of pyrimidine and purine **ribonucleotides** called **ribose nucleic acid,** or RNA. These instructions are then translated by means of other RNAs and special enzymes into the chains of amino acids that comprise proteins. It is the purpose of this appendix to outline briefly the structure of DNA and RNA, to show how the genetic instructions are encoded, and to describe the machinery whereby they are transcribed and translated. For a more extended treatment of these matters consult texts, such as Watson's (1976).

B.1 Structure of DNA

The deoxyribose nucleic acids are high molecular weight polymers composed of **nucleotides.** Each nucleotide is formed of a purine or pyrimidine base, a five-carbon sugar, **deoxyribose** (dR), to which the base is attached, and a **phosphate group** (P), to which the deoxyribose is attached. The nucleotides that form DNA are illustrated in Figure B.1. Nucleotides are linked by the formation of an ester bond between the phosphate group of one nucleotide and the 3' hydroxyl of the deoxyribose part of the adjacent one.

The bases found in deoxyribose nucleotides are the purines **adenine** and **guanine** and the pyrimidines **cytosine** and **thymine.** The nucleotides formed with these bases may be united in any linear order to form polynucleotide chains of great length. If we represent each base by its first letter, one such linear array might be the one illustrated in Figure B.2.

Analysis reveals that in the bases that are separated chemically from natural samples of DNA the amounts of A and T, as well as the amounts of G and C, occur in the ratio of 1:1. The ratios found for $(A + G)/(C + T)$, however, are quite variable from one organism to the next. These facts, together with data from x-ray diffraction studies and the analysis of hydrogen-bonding sites and covalent-bond angles, show that the nucleic acids of DNA are actually arranged as two parallel chains side by side in a ladder-like configuration (Figure B.3B). The sides of the ladder are formed by the dR–P–dR links, and the steps are formed by the purine and pyrimidine bases, which project inward from the sides and are held together by hydrogen bonds. Each step comprises either A and T, held together by two hydrogen bonds, or G and C, held by three hydrogen bonds (Figure B.3A). Thus, in any sample of double-stranded DNA, A and T always exist in equal numbers, and so do G and C. The entire double-chain molecule is twisted into a helix, so that the ladder is in the form of a spiral (Figure B.3C).

Clearly, the sequence of bases on either side of the ladder could form a kind of code or genetic language with an alphabet of four letters. The sequence of bases in DNA is such a code, and the con-

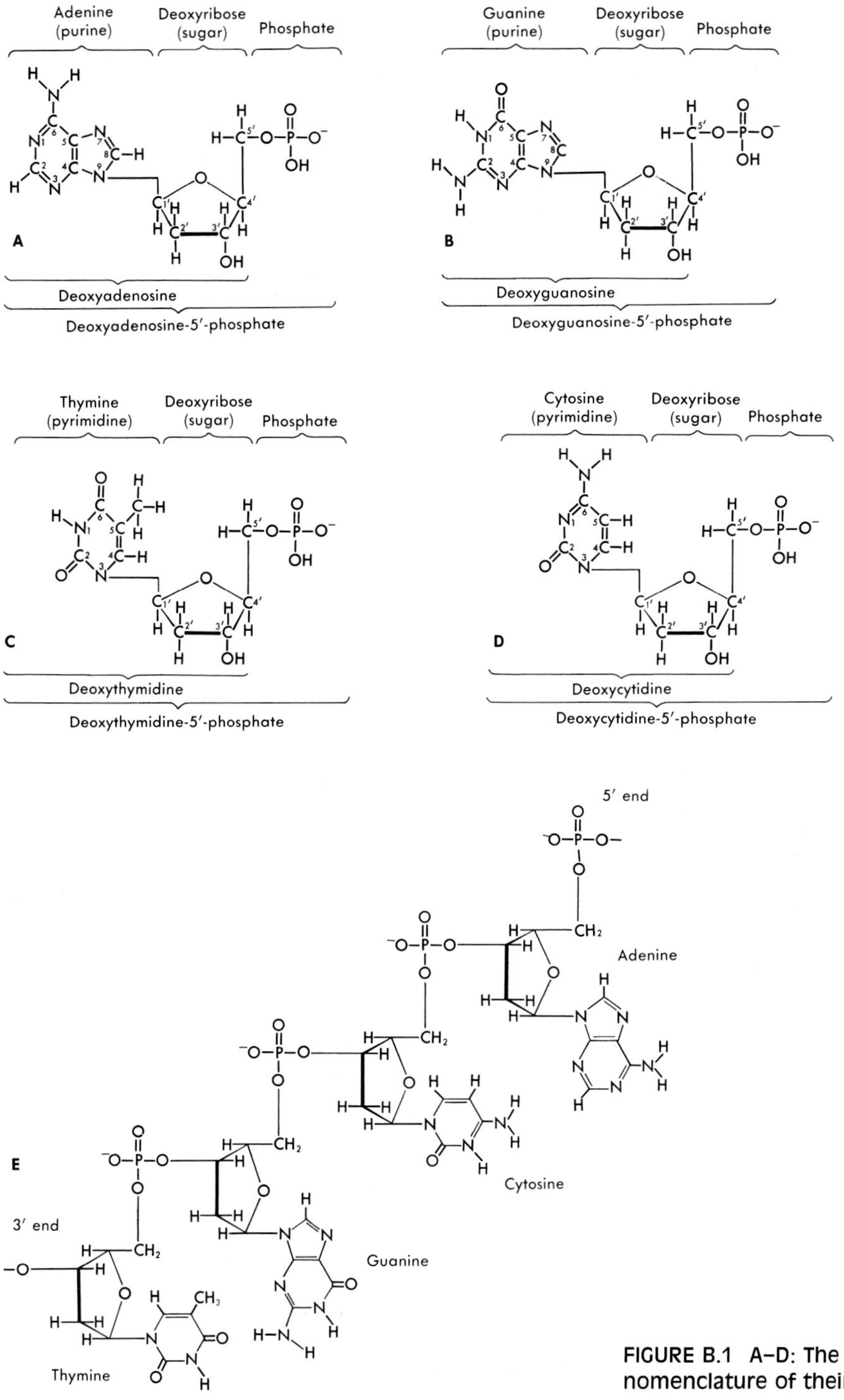

FIGURE B.1 A–D: The nucleotides of DNA and the nomenclature of their parts. E: The way they are linked.

FIGURE B.2 A linear array of purine and pyrimidine bases that might occur in a short DNA chain. The bases are represented by their first letters: A, adenine; C, cytosine; G, guanine; T, thymine. The backbone is a chain of alternating moieties of orthophosphate, P, and deoxyribose, dR.

struction manual for development is written in DNA language.

FIGURE B.3 The structure of DNA. A: Details of specific base pairing between thymine and adenine, which form two hydrogen bonds, and between guanine and cytosine, which form three hydrogen bonds. B: The DNA ladder. C: Coiling of the ladder into a helix.

B.2 Copying of Genes by RNA

The construction manual is in the nucleus, and protein, the basis of structure, is synthesized in the cytoplasm of the cell. Obviously, the cytoplasm must have copies of the nuclear instructions and the means of translating these copies into the amino acid sequences that occur in proteins. These needs

524

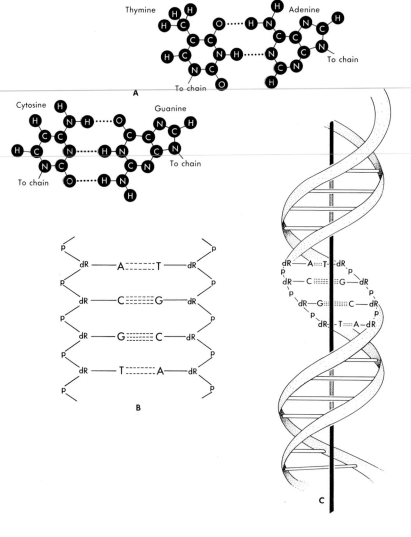

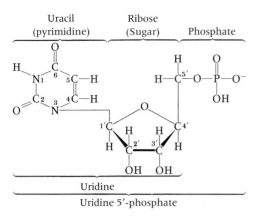

Uracil
(pyrimidine)   Ribose
(Sugar)   Phosphate

Uridine

Uridine 5'-phosphate

FIGURE B.4 The structure of uridine 5'-phosphate. Note the similarity of uracil to thymine. Compare the structure of ribose, shown here, with that of deoxyribose, shown in Figure B.1.

are provided for by other kinds of nucleic acids that we lump under the general heading of **ribose nucleic acid,** or RNA. RNA is much like DNA; it is constructed with ribose (R) in place of deoxyribose, however, and it has the pyrimidine uracil (U) in place of thymine (Figure B.4). Like DNA, RNA occurs in nucleotide chains, but unlike DNA, most RNA occurs in single strands (Figure B.5).

A glance at this chain shows that the base sequences in RNA, like those of DNA, could also form a genetic language written in an alphabet of four letters. This alphabet would be just like that of DNA except for the replacement of thymine (T) by uracil (U). If one were to make this substitution, the code could be rewritten or transcribed from a single strand of DNA in the form of RNA without changing the genetic words.

B.2.1 THE GENETIC COPY IS MESSENGER RNA. One of the forms of RNA that can be extracted from cells is, indeed, a copy of DNA words having to do with the construction of protein molecules; it is called

FIGURE B.5 A hypothetical RNA chain of ribose, R, and orthophosphate, P, with attached bases. U, uracil; other bases as in Figure B.2.

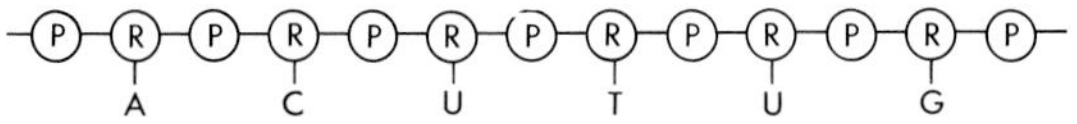

messenger RNA (mRNA) or **template RNA,** and it copies only one of the chains of the DNA helix. At the time of copying the individual RNA nucleotides align themselves, with the aid of the enzyme RNA polymerase, in register with the complementary bases of one strand of the DNA double helix—U of the incipient RNA molecule with A of the DNA, G with C, A with T, and C with G (Figure B.6). Copying of the genetic message occurs in longer or shorter segments of the genome, each describing in genetic language the way a protein molecule or portion thereof is to be constructed. The RNA units, of corresponding length, then detach from the DNA helix and move away. Thus it is that items of information about the sequence of amino acids in proteins encoded in the genome are copied.

B.2.2 GENETIC WORDS HAVE THREE LETTERS. The genetic information is conveyed by virtue of the arrangement of the four-letter alphabet of RNA into three-letter words called **codons.** Each codon denotes a particular amino acid. Does this provide a sufficiently large enough vocabulary for the construction manual? Yes, for if one arranges four letters in all possible sequences, three at a time, a total of 64 codons is possible. Because there are only 20 amino acids used in the construction of natural pro-

525

FIGURE B.6 Copying the genetic code. A segment of double-stranded DNA is shown opened up and a single strand of RNA is being transcribed from one strand of the DNA from right to left. The enzyme RNA polymerase catalyzes the transcription process. At the left is illustrated the attachment of a molecule of uridine phosphate that has been split from uridine triphosphate.

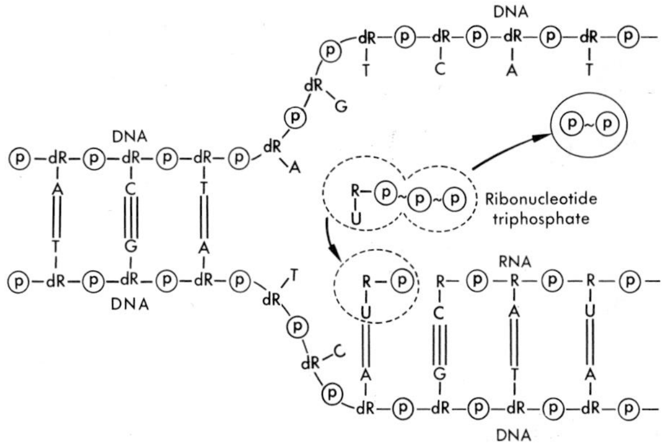

TABLE B.1   The Genetic Code

First Letter	Second	Letter			Third Letter
	U	C	A	G	
U	Phenylalanine	Serine	Tyrosine	Cysteine	U
	Phenylalanine	Serine	Tyrosine	Cysteine	C
	Leucine	Serine	STOP	STOP	A
	Leucine	Serine	STOP	Tryptophan	G
C	Leucine	Proline	Histidine	Arginine	U
	Leucine	Proline	Histidine	Arginine	C
	Leucine	Proline	Glutamine	Arginine	A
	Leucine	Proline	Glutamine	Arginine	G
A	Isoleucine	Threonine	Asparagine	Serine	U
	Isoleucine	Threonine	Asparagine	Serine	C
	Isoleucine	Threonine	Lysine	Serine	A
	Methionine START	Threonine	Lysine	Arginine	G
G	Valine	Alanine	Aspartic acid	Glycine	U
	Valine	Alanine	Aspartic acid	Glycine	C
	Valine	Alanine	Glutamic acid	Glycine	A
	Valine	Alanine	Glutamic acid	Glycine	G

526

teins, this vocabulary is clearly adequate. Actually, there are sufficient three-letter groups to permit the coding for the same amino acid by more than one group. Nucleotide sequences that code for the various amino acids are shown in Table B.1. Note that one codon AUG is the universal signal for the start of the synthesis of a protein; that is, it signals the point at which the amino acids that comprise the protein begin to be assembled in peptide linkage. The signal for terminating the peptide chain is UAA, UAG, or UGA.

B.3   Translation of the Genetic Message

Given a vocabulary that has words for each amino acid, the structure of a protein might be called for in the RNA language by a finite number of codons linearly arranged in the same order as the amino acids that compose the protein to be manufactured. The next requirement would be a mechanism for translating the RNA codons into the specific amino acids

represented by each. This requirement is met in the cell by another kind of RNA, of relatively low molecular weight, called **transfer RNA** (tRNA).

B.3.1   tRNAS ARE SPECIFIC.   For each kind of amino acid used in the building of a protein there are transcribed from the genome one or more specific kinds of tRNA corresponding to each of the codons designating that amino acid. Different tRNAs that correspond to the codons for a particular amino acid are called **isoaccepting tRNAs.** For example, there seem to be at least six tRNAs for leucine and some are preferentially used for the insertion of leucine into a particular protein or into a particular portion of a protein (see Chapter 18). tRNAs consist of chains of 75–90 ribonucleotides. All tRNAs have in common that they can be accommodated into the cloverleaf pattern illustrated in Figure 18.7. Each has the sequence cytosine, cytosine, adenine at the end that binds to amino acids (the so-called acceptor stem), and each has a series of loops in which certain invariable features are found, as shown in Figure 18.7. Of particular note is the fact that each kind of tRNA has a loop in which are exposed a series of three bases complementary to the mRNA codon of the amino acid for which it is

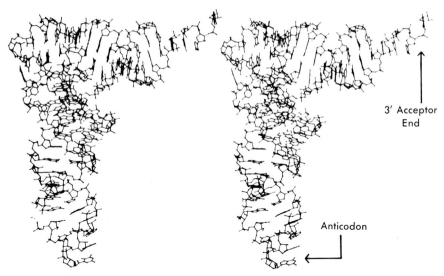

FIGURE B.7 A stereoscopic view of the predominant form of phenylalanine-tRNA from yeast. The structure of the molecule can be best viewed by means of stereoscopic glasses, but the two images can be fused without the use of glasses by simply relaxing the eye muscles until a three-dimensional image appears. The amino acid acceptor stem bearing CAA is located in the upper right projection, and the anticodon loop is at the bottom. The TΨC loop is at the upper left (cf. Figure 18.7). [From G. J. Quigley and A. Rich, *Science* 194:796–806 (1976). Copyright 1976 by the American Association for the Advancement of Science.]

527

In the figure: **3' Acceptor End** and **Anticodon**

specific. This sequence of bases is called an **anticodon**. The three-dimensional structure of phenylalanine tRNA from yeast is shown as a stereodiagram in Figure B.7.

Each kind of tRNA has the task of transporting the specific amino acid for which it has the anticodon to an assembly site where the amino acid will be joined with others in the sequence prescribed by the mRNA for a particular protein. This assembly site is provided by a special organelle, the **ribosome.**

B.3.2 SPECIFIC ACTIVATING ENZYMES ARE ALSO REQUIRED. Before describing the ribosomes and the assembly of amino acids to form protein, it is appropriate to examine the attachment of the amino acid to its tRNA (Figure B.8) and to recognize the remarkable group of enzymes that facilitate this attachment. For each kind of amino acid–tRNA pair there is a special enzyme called an **activating enzyme** or **amino acyl synthetase.** Each activating enzyme has two reactive sites, one for binding its amino acid and another for adapting to the amino

acid by catalyzing its reaction with **adenosine triphosphate (ATP)**, the high-energy compound generated principally in oxidative metabolism. This reaction results in the release of pyrophosphate (P~P) from ATP and the formation of a high-energy covalent bond between **adenosine monophosphate (AMP)** and the activated amino acid. The resulting molecule, AA~AMP, remains bound to the enzyme (E) until there is an effective collision with the appropriate tRNA molecule. The enzyme thereupon binds by its reactive site to the tRNA and then catalyzes the formation of a covalent bond between the amino acid and the ribose of the terminal adenylic nucleotide of the tRNA. AMP is thereupon released, and the activating enzyme detaches, free to activate and transport another molecule of the same amino acid.

AA + ATP + E → AA~AMP—E + P~P

AA~AMP—E + tRNA → AA~tRNA + AMP + E

The tRNA then unloads its amino acid at the ribosome.

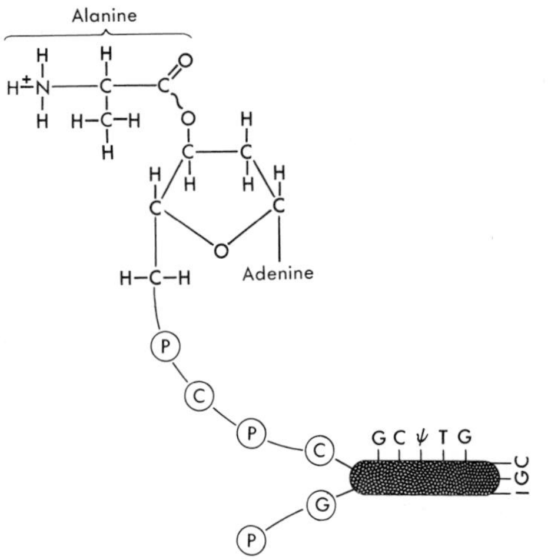

FIGURE B.8 Alanine-mRNA. A molecule of alanine is bound to the 3' position on the ribose moiety attached to the terminal adenine of the acceptor site. The TΨC sequence, flanked by Gs, is believed to be the site at which the tRNA is bound to 5S RNA and its associated ribosomal protein. In the anticodon the letter I stands for inosine. This base is formed post-transcriptionally by deamination of adenine at the 6 position and formation of a 6-keto group.

528

B.4 Assembly of Proteins on the Ribosome

The ribosome, as its name indicates, contains RNA. This RNA, conveniently called rRNA, is transcribed from the genome and processed into two subunits, each combined with protein. The smaller subunit from eukaryotic cells sediments in the analytical ultracentrifuge with a coefficient of 40 Svedberg units (the Svedberg unit is a measure of the velocity with which a substance moves under controlled conditions in a centrifugal field. It is a function of the size, shape, and composition of a substance), and is referred to as the 40 S subunit. Stripped of its protein, the RNA of the smaller subunit sediments at 18 S. The larger subunit sediments at 60 S, and its RNA consists of a 28 S segment to which a 5.8 S fragment is covalently bonded (Figure 18.6). A

molecule of 5 S RNA that is apparently complexed to a protein having a molecular weight of about 38,000 is also a part of the 60 S subunit. The larger and smaller subunits join as protein synthesis is initiated, forming a ribosome. A single ribosome is referred to as a **monosome.**

The process of protein synthesis involves three phases: (1) **initiation,** (2) **elongation,** and (3) **termination.** Initiation is concerned with the bringing together of the mRNA and the ribosome. Elongation is the process whereby successive amino acids are added to form a peptide chain. The elongating pepetide chain is referred to as **nascent protein.** Termination refers to the release of the completed protein from the ribosome. At this point the ribosome falls into its separate subunits.

Initiation involves one of the two isoaccepting tRNAs for methionine. For eukaryotes this species of tRNA is conveniently referred to as Met-tRNA$_i$ or as tRNA$_i^{Met}$. In either case "Met" refers to methionine and "i" to the term initiator. The other isoaccepting species for methionine is used only for inserting a methionine residue into the peptide chain.

For the initiation of protein synthesis proteins called **initiation factors** are required. One of these factors is named in what follows, but the list of currently known factors is given in Table 18.1 (Section 18.6). The initiation factors for eukaryotes, the kind of organisms with which this work is concerned, are called **eukaryotic initiation factors.** These factors are designated eIF, followed by a numerical designation. The first step in the initiation of protein synthesis is the assembly of eIF-2, Met-tRNA$_i$, and guanosine triphosphate (GTP) in a **ternary complex.** The binding of the ternary complex to the 40 S ribosomal subunit is then catalyzed by eIF-2. With this accomplished, mRNA is bound to the 40 S subunit in a reaction facilitated by the affinity of eIF-2 for double-stranded regions of the mRNA. (Double-stranded mRNA arises by the formation of hydrogen bonds between complementary bases of the mRNA.) Next, with the aid of the other initiation factors, the 40 S and 60 S ribosomal subunits unite to form an 80 S ribosomal **initiation complex.** In this reaction, GTP is hydrolyzed to GDP and inorganic phosphate.

The 80 S ribosome has two sites at which transfer RNA bearing an amino acid may be bound. These sites occupy space on both ribosomal sub-

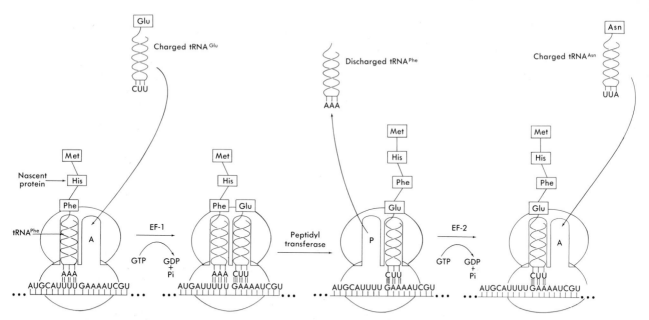

FIGURE B.9 Elongation of the peptide chain. The ribosomes are represented by outlines of their large and small subunits. The binding of mRNA to the smaller subunit is indicated, and the A and P sites (see text) are shown to involve both subunits. This illustration represents the translation of an internal segment of mRNA shifting from right to left and coding successively for methionine, histidine, phenylalanine, glutamic acid, asparagine, and arginine. The first three of these amino acids are already inserted into the nascent protein as shown at the left, and the incorporation of glutamic acid is illustrated as one follows the diagram from left to right. The same steps will be repeated when tRNA charged with asparagine binds to the A site (far right). After asparagine is incorporated, then arginine-tRNA will enter the A site as indicated by the codon CGU on the mRNA segment.

529

units (Figure B.9). One of these sites is called the **peptidyl site** (P), for it is in this site that the elongating peptide chain resides during the time required for the next amino acid to be brought to the ribosome. The next amino acid is brought to the ribosome at the **amino acid accepting site** (A). Currently it seems most likely that Met-tRNA$_i$, charged with methionine, is bound at the P site in the initiation complex before elongation begins. Here its anticodon is hydrogen bonded to the methionine codon AUG, which is apparently the only initiator codon utilized by eukaryotes. AUG also codes for internal methionine residues in the protein, but the isoaccepting Met-tRNA, rather than Met-tRNA$_i$, is used for inserting the methionine in such instances, as just noted.

Once initiation is completed, the initiation factors are released and elongation begins. Amino acids are added successively in the sequence called for by the mRNA. Addition of each amino acid calls for repetition of the steps illustrated in Figure B.9.

1. Binding of the next amino acid–tRNA (AA–tRNA) to the A site. The amino acid–tRNA complex bearing the correct anticodon to base-pair with the codon exposed at the A site, binds to the A site in a reaction catalyzed by **elongation factor 1** (EF-1). In this reaction there is expended energy released by the hydrolysis of one phosphate bond of GTP.
2. Formation of the peptide bond. As soon as the AA–tRNA is bound to the A site, the enzyme peptidyl transferase, one of the ribosomal proteins, catalyzes the formation of a peptide bond

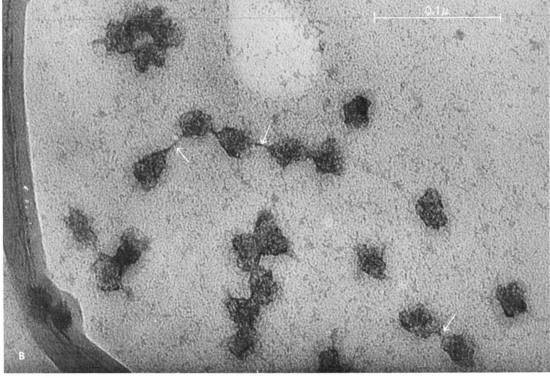

FIGURE B.10 (opposite) Polyribosomes translating hemoglobin mRNA. A: A preparation of polyribosomes isolated from reticulocytes deposited on an electron microscope grid, then washed, air dried and shadowed with platinum, and photographed (×80,000). Note that the ribosomes occur predominantly in clusters of 5. B: Similar polyribosomes fixed with formaldehyde, stained with uranyl acetate, and deposited on a grid for electron microscopy (×350,000). In this kind of preparation, the polyribosomes are often spread out, and individual ribosomes are sometimes seen to be connected by thin strands (arrows) 10–15 A in diameter, which are probably mRNA. [A from J. R. Warner, A. Rich, and C. D. Hall, Science 138:1398–1403 (1962) (cover photo, issue 3748). Copyright 1962 by the American Association for the Advancement of Science. B with permission from H. S. Slayter, J. R. Werner, A. Rich, and C. E. Hall, *J Mol Biol* 7:652–57 (1963). Copyright by Academic Press Inc. (London) Ltd. Both photographs courtesy of Dr. Alexander Rich.]

that links two amino acids in the peptide chain. This reaction involves removing the amino acid that had been attached to the tRNA in the P site and attaching it to the amino acid bound to the tRNA in the A site. (In the case in which Met-tRNA$_i$ occupies the P site, methionine is discharged without peptide bond formation.)

3. Release of discharged tRNA from the P site. The P site has little affinity for tRNA that is not charged with an amino acid.

4. Translocation of tRNA with its nascent peptide chain to the P site. This movement is accompanied by the shift of mRNA through the ribosome the distance of one codon. **Elongation factor 2** (EF-2) catalyzes this reaction, which also requires the expenditure of energy released by the hydrolysis of GTP to GDP + Pi. The A site is now open and ready to receive a new AA–tRNA.

The foregoing steps are repeated until a termination codon is brought into the A site. Thereupon another enzyme designated as the **releasing factor**, RF, catalyzes the release of the completed peptide chain with the concomitant hydrolysis of another molecule of GTP.

B.4.1 POLYRIBOSOMES. Molecules of mRNA are often relatively lengthy, and because the part of the molecule in contact with a single ribosome is relatively short, a single message can move through the mRNA binding sites of several ribosomes in succession, thus creating a chain of ribosomes held together by the mRNA strand (Figure B.10). Depending on the length of the messenger molecule, varying numbers of ribosomes may be moving in succession along it. The messenger for one of the polypeptide chains of hemoglobin, for example, is

about 150 nm in length. This length accommodates comfortably five or six ribosomes, each with a diameter of 23 nm. The messenger for the enzyme beta galactosidase has as many as 40 ribosomes strung along its length. When ribosomes of cells engaged in active protein synthesis are separated gently from other cellular components and placed on a coated grid for electron microscopy, they form clusters in which the individual ribosomes are apparently held in linear array by mRNA. In electron micrographs of sectioned cells, the clusters are often quite evident, but the connecting mRNA strand is not easily seen. These clusters are called **polyribosomes** or, more simply, **polysomes**. Figure B.11

531

FIGURE B.11 Diagrammatic representation of a polysome. A number of ribosomes are attached to mRNA. Each has progressed varying distances down the mRNA during translation. Each, therefore, has progressively longer protein chains attached. At the end of the message, the mRNA and ribosome separate and complete protein is released.

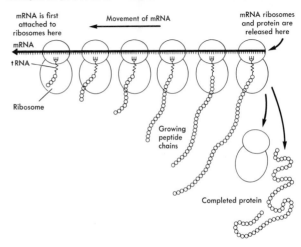

mRNA is first attached to ribosomes here

Movement of mRNA

mRNA ribosomes and protein are released here

mRNA

tRNA

Ribosome

Growing peptide chains

Completed protein

shows a diagrammatic representation of a polysome.

We have now presented the basic outline of the mechanism whereby proteins are synthesized in accordance with instructions coded in the nucleus. In sum: a segment of the genome coding for a particular protein is read out, or transcribed, in the form of messenger RNA. The message then feeds its codons sequentially through ribosomal binding sites; as it does so, transfer RNAs translate the message, discharging their amino acids in the order called for by the sequence of codons in the mRNA; peptide bonds form between successive amino acids; and when the sequence is assembled, the finished protein is released from the ribosome.

B.5 Suggestions for Further Readings

532

BAGLIONI, C., and B. COLOMBO. 1970. Protein synthesis. In D. M. Greenberg, ed. *Metabolic Pathways*, Vol 4, 3rd Ed. New York: Academic Press, pp. 278–351.

BARALLE, F. E., and G. G. BROWNLEE. 1978. AUG is the only recognizable signal in the 5′ non-coding regions of eukaryotic mRNA. *Nature* 274:84–87.

BARRIEUX, A., and M. G. ROSENFELD. 1977. Characterization of GTP-dependent Met-tRNA$_f$ binding protein. *J Biol Chem* 252:3843–47.

BRIMACOMBE, R., R. G. STOFFLER, and H. G. WITTMAN. 1978. *Ann Rev Biochem* 47:217–49.

DARNELL, J. E. 1976. mRNA structure and function. *Prog Nucleic Acid Res Mol Biol* 19:493–511.

EMANUILOV, I., D. D. SABATINI, J. A. LAKE, and C. FREIENSTEIN. 1978. Localization of eukaryotic initiation factor 3 on native small ribosomal subunits. *Proc Natl Acad Sci USA* 75:1389–93.

ERDMAN, V. A., M. SPRINZL, and O. PONGS. 1973. The involvement of 5 S RNA in the binding of tRNA to ribosomes. *Biochem Biophys Res Commun* 54:942–48.

FOX, J. W., and K.-P. WONG. 1978. Changes in the conformation and stability of 5 S RNA upon the binding of ribosomal proteins. *J Biol Chem* 253:18–20.

HENSHAW, E. C., G. G. GUINEY, and C. A. HIRSCH. 1973. The ribosome cycle in mammalian protein synthesis. I. The place of monomeric ribosomes and ribosomal subunits in the cycle. *J Biol Chem* 248:4367–76.

KAEMPFER, R., R. HOLLENDER, W. R. ABRAMS, and R. ISRAELI. 1978. Specific binding of messenger RNA and methionyl-tRNA$_f^{Met}$ by the same initiation factor for eukaryotic protein synthesis. *Proc Natl Acad Sci USA* 75:209–13.

KAEMFER, R., H. ROSEN, and R. ISRAELI. 1978. Translational control: recognition of the methylated 5′ end and an internal sequence in eukaroytic mRNA by the initiation factor that binds methionyl-tRNA$_f^{Met}$. *Proc Natl Acad Sci USA* 75:650–54.

LEADER, D. P., and G. C. MACHRAY. 1975. The binding sites for tRNA on eukaryotic ribosomes. *Nucleic Acids Res* 2:1177–88.

NAZAR, R. N. 1978. The release and reassociation of 5.8 S rRNA with yeast ribosomes. *J Biol Chem* 253:4505–07.

QUIGLEY, G. J., and A. RICH. 1976. Structural domains of transfer RNA molecules. *Science* 194:796–806.

RICH, A., and U. L. RAJBHANDARY. 1976. Transfer RNA: molecular structure, sequence, and properties. *Ann Rev Biochem* 45:805–60.

SERDYUK, I. N., and A. K. GRENADER. 1977. On the distribution and packing of RNA and protein in ribosomes. *Eur J Biochem* 79:495–504.

STENT, G. S., and R. CALENDAR. 1976. *Molecular Genetics: An Introductory Narrative*, 2nd Ed. San Francisco: Freeman.

WATSON, J. D. 1976. *The Molecular Biology of the Gene*, 3rd Ed. Menlo Park, Calif: Benjamin-Cummings.

WERTHEIMER, A. M., and M. S. KAULENAS. 1977. GDP kinase activity associated with salt-washed ribosomes. *Biochem Biophys Res Commun* 78:565–71.

Index

533

535

543

553

555